DIGITAL TRANSMISSION THEORY

SERGIO BENEDETTO
EZIO BIGLIERI
VALENTINO CASTELLANI

Politecnico di Torino
Turin, Italy

PRENTICE-HALL, INC.

Englewood Cliffs, New Jersey 07632

Library of Congress Cataloging-in-Publication Data

BENEDETTO, SERGIO.
   Digital transmission theory.

   Bibliography: p.
   Includes index.
   1. Digital communications.   I. Biglieri, Ezio.
II. Castellani, Valentino, 1940–   .   III. Title.
TK5103.7.B46   1988   621.398′1   86–25421
ISBN   0–13–214313–5

Editorial/production supervision and
   interior design: Reynold Rieger
Cover design: Ben Santora
Manufacturing buyer: S. Gordon Osbourne

A Note on the Front Cover

The pattern on the front cover, which closely resembles 64-QAM and 64-PSK signal constellations, is taken from the Chinese classic *Book of Changes* (*Yijing* or *I King*). Originally a book of divination, the *Yijing* was developed by Han dynasty scholars into a cosmological system in which the sixty-four exagrams represent, as mathematical formulas for modern scientists, all the possible mutations of creation, a universe in miniature. These exagrams inspired Leibniz in his invention of the system of binary number representation.

Printed in the United States of America

10   9   8   7   6   5   4   3   2   1

ISBN 0-13-214313-5   025

Prentice-Hall International (UK) Limited, *London*
Prentice-Hall of Australia Pty. Limited, *Sydney*
Prentice-Hall Canada Inc., *Toronto*
Prentice-Hall Hispanoamericana, S.A., *Mexico*
Prentice-Hall of India Private Limited, *New Delhi*
Prentice-Hall of Japan, Inc., *Tokyo*
Prentice-Hall of Southeast Asia Pte. Ltd., *Singapore*
Editora Prentice-Hall do Brasil, Ltda., *Rio de Janeiro*

Contents

10 DIGITAL TRANSMISSION OVER NONLINEAR CHANNELS 530

A USEFUL FORMULAS AND APPROXIMATIONS 577

Preface

*Yo conozco distritos en que los jóvenes se prosternan
ante los libros y besan con barbarie las páginas, pero no
saben descifrar una sola letra.*

(*J. L. Borges,* La Biblioteca de Babel)

This book is, in part, the outcome of some of the research activities of the authors in the field of digital communication systems and also the result of several lecture courses presented over the past years to the students of Politecnico di Torino. It is intended to serve as a graduate-level textbook and to cover the basic theory of digital communication with emphasis on its engineering applications.

During the last two decades or so, the field of digital transmission theory has been growing in importance, its rate of growth being partly motivated by the increasing relevance of the data communication business. Thus, it seemed useful to us to try to collect in a single book, in a comprehensive and unified manner, the most important material that has been published and is being published, for both theoretical and practical purposes, in scientific and technical journals throughout the world. However, it soon appeared that the area embraced by digital transmission theory is now too large to be covered adequately in a single book. Thus, a selective choice was made concerning both the topics to include and the depth of the coverage of each topic. We are aware that, as Lao Zi said, *"The wheel of a cart is not the cart,"* or, a selective choice of topics in digital transmission theory is not the whole theory. And, in fact, our selection leaves out a number of topics that could reasonably well be a part of a book like this. To name just one, a detailed treatment of synchronization problems is missing. But the size of the book had to be kept reasonable, and we prefer to provide a treatment of selected topics rather than to give the reader only a morsel of everything. As a result, this book is by no means intended to be encyclopedic. However, all the topics covered are developed in some depth, care is taken to include the most recent and useful results, and many references are given for further reading. Several examples are worked out in detail, and many exercises are included at the end of each chapter with the dual purpose of providing a better understanding of the material presented in the chapter and presenting new insights, applications, or results. As this work is also intended to serve as an introduction to research, our development of some topics has been shaped along the lines usually followed by current research papers, even at the price of a partial lack of coherence among some parts of the book (see, e.g., the different presentations of the Viterbi algorithm).

To cope with the ever-increasing demand for digital communication services, system planners and designers must look for more efficient systems: this is why the search for bandwidth- and power-efficient communication systems has been a very active research area. However, an important shift in perspective has recently occurred in the course of the search. This leads to the concept that the design trade-off is not only between bandwidth and power, because a third factor should be considered, the *complexity* of the communication system. In other words, an additional amount of complexity allowed at the transmitting and receiving ends of the system may prove beneficial to improve its performance without increasing either its bandwidth or its power. As Anderson and Lesh (1981) eloquently put it, ''Fortunately, we seem to be on the threshold of a third era in which engineers will find ways to trade both power and bandwidth for a third commodity, processing complexity. In years past, when active devices sold for five (1980) dollars apiece the word complexity was fraught with evil connotations. A new economics has now reversed this. Large-scale integration has reduced the cost of devices a million fold, and several future generations of such devices are waiting in the wings. It is clear that communications systems in the future will employ ever increasing complexity. Indeed, such complexity, if properly designed and implemented, will, in fact, be an economic necessity.''

From an instructor's point of view, a conclusion that can be drawn from these observations is that anybody wishing to design efficient systems should be able to put into action all the tools provided by a rather mature theory. This is why we deem it necessary for any serious student of digital communications to get a deep understanding of the theoretical material on which the techniques are based. Only when the basic concepts are firmly grasped can good designs be carried out, and only if the exact results are known can the quality of an approximation be fully evaluated. For these reasons, our goal in writing this book was more to illuminate the theory than to enumerate the practical results.

Our treatment is mainly theoretical, although we have tried to emphasize the points of view and the methods that are most likely to prove useful in applications. The level of mathematical rigor is intermediate so as not to obscure intuitive reasoning. However, no compromise in the level of mathematics has been made whenever the generality or the usefulness of the final results justifies the introduction of relatively sophisticated mathematical tools.

It is assumed that the reader has a basic understanding of Fourier transform techniques, probability theory, random variables, random processes (in particular, the Gaussian process), signal transmission through linear systems, the sampling theorem, linear modulation methods, matrix algebra, vector spaces and linear transformations. However, advanced knowledge of these topics is not required. Essentially, the reader is assumed to have taken courses in system theory, stochastic processes, and statistical communication theory.

English is not the mother tongue of any of the authors. However, we followed confidently the suggestion of Cato the Elder, who said *"Rem tene, verba sequentur"* (which is more or less the Duchess's remark to Alice: "Take care of the sense, and the sounds will take care of themselves.") Moreover, we are indebted to Dr. Stuart Wooding, who was kind enough to read the whole manuscript and did his best to improve the quality of our prose by removing from it most of its Italian flavor.

As this book is partly the outcome of our teaching activity, it should be apparent that the organization of the material and its presentation owe a great deal to some of our colleagues, too. We are indebted in particular to Professor Mario Pent for the development of some ideas about the treatment of the analytic signal in Chapter 2. We are also indebted to a number of colleagues who volunteered to read parts of the book, correct mistakes, and provide constructive criticism and suggestions for its improvement: Marco Ajmone Marsan, Tor Aulin, Michele Bellafemina, José Antonio Delgado Penin, Michele Elia, Umberto Mengali, Giovanni Vannucci, Kung Yao, and Zhongchen Zhang. We would like to thank them all for their time and effort. We also wish to thank the executive editor of Prentice-Hall, Inc., Bernard Goodwin, for his constant stimulus and encouragement, Luciano Brino for providing the line drawings, and Patrizia Vrenna, who, with great patience and skill, typed most of the manuscript.

Finally, we would like to express our great debt of gratitude to Professor Mario Boella, former chairman of our department, for inspiring us with his warm humanity and devotion to scholarship. This book is gratefully dedicated to him.

SERGIO BENEDETTO
EZIO BIGLIERI
VALENTINO CASTELLANI

Glossary of Acronyms

ACI   Adjacent channel interference
AM   Amplitude modulation
AM–PM   Amplitude and phase modulation
AMI   Alternate mark inversion
ARQ   Automatic repeat request
ASK   Amplitude-shift keying
AWGN   Additive white Gaussian noise
AZD   Ambiguity zone decoding
BCH   Bose–Chaudhuri–Hocquenghem
BEC   Binary erasure channel
BER   Bit error rate
BPF   Bandpass filter
BPSK   Binary phase-shift keying
BSC   Binary symmetric channel
B6ZS   Bipolar with 6-zero substitution
CAZAC   Constant amplitude zero autocorrelation
CCI   Cochannel interference
cdf   Cumulative distribution function
CHDB   Compatible high-density bipolar
CORPSK   Correlative phase-shift keying
CPFSK   Continuous-phase frequency-shift keying
CPM   Continuous-phase modulation
CPSK   Coherent phase-shift keying
dc   Direct current

DCPSK   Differentially coherent phase-shift keying
DECPSK   Differentially encoded coherent phase-shift keying
DFE   Decision-feedback equalizer
DM   Delay modulation
DS   Digital sum
DSB   Double sideband
DSV   Digital sum variation
FDM   Frequency-division multiplex
FDMA   Frequency-division multiple access
FEC   Forward error correction
FFSK   Fast frequency-shift keying
FIR   Finite impulse response
FM   Frequency modulation
FSK   Frequency-shift keying
FSOQ   Frequency-shift offset quadrature
GQR   Gauss quadrature rule
HDB   High-density bipolar
ICI   Interchannel interference
IF   Intermediate frequency
iid   Independent, identically distributed
IMUX   Input multiplexer
ISDN   Integrated services digital network
ISI   Intersymbol interference
LAN   Local area network
LCM   Least common multiple
LHS   Left-hand side
LS   Least squares
MAP   Maximum a posteriori
ML   Maximum likelihood
MLSE   Maximum-likelihood sequence estimation
MSE   Mean-square error
MSK   Minimum-shift keying
NRZ   Nonreturn to zero
OMUX   Output demultiplexer
OQPSK   Offset quaternary phase-shift keying
PAM   Pulse amplitude modulation
PCM   Pulse code modulation
pdf   Probability density function
PLL   Phase-locked loop
PM   Phase modulation
PN   Pseudonoise
PSK   Phase-shift keying
PST   Pair-selected ternary
PT   Pseudoternary
QAM   Quadrature amplitude modulation
QPSK   Quaternary phase-shift keying

RDS	Running digital sum
RF	Radio frequency
RHS	Right-hand side
RV	Random variable
RZ	Return to zero
SFSK	Sinusoidal frequency-shift keying
SNR	Signal-to-noise ratio
SQPSK	Staggered quaternary phase-shift keying
SSB	Single sideband
TDL	Tapped delay line
TDM	Time-division multiplex
TFM	Tamed frequency modulation
TWT	Traveling-wave tube
VCO	Voltage-controlled oscillator
VSB	Vestigial sideband
WS	Wide sense

About the Authors

Sergio Benedetto

Sergio Benedetto received the Laurea in Ingegneria Elettronica (summa cum laude) from Politecnico di Torino, Turin, Italy, in 1969. From 1970 to 1979 he was with the Istituto di Elettronica e Telecomunicazioni, first as Research Engineer, then as Associate Professor. In 1980, he was made Professor in Radio Communications at the Università di Bari. In 1981 he returned to Politecnico di Torino as Professor of Data Transmission Theory in the Dipartimento di Elettronica. He is coauthor of two books on signal theory and probability and random variables, and of more than 80 papers for leading engineering conferences and journals.

Ezio Biglieri

Ezio Biglieri had his training in Electrical Engineering from Politecnico di Torino, Turin, Italy, where he received the Laurea (summa cum laude) in 1967. From 1968 to 1975, he was with the Istituto di Elettronica e Telecomunicazioni, Politecnico di Torino, first as Research Engineer, then as Associate Professor. In 1975, he was made Professor of Electrical Engineering at the Università di Napoli, Naples. In 1977, he returned to Politecnico di Torino as Professor in the Department of Electrical Engineering. Since 1987, he has been Professor of Electrical Engineering at the University of California, Los Angeles. He has coauthored two books, edited one, and published more than 110 papers in scholarly journals or conference proceedings. He is a senior member of the IEEE.

Valentino Castellani

Valentino Castellani received the Laurea in Ingegneria Elettronica (summa cum laude) from Politecnico di Torino, Turin, Italy, in 1963 and the Master's Degree in Electrical Engineering from M.I.T. in 1966. Since 1966 he was with Istituto di Elettronica e Telecomunicazioni, Politecnico di Torino, first as Assistant Professor, then as Associate Professor. In 1980 he was made Professor of Electrical Communications. He is author of one book on noise and of some 60 papers for leading engineering conferences and journals.

CHAPTER 1

Introduction

In the information era that we are experiencing, communication networks are a vital part of society's infrastructure. The growing demand for the exchange of information in digital form and the need for sharing computation resources and for spreading the information contents of existing large data bases require ever increasing transmission capacity. The great variety of information to be transmitted (voice, images, alphanumeric characters, etc.) and the availability of highly reliable, low-cost, small-sized digital circuitry push irreversibly near the source the digital conversion of analog signals. From there, the information is transmitted in digital form through the integrated services digital networks (ISDN) (see Decina and Roveri, 1987). This leads to a large variety of digital signals of different formats and speeds that must be routed and switched to their proper destination.

The amount of data to be transmitted and the speed of transmission vary enormously from one system to another. When batches of data are sent for processing on a distant computer, a delivery time longer than 1 hour is sometimes acceptable. However, when a person–computer dialogue is taking place, the responses must be returned to the person quickly enough so as not to impede his or her train of thought. Response times between 100 milliseconds and 2 seconds are typical. In real-time systems, in which a machine or process is being controlled, response times can vary from a few milliseconds to many minutes. Figure 1.1 shows some of the common requirements for delivery time or response time and for amount of data transmitted. The block labeled "terminal dialogue systems," for example, indicates a response time requirement from 1 to 10 seconds and a message size ranging from 1 character (8 bits) to about 4000 characters (around 30,000 bits). The transmission speed required by the communication link equals the delivery time of one-way messages, as shown in Figure 1.1 (vertical axis), divided

1

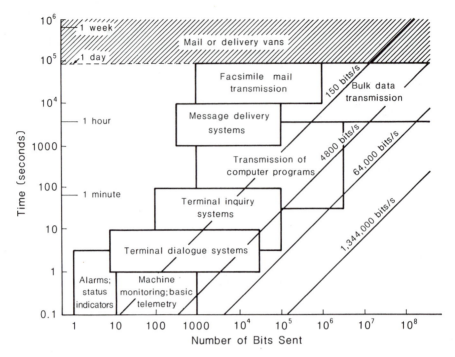

Figure 1.1 Desirable delivery time or response time and quantity of data for typical use of data transmission. (J. Martin, *System Analysis for Data Transmission*, © 1972. Reprinted by permission of Prentice-Hall, Inc., Englewood Cliffs, N.J.)

by the number of bits transmitted (horizontal axis). Straight lines corresponding to some available transmission speeds are also indicated. For most applications shown in the figure, the speeds allowed on telephone channels (up to, say, 9600 bits per second) are sufficient. Higher speeds are required when a very large bulk of data must be sent, as in the case of communication between computers.

The most common form of digital communication nowadays consists of people at terminals communicating with other terminals or a mainframe computer, as shown in Fig. 1.2. A community of users in a limited area is interconnected through a local area network (LAN) offering a variety of services such as computing resources, voice communication, teleconferencing, and electronic mail. These LANs can exchange informations over a packet-switched long-distance telecommunication network. They can normally reach one or several centralized computer systems for value-added services that require more powerful computing resources.

In a geographic computer network, the communication engineer must solve a great variety of global design problems, such as topological structure, line capacity allocation, routing, and flow-control procedures, and local design problems, such as the choice of the multiplexing scheme, the number of message sources per concentration point, the access technique (polling, random access), and buffer size. The final system choices will be a trade-off between costs and performance criteria, such as the specified response time and the specified reliability.

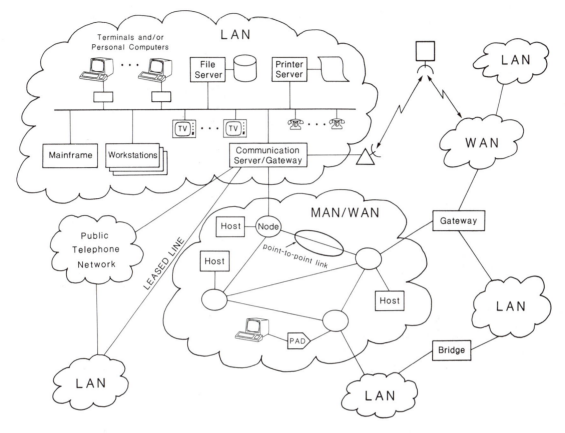

Figure 1.2 Local area networks, metropolitan area networks (MAN), and wide area networks (WAN) interconnected through various telecommunication links.

The exchange of information in a packet-switched network is governed by a layered protocol architecture, such as described, for example, in the ISO/OSI reference model (see Decina and Roveri, 1987). Level 1 of this layered architecture concerns the point-to-point communication between two nodes of the network. According to the physical medium that connects the nodes, different problems are encountered to establish a reliable link. Moreover, to access one of the nodes of the packet-switched network, the user may have to use the dialed public telephone network or a leased voice line.

Let us isolate and examine in greater detail one communication link in the system of Fig. 1.2, for example the one highlighted. It is shown magnified in Fig. 1.3. To be transmitted on a telephone channel, the bit stream emitted by the terminal must be converted into a sequence of waveforms using a special device known as the *modem*, short for modulator/demodulator. This converts the data into a signal whose range of frequencies matches the bandwidth of the telephone line. Besides data, the terminal and the modem exchange various line-control signals according to a standardized interface, as shown in Fig. 1.3. At the computer site, a modem converts the analog signals into a binary stream that is sent to the computer through a transmission control unit that supervises the communication with distant terminals.

Introduction   Chap. 1   **3**

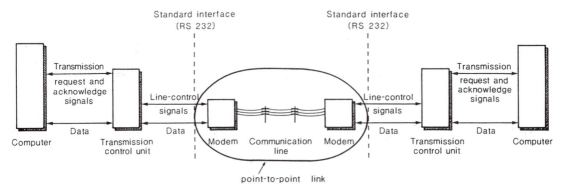

Figure 1.3 Point-to-point communication link between two computers.

The design of this *point-to-point* communication link is related to the choices made for the network in terms of available speeds, response times, multiplexing and access techniques, and transmission quality. In particular, matching the sources of information to the channel speeds may involve *source encoding, channel bandwidths* and the choice of *modulation schemes*. The response time and the access techniques pose constraints on the modem setup time, that is, on the choice of the *synchronization and adaptive equalization algorithms*. The transmission quality is usually given in terms of *bit error probability*, which in turn depends on *channel encoding* (*error correction or error detection and retransmission*), the *transmitted power* and the modulation schemes.

This book is primarily concerned with the theory of point-to-point digital communication. To resort to a more general and abstract context, let us expand the point-to-point connection of Fig 1.3 into the functional block diagram of Fig. 1.4. We only consider discrete information sources. When a source is analog in nature, such as a microphone actuated by speech or a TV camera scanning a scene, we assume that a process of sampling and quantizing takes place within the block so that the output is a sequence of discrete symbols or letters. Discrete information sources are characterized by a *source alphabet*, a *source rate* (symbols per second), and a probability law governing the emission of sequences of symbols, or *messages*. From these parameters we can construct a probabilistic model of the information source and define a source information rate (R_s) in bits (binary digits) per second. The input to the second block of Fig. 1.4, the *source encoder*, is a sequence of symbols occurring at a certain rate. The source encoder converts the symbol sequence into a binary sequence by assigning code words to the symbols of the input sequence according to a specified rule. This encoding process has the goal of reducing the *redundancy* of the source (i.e., of obtaining an output data rate approaching R_s). At the receiver, the source decoder converts the binary output of the channel decoder into a symbol sequence that is passed to the user.

Because the redundancy has been removed, the binary sequence at the output of the source encoder is very vulnerable to errors occurring during the process of transmitting the information to its destination. The *channel encoder* introduces a controlled redundancy into the binary sequence so as to achieve highly reliable transmission. At the receiver, the channel decoder recovers the information-bearing bits from the coded binary stream. Both the encoder and decoder can operate either in block mode or in a continuous sequential mode, depending on the type of code used in the system.

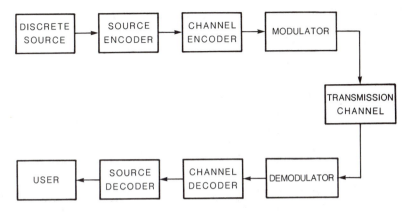

Figure 1.4 Block diagram of a point-to-point digital communication system.

The communication channel provides the electrical connection between the source and the destination. The channel may be a pair of wires, a telephone link, an optical fiber, or free space over which the signal is radiated in the form of electromagnetic waves. Communication channels introduce several impairments. Having finite bandwidths, they distort the signal in amplitude and phase. Moreover, the signal is attenuated and corrupted by unwanted random signals referred to as *noise*. For these reasons, an exact replica of the transmitted signal cannot be obtained at the receiver input. The primary objective of a good communication system design is to counteract the effects of noise and distortion.

The modulator converts the input binary stream into a waveform sequence suitable for transmission over the available channel. Being a powerful tool in the hands of the designer, modulation will receive considerable attention in this book. It involves a large number of choices, such as the number of waveforms, their shape, duration, and bandwidth, the power level, and more, allowing great flexibility in system design. At the receiver, the demodulator extracts the binary sequence from the received waveforms. Due to the impairments introduced by the channel, this process entails the possibility of errors between the input sequence to the modulator and the output sequence from the demodulator. A measure of this type of degradation is the *bit error probability*.

In practical communication systems, other functional blocks exist, which are not shown in Fig. 1.4 for reasons of simplicity. They are, for example, the adaptive equalizer, which reduces channel distortion, the carrier and symbol synchronizers, which allow coherent demodulation and proper sampling of the received signal to recover the information sequence, *scramblers* and *descramblers* used to prevent unwanted strings of symbols at the channel input, and *enciphering* and *deciphering* devices to ensure secure communication. Some of these blocks will be described in the rest of this book.

The book is organized as follows. Chapter 2 reviews the main results from the theory of random processes, spectral analysis, and detection theory, which can be considered as prerequisites to the book. In Chapter 3 we look at probabilistic models for discrete information sources and communication channels. The main results from classical information theory are introduced as conceptual background for the successive chapters. Chapter 4 is devoted to waveform transmission over additive Gaussian noise channels.

Introduction   Chap. 1 **5**

Optimum receiver structures are derived using results from detection theory, and their performance in terms of error probability is presented. A distinction is made between coherent and noncoherent transmission and between symbol-by-symbol and sequence receivers. The Viterbi algorithm, a crucial topic in modern digital communication, is used here for the first time; Appendix F is entirely devoted to it. In Chapter 5 the main modulation methods are described. Block diagrams of transmitters and receivers are given and their performance characteristics are presented in terms of bit error probability as a function of the signal-to-noise ratio. Both memoryless modulation and modulation schemes with memory are analyzed. The final part of the chapter is devoted to a comparison among different modulation schemes and a description of their practical fields of application.

In Chapter 6 we introduce the main impairments besides Gaussian noise that affect the transmitted signal. Great emphasis is given to techniques for the computation of the error probability in the presence of *intersymbol interference* due to linear channel distortion. Appendix E is devoted to the description of the most important methods used in this evaluation of the error probability. Also, in Chapter 6 a brief description of the problems of carrier and symbol synchronization can be found, together with the effects of imperfect carrier synchronization on system performance. The final part of the chapter deals with selective fading affecting digital radio relay links. Chapter 7 is devoted to system optimization and the analysis of optimum receivers (employing the Viterbi algorithm) for channels with intersymbol interference. In Chapter 8, adaptive equalization is analyzed in detail. Particular attention is given to the iterative algorithms used to control the equalizers, such as the stochastic gradient algorithm.

Chapter 9 deals with channel encoding. It describes *block codes* and *convolutional codes* as applied to both error detection and correction. The final part is devoted to the *line codes*, which are used mainly for spectral shaping of the transmitted signal. Finally, Chapter 10 presents some topics on digital transmission over nonlinear channels. Many previous examples are revisited in the context of systems using nonlinear devices.

CHAPTER 2

An Introduction to the Mathematical Techniques of Digital Communication Theory

Signal theory, system theory, probability, and stochastic processes are the essential mathematical tools for the analysis and design of digital communication systems. A comprehensive treatment of these topics would require several volumes. However, we assume that the reader has some familiarity with them. Thus, this chapter is selective. The topics selected and the depth of their presentation were decided according to two criteria. First, where possible, laborious and sophisticated mathematical apparatuses have been omitted. This entails a certain loss of rigor, but it should improve the presentation of the subject matter. Second, those topics most likely to be familiar to the reader are reviewed very quickly, whereas more attention is devoted to certain specialized points of particular relevance for applications.

The topics covered are deterministic and random signal theory for both discrete- and continuous-time models, linear and nonlinear system theory, and detection theory. The extensive bibliographical notes should guide the reader wishing to become more conversant with a specific topic.

2.1 SIGNALS AND SYSTEMS

In this section, we briefly present the basic concepts of the theory of linear and certain nonlinear systems. We shall begin with the time-discrete model for signals and systems and continue with the time-continuous model. To provide a higher level of generality to our presentation, we shall usually handle *complex* quantities. This is for mathematical convenience. The reasons for their use are explained in Section 2.4.

2.1.1 Discrete Signals and Systems

A *discrete-time signal* is a sequence of real or complex numbers, denoted by $(x_n)_{n=n_1}^{n_2}$, defined for every integer index n, $n_1 \le n \le n_2$. The index n is usually referred to as the *discrete time*. Whenever $n_1 = -\infty$ and $n_2 = \infty$, or when the upper and lower indexes need not be specified, we shall simply write (x_n). A *time-discrete system*, or for short a *discrete system*, is a mapping of a sequence (x_n), called the *input* of the system, into another sequence (y_n), called the *output* or *response*. We write

$$y_n = S[(x_n)] \tag{2.1}$$

for the general element of the sequence (y_n).

A discrete system is *linear* if, for any pair of input signals (x_n'), (x_n''), and for any pair of complex numbers A', A'', the following holds:

$$S[(A'x_n' + A''x_n'')] = A'S[(x_n')] + A''S[(x_n'')]. \tag{2.2}$$

Equation (2.2) means that if the system input is a linear combination of two signals, its output is the same linear combination of the two responses.

A discrete system is *time invariant* if the rule by which an input sequence is transformed into an output sequence does not change with time. Mathematically, this is expressed by the condition

$$S[(x_{n-k})] = y_{n-k} \tag{2.3}$$

for all integers k. This is equivalent to saying that, if the input is delayed by k time units, the output is delayed by the same quantity.

If (δ_n) denotes the sequence

$$\delta_n = \begin{cases} 1, & n = 0, \\ 0, & n \ne 0, \end{cases} \tag{2.4}$$

and S is a linear, time-invariant discrete system, its response (h_n) to the input (δ_n) is called the *(discrete) impulse response* of the system. Given a linear, time-invariant discrete system with impulse response (h_n), its response to any arbitrary input (x_n) can be computed via the *discrete convolution*

$$
\begin{aligned}
y_n &= \sum_{k=-\infty}^{\infty} x_k h_{n-k} \\
&= \sum_{k=-\infty}^{\infty} h_k x_{n-k}.
\end{aligned}
\tag{2.5}
$$

It may happen that the system output at time l, say y_l, depends only on a certain subset of the input sequence. In particular, the system is said to be *causal* if y_l depends only on $(x_n)_{n=-\infty}^{l}$. This means that the output at any given time depends only on the past and present values of the input, and not on its future values. In addition, the system is said to have a *finite memory L* if y_l depends only on the finite segment $(x_n)_{n=l-L}^{l}$ of the past input. When $L = 0$, and hence y_l depends only on x_l, the system is called *memoryless*. For a linear time-invariant system, causality implies $h_n = 0$ for all $n < 0$. A linear time-invariant system with finite memory L has an impulse response sequence (h_n) that may be nonzero only for $0 \le n \le L$. For this reason, a finite-memory system is often

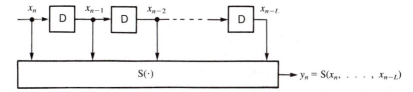

Figure 2.1 Transversal filter implementation of a time-invariant discrete system with memory L.

referred to also as a *finite impulse response* (FIR) system. A system with memory L can be implemented as in Fig. 2.1. The blocks labeled D denote unit-delay elements (i.e., systems that respond to the input x_n with the output $y_n = x_{n-1}$). A cascade of such unit-delay elements is called a *shift register*, and the resulting structure is called a *tapped delay line*, or *transversal*, filter. Here the function $S(\cdot)$ defining the input–output relationship has $L + 1$ arguments. When the system is linear, $S(\cdot)$ takes the form of a linear combination of its arguments:

$$S(x_n, x_{n-1}, \ldots, x_{n-L}) = \sum_{k=0}^{L} h_k x_{n-k}. \tag{2.6}$$

In this case, the structure of Fig. 2.1 becomes the linear transversal filter of Fig 2.2.

Discrete Volterra systems

Consider a time-invariant, nonlinear discrete system with memory L, and assume that the function $S(\cdot)$ is sufficiently regular to be expanded in a Taylor series in a neighborhood of the origin $x_n = 0, x_{n-1} = 0, \ldots, x_{n-L} = 0$. We have the representation

$$
\begin{aligned}
y_n = S(x_n, x_{n-1}, \ldots, x_{n-L}) &= h^{(0)} + \sum_{i=0}^{L} h_i^{(1)} x_{n-i} \\
&+ \sum_{i=0}^{L} \sum_{j=0}^{L} h_{ij}^{(2)} x_{n-i} x_{n-j} + \sum_{i=0}^{L} \sum_{j=0}^{L} \sum_{k=0}^{L} h_{ijk}^{(3)} x_{n-i} x_{n-j} x_{n-k} + \cdots
\end{aligned}
\tag{2.7}
$$

called a *Volterra series*. It is seen that the system is completely characterized by the coefficients of the expansion, say

$$h^{(0)}, h_i^{(1)}, h_{ij}^{(2)}, h_{ijk}^{(3)}, \ldots, \qquad i, j, k = 0, 1, 2, \cdots, L,$$

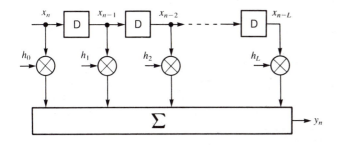

Figure 2.2 Linear discrete transversal filter.

which are proportional to the partial derivatives of the function $S(\cdot)$ at the origin. These are called the system's *Volterra coefficients*. The expansion (2.7) can be generalized to systems with infinite memory, although in the computational practice only a finite number of terms will be retained. Since in general the Volterra system representation involves an infinite number of infinite summations, if a truncation is not performed we must associate with the series suitable convergence conditions to guarantee that the representation is meaningful (see, e.g., Rugh, 1981).

Example 2.1

Consider the discrete system shown in Fig. 2.3 and obtained by cascading a linear, time-invariant, causal system with impulse response (h_n) to a memoryless nonlinear system with input–output relationship $y_n = g(w_n)$. Assume that $g(\cdot)$ is an analytic function, with a Taylor series expansion in the neighborhood of the origin

$$g(w) = \sum_{l=0}^{\infty} a_l w^l \tag{2.8}$$

The input–output relationship for the system of Fig 2.3 is then

$$y_n = g\left(\sum_{i=0}^{\infty} h_i x_{n-i}\right) \tag{2.9}$$

$$= a_0 + a_1 \sum_{i=0}^{\infty} h_i x_{n-i} + a_2 \sum_{i=0}^{\infty} \sum_{j=0}^{\infty} h_i h_j x_{n-i} x_{n-j} + \cdots$$

so that the Volterra coefficients for the system are:

$$h^{(0)} = a_0,$$
$$h_i^{(1)} = a_1 h_i$$
$$h_{ij}^{(2)} = a_2 h_i h_j . \ \ . \ \ . \ \ .$$

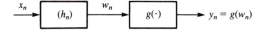

Figure 2.3 Discrete nonlinear system.

The following should be observed. First, if $g(\cdot)$ is a polynomial of degree K, the coefficients $a_{K+1}, a_{K+2}, \ldots$, in (2.8) are zero, so that only a finite number of summations will appear in (2.9). Second, if the impulse response sequence (h_n) is finite (i.e., the linear system of Fig. 2.3 has a finite memory), all the summations in (2.9) will include only a finite number of terms. $\square$

Discrete signals and systems in the transform domain

Given a sequence (x_n), we define its *Fourier transform* $\mathcal{F}[(x_n)]$ as the function of the frequency f defined as

$$X(f) \triangleq \sum_{n=-\infty}^{\infty} x_n e^{-jn2\pi f}, \tag{2.10}$$

where $j \triangleq \sqrt{-1}$. $X(f)$ is a periodic function of f with period 1, so it is customary to consider it only in the interval $-\frac{1}{2} \leq f \leq \frac{1}{2}$. The *inverse Fourier transform* gives the elements of the sequence (x_n) in terms of $X(f)$:

$$x_n = \int_{-1/2}^{1/2} X(f)e^{jn2\pi f} \, df. \tag{2.11}$$

The Fourier transform $H(f)$ of the impulse response (h_n) of a linear system is called the *frequency response*, or *transfer function*, of the system. We shall call $|H(f)|$ the *amplitude* and arg $[H(f)]$ the *phase* of the transfer function. The derivative of arg $[H(f)]$ taken with respect to f is called the *group delay* of the system. A basic property of the Fourier transform is that the response of a linear, time-invariant discrete system with transfer function $H(f)$ to a sequence with Fourier transform $X(f)$ has the Fourier transform $H(f)X(f)$.

Example 2.2

Consider the linear transversal filter of Fig 2.2. Its impulse response is the sequence of coefficients of the multipliers, so that its transfer function is

$$H(f) = \sum_{l=0}^{L} h_l \, e^{-jl2\pi f} \tag{2.12}$$

As for filters like this linearity of phase is often a desired goal, we shall now determine the constraints on the impulse response leading to a phase of the transfer function of the form

$$\vartheta(f) \stackrel{\triangle}{=} \arg [H(f)] = -\alpha 2\pi f,$$

where α is a constant phase delay measured in number of samples. Since

$$H(f) = |H(f)|e^{-j\alpha 2\pi f},$$

using (2.12) we get two equations:

$$|H(f)| \cos (\alpha 2\pi f) = \sum_{l=0}^{L} h_l \cos (l2\pi f), \tag{2.13a}$$

$$|H(f)| \sin (\alpha 2\pi f) = \sum_{l=0}^{L} h_l \sin (l2\pi f). \tag{2.13b}$$

By taking the ratio of (2.13b) to (2.13a), we get

$$\frac{\sin (\alpha 2\pi f)}{\cos (\alpha 2\pi f)} = \frac{\displaystyle\sum_{l=0}^{L} h_l \sin(l2\pi f)}{\displaystyle\sum_{l=0}^{L} h_l \cos(l2\pi f)}. \tag{2.14}$$

Discarding the trivial case in which $\alpha = 0$ (which leads to the solution $h_l = 0$ for $l > 0$), (2.14) can be rewritten in the form

$$\sum_{l=0}^{L} h_l \sin [(\alpha - l)2\pi f] = 0, \tag{2.15}$$

which is satisfied for all f if and only if the following conditions hold (assume L even):

$$\alpha = \frac{L}{2}$$
$$h_l = h_{L-l}, \qquad 0 \leq l \leq L. \;\; \square \tag{2.16}$$

2.1.2 Continuous Signals and Systems

A *continuous-time signal* is a real or complex function $x(t)$ of the real variable t (the *time*). Unless otherwise specified, the time is assumed to range from $-\infty$ to ∞. A *continuous-time* system, or for short a *continuous system*, is a mapping of a signal $x(t)$, the system *input*, into another signal $y(t)$, called the *output* or *response*. We write

$$y(t) = S[x(t)]. \tag{2.17}$$

A continuous system is *linear* if for any pair of input signals $x'(t)$, $x''(t)$ and for any pair of complex numbers A', A'', the following holds:

$$S[A'x'(t) + A''x''(t)] = A'S[x'(t)] + A''S[x''(t)]. \tag{2.18}$$

A continuous system is *time invariant* if (2.17) implies

$$S[x(t - \tau)] = y(t - \tau) \tag{2.19}$$

for all τ. Let $\delta(t)$ denote the delta function, characterized by the *sifting property*

$$\int_{-\infty}^{\infty} \delta(t)\phi(t)\,dt = \phi(0) \tag{2.20}$$

valid for every function $\phi(t)$ continuous at the origin. The response $h(t)$ of a linear, time-invariant continuous system to the input $\delta(t)$ is called the *impulse response* of the system. For a system with a known impulse response $h(t)$, the response $y(t)$ to any input signal $x(t)$ can be computed via the *convolution integral*

$$\begin{aligned} y(t) &= \int_{-\infty}^{\infty} x(\tau)h(t - \tau)d\tau \\ &= \int_{-\infty}^{\infty} h(\tau)x(t - \tau)\,d\tau \end{aligned} \tag{2.21}$$

It may happen that the system output $y(t)$ at time t depends on the input $x(t)$ only through the values taken by $x(t)$ in the time interval I. If $I = (-\infty, t]$, the system is said to be *causal*. If $I = (t - t_0, t]$, $0 < t_0 < \infty$, the system is said to have a *finite memory* t_0. If $I = \{t\}$ (i.e., the output at any given time depends only on the input at the same time), the system is called *memoryless*. It is easily seen from (2.21) that, for a linear time-invariant system, causality implies $h(t) = 0$ for all $t < 0$. A time function $h(t)$ with the latter property is sometimes called *causal*.

A linear system is said to be *stable* if its response to any bounded input is bounded. A linear, time-invariant system is stable if its impulse response is absolutely integrable.

Example 2.3

Figure 2.4 represents a linear, time-invariant continuous system with finite memory. The blocks labeled T are delay elements, that is, systems with impulse response $\delta(t - T)$. A cascade of such elements is called a *(continuous) tapped delay line* and the structure of Fig. 2.4 a linear *transversal filter*. The system has an impulse response

$$h(t) = \sum_{l=0}^{L} c_l \delta(t - lT) \tag{2.22}$$

and a memory LT. $\square$

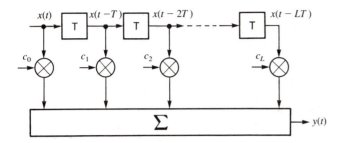

Figure 2.4 Linear continuous transversal filter.

Continuous Volterra systems

To motivate our general discussion of Volterra series, consider as an example the time-invariant, nonlinear continuous system shown in Fig. 2.5. Assume that the first block represents a linear time-invariant system with impulse response $h(t)$ and that $g(\cdot)$ is a function as in Example 2.1, so (2.8) holds. The input–output relationship for this system can thus be expanded in the form

$$y(t) = g\left[\int_{-\infty}^{\infty} h(\tau)x(t-\tau)\,d\tau\right]$$

$$= a_0 + a_1\int_{-\infty}^{\infty} h(\tau)x(t-\tau)d\tau \tag{2.23}$$

$$+ a_2\int_{-\infty}^{\infty}\int_{-\infty}^{\infty} h(\tau_1)h(\tau_2)x(t-\tau_1)x(t-\tau_2)\,d\tau_1\,d\tau_2 + \cdots.$$

By defining

$$h_0 \triangleq a_0,$$
$$h_1(t) \triangleq a_1 h(t), \tag{2.24}$$
$$h_2(t_1, t_2) \triangleq a_2 h(t_1)h(t_2),$$

$$\vdots$$

Eq. (2.23) can be rewritten as

$$y(t) = h_0 + \int_{-\infty}^{\infty} h_1(\tau)x(t-\tau)\,d\tau$$

$$+ \int_{-\infty}^{\infty}\int_{-\infty}^{\infty} h_2(\tau_1,\tau_2)x(t-\tau_1)x(t-\tau_2)d\tau_1\,t\tau_2 + \cdots \tag{2.25}$$

$$+ \int_{-\infty}^{\infty}\cdots\int_{-\infty}^{\infty} h_k(\tau_1,\tau_2,\ldots,\tau_k)\left[\prod_{i=1}^{k} x(t-\tau_i)\,d\tau_i\right] + \cdots.$$

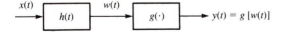

Figure 2.5 Continuous nonlinear system.

Equations (2.24) and (2.25) represent the input–output relationship of the system of Fig. 2.5. More generally, (2.25) without the definitions (2.24), that is, for a general set of functions h_0, $h_1(t)$, $h_2(t_1, t_2)$, . . . , provides an input–output relationship for nonlinear time-invariant continuous systems. The RHS of (2.25) is called a *Volterra series*, and the functions h_0, $h_1(t)$, $h_2(t_1, t_2)$, . . . , are called the *Volterra kernels* of the system. As a linear, time-invariant continuous system is completely characterized by its impulse response, so a nonlinear system whose input–output relationship can be expressed as a Volterra series is completely characterized by its Volterra kernels. It can be observed that the first-order kernel $h_1(t)$ is simply the impulse response of a linear system. The higher-order kernels can thus be viewed as higher-order impulse responses, which characterize the various orders of nonlinearity of the system. The zero-order term h_0 accounts for the response to a zero input.

It can be shown (see Problem 2.6) that a time-invariant system described by a Volterra series is causal if and only if, for all k,

$$h_k(t_1, t_2, \ldots, t_k) = 0 \qquad \text{for any } t_i < 0, \qquad i = 1, 2, \ldots, k. \qquad (2.26)$$

A Volterra series expansion can be made simpler if it is assumed that the system kernels are symmetric functions of their arguments. That is, for every $k \geqslant 2$ any of the $k!$ possible permutations of the k arguments of $h_k(t_1, t_2, \ldots, t_k)$ leaves the kernel unchanged. It can be proved (see Problem 2.5) that the assumption of symmetric kernels does not entail any loss of generality.

Volterra series can be viewed as "Taylor series with memory." As such they share with Taylor series some limitations, a major one being slow convergence. Moreover, the complexity in computation of the kth term of a Volterra series increases quickly with increasing k. Thus, it is expedient to use Volterra series only when the expansion (2.25) can be truncated to low-order terms. Applications of Volterra series and a discussion of the computational problems involved with their use are contained in Chapter 10.

Continuous signals and systems in the transform domain

With the notation $X(f) \triangleq \mathcal{F}[x(t)]$, we shall denote the Fourier transform of the signal $x(t)$; that is,

$$X(f) = \int_{-\infty}^{\infty} x(t)e^{-j2\pi ft}\, dt. \qquad (2.27)$$

Given its Fourier transform $X(f)$, the signal $x(t)$ can be recovered by computing the *inverse Fourier transform* $\mathcal{F}^{-1}[X(f)]$:

$$x(t) = \int_{-\infty}^{\infty} X(f)e^{j2\pi ft}\, df. \qquad (2.28)$$

The Fourier transform of a signal is also called the *amplitude spectrum* of the signal. If $h(t)$ denotes the impulse response of a linear, time-invariant system, its Fourier transform $H(f)$ is called the *frequency response*, or *transfer function*, of the system. We shall call $|H(f)|$ the *amplitude* and arg $[H(f)]$ the *phase* of the transfer function. The derivative of arg $[H(f)]$ taken with respect to f is called the *group delay* of the system. It is seen from (2.27) that when $x(t)$ is a real signal the real part of $X(f)$ is an even function of f, and the imaginary part is an odd function of f. It follows that for a real $x(t)$ the function $|X(f)|$ is even, and arg $[X(f)]$ is odd.

An important property of Fourier transform is that it relates products and convolutions of two signals $x(t)$, $y(t)$ with convolutions and products of their Fourier transforms $X(f)$ and $Y(f)$:

$$\mathcal{F}[x(t)y(t)] = \int_{-\infty}^{\infty} X(\alpha)Y(f - \alpha)\,d\alpha \qquad (2.29)$$

and

$$\mathcal{F}\left[\int_{-\infty}^{\infty} x(\tau)y(t - \tau)\,d\tau\right] = X(f)Y(f). \qquad (2.30)$$

In particular, (2.30) implies that the output $y(t)$ of a linear, time-invariant system with a transfer function $H(f)$ and an input signal $x(t)$ has the amplitude spectrum

$$Y(f) = H(f)X(f). \qquad (2.31)$$

Example 2.3 (continued)

The transfer function of the system shown in Fig. 2.4 is obtained by taking the Fourier transform of (2.22):

$$H(f) = \sum_{l=0}^{L} c_l\, e^{-jl2\pi ft}. \qquad (2.32)$$

It is left as an exercise for the reader to provide the conditions for which this system exhibits a linear phase. $\square$

Example 2.4

An important family of linear systems is provided by the *Butterworth filters*. The transfer function of the *nth-order low-pass Butterworth filter* with *cutoff frequency* f_c is

$$H(f) = \frac{1}{D_n(jf/f_c)}, \qquad (2.33)$$

where

$$D_n(s) \triangleq \prod_{i=1}^{n} [s - e^{j\pi(2i+n-1)/2n}] \qquad (2.34)$$

is an nth degree polynomial. Expressions of these polynomials for some values of n are

$$
\begin{aligned}
D_1(s) &= 1 + s, \\
D_2(s) &= 1 + \sqrt{2}\,s + s^2, \\
D_3(s) &= 1 + 2s + 2s^2 + s^3, \\
D_4(s) &= 1 + 2.613s + 3.414s^2 + 2.613s^3 + s^4.
\end{aligned}
\qquad (2.35)
$$

Figure 2.6 shows the amplitude $|H(f)|$ of the transfer function of the low-pass Butterworth filters for several values of the order n. It is seen that the curves of all orders pass through the 0.707 point at $f = f_c$. As $n \to \infty$, $|H(f)|$ approaches the ideal low-pass ("brickwall") characteristics:

$$|H(f)| = \begin{cases} 1, & |f| < f_c, \\ 0, & \text{elsewhere.} \end{cases} \square \qquad (2.36)$$

Sec. 2.1   Signals and Systems

15

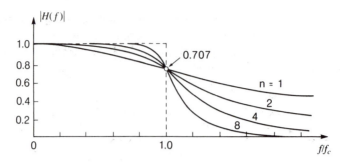

Figure 2.6 Amplitude of the transfer function of low-pass Butterworth filters of various orders.

2.2 RANDOM PROCESSES

2.2.1 Discrete-Time Processes

A *discrete-time random process*, or *random sequence*, is a sequence (ξ_n) of real or complex random variables (RV) defined on some sample space. The index n is usually referred to as the discrete time. A discrete-time process is completely characterized by providing the joint cumulative distribution functions (cdf) of the N-tuples ξ_{i+1}, ξ_{i+2}, . . . , ξ_{i+N} of RVs extracted from the sequence, for all integers i and N, $N > 0$. If the process is complex, these joint distributions are the joint $2N$-dimensional distributions of the real and imaginary components of ξ_{i+1}, . . . , ξ_{i+N}. The simplest possible case occurs when the RVs in the sequence are independent and identically distributed (iid). In this case the joint cdf of any N-tuple of RVs factors into the product of individual marginal cdfs. For a real process,

$$F_{\xi_{i+1}, \xi_{i+2}, \cdots, \xi_{i+N}}(x_{i+1}, x_{i+2}, \ldots, x_{i+N}) = \prod_{j=1}^{N} F_{\xi}(x_{i+j}), \qquad (2.37)$$

where $F_{\xi}(\cdot)$ is the common cdf of the RVs. Thus, a sequence of iid RVs is completely characterized by the single function $F_{\xi}(\cdot)$.

A random sequence is called *stationary* if for every N the joint distribution function of ξ_{i+1}, ξ_{i+2}, . . . , ξ_{i+N} does not depend on i. In other words, a stationary random sequence is one whose probabilistic properties do not depend on the time origin, so that for any given integer k the sequences (ξ_n) and (ξ_{n+k}) are identically distributed. An iid sequence extending from $n = -\infty$ to ∞ is an example of a stationary sequence.

The *mean* of a random sequence (ξ_n) is the sequence (μ_n) of mean values

$$\mu_n \triangleq E[\xi_n]. \qquad (2.38)$$

The *autocorrelation* of (ξ_n) is the two-index sequence $(r_{n,m})$ such that

$$r_{n,m} \triangleq E[\xi_n \xi_m^*]. \qquad (2.39)$$

For a stationary sequence,

(a) μ_n does not depend on n, and

(b) $r_{n,m}$ depends only on the difference $n - m$. Thus, the autocorrelation sequence is single index.

Conditions (a) and (b), which are necessary for the stationarity of the sequence (ξ_n), are generally not sufficient. If (a) and (b) hold true, we say that (ξ_n) is *wide-sense* (WS) *stationary*. Notice that wide-sense stationarity is exceedingly simpler to check for than stationarity. Thus, it is always expedient to verify whether wide-sense stationarity is enough to prove the properties that are needed. In practice, although stationarity is usually invoked, wide-sense stationarity is often sufficient.

Markov chains

For any real sequence $(\xi_n)_{n=0}^{\infty}$ of independent RVs, we have, for every n,

$$F_{\xi_n|\xi_{n-1},\xi_{n-2},\ldots,\xi_0}(x_n|x_{n-1}, x_{n-2}, \ldots, x_0) = F_{\xi_n}(x_n), \tag{2.40}$$

where $F_{\xi_n|\xi_{n-1},\xi_{n-2},\ldots,\xi_0}(\ldots)$ denotes the conditional cdf of the random variable ξ_n given all the "past" RVs $\xi_{n-1}, \xi_{n-2}, \ldots, \xi_0$. Equation (2.40) reflects the fact that ξ_n is independent of the past of the sequence. A first-step generalization of (2.40) can be obtained by considering a situation in which, for any n,

$$F_{\xi_n|\xi_{n-1},\xi_{n-2},\ldots,\xi_0}(x_n|x_{n-1},x_{n-2}, \ldots, x_0) = F_{\xi_n|\xi_{n-1}}(x_n|x_{n-1}); \tag{2.41}$$

that is, ξ_n depends on the past only through ξ_{n-1}.

When (2.41) holds, $(\xi_n)_{n=0}^{\infty}$ is called a *discrete-time (first-order) Markov* process. If in addition every ξ_n can take only a finite number of possible values, say the integers $1, 2, \ldots, q$, (ξ_n) is called a *(finite) Markov chain*, and the values of ξ_n are referred to as the *states* of the chain. To specify a Markov chain, it suffices to give, for all times $n \geq 0$ and $j, k = 1, 2, \ldots, q$, the probabilities $P\{\xi_n = j\}$ and $P\{\xi_{n+1} = k | \xi_n = j\}$. The latter quantity is the probability that the process will move to state k at time $n + 1$ given that it was in state j at time n. This probability is called the *one-step transition probability function* of the Markov chain.

A Markov chain is said to be *homogeneous* (or to have stationary transition probabilities) if the transition probabilities $P\{\xi_{l+m} = k | \xi_l = j\}$ depend only on the time difference m and not on l. We then call

$$p_{jk}^{(m)} \triangleq P\{\xi_{l+m} = k | \xi_l = j\}, \qquad l \geq 0, \qquad m \geq 1, \qquad j, k = 1, 2, \ldots, q, \tag{2.42}$$

the *m-step transition probability function* of the homogeneous Markov chain $(\xi_n)_{n=0}^{\infty}$. In other words, $p_{jk}^{(m)}$ is the conditional probability that the chain, being in state j at time l, will move to state k after m time instants. The one-step transition probabilities $p_{jk}^{(1)}$ are simply written p_{jk}:

$$p_{jk} \triangleq P\{\xi_{l+1} = k | \xi_l = j\}, \qquad l \geq 0, \qquad j,k = 1, 2, \ldots, q, \tag{2.43}$$

These transition probabilities can be arranged into a $q \times q$ *transition matrix* **P:**

$$\mathbf{P} \triangleq \begin{bmatrix} p_{11} & p_{12} & \cdots & p_{1q} \\ p_{21} & p_{22} & \cdots & p_{2q} \\ & & \cdots & \\ p_{q1} & p_{q2} & \cdots & p_{qq} \end{bmatrix}. \qquad (2.44)$$

The elements of $\mathbf{P}$ satisfy the conditions

$$p_{jk} \geq 0, \qquad j, k = 1, \ldots, q, \qquad (2.45)$$

and

$$\sum_{k=1}^{q} p_{jk} = 1, \qquad j = 1, 2, \ldots, q \qquad (2.46)$$

(i.e., the sum of the entries in each row of $\mathbf{P}$ equals 1). Any square matrix that satisfies the conditions (2.45) and (2.46) is called a *stochastic matrix* or a *Markov matrix*.

For a homogeneous Markov chain $(\xi_n)_{n=0}^{\infty}$, let $w_k^{(n)}$ denote the unconditional probability that state k occurs at time n; that is,

$$w_k^{(n)} \triangleq P\{\xi_n = k\}, \qquad k = 1, 2, \ldots q. \qquad (2.47)$$

The row q vector of probabilities $w_k^{(n)}$,

$$\mathbf{w}^{(n)} \triangleq [w_1^{(n)} \, w_2^{(n)} \, \ldots \, w_q^{(n)}], \qquad (2.48)$$

is called the *state distribution vector* at time n. With $\mathbf{w}^{(0)}$ denoting the initial state distribution vector, at time 1 we have

$$w_k^{(1)} = \sum_{j=1}^{q} w_j^{(0)} p_{jk}, \qquad k = 1, \ldots, q, \qquad (2.49)$$

or, in matrix notation,

$$\mathbf{w}^{(1)} = \mathbf{w}^{(0)}\mathbf{P}. \qquad (2.50)$$

Similarly, we obtain

$$\begin{aligned} \mathbf{w}^{(2)} &= \mathbf{w}^{(1)}\mathbf{P} \\ &= \mathbf{w}^{(0)}\mathbf{P}^2, \end{aligned} \qquad (2.51)$$

and, iterating the process,

$$\begin{aligned} \mathbf{w}^{(m)} &= \mathbf{w}^{(m-1)}\mathbf{P} \\ &= \mathbf{w}^{(0)}\mathbf{P}^m. \end{aligned} \qquad (2.52)$$

More generally, we have

$$\mathbf{w}^{(l+m)} = \mathbf{w}^{(l)}\mathbf{P}^m. \qquad (2.53)$$

Equation (2.53) shows that the elements of $\mathbf{P}^m$ are the m-step transition probabilities defined in (2.42). This proves in particular that a homogeneous Markov chain $(\xi_n)_{n=0}^{\infty}$ is completely described by the initial state distribution vector $\mathbf{w}^{(0)}$ and the transition probability matrix $\mathbf{P}$. In fact, these are sufficient to evaluate $P\{\xi_n = j\}$ for every $n \geq 0$ and $j = 1, 2, \ldots, q$, which, in addition to the elements of $\mathbf{P}$, characterize a Markov chain.

Consider now the behavior of the state distribution vector $\mathbf{w}^{(n)}$ as $n \to \infty$. If the limit

$$\mathbf{w} \triangleq \lim_{n \to \infty} \mathbf{w}^{(n)} \tag{2.54}$$

exists, the vector $\mathbf{w}$ is called the *stationary distribution vector*. A homogeneous Markov chain such that $\mathbf{w}$ exists is called *regular*. It can be proved that a homogeneous Markov chain is regular if and only if all eigenvalues of $\mathbf{P}$ having unit magnitude are identically 1. If, in addition, 1 is a simple eigenvalue of $\mathbf{P}$ (i.e., a simple root of the characteristic polynomial of $\mathbf{P}$), then the Markov chain is said to be *fully regular*. For a fully regular chain, the stationary state distribution vector is independent of the initial state distribution vector and can be evaluated by finding the unique solution of the system of homogeneous linear equations

$$\mathbf{w}\,\mathbf{P} = \mathbf{w} \tag{2.55}$$

subject to the constraints

$$\sum_{k=1}^{q} w_k = 1, \qquad w_k \geq 0, \qquad k = 1, 2, \ldots, q. \tag{2.56}$$

Also, for a fully regular chain the limiting transition probability matrix

$$\mathbf{P}^{\infty} \triangleq \lim_{n \to \infty} \mathbf{P}^n \tag{2.57}$$

exists and has identical rows, each row being the stationary distribution vector $\mathbf{w}$:

$$\mathbf{P}^{\infty} = \begin{bmatrix} \mathbf{w} \\ \mathbf{w} \\ \vdots \\ \mathbf{w} \end{bmatrix} q. \tag{2.58}$$

The existence of $\mathbf{P}^{\infty}$ in the form (2.58) is a sufficient, as well as necessary, condition for a homogeneous Markov chain to be fully regular.

Example 2.5

Consider a digital communication system transmitting the symbols 0 and 1. Each symbol passes through several blocks. At each block there is a probability $1 - p$, $p < \frac{1}{2}$, that the symbol at the output is equal to that at the input. Let ξ_0 denote the symbol entering the first block and ξ_n, $n \geq 1$, the symbol at the output of the nth block of the system. The sequence $\xi_0, \xi_1, \xi_2, \ldots$, is then a homogeneous Markov chain with transition probability matrix

$$\mathbf{P} = \begin{bmatrix} 1 - p & p \\ p & 1 - p \end{bmatrix}.$$

The n-step transition probability matrix is

$$\mathbf{P}^n = \begin{bmatrix} \frac{1}{2} + \frac{1}{2}(1 - 2p)^n & \frac{1}{2} - \frac{1}{2}(1 - 2p)^n \\ \frac{1}{2} - \frac{1}{2}(1 - 2p)^n & \frac{1}{2} + \frac{1}{2}(1 - 2p)^n \end{bmatrix}.$$

The eigenvalues of $\mathbf{P}$ are 1 and $1 + 2p$, so for $p \neq 0$ the chain is fully regular. Its stationary distribution vector is $\mathbf{w} = [\frac{1}{2}\ \frac{1}{2}]$, and

$$\mathbf{P}^{\infty} = \begin{bmatrix} \frac{1}{2} & \frac{1}{2} \\ \frac{1}{2} & \frac{1}{2} \end{bmatrix},$$

which shows that as $n \to \infty$ a symbol entering the system has the same probability $\frac{1}{2}$ of being received correctly or incorrectly. $\square$

Shift-register state sequences

An important special case of a Markov chain arises from the consideration of a stationary random sequence (α_n) of independent random variables, each taking on values in the set $\{a_1, a_2, \ldots, a_M\}$ with probabilities $p_k \triangleq P\{\alpha_n = a_k\}$, $k = 1, \ldots, M$, and of the sequence $(\sigma_n)_{n=0}^{\infty}$, with

$$\sigma_n \triangleq (\alpha_{n-1}, \ldots, \alpha_{n-L}). \tag{2.59}$$

If we consider an L-stage shift register fed with the sequence (α_n) (Fig. 2.7), σ_n represents the content (the "state") of the shift register at time n (i.e., when α_n is present at its input). For this reason, (σ_n) is called a *shift-register state sequence*. Each σ_n can take on M^L values, and it can be verified that (σ_n) forms a Markov chain. To derive its transition matrix, we shall first introduce a suitable ordering for the values of σ_n. This can be done in a natural way by first ordering the elements of the set $\{a_1, a_2, \ldots, a_M\}$ (a simple way to do this is to stipulate that a_i precedes a_j if and only if $i < j$) and then inducing the following "lexicographical" order among the L-tuples $a_{j_1}, a_{j_2}, \ldots, a_{j_L}$:

$$(a_{j_1}, a_{j_2}, \ldots, a_{j_L}) \text{ precedes } (a_{i_1}, a_{i_2}, \ldots, a_{i_L})$$

$$\text{if and only if} \begin{cases} j_1 < i_1, & \text{or} \\ j_1 = i_1 & \text{and} \quad j_2 < i_2, \quad \text{or} \\ j_1 = i_1, j_2 = i_2 & \text{and} \quad j_3 < i_3, \quad \text{etc.} \end{cases} \tag{2.60}$$

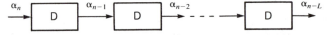

Figure 2.7 Generating a shift-register sequence.

Once the state set has been ordered according to the rule (2.60), each state can be represented by an integer number expressing its position in the ordered set. Thus, if i represents the state $(a_{i_1}, a_{i_2}, \ldots, a_{i_L})$ and j represents the state $(a_{j_1}, a_{j_2}, \ldots, a_{j_L})$, the one-step transition probability p_{ij} is given by

$$\begin{aligned} p_{ij} &\triangleq P\{\sigma_n = (a_{j_1}, \ldots, a_{j_L}) | \sigma_{n-1} = (a_{i_1}, \ldots, a_{i_L})\} \\ &= P\{\alpha_{n-1} = a_{j_1}, \ldots, \alpha_{n-L} = a_{j_L} | \alpha_{n-2} = a_{i_1}, \ldots, \alpha_{n-L-1} = a_{i_L}\} \\ &= p_{j_1} \delta_{i_1 j_2} \delta_{i_2 j_3} \ldots \delta_{i_{L-1} j_L}, \end{aligned} \tag{2.61}$$

where δ_{ij} denotes the Kronecker symbol ($\delta_{ii} = 1$ and $\delta_{ij} = 0$ for $i \neq j$).

Example 2.6

Assume $M = 2$, $a_1 = 0$, $a_2 = 1$, and $L = 3$. The shift register has eight states, whose lexicographically ordered set is $\{(000), (001), (010), (011), (100), (101), (110), (111)\}$. The transition probability matrix of the corresponding Markov chain is

$$\mathbf{P} = \begin{bmatrix} & (000) & (001) & (010) & (011) & (100) & (101) & (110) & (111) \\ & p_1 & 0 & 0 & 0 & p_2 & 0 & 0 & 0 \\ & p_1 & 0 & 0 & 0 & p_2 & 0 & 0 & 0 \\ & 0 & p_1 & 0 & 0 & 0 & p_2 & 0 & 0 \\ & 0 & p_1 & 0 & 0 & 0 & p_2 & 0 & 0 \\ & 0 & 0 & p_1 & 0 & 0 & 0 & p_2 & 0 \\ & 0 & 0 & p_1 & 0 & 0 & 0 & p_2 & 0 \\ & 0 & 0 & 0 & p_1 & 0 & 0 & 0 & p_2 \\ & 0 & 0 & 0 & p_1 & 0 & 0 & 0 & p_2 \end{bmatrix} \begin{matrix} (000) \\ (001) \\ (010) \\ (011) \\ (100) \\ (101) \\ (110) \\ (111) \end{matrix} \qquad (2.62)$$

As one can see, from state (xyz) the shift register can move only to states (wxy), with probability p_1 if $w = 0$ and p_2 if $w = 1$. $\square$

Consider now the m-step transition probabilities. These are the elements of the matrix $\mathbf{P}^m$. Since the shift register has L stages, its content after time $n + m$, $m \geq L$, is independent of its content at time n. Consequently, the states σ_{n+m}, $m \geq L$, are independent of σ_n; so

$$P\{\sigma_{n+m} = (a_{j_1}, \ldots, a_{j_L}) | \sigma_n = (a_{i_1}, \ldots, a_{i_L})\}$$
$$= \prod_{l=1}^{L} p_{j_l}, \qquad m \geq L. \qquad (2.63)$$

Thus, $\mathbf{P}^L = \mathbf{P}^{L+1} = \cdots$, and $\mathbf{P}^L$ has identical rows. We can write

$$\mathbf{P}^L = \mathbf{P}^\infty, \qquad (2.64)$$

which shows, in particular, that the shift-register state sequence defined in (2.59) is a fully regular Markov chain.

Example 2.6 (continued)

We have, by direct computation from (2.62) or using (2.63), that $\mathbf{P}^3$ has the structure (2.58), with $\mathbf{w}$, the stationary distribution vector, being equal to

$$\mathbf{w} = [p_1^3, \ p_1^2 p_2, \ p_1^2 p_2, \ p_1 p_2^2, \ p_1^2 p_2, \ p_1 p_2^2, \ p_1 p_2^2, \ p_2^3]. \ \square \qquad (2.65)$$

2.2.2 Continuous-Time Processes

A *continuous-time random process* (or *random continuous signal*) is a family of real or complex signals $\xi(t)$ defined on some probability space. At any N-tuple of times t_1, $t_2, \ldots, t_N$, the quantities $\xi(t_1), \xi(t_2), \ldots, \xi(t_N)$ are RVs. Consequently, a random process can be described by providing the joint distribution functions of the N RVs $\xi(t_1), \xi(t_2), \ldots, \xi(t_N)$ for all integers N and N-tuples of time instants.

A random process is called *stationary* if for every N, for any N-tuple t_1, t_2, $\ldots, t_N$, and for every real τ, the N-tuples of RVs $\xi(t_1), \xi(t_2), \ldots, \xi(t_N)$ and $\xi(t_1 + \tau), \xi(t_2 + \tau), \ldots, \xi(t_N + \tau)$ are identically distributed. Stated in another way, a stationary random process is one whose probabilistic properties do not depend on the time origin. Thus, for any given τ the processes $\xi(t)$ and $\xi(t + \tau)$ are identically distributed.

The *mean* of the process $\xi(t)$ is the deterministic signal

$$\mu(t) \triangleq E[\xi(t)]. \qquad (2.66)$$

The *autocorrelation* of $\xi(t)$ is the function

$$R_\xi(t_1, t_2) \triangleq E[\xi(t_1)\xi^*(t_2)]. \qquad (2.67)$$

For a stationary process,

(a) $\mu(t)$ does not depend on time, and

(b) $R_\xi(t_1, t_2)$ depends only on the difference $t_1 - t_2$. Consequently, we can write

$$R_\xi(t_1 - t_2) \triangleq E[\xi(t_1)\xi^*(t_2)]. \qquad (2.68)$$

Conditions (a) and (b) are generally not sufficient for the stationarity of $\xi(t)$. If (a) and (b) hold true, we say that $\xi(t)$ is *wide-sense* (WS) *stationary*. A random process $\xi(t)$ is called *cyclostationary with period T* if its probabilistic properties do not change when the time origin is shifted by a multiple of T; that is, we consider $\xi(t + kT)$, k an integer, instead of $\xi(t)$. *Wide-sense cyclostationarity* can also be defined as follows: $\xi(t)$ is WS cyclostationary if

(a) $\mu(t)$ is a periodic function of time with period T, and

(b) the autocorrelation of the process has the property

$$R_\xi(t + \tau, t) = R_\xi(t + \tau + kT, t + kT), \qquad (2.69)$$

k any integer. Equation (2.69) can be interpreted by saying that $R_\xi(t + \tau, t)$, when considered as a function of t, is periodic with period T.

Example 2.7

Consider a WS stationary sequence (α_n) of random variables with correlation (r_n) and the deterministic finite-energy signal $s(t)$. The random signal

$$\xi(t) \triangleq \sum_{l=-\infty}^{\infty} \alpha_l s(t - lT)$$

is WS cyclostationary with period T. In fact

$$\mu(t) = E[\alpha_l] \sum_{l=-\infty}^{\infty} s(t - lT)$$

is periodic with period T. Moreover,

$$R_\xi(t_1, t_2) = \sum_{l=-\infty}^{\infty} \sum_{m=-\infty}^{\infty} E[\alpha_l \alpha_m^*] s(t_1 - lT) s^*(t_2 - mT)$$

$$= \sum_{l=-\infty}^{\infty} \sum_{m=-\infty}^{\infty} r_{l-m} s(t_1 - lT) s^*(t_2 - mT)$$

and it can be verified that (2.69) holds. $\square$

Some important properties of stationary and cyclostationary processes are the following:

(a) If a stationary (cyclostationary) process is passed through a stable time-invariant system, it retains its stationarity (cyclostationarity).

(b) The sum of two stationary processes is a stationary process. The sum of a cyclostationary process and a stationary process is a cyclostationary process.

(c) Let $\xi(t)$ be a WS cyclostationary process with period T, and let $\eta(t)$ denote the randomly translated process

$$\eta(t) \triangleq \xi(t + \vartheta), \tag{2.70}$$

where ϑ is a random variable statistically independent of $\xi(t)$ and uniformly distributed in the interval $(0, T)$. Then the process $\eta(t)$ is WS stationary.

Gaussian processes

A real random process $\xi(t)$ is called *Gaussian* if, for any given time instant t, $\xi(t)$ is a Gaussian random variable. Formally, $\xi(t)$ is a Gaussian process if for any N-tuple $t_1, t_2, \ldots, t_N$ of time instants, N any integer ≥ 1, the row N vector of random variables $\boldsymbol{\xi} \triangleq [\xi(t_1), \xi(t_2), \ldots, \xi(t_N)]$ has a Gaussian distribution, that is, a probability density function of the form

$$f_{\boldsymbol{\xi}}(\mathbf{x}) = \frac{1}{(2\pi)^{N/2}(\det \boldsymbol{\Lambda}_{\boldsymbol{\xi}})^{1/2}} \exp\left[-\frac{1}{2}(\mathbf{x} - \boldsymbol{\mu}_{\boldsymbol{\xi}}) \boldsymbol{\Lambda}_{\boldsymbol{\xi}}^{-1}(\mathbf{x} - \boldsymbol{\mu}_{\boldsymbol{\xi}})'\right], \tag{2.71}$$

where $\boldsymbol{\mu}_{\boldsymbol{\xi}}$ is the mean vector

$$\boldsymbol{\mu}_{\boldsymbol{\xi}} \triangleq \mathrm{E}[\boldsymbol{\xi}] = (\mathrm{E}[\xi(t_1)], \mathrm{E}[\xi(t_2)], \cdots, \mathrm{E}[\xi(t_N)]) \tag{2.72}$$

and $\boldsymbol{\Lambda}_{\boldsymbol{\xi}}$ is the $N \times N$ *covariance matrix*

$$\boldsymbol{\Lambda}_{\boldsymbol{\xi}} \triangleq \mathrm{E}[(\boldsymbol{\xi} - \boldsymbol{\mu}_{\boldsymbol{\xi}})'(\boldsymbol{\xi} - \boldsymbol{\mu}_{\boldsymbol{\xi}})]. \tag{2.73}$$

Now let $\xi(t)$ be a *complex* random process, and let

$$\xi(t) = \xi_P(t) + j\xi_Q(t), \tag{2.74}$$

where $\xi_P(t)$, $\xi_Q(t)$ are real processes.

This process is called Gaussian if the joint distribution of $\xi_P(t_1)$, $\xi_P(t_2)$, . . . , $\xi_P(t_N)$, $\xi_Q(t_1)$, $\xi_Q(t_2)$, . . . , $\xi_Q(t_N)$ is $2N$-dimensional Gaussian for any N-tuple of time instants and for any integer $N \geq 1$.

Gaussian processes have the following properties:

(a) The output of any linear system whose input is a Gaussian process is still Gaussian.

(b) Let $\xi(t)$ be a WS stationary *real* Gaussian process. Then $\xi(t)$ is stationary.

(c) Let $\xi(t)$ be a WS stationary *complex* Gaussian process. Then $\xi(t)$ is stationary if and only if the average $\mathrm{E}[\xi(t_1)\xi(t_2)]$ is a function only of the time difference $t_1 - t_2$.

Property (c) deserves some comments. Wide-sense stationarity of $\xi(t)$ implies that $\mathrm{E}[\xi(t)\xi^*(s)]$ is a function of $t - s$, and $\mathrm{E}[\xi(t)]$ is a constant. For the stationarity, one must show that $\mathrm{E}[\xi_P(t)\xi_P(s)]$, $\mathrm{E}[\xi_P(t)\xi_Q(s)]$, $\mathrm{E}[\xi_Q(t)\xi_Q(s)]$ all depend only on the difference

$t - s$. But this is equivalent to showing that $E[\xi(t)\xi^*(s)]$ and $E[\xi(t)\xi(s)]$ depend only on $t - s$. To verify the latter property, it is sometimes useful to apply *Grettenberg's theorem* (Grettenberg, 1965). It states that for a complex Gaussian process $\xi(t)$ with mean zero we have $E[\xi(t)\xi(s)] = 0$ if and only if, for all $0 \leq \vartheta \leq 2\pi$, the processes $\xi(t)$ and $e^{j\vartheta}\xi(t)$ are identically distributed; that is, $\xi(t)$ is invariant under phase rotations.

2.3 SPECTRAL ANALYSIS OF DETERMINISTIC AND RANDOM SIGNALS

In the representation of signals in the Fourier transform domain, one associates with each frequency f a measure of its contribution to the signal. This representation is particularly useful when the signal is transformed by a linear time-invariant system, because in this case each of the frequency components of the signal is independently weighted by the system transfer function, according to the rule (2.31) (it holds for discrete and continuous signals). In this section we extend this concept to the spectral analysis of certain energetic quantities that one may want to associate to a given signal, such as its energy or its power (to be suitably defined). Specifically, assume that, for a given signal ξ, either discrete or continuous, deterministic or random, we have defined a nonnegative energetic quantity Π_ξ. The *density spectrum* of Π_ξ is a frequency function, say $V_\xi(f)$, carrying information regarding how much of Π_ξ is associated with each frequency f. The function $V_\xi(f)$ is nonnegative, and the two following properties hold:

(a) The integral of $V_\xi(f)$ gives Π_ξ:

$$\Pi_\xi = \int_I V_\xi(f)\, df. \tag{2.75}$$

(b) Let Π_η be the same energetic quantity defined at the output of a linear, time-invariant system with transfer function $H(f)$ and input $\xi(t)$. Then

$$\Pi_\eta = \int_I |H(f)|^2 V_\xi(f)\, df. \tag{2.76}$$

In (2.75) and (2.76), $I = (-\infty, \infty)$ if ξ is a continuous-time signal, and $I = (-\frac{1}{2}, \frac{1}{2})$ if ξ is a discrete-time signal.

Let us now specialize this general definition to some cases of practical interest.

Energy density spectrum: *Continuous deterministic signals*

Given a continuous deterministic signal $x(t)$, we define its *energy* as the quantity

$$\mathscr{E}_x \triangleq \int_{-\infty}^{\infty} |x(t)|^2\, dt, \tag{2.77}$$

provided that the integral in (2.77) is finite. In the transform domain, the energy of a signal $x(t)$ whose Fourier transform is $X(f)$ can be expressed in the form

$$\mathscr{E}_x = \int_{-\infty}^{\infty} |X(f)|^2\, df. \tag{2.78}$$

Equality (2.78) is a special case of *Parseval's theorem*. It states that for two signals $x_1(t)$, $x_2(t)$ with Fourier transforms $X_1(f)$, $X_2(f)$, respectively, the following holds:

$$\int_{-\infty}^{\infty} x_1(t)x_2^*(t)\,dt = \int_{-\infty}^{\infty} X_1(f)X_2^*(f)\,df. \tag{2.79}$$

The function

$$\mathcal{G}_x(f) \triangleq |X(f)|^2 \tag{2.80}$$

is the *energy (density) spectrum* of $x(t)$. It is easily seen that with this definition both (2.75) and (2.76) hold.

Power density spectrum: *Continuous deterministic signals*

For a continuous deterministic signal $x(t)$ whose energy is not finite, define its *average power* as the quantity

$$\mathcal{P}_x \triangleq \lim_{a \to \infty} \frac{1}{a} \int_{-a/2}^{a/2} |x(t)|^2\,dt, \tag{2.81}$$

provided that this limit exists. If we define the truncated signal

$$x_a(t) \triangleq \begin{cases} x(t), & -\dfrac{a}{2} < t < \dfrac{a}{2}, \\ 0, & \text{elsewhere}, \end{cases} \tag{2.82}$$

the average power of $x(t)$ can be written

$$\mathcal{P}_x = \lim_{a \to \infty} \frac{1}{a} \mathcal{E}_a, \tag{2.83}$$

where $\mathcal{E}_a$ denotes the energy of $x_a(t)$. Hence, for an aperiodic signal $x(t)$, we define its *power (density) spectrum* as the function

$$\mathcal{G}_x(f) \triangleq \lim_{a \to \infty} \frac{1}{a} |X_a(f)|^2, \tag{2.84}$$

where $|X_a(f)|^2$ is the energy spectrum of the truncated signal (2.82).

Average power density spectrum: *Discrete stationary random signals*

Consider a WS stationary random sequence (ξ_n) with autocorrelation (r_n). Its *average power* is defined as

$$\mathcal{P}_\xi \triangleq E\{|\xi_n|^2\}. \tag{2.85}$$

The *(average) power (density) spectrum* $\mathcal{G}_\xi(f)$ of (ξ_n) is the Fourier transform of the autocorrelation sequence (r_n); that is,

$$\mathcal{G}_\xi(f) = \sum_{n=-\infty}^{\infty} r_n e^{-j2\pi fn}, \qquad |f| \le \frac{1}{2}. \tag{2.86}$$

Let us show that with this definition (2.75) holds. We have

$$\int_{-1/2}^{1/2} \mathcal{G}_\xi(f)\, df = \sum_{n=-\infty}^{\infty} r_n \int_{-1/2}^{1/2} e^{-j2\pi fn}\, df = r_0, \tag{2.87}$$

and r_0 equals $E\{|\xi_n|^2\}$ because of (2.39) and the assumption of WS stationarity. Property (2.76) can be proved similarly.

Average power density spectrum: *Continuous stationary random signals*

Let $\xi(t)$ be a WS stationary continuous random process with autocorrelation $R_\xi(\tau)$. Its *average power* is defined as

$$\mathcal{P}_\xi \triangleq E\{|\xi(t)|^2\}. \tag{2.88}$$

The *(average) power (density) spectrum* $\mathcal{G}_\xi(f)$ of $\xi(t)$ is the Fourier transform of the autocorrelation function $R_\xi(\tau)$:

$$\mathcal{G}_\xi(f) = \int_{-\infty}^{\infty} R_\xi(\tau) e^{-j2\pi f\tau}\, d\tau. \tag{2.89}$$

In this situation, (2.76) takes the form

$$\mathcal{P}_\eta = \int_{-\infty}^{\infty} |H(f)|^2\, \mathcal{G}_\xi(f)\, df, \tag{2.90}$$

where $\eta(t)$ is the response of a linear time-invariant system with transfer function $H(f)$ to the input $\xi(t)$.

Example 2.8 White noise

A process with autocorrelation function

$$R_\xi(\tau) = \frac{N_0}{2} \delta(\tau) \tag{2.91}$$

has a power spectrum

$$\mathcal{G}_\xi(f) = \frac{N_0}{2}, \qquad -\infty < f < \infty. \tag{2.92}$$

Such a process is called a *white noise*. In practice, this process is not realizable, as its power $\mathcal{P}_\xi$ is not finite. However, this process can be very useful in instances where the actual process has an approximately constant spectral density over a frequency range wider than the bandwidth of the system under consideration. At the output of a linear time-invariant system with transfer function $H(f)$ we get the average power

$$\mathcal{P}_\eta = \frac{N_0}{2} \int_{-\infty}^{\infty} |H(f)|^2\, df, \tag{2.93}$$

which is finite provided that the integral in the RHS converges. In this situation, it is customary to define the *equivalent noise bandwidth* of the system as

$$B_{eq} \triangleq \frac{1}{2} \frac{\displaystyle\int_{-\infty}^{\infty} |H(f)|^2\, df}{\displaystyle\max_f |H(f)|^2}. \tag{2.94}$$

Notice the presence of the factor $\frac{1}{2}$ in (2.94), which can be interpreted by saying that the bandwidth is only defined for positive frequencies. This convention is assumed throughout the book for every possible definition of the bandwidth of a signal or a system. For linear systems with a real impulse response, $|H(f)|$ is an even function. Hence, the factor $\frac{1}{2}$ can be omitted in the RHS of (2.94) and the integration carried out from 0 to ∞. With definition (2.94), the power at the output of a linear, time-invariant system with equivalent noise bandwidth B_{eq} and whose input is a white noise with power spectral density $N_0/2$ turns out to be

$$\mathcal{P}_\eta = N_0 \cdot B_{eq} \cdot \max_f |H(f)|^2. \tag{2.95}$$

Equation (2.95) shows that B_{eq} can be interpreted as the bandwidth of a system with a rectangular transfer function, whose amplitude squared is $\max_f |H(f)|.^2$ Figure 2.8 illustrates this fact for a low-pass and a bandpass system.

For example, the low-pass Butterworth filters defined in Example 2.4 have an equivalent noise bandwidth

$$B_{eq} = f_c \frac{\pi/(2n)}{\sin[\pi/(2n)]}. \tag{2.96}$$

From (2.96) it is easily seen that, as $n \to \infty$, $B_{eq} \to f_c$, f_c being the cutoff frequency of the filter. $\square$

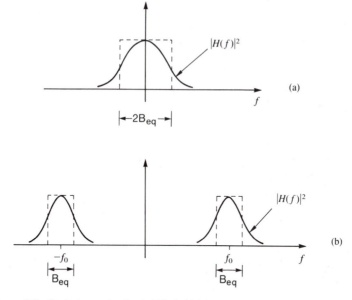

Figure 2.8 Equivalent noise bandwidth for (a) low-pass systems, and (b) bandpass systems.

Average power density spectrum: *Continuous nonstationary random signals*

Consider now a nonstationary continuous random process $\xi(t)$. Clearly, definition (2.88) cannot still be valid because in general $E\{|\xi(t)|^2\}$ varies with time. In this situation, the definition of average power that should be used is

$$\mathscr{P}_\xi \triangleq \lim_{a \to \infty} \frac{1}{a} \int_{-a/2}^{a/2} E\left\{|\xi(t)|^2\right\} dt, \tag{2.97}$$

that is, the time average of the mean value of the instantaneous power $|\xi(t)|^2$. With this definition a spectral density function that satisfies both (2.75) and (2.76) can also be defined for nonstationary processes, provided that we restrict our attention to an appropriate subclass of processes. This subclass is that of *harmonizable processes* (Loève, 1963, pp. 474–477). Roughly speaking, a process is harmonizable if we can define its Fourier transform:

$$\Xi(f) \triangleq \int_{-\infty}^{\infty} \xi(t) \, e^{-j2\pi ft} \, dt. \tag{2.98}$$

Equation (2.98) defines a new random process in the variable f. In certain cases, a proper interpretation of (2.98) requires some care. In fact, (2.98) is an equality in the sense of distribution theory (i.e., it becomes an equality if a linear operator is applied to both sides and the order of integrations is reversed in the RHS). Incidentally, this is the correct way to interpret equalities like

$$\delta(t) = \int_{-\infty}^{\infty} e^{j2\pi ft} \, df.$$

Harmonizable processes are a first-step generalization of WS stationary random processes. It has been shown (Cambanis and Liu, 1970) that, under some mild conditions, any random process obtained at the output of a linear system is harmonizable. The system may be randomly time variant and the input process need not be stationary, or even harmonizable.

For a harmonizable process $\xi(t)$, the power spectrum can be obtained as follows. Compute first the function

$$\Gamma_\xi(f_1, f_2) \triangleq E\left[\Xi(f_1) \, \Xi^*(f_2)\right]. \tag{2.99}$$

Consider then the bisector of the plane (f_1, f_2) and the line masses of $\Gamma_\xi(f_1, f_2)$ located on it. The distribution of these line masses provides us with a function $\mathscr{G}_\xi(f)$, the power spectrum of $\xi(t)$. Specifically, if $\Gamma_\xi(f_1, f_2)$ can be written in the form

$$\Gamma_\xi(f_1, f_2) = \mathscr{G}_\xi(f_1) \, \delta(f_1 - f_2) + \Delta_\xi(f_1, f_2), \tag{2.100}$$

where $\Delta_\xi(f_1, f_2)$ has no line masses located on the bisector $f_1 = f_2$, then $\mathscr{G}_\xi(f)$ is the required spectrum. [It may happen that $\mathscr{G}_\xi(f)$ is identically zero; in this case the process has finite energy.] Using (2.98), it can easily be seen that $\Gamma_\xi(f_1, f_2)$ can be written in a form equivalent to (2.99):

$$\Gamma_\xi(f_1, f_2) = \int_{-\infty}^{\infty} \int_{-\infty}^{\infty} R_\xi(t_1, t_2) \, e^{-j2\pi(f_1 t_1 - f_2 t_2)} \, dt_1 \, dt_2. \tag{2.101}$$

Equation (2.101) shows that $\Gamma_\xi(f_1, f_2)$ is the two-dimensional Fourier transform of the autocorrelation function of the process $\xi(t)$. This is equivalent to saying that $R_\xi(t_1, t_2)$ is the inverse Fourier transform of $\Gamma_\xi(f_1, f_2)$:

$$R_\xi(t_1, t_2) = \int_{-\infty}^{\infty}\int_{-\infty}^{\infty} \Gamma_\xi(f_1, f_2)\, e^{j2\pi(f_1 t_1 - f_2 t_2)}\, df_1\, df_2. \tag{2.102}$$

Example 2.9

Let $\xi(t)$ be WS stationary. Its autocorrelation function only depends on $t_1 - t_2$. Thus, from (2.101) we get

$$\begin{aligned}
\Gamma_\xi(f_1, f_2) &= \int_{-\infty}^{\infty}\int_{-\infty}^{\infty} R_\xi(t_1 - t_2)e^{-j2\pi[f_1(t_1-t_2)+(f_1-f_2)t_2]}\, dt_1\, dt_2 \\
&= \int_{-\infty}^{\infty} R_\xi(\tau)\, e^{-j2\pi f_1 \tau}\, d\tau \cdot \delta(f_1 - f_2),
\end{aligned} \tag{2.103}$$

which is consistent with (2.89) (as it should be). Also notice that, using (2.102), one sees that $R_\xi(t_1, t_2)$ depends on the difference $t_1 - t_2$ only if $\Gamma_\xi(f_1, f_2)$ has the form

$$\Gamma_\xi(f_1, f_2) = \mathcal{G}_\xi(f_1)\, \delta(f_1 - f_2) \tag{2.104}$$

(see Fig. 2.9). $\square$

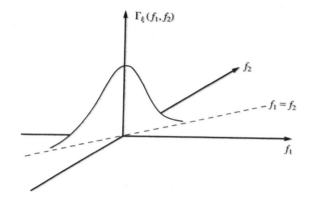

Figure 2.9 The function $\Gamma_\xi(f_1, f_2)$ for a wide-sense stationary process.

Example 2.10

Let $\xi(t)$ be a WS cyclostationary process with period T. Using the property (2.69), it is seen that $R_\xi(t_1, t_2)$ can be expanded in the Fourier series

$$R_\xi(t + \tau, t) = \sum_{n=-\infty}^{\infty} g_n(\tau)\, e^{jn2\pi t/T}, \tag{2.105}$$

where

$$g_n(\tau) \triangleq \frac{1}{T}\int_{-T/2}^{T/2} R_\xi(t + \tau, t)\, e^{-jn2\pi t/T}\, dt. \tag{2.106}$$

Using (2.101), we get

$$\begin{aligned}
\Gamma_\xi(f_1, f_2) &= \sum_{n=-\infty}^{\infty}\int_{-\infty}^{\infty}\int_{-\infty}^{\infty} g_n(\tau)\, e^{jn2\pi t/T}\, e^{-j2\pi[(f_1-f_2)t+f_1\tau]}\, dt\, d\tau \\
&= \sum_{n=-\infty}^{\infty} G_n(f_1)\, \delta\left(f_1 - f_2 - \frac{n}{T}\right),
\end{aligned} \tag{2.107}$$

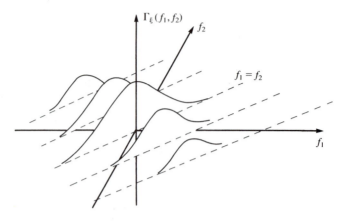

Figure 2.10 The function $\Gamma_\xi(f_1, f_2)$ for a wide-sense cyclostationary process.

where $G_n(\cdot)$ is the Fourier transform of $g_n(\cdot)$, $-\infty < n < \infty$. Equation (2.107) shows that $\Gamma_\xi(f_1, f_2)$ consists of line masses located on the lines $f_1 = f_2 + n/T$, $-\infty < n < \infty$, which are parallel to the bisector of the plane (f_1, f_2). This situation is shown qualitatively in Fig. 2.10.

The power spectrum of $\xi(t)$ is then

$$\mathcal{G}_\xi(f) = G_0(f). \tag{2.108}$$

It can also be shown that the power spectrum (2.108) can be obtained by considering the WS stationary process (2.70) and using (2.89). $\square$

2.3.1 Spectral Analysis of Random Digital Signals

In Chapter 4, devoted to the transmission of digital information using continuous signals, the following random process will be considered:

$$\xi(t) = \sum_{n=-\infty}^{\infty} s(t - nT; \alpha_n, \sigma_n). \tag{2.109}$$

It is called a *digitally modulated random signal*, or for short a *digital signal*. The random sequence (α_n) is stationary, made up of discrete RVs, and will be referred to as the sequence of *source symbols*. The sequence (σ_n) is a stationary sequence of discrete random variables referred to as the *states of the modulator*. The random waveforms $s(t; \alpha_n, \sigma_n)$ take values in a set $\{s_i(t)\}_{i=1}^{M}$ of deterministic, finite-energy signals. They are output sequentially by the modulator, one every T seconds, in accordance with the values of the source symbols and the modulator states.

Several special cases of (2.109) are of interest. If the modulator states σ_n do not appear in (2.109), the modulator is called *memoryless*, and we have

$$\xi(t) = \sum_{n=-\infty}^{\infty} s(t - nT; \alpha_n). \tag{2.110}$$

If in addition

$$s(t; \alpha_n) = \alpha_n s(t), \tag{2.111}$$

that is, the waveforms of the set $\{s_i(t)\}_{i=1}^M$ are scalar multiples of one and the same signal $s(t)$, the modulator is called *linear*, and we have

$$\xi(t) = \sum_{n=-\infty}^{\infty} \alpha_n s(t - nT). \tag{2.112}$$

Here we shall evaluate the power density spectrum of the signal (2.109), which is generally nonstationary. The Fourier transform of $\xi(t)$ is given by

$$\Xi(f) = \sum_{n=-\infty}^{\infty} S(f; \alpha_n, \sigma_n) e^{-j2\pi fnT}, \tag{2.113}$$

where $S(f; \alpha_n, \sigma_n)$, the Fourier transform of $s(t; \alpha_n, \sigma_n)$, takes values in the set $\{S_i(f)\}_{i=1}^M$, with $S_i(f) \triangleq \mathcal{F}[s_i(t)]$, $i = 1, 2, \ldots, M$. Thus, from (2.99) we get

$$\Gamma_\xi(f_1, f_2) = \sum_{m=-\infty}^{\infty} \sum_{n=-\infty}^{\infty} E\{S(f_1; \alpha_m, \sigma_m) S^*(f_2; \alpha_n, \sigma_n)\} e^{-j2\pi(f_1 m - f_2 n)T}$$

$$\tag{2.114}$$

$$= \sum_{l=-\infty}^{\infty} \sum_{n=-\infty}^{\infty} E\{S(f_1; \alpha_{n+l}, \sigma_{n+l}) S^*(f_2; \alpha_n, \sigma_n)\} e^{-j2\pi f_1 lT} e^{-j2\pi(f_1-f_2)nT}.$$

As the sequences (α_n), (σ_n) are stationary, the average in the last line of (2.114) depends only on l and not on n. Thus, recalling the equality (see, e.g., Jones, 1966, p. 135)

$$\sum_{m=-\infty}^{\infty} e^{-j2\pi mwz} = \frac{1}{w} \sum_{m=-\infty}^{\infty} \delta\left(z - \frac{m}{w}\right), \tag{2.115}$$

we obtain

$$\Gamma_\xi(f_1, f_2) = \frac{1}{T} \sum_{l=-\infty}^{\infty} E\{S(f_1; \alpha_{n+l}, \sigma_{n+l}) S^*(f_2; \alpha_n, \sigma_n)\} \cdot e^{-j2\pi f_1 lT}$$

$$\tag{2.116}$$

$$\cdot \sum_{m=-\infty}^{\infty} \delta\left(f_1 - f_2 - \frac{m}{T}\right).$$

Compare now (2.116) with (2.100). It is apparent that the power spectrum of $\xi(t)$ is given by

$$\mathcal{G}_\xi(f) = \frac{1}{T} \sum_{l=-\infty}^{\infty} G_l(f) e^{-j2\pi f lT}, \tag{2.117}$$

where

$$G_l(f) \triangleq E\{S(f; \alpha_{n+l}, \sigma_{n+l}) S^*(f; \alpha_n, \sigma_n)\}. \tag{2.118}$$

It is customary, in the computation of spectral densities, to separate their continuous part from their discrete part (line spectrum). This can be done in our situation by defining

$$G_\infty(f) \triangleq \lim_{l \to \infty} G_l(f)$$

$$\tag{2.119}$$

$$= |E\{S(f; \alpha_n, \sigma_n)\}|^2$$

(this does not depend on n because of stationarity) and rewriting (2.117) in the form

$$\mathcal{G}_\xi(f) = \frac{1}{T} \sum_{l=-\infty}^{\infty} [G_l(f) - G_\infty(f)] e^{-j2\pi f \, lT}$$

$$+ \frac{1}{T^2} G_\infty(f) \sum_{l=-\infty}^{\infty} \delta\left(f - \frac{l}{T}\right), \tag{2.120}$$

where (2.115) was used again. The second term in the RHS of (2.120) is a line spectrum with lines spaced $1/T$ hertz (Hz) apart. The first term is line-free if $G_l(f) - G_\infty(f)$ tends to zero fast enough as $l \to \infty$ for all f. We shall assume in the following that this is the case.

Equation (2.120) can be rewritten in a slightly different form by observing that, from definition (2.118), it follows that

$$G_{-l}(f) = G_l^*(f). \tag{2.121}$$

Thus, denoting by $\mathcal{G}_\xi^{(c)}(f)$ and $\mathcal{G}_\xi^{(d)}(f)$ the continuous and the discrete part of the power spectrum, respectively, we finally get

$$\mathcal{G}_\xi(f) = \mathcal{G}_\xi^{(c)}(f) + \mathcal{G}_\xi^{(d)}(f),$$

$$\mathcal{G}_\xi^{(c)}(f) = \frac{2}{T} \Re\left\{ \sum_{l=0}^{\infty} [G_l(f) - G_\infty(f)] e^{-j2\pi f \, lT} \right\}$$

$$- \frac{1}{T}[G_0(f) - G_\infty(f)], \tag{2.122}$$

$$\mathcal{G}_\xi^{(d)}(f) = \frac{1}{T^2} G_\infty(f) \sum_{l=-\infty}^{\infty} \delta\left(f - \frac{l}{T}\right).$$

We shall now proceed to specialize (2.122) to a number of cases of practical interest.

Linearly modulated digital signals

When the modulator is linear, that is, (2.109) reduces to (2.112), from (2.118) we get, with $S(f)$ denoting the Fourier transform of $s(t)$,

$$G_l(f) = \mathrm{E}\{\alpha_{n+l}\,\alpha_n^*\}\,|S(f)|^2. \tag{2.123}$$

If

$$\mathrm{E}\{\alpha_n\} = \mu \tag{2.124}$$

and

$$\mathrm{E}\{\alpha_l\,\alpha_m^*\} = \sigma_\alpha^2\,\rho_{l-m} + |\mu|^2 \tag{2.125}$$

with $\rho_0 = 1$ and $\rho_\infty = 0$, the power spectrum of $\xi(t)$ is given by

$$\mathcal{G}_\xi(f) = \mathcal{G}_\xi^{(c)}(f) + \mathcal{G}_\xi^{(d)}(f),$$

$$\mathcal{G}_\xi^{(c)}(f) = \frac{\sigma_\alpha^2}{T}|S(f)|^2\left\{2\,\mathcal{R}\sum_{l=0}^{\infty}\rho_l\,e^{-j2\pi f\,lT} - 1\right\}, \tag{2.126}$$

$$\mathcal{G}_\xi^{(d)}(f) = \frac{|\mu|^2}{T^2}|S(f)|^2\sum_{l=-\infty}^{\infty}\delta\left(f - \frac{l}{T}\right).$$

It is seen from (2.126) that $\mu = 0$ is a sufficient condition for $\mathcal{G}_\xi(f)$ to have no lines in its spectrum. This condition is not necessary, however, because when $\mu \neq 0$ a spectral line will show up at $f = l/T$ only if $S(l/T) \neq 0$.

When the random variables α_n are uncorrelated (i.e., $\rho_l = \delta_{0,l}$), we get from (2.126)

$$\mathcal{G}_\xi^{(c)}(f) = \frac{\sigma_\alpha^2}{T}|S(f)|^2. \tag{2.127}$$

Notice from (2.126) the two factors that separately influence the shape of $\mathcal{G}_\xi(f)$. The first is the waveform $s(t)$ through its energy spectrum. The second is the correlation of the sequence (α_n), which appears in the bracketed factor of (2.126). If this factor is rewritten as

$$2\mathcal{R}\sum_{l=0}^{\infty}\rho_l e^{-j2\pi f\,lT} - 1 = \sum_{l=\infty}^{\infty}\rho_l e^{-j2\pi f\,lT}$$

it is seen that it turns out to be the Fourier transform of the sequence (ρ_n). In practice, the fact that $\mathcal{G}_\xi(f)$ depends on two independent factors provides a degree of freedom that can be used to shape the signal spectrum. Indeed, a given spectrum can be obtained by choosing appropriately the waveform $s(t)$, or the correlation of (α_n), or both. This property will be exploited later in Chapter 9.

Example 2.11

Perhaps the simplest way to introduce correlation in a discrete sequence is to pass it through a linear system. Thus, let (β_n) denote a sequence of iid RVs with $E\beta_n = 0$ and $E|\beta_n|^2 = 1$, and let (α_n) denote a new sequence with

$$\alpha_n = \sum_m h_m \beta_{n-m}, \tag{2.128}$$

where (h_n) is the impulse response of a linear, time-invariant system. In this situation a simple computation shows that

$$E\{\alpha_{n+l}\alpha_n^*\} = \sum_m h_{m+l} h_m^*. \tag{2.129}$$

Thus, the power spectrum of (2.112), when (α_n) is as in (2.128), is

$$\mathcal{G}_\xi(f) = \frac{1}{T}|S(f)|^2|H(fT)|^2, \tag{2.130}$$

where $H(f)$ is the transfer function of the discrete linear system:

$$H(f) \triangleq \sum_m h_m e^{-jm2\pi f}. \tag{2.131}$$

It is immediately apparent from (2.130) that the same power spectrum for $\xi(t)$ could be obtained by using, instead of (α_n), the sequence (β_n) and a signal whose Fourier transform is $S(f)H(fT)$. $\square$

Nonlinearly modulated digital signals

We shall now consider the computation of the power spectrum of the digital signal $\xi(t)$ expressed by (2.109) when the sequence (σ_n) is assumed to have a special structure. In particular, we assume that (α_n) is an iid sequence, and that (σ_n) depends on (α_n) as follows:

$$\sigma_{n+1} = g(\alpha_n, \sigma_n), \tag{2.132}$$

where $g(\cdot, \cdot)$ is a completely known deterministic function. Equation (2.132) describes in which state the encoder is forced to move at time $n + 1$, when at time n it was in state σ_n, and the source symbol is α_n. The modulator uses the value of the pair α_n, σ_n to choose the waveform $s(t; \alpha_n, \sigma_n)$ from the set $\{s_i(t)\}_{i=1}^{M}$, which is then output sequentially.

For this model of a digital signal to be fully specified, it is sufficient to provide the function $g(\cdot, \cdot)$ and the mapping between pairs α_n, σ_n and waveforms of the set $\{s_i(t)\}_{i=1}^{M}$. We assume, hereafter, that σ_n takes on the q values $\Sigma_1, \Sigma_2, \ldots, \Sigma_q$ and α_n takes on the L values $a_1, a_2, \ldots, a_L$ (q and L both finite). Thus, our description of $\xi(t)$ can take a tabular form, by providing two $L \times q$ tables whose rows are labeled $a_1, a_2, \ldots, a_L$ and whose columns are labeled $\Sigma_1, \Sigma_2, \ldots, \Sigma_q$. In the first table we display the waveforms corresponding to the pairs (a_i, Σ_j), and in the second the values of $g(a_i, \Sigma_j)$. An equivalent representation is in the form of a *state diagram*. This is a directed graph consisting of q vertexes, each representing one state; an oriented branch is drawn from state Σ_i to state Σ_j if and only if there is a source symbol a_k such that $g(a_k, \Sigma_i) = \Sigma_j$. The branch is then labeled by a_k and by the waveform, say $s_l(t)$, corresponding to the pair (a_k, Σ_i) (see Fig. 2.11). Before proceeding further, we shall provide some examples of nonlinearly modulated digital signals and their representations.

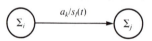

Figure 2.11 Element of the state-diagram representation of a modulated digital signal.

Example 2.12 "Bipolar-encoded" digital signal

The modulator has $q = 2$ states, say Σ_+ and Σ_-, and the source is binary; that is, $\alpha_n \in \{0, 1\}$. The modulator responds to a source symbol 0 with a zero waveform and to a source symbol 1 with the waveform $s(t)$ or $-s(t)$, according to whether its state is Σ_+ or Σ_-, respectively. Source symbols 1 make the modulator change its state. The tabular and state-diagram representations of this signal are provided in Fig. 2.12. $\square$

Example 2.13 "Miller-encoded" digital signal

The modulator has $M = 4$, $q = 4$ states, and the source is binary. Figure 2.13 describes this digital signal. $\square$

σ_n α_n	Σ_+	Σ_-
0	0	0
1	$s(t)$	$-s(t)$

σ_n α_n	Σ_+	Σ_-
0	Σ_+	Σ_-
1	Σ_-	Σ_+

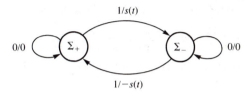

Figure 2.12 Representation of the bipolar-encoded digital signal: (a) Tabular form; (b) state diagram.

σ_n α_n	Σ_1	Σ_2	Σ_3	Σ_4
0	$s_4(t)$	$s_4(t)$	$s_1(t)$	$s_1(t)$
1	$s_2(t)$	$s_3(t)$	$s_2(t)$	$s_3(t)$

σ_n α_n	Σ_1	Σ_2	Σ_3	Σ_4
0	Σ_4	Σ_4	Σ_1	Σ_1
1	Σ_2	Σ_3	Σ_2	Σ_3

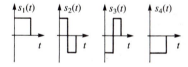

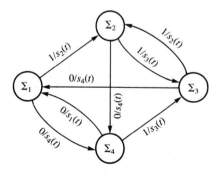

Figure 2.13 Representation of the Miller-encoded digital signal: (a) Tabular form; (b) state diagram.

Example 2.14 **"Ungerboeck-encoded" digital signal**

The modulator has $M = 8$, $q = 4$ states, and the source is quaternary. The available signals are

$$s_i(t) = \exp\left\{ j\left[2\pi f_0 t + (i - 1)\frac{\pi}{4} \right] \right\}, \qquad i = 1, \ldots ,8, \qquad 0 \le t < T, \qquad f_0 T >> 1.$$

Figure 2.14 describes the resulting digital signal. $\square$

For our future computations, the following quantities must be defined:

(a) The *state transition matrices* $\mathbf{E}_k$, $k = 1, 2, \ldots , L$, which are the $q \times q$ matrices whose entry $[\mathbf{E}_k]_{ij}$ is equal to 1 if $g(a_k, \Sigma_i) = \Sigma_j$, and zero otherwise. In clarification, the matrix $\mathbf{E}_k$ has a 1 in row i and column j if the source symbol a_k

σ_n / α_n	Σ_1	Σ_2	Σ_3	Σ_4
0	$s_1(t)$	$s_2(t)$	$s_3(t)$	$s_4(t)$
1	$s_5(t)$	$s_6(t)$	$s_7(t)$	$s_8(t)$
2	$s_3(t)$	$s_4(t)$	$s_1(t)$	$s_2(t)$
3	$s_7(t)$	$s_8(t)$	$s_5(t)$	$s_6(t)$

$$s_i(t) = \exp\left\{ j\left[2\pi f_0 t + (i - 1)\frac{\pi}{4} \right] \right\}$$
$$0 \le t < T$$
$$f_0 T >> 1$$

σ_n / α_n	Σ_1	Σ_2	Σ_3	Σ_4
0	Σ_1	Σ_3	Σ_1	Σ_3
1	Σ_1	Σ_3	Σ_1	Σ_3
2	Σ_2	Σ_4	Σ_2	Σ_4
3	Σ_2	Σ_4	Σ_2	Σ_4

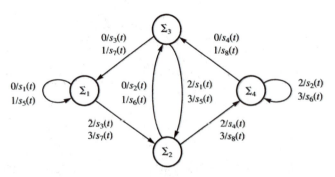

Figure 2.14 Representation of an Ungerboeck-encoded digital signal: (a) Tabular form; (b) state diagram.

forces a transition of the modulator from state Σ_i to state Σ_j. Otherwise, it has a zero.

(b) The row q vectors $\mathbf{s}_k(f)$, $k = 1, 2, \ldots, L$, whose q entries are the Fourier transforms of the waveforms of the set $\{s_i(t)\}_{i=1}^M$, according to the rule $[\mathbf{s}_k(f)]_i = \mathcal{F}[s(t; a_k, \Sigma_i)]$. That is, $\mathbf{s}_k(f)$ includes the amplitude spectra of the modulator waveforms corresponding to the source symbol a_k for the different modulator states.

Example 2.12 (continued)

In this case, letting $a_1 = 0$ and $a_2 = 1$, we have

$$\mathbf{E}_1 = \begin{bmatrix} 1 & 0 \\ 0 & 1 \end{bmatrix},$$

$$\mathbf{E}_2 = \begin{bmatrix} 0 & 1 \\ 1 & 0 \end{bmatrix},$$

and

$$\mathbf{s}_1(f) = [0 \quad 0],$$
$$\mathbf{s}_2(f) = S(f)[1 \quad -1],$$

where $S(f)$ is the Fourier transform of $s(t)$. $\square$

Example 2.13 (continued)

In this case, letting $a_1 = 0$, $a_2 = 1$, we have

$$\mathbf{E}_1 = \begin{bmatrix} 0 & 0 & 0 & 1 \\ 0 & 0 & 0 & 1 \\ 1 & 0 & 0 & 0 \\ 1 & 0 & 0 & 0 \end{bmatrix},$$

$$\mathbf{E}_2 = \begin{bmatrix} 0 & 1 & 0 & 0 \\ 0 & 0 & 1 & 0 \\ 0 & 1 & 0 & 0 \\ 0 & 0 & 1 & 0 \end{bmatrix},$$

and

$$\mathbf{s}_1(f) = S(f)[-1-z \quad -1-z \quad 1+z \quad 1+z],$$
$$\mathbf{s}_2(f) = S(f)[1-z \quad -1+z \quad 1-z \quad -1+z],$$

where

$$S(f) \triangleq \frac{T}{2} \frac{\sin \pi f T/2}{\pi f T/2}$$

and

$$z \triangleq e^{-j\pi f T}. \square$$

Example 2.14 (continued)

In this case, letting $a_i = i - 1$, $i = 1, 2, 3, 4$, we have

$$\mathbf{E}_1 = \mathbf{E}_2 = \begin{bmatrix} 1 & 0 & 0 & 0 \\ 0 & 0 & 1 & 0 \\ 1 & 0 & 0 & 0 \\ 0 & 0 & 1 & 0 \end{bmatrix},$$

$$\mathbf{E}_3 = \mathbf{E}_4 = \begin{bmatrix} 0 & 1 & 0 & 0 \\ 0 & 0 & 0 & 1 \\ 0 & 1 & 0 & 0 \\ 0 & 0 & 0 & 1 \end{bmatrix},$$

and

$$\mathbf{s}_1(f) = S(f) \begin{bmatrix} w^0 & w^1 & w^2 & w^3 \end{bmatrix},$$
$$\mathbf{s}_2(f) = S(f) \begin{bmatrix} w^4 & w^5 & w^6 & w^7 \end{bmatrix},$$
$$\mathbf{s}_3(f) = S(f) \begin{bmatrix} w^2 & w^3 & w^0 & w^1 \end{bmatrix},$$
$$\mathbf{s}_4(f) = S(f) \begin{bmatrix} w^6 & w^7 & w^4 & w^5 \end{bmatrix},$$

where

$$S(f) \triangleq T \frac{\sin \pi(f - f_0)T}{\pi(f - f_0)T},$$

and

$$w \triangleq e^{j\pi/4}. \quad \square$$

We want now to evaluate the power spectrum of the digital signal (2.109). The assumption that (α_n) is an iid sequence and relationship (2.132) imply that the state sequence (σ_n) is a *homogeneous Markov chain*. In fact, the probability that the encoder is in a given state at time $n + 1$ depends only on the state σ_n and the symbol α_n, and not on the preceding states $\sigma_{n-1}, \sigma_{n-2}, \ldots$ The transition matrix of this chain has entries

$$[\mathbf{P}]_{ij} \triangleq P\{\sigma_{n+1} = \Sigma_j | \sigma_n = \Sigma_i\}$$

$$= P\{g(\alpha_n, \sigma_n) = \Sigma_j | \sigma_n = \Sigma_i\}$$

$$= \sum_{k=1}^{L} P\{g(\alpha_n, \sigma_n) = \Sigma_j | \alpha_n = a_k, \sigma_n = \Sigma_i\} P\{\alpha_n = a_k\} \qquad (2.133)$$

$$= \sum_{k=1}^{L} p_k [\mathbf{E}_k]_{ij},$$

where, as already defined,

$$p_k \triangleq P\{\alpha_n = a_k\}, \qquad k = 1, 2, \ldots, L. \qquad (2.134)$$

We assume that the Markov chain is fully regular, and that its starting time is $n = -\infty$. This implies that, for any finite n, $\mathbf{w}^{(n)} = \mathbf{w}$. Thus, the transition matrix $\mathbf{P}$ provides a complete characterization of the sequence of modulator states; in particular, the stationary state probabilities

$$w_i \triangleq P\{\sigma_n = \Sigma_i\} \tag{2.135}$$

are obtained as the entries of vector $\mathbf{w}$ computed from (2.55) and (2.56).

Let us now define four quantities that play an important role in the expression of the power spectral density we are seeking. The first is the average value, taken over the source symbols, of the vectors $\mathbf{s}_k(f)$:

$$\mathbf{c}_2(f) \triangleq \sum_{k=1}^{L} p_k \mathbf{s}_k(f). \tag{2.136}$$

The ith component of $\mathbf{c}_2(f)$ is then the average amplitude spectrum of the waveforms available to the modulator when it is in state Σ_i.

The second is the q vector $\mathbf{c}_1(f)$ whose jth component is the average amplitude spectrum of the waveforms that, when output by the modulator, force it to state Σ_j. This jth component of $\mathbf{c}_1(f)$ is then given by

$$[\mathbf{c}_1(f)]_j \triangleq \sum_{k=1}^{L} \sum_{i=1}^{q} p_k w_i \, [\mathbf{E}_k]_{ij} \, \mathcal{F}[s(t; a_k, \Sigma_i)] \tag{2.137}$$

(recall from the definition of $\mathbf{E}_k$ that $[\mathbf{E}_k]_{ij} = 1$ only if the source symbol a_k takes the modulator from state Σ_i to state Σ_j). If we define the $q \times q$ diagonal matrix

$$\mathbf{D} \triangleq \text{diag}\,(w_1, w_2, \ldots, w_q) \tag{2.138}$$

we have from (2.137)

$$\mathbf{c}_1(f) = \sum_{k=1}^{L} p_k \mathbf{s}_k(f) \mathbf{D} \mathbf{E}_k. \tag{2.139}$$

Our third quantity is the average amplitude spectrum of the waveforms available from the modulator:

$$
\begin{aligned}
\mu(f) &\triangleq E\{\mathcal{F}[s(t; \alpha_n, \sigma_n)]\} \\
&= \sum_{k=1}^{L} \sum_{i=1}^{q} p_k w_i \mathcal{F}[s(t; a_k, \Sigma_i)] \\
&= \sum_{k=1}^{L} p_k \mathbf{w} \mathbf{s}'_k(f) \\
&= \mathbf{w} \mathbf{c}'_2(f).
\end{aligned}
\tag{2.140}
$$

Finally, the fourth quantity of interest is the average energy spectrum of the waveforms available from the modulator:

$$
\begin{aligned}
c_0(f) &\triangleq \sum_{k=1}^{L} \sum_{i=1}^{q} p_k w_i |\mathcal{F}[s(t; a_k, \Sigma_i)]|^2 \\
&= \sum_{k=1}^{L} p_k \mathbf{s}_k^*(f) \, \mathbf{D} \mathbf{s}'_k(f).
\end{aligned}
\tag{2.141}
$$

Before proceeding further, we evaluate these four quantities in a few examples.

Sec. 2.3   Spectral Analysis of Deterministic and Random Signals   **39**

Example 2.12 (continued)

Assuming that the source symbols 0 and 1 are equally likely, we have

$$\mathbf{P} = \mathbf{P}^\infty = \begin{bmatrix} \frac{1}{2} & \frac{1}{2} \\ \frac{1}{2} & \frac{1}{2} \end{bmatrix},$$

so that

$$\mathbf{w} = [\frac{1}{2} \quad \frac{1}{2}].$$

Moreover,

$$\mathbf{c}_2(f) = S(f) [\quad \frac{1}{2} \quad -\frac{1}{2}],$$
$$\mathbf{c}_1(f) = S(f) [-\frac{1}{4} \quad \frac{1}{4}],$$
$$\mu(f) = 0,$$

and

$$c_0(f) = \frac{1}{2}|S(f)|^2. \quad \square$$

Example 2.13 (continued)

Assuming that the source symbols 0 and 1 are equally likely, we have

$$\mathbf{P} = \frac{1}{2} \begin{bmatrix} 0 & 1 & 0 & 1 \\ 0 & 0 & 1 & 1 \\ 1 & 1 & 0 & 0 \\ 1 & 0 & 1 & 0 \end{bmatrix}$$

$$\mathbf{P}^\infty = \frac{1}{4} \begin{bmatrix} 1 & 1 & 1 & 1 \\ 1 & 1 & 1 & 1 \\ 1 & 1 & 1 & 1 \\ 1 & 1 & 1 & 1 \end{bmatrix}.$$

Thus, $\mathbf{w} = [\frac{1}{4} \ \frac{1}{4} \ \frac{1}{4} \ \frac{1}{4}]$, and

$$\mathbf{c}_2(f) = S(f) [-z \quad -1 \quad 1 \quad z],$$
$$\mathbf{c}_1(f) = \frac{1}{4} S(f) [1 + z \quad 1 - z \quad -1 + z \quad -1 - z],$$
$$\mu(f) = 0,$$
$$c_0(f) = 2|S(f)|^2. \quad \square$$

Example 2.14 (continued)

Assuming that the source symbols 0, 1, 2, 3, are equally likely, we have

$$\mathbf{P} = \frac{1}{2} \begin{bmatrix} 1 & 1 & 0 & 0 \\ 0 & 0 & 1 & 1 \\ 1 & 1 & 0 & 0 \\ 0 & 0 & 1 & 1 \end{bmatrix}$$

and

$$\mathbf{P}^\infty = \mathbf{P}^2 = \frac{1}{4} \begin{bmatrix} 1 & 1 & 1 & 1 \\ 1 & 1 & 1 & 1 \\ 1 & 1 & 1 & 1 \\ 1 & 1 & 1 & 1 \end{bmatrix}.$$

Moreover,

$$\mathbf{c}_2(f) = \mathbf{0},$$
$$\mathbf{c}_1(f) = \mathbf{0},$$
$$\mu(f) = 0,$$

and

$$c_0(f) = |S(f)|^2. \ \square$$

Consider now the computation of the power spectrum. This will be undertaken by applying (2.122). From (2.118) we have, for $l > 0$,

$$G_l(f) = \sum_{h=1}^{L} \sum_{k=1}^{L} \sum_{i=1}^{q} \sum_{j=1}^{q} S(f; a_h, \Sigma_j) S^*(f; a_k, \Sigma_i)$$
$$\cdot P\{\alpha_{n+l} = a_h, \ \alpha_n = a_k, \ \sigma_{n+l} = \Sigma_j, \ \sigma_n = \Sigma_i\}. \tag{2.142}$$

The probabilities appearing in (2.142) can be put in the form

$$P\{\alpha_{n+l} = a_h, \ \alpha_n = a_k, \ \sigma_{n+l} = \Sigma_j, \ \sigma_n = \Sigma_i\}$$
$$= P\{\alpha_{n+l} = a_h, \ \sigma_{n+l} = \Sigma_j \,|\, \alpha_n = a_k, \ \sigma_n = \Sigma_i\} \cdot p_k w_i. \tag{2.143}$$

As the source symbols are independent, we have

$$P\{\alpha_{n+l} = a_h, \ \sigma_{n+l} = \Sigma_j \,|\, \alpha_n = a_k, \ \sigma_n = \Sigma_i\}$$
$$= p_h P\{\sigma_{n+l} = \Sigma_j \,|\, \alpha_n = a_k, \ \sigma_n = \Sigma_i\}$$
$$= p_h \sum_{m=1}^{q} P\{\sigma_{n+l} = \Sigma_j \,|\, \sigma_{n+1} = \Sigma_m, \ \alpha_n = a_k, \ \sigma_n = \Sigma_i\}$$
$$\cdot P\{\sigma_{n+1} = \Sigma_m \,|\, \alpha_n = a_k, \ \sigma_n = \Sigma_i\} \tag{2.144}$$
$$= p_h \sum_{m=1}^{q} P\{\sigma_{n+l} = \Sigma_j \,|\, \sigma_{n+1} = \Sigma_m\}[\mathbf{E}_k]_{im}$$
$$= p_h \sum_{m=1}^{q} [\mathbf{P}^{l-1}]_{mj} [\mathbf{E}_k]_{im}.$$

For $l = 0$, we get instead

$$G_0(f) = \sum_{k=1}^{L} \sum_{i=1}^{q} |S(f; a_k, \Sigma_i)|^2 p_k w_i. \tag{2.145}$$

By combining (2.142) to (2.145), we have

$$G_l(f) = \begin{cases} \displaystyle\sum_{h=1}^{L} \sum_{k=1}^{L} p_h p_k \, \mathbf{s}_k^*(f) \mathbf{D} \mathbf{E}_k \mathbf{P}^{l-1} \mathbf{s}_h'(f), & l > 0, \\[3ex] \displaystyle\sum_{h=1}^{L} p_h \mathbf{s}_h^*(f) \mathbf{D} \mathbf{s}_h'(f), & l = 0, \end{cases} \tag{2.146}$$

and, using definitions (2.136) to (2.141),

Sec. 2.3   Spectral Analysis of Deterministic and Random Signals   **41**

$$G_l(f) = \begin{cases} \mathbf{c}_1^*(f)\mathbf{P}^{l-1}\mathbf{c}_2'(f), & l > 0, \\ c_0(f), & l = 0. \end{cases} \tag{2.147}$$

Also, from (2.119) and the definition (2.140) of $\mu(f)$, we get

$$G_\infty(f) = |\mu(f)|^2, \tag{2.148}$$

or, equivalently, if (2.147) is used,

$$G_\infty(f) = \mathbf{c}_1^*(f)\mathbf{P}^\infty \mathbf{c}_2'(f). \tag{2.149}$$

In conclusion, the continuous and discrete parts of the power spectrum of our digital signal are given by

$$\mathcal{G}_\xi^{(c)}(f) = \frac{1}{T}[c_0(f) - |\mu(f)|^2] + \frac{2}{T}\mathcal{R}[\mathbf{c}_1^*(f)\mathbf{\Lambda}(f)\mathbf{c}_2'(f)] \tag{2.150}$$

and

$$\mathcal{G}_\xi^{(d)}(f) = \frac{1}{T^2}|\mu(f)|^2 \sum_{l=-\infty}^{\infty} \delta\left(f - \frac{l}{T}\right), \tag{2.151}$$

where

$$\mathbf{\Lambda}(f) \triangleq \sum_{l=1}^{\infty} [\mathbf{P}^{l-1} - \mathbf{P}^\infty]e^{-j2\pi f \, lT}. \tag{2.152}$$

Whenever there exists a finite N such that $\mathbf{P}^N = \mathbf{P}^\infty$ [e.g., when (σ_n) is a shift-register state sequence], $\mathbf{\Lambda}(f)$ involves a finite number of terms, and its computation is straightforward. If such an N does not exist, we need a technique to evaluate the RHS of (2.152). Observe that, from the equality $\mathbf{P}^k\mathbf{P}^\infty = \mathbf{P}^\infty$, we have

$$\begin{aligned} \mathbf{P}^k - \mathbf{P}^\infty &= (\mathbf{I} - \mathbf{P}^\infty)(\mathbf{P}^k - \mathbf{P}^\infty) \\ &= (\mathbf{I} - \mathbf{P}^\infty)(\mathbf{P} - \mathbf{P}^\infty)^k \end{aligned} \tag{2.153}$$

for all $k > 0$. Thus

$$\begin{aligned} \mathbf{\Lambda}(f) &= e^{-j2\pi f T} \sum_{l=0}^{\infty} [\mathbf{P}^l - \mathbf{P}^\infty]e^{-j2\pi f \, lT} \\ &= e^{-j2\pi f T}(\mathbf{I} - \mathbf{P}^\infty)\sum_{l=0}^{\infty}(\mathbf{P} - \mathbf{P}^\infty)^l e^{-j2\pi f \, lT} \\ &= (\mathbf{I} - \mathbf{P}^\infty)[e^{j2\pi f T}\mathbf{I} - (\mathbf{P} - \mathbf{P}^\infty)]^{-1}, \end{aligned} \tag{2.154}$$

where the last equality holds because the matrix $(\mathbf{P} - \mathbf{P}^\infty)$ has all its eigenvalues with magnitude less than 1 (see Cariolaro and Tronca, 1974, for a proof).

It is seen from (2.154) that the matrix $\mathbf{\Lambda}(f)$, necessary to evaluate the RHS of (2.150), can be computed for each value of f by inverting a $q \times q$ matrix. This procedure is computationally inefficient because, if the spectrum value is needed for several f, many matrix inversions must be performed. For a more efficient technique, observe that $\mathbf{\Lambda}(f)$ is an analytic function of the matrix

$$\mathbf{A} \triangleq \mathbf{P} - \mathbf{P}^\infty, \tag{2.155}$$

so that $\Lambda(f)$ can be written in the form of a polynomial in $\mathbf{A}$ whose coefficients depend on f, say,

$$\Lambda(f) = (\mathbf{I} - \mathbf{P}^\infty) \sum_{i=0}^{K-1} \beta_i(f)\mathbf{A}^i. \tag{2.156}$$

The expansion (2.156) is not unique, unless we restrict K to take on its minimum possible value (i.e., the degree of the minimal polynomial of $\mathbf{A}$). Here we assume that the reader is familiar with the basic results of matrix calculus, as summarized in Appendix B. In this situation, equating the RHS of (2.154) and (2.156), we get

$$[e^{j2\pi f T}\mathbf{I} - \mathbf{A}] \sum_{i=0}^{K-1} \beta_i(f)\mathbf{A}^i - \mathbf{I} = \mathbf{0}. \tag{2.157}$$

As the LHS of (2.157) is a polynomial of $\mathbf{A}$ having degree K, its coefficients must be proportional to those of the minimal polynomial of $\mathbf{A}$. Denoting this minimal polynomial by

$$\Delta(\lambda) = \sum_{i=0}^{K} \delta_i \lambda^i, \qquad \delta_K = 1, \tag{2.158}$$

and equating the coefficients of $\mathbf{A}^i$, $i = 0, \ldots, K$, in (2.157) and in the identity

$$\sum_{i=0}^{K} \delta_i \mathbf{A}^i = \mathbf{0} \tag{2.159}$$

we get the coefficients $\beta_i(f)$, $i = 0, \ldots, K - 1$, needed to compute $\Lambda(f)$ according to (2.156). This procedure allows one to express $\Lambda(f)$ as a closed-form function of f, which can be computed for each value of f with modest computational effort.

Although the use of the minimal polynomial of $\mathbf{A}$ to obtain the representation (2.156) leads to the most economical way to compute the spectrum, every polynomial $\Delta(\lambda)$ such that (2.159) holds can be used instead of the minimal polynomial. In particular, the use of the characteristic polynomial of $\mathbf{A}$ (which has degree q) leads to a relatively simple computational algorithm (due to Faddeev and as first applied to this problem by Cariolaro and Tronca, 1974). According to this technique, $\Lambda(f)$ can be given the form

$$\Lambda(f) = (\mathbf{I} - \mathbf{P}^\infty) \frac{1}{\Delta(e^{j2\pi f T})} \mathbf{B}(e^{j2\pi f T}), \tag{2.160}$$

where $\Delta(\lambda)$ is now the characteristic polynomial of $\mathbf{A}$, and $\mathbf{B}(\cdot)$ is a $q \times q$ matrix polynomial:

$$\mathbf{B}(\lambda) = \lambda^{q-1}\mathbf{B}_0 + \lambda^{q-2}\mathbf{B}_1 + \cdots + \mathbf{B}_{q-1}. \tag{2.161}$$

The polynomials $\mathbf{B}(\cdot)$ and $\Delta(\cdot)$ can be computed simultaneously using the following recursive algorithm (Gantmacher, 1959). Starting with

$$\delta_q = 1,$$
$$\mathbf{B}_0 = \mathbf{I},$$

let

$$\mathbf{Q}_k = \mathbf{AB}_{k-1},$$

$$\delta_{q-k} = -\frac{1}{k}\operatorname{tr}\mathbf{Q}_k, \tag{2.162}$$

$$\mathbf{B}_k = \mathbf{Q}_k + \delta_{q-k}\mathbf{I},$$

for $k = 1, 2, \ldots, q$. At the final step, $\mathbf{B}_q$ must be equal to the null matrix, and $\delta_0 = 0$, because the matrix $\mathbf{A}$ has a zero eigenvalue.

Example 2.12 (continued)

Since $\mathbf{P} = \mathbf{P}^\infty$, from (2.152) we have

$$\Lambda(f) = (\mathbf{I} - \mathbf{P}^\infty)e^{-j2\pi f T}$$

$$= \begin{bmatrix} \frac{1}{2} & -\frac{1}{2} \\ -\frac{1}{2} & \frac{1}{2} \end{bmatrix} e^{-j2\pi f T}$$

so that

$$\mathcal{G}_\xi(f) \equiv \mathcal{G}_\xi^{(c)}(f) = \frac{1}{2T}|S(f)|^2(1 - \cos 2\pi f T). \quad \square \tag{2.163}$$

Example 2.13 (continued)

We have

$$\mathbf{A} \triangleq \mathbf{P} - \mathbf{P}^\infty = \tfrac{1}{4}\begin{bmatrix} -1 & 1 & -1 & 1 \\ -1 & -1 & 1 & 1 \\ 1 & 1 & -1 & -1 \\ 1 & -1 & 1 & -1 \end{bmatrix}.$$

Application of the Faddeev algorithm gives

$$\delta_4 = \delta_3 = 1, \qquad \delta_2 = \tfrac{1}{2}, \qquad \delta_1 = \delta_0 = 0,$$

$$\mathbf{B}_1 = \tfrac{1}{4}\begin{bmatrix} 3 & 1 & -1 & 1 \\ -1 & 3 & 1 & 1 \\ 1 & 1 & 3 & -1 \\ 1 & -1 & 1 & 3 \end{bmatrix}, \qquad \mathbf{B}_2 = \tfrac{1}{4}\begin{bmatrix} 1 & 0 & 0 & 1 \\ 0 & 1 & 1 & 0 \\ 0 & 1 & 1 & 0 \\ 1 & 0 & 0 & 1 \end{bmatrix},$$

and

$$\mathbf{B}_3 = \mathbf{0}.$$

Thus, using (2.150) and (2.160), we get

$$\mathcal{G}_\xi(f) \equiv \mathcal{G}_\xi^{(c)}(f)$$

$$-\frac{T}{2}\left(\frac{\sin \pi f T/2}{\pi f T/2}\right)^2 \frac{3 + \cos \pi f T + 2\cos 2\pi f T - \cos 3\pi f T}{9 + 12\cos 2\pi f T + 4\cos 4\pi f T}. \quad \square \tag{2.164}$$

Example 2.14 (continued)

From (2.150) we get

$$\mathcal{G}_\xi(f) \equiv \mathcal{G}_\xi^{(c)}(f) = \frac{1}{T}|S(f)|^2. \quad \square \tag{2.165}$$

We conclude this section by observing an important special case of the digital signal considered. If the modulator has *only one* state, or, equivalently, the waveform emitted at time nT depends, in a one-to-one way, only on the source symbol at the same instant, we have, from (2.150) and (2.151) and after some computations,

$$\mathscr{G}_{\xi}^{(c)}(f) = \frac{1}{T}\left[\sum_{i=1}^{M} p_i |S_i(f)|^2 - |\sum_{i=1}^{M} p_i S_i(f)|^2\right] \tag{2.166}$$

and

$$\mathscr{G}_{\xi}^{(d)}(f) = \frac{1}{T^2}|\sum_{i=1}^{M} p_i S_i(f)|^2 \sum_{l=-\infty}^{\infty} \delta\left(f - \frac{l}{T}\right), \tag{2.167}$$

where $\{S_i(f)\}_{i=1}^{M}$ are the Fourier transforms of the waveforms available from the modulator.

2.4 NARROWBAND SIGNALS AND BANDPASS SYSTEMS

For a real signal $x(t)$, its Fourier transform $X(f)$ shows certain symmetries around the zero frequency. In particular, the real part of $X(f)$ is an even function of f, and its imaginary part is odd. As a consequence, to determine it and be in a position to reconstruct $x(t)$, it is sufficient to specify $X(f)$ only for $f \geq 0$. Now suppose that $x(t)$ is passed through a linear, time-invariant system whose transfer function is the step function $\alpha u(f)$, α a constant. At the output of this system we observe a signal from which $x(t)$ can be recovered without information loss. This system has the impulse response

$$\frac{\alpha}{2}\left[\delta(t) + j\frac{1}{\pi t}\right],$$

so its response to $x(t)$ is $\alpha/2 \cdot [x(t) + j\hat{x}(t)]$, where

$$\hat{x}(t) \triangleq \frac{1}{\pi}\int_{-\infty}^{\infty}\frac{x(\vartheta)}{t - \vartheta}\,d\vartheta \tag{2.168}$$

and is called the *Hilbert transform* of $x(t)$. Notice that, because of the singularity in the integrand, the meaning of the RHS of (2.168) has to be made precise. Specifically, the integral is defined as the *Cauchy principal value*. By choosing $\alpha = 2$, we have

$$x(t) = \mathscr{R}[\mathring{x}(t)], \tag{2.169}$$

where $\mathring{x}(t)$, the system output, is

$$\mathring{x}(t) \triangleq x(t) + j\hat{x}(t). \tag{2.170}$$

Equation (2.169) shows that the original signal $x(t)$ can be recovered from the output of a system with transfer function $2u(f)$ by simply taking its real part. The complex signal $\mathring{x}(t)$ is called the *analytic signal* associated with $x(t)$.

Example 2.15

Let $x(t) = \cos(2\pi f_0 t + \varphi)$. Its Hilbert transform is $\hat{x}(t) = \sin(2\pi f_0 t + \varphi)$, so the corresponding analytic signal turns out to be $\mathring{x}(t) = \exp\{j(2\pi f_0 t + \varphi)\}$. We see from this simple example that the analytic signal representation is a generalization of the familiar complex representation of sinusoidal signals. $\square$

Sec. 2.4   Narrowband Signals and Bandpass Systems   **45**

Among the properties of analytic signals, two are worth mentioning here.

(a) The operation transforming the real signal $x(t)$ into the analytic signal $\mathring{x}(t)$ is linear and time invariant. In particular, if $x(t)$ is a Gaussian random process, $\mathring{x}(t)$ is a Gaussian random process.

(b) Consider two real signals $z(t)$ and $y(t)$, and their product

$$x(t) \triangleq z(t)y(t). \tag{2.171}$$

Assume that $z(t)$ is a baseband signal, that is, its (amplitude or energy or power) spectrum is zero for $|f| > f_1$, and $y(t)$ is a narrowband signal, that is, its spectrum is nonzero only for $f_2 < |f| < f_3, f_2 > f_1$ (see Fig. 2.15). With these assumptions, from our definition of an analytic signal, it follows that

$$\mathring{x}(t) = z(t)\mathring{y}(t), \tag{2.172}$$

that is, $\mathring{x}(t)$ is the product of the real signal $z(t)$ and the analytic signal associated with $y(t)$.

Example 2.16 Amplitude modulation of a sinusoidal carrier

Let $y(t) = \cos 2\pi f_0 t$, and let $z(t)$ be a deterministic baseband signal whose Fourier transform $Z(f)$ is confined to the interval $(-f_1, f_1), f_1 < f_0$. The analytic signal associated with their product is

$$\mathring{x}(t) = z(t)e^{j2\pi f_0 t}, \tag{2.173}$$

which shows that the amplitude spectrum of $\mathring{x}(t)$ is $Z(f - f_0)$, that is, is obtained by translating the amplitude spectrum of $z(t)$ around the frequency f_0. $\square$

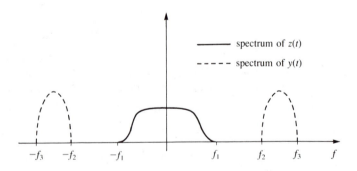

Figure 2.15 Spectra of a baseband signal $z(t)$ and of a narrowband signal $y(t)$.

2.4.1 Narrowband Signals: Complex Envelopes

A *narrowband signal* is one whose spectrum is to a certain extent concentrated around a nonzero frequency. We define a real signal to be narrowband if its (amplitude or energy or power) spectrum is zero for $|f| \notin (f_1, f_2)$, where (f_1, f_2) is a finite frequency

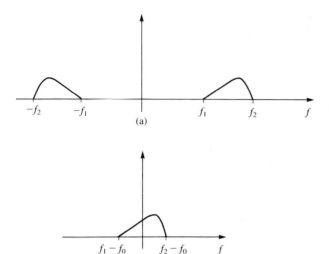

Figure 2.16 (a) Spectrum of a narrowband signal; (b) spectrum of its complex envelope.

interval not including the origin (see Fig. 2.16a). On the other hand, a signal whose spectrum is concentrated around the origin of the frequency axis is referred to as a *baseband signal*. For a given narrowband signal and a frequency $f_0 \in (f_1, f_2)$, the analytic signal $\mathring{x}(t)$ can be written, according to the result of Example 2.16, in the form

$$\mathring{x}(t) = \tilde{x}(t)e^{j2\pi f_0 t}, \tag{2.174}$$

where $\tilde{x}(t)$ is a (generally complex) signal whose spectrum is zero for $f > f_2 - f_0$ and $f < f_1 - f_0$ (see Fig. 2.16b).

The signal $\tilde{x}(t)$ is called the *complex envelope* associated with the real signal $x(t)$. From (2.174) we have the following representation for a narrowband $x(t)$:

$$\begin{aligned} x(t) &= \mathcal{R}[\mathring{x}(t)] \\ &= x_c(t) \cos 2\pi f_0 t - x_s(t) \sin 2\pi f_0 t, \end{aligned} \tag{2.175}$$

where

$$\begin{aligned} x_c(t) &\triangleq \mathcal{R}[\tilde{x}(t)] = \mathcal{R}\,[\mathring{x}(t)e^{-j2\pi f_0 t}] \\ &= x(t) \cos 2\pi f_0 t + \hat{x}(t) \sin 2\pi f_0 t \end{aligned} \tag{2.176a}$$

and

$$\begin{aligned} x_s(t) &\triangleq \mathcal{I}[\tilde{x}(t)] = \mathcal{I}[\mathring{x}(t)e^{-j2\pi f_0 t}] \\ &= -x(t) \sin 2\pi f_0 t + \hat{x}(t) \cos 2\pi f_0 t \end{aligned} \tag{2.176b}$$

are baseband signals. Equation (2.175) and direct computation prove that $x_c(t)$ and $x_s(t)$ can be obtained from $x(t)$ by using the circuitry shown in Fig. 2.17. There the filters are ideal low-pass.

From (2.174) it is also possible to derive a vector representation of the narrowband signal $x(t)$. To do this we define, at any time instant t, a two-dimensional vector whose

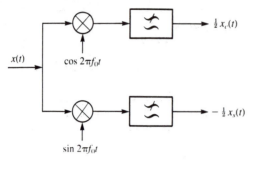

Figure 2.17 Obtaining the real and imaginary parts of the complex envelope of the narrowband signal $x(t)$.

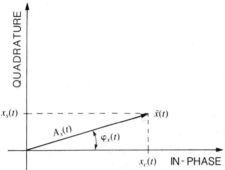

Figure 2.18 Vector representation of the narrowband signal $x(t)$.

components are the in-phase and quadrature components of $\tilde{x}(t)$, that is, $x_c(t)$ and $x_s(t)$ (see Fig. 2.18). The magnitude of this vector is

$$A_x(t) \triangleq |\tilde{x}(t)| = \sqrt{x_c^2(t) + x_s^2(t)} \qquad (2.177)$$

(see Fig. 2.19), and its phase is

$$\varphi_x(t) \triangleq \arg [\tilde{x}(t)] = \tan^{-1} \frac{x_s(t)}{x_c(t)} . \qquad (2.178)$$

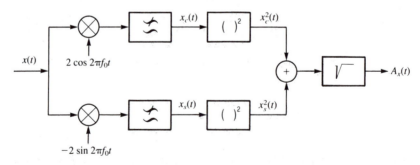

Figure 2.19 Obtaining the instantaneous envelope of the narrowband signal $x(t)$.

The time functions $A_x(t)$ and $\varphi_x(t) + 2\pi f_0 t$ are called, respectively, the *instantaneous envelope* and the *instantaneous phase* of $x(t)$. The *instantaneous frequency* of $x(t)$ is defined as $1/2\pi$ times the derivative of the instantaneous phase; that is,

$$f_x(t) \triangleq f_0 + \frac{1}{2\pi} \frac{\dot{x}_s(t)x_c(t) - x_s(t)\dot{x}_c(t)}{x_c^2(t) + x_s^2(t)}.$$ (2.179)

From (2.175) to (2.178) the following representation of the narrowband signal $x(t)$ can also be derived:

$$x(t) = A_x(t) \cos [2\pi f_0 t + \varphi_x(t)].$$ (2.180)

Narrowband random processes

Consider now a real narrowband, WS stationary random process $v(t)$, and the complex process

$$\mathring{v}(t) \triangleq v(t) + j\hat{v}(t).$$ (2.181)

The possible representations of $v(t)$ are

$$v(t) = \mathcal{R}[\tilde{v}(t)e^{j2\pi f_0 t}],$$ (2.182)

$$v(t) = v_c(t) \cos 2\pi f_0 t - v_s(t) \sin 2\pi f_0 t,$$ (2.183)

and

$$v(t) = A_v(t) \cos [2\pi f_0 t + \varphi_v(t)],$$ (2.184)

where

$$\begin{aligned} \tilde{v}(t) &= \mathring{v}(t)e^{-j2\pi f_0 t} \\ &= v_c(t) + jv_s(t) \end{aligned}$$ (2.185)

is the complex envelope of $v(t)$.

The power spectrum of $\mathring{v}(t)$ can be easily evaluated by observing that $\mathring{v}(t)$ can be thought of as the output of a linear, time-invariant system with transfer function $2u(f)$. Thus, its power spectrum equals the power spectrum of $v(t)$ times the squared magnitude of the transfer function:

$$\mathcal{G}_{\mathring{v}}(f) = 4\mathcal{G}_v(f)u^2(f) = 4\mathcal{G}_v(f)u(f).$$ (2.186)

Equation (2.186) shows that the spectral density of $\mathring{v}(t)$ is equal to four times the one-sided spectral density of $v(t)$. Consider then the complex envelope $\tilde{v}(t)$. From (2.185), its autocorrelation is

$$\begin{aligned} R_{\tilde{v}}(\tau) &= E[\mathring{v}(t + \tau)\mathring{v}^*(t)]e^{-j2\pi f_0 \tau} \\ &= R_{\mathring{v}}(\tau)e^{-j2\pi f_0 \tau}, \end{aligned}$$ (2.187)

and hence

$$\mathcal{G}_{\tilde{v}}(f) = \mathcal{G}_{\mathring{v}} (f + f_0),$$ (2.188)

which shows that the power spectral density of the complex envelope $\tilde{v}(t)$ is the version of $\mathcal{G}_{\mathring{v}}(f)$ translated around the origin (see Fig. 2.20). Consider finally $v_c(t)$ and $v_s(t)$, the real and imaginary parts of the complex envelope. It can be shown (see Problem 2.20) that the following equalities hold:

$$R_{v_c}(\tau) = R_{v_s}(\tau)$$ (2.189)

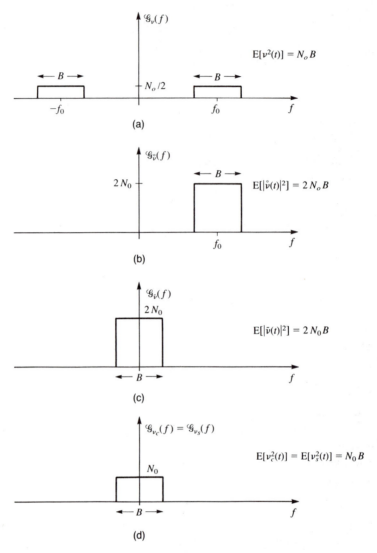

Figure 2.20 Representations of a narrowband white noise process $v(t)$: (a) Power spectrum of $v(t)$; (b) power spectrum of the analytic signal $\mathring{v}(t)$; (c) power spectrum of the complex envelope $\tilde{v}(t)$; (d) power spectra of the real and imaginary parts of $\tilde{v}(t)$.

and

$$E[v_c(t + \tau)v_s(t)] = -E[v_s(t + \tau)v_c(t)]. \tag{2.190}$$

Thus,

$$\begin{aligned}
R_{\tilde{v}}(\tau) &= E\{[v_c(t + \tau) + jv_s(t + \tau)][v_c(t) - jv_s(t)]\} \\
&= R_{v_c}(\tau) + R_{v_s}(\tau) + j\{E[v_s(t + \tau)v_c(t)] - E[v_c(t + \tau)v_s(t)]\} \tag{2.191} \\
&= 2[R_{v_c}(\tau) + jR_{v_sv_c}(\tau)],
\end{aligned}$$

where

$$R_{v_s v_c}(\tau) \triangleq \mathrm{E}[v_s(t + \tau)v_c(t)].$$ (2.192)

From (2.189) to (2.192) we can draw the following conclusions:

(a) As $R_{\tilde{v}}(0) = \mathrm{E}|\tilde{v}(t)|^2$ is a real quantity, Eqs. (2.191) and (2.192) show that

$$\mathrm{E}[v_s(t)v_c(t)] = 0.$$ (2.193)

That is, for any given t, $v_s(t)$ and $v_c(t)$ are uncorrelated RVs. As a special case, if $v(t)$ is a Gaussian process, for any given t, $v_s(t)$ and $v_c(t)$ are independent RVs.

(b) From (2.186) and (2.187) it follows that

$$\mathrm{E}|\tilde{v}(t)|^2 = \mathrm{E}|\mathring{v}(t)|^2 = 2\mathrm{E}[v^2(t)].$$ (2.194)

Similarly, from (2.189), (2.191), and result (a) we have

$$\mathrm{E}|\tilde{v}(t)|^2 = 2\mathrm{E}[v_c^2(t)] = 2\mathrm{E}[v_s^2(t)].$$ (2.195)

Thus,

$$\mathrm{E}[v_c^2(t)] = \mathrm{E}[v_s^2(t)] = \mathrm{E}[v^2(t)].$$ (2.196)

That is, the average power of $v_c(t)$ and $v_s(t)$ equals that of the original process $v(t)$.

(c) If the power spectrum of the process $v(t)$ is symmetric around the frequency f_0, from (2.188) it follows that the power spectrum of $\tilde{v}(f)$ is an even function. This implies that $R_{\tilde{v}}(\tau)$ is real for all τ, so (2.191) and (2.192) yield

$$\mathrm{E}[v_s(t + \tau)v_c(t)] = 0, \qquad \text{for all } \tau.$$ (2.197)

This means that the processes $v_c(t)$ and $v_s(t)$ are uncorrelated [or independent when $v(t)$ is Gaussian]. Thus, in this situation,

$$\mathcal{G}_{\tilde{v}}(f) = 2\mathcal{G}_{v_c}(f) = 2\mathcal{G}_{v_s}(f).$$ (2.198)

Example 2.17

Let $x(t)$ be a bandpass real signal, and let $\mathcal{G}_{\tilde{x}}(f)$ be the power density spectrum of its complex envelope. From (2.186) and (2.188) we have

$$\mathcal{G}_{\tilde{x}}(f) = \mathcal{G}_{\mathring{x}}(f + f_0)$$
$$= 4\mathcal{G}_x(f + f_0)u(f + f_0).$$

Recalling the fact that $\mathcal{G}_x(f)$ must be an even function of f, the last equality yields

$$\mathcal{G}_x(f) = \frac{1}{4}[\mathcal{G}_{\tilde{x}}(-f - f_0) + \mathcal{G}_{\tilde{x}}(f - f_0)].$$

As an example, consider the signal

$$x(t) = \Re\left[\sum_{n=-\infty}^{\infty} \alpha_n s(t - nT) \cdot e^{j2\pi f_0 t}\right],$$

where $E[\alpha_n] = 0$ and $E[\alpha_{n+m}\,\alpha_n^*] = \sigma_\alpha^2 \delta_{0,m}$.

From (2.126) we obtain the power spectrum of the complex envelope of $x(t)$:

$$\mathcal{G}_{\tilde{x}}(f) = \frac{\sigma_\alpha^2}{T}|S(f)|^2.$$

Hence, the power spectrum of the signal is

$$\mathcal{G}_x(f) = \frac{\sigma_\alpha^2}{4T}\{|S(-f - f_0)|^2 + |S(f - f_0)|^2\}. \quad \square$$

Narrowband white noise

As we shall see in later chapters, in problems concerning narrowband signals contaminated by additive noise it is usual to assume, as a model for the noise, a Gaussian process with a power density spectrum that is constant in a finite frequency interval and zero elsewhere. This occurs because a really white noise would have an infinite power (which is physically meaningless), and because any mixture of signal plus noise is always observed at the output of a bandpass filter that is usually not wider than the frequency occupation of the signal. Thus, in practice, we can assume that the noise has a finite bandwidth, an assumption entailing no loss of accuracy if the noise bandwidth is much wider than the filter bandwidth.

A *narrowband white noise* is a real, zero-mean, stationary random process whose power density spectrum is constant over a finite frequency interval not including the origin. In Fig. 2.20 we show the power spectrum of a narrowband white noise with a power spectral density $N_0/2$ in the bandwidth B centered at f_0.

2.4.2 Bandpass Systems

The complex envelope representation of narrowband signals can be extended to the consideration of bandpass systems (i.e., systems whose response to any input signal is a narrowband signal). We shall see in the following how to characterize the effects of a bandpass system directly in terms of complex envelopes. In other words, assume that $y(t)$ is the response of a bandpass system to the narrowband signal $x(t)$. We want to characterize a system whose response to $\tilde{x}(t)$, the complex envelope of $x(t)$, is exactly $\tilde{y}(t)$, the complex envelope of $y(t)$.

Bandpass linear systems

First, consider a bandpass linear, time-invariant system with impulse response $h(t)$ and transfer function $H(f)$. The analytic signal representation of $h(t)$ is $\overset{\circ}{h}(t) = h(t) + j\hat{h}(t)$, which corresponds to the transfer function $\overset{\circ}{H}(f) = 2H(f)u(f)$. If $x(t)$ is the narrowband input signal and $y(t)$ the response, the analytic signal $\overset{\circ}{y}(t)$ can be obtained by passing $x(t)$ into the cascade of the linear system under consideration and a filter with a transfer function $2u(f)$ (see Fig. 2.21a). In a cascade of linear transformations,

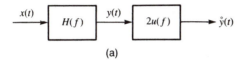

(a)

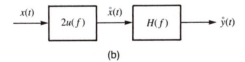

(b)

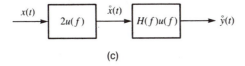

(c)

Figure 2.21 Three equivalent schemes to represent the analytic signal associated with the output of a linear system.

the order of the operations can be reversed without altering the final result, so we can substitute the scheme of Fig. 2.21b for that of Fig. 2.21a. Next, observe that $\mathring{x}(t)$ has a Fourier transform equal to zero for $f < 0$. Hence, we can substitute a system with transfer function $H(f)$ for another system having a transfer function $H(f)u(f)$ without altering the output. The latter system (see Fig. 2.21c) has an impulse response $\frac{1}{2}\mathring{h}(t)$, input $\mathring{x}(t)$, and output $\mathring{y}(t)$. These signals are related by the convolution integral

$$\mathring{y}(t) = \frac{1}{2} \int_{-\infty}^{\infty} \mathring{h}(\tau)\mathring{x}(t - \tau) \, d\tau. \tag{2.199}$$

This equation becomes particularly useful if both $\mathring{x}(t)$ and $\mathring{h}(t)$ are expressed in terms of complex envelopes. We get

$$\mathring{y}(t) = \frac{1}{2} e^{j2\pi f_0 t} \int_{-\infty}^{\infty} \tilde{h}(\tau)\tilde{x}(t - \tau) \, d\tau, \tag{2.200}$$

which shows that $\mathring{y}(t)$ is a narrowband signal, centered at f_0, with complex envelope

$$\tilde{y}(t) = \frac{1}{2} \int_{-\infty}^{\infty} \tilde{h}(\tau)\tilde{x}(t - \tau) \, d\tau. \tag{2.201}$$

In conclusion, the complex envelope of the response of a bandpass linear, time-invariant system with impulse response $h(t)$ to a given narrowband signal $x(t)$ can be obtained by passing the complex envelope $\tilde{x}(t)$ through the *low-pass equivalent system* whose impulse response is $\frac{1}{2}\tilde{h}(t)$ or, equivalently, whose transfer function is $H(f + f_0)u(f + f_0)$ (see Fig. 2.22). Notice that only if $H(f)$ is symmetric around f_0 will the low-pass equivalent system have a *real* impulse response. A nonreal impulse response will induce in the output signal a shift of the phase and a correlation between the in-phase and quadrature components. These effects are usually undesired.

Sec. 2.4   Narrowband Signals and Bandpass Systems   **53**

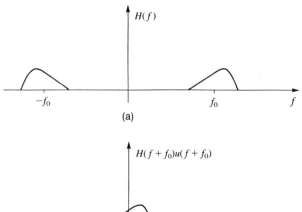

(a)

$H(f + f_0)u(f + f_0)$

(b)

Figure 2.22 (a) Transfer function of a bandpass linear system; (b) transfer function of the low-pass equivalent linear system.

Example 2.18

Let

$$x(t) = z(t)e^{j2\pi f_0 t}, \qquad (2.202)$$

where the Fourier transform of $z(t)$ is zero for $|f| > B$, $B < f_0$. Consider an LRC parallel resonator. Its transfer function is

$$H(f) = \frac{j2\pi f L}{R + j2\pi f L - 4\pi^2 f^2 LC}. \qquad (2.203)$$

The corresponding impulse response is, for $t \geq 0$,

$$h(t) = \frac{2\pi f_0}{Q} e^{-\pi f_0 t/2Q} \cos 2\pi f_0 t - \frac{\pi f_0}{Q^2} e^{-\pi f_0 t/2Q} \sin 2\pi f_0 t, \qquad (2.204)$$

where

$$Q \triangleq R\sqrt{\frac{C}{L}}$$

is the quality factor of the circuit, and

$$f_0 \triangleq \frac{1}{2\pi\sqrt{LC}} \left(1 - \frac{1}{4Q^2}\right)^{\frac{1}{2}}.$$

If $Q >> 1$, the computation of $\bar{h}(t)$ becomes very easy. In fact, the second term in the RHS of (2.204) can be disregarded. Additionally, we can safely assume that the exponential factor $e^{-\pi f_0 t/2Q}$ is a bandlimited signal. Thus, from (2.204), we have

$$\overset{\circ}{h}(t) \cong \frac{2\pi f_0}{Q} e^{-\pi f_0 t/2Q} e^{j2\pi f_0 t}, \qquad t \geq 0, \qquad (2.205)$$

where

$$f_0 = \frac{1}{2\pi\sqrt{LC}}.$$

In conclusion,

$$\bar{h}(t) \cong \frac{2\pi f_0}{Q} e^{-\pi f_0 t/2Q}, \qquad t \geq 0. \tag{2.206}$$

Notice that the approximations in the computation of $\bar{h}(t)$ make the Fourier transform of (2.206) symmetrical around the origin of the frequency axis. $\square$

Bandpass memoryless nonlinear systems

We shall now examine a class of nonlinear systems that are often encountered in radio-frequency transmission. Some examples of these will be described in Chapter 10. We want to consider a nonlinear time-invariant system whose input signal bandwidth is so narrow that the system's behavior is essentially frequency independent. Moreover, the system is assumed to be bandpass. This in turn means that it can be thought of as being followed by a *zonal filter* whose aim is to stop all the frequency components of the output not close to the center frequency of the input signal. For a simple example of such a system, consider a sinusoidal signal $x(t) = A \cos 2\pi f_0 t$ sent into a time-invariant nonlinear system. Its output includes a sum of several harmonics centered at frequency $0, f_0, 2f_0, \ldots$. If only the harmonic at f_0 is retained at the output, the observed output signal is a sinusoid $y(t) = F(A) \cos [2\pi f_0 t + \phi(A)]$. If we consider the complex envelopes $\tilde{x}(t) = A$ and $\tilde{y}(t) = F(A) \exp[j\phi(A)]$, we see that the system operation for sinusoidal inputs can be characterized by the two functions $F(\cdot)$ and $\phi(\cdot)$. In the following we shall show that this result holds true even when the input signal is a more general narrowband signal.

Consider a narrowband signal $x(t)$, with a spectrum centered at f_0. Its analytic signal representation can be given the form

$$\overset{\circ}{x}(t) = A_x(t)e^{j[2\pi f_0 t + \varphi_x(t)]}, \tag{2.207}$$

where $A_x(t)$ and $\varphi_x(t)$ are baseband signals. Letting

$$\psi_x(t) \triangleq 2\pi f_0 t + \varphi_x(t), \tag{2.208}$$

we rewrite (2.207) as

$$\overset{\circ}{x}(t) = A_x(t)e^{j\psi_x(t)}. \tag{2.209}$$

Consider then the effect of a nonlinear memoryless system whose input–output relationship is assumed to have the form

$$y(t) = S_e[A_x(t) \cos \psi_x(t)] + S_o[A_x(t) \sin \psi_x(t)], \tag{2.210}$$

where $S_e[\cdot]$ is an even function of $\psi_x(t)$, and $S_o[\cdot]$ is an odd function. It is seen that $y(t)$, when expressed as a function of $\psi_x(t)$, is periodic with period 2π. Thus, we can expand $y(t)$ in a Fourier series:

$$y(t) = y[\psi_x(t)] = \sum_{l=-\infty}^{\infty} c_l[A_x(t)]e^{jl\psi_x(t)}, \tag{2.211}$$

where

Sec. 2.4   Narrowband Signals and Bandpass Systems   **55**

$$c_l(A) \triangleq \frac{1}{2\pi} \int_0^{2\pi} \{S_e[A \cos \psi] + S_o[A \sin \psi]\}e^{-jl\psi}d\psi. \tag{2.212}$$

The quantity $c_l(A)$ is generally complex. In particular, we have

$$\Re[c_l(A)] = \frac{1}{2\pi} \int_0^{2\pi} S_e[A \cos \psi] \cos l\psi \, d\psi \tag{2.213}$$

and

$$\mathscr{I}[c_l(A)] = -\frac{1}{2\pi} \int_0^{2\pi} S_o[A \sin \psi] \sin l\psi \, d\psi. \tag{2.214}$$

From the definition (2.208) of $\psi_x(t)$, we see how (2.211) expresses the fact that the spectrum of $y(t)$ includes several spectral components, each centered around the frequencies $\pm lf_0$, $l = 0, 1, \ldots$. Figure 2.23 illustrates qualitatively this situation. Notice that we must assume that the signals $c_l[A_x(t)]$ have spectra that do not significantly extend beyond the interval $(-f_0/2, f_0/2)$.

The assumption that the memoryless system is bandpass implies that only one of the spectral components of $y(t)$ can survive at the system output (i.e., that centered at $\pm f_0$). The analytic signal representation of the output of such a bandpass memoryless system is then

$$\mathring{y}(t) = c[A_x(t)]e^{j\psi_x(t)}, \tag{2.215}$$

where

$$c(A) \triangleq 2c_1(A). \tag{2.216}$$

As $c(A)$ is generally a complex number, we can put it in the form

$$c(A) = F(A)e^{j\phi(A)}, \tag{2.217}$$

so

$$\mathring{y}(t) = F[A_x(t)]e^{j\{\psi_x(t)+\phi[A_x(t)]\}}, \tag{2.218}$$

or, in terms of complex envelopes,

$$\tilde{y}(t) = F[A_x(t)]e^{j\{\varphi_x(t)+\phi[A_x(t)]\}}. \tag{2.219}$$

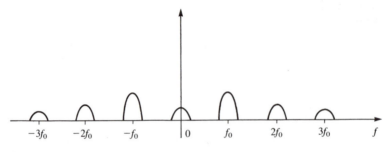

Figure 2.23 Spectrum of the output of a memoryless nonlinear system when the input is a narrowband signal centered at frequency f_0.

Comparing the last equation with the complex envelope of the input signal,

$$\tilde{x}(t) = A_x(t)e^{j\varphi_x(t)}, \tag{2.220}$$

we can see that the effect of a bandpass memoryless nonlinear system is to alter the amplitude and to shift the phase of the input signal according to a law that depends only on the values of its instantaneous envelope. This shows, in particular, that the system can be characterized by assigning the two functions $F[\cdot]$, $\phi[\cdot]$, which describe the so-called AM/AM conversion and AM/PM conversion effects of the system (AM denotes *amplitude modulation* and PM *phase modulation*). These functions can be determined experimentally by taking as an input signal a single sinusoid with a frequency close to f_0 and an envelope A, and by measuring, for different values of A, the output envelope $F(A)$ and the output phase shift $\phi(A)$. Notice that, for the validity of this nonlinear system model, the functions $F(A)$ and $\phi(A)$ should not depend appreciably on the frequency of the test sinusoid as it varies within the range of interest.

Finally, notice that the system we are dealing with can be represented as in Fig. 2.24, where S_e and S_o denote memoryless nonlinear devices as defined in Section 2.1.2. From this scheme it is seen that only if S_o is present can the system show an AM/PM conversion effect.

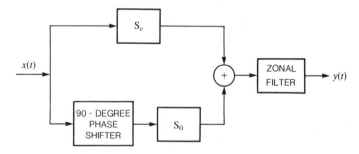

Figure 2.24 Representation of a bandpass memoryless nonlinear system.

Example 2.19 Polynomial-law devices

Consider a nonlinear system whose input–output relationship is

$$y(t) = Cx^l(t), \tag{2.221}$$

l an integer greater than 1. If $x(t)$ is written in the form

$$\begin{aligned} x(t) &= \Re[\tilde{x}(t)e^{j2\pi f_0 t}] \\ &= \frac{1}{2}[\tilde{x}(t)e^{j2\pi f_0 t} + \tilde{x}^*(t)e^{-j2\pi f_0 t}], \end{aligned} \tag{2.222}$$

we get

$$x^l(t) = \frac{1}{2^l}\sum_{k=0}^{l}\binom{l}{k}[\tilde{x}(t)]^k[\tilde{x}^*(t)]^{l-k}e^{j2\pi(2k-l)f_0 t}. \tag{2.223}$$

When $y(t)$ is filtered through a zonal filter, all its frequency components other than those centered at $\pm f_0$ will be removed. Thus, only the terms with $2k - l = \pm 1$ will contribute to the system output. This shows, in particular, that only when l is *odd* can the output of the zonal filter be nonzero. For l odd, the complex envelope of the system output is then

$$\tilde{y}(t) = \frac{C}{2^{l-1}} \binom{l}{\frac{l+1}{2}} \tilde{x}(t) \, |\tilde{x}(t)|^{l-1}. \tag{2.224}$$

More generally, if the system is polynomial,

$$y(t) = \sum_{l=1}^{L} a_l \, x^l(t), \tag{2.225}$$

we shall get, for L odd,

$$\tilde{y}(t) = \tilde{x}(t) \sum_{m=0}^{(L-1)/2} \frac{a_{2m+1}}{2^{2m}} \binom{2m+1}{m+1} |\tilde{x}(t)|^{2m}. \; \square \tag{2.226}$$

Bandpass Volterra systems

Consider Volterra systems, as defined in Section 2.1.2. These are systems whose input–output relationship is of the form

$$y(t) = \sum_{k=1}^{\infty} y_k(t), \tag{2.227}$$

where

$$y_k(t) \triangleq \int_{-\infty}^{\infty} \cdots \int_{-\infty}^{\infty} h_k(\tau_1, \ldots, \tau_k) \left[\prod_{i=1}^{k} x(t - \tau_i) d\tau_i \right], \tag{2.228}$$

and the functions $h_k(\tau_1, \ldots, \tau_k)$, $k = 1, 2, \ldots$, are the kernels of the system [for simplicity, the constant term of (2.25) has been disregarded in the RHS of (2.227)]. Here we want to derive an input–output relationship for a Volterra system in the case where both the input signal and the system are bandpass. We assume, as is usual, that the kernels are symmetric.

Our first step is to express $x(t)$ as in (2.222) and the Volterra kernels in a similar way. To do this, we assume that the multidimensional Fourier transforms of the kernels, that is,

$$H_k(f_1, \ldots, f_k) \triangleq \int_{-\infty}^{\infty} \cdots \int_{-\infty}^{\infty} h_k(t_1, \ldots, t_k) \left[\prod_{i=1}^{k} e^{-j2\pi f_i t_i} \, dt_i \right], \tag{2.229}$$

differ significantly from zero only in small neighborhoods of the 2^k points ($\pm f_0$, $\pm f_0$, $\ldots$, $\pm f_0$) of the k-dimensional space in which they are defined (see Fig. 2.25). This assumption expresses mathematically the fact that the system is bandpass and is an extension of a similar assumption for transfer functions of linear systems.

Example 2.20

If the input signal is the sinusoid $A \cos 2\pi f't$, simple calculations show that $y_k(t)$ is made up by a linear combination of terms of the type $H_k(\pm f', \pm f', \ldots, \pm f')$. Due to the

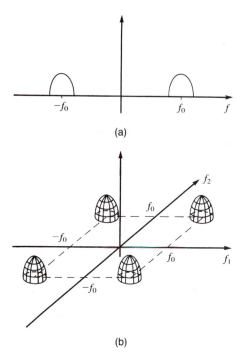

(a)

(b)

Figure 2.25 Fourier transform $H_k(f_1, f_2, \ldots, f_k)$ of the Volterra kernels of a bandpass system: (a) $k = 1$; (b) $k = 2$.

bandpass hypothesis, we expect these terms to be significantly nonzero only for $f' \cong f_0$, which is achieved if $H_k (\cdots)$ satisfies our assumption. $\square$

From our previous discussion, it turns out that $H_k(f_1, \ldots, f_k)$ can be expressed as a sum of 2^k "low-pass" functions, whose arguments are $f_1 \pm f_0, f_2 \pm f_0, \ldots,$ $f_k \pm f_0$, with all the possible combinations of signs. Each of these functions is significantly nonzero only in the neighborhood of the origin $(0, 0, \ldots, 0)$. By taking the inverse Fourier transform of these functions, it results that $h_k(t_1, \ldots, t_k)$ is the sum of 2^k "baseband" kernels, each one being multiplied by a factor $\exp[j2\pi f_0(\pm t_1 \pm t_2 \pm \cdots \pm t_k)]$.

Example 2.21

The Volterra system of Fig. 2.26 has only one kernel, given by

$$h_3(t_1, t_2, t_3) = h(t_1)h(t_2)h(t_3). \tag{2.230}$$

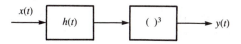

Figure 2.26 Volterra system.

If $h(t)$ is the impulse response of a bandpass filter, the Volterra system is bandpass. Letting

$$h(t) = \frac{1}{2}[\bar{h}(t)e^{j2\pi f_0 t} + \bar{h}^*(t)e^{-j2\pi f_0 t}], \tag{2.231}$$

Sec. 2.4   Narrowband Signals and Bandpass Systems   **59**

we get

$$
\begin{aligned}
h_3(t_1, t_2, t_3) = \frac{1}{8} [& \tilde{h}(t_1)\tilde{h}(t_2)\tilde{h}(t_3)e^{j2\pi f_0(t_1+t_2+t_3)} \\
& + \tilde{h}(t_1)\tilde{h}(t_2)\tilde{h}*(t_3)e^{j2\pi f_0(t_1+t_2-t_3)} \\
& + \tilde{h}(t_1)\tilde{h}*(t_2)\tilde{h}(t_3)e^{j2\pi f_0(t_1-t_2+t_3)} \\
& + \tilde{h}(t_1)\tilde{h}*(t_2)\tilde{h}*(t_3)e^{j2\pi f_0(t_1-t_2-t_3)} \\
& + \tilde{h}*(t_1)\tilde{h}(t_2)\tilde{h}(t_3)e^{j2\pi f_0(-t_1+t_2+t_3)} \\
& + \tilde{h}*(t_1)\tilde{h}(t_2)\tilde{h}*(t_3)e^{j2\pi f_0(-t_1+t_2-t_3)} \\
& + \tilde{h}*(t_1)\tilde{h}*(t_2)\tilde{h}(t_3)e^{j2\pi f_0(-t_1-t_2+t_3)} \\
& + \tilde{h}*(t_1)\tilde{h}*(t_2)\tilde{h}*(t_3)e^{j2\pi f_0(-t_1-t_2-t_3)}]. \quad \square
\end{aligned}
\tag{2.232}
$$

Let us now insert into (2.228) our representation of $h_k(\cdots)$ in terms of baseband kernels, and express $x(t)$ as in (2.222). This produces a sum of numerous integrals the majority of which do not contribute to the final result. In fact, a complete tabulation of these terms will show that, for many of them, one of the following situations occurs:

(a) The integrand is a baseband function multiplied by a factor like $e^{\pm j2\pi n f_0 \tau_i}$, $n \neq 0$. For f_0 much larger than the bandwidth of the input signal, this term will provide a vanishing contribution to $y_k(t)$.

(b) The resulting integral is the product of a baseband function and of a factor $e^{\pm j2\pi m f_0 t}$, $m \neq 1$. This term will not survive given the assumption that the system is bandpass.

In conclusion, lengthy but straightforward computations will show that if k is even there is no surviving term [so $y_k(t) \equiv 0$], whereas for $k = 2m + 1$ we have

$$
\begin{aligned}
y_{2m+1}(t) = \frac{1}{2^{2m+2}} \binom{2m+1}{m} \Bigg\{ & e^{j2\pi f_0 t} \int_{-\infty}^{\infty} \cdots \int_{-\infty}^{\infty} \tilde{h}_{2m+1}(\tau_1, \ldots, \tau_{2m+1}) \cdot \prod_{i=1}^{m+1} \tilde{x}(t-\tau_i) \\
& \cdot \prod_{l=m+2}^{2m+1} \tilde{x}*(t-\tau_l) d\tau_1 d\tau_2 \cdots d\tau_{2m+1} + e^{-j2\pi f_0 t} \int_{-\infty}^{\infty} \cdots \int_{-\infty}^{\infty} \\
& \tilde{h}_{2m+1}^*(\tau_1, \ldots, \tau_{2m+1}) \cdot \prod_{i=1}^{m+1} \tilde{x}*(t-\tau_i) \prod_{l=m+2}^{2m+1} \tilde{x}(t-\tau_l) d\tau_1 \cdots d\tau_{2m+1} \Bigg\}
\end{aligned}
\tag{2.233}
$$

where the function

$$
\frac{1}{2^{2m+1}} \binom{2m+1}{m} \tilde{h}_{2m+1}(\cdots),
$$

the *low-pass equivalent kernel*, is defined as follows. Consider the component of $H_{2m+1}(\cdots)$ that is zero in the neighborhood of $(\underbrace{f_0, f_0, \ldots, f_0}_{m+1}, \underbrace{-f_0, -f_0, \ldots, -f_0}_{m})$, and take its inverse Fourier transform. Then multiply it by $2 \exp\{-j2\pi f_0(t_1 + t_2 + \cdots + t_{m+1} - t_{m+2} - \cdots - t_{2m+1})\}$. The resulting function is $\tilde{h}_{2m+1}(t_1, \ldots, t_{2m+1})$. The factor $\binom{2m+1}{m}$ accounts for the fact that the only surviving terms are those whose integrands include $m + 1$ factors of the type $\tilde{x}(t - \tau_i)$ and m factors of the type $\tilde{x}^*(t - \tau_i)$, as well as their conjugates. There are exactly $\binom{2m+1}{m}$ such conjugate pairs. They give equal contributions to $y_{2m+1}(t)$ because of the assumed symmetry of the kernels $h_k(\cdots)$. Furthermore, in (2.233) the integrand functions are baseband, so we finally get

$$\tilde{y}(t) = \sum_{m=0}^{\infty} \tilde{y}_{2m+1}(t). \tag{2.234}$$

where

$$\tilde{y}_{2m+1}(t) \triangleq \frac{1}{2^{2m+1}} \binom{2m+1}{m} \int_{-\infty}^{\infty} \cdots \int_{-\infty}^{\infty} \tilde{h}_{2m+1}(\tau_1, \ldots, \tau_{2m+1})$$
$$\cdot \prod_{i=1}^{m+1} \tilde{x}(t - \tau_i) \prod_{l=m+2}^{2m+1} \tilde{x}^*(t - \tau_l) d\tau_1 \cdots d\tau_{2m+1}. \tag{2.235}$$

Example 2.21 (continued)

Using (2.232) and (2.222) in (2.228), it is seen that $y_3(t)$ includes $2^6 = 64$ terms. Some examples of integrand functions are

$$\frac{1}{64} \tilde{h}(\tau_1)\tilde{h}(\tau_2)\tilde{h}(\tau_3)\tilde{x}(t - \tau_1)\tilde{x}(t - \tau_2)\tilde{x}(t - \tau_3)e^{j6\pi f_0 t},$$

which is screened out because the system is bandpass;

$$\frac{1}{64} \tilde{h}(\tau_1)\tilde{h}(\tau_2)\tilde{h}(\tau_3)\tilde{x}(t - \tau_1)\tilde{x}(t - \tau_2)\tilde{x}^*(t - \tau_3)e^{j2\pi f_0 t} e^{j4\pi f_0 \tau_3},$$

which is screened out because it gives rise to a vanishing contribution to $y_3(t)$;

$$\frac{1}{64} \tilde{h}(\tau_1)\tilde{h}(\tau_2)\tilde{h}^*(\tau_3) \tilde{x}(t - \tau_1)\tilde{x}(t - \tau_2)\tilde{x}^*(t - \tau_3) e^{j2\pi f_0 t},$$

$$\frac{1}{64} \tilde{h}(\tau_1)\tilde{h}^*(\tau_2)\tilde{h}(\tau_3) \tilde{x}(t - \tau_1)\tilde{x}^*(t - \tau_2)\tilde{x}(t - \tau_3) e^{j2\pi f_0 t},$$

$$\frac{1}{64} \tilde{h}^*(\tau_1)\tilde{h}(\tau_2)\tilde{h}(\tau_3) \tilde{x}^*(t - \tau_1)\tilde{x}(t - \tau_2)\tilde{x}(t - \tau_3) e^{j2\pi f_0 t},$$

which are retained (and give equal contributions); and so forth.

The resulting expression for $\tilde{y}_3(t)$ is

$$\tilde{y}_3(t) = \frac{3}{32} \int_{-\infty}^{\infty}\int_{-\infty}^{\infty}\int_{-\infty}^{\infty} \tilde{h}(\tau_1)\tilde{h}(\tau_2)\tilde{h}^*(\tau_3)\tilde{x}(t - \tau_1)\tilde{x}(t - \tau_2)\tilde{x}^*(t - \tau_3) \, d\tau_1 \, d\tau_2 \, d\tau_3, \tag{2.236}$$

which corresponds to (2.235) if we let

$$\tilde{h}_3(\tau_1, \tau_2, \tau_3) \triangleq \frac{1}{4} \tilde{h}(\tau_1)\tilde{h}(\tau_2)\tilde{h}^*(\tau_3), \tag{2.237}$$

according to the definition of low-pass equivalent kernels. $\square$

It is left as an exercise for the interested reader to show that (2.235) can be specialized, under the appropriate assumptions, to (2.201) and (2.226).

Sec. 2.4   Narrowband Signals and Bandpass Systems   **61**

2.5 DISCRETE REPRESENTATION OF CONTINUOUS SIGNALS

In this section we consider the problem of associating a discrete representation with a continuous signal. In other words, we wish to represent a given continuous signal in terms of a (possibly finite) sequence. The representation may be exact or only approximate, in which case it will be chosen on the basis of a compromise between accuracy and simplicity.

2.5.1 Orthonormal Expansions of Finite-Energy Signals

A fundamental type of discrete representation is based on sets of signals called *orthonormal*. To define these sets, consider first the notion of *scalar product* between two finite-energy signals $x(t)$ and $y(t)$: it is denoted by (x, y) and defined as the value of the integral

$$(x, y) \triangleq \int_{-\infty}^{\infty} x(t)y^*(t) \, dt. \qquad (2.238)$$

If $X(f)$ and $Y(f)$ denote the Fourier transforms of $x(t)$ and $y(t)$, respectively, and we let

$$(X, Y) = \int_{-\infty}^{\infty} X(f)Y^*(f) \, df, \qquad (2.239)$$

Parseval's equality relates the scalar products defined in the time and in the frequency domain:

$$(x, y) = (X, Y). \qquad (2.240)$$

If $(x, y) = 0$, or equivalently $(X, Y) = 0$, the signals $x(t)$ and $y(t)$ are called *orthogonal*. From the definitions of scalar product and of orthogonality, it immediately follows that $(x, x) = \mathcal{E}_x$, the energy of $x(t)$, and that the energy of the sum of two orthogonal signals equals the sum of their energies.

Suppose now that we have a sequence $(\psi_i(t))_{i \in I}$ of orthogonal signals; that is,

$$(\psi_i, \psi_j) = \begin{cases} \mathcal{E}_i, & i = j, \\ 0, & i \neq j, \end{cases} \qquad (2.241)$$

where I is a finite or countable index set.

If $\mathcal{E}_i = 1$ for all $i \in I$, the signals of this sequence are called *orthonormal*. Obviously, an orthonormal sequence can be obtained from an orthogonal one by dividing each $\psi_i(t)$ by $\sqrt{\mathcal{E}_i}$. Given an orthonormal sequence, we wish to approximate a given finite-energy signal $x(t)$ with a linear combination of signals belonging to this sequence, that is, with the signal

$$\hat{x}(t) \triangleq \sum_{i \in I} c_i \psi_i(t). \qquad (2.242)$$

A suitable criterion for the choice of the constants c_i appearing in (2.242), and hence of the approximation $\hat{x}(t)$, is to minimize the energy of the error signal

$$e(t) \triangleq x(t) - \hat{x}(t). \qquad (2.243)$$

Thus, the task is to minimize

$$\mathcal{E}_e \triangleq \int_{-\infty}^{\infty} |x(t) - \hat{x}(t)|^2 \, dt$$

$$= \mathcal{E}_x + \mathcal{E}_{\hat{x}} - 2\Re(x, \hat{x}) \tag{2.244}$$

$$= \mathcal{E}_x + \sum_{i \in I} |c_i|^2 - 2\Re \sum_{i \in I} c_i(x, \psi_i)$$

with respect to c_i, $i \in I$. By completing the square, we can also write

$$\mathcal{E}_e = \mathcal{E}_x + \sum_{i \in I} |c_i - (x, \psi_i)|^2 - \sum_{i \in I} |(x, \psi_i)|^2. \tag{2.245}$$

As the middle term in the RHS of (2.245) is nonnegative, $\mathcal{E}_e$ is minimized if the c_i are chosen such as to render it equal to zero. This is achieved for

$$c_i = (x, \psi_i) \triangleq \int_{-\infty}^{\infty} x(t)\psi_i^*(t) \, dt, \qquad i \in I. \tag{2.246}$$

The minimum value of $\mathcal{E}_e$ is then given by

$$(\mathcal{E}_e)_{\min} = \mathcal{E}_x - \sum_{i \in I} |c_i|^2. \tag{2.247}$$

When c_i, $i \in I$, are computed using (2.246), the signal $\hat{x}(t)$ of (2.242) is called the *projection* of $x(t)$ onto the space spanned by the signals of the sequence $(\psi_i(t))_{i \in I}$, that is, on the set of signals that can be expressed as linear combinations of the $\psi_i(t)$. This denomination stems from the fact that, if (2.246) holds, the error $e(t)$ is orthogonal to every $\psi_i(t)$, $i \in I$, and hence to $\hat{x}(t)$. In fact,

$$(e, \psi_i) = (x - \hat{x}, \psi_i)$$

$$= (x, \psi_i) - (\hat{x}, \psi_i)$$

$$= c_i - c_i = 0, \qquad i \in I.$$

(See Fig. 2.27 for a pictorial interpretation of this property in the case $I = \{1, 2\}$.)

An important issue with this theory is the investigation of the conditions under which $(\mathcal{E}_e)_{\min} = 0$. When this happens, the sequence $(\psi_i(t))_{i \in I}$ is said to be *complete* for the signal $x(t)$, and from (2.247) we have the equality

$$\mathcal{E}_x = \sum_{i \in I} |c_i|^2. \tag{2.248}$$

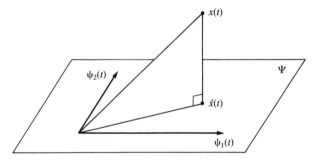

Figure 2.27 $\hat{x}(t)$ is the projection of $x(t)$ on Ψ, the signal space spanned by $\psi_1(t)$ and $\psi_2(t)$.

Sec. 2.5 Discrete Representation of Continuous Signals **63**

In this case we write

$$x(t) = \sum_{i \in I} c_i \psi_i(t), \qquad (2.249)$$

although this equality is not to be interpreted in the sense that its RHS and LHS are equal for every t, but rather in the sense that the energy of their difference is zero. This fact is sometimes expressed by saying that the RHS and LHS of (2.249) are equal *almost everywhere*.

In conclusion, once an orthonormal signal set has been chosen, a signal $x(t)$ can be represented by the sequence $(c_i)_{i \in I}$ defined by (2.246). This representation is exact (in the sense just specified) if the orthonormal set is complete with respect to $x(t)$.

Example 2.22   The complex Fourier series

The orthonormal sequence

$$\left(\frac{1}{\sqrt{T}} e^{ji2\pi t/T} \right)_{i=-\infty}^{\infty}$$

is complete for every complex signal $x(t)$ defined in the interval $(-T/2, \ T/2)$ and having bounded variation with finitely many discontinuity points. The expansion

$$x(t) = \frac{1}{\sqrt{T}} \sum_{i=-\infty}^{\infty} c_i e^{ji2\pi t/T}, \qquad t \in (-T/2, T/2), \qquad (2.250)$$

with

$$c_i = \frac{1}{\sqrt{T}} \int_{-T/2}^{T/2} x(t) e^{-ji2\pi t/T} \, dt, \qquad (2.251)$$

is the familiar complex Fourier-series representation. $\square$

Gram–Schmidt procedure

Because of the importance of orthonormal signal sequences, algorithms for constructing these sequences are of interest. One such algorithm, which is computationally convenient because of its iterative nature, is called the *Gram–Schmidt orthogonalization procedure*. Assume it is given a sequence $(\phi_i(t))_{i=1}^{N}$ of finite-energy, linearly independent signals, that is, such that any linear combination $\sum_{i=1}^{N} c_i \phi_i(t)$ is zero almost everywhere only if all the c_i are zero. An orthonormal sequence $(\psi_i(t))_{i=1}^{N}$ is generated using the following algorithm (see Problem 2.21):

$$\psi_1'(t) = \phi_1(t),$$

$$\psi_1(t) = \frac{\psi_1'(t)}{\sqrt{(\psi_1', \psi_1')}},$$

$$\psi_2'(t) = \phi_2(t) - (\phi_2, \psi_1)\psi_1(t),$$

$$\psi_2(t) = \frac{\psi_2'(t)}{\sqrt{(\psi_2', \psi_2')}},$$

$$\vdots$$

$$\psi_i'(t) = \phi_i(t) - \sum_{l=1}^{i-1} (\phi_i, \psi_l)\psi_l(t),$$

$$\psi_i(t) = \frac{\psi_i'(t)}{\sqrt{(\psi_i', \psi_i')}}, \qquad i = 3, \dots, N.$$

(2.252)

Example 2.23 The Legendre polynomials

If the Gram–Schmidt procedure is applied to the sequence of signals $(t^i)_{i=0}^N$ defined in the interval $(-1, 1)$, it results in the orthonormal polynomials

$$\psi_0(t) = \frac{1}{\sqrt{2}},$$

$$\psi_1(t) = \sqrt{\frac{3}{2}}\, t,$$

$$\psi_2(t) = \sqrt{\frac{5}{2}} \left(\frac{3}{2} t^2 - \frac{1}{2} \right),$$

$$\vdots$$

$$\psi_i(t) = \sqrt{\frac{2i+1}{2}}\, P_i(t),$$

where $P_i(t)$, $i = 1, \dots, N$, is the ith-degree Legendre polynomial, defined as

$$P_i(t) \triangleq \frac{1}{2^i i!} \frac{d^i}{dt^i} (t^2 - 1)^i. \quad \square$$

Geometrical representation of a set of signals

The theory of orthonormal expansions of finite-energy signals shows that a signal $x(t)$ can be represented by the (generally complex) sequence $(c_i)_{i \in I}$ of scalar products (2.246), once an orthonormal sequence that is *complete* for $x(t)$ has been provided. Now, if we consider a given sequence $(\psi_i(t))_{i=1}^N$ of N orthonormal *real* signals, it will be complete for any real $x(t)$ that can be written as a linear combination of the $\psi_i(t)$, that is, in the form

$$x(t) = \sum_{i=1}^{N} x_i \, \psi_i(t). \tag{2.253}$$

Thus, every such signal can be represented by the real N vector $\mathbf{x} \triangleq (x_1, x_2, \ldots, x_N)$ or, equivalently, by a point in the N-dimensional Euclidean space whose coordinate axes correspond to the signals $\psi_i(t)$, $i = 1, \ldots, N$.

Consider now a set $\{x_i(t)\}_{i=1}^{M}$ of real signals. Can we find an orthonormal sequence that is complete for these M signals? If so, we can represent $x_1(t), x_2(t), \ldots, x_M(t)$ as M vectors or as M points in a Euclidean space of suitable dimensionality. If the signals in the set $\{x_i(t)\}_{i=1}^{M}$ are linearly independent, it suffices to apply to it the Gram–Schmidt procedure to find such an orthonormal sequence. In fact, (2.252) shows that each of the $\psi_i(t)$ is expressed as a linear combination of signals in $\{x_i(t)\}_{i=1}^{M}$; hence, each of the $x_i(t)$ can be expressed as a linear combination of the $\psi_i(t)$. Suppose, instead, that only N signals in $\{x_i(t)\}_{i=1}^{M}$ are linearly independent, and hence $M - N$ of them can be expressed as linear combinations of the remaining signals. In this case, the Gram–Schmidt procedure can still be used, but it will produce *only $N < M$ nonzero* orthonormal signals. Every $x_i(t)$ is then represented by the N vector

$$\mathbf{x}_i \triangleq (x_{i_1}, x_{i_2}, \ldots, x_{i_N}), \tag{2.254}$$

where

$$x_{ij} \triangleq (x_i, \psi_j), \qquad i = 1, \ldots, M, \qquad j = 1, \ldots, N, \tag{2.255}$$

or, equivalently, as a point in the N-dimensional Euclidean space whose coordinate axes correspond to the nonzero orthonormal signals found through the Gram–Schmidt procedure. In this situation, we say that the signal set $\{x_i(t)\}_{i=1}^{M}$ *has dimensionality N.*

Example 2.24

Consider the four signals

$$x_i(t) = \cos\left[\frac{2\pi}{T}t + (i-1)\frac{\pi}{2}\right], \qquad t \in (0, T), \qquad i = 1, 2, 3, 4. \tag{2.256}$$

Using the Gram–Schmidt procedure, we get

$$\psi_1'(t) = \cos\frac{2\pi}{T}t,$$

$$\psi_1(t) = \sqrt{\frac{2}{T}}\cos\frac{2\pi}{T}t,$$

$$\psi_2'(t) = -\sin\frac{2\pi}{T}t, \tag{2.257}$$

$$\psi_2(t) = -\sqrt{\frac{2}{T}}\sin\frac{2\pi}{T}t,$$

$$\psi_3(t) = \psi_4(t) = 0,$$

which shows that the signal set (2.256) has dimensionality 2 and is represented by the four vectors:

$$\mathbf{x}_1 = \left(\sqrt{\frac{T}{2}}, 0\right),$$

$$\mathbf{x}_2 = \left(0, \sqrt{\frac{T}{2}}\right),$$

$$\mathbf{x}_3 = \left(-\sqrt{\frac{T}{2}}, 0\right), \tag{2.258}$$

$$\mathbf{x}_4 = \left(0, -\sqrt{\frac{T}{2}}\right),$$

The reader should observe that the signals

$$x_i(t) = \cos\left[\frac{2\pi}{T}t + (2i-1)\frac{\pi}{4}\right], \qquad t \in (0, T), \qquad i = 1, 2, 3, 4, \tag{2.259}$$

can also be represented using the same orthonormal basis. $\square$

Sampling expansion of bandlimited signals

Consider now the set of signals $x(t)$ strictly bandlimited in the frequency interval $(-B, B)$, that is, such that their Fourier transform $X(f)$ is identically zero for $|f| \geq B$. An orthonormal basis for any such $x(t)$ can be found as follows. Expand $X(f)$ in a Fourier series according to (2.250) and (2.251). Then take the inverse Fourier transform to get an expansion for $x(t)$. This procedure yields

$$X(f) = \frac{1}{\sqrt{2B}} \sum_{i=-\infty}^{\infty} c_i e^{ji\pi f/B}, \qquad f \in (-B, B), \tag{2.260}$$

$$c_i = \frac{1}{\sqrt{2B}} \int_{-B}^{B} X(f) e^{-ji\pi f/B} df, \tag{2.261}$$

and finally

$$x(t) = \frac{1}{\sqrt{2B}} \sum_{i=-\infty}^{\infty} c_i \int_{-B}^{B} e^{-ji\pi f/B} e^{j2\pi ft} df$$

$$= \sqrt{2B} \sum_{i=-\infty}^{\infty} c_i \frac{\sin 2\pi B(t - i/2B)}{2\pi B(t - i/2B)}, \tag{2.262}$$

which is an expansion valid for every $x(t)$ with bandwidth B. Observing further that the integral in the RHS of (2.261) is proportional to the inverse Fourier transform of $X(f)$ computed for $t = i/2B$, c_i can be put in the form

$$c_i = \frac{1}{\sqrt{2B}} x\left(\frac{i}{2B}\right). \tag{2.263}$$

This shows that the coefficients of the series expansion (2.262) are the *samples* of the signal $x(t)$ taken at the time instants $i/2B$, $-\infty < i < \infty$. In conclusion,

$$x(t) = \sum_{i=-\infty}^{\infty} x\left(\frac{i}{2B}\right) \frac{\sin 2\pi B(t - i/2B)}{2\pi B(t - i/2B)}. \tag{2.264}$$

Equation (2.264) shows that every finite-energy signal with bandwidth B hertz can be completely recovered from the knowledge of its samples taken at the rate of $2B$ samples per second. More generally, as any signal bandlimited in $(-B, B)$ is also bandlimited in $(-B', B')$, where $B' > B$, we can say that any finite-energy bandlimited signal can be represented using the sequence of its samples, provided that they are taken at a rate *not less than* $2B$. This minimum sampling rate of $2B$ is usually called the *Nyquist sampling rate* for $x(t)$. If $x(t)$ is a narrowband signal, it should be observed that it is convenient to apply the sampling expansion (2.264) to its complex envelope instead of the signal itself. This results in a much lower Nyquist frequency and, hence, to a more economical representation.

Observe now that (2.264) can also be written in the form

$$x(t) = \left\{ \sum_{i=-\infty}^{\infty} x\left(\frac{i}{2B}\right) \delta\left(t - \frac{i}{2B}\right) \right\} * \frac{\sin 2\pi Bt}{2\pi Bt}. \tag{2.265}$$

Now $\sin(2\pi Bt)/(2\pi Bt)$ can be interpreted as the impulse response of a linear, time-invariant system with frequency response

$$H(f) = \begin{cases} 1/2B, & |f| < B, \\ 0, & \text{elsewhere,} \end{cases} \tag{2.266}$$

that is, an ideal low-pass filter with cutoff frequency B. Thus, (2.265) suggests how to implement a system that recovers $x(t)$ from its samples. The sequence of samples is used to modulate linearly a train of impulses, which is then passed through an ideal low-pass filter (see Fig. 2.28).

A frequency-domain interpretation of the reconstruction of a sampled signal can also be provided. Let the signal $x(t)$ be sampled every T_s seconds, and observe that we can write

$$\sum_{i=-\infty}^{\infty} x(iT_s)\delta(t - iT_s) = x(t) \cdot \sum_{i=-\infty}^{\infty} \delta(t - iT_s). \tag{2.267}$$

The spectrum of this signal is obtained by taking the convolution of $X(f)$ and the Fourier transform of a train of impulses with period T_s. This is given by

$$\frac{1}{T_s} \sum_{i=-\infty}^{\infty} \delta(f - i/T_s)$$

[use (2.115)]. Thus, the spectrum of (2.267) is

$$X_s(f) \triangleq \frac{1}{T_s} \sum_{i=-\infty}^{\infty} X\left(f - \frac{i}{T_s}\right), \tag{2.268}$$

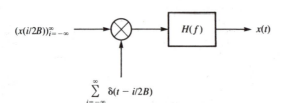

Figure 2.28 Recovering a bandlimited signal $x(t)$ from its samples. $H(f)$ is as in (2.266).

which is periodic with period $1/T_s$ (see Fig. 2.29). The original signal can be recovered from $X_s(f)$ by using the ideal low-pass filter whose transfer function $H(f)$ is shown in Fig. 2.29, provided that the translated copies of $X(f)$ forming $X_s(f)$ do not overlap. This condition holds only if $B \leq \dfrac{1}{T_s} - B$, that is,

$$f_s \geq 2B, \qquad (2.269)$$

where $f_s \triangleq 1/T_s$ is the sampling rate.

If (2.269) does not hold (i.e., the signal is sampled at a rate lower than Nyquist's), $x(t)$ cannot be recovered exactly from its samples. The signal obtained at the output of the ideal low-pass filter has the Fourier transform

$$H(f)X_s(f) = \begin{cases} \displaystyle\sum_{i=-\infty}^{\infty} X\left(f - \frac{i}{T_s}\right), & |f| < \dfrac{1}{2T_s}, \\ 0, & \text{elsewhere.} \end{cases} \qquad (2.270)$$

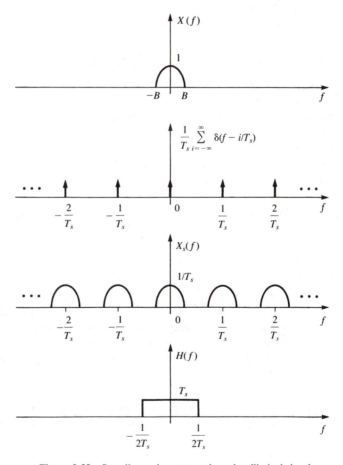

Figure 2.29 Sampling and reconstructing a bandlimited signal.

Sec. 2.5   Discrete Representation of Continuous Signals   **69**

It is important to observe that if this situation occurs (i.e., the signal is "undersampled"), even the *phase of the sampling process* affects the shape of the reconstructed signal. Specifically, if the sampled signal is

$$x(t) \sum_{i=-\infty}^{\infty} \delta(t - iT_s + \Theta),$$

where Θ is a constant smaller than T_s, at the output of the low-pass filter we get a signal whose Fourier transform is

$$\sum_{i=-\infty}^{\infty} X\left(f - \frac{i}{T_s}\right) e^{ji2\pi\Theta/T_s}, \qquad |f| < \frac{1}{2T_s}. \tag{2.271}$$

If the bandwidth of $x(t)$ does not exceed $1/(2T_s)$, (2.271) gives the spectrum of $x(t)$ (as it should). Otherwise, the shape of the signal recovered will also depend on the value of Θ.

Example 2.25

If the amplitude spectrum of $x(t)$ is as in Fig. 2.30a, the signal is undersampled as the spectrum $X(f)$ extends beyond $fT_s = 1$. Figure 2.30b shows the amplitude spectrum of the reconstructed signal for different values of the ratio Θ/T_s. $\square$

$2BT$ theorem and the uncertainty principle

The sampling expansion (2.264), which is valid for any $x(t)$ bandlimited in the interval $(-B, B)$, when applied to a signal vanishing outside the time interval $(0, T)$ has nonzero terms occurring only for $0 \le i/2B \le T$ (i.e., for $i = 0, 1, 2, \ldots,$ $2BT$). Thus, any *bandlimited and time limited* $x(t)$ is specified by $2BT + 1 \sim 2BT$ constants. For real signals, this can be interpreted as "the space of real signals of duration T and bandwidth B has dimension $2BT$."

However, these arguments are fallacious, because *no bandlimited signal* (besides the trivial null signal) *can have a finite duration.* The proof of this property is based on the fact that a signal $x(t)$ whose amplitude spectrum vanishes for $|f| > B$ can be written as

$$x(t) = \int_{-B}^{B} X(f)e^{j2\pi ft}\, df. \tag{2.272}$$

Now, if we allow t in (2.272) to be a complex variable, this extended $x(t)$ is an *entire function* of t. In other words, $x(t)$ has no singularities in the finite t plane, and its Taylor series expansion about every point has an infinite radius of convergence. Thus, any $x(t)$ vanishing on any interval of the time axis would have all its derivatives zero at some interior point of the interval. Hence, its Taylor series expansion would require it to be identically zero.

This impossibility for a signal to be simultaneously bandlimited and time limited is a special case of the *uncertainty principle* for a signal and its Fourier transform. One way to express this principle is the following (stated without proof). Define

$$\alpha \triangleq \frac{1}{\mathcal{E}_x} \int_{-T/2}^{T/2} |x(t)|^2\, dt \tag{2.273}$$

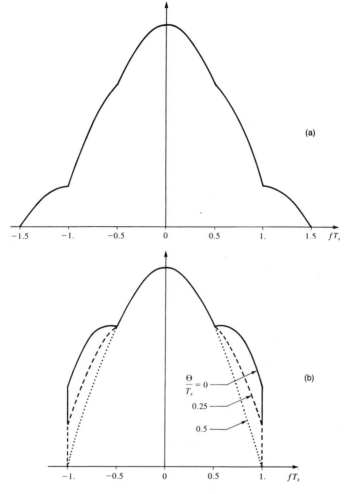

Figure 2.30 The influence of the sampling phase Θ in the reconstruction of an under-sampled signal: (a) Amplitude spectrum of the original signal; (b) amplitude spectrum of the reconstructed signal.

and

$$\beta \triangleq \frac{1}{\mathcal{E}_x} \int_{-B}^{B} |X(f)|^2 \, df, \tag{2.274}$$

where $\mathcal{E}_x$ denotes the energy of the signal $x(t)$. The quantities $\alpha \in [0, 1]$ and $\beta \in [0, 1]$ are, respectively, two measures of the concentration of the signal $x(t)$ and of its spectrum. The uncertainty principle states that

$$\pi BT \geq c(\lambda_0), \tag{2.275}$$

where

$$\lambda_0 = \cos^2(\vartheta_1 + \vartheta_2), \qquad \cos^2 \vartheta_1 = \alpha, \qquad \cos^2 \vartheta_2 = \beta, \tag{2.276}$$

and the function $c(\cdot)$ is shown in Fig. 2.31. Notice that $c(\lambda_0) \to \infty$ as $\lambda_0 \to 1$.

Sec. 2.5   Discrete Representation of Continuous Signals   **71**

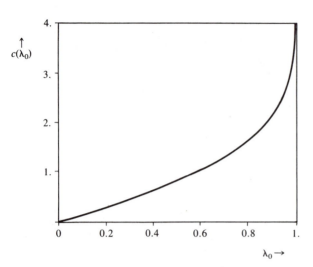

Figure 2.31 The function $c(\lambda_0)$ of the uncertainty principle (2.275).

With a signal both time limited and bandlimited, we should have $\alpha = \beta = 1$ for a finite product BT. But this would be in conflict with (2.275), as in this case $\lambda_0 = 1$, and hence $c(\lambda_0) = \infty$.

Example 2.26

For an example of the application of the uncertainty principle, determine the minimum value of the product πBT for $\alpha = \beta = 0.95$. From (2.276) we get $\vartheta_1 = \vartheta_2 = 0.2255$ and $\lambda_0 = \cos^2 0.4510 = 0.81$. The curve in Fig. 2.31 yields $c(\lambda_0) \cong 1.6$. $\square$

Let us now return to our $2BT$ theorem. Although it is not strictly true in the form stated at the beginning of this section, it can be reformulated in a more rigorous manner. To this end, assume that there is a smallest amount, say ϵ, of energy that we could measure by any means. We say that a real signal $x(t)$ is time limited to $(-T/2, T/2)$ *at level* ϵ if

$$\int_{|t|>T/2} x^2(t)\, dt < \epsilon, \tag{2.277}$$

and is bandlimited with bandwidth B *at level* ϵ if

$$\int_{|f|>B} |X(f)|^2\, df < \epsilon. \tag{2.278}$$

Conditions (2.277) and (2.278) indicate that the energy lying outside the time interval $(-T/2, T/2)$ and the frequency range $(-B, B)$ is less than we can measure. Furthermore, a set S of real signals is said to have dimension N *at level* ϵ if there is a set of N signals $\{\phi_i(t)\}_{i=1}^{N}$ such that, for each $x(t) \in$ S, there exist $a_1, a_2, \ldots, a_N$ such that

$$\int_{-T/2}^{T/2} [x(t) - \sum_{i=1}^{N} a_i\phi_i(t)]^2\, dt < \epsilon, \tag{2.279}$$

and there is no set of $N - 1$ functions that will approximate every $x(t) \in$ S in this manner. In words, every signal in S can be so well approximated in $(-T/2, T/2)$ by a linear combination of $\phi_1(t), \ldots, \phi_N(t)$ that we could not measure the energy of the difference between the signal and its approximation. With these definitions, we have the following theorem, due to Slepian (1976):

THEOREM. Let S_ϵ be the set of real signals time limited to $(-T/2, T/2)$ at level ϵ and bandlimited to $(-B, B)$ at level ϵ. Let $N = N(B, T, \epsilon, \epsilon')$ be the approximate dimension of S_ϵ at level ϵ'. Then, for every $\epsilon' > \epsilon$,

$$\lim_{T \to \infty} \frac{N(B, T, \epsilon, \epsilon')}{T} = 2B, \qquad \lim_{B \to \infty} \frac{N(B, T, \epsilon, \epsilon')}{2B} = T. \tag{2.280}$$

This $2BT$ theorem renders precise the concept that for large BT the space of signals of approximate duration T and approximate bandwidth B has approximate dimension $2BT$. The proof of the theorem will not be reported here. The interested reader is referred to Slepian (1976).

2.5.2 Orthonormal Expansions of Random Signals

We shall now briefly consider the problem of associating a discrete representation with a random signal $\xi(t)$. Quite generally, we look for a series expansion of the form

$$\xi(t) = \sum_{i=-\infty}^{\infty} \gamma_i \psi_i(t), \tag{2.281}$$

where $\psi_i(t)$, $-\infty < i < \infty$, are deterministic random functions, γ_i are random variables, and the equality is to be interpreted in the sense that

$$\lim_{k \to \infty} E\left| \xi(t) - \sum_{i=-k}^{k} \gamma_i \psi_i(t) \right|^2 = 0. \tag{2.282}$$

Various constraints may be imposed on $\xi(t)$, $\psi_i(t)$, and the random sequence (γ_i). In the *Karhunen–Loève expansion*, $\xi(t)$ is a WS stationary process defined in the finite interval $(0, T)$, $\{\psi_i(t)\}_{i=0}^{\infty}$ is a set of finite-energy orthogonal signals, and the coefficients γ_i are uncorrelated random variables.

If the process $\xi(t)$ is bandlimited, in the sense that its power spectrum $\mathcal{G}_\xi(f)$ vanishes outside the interval $(-B, B)$, we have the sampling expansion

$$\xi(t) = \sum_{i=-\infty}^{\infty} \xi\left(\frac{i}{2B}\right) \frac{\sin 2\pi B(t - i/2B)}{2\pi B(t - i/2B)}. \tag{2.283}$$

The coefficients $\xi(i/2B)$ are uncorrelated if and only if $\mathcal{G}_\xi(f)$ is constant over $(-B, B)$.

More general classes of series representations of WS stationary random processes were derived by Masry, Liu, and Steiglitz (1968) and Campbell (1969). Similar results were obtained for harmonizable processes by Cambanis and Liu (1970) and for an even more general class of processes (the "weakly continuous" processes) by Cambanis and Masry (1971).

2.6 ELEMENTS OF DETECTION THEORY

In this section we shall consider the problem of recognizing a signal chosen in a finite known set $\{s_i(t)\}_{i=1}^M$ once it has been perturbed by a random disturbance in the form of a noise process $v(t)$ independent of the signal and added to it. More specifically, the problem is to decide which one, among the signals $s_1(t)$, $s_2(t)$, . . . , $s_M(t)$, has given rise to the observed signal $y(t)$, when it is known that $y(t)$ has the form

$$y(t) = s_j(t) + v(t) \tag{2.284}$$

for some j, $1 \le j \le M$. Signals and noise may be either real or complex (i.e., complex envelopes of narrowband time functions). It will be assumed here that $v(t)$ is a white Gaussian process, with power spectral density $N_0/2$ (real signals) or $2N_0$ (complex envelopes). The signals dealt with have a finite energy and a finite duration. Also, their starting and ending times are known to the observer. We assume that $s_i(t)$, $1 \le i \le M$, are defined in the interval $0 \le t < T$ and that $y(t)$ is observed in the same interval.

This problem, called a *detection* problem, is central in digital transmission theory. It will be provided further motivation in Chapter 4, which includes a number of applications.

2.6.1 Optimum Detector: One Real Signal in Noise

We shall consider first, for simplicity's sake, the case in which there are only two signals, one of which is zero. Thus, the task is to decide between the two hypotheses

$$\begin{aligned} H_0&: y(t) = v(t), \\ H_1&: y(t) = s(t) + v(t), \end{aligned} \tag{2.285}$$

where $s(t)$ is a finite-energy real signal. The decision is based on the observation of $y(t)$ for $0 \le t < T$, and we want it to be made in such a way that the probability of a wrong decision is minimized. In words, we say that $H_1(H_0)$ is true when the observed signal contains (does not contain) $s(t)$.

A basic step in our derivation of the optimum detector is the discrete representation of the signals involved. This allows us to avoid further consideration of time functions. To do this, we expand $y(t)$ in an orthonormal series and represent it using the sequence of its coefficients. As a basis for this expansion, we choose any complete sequence of real signals $(\psi_i(t))_{i=1}^\infty$ orthonormal in the interval $(0, T)$ and such that $\psi_1(t) = s(t)/\sqrt{\mathcal{E}_s}$ (see Problem 2.23). Hence, $s(t)$ will be represented by the sequence $(\sqrt{\mathcal{E}_s}, 0, 0, . . .)$ and $v(t)$ by the sequence $(v_1, v_2, v_3, . . .)$, where

$$v_i \triangleq \int_0^T v(t)\psi_i(t)\, dt, \qquad i = 1, 2, ... \tag{2.286}$$

By direct calculation it can be shown that $E\{v_i\} = 0$, $i = 1, 2, . . .$, and

$$E\{v_i v_j\} = \int_0^T \int_0^T E\{v(t)v(\tau)\}\psi_i(t)\psi_j(\tau)\, dt\, d\tau$$

$$= \frac{N_0}{2}\int_0^T \int_0^T \delta(t-\tau)\psi_i(t)\psi_j(\tau)\, dt\, d\tau \qquad (2.287)$$

$$= \frac{N_0}{2}\int_0^T \psi_i(t)\psi_j(t)\, dt = \frac{N_0}{2}\delta_{ij}, \qquad i,j = 1, 2, \ldots$$

Since v_i, $i = 1, 2, \ldots$, are Gaussian RVs, (2.287) shows that they are independent.

In terms of these discrete representations, we can formulate our decision problem as follows. Decide among the hypotheses

$$H_0\colon (Y_i)_{i=1}^{\infty} = (v_1, v_2, \ldots),$$
$$H_1\colon (Y_i)_{i=1}^{\infty} = (v_1 + \sqrt{\mathscr{E}_s}, v_2, \ldots), \qquad (2.288)$$

on the basis of the observation of the quantities

$$Y_i \triangleq \int_0^T y(t)\psi_i(t)\, dt, \qquad i = 1, 2, \ldots \qquad (2.289)$$

Consider now a crucial point. Under both hypotheses H_0 and H_1, the observed quantities $Y_2, Y_3, \ldots$, are equal to $v_2, v_3, \ldots$, respectively, and these are independent of each other and Y_1. Thus, observation of $Y_2, Y_3, \ldots$, *does not add* any information to the decision process. Hence, it can be based solely on the observation of

$$Y_1 \triangleq \frac{1}{\sqrt{\mathscr{E}_s}}\int_0^T y(t)s(t)\, dt. \qquad (2.290)$$

In conclusion, the problem is reduced to the decision between the two hypotheses

$$H_0\colon Y_1 = v_1,$$
$$H_1\colon Y_1 = v_1 + \sqrt{\mathscr{E}_s}, \qquad (2.291)$$

upon observation of Y_1 as defined in (2.290). The quantity Y_1 is called the *sufficient statistics* for deciding between H_0 and H_1, because it extracts from the observed signal $y(t)$ all that is required to perform the decision. All other information about $y(t)$ is *irrelevant* to the decision process.

As the decision is based only on the observation of the scalar quantity Y_1, the optimum detector will first compute the scalar product (2.290) of the observed signal $y(t)$ and $s(t)$. Then it will choose either H_0 or H_1 according to the value taken by Y_1. If we denote by S_1 and $S_0 = R - S_1$ two subsets of the real line R, the decision rule is

$$\text{choose } H_0 \text{ if } y_1 \in S_0;$$
$$\text{choose } H_1 \text{ if } y_1 \in S_1; \qquad (2.292)$$

where y_1 is the observed value of Y_1. Hence, the optimum decision rule can be specified by choosing S_0 and S_1 in such a way that the average error probability is minimized. The error probability is given by

$$P(e) = \int_{S_0} p_1 f_{Y_1|H_1}(y|H_1)\, dy + \int_{S_1} p_0 f_{Y_1|H_0}(y|H_0)\, dy$$

$$= \int_{R} p_1 f_{Y_1|H_1}(y|H_1)\, dy - \int_{S_1} p_1 f_{Y_1|H_1}(y|H_1)\, dy \qquad (2.293)$$

$$+ \int_{S_1} p_0 f_{Y_1|H_0}(y|H_0)\, dy$$

$$= p_1 - \int_{S_1} [p_1 f_{Y_1|H_1}(y|H_1) - p_0 f_{Y_1|H_0}(y|H_0)]\, dy,$$

where $p_0 \triangleq P\{H_0\}$, $p_1 \triangleq P\{H_1\}$ are the a priori probabilities that H_0 is true [i.e., the observed signal does not contain $s(t)$] and H_1 is true [i.e., the observed signal contains $s(t)$], respectively.

To minimize $P(e)$, we should maximize the contribution to the integral of the term in brackets in the last expression of (2.293). This can be done by including in S_1 all the values y assumed by Y_1 such that $p_1 f_{Y_1|H_1}(y|H_1) > p_0 f_{Y_1|H_0}(y|H_0)$ and in S_0 the remaining values. Values of Y_1 such that the integrand is zero do not affect the value of $P(e)$. They may be included in either S_0 or S_1 arbitrarily. Hence, if we define the *likelihood ratio* among the hypotheses H_0 and H_1 as

$$\Lambda(y) \triangleq \frac{f_{Y_1|H_1}(y|H_1)}{f_{Y_1|H_0}(y|H_0)}, \qquad (2.294)$$

the decision rule becomes

$$\text{choose } H_0 \text{ if } \Lambda(y_1) \leq \frac{p_0}{p_1};$$

$$\qquad (2.295)$$

$$\text{choose } H_1 \text{ if } \Lambda(y_1) > \frac{p_0}{p_1}.$$

In conclusion, the optimum detector consists of a device that computes the likelihood ratio $\Lambda(y_1)$ and compares its value with the threshold p_0/p_1. Explicitly, we have

$$\Lambda(y) = \frac{e^{-(y-\sqrt{\mathscr{E}_s})^2/N_0}}{e^{-y^2/N_0}}$$

$$\qquad (2.296)$$

$$= \exp\left\{\frac{2}{N_0} y\sqrt{\mathscr{E}_s} - \frac{1}{N_0}\mathscr{E}_s\right\},$$

so that, using (2.290),

$$\Lambda(Y_1) = \exp\left\{\frac{2}{N_0}\int_0^T y(t)s(t)\, dt - \frac{1}{N_0}\int_0^T s^2(t)\, dt\right\}. \qquad (2.297)$$

Because of the likelihood ratio's structure, it is customary to define the *log-likelihood ratio* as the logarithm of $\Lambda(\cdot)$:

$$\lambda(y) \triangleq \ln \Lambda(y), \qquad (2.298)$$

so (2.297) becomes

76   Mathematical Techniques of Digital Communication Theory   Chap. 2

$$\lambda(Y_1) = \frac{2}{N_0} \int_0^T y(t)s(t)\, dt - \frac{1}{N_0} \int_0^T s^2(t)\, dt, \tag{2.299}$$

and the decision rule becomes

$$\text{choose } H_0 \text{ if } \lambda(y_1) \leq \ln \frac{p_0}{p_1}; \tag{2.300}$$

$$\text{choose } H_1 \text{ if } \lambda(y_1) > \ln \frac{p_0}{p_1}.$$

An important special case occurs when $p_0 = p_1$ (i.e., the two hypotheses are equally likely). In such a case, the decision is made by comparing $\lambda(y_1)$ against a zero threshold. Moreover, from (2.299) it is seen that the value of the constant N_0 is not relevant to the decision. Hence, when $p_0 = p_1$ the decision procedure does not depend on the spectral density of the noise. This simplification and the fact that the a priori probabilities p_0 and p_1 might be unknown justify the frequent use of the simplified decision rule (called the *maximum likelihood*, or ML, rule):

$$\text{choose } H_0 \text{ if } \lambda(y_1) \leq 0; \tag{2.301}$$
$$\text{choose } H_1 \text{ if } \lambda(y_1) > 0,$$

although it gives minimum error probability only when $p_0 = p_1$. [The rule (2.300) is referred to as the *maximum a posteriori probability*, or MAP, rule.] The structure of the ML detector is shown in Fig. 2.32.

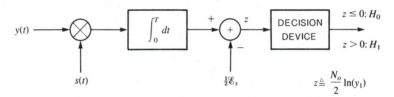

Figure 2.32 ML detection of a real signal $s(t)$ in Gaussian noise.

Example 2.27   The integrate-and-dump receiver

A simple special case of the general ML detector previously considered arises when the signal $s(t)$ has a constant amplitude A in the interval $0 \leq t < T$. The task is then to decide between the two hypotheses

$$H_0: y(t) = v(t), \tag{2.302}$$
$$H_1: y(t) = A + v(t),$$

upon observation of $y(t)$ for $0 \leq t < T$. In this case, $\mathcal{E}_s = A^2 T$, and from (2.290) we have

$$Y_1 = \frac{1}{\sqrt{T}} \int_0^T y(t)\, dt. \tag{2.303}$$

Equation (2.303) shows that the sufficient statistics for the detection are computed by averaging out the noise from the observed signal. This is obtained by integrating $y(t)$ over the observation interval.

Consider the performance of this detector when $p_0 = p_1$. The RV

$$v_1 \triangleq \frac{1}{\sqrt{T}} \int_0^T v(t)\, dt \qquad (2.304)$$

is Gaussian, with mean zero and variance $N_0/2$. Thus, the error probability under H_0 (i.e., the probability of choosing H_1 when H_0 is true) is

$$P(e|H_0) = P\{\lambda(Y_1) > 0|H_0\}$$
$$= P\left\{ v_1 > \frac{A\sqrt{T}}{2} \right\} \qquad (2.305)$$
$$= \frac{1}{2}\, erfc\left(\frac{1}{2} \sqrt{\frac{A^2 T}{N_0}} \right),$$

where $erfc(\cdot)$ is the complementary error function (see Appendix A). Similarly, the error probability under H_1 is

$$P(e|H_1) = P\{\lambda(Y_1) \le 0|H_1\}$$
$$= P\left\{ v_1 < -\frac{A\sqrt{T}}{2} \right\} \qquad (2.306)$$
$$= \frac{1}{2}\, erfc\left(\frac{1}{2} \sqrt{\frac{A^2 T}{N_0}} \right),$$

so that

$$P(e) = P(e|H_0)p_0 + P(e|H_1)p_1 = \frac{1}{2}\, erfc\left(\frac{1}{2} \sqrt{\frac{A^2 T}{N_0}} \right). \qquad (2.307)$$

If we define the *signal-to-noise ratio*,

$$\eta \triangleq \frac{A^2 T}{N_0}, \qquad (2.308)$$

it is seen that, as $P(e)$ is a monotone decreasing function of η, the error probability will decrease by increasing the level A, or by increasing the duration T of the observation interval, or by decreasing the noise spectral density. $\square$

Matched filter

Consider again (2.290). This equation shows that the sufficient statistics can be obtained, apart from a constant factor, as the output at time $t = T$ of a linear, time-invariant filter whose impulse response is

$$h(t) \triangleq s(T - t). \qquad (2.309)$$

In fact, with this definition we have

$$y(t) * h(t) \Big|_{t=T} = \int_0^T y(\tau)h(T - \tau)\, d\tau = \int_0^T y(\tau)s(\tau)\, d\tau. \qquad (2.310)$$

A filter whose impulse response is (2.309), or, equivalently, whose tranfer function is

$$H(f) \triangleq S^*(f)e^{-j2\pi f T}, \qquad (2.311)$$

where $S(f) \triangleq \mathcal{F}[s(t)])$, is called the filter *matched to the signal $s(t)$*. Thus, we can say that a matched filter whose output is sampled at $t = T$ extracts from the observed signal $y(t)$ the sufficient statistics for our decision problem.

An important property of the matched filter is that it maximizes the signal-to-noise ratio at its output, in the following sense. When a filter input is the sum of the signal $s(t)$ plus white noise $v(t)$, at time $t = T$ its output will be made up of two terms. The first is the signal part $\int_{-\infty}^{\infty} H(f)S(f)e^{j2\pi f T} df$, where $H(f)$ is the transfer function of the filter. The second is the noise part, a Gaussian RV with mean zero and variance $(N_0/2) \int_{-\infty}^{\infty} |H(f)|^2 df$. If we define the signal-to-noise ratio at the filter output as

$$z^2 \triangleq \frac{\left[\int_{-\infty}^{\infty} H(f)S(f)e^{j2\pi f T} df\right]^2}{(N_0/2) \int_{-\infty}^{\infty} |H(f)|^2 df},$$
(2.312)

(i.e., the ratio between the instantaneous power of the signal part and the variance of the noise part), we can show that z^2 is maximized if $H(f)$ has the form (2.311); that is, the filter is matched to the signal $s(t)$. The proof is based on Schwarz's inequality. It states that if $A(\cdot)$ and $B(\cdot)$ are two complex functions, then

$$|\int AB^*|^2 \leq \int |A|^2 \int |B|^2,$$
(2.313)

with equality if and only if $A = \alpha B^*$, where α is any complex constant. Using (2.313) in (2.312), we get

$$z^2 \leq \frac{\int_{-\infty}^{\infty} |H(f)|^2 df \int_{-\infty}^{\infty} |S(f)|^2 df}{(N_0/2) \int_{-\infty}^{\infty} |H(f)|^2 df}$$
(2.314)

$$= \frac{2\mathcal{E}_s}{N_0}.$$

Thus, the maximum value of the signal-to-noise ratio z^2 is obtained for

$$H(f) = \alpha S^*(f)e^{-j2\pi f T}.$$
(2.315)

Since α can be any constant, we can set $\alpha = 1$ without loss of optimality, so that the filter sought is indeed the matched filter as defined by (2.311). Notice that this filter is generally unrealizable. Also, its response to the input $s(t)$ is, at time $t = T$:

$$\int_{-\infty}^{\infty} H(f)S(f)e^{j2\pi f T} df = \int_{-\infty}^{\infty} |S(f)|^2 df = \mathcal{E}_s,$$
(2.316)

that is, the energy of $s(t)$.

Sec. 2.6   Elements of Detection Theory

2.6.2 Optimum Detector: M Real Signals in Noise

We now want to solve the most general problem stated at the beginning of this section, that is, to decide among the M hypotheses

$$H_j: y(t) = s_j(t) + v(t), \qquad j = 1, 2, \ldots, M, \qquad (2.317)$$

upon observation of $y(t)$ in the time interval $(0, T)$. The M real signals $s_j(t)$, $j = 1$, $\ldots, M$, are known and have a finite duration and a finite energy. Using the Gram–Schmidt procedure, we can determine an orthonormal signal set $\{\psi_i(t)\}_{i=1}^{N}$, $N \leq M$, such that each $s_j(t)$, $j = 1, 2, \ldots, M$, can be expressed as a linear combination of these signals. Also, consider a complete orthonormal signal sequence such that its first N signals are $\psi_1(t), \ldots, \psi_N(t)$ (see Problem 2.23). Denote with $(\psi_i(t))_{i=1}^{\infty}$ this sequence, and define

$$s_{ji} \triangleq \int_0^T s_j(t)\psi_i(t)\, dt, \qquad j = 1, \ldots, M, \qquad i = 1, 2, \ldots, \qquad (2.318)$$

and v_i, Y_i as in (2.286) and (2.289), respectively. The decision problem can be formulated in a discrete form as follows. Choose among the M hypotheses

$$H_j: (Y_i)_{i=1}^{\infty} = (s_{j1} + v_1, s_{j2} + v_2, \ldots, s_{jN} + v_N,$$
$$v_{N+1}, v_{N+2}, \ldots), \qquad j = 1, 2, \ldots, M \qquad (2.319)$$

on the basis of the observation of the values taken by the RVs, $Y_1, Y_2, \ldots$. As the noise components $v_{N+1}, v_{N+2}, \ldots$, are independent of $v_1, \ldots, v_N$ and of the hypothesis, observation of $Y_{N+1}, Y_{N+2}, \ldots$, does not add any information to the decision process. Thus, it can be based solely on the observation of $Y_1, Y_2, \ldots, Y_N$. By defining the row N vectors $\mathbf{Y} \triangleq [Y_1, Y_2, \ldots, Y_N]$, $\mathbf{v} \triangleq [v_1, v_2, \ldots, v_N]$, and $\mathbf{s}_j \triangleq [s_{j1}, s_{j2}, \ldots, s_{jN}]$, $j = 1, 2, \ldots, M$, (2.319) can be reduced to the vector form

$$H_j: \mathbf{Y} = \mathbf{s}_j + \mathbf{v}, \qquad j = 1, 2, \ldots, M. \qquad (2.320)$$

Thus, the optimum detector sought for will operate as follows:

$$\text{choose } H_j \text{ if } \quad \mathbf{y} \in S_j, \qquad (2.321)$$

where $\mathbf{y}$ denotes the observed value of the random vector $\mathbf{Y}$, and $S_1, S_2, \ldots, S_M$ is a partition of the N-dimensional vector space such that the rule (2.321) gives a minimum of the average error probability

$$P(e) = 1 - \sum_{j=1}^{M} p_j \int_{S_j} f_{\mathbf{Y}|H_j}(\mathbf{z}|H_j)\, d\mathbf{z}, \qquad (2.322)$$

where $p_j \triangleq P\{H_j\}$, $j = 1, 2, \ldots, M$. It is seen from (2.322) that $P(e)$ is minimized if every S_j is chosen in such a way that

$$\mathbf{z} \in S_j \quad \text{if and only if} \quad p_j f_{\mathbf{Y}|H_j}(\mathbf{z}|H_j) = \max_i \{p_i f_{\mathbf{Y}|H_i}(\mathbf{z}|H_i)\}. \qquad (2.323)$$

By combining (2.321) and (2.323), we obtain the MAP decision rule. In this situation the M-dimensional regions S_j are called the MAP *decision regions*. In the special case where the hypotheses H_j are equally likely, that is, $p_j = 1/M$, $j = 1, 2, \ldots, M$, (2.323) becomes

$$\mathbf{z} \in S_j \quad \text{if and only if} \quad f_{\mathbf{Y}|H_j}(\mathbf{z}|H_j) = \max_i f_{\mathbf{Y}|H_i}(\mathbf{z}|H_i), \tag{2.324}$$

which corresponds to the *maximum-likelihood* (ML) decision rule (accordingly, the S_j are called the ML decision regions). Although it minimizes the average error probability only for equally likely H_j, (2.324) is the most used detection rule, so in the following we shall mostly confine our attention to ML detection.

By defining the auxiliary hypothesis

$$H_0: \mathbf{Y} = \mathbf{\nu}, \tag{2.325}$$

(2.324) can also be written in the form

$$\mathbf{z} \in S_j \quad \text{if and only if} \quad \Lambda_j(\mathbf{z}) = \max_i \Lambda_i(\mathbf{z}), \tag{2.326}$$

where we define the *likelihood ratios*

$$\Lambda_j(\mathbf{y}) \triangleq \frac{f_{\mathbf{Y}|H_j}(\mathbf{y}|H_j)}{f_{\mathbf{Y}|H_0}(\mathbf{y}|H_0)}. \tag{2.327}$$

Thus, the ML decision rule is

$$\text{choose } H_j \quad \text{if } \Lambda_j(\mathbf{y}) = \max_i \Lambda_i(\mathbf{y}), \tag{2.328}$$

where, as usual, $\mathbf{y}$ denotes the observed value of $\mathbf{Y}$. That is, the ML detector operates by computing the M likelihood ratios $\Lambda_1(\mathbf{y}), \Lambda_2(\mathbf{y}), \ldots, \Lambda_M(\mathbf{y})$, and then choosing the hypothesis that corresponds to the largest among them. Let us now compute explicitly the likelihood ratios (2.327). By observing that, under hypothesis H_j, $j = 0, 1, \ldots, M$, $\mathbf{Y}$ is a Gaussian random vector with mean $\mathbf{s}_j$ (or zero for $j = 0$), independent components, and variance $N_0/2$ for each component, we have

$$\Lambda_j(\mathbf{y}) = \frac{\exp[-(1/N_0)|\mathbf{y} - \mathbf{s}_j|^2]}{\exp[-(1/N_0)|\mathbf{y}|^2]} = \exp\left\{\frac{2}{N_0}\mathbf{y}\mathbf{s}_j' - \frac{1}{N_0}|\mathbf{s}_j|^2\right\}, j = 1, \ldots, M, \tag{2.329}$$

where $|\mathbf{x}|^2 = \mathbf{x}\mathbf{x}' = \sum_{i=1}^{N} x_i^2$ denotes the squared modulus of the row vector $\mathbf{x}$. Consideration of the log-likelihood ratios

$$\lambda_j(\mathbf{y}) \triangleq \ln \Lambda_j(\mathbf{y}) \tag{2.330}$$

allows us to rewrite (2.328) in the following simple form:

$$\text{choose } H_j \quad \text{if } \mathbf{y}\mathbf{s}_j' - \frac{1}{2}|\mathbf{s}_j|^2 = \max_i \left\{\mathbf{y}\mathbf{s}_i' - \frac{1}{2}|\mathbf{s}_i|^2\right\}. \tag{2.331}$$

A different expression for the log-likelihood ratio can be derived as follows. Because

$$y(t) = \sum_{i=1}^{\infty} Y_i\psi_i(t) \tag{2.332}$$

and

$$s_j(t) = \sum_{l=1}^{N} s_{jl}\psi_l(t), \tag{2.333}$$

we have

$$\int_0^T y(t)s_j(t)\, dt = \sum_{i=1}^{\infty} \sum_{l=1}^{N} Y_i s_{jl} \int_0^T \psi_i(t)\psi_l(t)\, dt$$

$$= \sum_{i=1}^{N} Y_i s_{ji} \tag{2.334}$$

$$\triangleq \mathbf{Y}\mathbf{s}_j'$$

and

$$\int_0^T s_j^2(t)dt = \sum_{l=1}^{N} \sum_{k=1}^{N} s_{jl}s_{jk} \int_0^T \psi_l(t)\psi_k(t)\, dt$$

$$= \sum_{l=1}^{N} s_{jl}^2 \tag{2.335}$$

$$\triangleq |\mathbf{s}_j|^2$$

so that

$$\lambda_j(y) = \frac{2}{N_0} \int_0^T y(t)s_j(t)\, dt - \frac{1}{N_0} \int_0^T s_j^2(t)\, dt, \qquad j = 1, \ldots, M. \tag{2.336}$$

In Chapter 4 the structures of the optimum detectors based on (2.329) and (2.336) will be reexamined and discussed.

2.6.3 Detection Problem for Complex Signals

We shall now focus our attention on the problem of detecting *complex* signals in Gaussian noise. This situation occurs when we are dealing with narrowband signals that we want to describe using complex envelopes. Let us first consider the detection of a single complex signal in noise, that is, the decision among the hypotheses

$$H_0: y(t) = \frac{1}{\sqrt{2}} v(t), \tag{2.337}$$

$$H_1: y(t) = \frac{1}{\sqrt{2}} s(t) + \frac{1}{\sqrt{2}} v(t),$$

where $t \in (0, T)$, and $y(t)$, $v(t)$, and $s(t)$ are complex envelopes of narrowband signals (for notational simplicity, we omit the tilde). In particular, we have

$$s(t) = s_c(t) + js_s(t), \tag{2.338a}$$

$$v(t) = v_c(t) + jv_s(t), \tag{2.338b}$$

$$y(t) = y_c(t) + jy_s(t), \tag{2.338c}$$

where $v(t)$ is a complex Gaussian noise process with power spectral density $2N_0$, and $v_c(t)$, $v_s(t)$ are independent, white Gaussian baseband processes with power spectral density N_0 (see Fig. 2.20). Hence, $(1/\sqrt{2})\, v_c(t)$ and $(1/\sqrt{2})\, v_s(t)$ have spectral density $N_0/2$, and the energy of $(1/\sqrt{2})\, s(t)$ is the same as the real signal with which it is associated. Choose now a real orthonormal sequence $(\psi_{c1}(t),\ \psi_{s1}(t),\ \psi_{c2}(t),\ \psi_{s2}(t),\ \ldots)$ complete for any real signal in the time interval $(0,\ T)$, with $\psi_{c1}(t) = s_c(t)$ and $\psi_{s1}(t) = s_s(t)$. By formulating our detection problem in a discrete form, we have

$$H_0: (Y_{c1}, Y_{s1}, Y_{c2}, Y_{s2}, \ldots) = (v_{c1}, v_{s1}, v_{c2}, v_{s2}, \ldots),$$
$$H_1: (Y_{c1}, Y_{s1}, Y_{c2}, Y_{s2}, \ldots) = (s_{c1} + v_{c1}, s_{s1} + v_{s1}, v_{c2}, v_{s2}, \ldots),$$

(2.339)

where

$$s_{ci} \triangleq \frac{1}{\sqrt{2}} \int_0^T s_c(t)\psi_{ci}(t)\, dt, \tag{2.340a}$$

$$s_{si} \triangleq \frac{1}{\sqrt{2}} \int_0^T s_s(t)\psi_{si}(t)\, dt, \tag{2.340b}$$

$$v_{ci} \triangleq \frac{1}{\sqrt{2}} \int_0^T v_c(t)\psi_{ci}(t)\, dt, \tag{2.340c}$$

$$v_{si} \triangleq \frac{1}{\sqrt{2}} \int_0^T v_s(t)\psi_{si}(t)\, dt, \tag{2.340d}$$

$$Y_{ci} \triangleq \frac{1}{\sqrt{2}} \int_0^T y_c(t)\psi_{ci}(t)\, dt, \tag{2.340e}$$

$$Y_{si} \triangleq \frac{1}{\sqrt{2}} \int_0^T y_s(t)\psi_{si}(t)\, dt. \tag{2.340f}$$

Discarding the data irrelevant to the decision process, (2.339) can be put in the equivalent form

$$H_0: Y_1 = v_1,$$
$$H_1: Y_1 = s_1 + v_1,$$

(2.341)

where $Y_1 \triangleq Y_{c1} + jY_{s1}$, $v_1 \triangleq v_{c1} + jv_{s1}$, and $s_1 \triangleq s_{c1} + js_{s1}$. In this situation, the decision regions S_0 and S_1 are two dimensional, and the likelihood ratio

$$\Lambda(y_c, y_s) \triangleq \frac{f_{Y_{c1}, Y_{s1}|H_1}(y_c, y_s|H_1)}{f_{Y_{c1}, Y_{s1}|H_0}(y_c, y_s|H_0)} \tag{2.342}$$

is equal to

$$\Lambda(y_c, y_s) = \exp\left\{\frac{2}{N_0}(y_c s_{c1} + y_s s_{s1}) - \frac{1}{N_0}(s_{c1}^2 + s_{s1}^2)\right\}$$

$$= \exp\left\{\frac{2}{N_0}\mathscr{R}(y^* s_1) - \frac{1}{N_0}|s_1|^2\right\},$$

(2.343)

Sec. 2.6 Elements of Detection Theory

83

where $y \triangleq y_c + jy_s$. Through computations similar to those that led to (2.336), it can be shown that the likelihood ratio among H_0 and H_1 can also be written in the form

$$\Lambda(Y_1) = \exp\left\{\frac{2}{N_0}\mathcal{R}\int_0^T y^*(t)s(t)\,dt - \frac{1}{N_0}\int_0^T |s(t)|^2\,dt\right\}. \qquad (2.344)$$

This result can be extended to the problem of detecting one out of M complex signals in noise. In this case the likelihood ratio among the hypotheses

$$H_j\colon y(t) = \frac{1}{\sqrt{2}}s_j(t) + \frac{1}{\sqrt{2}}v(t), \qquad j = 1, 2, \ldots, M,$$

and

$$H_0\colon y(t) = \frac{1}{\sqrt{2}}v(t)$$

(where all the signals are complex) is given by

$$\Lambda_j(y) = \exp\left\{\frac{2}{N_0}\mathcal{R}\int_0^T y^*(t)s_j(t)\,dt - \frac{1}{N_0}\int_0^T |s_j(t)|^2\,dt\right\}. \qquad (2.345)$$

The proof of (2.345) is left to the reader.

BIBLIOGRAPHICAL NOTES

There are several excellent books covering the area of signal and system theory in which the reader can find further details regarding the topics covered in this chapter. Continuous-time and discrete-time deterministic signals and systems are treated extensively by Oppenheim, Willsky, and Young (1983). Discrete-time signals and systems are studied in Schwartz and Shaw (1975), whereas Papoulis (1977) covers both continuous- and discrete-time systems and deterministic and random signals.

Volterra series were first studied by the Italian mathematician Vito Volterra around 1880 as a generalization of the Taylor series of a function. His work in this area is summarized in Volterra (1959). The application of Volterra series to the analysis of nonlinear systems with memory was suggested by Norbert Wiener. Extensive treatments of Volterra series as applied to the description of nonlinear systems can be found in Schetzen (1980) and Rugh (1981). Applications are covered by Weiner and Spina (1980), and, among others, in the papers by Bedrosian and Rice (1971) and Benedetto, Biglieri, and Daffara (1976 and 1979).

Probability theory and random processes, at the level needed for this book, are covered by Parzen (1962), Papoulis (1965), and Davenport (1970). A comprehensive treatment of cyclostationary processes can be found in the dissertation by Hurd (1969) and in the papers by Gardner and Franks (1975) and Gardner (1978). Complex random processes are covered extensively by Miller (1974). Further details on Markov chains can be found in the classic by Feller (1968) or in Kemeny and Snell (1960). The two volumes by Gantmacher (1959) on matrix theory include a treatment of Markov chains based on their matrix description. The reader is warned, however, that the nomenclature in Markov chain theory varies in the literature.

Fourier series and Fourier transforms are covered by Bracewell (1978), Dym and McKean (1972), and Papoulis (1962). Arsac (1966) emphasizes generalized functions. The approach to the computation of the power density spectrum of a random process $\xi(t)$ based on the function $\Gamma_\xi (f_1, f_2)$ is described in some detail in Blanc-Lapierre and Fortet (1968) and in Papoulis (1965). Spectral analysis of digital signals based on a Markov chain model was first discussed by Huggins (1957) and Zadeh (1957). Since then, several authors have expanded on the basic results. For a comprehensive and detailed discussion of this topic, see Cariolaro, Pierobon, and Tronca (1983) and Galko and Pasupathy (1981), where the whole treatment is given a firm mathematical basis.

For a more detailed treatment of narrowband signals and bandpass systems than was possible here, the reader is referred to Schwartz, Bennett, and Stein (1966, pp. 29–45) and Franks (1969, pp. 79–97, 195–200), or to the papers of Arens (1957), Dugundji (1958), and Bedrosian (1962). Different possible definitions for the envelope of a narrowband signal are discussed and compared in Rice (1982). Bandpass nonlinear systems are introduced in Blachman (1971) (see also Blachman, 1982).

Orthonormal expansions of finite-energy signals and the Gram–Schmidt orthogonalization procedure are dealt with by Franks (1969), including an introduction to the Karhunen–Loève expansion and to the sampling theorem for random processes.

A profound treatment of detection theory can be found in the classics by Van Trees (1968) and Helstrom (1968). For the computation of the likelihood ratio in signal-detection problems, see also Turin (1969) and Kailath (1971). In the latter paper the case of nonwhite Gaussian noise is treated using the techniques of "reproducing-kernel Hilbert spaces."

PROBLEMS

2.1 A given (discrete or continuous) system may or may not be linear, time-invariant, memoryless, or causal. Determine which of these properties hold and which do not for each system with the given input–output relationship. In particular, when a system is not memoryless, determine the length of its memory.

(a) $y_n = 2x_n + 1$

(b) $y_n = nx_n$

(c) $y_n = 1 + \sum_{i=0}^{L} a_i x_{n-i}$

(d) $y_n = x_{[n/2]}$ ($[z] \triangleq$ integer part of z)

(e) $y_n = x_n[1 - \delta_n]$ [δ_n is defined in (2.4)]

(f) $y_n = x_n^*$

(g) $y(t) = 1 + \int_{-\infty}^{t} h(t - \tau)x(\tau)d\tau$

(h) $y(t) = \dfrac{dx(t)}{dt}$

(i) $y(t) = x(t - T) - x(t + T)$

(j) $y(t) = \int_{t}^{t+T} x(\tau)\, d\tau$

(k) $y(t) = x(t)e^{j2\pi f_0 t}$

(l) $y(t) = \displaystyle\int_{-\infty}^{\infty} x(\tau)e^{-j2\pi\tau t}\, d\tau$

2.2 Find the Fourier transform of the sequence (x_n), for

    (a) $x_n = \delta_n$ [δ_n is defined in (2.4)]

    (b) $x_n = \begin{cases} a^n, & n \geq 0 \\ 0, & n < 0 \end{cases}$ $(a < 1)$

    (c) $x_n = a^{|n|}$, $(a < 1)$

    (d) $x_n = \begin{cases} 1, & 0 \leq n \leq N - 1 \\ 0, & \text{elsewhere} \end{cases}$

    (e) $x_n = (-1)^n$

2.3 Given the discrete linear time-invariant system whose input–output relationship is described by the difference equation

$$y_n - \frac{5}{6}y_{n-1} + \frac{1}{6}y_{n-2} = x_n,$$

compute its transfer function $H(f)$ and its impulse response (h_n). Determine the output (y_n) when the input sequence is (x_n), with

$$x_n = \begin{cases} (\tfrac{1}{2})^n, & n \geq 0, \\ 0, & n < 0. \end{cases}$$

2.4 *Parseval equality:* Prove that

$$\sum_{n=-\infty}^{\infty} |x_n|^2 = \int_{-1/2}^{1/2} |X(f)|^2\, df,$$

where $X(f)$ denotes the Fourier transform of the sequence (x_n).

2.5 Prove that for a continuous or discrete time-invariant Volterra system there is no loss of generality if it is assumed that the kernels describing the system are symmetric (i.e., any permutation of their arguments leaves the kernels unchanged).

2.6 Prove that a continuous time-invariant Volterra system is causal if and only if (2.26) holds for all k. Give a corresponding condition for the kernels of a discrete time-invariant Volterra system.

2.7 Find the Volterra kernels for the continuous nonlinear system obtained by cascading a memoryless nonlinearity and a linear time-invariant system with impulse response $h(t)$ (Fig. P2.1).

Figure P2.1

It is assumed that

$$g(x) = \sum_{i=0}^{\infty} a_i x^i.$$

2.8 Find the input–output relationship of the system of Fig. P2.2, where $x(t)$ is a low-pass signal whose spectrum is zero for $|f| > f_1$, and $H(f)$ denotes an ideal bandpass filter centered at f_0, $f_0 \gg f_1$.

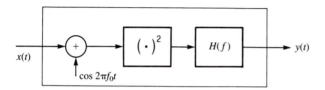

Figure P2.2

2.9 Consider a finite-energy signal $s(t)$ defined for $|t| \leqslant T/2$, and its Fourier transform $S(f)$. Denoting by $s^{(k)}(t)$ the kth derivative of $s(t)$, $[s^{(0)}(t) \triangleq s(t)]$, show that if

$$s^{(k)}\left(\frac{T}{2}\right) = s^{(k)}\left(\frac{-T}{2}\right) = 0, \qquad 0 \leqslant k \leqslant K,$$

then, as $f \rightarrow \infty$

$$|S(f)|^2 = O(f^{-2K-4}).$$

Consider then the signal

$$s(t) = \cos\left(\frac{\pi}{T}t + \alpha \sin\frac{4\pi}{T}t\right)$$

and find α so as to get $|S(f)|^2 = O(f^{-6})$.

2.10 *Discrete matched filter:* The input of a discrete linear time-invariant system with transfer function $H(f)$ is the real sequence $(x_n) = (s_n + w_n)$, where (s_n) is a deterministic sequence with Fourier transform $S(f)$, and (w_n) is a sequence of independent, identically distributed random variables. If (y_n) denotes the system response to (s_n) and (v_n) the system response to (w_n), find $H(f)$ such that the ratio $y_0^2/\mathrm{E}[v_n^2]$ is a maximum.

2.11 For a homogeneous Markov chain (ξ_n) with transition matrix $\mathbf{P}$, given an integer N and an N-tuple of integers $k_1 < k_2 < \cdots < k_N$, express the probability $P\{\xi_{k_1} = i_1, \xi_{k_2} = i_2, \ldots, \xi_{k_N} = i_N\}$ in terms of the entries of $\mathbf{P}$ and of the initial state distribution vector $\mathbf{w}^{(0)}$.

2.12 Prove that, if the transition matrix $\mathbf{P}$ of a fully regular homogeneous Markov chain is *doubly stochastic* (i.e., the sum of its entries in each row and column equals 1), then its stationary distribution vector $\mathbf{w}$ has equal components.

2.13 Let $\xi(t)$ be a WS stationary random process and $R_\xi(\tau)$ its autocorrelation function. Prove that, given $n \geqslant 1$, for any n-tuple of time instants $\tau_1, \tau_2, \ldots, \tau_n$ and for any n-tuple of complex numbers $a_1, a_2, \ldots, a_n$, the following inequality holds:

$$\sum_{i=1}^{n}\sum_{j=1}^{n} a_i^* a_j R_\xi(\tau_i - \tau_j) \geqslant 0.$$

2.14 Consider the linearly modulated digital signal

$$\xi(t) = \sum_{n=-\infty}^{\infty} \alpha_n s(t - nT),$$

where (α_n) is a sequence of independent, identically distributed random variables taking on values ± 1 with equal probabilities. Compute $\mathrm{E}[\xi^2(t)]$ and show that it is a periodic signal with period T.

2.15 Let $\xi(t)$ be a WS cyclostationary process with period T. Consider a random linear system whose output is $\eta(t) = \xi(t - \Theta)$, Θ an RV independent of $\xi(t)$ and uniformly distributed in the interval $(0, T)$. Prove that $\eta(t)$ is WS stationary. *Hint:* Consider $\Gamma_\eta(f_1, f_2)$.

2.16 Evaluate the power density spectrum of the digital signal

Chap. 2   Problems

87

$$\xi(t) = \sum_{n=-\infty}^{\infty} s(t - nT; \alpha_n),$$

where (α_n) is a sequence of independent, identically distributed random variables taking on values 1, 2, 3, 4 with equal probabilities,

$$s(t; \alpha_n) \triangleq \beta'_n r(t) + j\beta''_n r\left(t - \frac{T}{2}\right)$$

with $r(t)$ defined for $t \in (-T/2, T/2)$, and β'_n, β''_n are obtained from α_n according to the following table:

α_n	β'_n	β''_n
1	1	1
2	-1	1
3	1	-1
4	-1	-1

Specialize the result to the cases (which will be treated in detail in Chapter 5)

$$r(t) = 1 \quad \text{(offset PSK)}$$

$$r(t) = \cos\frac{\pi}{T}t \quad \text{(MSK)},$$

$$r(t) = \cos\left(\frac{\pi}{T}t - \frac{1}{4}\sin\frac{4\pi}{T}t\right) \quad \text{(SFSK)},$$

and plot the resulting power spectra for $|fT| \leq 10$.

2.17 *FSK digital signals*: Find the power density spectrum of the signal

$$\xi(t) = \exp\left\{j2\pi f_d \int_0^t q(\tau)\, d\tau\right\},$$

where

$$q(t) = \sum_{n=0}^{\infty} \alpha_n s(t - nT),$$

(α_n) is a sequence of independent, identically distributed random variables taking on values ± 1 with equal probabilities, and

$$s(t) = \begin{cases} 1, & 0 \leq t < T, \\ 0, & \text{elsewhere.} \end{cases}$$

(This refers to CPFSK modulated signals. They will be treated in Chapter 5).

2.18 Evaluate the power density spectrum of the digital signal

$$\xi(t) = \exp\left\{j\sum_{n=-\infty}^{\infty} \alpha_n g(t - nT)\right\},$$

where $g(t)$ is a signal with duration $\tau \leq T$, and (α_n) is a sequence of iid RVs taking on the M values $(\pi/M)(2i - 1)$, $i = 1, \ldots, M$, with equal probabilities.

2.19 Prove the following properties of the Hilbert transform $\hat{x}(t)$ of the signal $x(t)$:

(a) If $x(t)$ is an even function of t, $\hat{x}(t)$ is an odd function.

(b) If $x(t)$ is an odd function of t, $\hat{x}(t)$ is an even function.

(c) The Hilbert transform of $\hat{x}(t)$ is equal to $-x(t)$.

(d) The energy of $\hat{x}(t)$ is equal to the energy of $x(t)$.

(e) The energy of $x(t) + j\hat{x}(t)$ is equal to twice the energy of $x(t)$.

(f) $R_{\hat{x}\hat{x}}(\tau) = R_{xx}(\tau)$.

(g) $E\{\hat{x}(t + \tau)x(t)\}$ is equal to $\hat{R}_{xx}(\tau)$, the Hilbert transform of the autocorrelation function of $x(t)$, and $E\{x(t + \tau)\hat{x}(t)\}$ is equal to $-\hat{R}_{xx}(\tau)$.

2.20 Prove the equalities (2.189) and (2.190) *Hint*: Compute $R_{vv}(\tau)$ using (2.183).

2.21 Show that the Gram–Schmidt procedure (2.252) generates an orthonormal set of signals.

2.22 Consider the M signals

$$x_i(t) = \cos\left[2\pi f_0 t + (2i - 1)\frac{\pi}{M}\right], \qquad i = 1, 2, \ldots , M,$$

defined for $t \in (0, T)$, $f_0 T \gg 1$. Find the dimensionality of this signal set, an orthonormal basis, and a vector representation. Show that if the phase of the signals $x_i(t)$, $i = 1, \ldots ,$ M, is shifted by one and the same amount φ, they can still be represented using the same orthonormal basis.

2.23 Consider a given orthonormal signal set $\{\psi_i(t)\}_{i=1}^{N}$. Prove that it is possible to find a *complete* orthonormal signal sequence such that its first elements are $\psi_1(t), \ldots , \psi_N(t)$.

2.24 Let $x_c(t)$ be a continuous-time signal and (x_n) the sequence of its samples taken every T s; that is, $x_n \triangleq x_c(nT)$. If $X(f)$ denotes the Fourier transform of the sequence (x_n), and $X_p(f)$ denotes the Fourier transform of the signal

$$x_p(t) \triangleq x_c(t) \sum_{n=-\infty}^{\infty} \delta(t - nT),$$

prove that

$$X_p(f) = X(fT).$$

2.25 Generalize Eq. (2.268) to the case in which the sampling waveform, instead of being a train of ideal impulses (delta functions), is the periodic signal $\sum_{i=-\infty}^{\infty} p(t - iT)$, where $p(t)$ is a rectangular pulse with duration $\tau < T$. Can the original signal $x(t)$ still be recovered exactly from the product signal $x(t) \sum_{i=-\infty}^{\infty} p(t - iT)$?

2.26 *Matched filter for nonwhite noise*: Consider a continuous linear time-invariant system whose input is the sum of a deterministic signal $s(t)$ and a WS stationary noise $v(t)$ whose power spectral density $\mathcal{G}_v(f)$ is nonzero for all f. Find the transfer function of the system that maximizes the ratio between the instantaneous power of the signal part and the variance of the noise part at its output.

CHAPTER 3

Basic Results from Information Theory

In this chapter we deal with information sources and communication channels. Apart from a brief description of continuous sources and channels at the end of the chapter, aimed at obtaining the capacity of the bandlimited Gaussian channel, the main part of the treatment is devoted to the discrete case.

The first part of the chapter defines a discrete stationary source and shows how the quantity of information that is emitted from the source can be measured. In general, the source output (the *message*) consists of a sequence of *symbols* chosen from a finite set, the *alphabet* of the source. A probability distribution is associated with the source alphabet, and a probabilistic mechanism governs the emission of successive symbols in the message. Generally, different messages convey different quantities of information; thus an average information quantity, or *entropy*, must be defined for the source. The unit of measure for the information is taken to be the *bit* (i.e., the information provided by the emission of one among two equally likely symbols). The entropy of the source gives the minimum average number of binary symbols (*digits*) that is necessary to represent each symbol in the message. The source output can thus be replaced by a string of binary symbols conveying the same quantity of information and having an average number of digits per symbol of the original source as close as desired to the source entropy. The block in the system that implements this function is called *source encoder* and is described in this chapter.

The communication channel is the physical medium used to connect the source of information with its user. In the second part of the chapter, discrete memoryless channels are defined and their properties are studied. They are specified by a probability law that connects symbols of the channel input alphabet to symbols of the channel output alphabet. A basic point is the knowledge of the maximum average information flow that can reliably pass through the channel. This leads to the definition of the channel *capacity* and to the problem of computing it. Both topics are addressed in this chapter.

The final part of the chapter is devoted to a presentation of the channel coding theorem and its converse. They provide a link between the concepts of entropy of a source and capacity of a channel and assess precisely what has to be meant by reliable transmission.

The main goal of this chapter is to provide a general frame to the subsequent material, which deals with specific aspects of data transmission systems. It also assesses the theoretical limits in performance that can be obtained over a binary channel and an additive Gaussian channel. The communication sytems used in the practice will be compared in Chapter 5, with these limits in mind.

3.1 INTRODUCTION

The goal of every communication system is the reproduction of a message emitted from a source to a place where the user of the information is located. The distance between the source and the user may be either considerable, as in the case of intercontinental transmission, or very small, as in the storage and retrieval of data using the disk unit of a digital computer. However, irrespective of distance, there exists between the source and the user a communicating channel affected by noise.

The presence of noise means that the exact reproduction of the message emitted from the source, at the user's premises, is an impossible achievement. Nevertheless, the designer of a communication system will always be asked to provide the user with an "as close as possible" replica of the original message. A closer insight into the characteristics of the user better specifies, case by case, the meaning of "as close as possible," that is, the specification of a *user-oriented* criterion of acceptability. For example, in the case of speech communication in the area of service communications, one can be satisfied when the listener can understand the semantic content of what the speaker is saying. However, the listener often wishes to recognize the identity and mood of the speaker through the timbre and inflection of his or her voice, and this gives rise to a more rigorous criterion of acceptability. Hence, as illustrated in these examples, different user requirements lead to different criteria of acceptability and, consequently, to different bandwidth requirements for speech transmission.

As explained, the problem of noise in the communication channel creates the need for user-sensitive specifications of criteria of acceptability in the design of communication systems. This can be accomplished in the following fashion. The source outputs, for a given time interval, are divided into equivalence classes on the basis of a certain criterion of acceptability. This permits one to regard such source outputs as a set of equivalence classes where those source outputs residing in the same equivalence class are indistinguishable with respect to the acceptability criterion. Thus, the communication system, in this regard, is reduced to the transmission of an indication of the specific class to which the source output belongs in each successive time interval.

In Fig. 3.1, one possible class of equivalence is depicted for the transmission of written texts, where the criterion of acceptability is merely the semantic intelligibility

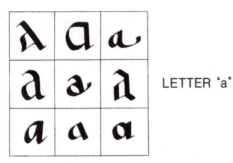

LETTER "a"

Figure 3.1 Class of equivalence relative to the letter ''a'' in handwritten texts.

of the message. The class represents different ways of writing the letter a. Another well-known example is the quantization process performed in connection with *pulse-code modulation* (PCM). In Fig. 3.2, the process is schematically outlined. The source voltage is first sampled every T seconds. Each voltage sample is then quantized, (i.e., the closest value in a discrete preselected set of voltages is substituted for it), and kept constant for T seconds.

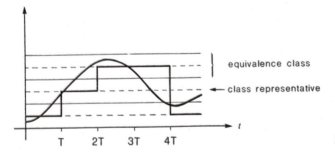

equivalence class

class representative

Figure 3.2 Quantization process in PCM.

From here on we shall assume that the criterion of acceptability gives rise to a finite number of equivalence classes, say M, in a specified time interval. The transmission of information is then equivalent to the communication of a sequence of integer numbers chosen in the set $\{1, 2, \ldots, M\}$ from the source to the user, which, in turn, generates the source output representative of each indicated class.

3.2 DISCRETE STATIONARY SOURCES

Consider a finite *alphabet* X formed by the M symbols $\{x_i\}_{i=1}^{M}$ and define a *message* as a sequence of symbols such as $(x_n)_{n=0}^{\infty}$. A discrete stochastic source is a device emitting messages through the selection of symbols from the alphabet X according to a probability distribution $\{p_i\}_{i=1}^{M}$, where $p_i \triangleq P(x_i)$. From a probabilistic point of view, one can regard the whole set of messages as a discrete random process, that is, a sequence $(\xi_n)_{n=0}^{\infty}$ of random variables (RVs), each taking values in the set X with the probability distribution $\{p_i\}$.

We shall assume that the source is *stationary*; that is,

$$P\{\xi_{i_1} = x_1, \ldots, \xi_{i_k} = x_k\} = P\{\xi_{i_1+h} = x_1, \ldots, \xi_{i_k+h} = x_k\} \tag{3.1}$$

for all nonnegative integers $i_1, \ldots, i_k, h$ and all $x_1, \ldots, x_k \in X$. In this case the message set forms a discrete-time stationary random process.

3.2.1 A Measure of Information: Entropy of the Source Alphabet

The quantity of information carried by one particular symbol of the source alphabet is strictly related to its uncertainty. Increased uncertainty should correspond to more information. As an example, the letter size in a newspaper headline is larger when the news is unexpected, like "Found Life on Mars!" It is then fairly natural that the information content of the ith symbol, denoted by $I(x_i)$, be a decreasing function of its probability

$$I(x_j) > I(x_i), \qquad \text{if } p_j < p_i, \tag{3.2}$$

and that the information content associated with the emission of two independent symbols be the sum of the two individual informations

$$\text{If } P(x_i, x_j) = P(x_i)P(x_j) \quad \text{then} \quad I(x_i, x_j) = I(x_i) + I(x_j). \tag{3.3}$$

A definition of the information content satisfying both (3.2) and (3.3) is

$$I(x_i) \triangleq \log_a \frac{1}{p_i}. \tag{3.4}$$

In (3.4) the base of the logarithm (indicated with a) is unspecified. Its choice determines the unit of measurement assigned to the information content. If the natural (base e) logarithm is used, then the unit is called *nat*, and if the base is 2, then the unit is widely known as *bit* (binary digit). The use of bit is based on the fact that the correct identification of one out of two equally likely symbols conveys an amount of information equal to $I(x_1) = I(x_2) = \log_2 2 = 1$ bit. Unless otherwise specified, we shall use the base 2 in this chapter and write log to mean $\log_2$.

The definition (3.4) allows one to associate with each symbol of the source alphabet its information content. A characterization of the whole alphabet can be obtained by defining the *average information content* of X:

$$H(X) \triangleq \sum_{i=1}^{M} p_i I(x_i) = \sum_{i=1}^{M} p_i \log \frac{1}{p_i}, \tag{3.5}$$

which is called the *entropy of the source alphabet* and is measured in *bits/symbol*.

Example 3.1

The source alphabet consists of four possible symbols with probabilities $p_1 = \frac{1}{2}$, $p_2 = \frac{1}{4}$, $p_3 = p_4 = \frac{1}{8}$. To compute the entropy of the source alphabet, we apply definition (3.5):

$$H(X) = \tfrac{1}{2} \log 2 + \tfrac{1}{4} \log 4 + 2 \cdot \tfrac{1}{8} \log 8 = 1.75 \text{ bits/symbol}.$$

If the source alphabet consists of M equally likely symbols, we have

$$H(X) = \sum_{i=1}^{M} \frac{1}{M} \log M = \log M \text{ bits/symbol}.$$

When the source alphabet consists of two symbols with probabilities p and $q = 1 - p$, the alphabet entropy is

Sec. 3.2   Discrete Stationary Sources   **93**

$$H(X) = p \log \frac{1}{p} + (1 - p) \log \frac{1}{1 - p} \triangleq H(p). \tag{3.6}$$

In Fig. 3.3, the function $H(p)$ is plotted. It can be seen that the maximum occurs for $p = 0.5$, that is, when the two symbols are equally likely. $\square$

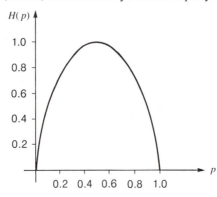

Figure 3.3 Plot of the entropy $H(p)$ of a binary source with $P(x_1) = p$ and $P(x_2) = 1 - p$.

The last result of Example 3.1 is fairly general, as will be stated in the following theorem.

THEOREM 3.1. The entropy $H(X)$ of a source alphabet with M symbols satisfies the inequality

$$H(X) \leq \log M \tag{3.7}$$

with equality when the symbols are equally likely.

To prove the theorem, consider the difference

$$H(X) - \log M = \sum_{i=1}^{M} p_i \log \frac{1}{p_i} - \sum_{i=1}^{M} p_i \log M \tag{3.8}$$

$$= \sum_{i=1}^{M} p_i \log \frac{1}{p_i M}.$$

Making use of the inequality

$$\ln y \leq y - 1 \tag{3.9}$$

in the RHS of (3.8), we obtain

$$H(X) - \log M \leq \log e \sum_{i=1}^{M} \left(\frac{1}{M} - p_i \right) = 0. \quad \blacksquare$$

3.2.2 Coding of the Source Alphabet

For a given source, we are now able to compute the information content of each symbol in the source alphabet and the entropy of the alphabet itself. Suppose now that we want to transmit each symbol using a binary channel. Before being delivered to the channel, each symbol must be represented by a finite sequence of digits, called *code word*. Leaving aside the problem of possible channel errors, efficient communication would involve transmitting a symbol in the shortest possible time, which, in turn, means representing it with a code word as short as possible. As usual, we are interested in average quantities, so our goal will be that of minimizing the *average length* of a code word

$$\bar{n} \triangleq \mathrm{E}\{n\} = \sum_{i=1}^{M} p_i n_i, \qquad (3.10)$$

where n_i is the length (number of digits) of the code word representing the symbol x_i, and n is the random variable representing its length (i.e., assuming the value n_i with probability p_i, $i = 1, 2, \ldots, M$).

The minimization of (3.10) has to be accomplished according to an important constraint placed on the assignment of code words to the alphabet symbols. For example, consider the following code:

Symbols	Code words
x_1	0
x_2	01
x_3	10
x_4	100

the binary sequence 010010 could correspond to any one of the five messages $x_1 x_3 x_2 x_1$, $x_2 x_1 x_1 x_3$, $x_1 x_3 x_1 x_3$, $x_2 x_1 x_2 x_1$, or $x_1 x_4 x_3$. The code is *ambiguous* or *not uniquely decipherable*. It then seems natural to require that the code be *uniquely decipherable*, which means that every finite sequence of binary digits corresponds to, at most, one message. A condition that ensures unique decipherability is to require that no code word be a prefix of a longer code word. The codes described in the sequel satisfy this constraint.

A very useful graphical representation for a code satisfying the prefix constraint is that which associates to each code word a terminal node in a binary tree, like the one of Fig. 3.4. Starting from the root of the tree, the two branches leading to the first-order nodes correspond to the choice between 0 and 1 as the first digit in the code words. The two branches stemming from each of the first-order nodes correspond to

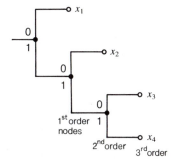

Symbols	Code words
x_1	0
x_2	10
x_3	110
x_4	111

Figure 3.4 Binary tree associated with a binary source code.

the second digit of the code words, and so on. Since code words are assigned only to terminal nodes, no code word can be a prefix of another code word. A tree is said to be of order n if it contains nodes up to the nth order.

A necessary and sufficient condition for a given code to satisfy the prefix constraint is given in the following theorem.

THEOREM 3.2 KRAFT INEQUALITY. A binary code satisfying the prefix constraint with word lengths $n_1, n_2, \ldots, n_M$ exists if and only if

$$\sum_{i=1}^{M} 2^{-n_i} \leq 1. \tag{3.11}$$

We shall prove first that (3.11) is a necessary condition. As the code satisfies the prefix constraint, it is imbedded in a tree of order $n = \max (n_1, n_2, \ldots, n_M)$. The presence in the tree of a terminal node of order n_i eliminates 2^{n-n_i} of the possible nodes of order n. Thus, for the code to be embedded in the tree, the sum of all nodes of order n eliminated by terminal nodes associated with code words must be less than or equal to the number of nodes of order n in the tree; that is,

$$\sum_{i=1}^{M} 2^{n-n_i} \leq 2^n.$$

Dividing both sides of the last inequality by 2^n yields (3.11). To prove that (3.11) is a sufficient condition for a code to satisfy the prefix constraint, let us assume that the n_i's are arranged in nondecreasing order, $n_1 \leq n_2 \leq \cdots \leq n_M$. Choose as the first terminal node in the code tree any node of order n_1 in a tree of order n_M containing all branches. All nodes on the tree of each order greater than or equal to n_1 are still available for use as terminal nodes in the code tree, except for the fraction 2^{-n_1} that stems from the chosen node. Next, choose any available node of order n_2 as the next terminal node in the code tree. All nodes in the tree of each order greater than or equal to n_2 are still available except for the fraction $2^{-n_1} + 2^{-n_2}$ that stem from either of the two chosen nodes. Continuing in this way, after the assignment of the jth terminal node in the code tree, the fraction of nodes eliminated by previous choices is $\sum_{i=1}^{j} 2^{-n_i}$. From (3.11), this fraction is always strictly less than 1 for $j < M$, and thus there is always a node available to be assigned to the next code word. ∎

Since we are using a binary code, the maximum information content of each digit in the code word is 1 bit. So the average information content in each code word is, at most, equal to $\bar{n}$. On the other hand, to specify uniquely a symbol of the source alphabet, we need an average amount of information equal to H(X) bits. Hence we can intuitively conclude that

$$\bar{n} \geq \text{II(X)}. \tag{3.12}$$

Comparing the definitions (3.10) and (3.5) of $\bar{n}$ and H(X), it can be seen that the condition (3.12) can be satisfied with the equal sign if and only if

$$p_i = 2^{-n_i}, \qquad i = 1, 2, \ldots, M. \tag{3.13}$$

In this case, (3.11) also becomes an equality.

Example 3.2

The following is an example of a code satisfying (3.12) with the equal sign and subject to the prefix constraint.

Symbols	p_i	Code words
x_1	$\frac{1}{2}$	1
x_2	$\frac{1}{4}$	00
x_3	$\frac{1}{8}$	010
x_4	$\frac{1}{16}$	0110
x_5	$\frac{1}{16}$	0111

Computing the value of $\bar{n}$ defined in (3.10), one obtains

$$\bar{n} = H(X) = \tfrac{15}{8}. \quad \square$$

In general, condition (3.13) with n_i integers is not satisfied. So we cannot hope to attain the lower bound for $\bar{n}$ as in the previous example. However, a code satisfying the prefix constraint can be found whose $\bar{n}$ obeys the following theorem.

THEOREM 3.3. A binary code satisfying the prefix constraint can be found for any source alphabet of entropy H(X) whose average code word length $\bar{n}$ satifies the inequality

$$H(X) \leq \bar{n} < H(X) + 1. \tag{3.14}$$

An intuitive proof of the lower bound has already been given when introducing (3.12). Let us now choose for the code word representing the symbol x_i a number of bits n_i corresponding to the smallest integer greater than or equal to $I(x_i)$. So we have

$$I(X_i) \leq n_i < I(x_i) + 1. \tag{3.15}$$

Multiplying (3.15) by p_i and summing over i, we obtain

$$H(X) \leq \bar{n} < H(X) + 1.$$

To complete the proof of the theorem, we still have to show that the code satisfies the prefix constraint, that is the lengths n_i's of the code words obey the Kraft inequality (3.11). Recalling the definition (3.4) of $I(x_i)$, the left-hand inequality of (3.15) leads to $p_i \geq 2^{-n_i}$; so, summing with respect to i, we get

$$\sum_{i=1}^{M} 2^{-n_i} \leq \sum_{i=1}^{M} p_i = 1. \quad \blacksquare$$

The last step in our description of the source alphabet coding is the *construction* of a code uniquely decipherable that minimizes the average code word length. We shall present a method for the construction of such optimal codes due to Huffman; the proof of optimality will be omitted. The Huffman procedure will be described step by step. The reader is referred to Fig. 3.5, in which the steps can be spotted in the tree originated by the encoding procedure.

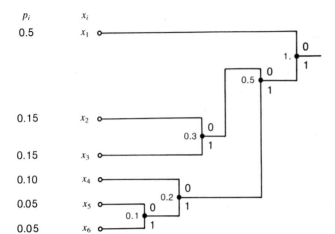

Figure 3.5 Tree generated by the Huffman encoding procedure for a source with six symbols.

Step 1 Have the M symbols ordered according to nonincreasing values of their probabilities.

Step 2 Group the last two symbols x_{M-1} and x_M into an equivalent "symbol," with probability $p_{M-1} + p_M$.

Step 3 Repeat steps 1 and 2 until only one "symbol" is left.

Step 4 Looking at the tree originated by the preceding steps (see Fig. 3.5), associate the binary symbols 0 and 1 to each pair of branches departing from intermediate nodes. The code word of each symbol can be read as the binary sequence found when starting from the root of the tree and reaching the terminal node associated with the symbol at hand.

For the example of Fig. 3.5, the code words obtained using the Huffman procedure are

Symbols	Code words
x_1	0
x_2	100
x_3	101
x_4	110
x_5	1110
x_6	1111

The average length $\bar{n}$ and the entropy $H(X)$ are 2.1 digits/symbol and 2.09 bits/symbol, respectively; they satisfy (3.14), and no other code can do better.

Example 3.3

For the code shown in Fig. 3.5, use the tree to decode the received sequence 1100101100110. Starting from the root of the tree, we follow the branches at each intermediate node according to the binary digits in the received sequence, until a terminal node (and hence a symbol) is reached. Then we restart the procedure. The decoded sequence is $x_4 x_1 x_3 x_2 x_4$. □

The reader is invited to repeat the decoding procedure of Example 3.3, assuming that an error had been introduced by the channel in the first position. This permits us to verify the catastrophic effect of the error's propagation in these variable-length codes. On the other hand, the goal of the source coding is the *reduction of redundancy* of the source alphabet, and not the protection against channel errors. This is the scope of *channel encoding*, as we will see in Chapter 9.

So far, we have seen how a code word can be efficiently assigned to each symbol x_i of the source alphabet X. The main result is represented by (3.14). In fact, the lower bound of (3.14) can be approached as closely as desired if we are allowed to encode a *block of symbols* instead of single symbols. Suppose we take a sequence of independent observations of X and assign a code word to the resulting group of symbols. In other words, we construct a code for a new alphabet $Y \equiv X^\nu$ containing M^ν symbols y_i. The probability of y_i is then given by the product of the probabilities corresponding to the ν symbols of X that specify y_i. By Theorem 3.3, we may construct a code for Y whose average code word length $\bar{n}_\nu$ satisfies

$$H(Y) \le \bar{n}_\nu < H(Y) + 1. \tag{3.16}$$

But every symbol in Y is made by ν independent symbols of the original alphabet X; so the entropy of Y is $H(Y) = \nu H(X)$ (see Problem 3.2). Thus, from (3.16) we get

$$H(X) \le \frac{\bar{n}_\nu}{\nu} \le H(X) + \frac{1}{\nu}. \tag{3.17}$$

But $\bar{n}_\nu/\nu$ is the average number of digits/symbol of X; so from (3.17) it follows that it can be made arbitrarily close to H(X) by choosing ν sufficiently large.

The *efficiency* ϵ of a code is defined as

$$\epsilon \triangleq \frac{\nu H(X)}{\bar{n}_\nu},$$

and its *redundancy* is $(1 - \epsilon)$.

Example 3.4

Given the source alphabet $X = \{x_1, x_2, x_3\}$, with $p_1 = 0.5$, $p_2 = 0.3$, and $p_3 = 0.2$, we want to construct the new alphabet $Y = X^2 = \{y_1, y_2, \ldots, y_9\}$, obtained by grouping the symbols x_i two by two.

Pair of symbols	Probability
$y_1 = x_1 x_1$	$P(y_1) = P(x_1 x_1) = P(x_1)P(x_1) = 0.25$
$y_2 = x_1 x_2$	$P(y_2) = 0.15$
$y_3 = x_2 x_1$	$P(y_3) = 0.15$
$y_4 = x_2 x_2$	$P(y_4) = 0.09$
$y_5 = x_1 x_3$	$P(y_5) = 0.1$
$y_6 = x_3 x_1$	$P(y_6) = 0.1$
$y_7 = x_3 x_3$	$P(y_7) = 0.04$
$y_8 = x_2 x_3$	$P(y_8) = 0.06$
$y_9 = x_3 x_2$	$P(y_9) = 0.06$

The reader is invited to construct the Huffman codes for block lengths $\nu = 1$ and $\nu = 2$ and compare the average numbers of digits/symbol obtained in both cases, using the preceding definition of code efficiency. $\square$

3.2.3 Entropy of Stationary Sources

Although our definition of a discrete stationary source is fairly general, we have so far considered in detail only the information content and the encoding of the source alphabet. Even when we have described the achievement of the block encoding of the source, we made the assumption of independence between the symbols forming each block. Of course, when the messages emitted by the source are actually a sequence of independent random variables, then the results obtained for the source alphabet also hold true for the source message. But, in practice, this is rarely the case. Thus we need to extend our definition of the information content of the source alphabet to the information content of the source, which will involve consideration of the statistical dependence between symbols in a message.

Let us consider a message emitted by the source, like $(x_n)_{n=0}^{\infty}$, and try to compute the average information needed to specify each symbol x_n in the message. The information content of the first symbol x_0 is, of course, the entropy $H(X_0)$†:

$$H(X_0) = \sum_{i=1}^{M} p_i \log \frac{1}{p_i}.$$

The information content of the second symbol x_1, having specified x_0, is the *conditional entropy* $H(X_1|X_0)$ based on the conditional information $I(x|y) \triangleq \log (1/P(x|y))$:

$$H(X_1|X_0) \triangleq \sum_{X_0} \sum_{X_1} P(x_0, x_1) I(x_1|x_0) = \sum_{X_0} \sum_{X_1} P(x_0, x_1) \log \frac{1}{P(x_1|x_0)}. \qquad (3.18)$$

In general, the information content of the *i*th symbol, given the previous *h* symbols in the message, is obtained as

$$H(X_i|X_{i-1}, \ldots, X_{i-h}) = \sum_{X_{i-h}} \cdots \sum_{X_i} P(x_{i-h}, x_{i-h+1}, \ldots, x_i)$$
$$\cdot \log \frac{1}{P(x_i|x_{i-1}, \ldots, x_{i-h})}, \qquad 1 \leq h \leq i. \qquad (3.19)$$

It thus seems fairly intuitive to define the information content of the source, or its *entropy* $H_{\infty}(X)$, as the information content of any symbol produced by the source, given that we have observed all previous symbols. Given a stationary information source $(\xi_n)_{n=0}^{\infty}$, its entropy, denoted by $H_{\infty}(X)$, is defined as

$$H_{\infty}(X) \triangleq \lim_{n \to \infty} H(X_n|X_{n-1}, X_{n-2}, \ldots, X_0).$$

To gain a deeper insight into the definition of $H_{\infty}(X)$, we shall prove the following theorem.

---

† We are using the notation X_i to indicate the alphabet pertaining to the *i*th symbol in the message; usually all X_i's refer to the same set X, but it is notationally convenient to keep them distinct.

THEOREM 3.4. The conditional entropy $H(X_1|X_0)$ satisfies the inequality

$$H(X_1|X_0) \leq H(X_1). \tag{3.20}$$

To prove the theorem, consider the difference

$$H(X_1|X_0) - H(X_1) = \sum_{X_0} \sum_{X_1} P(x_0, x_1) \log \frac{P(x_1)}{P(x_1|x_0)}$$

and use in the RHS the inequality (3.9) so as to get

$$H(X_1|X_0) - H(X_1) \leq \log e \sum_{X_0} \sum_{X_1} P(x_0, x_1) \left[\frac{P(x_1)}{P(x_1|x_0)} - 1 \right] = 0. \quad \blacksquare$$

The relationship (3.20) becomes an equality when ξ_1 and ξ_0 are independent random variables; in this case, in fact, $P(x_1|x_0) = P(x_1)$. A shrewd extension of Theorem 3.4 and the exploitation of the stationarity of the sequence $(\xi_n)_{n=0}^{\infty}$ allow one to write

$$H(X_n|X_{n-1}, \ldots , X_0) \leq H(X_n|X_{n-1}, \ldots , X_1)$$
$$= H(X_{n-1}|X_{n-2}, \ldots , X_0). \tag{3.21}$$

So the sequence $H(X_n|X_{n-1}, \ldots , X_0)$, $n = 1, 2, \ldots$, is nonincreasing, and since the terms of the sequence are nonnegative, the limit $H_\infty(X)$ exists. Moreover, it satisfies the following inequality:

$$0 \leq H_\infty(X) \leq H(X), \tag{3.22}$$

where the RHS inequality becomes an equality when the symbols in the sequence are independent.

The entropy of an information source is difficult to compute in most cases. We shall describe how this is achieved for a particular class of sources, the *stationary Markov sources*. A stationary Markov source is an information source whose output can be modeled as a finite-state, fully regular Markov chain (see Section 2.2.1). The properties of a stationary Markov source can be described as follows:

(i) At the beginning of each symbol interval, the source is in one of q possible states $\{S_j\}_{j=1}^{q}$. During each symbol interval, the source changes state, say from S_j to S_k, according to a *transition probability* p_{jk} whose value is independent of the particular symbol interval.

(ii) The change of the state is accompanied by the emission of a symbol x_i, chosen from the source alphabet X, which depends only on the initial state S_j and the new state S_k.

(iii) The state S_j and the emitted symbol x_i uniquely determine the new state S_k.

In other words, the current symbol emitted by the source depends on the past symbols only through the state of the source. The stationary Markov model for information sources is a useful approximation in many physical situations. The interested reader is referred to Ash (1967, Chapter 6) for a detailed exposition of the subject.

Example 3.5

Let a stationary information source $(\xi_n)_{n=0}^{\infty}$ be characterized by the property $P(x_n|x_{n-1},$
$\ldots, x_0) = P(x_n|x_{n-1})$; that is, each symbol in the sequence depends only on the previous
one. We assume that the alphabet X is formed by three symbols, say the letters A, B,
and C. The probabilities $P(x_n|x_{n-1})$ are given as follows:

x_{n-1} \ x_n	A	B	C
A	0.2	0.4	0.4
B	0.3	0.5	0.2
C	0.6	0.1	0.3

This source can be represented using the directed graph of Fig. 3.6, where each state
represents the last emitted symbol and the transitions are identified by their probabilities
and the presently emitted symbols. It can be verified that this source satisfies properties
(i), (ii) and (iii); thus it is a stationary Markov source. $\square$

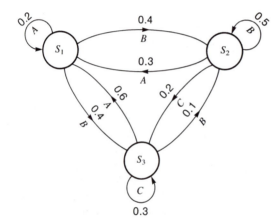

Figure 3.6 Graph representation of a
stationary Markov source with alphabet X
$= \{A,B,C\}$.

Let us compute now the source entropy $H_\infty(X)$. Defining the *entropy of the state*
S_j as

$$H(S_j) \triangleq \sum_{k=1}^{M_j} p_{jk} \log \frac{1}{p_{jk}}, \tag{3.23}$$

where M_j represents the number of symbols available at the state S_j, and denoting by
$\{w_i\}_{i-1}^q$ the components of the stationary distribution vector (2.54), the following basic
theorem can be proved (see, e.g., Ash, 1967, Chapter 6).

THEOREM 3.5. The entropy $H_\infty(X)$ of a stationary Markov source is given by

$$H_\infty(X) = \sum_{j=1}^{q} w_j H(S_j). \tag{3.24}$$

Example 3.5 (continued)

The entropy $H_\infty(X)$ of the source of Fig. 3.6 is 1.441. $\square$

Encoding of stationary Markov sources

The inequalities (3.21) and (3.22) show that the average information content of a source emitting nonindependent symbols decreases as the message length increases. If we define the entropy of a block of subsequent symbols of a source message as

$$H(X^\nu) \triangleq H(X_0, X_1, \ldots, X_{\nu-1})$$

$$= \sum_{X_0} \cdots \sum_{X_{\nu-1}} P(x_0, \ldots, x_{\nu-1}) \log \frac{1}{P(x_0, \ldots, x_{\nu-1})} \quad (3.25)$$

and apply the way of reasoning that led to (3.17), we find that a code can be devised for blocks of n consecutive source symbols whose average number of bits per symbol satisfies the inequality

$$\frac{H(X^\nu)}{\nu} \leqslant \frac{\bar{n}_\nu}{\nu} < \frac{H(X^\nu)}{\nu} + \frac{1}{\nu} \cdot \quad (3.26)$$

Moreover, using (3.21) and some easy algebra, it can be proved that the sequence $H(X^\nu)/\nu$, $\nu = 1, 2, \ldots$, is nonincreasing, and its limit is $H_\infty(X)$; that is,

$$\lim_{\nu \to \infty} \frac{H(X^\nu)}{\nu} = H_\infty(X). \quad (3.27)$$

Thus we can see that increasing ν (the block length) makes the code more efficient at each step, and, as ν goes to infinity, the average length $\bar{n}_\nu/\nu$ approaches the *source entropy* $H_\infty(X)$ as close as desired; that is,

$$H_\infty(X) \leqslant \frac{\bar{n}_\nu}{\nu} < H_\infty(X) + O(\nu^{-1}), \qquad \nu \to \infty. \quad (3.28)$$

The price that must be paid for this increased efficiency lies in the complexity of the encoder, whose input alphabet size increases exponentially with ν, and in the delay suffered in the reconstructed sequence. As a matter of fact, before obtaining the first symbol in every block one must wait for the decoding of the entire block of ν symbols at the output of the source decoder.

Turning our attention to the particular case of Markov sources, we can apply the Huffman procedure to encode the symbols of the alphabet for *each state S_j*. This requires using a different set of code words for each state of the source. The performance of such a coding procedure is easily obtained. Using Theorem 3.3 and denoting by $\bar{n}(S_j)$ the average number of digits/symbol in the code words for state S_j, we obtain

$$H(S_j) \leqslant \bar{n}(S_j) < H(S_j) + 1. \quad (3.29)$$

Thus, the average length of a code word is

$$\bar{n} = \sum_{j=1}^{q} w_j \, \bar{n}(S_j) \quad (3.30)$$

and satisfies the inequality

$$H_\infty(X) \leq \bar{n} < H_\infty(X) + 1, \tag{3.31}$$

where (3.24) has been taken into account.

Example 3.6

Let us use the state-dependent Huffman procedure to encode the Markov source of Example 3.5 and compute its efficiency. Using the tree-encoding procedure for the three symbols that can be emitted from the source in any state, we obtain the following code

	S_1	S_2	S_3
A	11	10	0
B	10	0	11
C	0	11	10

The average number of digits/symbol is

$$\bar{n} = \sum_{j=1}^{3} w_j \, \bar{n}(S_j) \cong 1.5054.$$

Using the result of Example 3.5, we can compute the efficiency ε_∞ of the code, which is defined as

$$\varepsilon_\infty \triangleq \frac{H_\infty(X)}{\bar{n}}$$

and is equal to 0.957 in this case. $\square$

Information rate of a stationary source

In our definition of a discrete stationary source at the beginning of Section 3.2, *time* was not taken into account. To overcome this, we need to place the events forming a source message in correspondence with a sequence of points on the time axis. In particular, let us assume that the source emits the symbols forming a message at equally spaced time instants, and that the time period between two consecutive emissions is T_s. Thus we can define the *average information rate* of the source, R_s, as

$$R_s \triangleq \frac{H_\infty(X)}{T_s} \text{ bits/s.} \tag{3.32}$$

As we shall see later, the appearance of time in our paradigm is strictly related to the *bandwidth* of the channel that will be used to convey the information.

3.3 COMMUNICATION CHANNELS

The communication channel is the physical medium used to connect the source of information (in general, the transmitter) and its user (the receiver). As we saw in Chapter 1, according to the block diagram of Fig. 1.4, different kinds of channels can be specified, depending on the points of the system we are observing. Between the output of the modulator and the input of the demodulator, for example, we have a *continuous channel*, which can be modeled in its simplest form by the additive channel shown in Fig. 3.7; in it, $x(t)$ is the information signal emitted by the modulator, $n(t)$ represents the *noise* added to the signal on the channel, and $y(t)$ is the received signal. The channel is completely characterized by the probability distribution of the noise. If we now observe the block diagram of Fig. 1.4 between the channel encoder output and the decoder input, we have a *discrete channel*, which accepts symbols x_i belonging to the input alphabet X of the channel encoder and returns symbols y_j belonging to its own output alphabet Y. When X and Y contain the same symbols, y_j is an estimate of the j th transmitted symbol x_j.

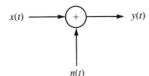

Figure 3.7 Additive noise channel.

In the following, we shall see how to characterize a communication channel and how to compute the rate at which the information can be reliably transmitted through it.

3.3.1 Discrete Memoryless Channel

A *discrete channel* is characterized by an input alphabet $X = \{x_i\}_{i=1}^{N_X}$, an output alphabet $Y = \{y_j\}_{j=1}^{N_Y}$, and by a set of *conditional probabilities* p_{ij}, $i = 1, 2, \ldots, N_X$, $j = 1, 2, \ldots, N_Y$, where $p_{ij} \triangleq P(y_j|x_i)$ represents the probability of receiving the symbol y_j, given that the symbol x_i has been transmitted. We assume that the channel is *memoryless*; that is,

$$P(y_1, \ldots, y_n|x_1, \ldots, x_n) = \prod_{i=1}^{n} P(y_i|x_i),$$

where $x_1, \ldots, x_n$ and $y_1, \ldots, y_n$ represent n consecutive transmitted and received symbols, respectively.

A graphical model of the discrete memoryless channel is shown in Fig. 3.8. Each arrow represents a transition from one of the symbols of the input alphabet to

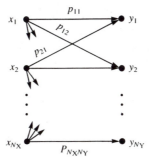

Figure 3.8 Model of a discrete memoryless channel.

one of the symbols of the output alphabet, and it is labeled with the conditional probability of that transition. The sum of all the transition probabilities labeling the arrows stemming from the same input symbol is equal to 1.

Example 3.7 The binary channel

It is a special case of the discrete channel when $N_X = N_Y = 2$, as depicted in Fig. 3.9. The average error probability $P(e)$ is defined as

$$P(e) \triangleq P(x_1, y_2) + P(x_2, y_1) \tag{3.33}$$

and can be computed as

$$P(e) = P(x_1)P(y_2|x_1) + P(x_2)P(y_1|x_2) = P(x_1)p_{12} + P(x_2)p_{21}. \tag{3.34}$$

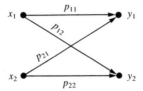

Figure 3.9 Binary channel.

If the two transition probabilities p_{12} and p_{21} are equal, say, to p, the channel is called *binary symmetric channel* (BSC) and (3.34) becomes

$$P(e) = p[P(x_1) + P(x_2)] = p. \quad \Box \tag{3.35}$$

It is customary to arrange the conditional probabilities $\{p_{ij}\}$ into the *channel matrix*

$$\mathbf{P} \triangleq \begin{bmatrix} p_{11} & p_{12} & \cdots & p_{1N_Y} \\ p_{12} & p_{22} & \cdots & p_{2N_Y} \\ \cdot & & & \cdot \\ \cdot & & & \cdot \\ \cdot & & & \cdot \\ p_{N_X1} & p_{N_X2} & \cdots & p_{N_XN_Y} \end{bmatrix}. \tag{3.36}$$

In (3.36) the numbers p_{ij}'s represent probabilities, so they satisfy the inequality $0 \leq p_{ij} \leq 1$, and, obviously, the relationship

$$\sum_{j=1}^{N_Y} p_{ij} = 1, \quad i = 1, 2, \ldots, N_X.$$

That is, the sum of the elements in each row of $\mathbf{P}$ is 1. The *average error probability* is defined by extension of (3.33) for $N_Y = N_X$, as

$$P(e) \triangleq \sum_{i=1}^{N_X} \sum_{\substack{j=1 \\ j \neq i}}^{N_X} P(x_i, y_j) = \sum_{i=1}^{N_X} P(x_i) \sum_{\substack{j=1 \\ j \neq i}}^{N_X} P(y_j|x_i) \tag{3.37}$$

$$= \sum_{i=1}^{N_X} P(x_i) \sum_{\substack{j=1 \\ j \neq i}}^{N_X} p_{ij},$$

whereas the probability of correctly receiving the symbol transmitted over the channel is given by

$$P(c) \triangleq 1 - P(e) = \sum_{i=1}^{N_X} P(x_i)P(y_i|x_i) = \sum_{i=1}^{N_X} P(x_i)p_{ii}. \tag{3.38}$$

Particular forms of $\mathbf{P}$ lead to some cases of interest.

Example 3.8   The noiseless channel

For this channel, we have $N_X = N_Y = N$, and the conditional probabilities p_{ij} satisfy the relationship

$$p_{ij} = \begin{cases} 1, & i = j, \\ 0, & i \neq j. \end{cases} \tag{3.39}$$

In words, the symbols of the input alphabet are in a one-to-one correspondence with the symbols of the output alphabet. It can be easily verified that in this case $P(e) = 0$, as it is intuitive. $\square$

Example 3.9   The useless channel

For this channel, we have $N_X = N_Y = N$, and the output symbols are independent from the input ones, or

$$P(y_j|x_i) = P(y_j), \qquad \text{all } j, i. \tag{3.40}$$

Regarding matrix $\mathbf{P}$, it is evident that (3.40) is verified if and only if $\mathbf{P}$ has identical rows. $\square$

A noiseless channel and a useless channel represent extremes of possible channel behavior. The output symbol of a noiseless channel uniquely specifies the input symbol, whereas a useless channel completely scrambles all input symbols.

Example 3.10   The symmetric channel

For this channel, each row of the matrix $\mathbf{P}$ contains the same set of numbers $\{p_j\}_{j=1}^{N_Y}$, and each column contains the same set of numbers $\{q_i\}_{i=1}^{N_X}$. The following matrices provide examples of symmetric channels:

$$\mathbf{P} = \begin{bmatrix} \frac{1}{2} & \frac{1}{3} & \frac{1}{6} \\ \frac{1}{6} & \frac{1}{2} & \frac{1}{3} \\ \frac{1}{3} & \frac{1}{6} & \frac{1}{2} \end{bmatrix}, \qquad \mathbf{P} = \begin{bmatrix} \frac{1}{3} & \frac{1}{3} & \frac{1}{6} & \frac{1}{6} \\ \frac{1}{6} & \frac{1}{6} & \frac{1}{3} & \frac{1}{3} \end{bmatrix}. \quad \square$$

According to the input and output channel alphabets X and Y and to their probabilistic dependence specified by the channel matrix $\mathbf{P}$, we can define five entropies.

(i) The *input entropy* H(X),

$$H(X) \triangleq \sum_{i=1}^{N_X} P(x_i) \log \frac{1}{P(x_i)} \quad \text{bits/symbol}, \tag{3.41}$$

which measures the average information content of the input alphabet.

(ii) The *output entropy* H(Y),

$$H(Y) \triangleq \sum_{j=1}^{N_Y} P(y_j) \log \frac{1}{P(y_j)} \quad \text{bits/symbol}, \tag{3.42}$$

which measures the average information content of the output alphabet.

(iii) The *joint entropy* H(X, Y),

$$H(X, Y) \triangleq \sum_{i=1}^{N_X} \sum_{j=1}^{N_Y} P(x_i, y_j) \log \frac{1}{P(x_i, y_j)} \quad \text{bits/symbol,} \tag{3.43}$$

which measures the average information content of a pair of input and output symbols, or the average uncertainty of the communication system formed by the input alphabet, the channel, and the output alphabet as a whole.

(iv) The *conditional entropy* H(Y|X),

$$H(Y|X) \triangleq \sum_{i=1}^{N_X} \sum_{j=1}^{N_Y} P(x_i, y_j) \log \frac{1}{P(y_j|x_i)} \quad \text{bits/symbol,} \tag{3.44}$$

which measures the average information quantity needed to specify the output symbol y when the input symbol x is known.

(v) The *conditional entropy* H(X|Y),

$$H(X|Y) \triangleq \sum_{i=1}^{N_X} \sum_{j=1}^{N_Y} P(x_i, y_j) \log \frac{1}{P(x_i|y_j)} \quad \text{bits/symbol,} \tag{3.45}$$

which measures the average information quantity needed to specify the input symbol x when the output (or received) symbol y is known. This conditional entropy represents the average amount of information that has been lost on the channel, and it is called *equivocation*. The term equivocation seems appropriate if one realizes that for a noiseless channel $H(X|Y) = 0$ (the received symbol uniquely determines the transmitted one), whereas for a useless channel we can find that $H(X|Y) = H(X)$. In this case the uncertainty about the transmitted symbol remains unaffected by the reception of an output symbol (all the information has been lost on the channel).

Using these definitions and (3.20), it is seen that the following relationships between the entropies just defined hold true:

$$H(X, Y) = H(Y, X) = H(X) + H(Y|X) = H(Y) + H(X|Y), \tag{3.46}$$

$$H(X|Y) \leq H(X), \tag{3.47}$$

$$H(Y|X) \leq H(Y). \tag{3.48}$$

The reader is invited to verify some of the following results by applying the definitions and properties (3.41) to (3.48) (see Problem 3.17).

Example 3.8 (continued)

$$H(X|Y) = 0, \tag{3.49}$$

$$H(Y|X) = 0, \tag{3.50}$$

and

$$H(X, Y) = H(X) = H(Y). \ \square \tag{3.51}$$

Example 3.9 (continued)

$$H(X|Y) = H(X), \qquad (3.52)$$

$$H(Y|X) = H(Y), \qquad (3.53)$$

$$H(X, Y) = H(X) + H(Y). \qquad (3.54)$$

Equation (3.52) says that *all* the transmitted information is lost in the channel. $\square$

Example 3.10 (continued)

An important property of the symmetric channel is that $H(Y|X)$ is independent of the input probabilities $P(x_i)$ and depends only on the channel matrix $\mathbf{P}$. To show this, let us write

$$H(Y|X) = \sum_{i=1}^{N_X} P(x_i)\, H(Y|x_i), \qquad (3.55)$$

where

$$H(Y|x_i) \triangleq \sum_{j=1}^{N_Y} p_{ij} \log \frac{1}{p_{ij}}. \qquad (3.56)$$

According to the definition of symmetry, all the rows of $\mathbf{P}$ are permutations of the same set of numbers $\{p_j\}_{j=1}^{N_Y}$. Thus

$$H(Y|x_i) = \sum_{j=1}^{N_Y} p_j \log \frac{1}{p_j}, \qquad i = 1, \ldots, N_X, \qquad (3.57)$$

and inserting (3.57) into (3.55) gives

$$H(Y|X) = \sum_{i=1}^{N_X} P(x_i) \sum_{j=1}^{N_Y} p_j \log \frac{1}{p_j} = \sum_{j=1}^{N_Y} p_j \log \frac{1}{p_j}, \qquad (3.58)$$

which does not depend on the input probabilities $P(x_i)$, $i = 1, \ldots, N_X$. $\square$

3.3.2 Capacity of the Discrete Memoryless Channel

We have seen that a part of the information $H(X)$ that must be transmitted over the channel is lost because of the noise present in the channel itself. This part is measured by the channel equivocation $H(X|Y)$. Thus it seems natural to define the *average information flow* $I(X; Y)$ through the channel as

$$I(X; Y) \triangleq H(X) - H(X|Y) \quad \text{bits/symbol.} \qquad (3.59)$$

Using (3.46), the following alternative forms can be derived:

$$I(X; Y) = H(Y) - H(Y|X) = H(X) + H(Y) - H(X, Y). \qquad (3.60)$$

Comparing (3.59) and the first equality of (3.60), it is apparent that $I(X; Y) = I(Y; X)$.

Example 3.11

Let us compute $I(X; Y)$ for the BSC with error probability $p = 0.1$ and equally likely input symbols. Because $P(x_1) = P(x_2) = 0.5$, the output symbols y_1 and y_2 are also equally likely. Thus we have

$$H(X) = H(Y) = 1 \text{ bit/symbol.}$$

To compute I(X; Y) using (3.60), we need the joint entropy H(X, Y) given by (3.43). The joint probabilities $P(x_i, y_j)$ are easily computed as

$$P(x_1, y_1) = P(x_1)P(y_1|x_1) = 0.5 \cdot 0.9 = 0.45,$$
$$P(x_2, y_1) = 0.05,$$
$$P(x_2, y_2) = 0.45,$$
$$P(x_1, y_2) = 0.05.$$

Thus we have

$$H(X, Y) \cong 1.469,$$

and in conclusion

$$I(X; Y) = 1 + 1 - 1.469 = 0.531 \text{ bits/symbol.}$$

The results show that almost one-half of the information is lost on the channel. How does this result compare with the intuitive remark that, on the average, only 10 percent of the received bits are in error if $p = 0.1$? □

Let us consider again the BSC and see how I(X; Y) depends on the probability distribution of the input symbols. Using the form

$$I(X; Y) = H(Y) - H(Y|X)$$

and computing H(Y|X) using (3.55), we get

$$I(X; Y) = H(Y) - H(p), \tag{3.61}$$

where $H(p)$ was defined in (3.6). In Fig. 3.10 the plot of I(X; Y) versus $P(x_1)$ for different values of p is shown. It can be observed that the maximum value of I(X; Y), no matter what the value of p is, is obtained when $P(x_1) = 0.5$, that is, when the in-

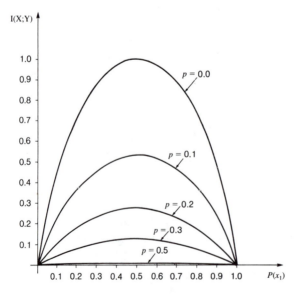

Figure 3.10 Plot of the average information flow through a BSC as a function of the probability $P(x_1)$. The error probability p of the channel is the parameter.

put symbols are equally likely. Then, fixing $P(x_1) = 0.5$, we obtain a value for I(X; Y) that depends only on the channel and represents the *maximum information flow* through a BSC. It is given by

$$\max_{P(x)} I(X; Y) = 1 - H(p) = 1 + p \log p + (1 - p) \log (1 - p), \quad (3.62)$$

where P(x) is the set of all possible probability distributions of the input symbols. The maximum value of I(X; Y) is called *channel capacity C* and is plotted in Fig. 3.11. Notice that the capacity is maximum when p is equal to 0 or equal to 1, since both these situations lead to a noiseless channel. For $p = 0.5$, the capacity is zero, since the output symbols are independent from the input symbols, and no information can flow through the channel. Note that, due to the symmetry of the channel, $H(p) = H(1 - p)$.

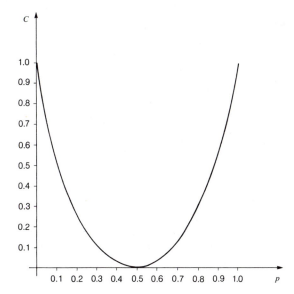

Figure 3.11 Capacity of the BSC as a function of the error probability p.

The *capacity C* of a discrete memoryless channel is defined as the maximum information I(X; Y) that can be transmitted through the channel. Recalling the first equality of (3.60), we obtain

$$C \triangleq \max_{P(x)} I(X; Y) = \max_{P(x)} \sum_{i=1}^{N_X} \sum_{j=1}^{N_Y} P(x_i)p_{ij} \log \frac{p_{ij}}{\sum_{k=1}^{N_X} P(x_k)p_{kj}}. \quad (3.63)$$

The meaning of the channel capacity and its significance are not completely apparent so far. However, it will be proved later in this chapter that reliable transmission through the channel is not possible when the average number of bits per channel symbol is greater than the channel capacity. The analytical computation of the channel capacity is difficult in most cases. However, numerical algorithms are available, such as those due to Arimoto and Blahut (see Viterbi and Omura, 1979, Appendix 3C). It becomes simpler in some particular cases, like those described in Examples 3.8 to 3.10.

Example 3.8 (continued)

Using the result (3.49), we can write

$$C = \max_{P(x)} [H(X) - H(X|Y)] = \max_{P(x)} H(X) = \log N_X. \tag{3.64}$$

Hence no information is lost on the channel. As a matter of fact, the information flow through the channel equals the average quantity of information $H(X)$ that is needed to specify an input symbol. $\square$

Example 3.9 (continued)

Using (3.52), we get

$$C = H(X) - H(X|Y) = 0. \tag{3.65}$$

Hence no information can flow through the channel. $\square$

Example 3.10 (continued)

We have proved that the conditional entropy $H(Y|X)$ does not depend on the input probability distribution. Thus, the problem of maximizing $I(X; Y) = H(Y) - H(Y|X)$ reduces to the problem of maximizing the output entropy $H(Y)$. We know that the following inequality holds:

$$H(Y) \leq \log N_Y,$$

where the equal sign refers to the case of equally likely outputs; that is, $P(y_j) = \dfrac{1}{N_Y}$, $j = 1, 2, \ldots , N_Y$. We shall prove that the output symbols are equally likely when the inputs are equally likely. In fact, if $P(x_i) = 1/N_X$, $i = 1, 2, \ldots , N_X$, we have

$$P(y_j) = \sum_{i=1}^{N_X} P(x_i, y_j) = \sum_{i=1}^{N_X} P(x_i)p_{ij} = \frac{1}{N_X} \sum_{i=1}^{N_X} p_{ij}.$$

But the term $\sum_{i=1}^{N_X} p_{ij}$ is the sum of the entries of the jth column of the channel matrix $\mathbf{P}$, and, by definition of symmetric channels, it does not depend on j. Thus all symbols $y_j \in Y$ have the same probability, and the capacity of a symmetric channel is given by

$$C = \log N_Y + \sum_{j=1}^{N_Y} p_j \log p_j \quad \text{bits/symbol.} \ \square \tag{3.66}$$

Example 3.12

The capacity of the symmetric channel whose matrix is

$$\mathbf{P} = \begin{bmatrix} \frac{1}{3} & \frac{1}{3} & \frac{1}{6} & \frac{1}{6} \\ \frac{1}{6} & \frac{1}{6} & \frac{1}{3} & \frac{1}{3} \end{bmatrix}$$

can be computed using (3.66) with $N_Y = 4$, and gives

$$C = 2 + 2(\tfrac{1}{3} \log \tfrac{1}{3} + \tfrac{1}{6} \log \tfrac{1}{6}) \cong 0.082 \text{ bits/symbol.} \ \square$$

Example 3.13

Consider a channel with $N_X = N_Y = N$ and probabilities $p_{ij} \in \mathbf{P}$ given by

$$p_{ij} = \begin{cases} 1 - p, & i = j, \\ \dfrac{p}{N - 1}, & i \neq j. \end{cases}$$

The rows and columns of $\mathbf{P}$ are in this case permutations of the N numbers

$$\left(1 - p, \frac{p}{N-1}, \ldots, \frac{p}{N-1}\right).$$

Thus the channel is symmetric, and its capacity C is given by

$$C = \log N + (1 - p) \log (1 - p) + p \log \frac{p}{N-1}. \tag{3.67}$$

The capacity of the BSC is obtained as a particular case of (3.67) with $N = 2$. $\square$

Example 3.14   The binary erasure channel (BEC)

Consider the channel of Fig. 3.12. The outputs y_1 and y_2 correspond to the input symbols x_1 and x_2, whereas y_3 refers to an ambiguous output for which no decision about the transmitted symbol will be taken. This model represents a system that is used in data transmission (see Problem 3.21). Let us compute the capacity of this channel. The channel matrix is the following:

$$\mathbf{P} = \begin{bmatrix} 1 - p & 0 & p \\ 0 & 1 - p & p \end{bmatrix}. \tag{3.68}$$

It does not satisfy the symmetry conditions, so (3.66) cannot be applied. Starting from the first equality of (3.60), it is straightforward to show that I(X; Y) is given by (3.61) also in this case. Therefore, the capacity C is obtained for the input distribution that maximizes H(Y). Denoting $P(x_1)$ by α and computing H(Y), we get

$$H(Y) = -p \log p - (1 - p) \log(1 - p)$$
$$- \alpha(1 - p) \log \alpha - (1 - \alpha)(1 - p) \log(1 - \alpha), \tag{3.69}$$

which is seen to have a maximum when $\alpha = \frac{1}{2}$. So the capacity is obtained for equally likely input symbols, and it is given by

$$C = 1 - p. \tag{3.70}$$

Comparing the plot of C shown in Fig. 3.13 with the capacity of the BSC of Fig. 3.11, we can see that erasing the received symbol when the information is not reliable can improve the information flow through the channel. This is true also in a more realistic situation when the probabilities $P(y_2|x_1)$ and $P(y_1|x_2)$ are different from zero (see Problem 3.21). $\square$

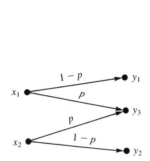

Figure 3.12   Binary erasure channel (BEC).

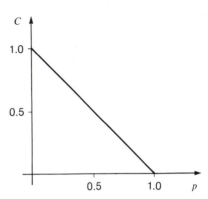

Figure 3.13   Capacity of the BEC.

We have seen that the channel capacity of the BEC is achieved with equally likely input symbols, although the channel is not symmetric, according to our definition. However, by inspection of Fig. 3.12, we can see that the structure of the channel exhibits a clear symmetry with respect to the inputs. As a matter of fact, it is possible to generalize the definition of symmetric channels in such a way as to handle situations like the one of Fig. 3.12. The reader is referred to Gallager (1968, Chapter 4) for useful theorems on the computation of channel capacity in some particular cases of interest.

3.3.3 Equivocation and Error Probability

In Section 3.3.2 we defined the average error probability $P(e)$ of a discrete channel and its equivocation $H(X|Y)$. Both may be used as measures of channel quality, and certainly they are not independent quantities. In the following, we shall derive a relationship between them. Let us refer to a channel matrix $\mathbf{P}$, with $N_X = N_Y = N$. Applying the definition of error probability given in (3.37) to this case, we can write

$$P(e) = \sum_{i=1}^{N} \sum_{\substack{j=1 \\ j \neq i}}^{N} P(x_i, y_j). \tag{3.71}$$

Let us define now the entropy $H(e)$ as

$$H(e) \triangleq -P(e) \log P(e) - [1 - P(e)] \log [1 - P(e)], \tag{3.72}$$

that is, consider $H(e)$ like the entropy of a binary alphabet with probabilities $P(e)$ and $1 - P(e)$, which corresponds to the amount of information needed to specify if an error has occurred during the transmission. We can prove the following theorem.

THEOREM 3.6 FANO'S INEQUALITY. Given a discrete memoryless channel whose input and output alphabets X and Y have the same number of symbols N, and whose error probability is $P(e)$, the following inequality holds:

$$H(X|Y) \leq H(e) + P(e) \log (N - 1). \tag{3.73}$$

To prove the theorem, we use the definition (3.45) of the equivocation $H(X|Y)$ to write

$$H(X|Y) = \sum_{i=1}^{N} \sum_{\substack{j=1 \\ j \neq i}}^{N} P(x_i, y_j) \log \frac{1}{P(x_i|y_j)} + \sum_{i=1}^{N} P(x_i, y_i) \log \frac{1}{P(x_i|y_i)}$$

and the definition (3.37) of $P(e)$ to get

$$H(X|Y) - P(e) \log (N - 1) - H(e) = \sum_{i=1}^{N} \sum_{\substack{j=1 \\ j \neq i}}^{N} P(x_i, y_j) \log \frac{P(e)}{(N - 1)P(x_i|y_j)} \tag{3.74}$$

$$+ \sum_{i=1}^{N} P(x_i, y_i) \log \frac{1 - P(e)}{P(x_i|y_i)}.$$

Applying now the inequality (3.9) to the RHS of (3.74), we obtain

$$H(X|Y) - P(e) \log (N - 1) - H(e) \leq \log e \left\{ \sum_{i=1}^{N} \sum_{\substack{j=1 \\ j \neq i}}^{N} P(x_i, y_j) \left[\frac{P(e)}{(N-1)P(x_i|y_j)} - 1 \right] \right.$$

$$+ \sum_{i=1}^{N} P(x_i, y_i) \left[\frac{1 - P(e)}{P(x_i|y_i)} - 1 \right] \right\}$$

$$= \log e \left\{ \frac{P(e)}{N-1} \sum_{i=1}^{N} \sum_{\substack{j=1 \\ j \neq i}}^{N} P(y_j) - \sum_{i=1}^{N} \sum_{\substack{j=1 \\ j \neq i}}^{N} P(x_i, y_j) \right.$$

$$+ [1 - P(e)] \sum_{i=1}^{N} P(y_i) - \sum_{i=1}^{N} P(x_i, y_i) \right\}$$

$$= \log e \left\{ P(e) - P(e) + [1 - P(e)] - [1 - P(e)] \right\} = 0.$$

And the theorem is proved. ∎

The inequality (3.73) can also be given this intuitive interpretation. Having received a symbol $y \in Y$, if we detect whether or not an error has occurred, we remove an uncertainty equal to $H(e)$. If no error occurred, the remaining uncertainty about the transmitted symbol is zero. If an error occurred, with probability $P(e)$, we have to decide which of the remaining $N - 1$ symbols has been transmitted. The uncertainty about this choice cannot exceed $\log (N - 1)$.

In Fig. 3.14 the function $P(e) \log (N - 1) + H(e)$ is plotted. Theorem 3.6 states that the allowed pairs of values $H(X|Y)$, $P(e)$ are represented by the points of the shaded region in Fig. 3.14. Moreover, being $H(X|Y) = H(X) - I(X; Y)$, the theorem

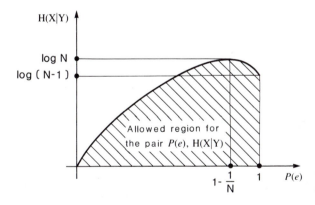

Figure 3.14 Plot of the function $H(e) + P(e) \log (N - 1)$ versus $P(e)$.

gives a lower bound to the error probability in terms of the excess of entropy of the input alphabet X with respect to the information flow through the channel. Considering now that $I(X; Y) \leq C$, (3.73) can be written as

$$H(X) - C \leq H(e) + P(e) \log(N - 1). \tag{3.75}$$

The curve $C + H(e) + P(e) \log (N - 1)$ is reported in Fig. 3.15. It is seen by inspection that the region of the allowed pairs $P(e)$, $H(X)$ contains points with $P(e) = 0$ only if $H(X) \leq C$. In other words, if *the entropy of the input alphabet exceeds the channel*

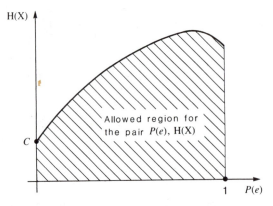

Figure 3.15 Plot of the function $C + H(e) + P(e) \log (N - 1)$ versus $P(e)$.

capacity, it is impossible to transmit the information through the channel with arbitrarily small error probability. This result is a simplified version of the converse to the fundamental theorem of information theory. If we identify the input alphabet of the channel with the output alphabet of the source encoder, the previously described situation refers to a communication system in which the symbols at the output of the source encoder are sent directly through the channel: no channel encoding is performed. We shall now include the channel encoder into our system and try to extend the previous results. Let us consider the system shown in Fig. 3.16 in which the block labeled ''source'' represents the cascade of the source itself and the source encoder. Suppose for simplicity that the output of the source is a sequence of binary symbols emitted every T_s seconds. The channel encoder is a *block encoder*; that is, it transforms blocks of k consecutive source digits (a *word*) into blocks of n symbols belonging to the input channel alphabet X. An encoding rate R_c can be defined as

$$R_c \triangleq \frac{k}{n} \cdot \tag{3.76}$$

Since n symbols must be transmitted over the channel every kT_s seconds, the channel must be used every $T_c = R_c T_s$ seconds. Denoting with W the set of 2^k messages at the

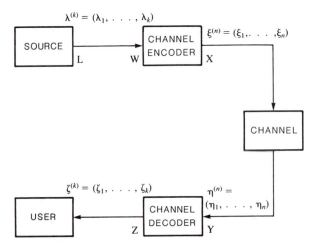

Figure 3.16 Block diagram of a discrete communication system.

input of the channel encoder and with Z the set of 2^k messages at the output of the channel decoder, we can apply Theorem 3.6 to these two sets, obtaining

$$H(W|Z) \leq H(e) + P_w(e) \log(2^k - 1), \qquad (3.77)$$

where the subscript w in $P_w(e)$ denotes "word" and $P_w(e)$ represents the average probability of a decoding word error. Moreover, since $H(W|Z) = H(W) - I(W; Z)$ and taking into account that the following inequality holds true (*data-processing theorem*; see Viterbi and Omura, 1979, Chapter 1 for a proof),

$$I(W; Z) \leq I(X; Y),$$

we get

$$H(W|Z) \geq H(W) - I(X; Y). \qquad (3.78)$$

Since the transmission of each block of k bits involves using n times the channel, we can also write

$$I(X; Y) \leq nC \qquad (3.79)$$

so that, inserting (3.79) into (3.78) and the result into (3.77), we obtain

$$H(W) - nC \leq H(e) + P_w(e) \log(2^k - 1). \qquad (3.80)$$

The inequality (3.80) is the *converse to the coding theorem*. Since the alphabet W is made by grouping k consecutive symbols at the output of the source, the entropy $H(W)$ is given by

$$H(W) = kH_\infty(L), \qquad (3.81)$$

where $H_\infty(L)$ is the entropy of the source.

Thus, inequality (3.80) states that the probability of erroneously decoding a sequence of k source symbols cannot be made arbitrarily small when the encoding rate R_c is greater than the ratio $C/H_\infty(L)$. A lower bound to the error probability can be derived from (3.80) and (3.81) as follows:

$$P_w(e) \geq \frac{kH_\infty(L) - nC - H(e)}{\log(2^k - 1)} > \frac{kH_\infty(L) - nC - 1}{k} = H_\infty(L) - \frac{C}{R_c} - \frac{1}{k} \cdot$$

Now, letting k and n go to infinity and keeping constant their ratio R_c leads to

$$P_w(e) > H_\infty(L) - \frac{C}{R_c} \cdot \qquad (3.82)$$

Channel coding theorem

We have seen in the previous subsection that there exists a lower bound to the error probability, different from zero, when the encoding rate R_c is greater than the channel capacity C. What happens when R_c is smaller than C? The answer is given by the *channel coding theorem*, which will be stated without proof in the following. It was proved in 1948 by C. E. Shannon, and the interested reader is referred to his original paper (Shannon, 1948) or to one of the many books available, for example, Gallager (1968, Chapter 5).

THEOREM 3.7. Given a binary information source, with entropy $H_\infty(L)$ bits/symbol and a discrete, memoryless channel with capacity C bits/symbol, there exists a code of rate $R_c = k/n$ for which the word error probability is bounded by

$$P_w(e) < e^{-nE(R)}, \qquad R = R_c H_\infty(L) \tag{3.83}$$

where $E(R)$ is a convex $\cup$, decreasing, positive function of R for $0 \le R \le C$.†

A typical behavior of the function $E(R)$ is shown in Fig. 3.17. Based on (3.83), we can undertake three ways to improve the performance of a data transmission system.

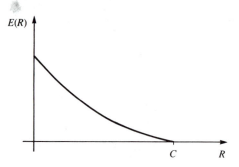

Figure 3.17 Typical behavior of the function $E(R)$.

(i) Decrease R by decreasing $R_c = k/n$. This means increasing the redundancy of the code and, for a given source emission rate, using the channel more often. In other words, we need a channel with a larger bandwidth. What happens is shown in Fig. 3.18. We move from R_1 to R_2, so that $E(R)$ increases and the bound in (3.83) decreases.

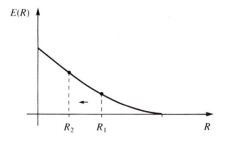

Figure 3.18 Increasing the value of $E(R)$ by decreasing $R_c = k/n$.

(ii) Increase the channel capacity C by increasing the signal-to-noise ratio over the channel. The situation is depicted in Fig. 3.19. The operating point moves from $E_1(R)$ to a larger $E_2(R)$, thus improving the error bound.

(iii) *Keeping the ratio $R_c = k/n$ fixed, increase n.* This third approach does not require any intervention on the bandwidth and signal-to-noise ratio of the channel. It

† Notice that, when the whole redundancy of the source has been removed, $H_\infty(L) = 1$ and, therefore, $R = R_c$.

118   Basic Results from Information Theory   Chap. 3

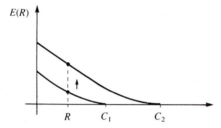

Figure 3.19 Increasing the value of $E(R)$ by increasing the capacity of the channel.

allows one to improve the performance of the communication system by simply increasing the length of the block entering the channel encoder, and thus at the expense of a greater complexity of the encoder–decoder pair and of a longer delay in reconstructing the decoded sequence.

While (i) and (ii) were well-known remedies to counteract the disturbances in a communication system, the use of the third way is one of the major achievements of Shannon's theory.

3.3.4 Additive Gaussian Channel

In this section we shall consider a communication channel that is *amplitude continuous* and *time continuous*. This situation will be approached in two steps. First, we shall examine the *time-discrete, amplitude-continuous* Gaussian channel shown in Fig. 3.20.

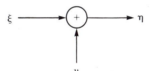

Figure 3.20 Block diagram of a discrete-time additive Gaussian channel.

The source, every T_s seconds, transmits a symbol chosen from a possibly uncountable alphabet. The channel disturbance has the form of an unwanted noise superimposed on the signal to be transmitted. The assumption that the noise is Gaussian, which is highly desirable from a mathematical point of view, is reasonable in a wide variety of physical settings. After the analysis of this simplified case has been completed, we shall extend the results to the time-continuous channel, in which the transmission of information will be allowed to be continuous in time.

The derivation of the main result will be done in a heuristic manner, avoiding all the mathematical subtleties. The unsatisfied reader is invited to quench the thirst for rigor in one of the books already quoted (see, e.g., Ash, 1967, Chapter 3). For the channel of Fig. 3.20, we have $\eta = \xi + v$, where ξ and η are RVs representing the input and output symbols and taking values $x \in X$, $y \in Y$ with the probability density functions (pdf) $f_\xi(x)$ and $f_\eta(y)$, respectively, and v is the zero-mean Gaussian RV representing the noise. Denoting the noise variance with σ_v^2, we have $v \sim \mathcal{N}(0, \sigma_v^2)$, and the output η, given $\xi = x$, is $\mathcal{N}(x, \sigma_v^2)$. Our goal is the evaluation of the channel capacity; thus we need to extend our definition of the information measure to the continuous case.

Measure of information in the continuous case

Let ξ be a continuous RV taking values in X with pdf $f_\xi(x)$. We define its *entropy* H(X) as

$$H(X) = -\int_{-\infty}^{\infty} f_\xi(x) \log f_\xi(x)\, dx. \tag{3.84}$$

Although (3.84) seems a straightforward extension of the definition (3.5) used in the discrete case, some differences arise. The main one consists in the fact that H(X) defined in (3.84) may be arbitrarily large, positive, or negative (see Problem 3.30). In the same way as for (3.84), we can define for two random variables ξ and η having a joint probability density function $f_{\xi\eta}(x, y)$, the joint entropy H(X, Y), and the conditional entropies H(X|Y) and H(Y|X) as

$$H(X, Y) \triangleq -\int_{-\infty}^{\infty}\int_{-\infty}^{\infty} f_{\xi\eta}(x, y) \log f_{\xi\eta}(x, y)\, dx\, dy,$$

$$H(X|Y) \triangleq -\int_{-\infty}^{\infty}\int_{-\infty}^{\infty} f_{\xi\eta}(x, y) \log f_{\xi|\eta}(x|y)\, dx\, dy,$$

$$H(Y|X) \triangleq -\int_{-\infty}^{\infty}\int_{-\infty}^{\infty} f_{\xi\eta}(x, y) \log f_{\xi|\eta}(y|x)\, dx\, dy.$$

Assuming now that both H(X) and H(Y) are finite, the following relationships hold true, as in the discrete case:

$$H(X, Y) \leq H(X) + H(Y),$$
$$H(X, Y) = H(X) + H(Y|X) = H(Y) + H(X|Y),$$
$$H(Y|X) \leq H(Y), \qquad H(X|Y) \leq H(X).$$

In all the preceding, inequalities become equalities if ξ and η are statistically independent. In the discrete case, we have proved (Theorem 3.1) that the entropy is maximized by equally likely symbols. In the continuous case, the following theorem holds.

THEOREM 3.8   Let ξ be a continuous RV with pdf $f_\xi(x)$. If ξ has finite variance σ_ξ^2, then H(X) exists and satisfies the inequality

$$H(X) \leq \frac{1}{2} \log 2\pi e \sigma_\xi^2 \tag{3.85}$$

with equality if and only if $\xi \sim \mathcal{N}(\mu, \sigma_\xi^2)$.

For the proof, see Problem 3.31. ■

Capacity of the discrete-time Gaussian channel

Suppose that the continuous RVs ξ and η represent the input and output symbols for the channel of Fig. 3.20. As we did for the discrete channel, we define the average information flow through the channel as

$$I(X; Y) \triangleq H(X) - H(X|Y) = I(Y; X) = H(Y) - H(Y|X)$$

and the channel capacity C as

$$C \triangleq \max_{f_\xi(x)} I(X; Y). \tag{3.86}$$

We know that, given $\xi = x$, $\eta \sim \mathcal{N}(x, \sigma_\nu^2)$. Thus $H(Y|X) = \frac{1}{2} \log 2\pi e \sigma_\nu^2$, and

$$C = \max_{f_\xi(x)} H(Y) - \frac{1}{2} \log 2\pi e \sigma_\nu^2. \tag{3.87}$$

By Theorem 3.8, $H(Y)$ is maximum when η is Gaussian, and this in turn happens if and only if ξ is Gaussian. Therefore, the capacity C is attained for a Gaussian input ξ, say $\xi \sim \mathcal{N}(0, \sigma_\xi^2)$, and its value is given by

$$C = \frac{1}{2} \log 2\pi e (\sigma_\xi^2 + \sigma_\nu^2) - \frac{1}{2} \log 2\pi e \sigma_\nu^2 = \frac{1}{2} \log \left(1 + \frac{\sigma_\xi^2}{\sigma_\nu^2} \right) \quad \text{bits/symbol.} \tag{3.88}$$

Capacity of the bandlimited Gaussian channel

We have discussed up to this point only time-discrete channels. On the other hand, many channels of practical interest are time continuous, in the sense that their inputs and outputs are time-continuous functions. To extend the result (3.88) to this new situation, let us suppose that the input signals and the noise are strictly bandlimited to the interval $(-B, B)$. Then, by the sampling theorem (see Section 2.5.1), we can represent each signal using at least $2B$ samples per second, each having a variance σ_ξ^2 equal to the signal power $\mathcal{P}$. Moreover, the noise is assumed to be a white Gaussian random process, with two-sided power spectral density $N_0/2$ sampled every $1/2B$ seconds. Hence its power is $\sigma_\nu^2 = (N_0/2) \cdot (2B) = N_0 B$. We can now use the result (3.88), taking into account that we are using $2B$ times per second a discrete-time Gaussian channel with capacity C given by (3.88). So, finally, we obtain the *capacity C* of a *bandlimited white Gaussian channel* as

$$C = B \log \left(1 + \frac{\mathcal{P}}{N_0 B} \right) \quad \text{bits/s.} \tag{3.89}$$

The result (3.89) is of fundamental importance, since it gives the upper limit that can be reached when information is to be reliably transmitted over Gaussian channels. The designer of a data communication system tries to choose the system parameters in such a way as to approach the capacity C as closely as possible with a preassigned error rate. In Chapter 5 the modulation schemes most widely used for data transmission will be compared, taking into account the limit given by C in (3.89).

Example 3.15

Compute the capacity C of a Gaussian channel with bandwidth $B = 2500$ Hz and signal-to-noise ratio equal to 30 dB. Using (3.89), with $B = 2500$ and $\mathcal{P}/N_0 B = 10^3$, we obtain

$$C = 2500 \log (1 + 1000) = 24,918 \text{ bits/s.}$$

The values of the parameters in this example reflect a typical situation of standard voice-grade telephone lines. On these channels, however, the maximum data rate commercially achievable so far is 9600 bits/s. The gap between the two figures is due to the fact that by no means can a telephone line be accurately modeled as a Gaussian channel. As we shall see in Chapters 6 and 10, the additive Gaussian noise is only one among the numerous disturbances that dwell in a telephone line. □

BIBLIOGRAPHICAL NOTES

This chapter has no presumptions of originality, and its material results from the many excellent textbooks available on the subject of information theory. All of them stem from the pioneering work of C. E. Shannon that was published in his fundamental paper of 1948 (Shannon 1948); see also the collection of fundamental papers by Shannon in Slepian (1974).

As students first, as researchers and teachers later, the authors have been familiar with the classical books by Fano (1961), Ash (1967), and Gallager (1968). We are indebted to these books for the development of topics in this chapter. In the following, we give some suggestions to the reader wishing to go deeper into the subject.

Berger (1971) wrote an advanced book dealing wholly with the source coding theorem, its generalizations, and its practical applications. Chapter 3 of McEliece (1977) is devoted to a modern and original presentation of discrete memoryless sources and their rate-distortion functions. Source coding and recent advances in rate-distortion theory are also treated extensively in Viterbi and Omura (1979, Chapters 7 and 8).

For the subject of discrete channels with memory (not covered in this chapter), see Ash (1967, Chapter 7), Gallager (1968, Chapter 4), and Viterbi and Omura (1979, Chapter 2). The continuous-time Gaussian channel, briefly mentioned here, is treated in detail in Fano (1961, Chapter 5), Gallager (1968, Chapter 8), and Ash (1967, Chapter 8).

PROBLEMS†

3.1. For the third source alphabet of Example 3.1, show by direct differentiation that the entropy has a maximum when $p = 0.5$.

3.2. For the source of Problem 3.1, consider sequences of two outputs as a single output of an extended source with alphabet $X^2 = \{(x_1, x_1), (x_1, x_2), (x_2, x_1), (x_2, x_2)\}$. Under the hypothesis that consecutive outputs from the source are statistically independent, show directly that $H(X^2) = 2 \cdot H(X)$. Extend to the case of an extended source X^ν.

3.3. A source emits a sequence of independent symbols from an alphabet X consisting of five symbols $x_1, \ldots, x_5$, with probabilities $\frac{1}{4}, \frac{1}{8}, \frac{1}{8}, \frac{3}{16}, \frac{5}{16}$, respectively. Find the entropy of the source alphabet.

3.4. A black and white TV picture consists of 525 lines of picture information. Assume that each line consists of 525 picture elements and that each element can have 256 different brightness levels. Pictures are repeated at the rate of 30 per second. What is the average rate of information conveyed by a TV set assuming independence?

† Problems marked with an asterisk should be solved with the aid of a computer.

3.5. Consider two discrete sources with alphabets X_1 and X_2, having M_1 and M_2 symbols, respectively, and probability distributions $\{p_i\}_{i=1}^{M_1}$ and $\{q_i\}_{i=1}^{M_2}$. From these sources a new source is formed, with $M_1 + M_2$ symbols: the first M_1 symbols have probability distribution $\{\lambda p_i\}_{i=1}^{M_1}$, while the last M_2 symbols have probability distribtuion $\{(1 - \lambda)q_i\}_{i=1}^{M_2}$, $0 \leq \lambda \leq 1$. Find the entropy of the new source and the value of λ that maximizes it.

3.6. Given a source with alphabet X with M symbols and probability distribution $\{p_i\}_{i=1}^{M}$, group the last m symbols to form a new source X' with $M' = M - m + 1$ symbols and probability distribution $\{q_i\}_{i=1}^{M'}$:

$$q_i = p_i, \qquad i = 1, 2, \ldots, M' - 1,$$
$$q_{M'} = p_{M'} + \cdots + p_M.$$

Show that the entropy of the new source satisfies the inequality $H(X') \leq H(X)$.

3.7. A binary source with alphabet X and symbols $\{x_1, x_2\}$ has probabilities $P\{x_1\} = 0.1$ and $P\{x_2\} = 0.9$. Construct the Huffman codes corresponding to the sources X^ν ($\nu = 1, 2, 3, 4$) obtained by grouping the outputs of X in words of length ν. For every value of ν, find the efficiency ϵ of the encoding scheme.

3.8. Given the source with alphabet X having symbols $\{x_i\}_{i=1}^{8}$ and probability distribution $\{\frac{1}{2}, \frac{1}{4}, \frac{1}{8}, \frac{1}{16}, \frac{1}{64}, \frac{1}{64}, \frac{1}{64}, \frac{1}{64}\}$, find three binary codes satisfying the prefix constraint such that:
   (a) The average number of digits/symbol $\bar{n}$ is minimized;
   (b) The maximum number of digits in every code word is minimized;
   (c) The average number of digits/symbol $\bar{n}$ is minimized subject to the constraint that the maximum length of the code words is 4 digits.

3.9. Consider a source alphabet with N symbols, and two probability distributions $\{p_1, p_2, p_3, \ldots, p_N\}$ and $\{p'_1, p'_2, p'_3, \ldots, p_N\}$, where

$$\left.\begin{array}{l} p'_1 = p_1 - \Delta p \\ p'_2 = p_2 + \Delta p \end{array}\right\} \Delta p > 0, \qquad p_1 > p_2.$$

Show that the entropy of the alphabet is greater for the second probability distribution provided that $p'_1 > p'_2$. *Hint*: Applying the inequality $\ln x \leq x - 1$, with $x = q_i/p_i$, show that the following inequality holds true:

$$- \sum_i p_i \log p_i \leq - \sum_i p_i \log q_i,$$

where $\{p_i\}_{i=1}^{N}$ and $\{q_i\}_{i=1}^{N}$ are probability distributions.

3.10. In Fig. P3.1, the representations of symbols 0, 1, 2, 3, $\ldots$, 9 are shown. A two-

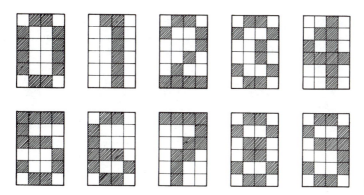

Figure P3.1

dimensional field with 4×6 positions (white or black) is used. Find the redundancy of this code under the hypothesis that the symbols are equally likely.

3.11. A binary source with alphabet X having symbols $\{x_1, x_2\}$ has a probability distribution $\{p_1 = 0.005, p_2 = 0.995\}$. The outputs from the source are grouped in blocks of 100 each, and a code word is associated only with those blocks containing no more than three symbols x_1. Assuming that the symbols from the source are statistically independent:
(a) Find the minimum code word length for a fixed-length code;
(b) Find the probability that a block is not encoded.

3.12. A source has an alphabet of four symbols. The probabilities of the symbols and two possible sets of binary code words for the source are as follows:

Symbols	$P(x_i)$	Code I	Code II
x_1	0.4	1	1
x_2	0.3	01	10
x_3	0.2	001	100
x_4	0.1	000	1000

For each code, answer the following questions:
(a) Does the code satisfy the prefix condition?
(b) Is the code uniquely decipherable?
(c) What is the mutual information provided about the source symbol by the specification of the first digit of the code word?

3.13. For the source of Example 3.5, compute the entropy and check the result of Example 3.6.

3.14. It is desired to reencode more efficiently the output of the Markov source illustrated in Figure P3.2.
(a) Find the entropy $H_\infty(X)$ of the source.
(b) Construct the Markov diagram for pairs of symbols.
(c) Construct the optimum binary code for all pairs of symbols and evaluate the resulting code efficiency.

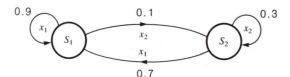

Figure P3.2

3.15. Consider a stationary source with a ternary alphabet $X = \{x_1, x_2, x_3\}$, for which the probability of each symbol depends only on the preceding symbol. The probabilities of the possible ordered symbol pairs are given in the following table:

n \\ $n-1$	x_1	x_2	x_3
x_1	0.2	0.05	0.15
x_2	0.15	0.05	0.1
x_3	0.05	0.2	0.05

Determine the optimum binary code words and the resulting code efficiency for the following encoding schemes:

(a) The sequence is divided into successive pairs of symbols and each pair is represented by a code word;

(b) A Markov model for the source is devised, and a state-dependent code is used to encode each symbol.

3.16. The state diagram of a Markov source is given in Fig. P3.3, and the symbol probabilities for each state are as follows:

	S_1	S_2	S_3	S_4
x_1	0.7	0.3	0.5	0.3
x_2	0.125	0.5	0.1	0.5
x_3	0.075	0	0.1	0
x_4	0.1	0.2	0.3	0.2

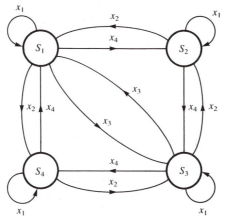

Figure P3.3

(a) Evaluate the source entropy $H_\infty(X)$.

(b) Construct an optimum set of binary code words for each state and evaluate the resulting code efficiency.

3.17. Prove the results (3.49) to (3.54).

3.18. Outputs from a source with alphabet $X = \{x_1, x_2, x_3\}$ and probability distribution $\{\frac{1}{4}, \frac{1}{4}, \frac{1}{2}\}$ are sent independently through the channels shown in the Figure P3.4. Evaluate $H(X)$, $H(Y)$, $H(Z)$, $H(X|Y)$, and $H(X|Z)$.

Figure P3.4

3.19. A channel with input alphabet X, output alphabet Y, and channel matrix P_1 is cascaded with a channel with input alphabet Y, output alphabet Z, and channel matrix P_2 (Fig.

Chap. 3   Problems

125

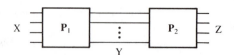

Figure P3.5

P3.5). If transmissions over the two channels are independent, find the channel matrix **P** of the equivalent channel with input X and output Z.

3.20. Prove that the cascade of n BSCs is still a BSC. Under the hypothesis that the n channels are equal, evaluate the channel matrix and the error probability of the equivalent channel and let $n \to \infty$.

3.21. Consider the transmission system shown in Fig. P3.6. A modulator associates with the two symbols x_1, x_2 emitted from the source the voltages of $+1$ and -1 volt, respectively.

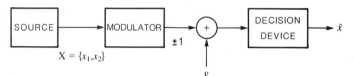

Figure P3.6

A Gaussian noise represented by the RV v is added to the modulator output, with $v \sim \mathcal{N}(0, 1)$. The decision device can operate in two ways:

(i) It compares the received voltage with the threshold zero and decides that x_1 has been transmitted when the threshold value is exceeded or x_2 has been transmitted when the threshold value is not exceeded.

(ii) It compares the received voltage with two thresholds, $+\delta$ and $-\delta$. When $+\delta$ is exceeded, it decides for x_1; when $-\delta$ is not exceeded, it decides for x_2; if the voltage lies between $-\delta$ and $+\delta$, it does not decide and erases the symbol.

Compute and plot the capacity of the two discrete channels resulting from the application of decision schemes (i) and (ii). Plot the second one as a function of δ.

3.22. For two situations depicted in Fig. P3.7, show that $I(X; Y) = I(X; Z)$. Verify also that in the first case $H(Y) > H(Z)$, whereas in the second case $H(Y) < H(Z)$.

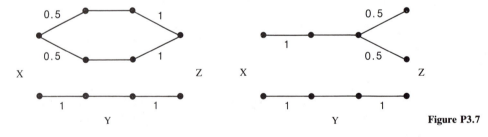

Figure P3.7

3.23. Compute the capacity C of the channel shown in Fig. P3.8. Cascading n such channels, compute the capacity C_n of the equivalent channel and let $n \to \infty$.

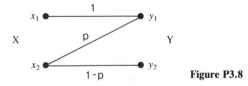

Figure P3.8

3.24. Consider the cascade of two BSC channels. Prove that the capacity of the equivalent channel cannot exceed the capacity of each single channel.

3.25. Consider the BEC channel with channel matrix given in (3.68). Compute and plot H(X|Y) as a function of $P(x_1)$.

*3.26. Write a computer program implementing the Arimoto–Blahut algorithm (see Viterbi and Omura, 1979, Appendix 3C) to find the capacity of a discrete channel.

3.27. Find the capacity and an optimizing input probability assignment for each of the discrete channels shown in Fig. P3.9 (Gallager, 1968).

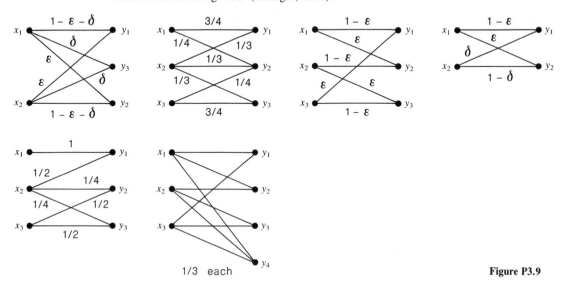

1/3 each

Figure P3.9

*3.28. Compute the capacity of the channels of Problem 3.27 through the use of the program developed in Problem 3.26 and compare the results with those obtained analytically in Problem 3.27.

3.29. Let ξ be a continuous RV uniformly distributed in the interval $X = (0, b)$. Compute the entropy H(X).

3.30. Show that the entropy H(X) of the RV ξ having probability density function:

$$f_\xi(x) = \begin{cases} \dfrac{1}{x} \cdot \dfrac{1}{(\log x)^2}, & x \geqslant e, \\ 0, & x < e, \end{cases}$$

is infinite.

3.31. Prove Theorem 3.8. *Hint*: Prove first that the following inequality holds:

$$-\int_{-\infty}^{\infty} f_\xi(x) \log f_\xi(x) \, dx \leqslant -\int_{-\infty}^{\infty} f_\xi(x) \log f_\eta(x) \, dx,$$

where $f_\xi(x)$ and $f_\eta(y)$ are arbitrary probability density functions. Then apply it by considering an arbitrary probability density function $f_\xi(x)$ with finite variance σ_ξ^2, and $f_\eta(y) = \dfrac{1}{\sqrt{2\pi}\sigma_\xi}$ exp $[-(x - \mu)^2/2\sigma_\xi^2]$.

3.32. Compute the upper limit of the capacity (3.89) of a bandlimited Gaussian channel as the bandwidth $B \to \infty$.

Chap. 3   Problems

127

CHAPTER 4

Waveform Transmission Over the Gaussian Channel

As we saw in Chapter 3, the digital information to be delivered to a user through a communicating channel can be represented by a sequence of binary digits corresponding to the messages emitted by the source. These digits are either generated directly by the source or represent the output of a source encoder or, more generally, are produced by a channel encoder (see Fig. 3.16). A physical channel may consist of a pair of wires, a coaxial cable, an optical fiber, a radio link or a combination of these. It is, therefore, necessary to convert the sequence of binary digits into waveforms that match the physical properties of the transmission medium. The mapping of digital sequences into a set of waveforms is called *digital modulation*.

The *digital modulator* is the functional device that achieves such mapping. The modulator may simply map binary digits into a set of two waveforms. This type of transmission is called either *binary modulation* or *binary signaling*. Alternatively, the modulator may use $M = 2^h$ different waveforms by mapping h digits at a time, into a set of M waveforms. This type of transmission is called *M-ary modulation* or *M-ary signaling* or, more generally, *multilevel modulation* or *multilevel signaling*.

All real channels corrupt the transmitted waveforms with different impairments such as distortions, interferences, and noise. However, throughout this chapter, we shall assume that the only impairment introduced by the channel is additive Gaussian noise. Other impairments will be considered in Chapters 6 and 10. At the receiving side, the noise-corrupted waveform is processed by the *digital demodulator*. Its task is that of estimating which particular waveform was actually transmitted by the modulator.

This chapter deals with the performance evaluation of the modulator–demodulator pair on the additive Gaussian noise channel. The performance of each transmission scheme will be evaluated with respect to the *symbol error probability*, that is, the probability that the demodulator chooses a waveform different from the transmitted one. One can argue that this performance parameter is irrelevant to the user, as he or she is interested in receiving the binary sequences associated with the waveforms without errors. This observation is true and deserves some attention. In the absence of a channel encoder, no redundancy is added to the information sequence. Therefore, the demodulator, after the selection of a waveform, can perform the inverse mapping to the digital sequence. It is not a difficult task to relate the symbol error probability to the *bit error probability* affecting the binary sequence delivered to the user. When a channel encoder is present, the symbol error probability at the demodulator output is not a meaningful parameter and the problem requires a more careful discussion. We shall treat this issue in Chapter 9. In this chapter, only uncoded transmission will be considered. After a heuristic introduction of different signaling schemes, we shall emphasize the general approach to the study of the modulator–demodulator pair.

4.1 INTRODUCTION

We assume that the analog channel connecting the modulator output to the demodulator input is the additive white Gaussian noise (AWGN) channel discussed in Section 3.3. To allow distortionless transmission of any possible waveform, an infinite bandwidth is assigned to this channel. The model is reproduced in Fig. 4.1 with some additional details. A source produces an independent identically distributed binary sequence with

Figure 4.1 Transmission systems considered in this chapter.

an average information rate of R_s bits/s. No channel encoding is performed. Therefore, the binary digits are presented for transmission to the channel every $T_c = T_s$ seconds, where the subscripts c and s refer to the channel and source, respectively. The binary digits are grouped in sequences of length h called symbols. The statistical occurrence of the $M = 2^h$ symbols is described by the probabilities $p_i = 1/M$ assigned to the set of symbols $\{m_i\}_{i=1}^M$. So a time $T = hT_c$ is available for the transmission of one symbol and the modulator maps the symbols into a set of M waveforms. The sequence of waveforms generated by the modulator forms the transmitted signal $v_\xi(t)$. The rate $1/T$ at which the waveforms are transmitted over the channel is called the *signaling rate* and is given by $R_s/\log_2 M$.

A great variety of digital modulation schemes (or mapping rules) is available, and an effort will be made to present them in a unifying conceptual frame. To this purpose, some practical examples will now be introduced.

Example 4.1

Given a set of $M = 2^h$ symbols, the digital modulator maps them into a set of M different waveforms

$$\{s_i(t)\}_{i=1}^M$$

of duration $T = hT_c = hT_s$. This signaling scheme uses a *memoryless* correspondence between each symbol of h digits and one of the available waveforms. Different schemes are characterized by the properties of the signal set $\{s_i(t)\}_{i=1}^{M}$. For instance, baseband or bandpass signals can be chosen.

Let us examine some particular cases (Fig. 4.2).

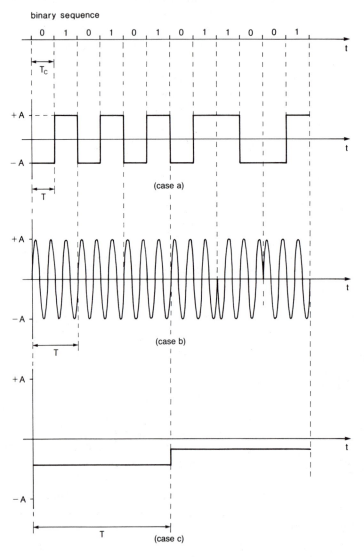

Figure 4.2 Example of memoryless modulation schemes: (a) Baseband binary modulation; (b) bandpass quaternary modulation; (c) baseband multilevel modulation.

Case (a) *Baseband binary modulation*

In this case, the modulator uses only two waveforms ($M = 2$) consisting of a positive or negative rectangular pulse of the same magnitude and of duration $T = T_c$. Each waveform corresponds to a symbol of one digit. We have

$$s_1(t) = +A, \qquad s_2(t) = -A, \qquad 0 \leqslant t < T = T_c. \tag{4.1}$$

Case (b)   *Bandpass quaternary modulation*

The modulator uses four sinusoidal waveforms ($M = 4$) with different initial phases. Each symbol of two digits is associated to the values of these initial phases. The signals are bandpass and the frequency f_0 determines their spectral allocation. We have, for this case,

$$s_i(t) = A \cos (2\pi f_0 t + \varphi_i), \qquad 0 \leqslant t < T = 2T_c,$$

$$\varphi_i = \frac{(2i - 1)\pi}{4}, \qquad i = 1, 2, 3, 4. \tag{4.2}$$

Notice the lower signaling rate with respect to the previous case and the possible phase discontinuities in the transmitted waveform.

Case (c)   *Baseband multilevel modulation*

The modulator uses 64 different waveforms consisting of rectangular pulses of different amplitudes; that is,

$$s_i(t) = \frac{2i - 65}{63} A, \qquad 0 \leqslant t < T = 6T_c, \qquad i = 1, 2, 3, \ldots, 64. \tag{4.3}$$

In this case each waveform corresponds to a message of six digits and its duration is, therefore, six times longer than in case (a). ☐

A crucial question is raised when considering these examples: what are the substantial differences among the various choices? Intuitively, we can affirm that, when the channel rate $1/T_c$ is kept constant, the waveforms of case (c) require less bandwidth than those of case (a), since the pulse duration in the former case is longer. For the waveforms of case (b), one obvious observation is that the center frequency f_0 of the bandpass signals allows one to choose among different frequency ranges and thus to match the physical properties of the transmission channel. Some additional considerations may give a better insight into this case. Notice that these waveforms are characterized by a constant envelope. In fact, the messages are associated with the phase discontinuities and do not affect the envelope of the waveform. Consequently, these waveforms appear to be good candidates for those applications for which a constant envelope is required. Typically, this is the case of a satellite communication channel in which it is common practice to use bandpass nonlinear amplifiers that require an input signal with nearly constant envelope to avoid AM/AM and AM/PM conversions (see Section 2.4.2). However, unfortunately, when the waveforms are passed through a bandlimiting filter their phase discontinuities give rise to envelope variations. In particular, a phase reversal causes the envelope to approach zero. As filtering cannot be avoided, the desired advantages are lost. The way out of this drawback seems to be the use of constant envelope waveforms without phase discontinuities. In part, Example 4.2 is motivated by this possibility. But, before introducing it, let us reconsider the cases of Example 4.1 from a more general viewpoint. In fact, it has already been noted that in all cases the modulator establishes a memoryless correspondence between a sequence of symbols and a sequence of waveforms. But why should the waveforms be chosen in this way? Introducing correlation among successive waveforms would necessitate more complex signaling schemes. But, at least in principle, it could allow one to achieve new desirable properties with regard to the transmitted signal. It is fairly intuitive that correlation in the time domain could be used to shape the power spectrum and to reduce the transmission bandwidth. Let us see in Example 4.2 what kind of signals can be designed when following this more general approach to obtain constant envelope signals without phase discontinuities.

Example 4.2

Assume a binary modulation scheme in which the modulator can choose two different frequencies for transmission and the modulated signal is constrained to be phase continuous. Each transmitted frequency corresponds to a symbol of one digit. The signals are bandpass and have the form

$$s_i(t) = A \cos \{2\pi f_0 t + \Phi_{ik}(t)\},$$

$$\Phi_{ik}(t) = \frac{(2i - 3)\pi}{2T}(t - kT) + \varphi_k, \qquad i = 1, 2, \qquad k = 0, 1, 2, \ldots, \qquad (4.4)$$

$$kT \leqslant t < (k + 1)T, \qquad T = T_c.$$

The two transmitted frequencies are $f_0 \pm 1/4T$ and the phase φ_k is chosen in each signaling period to meet the requirement of phase continuity at the transition times $t = kT$. Notice that, as a consequence of the phase-continuity constraint, this modulator has memory: the frequency depends on the transmitted symbol and the phase on the whole transmitted sequence.

A particular example of this signaling scheme is shown in Fig. 4.3. Notice the phase trajectory as a function of time; this trajectory represents the binary sequence. Other possible phase paths are shown as dotted lines on the graph. □

The main feature of Example 4.2 is that there is correlation among the successively transmitted waveforms. Therefore, the demodulator must take it into account when processing the received signal. At least in principle, the transmitted information should be recovered only after observing the whole received sequence. On the other hand, with respect to the previous discussion regarding phase discontinuities, the transmitted signal should exhibit a narrower spectrum, and hence a smaller amount of envelope fluctuations, even in the presence of narrowband filtering. We shall discuss this problem in greater detail in Chapter 5, when considering practical applications.

Let us now take a more formal approach to the problem considered in this chapter. As already stated, a time T is assigned to the transmission of one symbol m_k of h digits. The modulator generates a particular waveform in such a way that there is a one-to-one correspondence between the symbol m_k and the value of a RV ξ_k associated with the waveform. This RV can be the amplitude, the phase, the frequency, or some other parameter that defines the waveform.

Notice that, in general, the generated waveform can have a duration longer than T. The choice of the waveform depends on the value of the RV ξ_k to be transmitted in the kth time interval and on the *state* σ_k of the modulator, which takes into account the memory of the modulation process, if any. Therefore, when a sequence ξ of RVs ξ_k, $k = 0, 1, \ldots, K - 1$, is transmitted, the modulated signal $v_\xi(t)$ can be written as follows:

$$v_\xi(t) = \sum_{k=0}^{K-1} s(t - kT; \xi_k, \sigma_k), \qquad 0 \leqslant t < KT, \qquad (4.5)$$

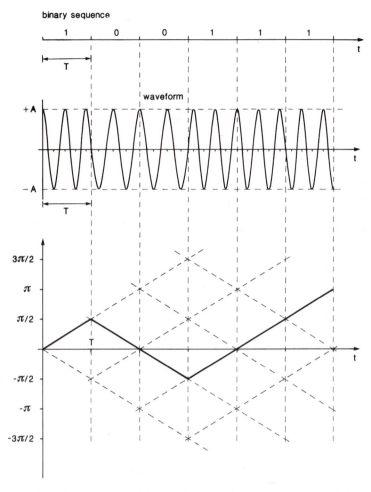

Figure 4.3 Example of a modulation scheme with memory. The sinusoidal signals are constrained to be phase continuous. The phase trajectory corresponding to the binary sequence (100111) is shown as a solid path in the graph.

where $s(t - kT; \xi_k, \sigma_k)$ is the signal transmitted in the kth interval of duration T. For simplicity, only real signals will be considered. The received signal at the output of the channel is corrupted by an AWGN $n(t)$ with power spectral density $N_0/2$. Two different cases will be considered. In the first, the receiver has complete knowledge of the set of possible transmitted signals. We call this receiver a *coherent receiver*. We can write the received signal in the form

$$r(t) = v_\xi(t) + n(t). \tag{4.6}$$

In the second case, we shall consider a typical situation arising in bandpass communication systems where the signals, in addition to the additive noise, are characterized by an uncertainty due to a random phase angle ϑ. The received signal will be written in the form

$$r(t) = v_{\xi}(t; \vartheta) + n(t). \qquad (4.7)$$

Another important difference between demodulator structures is the presence of memory in the modulation process. It is intuitive that much simpler demodulators will be allowed when the modulator is memoryless.

The purpose of the demodulator is to process the received signal $r(t)$ to produce an estimate $\hat{\xi}$ of the transmitted sequence ξ, and hence an estimate $\hat{m}_k$ of each transmitted symbol. The performance of the signaling scheme will be evaluated through the *symbol error probability*

$$P(e) \triangleq P\{\hat{\xi}_k \neq \xi_k\} = P\{\hat{m}_k \neq m_k\}. \qquad (4.8)$$

We are interested in the demodulator that achieves the minimum value of $P(e)$. We call it *optimum* in this sense. We shall assume throughout the chapter that the transmitted messages are equally likely. For this case, it was shown in Section 2.6 that minimum error probability is achieved with a maximum likelihood (ML) receiver. The material of the following sections deals with three different cases:

(i) *Memoryless modulators and coherent receivers.* Each generated waveform has a duration strictly limited to the time interval T and the modulator is memoryless. That is, it presents only one possible state (see the cases of Example 4.1).

(ii) *Memoryless modulators and incoherent receivers.* Each generated waveform still has a duration limited to the interval T and the modulator is again memoryless. But the receiver has an uncertainty due to a random phase angle as in (4.7).

(iii) *Modulators with memory and coherent receivers.* The modulation process implies different states of the modulator, as in the simple case of Example 4.2.

4.2 MEMORYLESS MODULATION AND COHERENT DEMODULATION

The complete knowledge of the set of possible transmitted signals (coherent receiver) implies perfect synchronization of the receiver. In other words, if the signal is bandpass, its carrier is perfectly recovered at the receiver end. Furthermore, an ideal clock, at the signaling rate, is also available (see Section 6.3 for the problems related to this topic).

The memoryless nature of the modulation process implies that the waveforms available at the modulator are strictly limited to the time interval T and that the modulator has only one state, which therefore does not affect the generated signals. These can now be written as

$$s(t - kT; \xi_k, \sigma_k) = s(t - kT; \xi_k), \qquad kT \leq t < (k + 1)T. \qquad (4.9)$$

In principle, the demodulator forms the set of signals like

$$v_{\xi}(t) = \sum_{k=0}^{K-1} s(t - kT; \xi_k), \qquad 0 \leq t < KT, \qquad (4.10)$$

for all possible sequences ξ. From these signals, it recovers an estimate $\hat{\xi}$ of the transmitted symbol sequence, and hence of the transmitted symbols. The log-likelihood ratio for a sequence ξ, based on the observation of $r(t)$, is given by (see Section 2.6)

$$\lambda_\xi = \frac{2}{N_0} \int_0^{KT} r(t)v_\xi(t)\,dt - \frac{1}{N_0} \int_0^{KT} v_\xi^2(t)\,dt. \qquad (4.11)$$

The optimum demodulator chooses the sequence $\hat{\xi}$ that maximizes λ_ξ in (4.11); that is,

$$\hat{\xi}: \quad \lambda_{\hat{\xi}} = \max_{\xi} \lambda_\xi. \qquad (4.12)$$

Inserting (4.10) in (4.11) and multiplying by $N_0/2$, the ML sequence maximizes the quantity

$$l_\xi \triangleq \int_0^{KT} r(t) \sum_{k=0}^{K-1} s(t - kT; \xi_k)\,dt - \frac{1}{2} \int_0^{KT} \left[\sum_{k=0}^{K-1} s(t - kT; \xi_k) \right]^2 dt. \qquad (4.13)$$

Taking advantage of the memoryless nature of the modulation and of the duration T of $s(t; \xi_k)$, (4.13) can be rewritten in the form

$$l_\xi = \sum_{k=0}^{K-1} l_{\xi_k} \qquad (4.14)$$

where

$$l_{\xi_k} \triangleq \int_{kT}^{(k+1)T} r(t)s(t - kT; \xi_k)\,dt - \frac{1}{2} \int_{kT}^{(k+1)T} s^2(t; \xi_k)\,dt. \qquad (4.15)$$

From (4.14) we can conclude that the ML sequence $\hat{\xi}$ is obtained as an ML symbol-by-symbol decision by maximizing in each time interval T the quantities l_{ξ_k} of (4.15). In fact, in our case the maximum value of the sum (4.14) corresponds to the sum of the maximum values of its components (see Appendix F). Considering this fact, from now on we shall restrict ourselves to the time interval $(0, T)$ corresponding to $k = 0$.

The quantities in (4.15) are said to form a *sufficient statistics* of the received signal $r(t)$. This is a technical term meaning that all we need to know about the received signal $r(t)$ to allow an ML decision is contained in these quantities. Since the RV ξ_0 can take M different values, each signal $s(t; \xi_0)$ is chosen from a set $\{s_i(t)\}_{i=1}^M$ of M different waveforms of duration T. Therefore, the RV l_{ξ_0} of (4.15) becomes

$$l_i = \int_0^T r(t)s_i(t)\,dt + c_i,$$

$$c_i \triangleq -\frac{1}{2} \int_0^T s_i^2(t)\,dt = -\frac{1}{2}\mathscr{E}_i, \qquad (4.16)$$

$$i = 1, 2, \ldots, M,$$

where $\mathscr{E}_i$ is the energy of the ith waveform.

The block diagram of the ML demodulator is shown in Fig. 4.4 and is usually referred to as a *correlation demodulator*. For simplicity the figure also refers to the detection of the first symbol, that is, $k = 0$.

Sec. 4.2   Memoryless Modulation and Coherent Demodulation   **135**

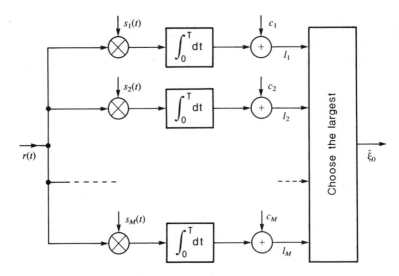

Figure 4.4 Block diagram of the correlation receiver for the case of memoryless modulation and coherent demodulation.

Another method to get the quantities (4.16) is to replace the bank of correlators with a bank of filters matched to the signals $\{s_i(t)\}$. The filter matched to $s_i(t)$ has an impulse response $h_i(t) = s_i(T - t)$. Therefore, the output of this filter at $t = T$, when the input is $r(t)$, is exactly that required in (4.16). The block diagram of this *matched filter demodulator* is shown in Fig. 4.5. This version of the optimum demodulator shows that a bank of M matched filters supplies the sufficient statistics to solve our decision problem.

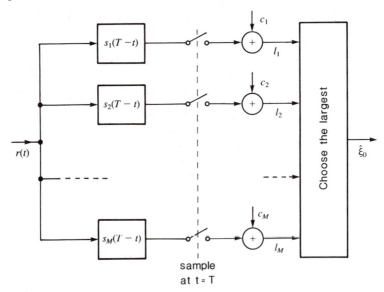

Figure 4.5 Block diagram of the matched filter receiver for the case of memoryless modulation and coherent demodulation.

A different version of the optimum demodulator can be obtained by representing the signals $\{s_i(t)\}_{i=1}^{M}$ in the orthonormal basis $\{\psi_j(t)\}_{j=1}^{N}$, $N \leq M$, by using the Gram–Schmidt procedure (see Section 2.5). We get

$$s_i(t) = \sum_{j=1}^{N} s_{ij}\psi_j(t), \qquad i = 1, 2, \ldots, M, \qquad 0 \leq t < T. \tag{4.17}$$

By inserting (4.17) in (4.16) and after some easy algebra, we get

$$l_i = \sum_{j=1}^{N} s_{ij}r_j - \frac{1}{2}\sum_{j=1}^{N} s_{ij}^2, \qquad i = 1, 2, \ldots, M, \tag{4.18}$$

where

$$r_j \triangleq \int_0^T r(t)\,\psi_j(t)\,dt. \tag{4.19}$$

Due to the complete knowledge of the signal set $\{s_i(t)\}$, we can conclude from (4.18) that a sufficient statistics for the received signal $r(t)$ is the set $\{r_j\}_{j=1}^{N}$ of the received signal components in the N-dimensional space spanned by the orthonormal basis $\{\psi_j(t)\}_{j=1}^{N}$. In fact, (4.18) shows that only the projections of $r(t)$ onto the signal space are relevant for the decision. The demodulator ignores as irrelevant data any component of the received signal that is orthogonal to the N-dimensional space of the transmitted signals. The block diagram of the demodulator based on (4.18) is shown in Fig. 4.6. There is no conceptual difference between this demodulator and that of Fig. 4.4. The only practical difference depends on the number N of correlators. Sometimes these can be much less than M. On the other hand, some additional processing is present in the demodulator of Fig. 4.6. This must be traded off for the reduced number of correlators.

Preferably, we shall refer to the scheme of Fig. 4.6 since it is usually less redundant.

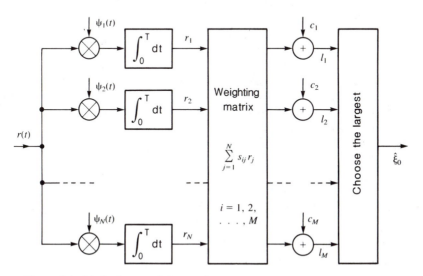

Figure 4.6 Block diagram of the correlation receiver with signals expanded in an N-dimensional orthonormal basis $\{\psi_j(t)\}_{j=1}^{N}$.

Furthermore, it implies that the geometric viewpoint be emphasized, and this is very useful to calculate the symbol error probability.

Example 4.3

Let us reconsider in this example the baseband binary modulation of case (a) of Example 4.1. The two transmitted signals are written as

$$s_1(t) = A\sqrt{T}\psi(t),$$
$$s_2(t) = -A\sqrt{T}\psi(t), \qquad 0 \le t < T, \tag{4.20}$$

with

$$\psi(t) \triangleq \frac{1}{\sqrt{T}} u_T(t), \qquad u_T(t) \triangleq u(t) - u(t-T).$$

In this case the orthonormal basis consists of only one function $\psi(t)$.

The computation of the quantities in (4.18) gives

$$l_1 = A\sqrt{T}r - \frac{1}{2}A^2 T, \qquad l_2 = -A\sqrt{T}r - \frac{1}{2}A^2 T, \tag{4.21}$$

where

$$r = \frac{1}{\sqrt{T}} \int_0^T r(t)\, dt.$$

The sufficient statistics is represented now by the RV r. That is, the component of $r(t)$ along $\psi(t)$. By inspection of (4.21), we can conclude that the decision can be taken by simply considering the sign of the RV r: for positive r we have $l_1 > l_2$, whereas, for negative r, we have $l_2 > l_1$. The block diagram of the optimum demodulator is shown in Fig. 4.7. $\square$

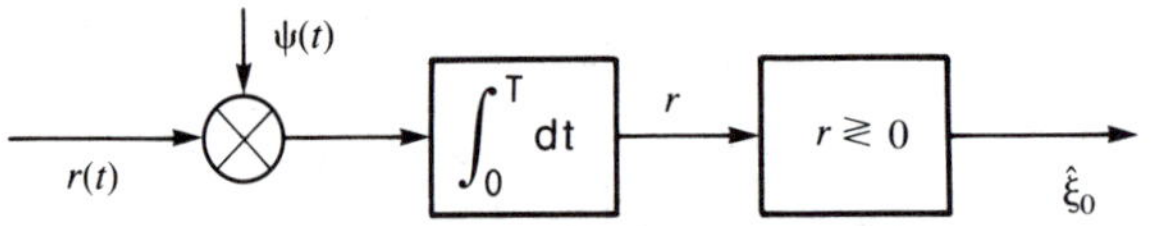

Figure 4.7 Optimum receiver in the binary case and coherent demodulation.

Now we want to consider the optimum demodulator of Fig. 4.6 from a different point of view, with the aim of enhancing a geometric interpretation of the optimum decision rule. First, the N-dimensional space used to represent the signal set also contains the sufficient statistics for the received signal $r(t)$. Therefore, using (4.17) and (4.19), the received signal can be put in vector form:

$$\mathbf{r} = \mathbf{s}_j + \mathbf{n}, \qquad j = 1, 2, \ldots, M, \tag{4.22}$$

where

$$\mathbf{r} \triangleq (r_1, r_2, \ldots, r_N),$$
$$\mathbf{s}_j \triangleq (s_{j1}, s_{j2}, \ldots, s_{jN}),$$
$$\mathbf{n} \triangleq (n_1, n_2, \ldots, n_N),$$

and

$$n_j \triangleq \int_0^T n(t)\psi_j(t)\, dt \tag{4.23}$$

are the components of the noise vector in the N-dimensional signal space. Second, we can take the likelihood ratios defined in Section 2.6 and complete the square in the exponent by adding a term that does not depend on ξ and therefore is irrelevant in the decision. We define the likelihood function

$$\Lambda'_\xi \triangleq \exp\left\{-\frac{2}{N_0}\int_0^{KT}[r(t)-v_\xi(t)]^2\,dt\right\}. \tag{4.24}$$

Following the discussion that led to (4.14) and (4.16), we can conclude that the optimum demodulator takes its symbol by symbol decision by looking for the maximum of the likelihood function:

$$\Lambda'_i = \exp\left\{-\frac{2}{N_0}\int_0^{T}[r(t)-s_i(t)]^2\,dt\right\}. \tag{4.25}$$

The maximization of (4.25) is achieved by looking for the minimum of the exponent. Using the vector representation (4.22), we can write

$$\int_0^{T}[r(t)-s_i(t)]^2\,dt = \sum_{j=1}^{N}(r_j-s_{ij})^2 = |\mathbf{r}-\mathbf{s}_i|^2. \tag{4.26}$$

If we now define the Euclidean distance between the two vectors $\mathbf{r}$ and $\mathbf{s}_i$ as

$$d(\mathbf{r},\mathbf{s}_i) \triangleq |\mathbf{r}-\mathbf{s}_i|, \tag{4.27}$$

we can restate the optimum decision rule in a geometrical parlance. The optimum demodulator chooses the signal vector $\mathbf{s}_i$ that is closer to the received vector $\mathbf{r}$. In conclusion, *the ML decision is implemented by a minimum distance demodulator.*

This interpretation is shown as an example in Fig. 4.8 for a two-dimensional case. The receiver observes the vector $\mathbf{r}$ obtained by addition of the noise vector $\mathbf{n}$ to the transmitted vector $\mathbf{s}_i$. The decision rule is intuitively understood because the most ''reasonable'' choice (which is indeed optimum for the Gaussian case) is that of deciding for the signal that is closer to $\mathbf{r}$. With this geometric approach, the implementation of

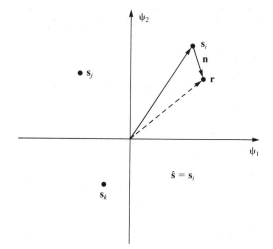

Figure 4.8 Geometry of the minimum-distance decision rule.

the optimum decision rule can be easily described by considering again a two-dimensional case as in Fig. 4.9. Each point of the plane represents a possible received vector $\mathbf{r}$; the demodulator computes the distances from each possible transmitted vector. The plane is thus partitioned into decision regions R_i that are the loci of the points closer to the vector $\mathbf{s}_i$ than to any other one. More formally,

$$R_i \triangleq \{\mathbf{r}: d(\mathbf{r}, \mathbf{s}_i) = \min_j d(\mathbf{r}, \mathbf{s}_j)\}. \tag{4.28}$$

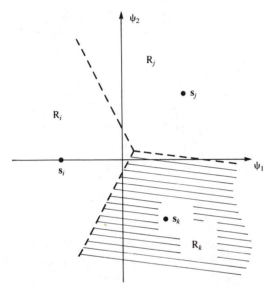

Figure 4.9 ML decision regions in a two-dimensional case.

The ML decision rule is therefore

$$\text{choose} \quad \hat{\mathbf{s}} = \mathbf{s}_i \quad \text{whenever} \quad \mathbf{r} \in R_i. \tag{4.29}$$

In the more general case, the decision rule implemented by the optimum demodulator entails partitioning the N-dimensional signal space into M disjoint decision regions by applying the minimum-distance concept. The decision regions R_i are always delimited by hyperplanes, which are the loci of the points equidistant from two neighbor signals.

4.2.1 Error Probability Evaluation

Having equally likely symbols, the symbol error probability (4.8) can be written as

$$P(e) = 1 - P(c)$$
$$= 1 - \frac{1}{M} \sum_{j=1}^{M} P(c|\mathbf{s}_j), \tag{4.30}$$

where $P(c|\mathbf{s}_j)$ is the probability of a correct decision given that the signal vector $\mathbf{s}_j$, corresponding to the symbol m_j, was transmitted. Therefore, the computation of $P(e)$ requires the computation of the set of probabilities $\{P(c|\mathbf{s}_j)\}_{j=1}^{M}$.

We have shown that the decision process implemented by the optimum demodulator is based on the RVs l_i of (4.18) and that the decision consists of choosing the largest among them. Assuming that the signal vector $\mathbf{s}_j$ is transmitted, we obtain from (4.18) and (4.22)

$$l_i = (\mathbf{s}_j + \mathbf{n})\,\mathbf{s}_i' - \frac{1}{2}|\mathbf{s}_i|^2, \qquad i, j = 1, 2, \ldots, M, \tag{4.31}$$

where

$$\mathbf{s}_j\,\mathbf{s}_i' = \int_0^T s_j(t)s_i(t)\,dt.$$

It is quite straightforward to show that, given $\mathbf{s}_j$, the RVs l_i are jointly Gaussian with means and covariances depending only on the two-sided noise power spectral density $N_0/2$ and on the scalar products $\mathbf{s}_j\,\mathbf{s}_i'$. The problem of the evaluation of (4.30) is thus reduced to a classical Gaussian problem in detection theory.

Instead of solving this problem with this approach, we prefer a more intuitive geometric viewpoint. Recalling the optimum decision rule in the form of (4.29), we can assign to (4.30) the expression

$$P(e) = 1 - \frac{1}{M}\sum_{j=1}^{M} P\{\mathbf{r} \in R_j | \mathbf{s}_j\}. \tag{4.32}$$

Equation (4.32) means that the demodulator makes an error whenever the received signal vector $\mathbf{r} = \mathbf{s}_j + \mathbf{n}$ does not belong to the decision region R_j of the transmitted signal vector $\mathbf{s}_j$.

Let us now fully exploit the geometric implications of the result (4.32). The error probability depends only on the geometrical configuration of the decision regions that are a direct consequence, through (4.28), of the given set of transmitted signal vectors. Therefore, we can conclude that the choice of the orthonormal basis $\{\psi_j(t)\}$ in (4.17) does not affect the error probability. In other words, changing the waveforms used for transmission does not change the error probability, provided that the geometrical configuration of the set of signal vectors is left unchanged. This important result is also due to the distortionless nature of the channel. We shall see in Chapter 6 that, when distortion is present, the waveforms are also relevant for the computation of the error probability. In conclusion, signal sets having identical geometrical configurations are equivalent insofar as error probability is concerned.

This concept of equivalence among different signal sets can be extended after a more careful consideration of the geometry of the decision regions. If we denote as *signal points* in the signal space the tips of the signal vectors, we can say that the configuration of the decision regions depends only on the set of signal points. Therefore, any rotation or translation of the signal points as a whole does not change the value of the error probability. Notice, however, that although the error probability is invariant with respect to translations of the signal points such a modification does change the energy required to transmit each signal. Since the average energy required for transmitting a given signal set is given by

$$\mathcal{E} = \frac{1}{M}\sum_{i=1}^{M}|\mathbf{s}_i|^2, \tag{4.33}$$

one meaningful question can be raised. Given a particular configuration of the signal points, and therefore a value of the error probability, what choice of the origin requires the minimum average energy for transmission?

After recognizing that (4.33) is precisely the definition of the moment of inertia around the origin for a set of M equal point masses located at the signal points, the answer is that their center of gravity must be the origin. This condition can be stated mathematically as

$$\sum_{i=1}^{m} \mathbf{s}_i \triangleq \mathbf{0}. \tag{4.34}$$

Therefore, a signal set satisfying (4.34) for a given error probability requires the minimum average energy for transmission.

To conclude these geometrical considerations about (4.32), we define a signal set that presents a *complete symmetry*. A geometrical configuration of signal points presents a complete symmetry when it has congruent decision regions. In this situation, the conditional probability of a correct decision is independent of the transmitted signal. As a consequence, (4.32) simplifies for this case to

$$P(e) = 1 - P\{\mathbf{r} \in \mathrm{R}_j | \mathbf{s}_j\}, \qquad \text{any } j. \tag{4.35}$$

To proceed further, we need to know the conditional pdf of the received vector $\mathbf{r}$ for a given transmitted signal $\mathbf{s}_j$. Recalling (4.22) and (4.23), first notice (see Problem 4.1) that the RVs n_i are statistically independent zero-mean Gaussian RVs with variance $N_0/2$. Thus, the pdf of the noise vector $\mathbf{n}$ is given by

$$f_{\mathbf{n}}(\boldsymbol{\alpha}) = \frac{1}{(\pi N_0)^{N/2}} \exp\left\{ -\frac{|\boldsymbol{\alpha}|^2}{N_0} \right\}. \tag{4.36}$$

Since the noise vector $\mathbf{n}$ is independent of the signal $\mathbf{s}_j$, the conditional pdf of the received vector $\mathbf{r}$ is still Gaussian with mean $\mathbf{s}_j$ and is given by

$$f_{\mathbf{r}|\mathbf{s}_j}(\boldsymbol{\alpha}|\mathbf{s}_j) = f_{\mathbf{n}}(\boldsymbol{\alpha} - \mathbf{s}_j). \tag{4.37}$$

Finally, introducing (4.37) into (4.32), we get

$$P(e) = 1 - \frac{1}{M} \sum_{j=1}^{M} \int_{\mathrm{R}_j} f_{\mathbf{n}}(\boldsymbol{\alpha} - \mathbf{s}_j) d\boldsymbol{\alpha}. \tag{4.38}$$

This simple but general result will now be applied to some significant signal configurations.

Binary signals

The general case of two signals is shown in Fig. 4.10a. An equivalent signal set with the same error probability, but minimum energy, is shown in Fig. 4.10b. In this second case, the vertical axis is the boundary between the two decision regions R_1 and R_2. The situation was already encountered in Example 4.3. Notice that only the noise component n_1 can cause errors by forcing the received signal point to cross the boundary of the decision region assigned to the transmitted signal.

Due to the complete symmetry of the signal set, we can use (4.35) and Fig.

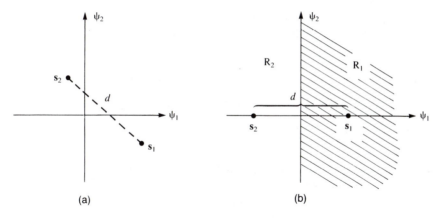

(a)   (b)

Figure 4.10 Geometric representation of the binary signaling scheme: (a) General case; (b) minimum-energy configuration: antipodal signals (Example 4.4).

4.10b to get

$$P(e) = 1 - P\{\mathbf{r} \in R_1|\mathbf{s}_1\} = P\{r_1 < 0|\mathbf{s}_1\}. \tag{4.39}$$

Recalling (4.22), we have

$$P(e) = P\left\{n_1 < -\frac{d}{2}\right\}, \tag{4.40}$$

where $d \triangleq d(\mathbf{s}_1, \mathbf{s}_2)$ is the Euclidean distance between the two signals. Finally, using a result of Appendix A, we get

$$P(e) = \frac{1}{2} erfc\left(\frac{d}{2\sqrt{N_0}}\right). \tag{4.41}$$

Interpreting this result, we can see that in the coherent demodulation of two equally likely signals transmitted on the AWGN channel the error probability depends only on the Euclidean distance between the two signals. Using (4.26) and (4.27), we can relate the geometric parameter d to the signal energies. In fact, we have

$$d^2 = \mathscr{E}_1 + \mathscr{E}_2 - 2\rho\sqrt{\mathscr{E}_1\mathscr{E}_2}, \tag{4.42}$$

where $\mathscr{E}_1$ and $\mathscr{E}_2$ are the energies of the two signals and

$$\rho \triangleq \frac{1}{\sqrt{\mathscr{E}_1\mathscr{E}_2}} \int_0^T s_1(t)s_2(t)\, dt \tag{4.43}$$

is the *correlation coefficient* between the signals. Notice that we always have $-1 \leq \rho \leq 1$. For the case of equal energy signals ($\mathscr{E}_1 = \mathscr{E}_2 = \mathscr{E}$), the error probability of (4.41) can be expressed in terms of the energy $\mathscr{E}$ as follows:

$$P(e) = \frac{1}{2} erfc\left(\sqrt{\frac{\mathscr{E}(1-\rho)}{2N_0}}\right). \tag{4.44}$$

Example 4.4 Antipodal signals

The minimum energy in the binary case is achieved when $s_1(t) = -s_2(t)$; for this case $\rho = -1$. This is again the situation dealt with in Example 4.3. The error probability is now, from (4.44),

$$P(e) = \frac{1}{2} erfc \left(\sqrt{\frac{\mathcal{E}}{N_0}} \right) \tag{4.45}$$

and its curve is shown in Fig. 4.12. □

Example 4.5 Orthogonal signals

Two orthogonal signals are shown in Figure 4.11. In this case $\rho = 0$. Therefore, the error probability is given by

$$P(e) = \frac{1}{2} erfc \left(\sqrt{\frac{\mathcal{E}}{2N_0}} \right) \tag{4.46}$$

and its curve is shown in Fig. 4.12. There is a 3-dB penalty in the signal energy to be paid, with respect to the antipodal case. □

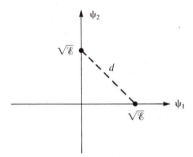

Figure 4.11 Binary orthogonal signals (Example 4.5).

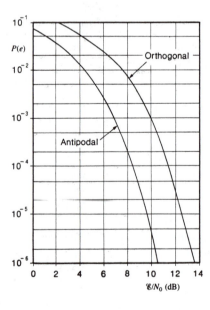

Figure 4.12 Error probability as a function of $\mathcal{E}/N_0$ for the cases of binary antipodal and orthogonal signals.

Let us now consider a slightly more complex example with two bandpass signals of different frequencies.

Example 4.6

Let the signals be given by

$$s_1(t) = A \cos \{2\pi(f_0 - f_d)t\}, \qquad 0 \le t < T. \tag{4.47}$$
$$s_2(t) = A \cos \{2\pi(f_0 + f_d)t\},$$

We assume that $(f_0 \pm f_d)T \gg 1$ so that we have approximately

$$\mathscr{E} = \int_0^T s_1^2(t)\, dt = \int_0^T s_2^2(t)\, dt = \frac{A^2 T}{2}. \tag{4.48}$$

Under the same conditions, we get from (4.43)

$$\rho = \frac{\sin 4\pi f_d T}{4\pi f_d T}. \tag{4.49}$$

The behavior of ρ as a function of $2\pi f_d T$ is shown as a solid line in Fig. 4.13.

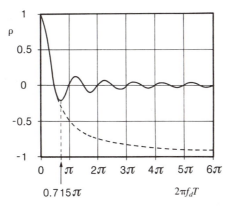

Figure 4.13 Correlation coefficient of the signals in Example 4.6.

Recalling (4.44), we get the minimum value for the error probability when ρ achieves its minimum; that is, $\rho \cong -0.22$. In this case,

$$P(e) \cong \frac{1}{2} erfc \left(\sqrt{0.61 \frac{\mathscr{E}}{N_0}} \right). \tag{4.50}$$

This value is greater than the result obtained in (4.45) for antipodal signals. An improvement toward the absolute minimum (4.45) can be obtained by using the frequencies $f_0 - f_d$ and $f_0 + f_d$ only for a duration $\tau < T$ and the frequency f_0 for the remaining part of the transmission time. The signals in this case are

$$s_1(t) = \begin{cases} A \cos \{2\pi(f_0 - f_d)t\}, & 0 \le t < \tau, \\ A \cos \{2\pi f_0 t - 2\pi f_d \tau\}, & \tau \le t < T, \end{cases}$$

$$s_2(t) = \begin{cases} A \cos \{2\pi(f_0 + f_d)t\}, & 0 \le t < \tau, \\ A \cos \{2\pi f_0 t + 2\pi f_d \tau\}, & \tau \le t < T. \end{cases} \tag{4.51}$$

Under the same assumptions as before, we get

$$\rho = \frac{\sin 4\pi f_d \tau}{4\pi f_d T} + \frac{T - \tau}{T} \cos 4\pi f_d \tau. \tag{4.52}$$

For a given value of f_d, the minimum value of ρ is achieved when

$$\tau = \frac{1}{4f_d}, \tag{4.53}$$

provided that $f_d \geq 1/4T$. Using (4.52) and (4.53) in (4.44), we get

$$P(e) = \frac{1}{2} erfc \left(\sqrt{\frac{\mathscr{E}}{N_0} \left(1 - \frac{1}{8f_d T} \right)} \right). \tag{4.54}$$

The behavior of the minimum value of ρ is also shown as a dotted line in Fig. 4.13. The conclusion is that by choosing τ according to (4.53) we can obtain values of ρ smaller than (4.49). Therefore, the error probability (4.54) can be smaller than the corresponding (4.50). We could actually approach asymptotically the value of (4.45) by letting $f_d \to \infty$. However, in this case, τ goes to zero and the physical implementation of the signal set (4.51) is, of course, questionable. $\square$

Rectangular signal sets

A straightforward extension of the binary case calculations can be applied to signal sets that have a "rectangular" configuration. By this term we mean the cases in which the boundaries of the decision regions are, in the general case, hyperplanes parallel to the coordinate hyperplanes. A two-dimensional example is shown in Fig. 4.14. Notice

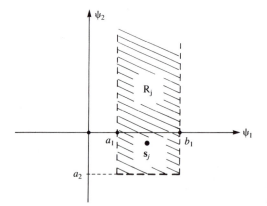

Figure 4.14 Decision region for one signal of a rectangular signal set.

that the signal set has the minimum energy configuration. To evaluate the error probability for such a signal set, we must compute the RHS of (4.32). The computational steps can be followed by looking at Fig. 4.14. In two dimensions, the boundaries of the decision regions are straight lines parallel to the coordinate axes. Hence, from Fig. 4.14 we get

$$\begin{aligned} P(c|\mathbf{s}_j) &= P\{\mathbf{r} \in R_j | \mathbf{s}_j\} \\ &= P\{a_1 \leq r_1 < b_1, a_2 \leq r_2 < \infty | \mathbf{s}_j\}. \end{aligned} \tag{4.55}$$

Recalling (4.22) and the independence of the noise components, we can write (4.55) as follows:

$$P(c|\mathbf{s}_j) = P\{a_1 - s_{j1} \leq n_1 < b_1 - s_{j1}\} \\ \cdot P\{a_2 - s_{j2} \leq n_2 < \infty\}. \tag{4.56}$$

Finally, since the noise components have the same Gaussian pdf, we have

$$P(c|\mathbf{s}_j) = \int_{a_1-s_{j1}}^{b_1-s_{j1}} f_n(\alpha)d\alpha \int_{a_2-s_{j2}}^{\infty} f_n(\alpha)d\alpha. \tag{4.57}$$

Similar expressions can be obtained for other rectangular regions. These can also be extended to the N-dimensional case.

Example 4.7

Consider the rectangular signal set of 16 equally likely signals of Fig. 4.15. There are only three noncongruent types of rectangular decision regions concerning the signal subsets: $\{\mathbf{s}_1, \mathbf{s}_4, \mathbf{s}_{13}, \mathbf{s}_{16}\}$, $\{\mathbf{s}_2, \mathbf{s}_3, \mathbf{s}_5, \mathbf{s}_8, \mathbf{s}_9, \mathbf{s}_{12}, \mathbf{s}_{14}, \mathbf{s}_{15}\}$, and $\{\mathbf{s}_6, \mathbf{s}_7, \mathbf{s}_{10}, \mathbf{s}_{11}\}$. Therefore, we need to compute only the following probabilities:

$$p_1 \overset{\triangle}{=} P(c|\mathbf{s}_1), \\ p_2 \overset{\triangle}{=} P(c|\mathbf{s}_2), \\ p_6 \overset{\triangle}{=} P(c|\mathbf{s}_6). \tag{4.58}$$

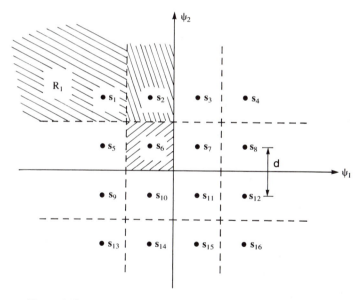

Figure 4.15 Two-dimensional signal set with 16 signals (Example 4.7).

Using (4.57), we have for p_1

$$p_1 = P\left\{n_1 < \frac{d}{2}\right\} P\left\{n_2 > -\frac{d}{2}\right\} = (1-p)^2, \tag{4.59}$$

where we have defined

$$p \triangleq \frac{1}{2} erfc \left(\frac{d}{2\sqrt{N_0}} \right)$$

and d is the distance between two adjacent signals shown in Fig. 4.15. With similar computations, we get

$$p_2 = (1 - 2p)(1 - p),$$
$$p_6 = (1 - 2p)^2. \tag{4.60}$$

Using (4.58) and (4.60) in (4.30), we have finally

$$P(e) = 1 - \frac{1}{4}(2 - 3p)^2. \tag{4.61}$$

$\square$

Orthogonal signal sets

Another important signal configuration is the set of M equal energy orthogonal signals. The two-dimensional case was shown in Fig. 4.11, whereas the case $M = 3$ is shown in Fig. 4.16. This signal set is completely symmetric. Thus, equation (4.35)

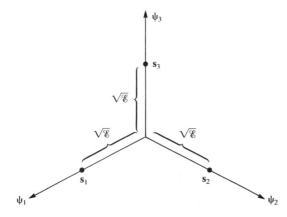

Figure 4.16 Orthogonal signal set in three dimensions.

can be used to compute the error probability. The decision regions are rather difficult to visualize, but the decision rule can be given a simple formulation. Assuming that the signal $\mathbf{s}_1$ is transmitted, the received signal point $\mathbf{r}$ is assigned to the decision region R_1 provided that the received signal component r_1 is greater than any other. This property can be verified by inspection of Fig. 4.17, where only the two axes ψ_1 and ψ_j are shown. Therefore, we have

$$P(c|\mathbf{s}_1) = P\{r_1 > r_2, r_1 > r_3, \ldots, r_1 > r_M|\mathbf{s}_1\}. \tag{4.62}$$

When $\mathbf{s}_1$ is transmitted, the received signal components are independent Gaussian RVs given by

$$r_i = \begin{cases} n_1 + \sqrt{\mathcal{E}}, & i = 1, \\ n_i, & i > 1. \end{cases} \tag{4.63}$$

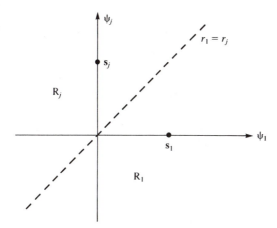

Figure 4.17 Boundary of decision regions for a pair of orthogonal signals.

Using (4.62), we then get:

$$P(c|\mathbf{s}_1) = \frac{1}{\sqrt{\pi N_0}} \int_{-\infty}^{\infty} \exp\left[-\frac{(\alpha - \sqrt{\mathscr{E}})^2}{N_0}\right] \cdot \left\{1 - \frac{1}{2} erfc\left(\frac{\alpha}{\sqrt{N_0}}\right)\right\}^{M-1} d\alpha. \tag{4.64}$$

Finally, introducing (4.64) into (4.35), one obtains the error probability. The integral (4.64) cannot be further simplified, but due to its importance it has been tabulated and can be found, for example, in Lindsey and Simon (1973). The resulting error probability is plotted in Fig. 4.18 as a function of the commonly used parameter $\mathscr{E}/N_0$. A brief discussion of this result should allow a first insight into the problem of comparing different transmission schemes. Apparently, looking at the curves of Fig. 4.18, one could conclude that a small value of M is better, since it gives a smaller error probability for the same value of $\mathscr{E}/N_0$. However, notice that the greater M is, the higher is the information content of each signal, which in fact conveys $h = \log_2 M$ bits. As a consequence, the transmission of one binary digit requires an energy $\mathscr{E}/\log_2 M$. Given this, a reasonable question is: given the noise power spectral density $N_0/2$, what happens to the error probability $P(e)$ when M is increased but the energy per bit, $\mathscr{E}/\log_2 M$, is kept constant? The answer can be obtained from Fig. 4.19, where the same curves of the previous figure are plotted as a function of $\mathscr{E}/(hN_0)$. It can be seen that, at least for low error probabilities, increasing the size M of the signal set requires less energy per bit to obtain the same error probability.

Other problems deserve consideration. The first is the complexity of the demodulator, which increases with M. The second is the required bandwidth. Throughout the chapter a channel with infinite bandwidth is assumed, but in practice the channel bandwidth is always an expensive resource. Therefore, much more careful arguments must be employed in drawing conclusions such as those based on Fig. 4.19. This issue will be discussed in detail in Chapter 5.

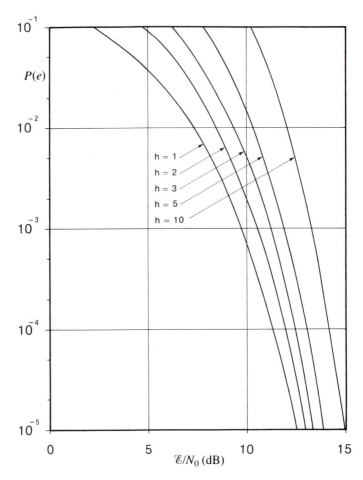

Figure 4.18 Error probability for coherent demodulation of an orthogonal set of $M = 2^h$ signals. Curves are plotted as a function of $\mathscr{E}/N_0$.

4.2.2 Upper and Lower Bounds to the Error Probability

In many practical cases the probability of error cannot be computed in closed form. An example was already encountered in the case of orthogonal signals. It is common engineering practice to try to upper and lower bound the true error probability. These bounds are of practical interest only if they require simple computations and if they are "tight," that is, if the interval of possible values is reasonably small around the true unknown value of error probability.

Let us start by deriving a useful upper bound. It is based on this definition: an *error event* e_{ij} occurs whenever the transmitted signal vector $\mathbf{s}_i$ is transformed by the channel into a received signal vector $\mathbf{r}$ that lies closer to the vector $\mathbf{s}_j$ different from $\mathbf{s}_i$:

$$e_{ij} \triangleq \{\mathbf{r}: d(\mathbf{r}, \mathbf{s}_j) < d(\mathbf{r}, \mathbf{s}_i)|\mathbf{s}_i\}. \tag{4.65}$$

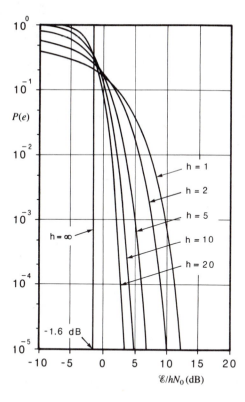

Figure 4.19 Same curves as in Fig. 4.18, but plotted as a function of $\mathcal{E}/hN_0$.

When $\mathbf{s}_i$ is transmitted, the error event is the union of all the events e_{ij} just defined. Thus, we can write

$$P(e|\mathbf{s}_i) = P\left\{\bigcup_{\substack{j=1\\j\neq i}}^{M} e_{ij}\right\} \leq \sum_{\substack{j=1\\j\neq i}}^{M} P(e_{ij}). \tag{4.66}$$

This is called the *union bound* of the conditional error probability $P(e|\mathbf{s}_i)$. The result is a direct consequence of the property that the probability of a union of events is upper bounded by the sum of the probabilities of these events. Computing $P(e_{ij})$ in (4.66) is easy because only two signals are implied by this event and we are in the same situation that led to (4.41). Using that result, we can rewrite (4.66) in the following way:

$$P(e|\mathbf{s}_i) \leq \sum_{\substack{j=1\\j\neq i}}^{M} \frac{1}{2} erfc\left(\frac{d_{ij}}{2\sqrt{N_0}}\right), \tag{4.67}$$

where $d_{ij} \triangleq d(\mathbf{s}_i, \mathbf{s}_j)$. Finally, averaging (4.67) over the signal set, we get

$$P(e) \leq \frac{1}{2M} \sum_{i=1}^{M} \sum_{\substack{j=1\\j\neq i}}^{M} erfc\left(\frac{d_{ij}}{2\sqrt{N_0}}\right). \tag{4.68}$$

This is the most general formulation of the union bound.

Sec. 4.2   Memoryless Modulation and Coherent Demodulation   **151**

A simpler bound can be obtained from (4.68) by defining the minimum distance d_{min} of the signal set as

$$d_{min} \triangleq \min_{\substack{i,j \\ i \neq j}} d_{ij}. \tag{4.69}$$

Observing that $erfc(\cdot)$ is a monotone decreasing function of its argument, each term of the sum (4.68) can be bounded from above by using the definition (4.69), and we finish with

$$P(e) \leqslant \frac{M-1}{2} erfc\left(\frac{d_{min}}{2\sqrt{N_0}}\right). \tag{4.70}$$

The derivation of a lower bound starts from the property that the function $erfc(\cdot)$ is convex upward for nonnegative arguments. That is, it has a nonnegative second derivative.

Let us define the distance

$$d_i \triangleq \min_{\substack{j \\ j \neq i}} (d_{ij}). \tag{4.71}$$

It is the minimum distance of the signal s_i from any other signal. We then have:

$$P(e|s_i) \geqslant \frac{1}{2} erfc\left(\frac{d_i}{2\sqrt{N_0}}\right). \tag{4.72}$$

An explanation of (4.72) based on information theory arguments can also be given (Forney, 1972a; see also Section 7.5.2). Averaging over the signal set, we get

$$P(e) \geqslant \frac{1}{2M} \sum_{i=1}^{M} erfc\left(\frac{d_i}{2\sqrt{N_0}}\right). \tag{4.73}$$

If we now define an average distance as

$$\bar{d} \triangleq \frac{1}{M} \sum_{i=1}^{M} d_i \tag{4.74}$$

and recall the Jensen's inequality, valid for any convex upward function $g(\cdot)$,

$$\frac{1}{M} \sum_{i=1}^{M} g(z_i) \geqslant g\left(\frac{1}{M} \sum_{i=1}^{M} z_i\right),$$

we can get from (4.73) the final result:

$$P(e) \geqslant \frac{1}{2} erfc\left(\frac{\bar{d}}{2\sqrt{N_0}}\right). \tag{4.75}$$

A looser bound can be obtained by noticing that from (4.73) we have

$$P(e) \geqslant \frac{1}{2M} \sum_{i \in D_m} erfc\left(\frac{d_i}{2\sqrt{N_0}}\right) \tag{4.76}$$

$$= \frac{N_m}{2M} erfc\left(\frac{d_{min}}{2\sqrt{N_0}}\right),$$

where D_m is the subset of indexes i such that $d_i = d_{\min}$, and N_m is the cardinality of D_m. Since $N_m \geq 2$, we get from (4.76)

$$P(e) \geq \frac{1}{M} \, erfc \left(\frac{d_{\min}}{2\sqrt{N_0}} \right).$$ (4.77)

Combining (4.77) with (4.70), we can finally write

$$\frac{1}{M} \, erfc \left(\frac{d_{\min}}{2\sqrt{N_0}} \right) \leq P(e) \leq \frac{M-1}{2} \, erfc \left(\frac{d_{\min}}{2\sqrt{N_0}} \right).$$ (4.78)

Equation (4.78) enhances the importance of the parameter $d_{\min}$ as defined in (4.69). In fact, it allows us to both upper and lower bound the error probability of a given signal set, and the bounds differ only for a multiplicative constant.

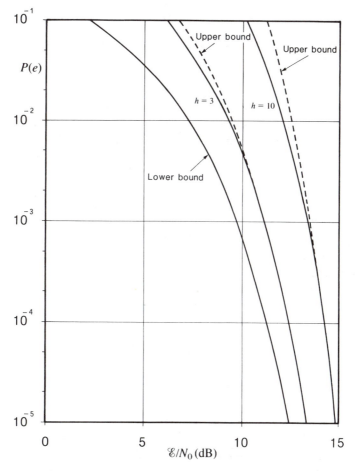

Figure 4.20 Error probability for coherent demodulation of orthogonal signals. Upper and lower bounds of (4.79) for the two cases of $M = 8$ and $M = 1024$. The continuous curves refer to the exact values.

Sec. 4.2   Memoryless Modulation and Coherent Demodulation   **153**

Example 4.8

Let us apply the bounds of (4.70) and (4.75) to the case of M orthogonal signals with equal energy $\mathcal{E}$. In this case, all the distances d_{ij} in (4.68) are equal, so

$$d_{ij} = d_{min} = \sqrt{2\mathcal{E}}.$$

Therefore, we get the following bounds:

$$\frac{1}{2} erfc \left(\sqrt{\frac{\mathcal{E}}{2N_0}} \right) \leq P(e) \leq \frac{M-1}{2} erfc \left(\sqrt{\frac{\mathcal{E}}{2N_0}} \right). \tag{4.79}$$

Notice that, for the binary case, both the upper and lower bounds coincide with the exact result given by (4.46). The two bounds are compared with the exact result for two different values of M in Fig. 4.20. Notice that the lower bound is independent of M, whereas the upper bound (union bound) becomes increasingly tight for a given M as $\mathcal{E}/N_0$ is increased. This property is of practical importance, since for applications we are usually interested in low error probability values. $\square$

4.3 BANDPASS MEMORYLESS MODULATION AND INCOHERENT DEMODULATION

The case that will be addressed in this section arises in all practical situations where bandpass signals are transmitted, but the demodulator does not know the phase of the oscillator at the transmitter side. Consequently, there is an uncertainty at the receiver side due to a random phase ϑ whose pdf is assumed to be known.

The modulator is memoryless and the waveforms used to convey the RVs ξ_k are limited to the time interval T. The transmitted waveform in the kth time interval can therefore be written as

$$s(t - kT; \xi_k) = \mathcal{R}\{\tilde{s}(t - kT; \xi_k)e^{j(2\pi f_0 t + \vartheta)}\}, \qquad kT \leq t < (k+1)T, \tag{4.80}$$

where the complex envelope $\tilde{s}(t - kT; \xi_k)$ bears the transmitted symbol m_k associated with the RV ξ_k. [In this section we are using the analytic signal representation (see Section 2.4.1) because it makes easier to account for the random phase ϑ.]

In principle, a coherent demodulator should form, for all possible values of $\boldsymbol{\xi}$, the set of signals like

$$v_{\boldsymbol{\xi}}(t; \vartheta) = \sum_{k=0}^{K-1} s(t - kT; \xi_k, \vartheta), \qquad 0 \leq t < KT, \tag{4.81}$$

where

$$s(t - kT; \xi_k, \vartheta) \triangleq \mathcal{R}\{\tilde{s}(t - kT; \xi_k)e^{j\vartheta}e^{j2\pi f_0 t}\}. \tag{4.82}$$

Doing this, it would be possible to recover an estimate of the transmitted symbol sequence. But the value of ϑ is unknown and hence we want to get rid of it in the ML decision. To this purpose, we first evaluate the likelihood ratios of the transmitted sequences conditioned on a given value of ϑ and then average over its known pdf. The first step leads to the case already solved in Section 4.2. Therefore, the optimum demodulator is based on symbol-by-symbol ML decisions.

Let us now go through the steps that lead to the optimum demodulator structure by restricting ourselves to the time interval $(0, T)$ corresponding to $k = 0$. In this time interval, the conditional likelihood ratio can be written as

$$\Lambda_i(\vartheta) = \exp\left\{\frac{2}{N_0}\int_0^T r(t)s_i(t; \vartheta)dt - \frac{1}{N_0}\int_0^T s_i^2(t; \vartheta)dt\right\}, \qquad i = 1, 2, \ldots, M, \qquad (4.83)$$

where

$$s_i(t; \vartheta) \triangleq \mathcal{R}\{\tilde{s}_i(t)e^{j\vartheta}e^{j2\pi f_0 t}\}, \qquad (4.84)$$

and $\{\tilde{s}_i(t)\}_{i=1}^M$ is the set of waveforms in which the signals $\tilde{s}(t; \xi_0)$ are chosen by the modulator. The conditioning on ϑ can be now removed by averaging over the pdf $f_\vartheta(\cdot)$; that is,

$$\Lambda_i = \int_{-\pi}^{\pi} \Lambda_i(z)f_\vartheta(z)dz, \qquad i = 1, 2, \ldots, M. \qquad (4.85)$$

Let us now define the complex quantity L_i as

$$L_i \triangleq \frac{1}{\sqrt{\mathcal{E}_i}}\int_0^T r(t)\tilde{s}_i^*(t)e^{-j2\pi f_0 t}dt \qquad (4.86)$$

and notice that the second integral in (4.83) represents the energy $\mathcal{E}_i$ of the signal $s_i(t; \vartheta)$ and that this does not depend on the random phase ϑ. Using (4.86) and (4.84) in (4.83), we get from (4.85)

$$\Lambda_i = \exp\left(-\frac{\mathcal{E}_i}{N_0}\right)\int_{-\pi}^{\pi} \exp\left\{\mathcal{R}\left[\frac{2\sqrt{\mathcal{E}_i}}{N_0}L_i^* e^{jz}\right]\right\}f_\vartheta(z)\, dz,$$
$$i = 1, 2, \ldots, M. \qquad (4.87)$$

When no a priori information is available about the random phase ϑ, the most sensible assumption is that its distribution is uniform. That is,

$$f_\vartheta(z) = \frac{1}{2\pi}, \qquad -\pi \leq z \leq \pi. \qquad (4.88)$$

Introducing (4.88) into (4.87) and recalling the definition of the modified Bessel function of the first kind, $I_0(\cdot)$ (see Appendix A), we get

$$\Lambda_i = \exp\left(-\frac{\mathcal{E}_i}{N_0}\right)I_0\left(\frac{2\sqrt{\mathcal{E}_i}}{N_0}|L_i|\right), \qquad i = 1, 2, \ldots, M, \qquad (4.89)$$

and the log-likelihood ratio for each decision becomes

$$\lambda_i = \ln I_0\left(\frac{2\sqrt{\mathcal{E}_i}}{N_0}|L_i|\right) + c_i, \qquad i = 1, 2, \ldots, M, \qquad (4.90)$$

where $c_i \triangleq -\mathcal{E}_i/N_0$.

An interesting and practical case is encountered when a set of equal energy signals is transmitted. In this case the constants c_i are all equal and can be omitted from (4.90). Also taking advantage of the monotone behavior of the function $\ln I_0(\cdot)$ for nonnegative arguments, the ML decisions can be based on the quantities

$$l_i \triangleq |L_i|^2, \qquad i = 1, 2, \ldots, M. \tag{4.91}$$

The block diagram of the optimum demodulator in this case is shown in Fig. 4.21. It is called a *correlation demodulator*. The extension to the more general case of (4.90) is quite straightforward. Also in this case, an alternative structure of the optimum demodulator can be derived. Let us define a *bandpass matched filter* having the following impulse response:

$$h_i(t) \triangleq \Re\{\tilde{s}_i^*(T - t)e^{j2\pi f_0 t}\}. \tag{4.92}$$

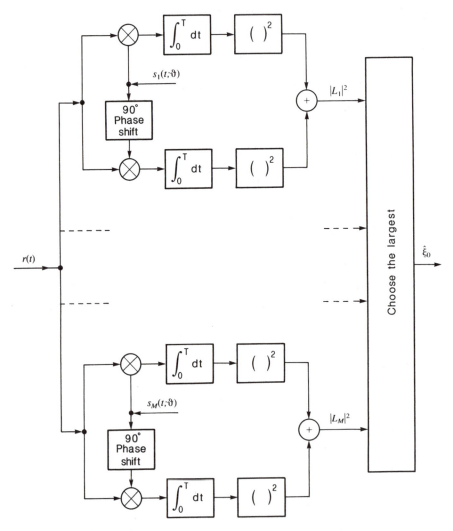

Figure 4.21 Block diagram of the correlation receiver for the case of incoherent demodulation of bandpass signals with equal energy.

It can be verified that the output $y(t)$ from this filter at $t = T$, when the input is $r(t)$, can be put in the form

$$y(T) = \Re\{L_i e^{j2\pi f_0 T}\}$$
$$= |L_i| \cos\{2\pi f_0 T + \arg(L_i)\}. \tag{4.93}$$

Therefore, an envelope detector can recover the required sufficient statistics defined by (4.91). A *matched-filter* envelope demodulator for the case of equal-energy signals is shown in Fig. 4.22.

Example 4.9 On–off signaling scheme

Let us consider a binary case, with the two transmitted signals given by

$$s_1(t) \equiv 0, \tag{4.94}$$

$$s_2(t) = \sqrt{\frac{2\mathcal{E}}{T}} \cos 2\pi f_0 t, \qquad 0 \le t < T.$$

The decision rule based on (4.90) requires the computation of

$$L_2 = \sqrt{\frac{2}{T}} \int_0^T r(t) e^{-j2\pi f_0 t} \, dt. \tag{4.95}$$

Furthermore,

$$\lambda_1 = 0, \qquad \lambda_2 = \ln I_0 \left(\frac{2\sqrt{\mathcal{E}}}{N_0} |L_2| \right) - \frac{\mathcal{E}}{N_0}. \tag{4.96}$$

Thus, the ML decision rule can be reduced to the comparison of $|L_2|^2$ against a suitable threshold ν. The block diagram of the demodulator is shown in Fig. 4.23. $\square$

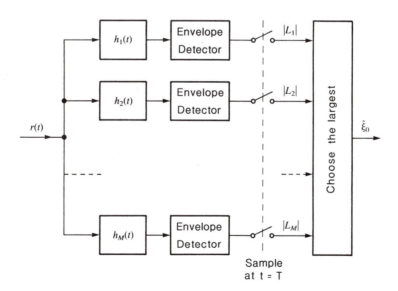

Figure 4.22 Block diagram of the matched-filter envelope receiver for the case of incoherent demodulation of bandpass signals with equal energy.

Sec. 4.3   Bandpass Memoryless Modulation and Incoherent Demodulation   **157**

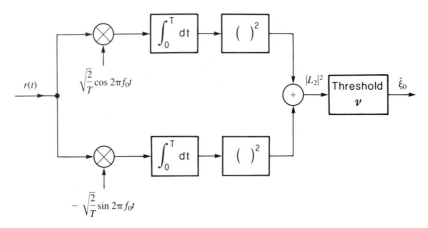

Figure 4.23 Optimum receiver for the case of incoherent demodulation of binary signals. On–off signaling scheme (Example 4.9).

4.3.1 Error Probability Evaluation

The sufficient statistics for the received signal $r(t)$ in the case of incoherent demodulation is given by the RVs (4.90) or, in the special case of equal-energy signals, by the RVs (4.91). Therefore, the ML demodulator implements a partitioning of the N-dimensional signal space into M disjoint decision regions defined by

$$R_i \triangleq \{\mathbf{r}: l_i = \max_j l_j\}, \qquad i, j = 1, 2, \ldots, M, \qquad (4.97)$$

and the general expression for the error probability is again given by (4.32). However, since the sufficient statistics is based on the envelope of the received signal, the optimum demodulator is no longer a minimum-distance demodulator. A geometric approach can also be tried for the incoherent demodulator, but the general solution to this problem is not so neat and useful as it was for the coherent case. Therefore, we prefer to consider some particular cases and discuss them in the context of the error performance evaluation.

On–off binary signals

This case was already considered in Example 4.9. The decision variable (4.95) requires the computation of

$$L_2 = L_c + jL_s \qquad (4.98)$$

with

$$L_c \triangleq \sqrt{\frac{2}{T}} \int_0^T r(t) \cos 2\pi f_0 t \, dt = r_1,$$

$$\qquad (4.99)$$

$$L_s \triangleq -\sqrt{\frac{2}{T}} \int_0^T r(t) \sin 2\pi f_0 t \, dt = r_2.$$

These equations represent the received signal in the two-dimensional space defined by the basis

$$\psi_1(t) = \sqrt{\frac{2}{T}} \cos 2\pi f_0 t,$$

$$0 \leq t < T. \qquad (4.100)$$

$$\psi_2(t) = \sqrt{\frac{2}{T}} \sin 2\pi f_0 t,$$

Because the decision rule is based on the envelope of the received signal, it is immediately realized that the boundary between the two decision regions is a circle of radius $\sqrt{\nu}$, ν being the decision threshold (see Fig. 4.24). Extending to this case the general arguments that led to (4.38), we can write the expression of the conditional error probability as follows:

$$P(e|\mathbf{s}_i) = \int_{-\pi}^{\pi} f_{\vartheta}(z)\, dz \int_{\overline{R}_i} f_{\mathbf{n}}[\boldsymbol{\alpha} - \mathbf{s}_i(z)]\, d\boldsymbol{\alpha}, \qquad i = 1, 2, \qquad (4.101)$$

where $\mathbf{s}_i(\vartheta)$ is the vector corresponding to the signal $s_i(t; \vartheta)$, and $\overline{R}_i$ is the complement of the region R_i. When the signal $s_1(t)$ of (4.94) is transmitted, r_1 and r_2 of (4.99) are zero-mean independent Gaussian RVs. Therefore,

$$P(e|\mathbf{s}_1) = \frac{1}{\pi N_0} \iint_{R_2} \exp\left(-\frac{\alpha_1^2 + \alpha_2^2}{N_0}\right) d\alpha_1\, d\alpha_2. \qquad (4.102)$$

Shifting to polar coordinates and solving the integral, we get

$$P(e|\mathbf{s}_1) = \exp\left(-\frac{\nu}{N_0}\right). \qquad (4.103)$$

When the signal $s_2(t)$ is transmitted, considering that $r(t) = s_2(t; \vartheta) + n(t)$, we get from (4.99)

$$r_1 = \sqrt{\mathscr{E}} \cos \vartheta + n_1,$$

$$r_2 = \sqrt{\mathscr{E}} \sin \vartheta + n_2. \qquad (4.104)$$

For a given value of ϑ, r_1 and r_2 are again independent Gaussian RVs. Therefore,

$$P(e|\mathbf{s}_2, \vartheta) =$$

$$\frac{1}{\pi N_0} \iint_{R_1} \exp\left\{-\frac{(\alpha_1 - \sqrt{\mathscr{E}} \cos \vartheta)^2 + (\alpha_2 - \sqrt{\mathscr{E}} \sin \vartheta)^2}{N_0}\right\} d\alpha_1\, d\alpha_2. \qquad (4.105)$$

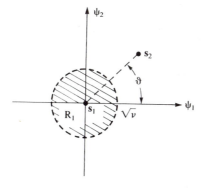

Figure 4.24 Optimum decision regions for the case of on–off signaling and incoherent demodulation.

Shifting to polar coordinates and integrating with respect to the angular coordinate, we get

$$P(e|\mathbf{s}_2, \vartheta) = \frac{2}{N_0} \int_0^{\sqrt{v}} z \exp\left\{ -\frac{z^2 + \mathscr{E}}{N_0} \right\} I_0\left(\frac{2z\sqrt{\mathscr{E}}}{N_0} \right) dz. \qquad (4.106)$$

As expected from the circular symmetry of the decision region R_1, this conditional probability is actually independent of ϑ. The integral cannot be evaluated in closed form but can be expressed in terms of the Marcum's Q function (see Appendix A). We get

$$P(e|\mathbf{s}_2) = 1 - Q\left(\sqrt{\frac{2\mathscr{E}}{N_0}}, \sqrt{\frac{2v}{N_0}} \right). \qquad (4.107)$$

Combining (4.103) and (4.107), we get the final expression for the error probability:

$$P(e) = \frac{1}{2}\left\{ \exp\left(-\frac{v}{N_0} \right) + 1 - Q\left(\sqrt{\frac{2\mathscr{E}}{N_0}}, \sqrt{\frac{2v}{N_0}} \right) \right\}. \qquad (4.108)$$

The minimum value of $P(e)$ is achieved by optimizing the choice of the threshold v in (4.108). We get for v_{opt} the implicit solution

$$I_0\left(\frac{\sqrt{4v_{opt}\mathscr{E}}}{N_0} \right) = \exp\left(\frac{\mathscr{E}}{N_0} \right). \qquad (4.109)$$

Equation (4.109) shows that v_{opt} is a function of the ratio $\mathscr{E}/N_0$. In other words, the optimum threshold must be "adapted" to the signal-to-noise ratio. For very high values of $\mathscr{E}/N_0$, using for $I_0(\cdot)$ the crude approximation $I_0(x) \sim e^x$, we get

$$\sqrt{v_{opt}} \sim \frac{\sqrt{\mathscr{E}}}{2}. \qquad (4.110)$$

Since the distance between the two transmitted signal is $\sqrt{\mathscr{E}}$, (4.110) shows that for high values of $\mathscr{E}/N_0$ the optimum demodulator is again a minimum-distance demodulator. Consequently, the asymptotic performance of the incoherent demodulator is the same as for the coherent case. This result can be explained by considering that in a coherent demodulator the quadrature noise component of the received signal is irrelevant, while in the incoherent demodulator this noise component becomes asymptotically irrelevant for the envelope of the received signal. Hence, when the signal-to-noise ratio is very high, the only significant noise component is the in-phase one and, as a consequence, the performance is the same for both cases.

Equal-energy binary signals

Assume two signals with equal energy $\mathscr{E}$ and correlation coefficient

$$\rho \triangleq \frac{1}{2\mathscr{E}} \int_0^T \tilde{s}_1(t)\tilde{s}_2^*(t)\, dt. \qquad (4.111)$$

The configuration is completely symmetric and therefore $P(e) = P(e|\mathbf{s}_1)$. The sufficient statistics for the received signal vector $\mathbf{r}$ is given by (4.91). Given the transmitted signal vector $\mathbf{s}_1$, we have

$$r(t) = \mathcal{R}\{[\tilde{s}_1(t)e^{j\vartheta} + \tilde{n}(t)]e^{j2\pi f_0 t}\}$$
$$= \mathcal{R}\{\tilde{r}(t)e^{j2\pi f_0 t}\}, \qquad (4.112)$$

where

$$\tilde{r}(t) \triangleq \tilde{s}_1(t)e^{j\vartheta} + \tilde{n}(t). \qquad (4.113)$$

Let us use in (4.86) the equality

$$r(t) = \frac{1}{2}[\tilde{r}(t)e^{j2\pi f_0 t} + \tilde{r}^*(t)e^{-j2\pi f_0 t}];$$

assuming that $f_0 T \gg 1$, we can drop the high-frequency terms and get

$$L_i = \frac{1}{2\sqrt{\mathscr{E}}} \int_0^T \tilde{r}(t)\tilde{s}_i^*(t)\, dt. \qquad (4.114)$$

Notice that, since the sufficient statistics (4.91) is based on the squared envelope of L_i, we can use instead the quantities $L_i e^{-j\vartheta}$ without affecting the decision. Using (4.113) and (4.111) in (4.114), we get

$$L_1 e^{-j\vartheta} = \sqrt{\mathscr{E}} + n_1,$$
$$L_2 e^{-j\vartheta} = \rho\sqrt{\mathscr{E}} + n_2, \qquad (4.115)$$

where n_1 and n_2, conditioned on ϑ, are Gaussian RVs defined as

$$n_i \triangleq \frac{1}{2\sqrt{\mathscr{E}}} \int_0^T \tilde{n}(t)e^{-j\vartheta}\tilde{s}_i^*(t)\, dt. \qquad (4.116)$$

We have

$$\mathrm{E}\{n_i|\vartheta\} = 0, \qquad \mathrm{E}\{|n_i|^2|\vartheta\} = N_0, \qquad i = 1, 2,$$
$$\mathrm{E}\{n_1^* n_2|\vartheta\} = N_0\rho. \qquad (4.117)$$

The error probability is given by

$$P(e) = P\{|L_2|^2 > |L_1|^2\}$$
$$= P\{|L_2 e^{-j\vartheta}| > |L_1 e^{-j\vartheta}|\}. \qquad (4.118)$$

The envelopes have a Rician distribution, but the RVs n_1 and n_2 are not independent. The evaluation of (4.118) is, however, a classical problem in probability theory, and the result (see Problem 4.15) is the following:

$$P(e) = Q\left(\sqrt{a\frac{\mathscr{E}}{N_0}}, \sqrt{b\frac{\mathscr{E}}{N_0}}\right) - \frac{1}{2}\exp\left(-\frac{\mathscr{E}}{N_0}\frac{a+b}{2}\right) I_0\left(\frac{\mathscr{E}}{N_0}\sqrt{ab}\right), \qquad (4.119)$$

where

$$a \triangleq \frac{1}{2}(1 - \sqrt{1 - |\rho|^2}),$$

$$b \triangleq \frac{1}{2}(1 + \sqrt{1 - |\rho|^2}).$$

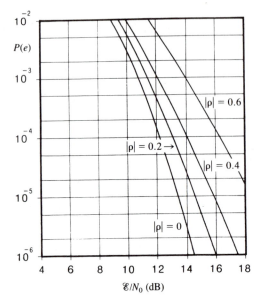

Figure 4.25 Error probability for incoherent demodulation of two equal energy signals with arbitrary correlation ρ. (Reprinted with permission from Proakis, 1983.)

The error probability (4.119) for some values of $|\rho|$ is shown in Fig. 4.25. The minimum value of $P(e)$ is achieved when $|\rho| = 0$, that is, when the two signals are orthogonal. For this case, we get

$$P(e) = Q\left(0, \sqrt{\frac{\mathcal{E}}{N_0}}\right) - \frac{1}{2}e^{-\mathcal{E}/2N_0} = \frac{1}{2}e^{-\mathcal{E}/2N_0}. \tag{4.120}$$

Equal-energy M-ary orthogonal signals

In this case, we assume M equal-energy orthogonal signals expressed as

$$s_i(t; \vartheta) = \mathcal{R}\left\{\sqrt{\frac{2\mathcal{E}}{T}}\,e^{j(2\pi f_i t + \vartheta)}\right\}, \qquad i = 1, 2, \ldots, M, \qquad 0 \leq t < T, \tag{4.121}$$

where $f_i T = K_i$, K_i integer, so that

$$\int_0^T s_i(t; \vartheta) s_j(t; \vartheta)\, dt = \begin{cases} \mathcal{E}, & i = j, \\ 0, & i \neq j. \end{cases} \tag{4.122}$$

This signal set can be represented using an orthonormal basis of dimensionality $2M$ with coordinates

$$\psi_{ci}(t) \triangleq \sqrt{\frac{2}{T}}\cos 2\pi f_i t,$$
$$\qquad\qquad\qquad\qquad i = 1, 2, \ldots, M, \qquad 0 \leq t < T. \tag{4.123}$$
$$\psi_{si}(t) \triangleq \sqrt{\frac{2}{T}}\sin 2\pi f_i t,$$

The sufficient statistics for the received signal is again given by (4.91). We compute the decision complex quantity L_j under the hypothesis that the signal $s_i(t; \vartheta)$ is transmitted. Expanding the received signal components using the basis (4.123), we get:

$$L_j = \begin{cases} (\sqrt{\mathcal{E}} \cos \vartheta + n_{ci}) + j(\sqrt{\mathcal{E}} \sin \vartheta + n_{si}), & j = i, \\ n_{cj} + jn_{sj}, & j \neq i. \end{cases} \tag{4.124}$$

In this case, given ϑ, all the noise components are independent Gaussian RVs. Defining the normalized squared envelope $|R_j|^2$ as

$$|R_j|^2 \triangleq \frac{2}{N_0} |L_j|^2, \tag{4.125}$$

it is straightforward to get the pdf of the envelope $|R_j|$ conditioned on the transmitted signal $s_i(t; \vartheta)$. We obtain

$$f_{|R_j| \, |s_i}(\alpha|s_i) = \begin{cases} \alpha \exp\left\{ -\frac{1}{2}\left(\alpha^2 + \frac{2\mathcal{E}}{N_0} \right) \right\} I_0\left(\alpha \sqrt{\frac{2\mathcal{E}}{N_0}} \right), & j = i, \\ \alpha \exp\left(-\frac{\alpha^2}{2} \right), & j \neq i, \alpha \geq 0. \end{cases} \tag{4.126}$$

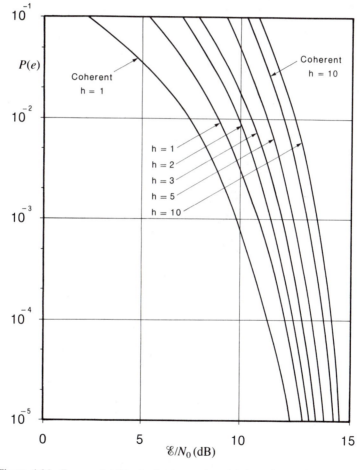

Figure 4.26 Error probability for incoherent demodulation of an orthogonal set of $M = 2^h$ signals. Curves are plotted as a function of $\mathcal{E}/N_0$. Two curves for coherent demodulation are also shown for comparison.

Recalling (4.97), we now have the result:

$$P(c|\mathbf{s}_i) = P\{|R_i| = \max_j |R_j| \, |\mathbf{s}_i\}. \tag{4.127}$$

A consequence of the independence of the noise components in (4.124) is that the envelopes $|R_j|$ are independent as well. Then, using (4.126), we get from (4.127)

$$P(c|\mathbf{s}_i) = \int_0^\infty \alpha \exp\left[-\frac{1}{2}\left(\alpha^2 + \frac{2\mathscr{E}}{N_0}\right)\right]$$

$$\cdot I_0\left(\alpha \sqrt{\frac{2\mathscr{E}}{N_0}}\right)\left[1 - \exp\left(-\frac{\alpha^2}{2}\right)\right]^{M-1} d\alpha. \tag{4.128}$$

Notice that the RHS of (4.128) is independent of the transmitted vector $\mathbf{s}_i$. This is because the configuration of the signal space is completely symmetric. Therefore, the integral (4.128) gives the probability $P(c)$. It can be solved by using the binomial expansion for the bracketed term raised to the $(M-1)$th power and then integrating termwise. The final expression for the error probability is found to be

$$P(e) = \frac{1}{M} \exp\left(-\frac{\mathscr{E}}{2N_0}\right) \sum_{i=2}^M \binom{M}{i} (-1)^i \exp\left[\frac{(2-i)\mathscr{E}}{2iN_0}\right]. \tag{4.129}$$

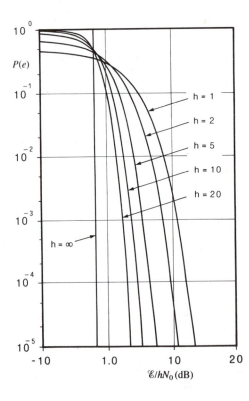

Figure 4.27 Same curves as in Fig. 4.26, but plotted as a function of $\mathscr{E}/hN_0$.

Notice that for $M = 2$ this result checks with (4.120). The curves of (4.129) are shown in Fig. 4.26 as a function of $\mathscr{E}/N_0$. On the same graph, two cases of coherent demodulation are also redrawn for comparison. The penalty due to the presence of the unknown random phase ϑ is thus apparent. In particular, for the binary case, this penalty is monotonically smaller as the ratio $\mathscr{E}/N_0$ increases. This feature can be given the same intuitive interpretation as for the simple on–off signaling case previously examined. In analogy with the case of coherent demodulation of orthogonal signals, the error probability of (4.129) is shown in Fig. 4.27 as a function of the ratio $\mathscr{E}/hN_0$. Similar arguments also hold for the present case of incoherent demodulation.

4.4 MODULATION WITH MEMORY AND COHERENT DEMODULATION

The main feature of the issue addressed in this section is the fact that the signals transmitted in each time interval T depend not only on the RV ξ_k, but also on the modulator state σ_k. This can be the consequence of two different situations, which can be present either separately or together. The first arises when the signals $s(t - kT; \xi_k, \sigma_k)$ are formed using waveforms with a duration longer than T, typically, a multiple of this time interval.† But even if the waveforms last only T, the modulator can choose the waveform to be transmitted depending on the signal in the previous intervals. A typical example of this is the requirement of phase continuity as in Example 4.2. In the second situation, the modulator chooses the waveform to be transmitted from a set that depends on a sequence of previous symbols. Typical examples arise from combined modulation and coding considered in Section 9.7. In these situations a receiver based on symbol-by-symbol decisions is no longer optimum.

We shall address only the problem of coherent reception. This means that the possible transmitted signals are completely known by the demodulator, whose task is to make an ML estimate of the transmitted symbol sequence $\boldsymbol{\xi}$. To this purpose, the demodulator should form the set of signals like

$$v_{\boldsymbol{\xi}}(t) = \sum_{k=0}^{K-1} s(t - kT; \xi_k, \sigma_k), \qquad 0 \leqslant t < KT, \tag{4.130}$$

for all possible values of $\boldsymbol{\xi}$ and choose the sequence $\hat{\boldsymbol{\xi}}$ that maximizes the likelihood ratio. Recalling the derivation that led to (4.14), this maximization is equivalent to the maximization of the quantities

$$\begin{aligned} l_{\boldsymbol{\xi}} &\triangleq \int_0^{KT} r(t)v_{\boldsymbol{\xi}}(t)\, dt - \frac{1}{2}\int_0^{KT} v_{\boldsymbol{\xi}}^2(t)\, dt \\ &= \sum_{k=0}^{K-1} l_{\xi_k}, \end{aligned} \tag{4.131}$$

---

† Notice that the signals $s(t - kT; \xi_k, \sigma_k)$ are defined only in an interval of duration T and they take into account into this interval the eventual longer duration of the modulator waveforms.

where

$$l_{\xi_k} \triangleq \int_{kT}^{(k+1)T} r(t)s(t - kT; \xi_k, \sigma_k)\, dt \qquad (4.132)$$

$$-\frac{1}{2} \int_{kT}^{(k+1)T} s^2(t - kT; \xi_k, \sigma_k)\, dt, \qquad k = 0, 1, \ldots, K - 1.$$

Therefore, the maximization of (4.131) can be performed by accumulating the quantities of (4.132) obtained by the receiver after each interval of duration T. Introducing the definition of the partial sum over the first k intervals of duration T,

$$l_{\xi(k)} \triangleq \sum_{i=0}^{k-1} l_{\xi_i}, \qquad (4.133)$$

we can obtain the recurrent form

$$l_{\xi(k+1)} = l_{\xi(k)} + l_{\xi_k}, \qquad (4.134)$$

where of course $l_{\xi(K)} = l_\xi$.

Equation (4.134) will be very useful for computational purposes later. The ML demodulator requires that (4.131) be maximized over the set of possible sequences ξ. The brute-force approach of computing all the possible values of l_ξ is not feasible because of the exponential growth of the computational effort with the sequence length. On the other hand, a termwise maximization of (4.131) does not lead to the optimum, because the possible values of the l_{ξ_k}'s in different time intervals are not independent (see Appendix F).

To perform such a maximization, a sequential algorithm (the *Viterbi algorithm* described in Appendix F) based on (4.134) is available. It allows a greatly reduced computational effort. From now on we shall assume that the reader is familiar with Appendix F. The description of the Viterbi algorithm is easier if the problem is formulated in terms of the search of the optimum path into a trellis. This search is equivalent to tracking in the trellis the ML symbols sequence or the corresponding sequence of the modulator states σ_k.

Instead of continuing this analysis in the more general and abstract case, we shall restrict the discussion to a class of transmitted signals that are a generalization of those presented in Example 4.2. These signals have a constant envelope and a phase associated with the information symbols. Moreover, the phase continuity is imposed. The important properties of these signals in the applications will be examined in Chapter 5.

4.4.1 Continuous Phase Modulated (CPM) Signals

The transmitted signal is a sequence of bandpass waveforms of the form

$$s(t - kT; \xi_k, \sigma_k) = \Re\left\{\bar{s}(t - kT; \xi_k, \sigma_k)e^{j2\pi f_0 t}\right\}, \qquad kT \leq t < (k + 1)T, \qquad (4.135)$$

where

$$\bar{s}(t - kT; \xi_k, \sigma_k) \triangleq \sqrt{\frac{2\mathcal{E}}{T}}\, e^{j\Phi(t; \xi_k, \sigma_k)}, \qquad kT \leq t < (k + 1)T, \qquad (4.136)$$

with $\mathscr{E}$ being the energy transmitted per symbol.

The name of CPM signals comes from the property of the information-carrying phase $\Phi(t; \xi_k, \sigma_k)$, which is a continuous function of time. This phase function is a linear function of the RVs ξ_k, which we assume to take values in the set $\{\pm 1, \pm 3, \ldots, \pm(M-1)\}$, M being an even integer. The information-carrying phase has the following general expression:

$$\Phi(t; \xi_k, \sigma_k) = 2\pi h \int_0^t \sum_{i=0}^k \xi_i g(\tau - iT) \, d\tau + \psi_0$$

$$= 2\pi h \sum_{i=0}^k \xi_i q(t - iT) + \psi_0, \qquad kT \le t < (k+1)T, \tag{4.137}$$

where the definition $q(t) \triangleq \int_0^t g(\tau) d\tau$ has been introduced, and ψ_0 is an arbitrary initial phase.

Looking at (4.137), we see that the parameters that can be varied to control the properties of a specific CPM signaling scheme are

(i) the number M of possible values of ξ_k;

(ii) the variable h, which is called the *modulation index*; and

(iii) the shape and duration of the baseband ''frequency'' pulse $g(t)$, or equivalently of $q(t)$.†

A deeper insight into (4.137) leads to the conclusion that, in general, the phase over a symbol interval is due to contributions from several preceding input symbols. Stated in another way, the information carried by one input symbol can be stretched over more symbol intervals in the phase trajectory. As a consequence, successive phase changes are correlated. For this reason, these modulation schemes are also named *correlative phase modulations*.

Let us first define the choice of the baseband frequency pulse $g(t)$. In general, we have

$$g(t) = 0, \qquad t < 0, \qquad t > LT; \tag{4.138}$$

where L is called the *correlation length* of the pulse. Moreover, the following normalizing convention is used:

$$\int_0^\infty g(t) \, dt = \int_0^{LT} g(t) \, dt = \frac{1}{2}. \tag{4.139}$$

With this choice, the accumulated phase values due to completed $g(t)$ pulses are always multiples of $h\pi$. Moreover, using (4.137), the maximum absolute phase change over one symbol interval is $(M-1)h\pi$. Finally, we have, from (4.138) and (4.139),

$$q(t) = 0, \qquad t \le 0,$$

$$q(t) = \frac{1}{2}, \qquad t \ge LT. \tag{4.140}$$

† Notice that the continuity of the phase function implies that the baseband pulse $g(t)$ does not contain any impulse.

To define the state σ_k of the modulator at time $t = kT$, we need to examine in detail the structure of the phase function (4.137). At this time, the RV ξ_k is presented for transmission and the state σ_k represents the memory of the past input symbols. Let us rewrite (4.137) by dividing the summation into three parts, as follows:

$$\Phi(t; \xi_k, \sigma_k) = 2\pi h \xi_k q(t - kT) \tag{4.141}$$

$$+ 2\pi h \sum_{i=k-L+1}^{k-1} \xi_i q(t - iT) + \pi h \sum_{i=0}^{k-L} \xi_i + \psi_0, \qquad kT \leq t < (k + 1)T.$$

The first term in the RHS of (4.141) is the contribution of the present symbol ξ_k, which, as expected, starts at $t = kT$. The second and third terms define the state σ_k that we are looking for. The second term of (4.141) is the contribution to σ_k of the more recent $(L - 1)$ symbols whose corresponding $q(t)$ pulses have not yet reached their final values. Instead, the third term of (4.141) is the contribution to σ_k of the oldest symbols whose corresponding $q(t)$ pulses have reached their final value. This term is a constant phase inside the present symbol interval, given by

$$\varphi_k \triangleq \pi h \sum_{i=0}^{k-L} \xi_i. \tag{4.142}$$

Therefore, in conclusion, we can define the state σ_k of the modulator by the L-tuple

$$\sigma_k \triangleq (\varphi_k, \xi_{k-1}, \xi_{k-2}, \ldots, \xi_{k-L+1}). \tag{4.143}$$

At the end of the present symbol interval, say at $t = (k + 1)T$, the RV ξ_k has been transmitted and the modulator goes to the next state σ_{k+1}, given by

$$\sigma_{k+1} \triangleq (\varphi_{k+1}, \xi_k, \xi_{k-1}, \ldots, \xi_{k-L+2}). \tag{4.144}$$

The constant phase φ_{k+1} is obtained from φ_k by accounting for the contribution of the oldest symbol of the second term of (4.141) as the corresponding $q(t)$ pulse reaches its final value. That is,

$$\varphi_{k+1} = \varphi_k + \pi h \xi_{k-L+1}. \tag{4.145}$$

The notations involved with the transitions between states are shown in Fig. 4.28. The graph describing the possible transitions among the states as a function of time is called the *state trellis* of the modulator.

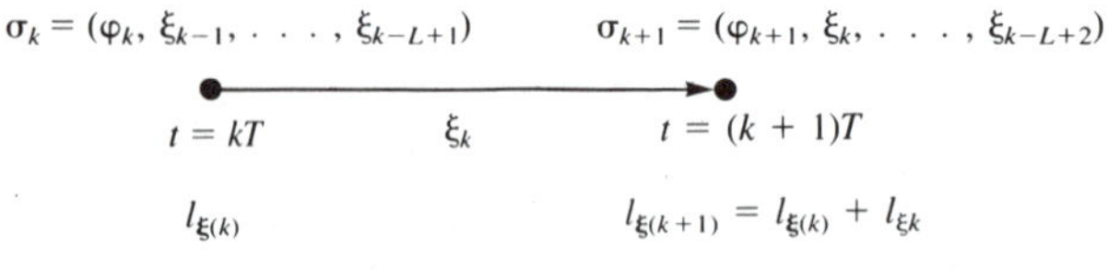

$\sigma_k = (\varphi_k, \xi_{k-1}, \ldots, \xi_{k-L+1})$ $\qquad$ $\sigma_{k+1} = (\varphi_{k+1}, \xi_k, \ldots, \xi_{k-L+2})$

$t = kT$ $\qquad$ ξ_k $\qquad$ $t = (k + 1)T$

$l_{\xi(k)}$ $\qquad$ $l_{\xi(k+1)} = l_{\xi(k)} + l_{\xi k}$

Figure 4.28 Transition between two successive states σ_k and σ_{k+1} in CPM.

A problem to be solved is to count the number of possible states σ_k of the modulator. This number, say N_σ, depends on the set of values assumed by the second and third term of the RHS of (4.141). The application of the Viterbi algorithm, to get ML decisions, requires that N_σ be finite and, for practical purposes, small.

The third term of the RHS of (4.141), redefined as φ_k in (4.142), is called the *phase state* at time kT. A restricted number of such phase states with the normalization

(4.139) can be achieved only for rational values of the modulation index h. Let us therefore write

$$h = \frac{i}{m} \tag{4.146}$$

with i and m being relatively prime integers. Recalling (4.142) and the set of values assumed by ξ_k, we have the following two cases:

(a) The integer i of (4.146) is even. The possible distinct (mod 2π) values for the phase states φ_k are

$$0, \, 2\pi \frac{i}{2m}, \, 2\pi \frac{2i}{2m}, \, \ldots, \, 2\pi \frac{(m-1)i}{2m},$$

and the total number p of phase states is equal to m.

(b) The integer i of (4.146) is odd. Now the phase states are

$$0, \, 2\pi \frac{i}{2m}, \, 2\pi \frac{2i}{2m}, \, \ldots, \, 2\pi \frac{(2m-1)i}{2m},$$

and the total number of phase states is $p = 2m$.

Example 4.10

Assume that $h = \frac{2}{3}$. Then the number of phase states is $p = 3$ and they take values in the set

$$\left\{ 0, \frac{2\pi}{3}, \frac{4\pi}{3} \right\}.$$

If, instead, $h = \frac{3}{4}$, the number of phase states is $p = 8$ and they take values in the set

$$\left\{ 0, \frac{\pi}{4}, \frac{\pi}{2}, \frac{3\pi}{4}, \pi, \frac{5\pi}{4}, \frac{3\pi}{2}, \frac{7\pi}{4} \right\}. \quad \square$$

Let us calculate how many different values the second term of the RHS of (4.141) can assume at time kT. As said before, this second term depends on the $(L-1)$ RVs preceding ξ_k and assumes $M^{(L-1)}$ different values. The $(L-1)$ RVs $\xi_{k-1}, \xi_{k-2}, \ldots, \xi_{k-L+1}$ are said to form the *correlative state* at time $t = kT$.

As a conclusion, each modulator state σ_k is one combination of a phase state with a correlative state. We get therefore, for the number of states σ_k,

$$N_\sigma = pM^{(L-1)} \tag{4.147}$$

where p is the number of phase states, M the dimensionality of the symbol alphabet, and L the correlation length of the baseband pulse $g(t)$.

To conclude the description of the modulator, it is very useful to examine the evolution of the phase function $\Phi(t; \xi_k, \sigma_k)$ in time. The phase trajectory in the kth time interval is given by (4.141). Let us define the *phase branch* as

$$\vartheta_k(t) \triangleq 2\pi h \sum_{i=k-L+1}^{k} \xi_i q(t - iT), \qquad kT \leq t < (k+1)T. \tag{4.148}$$

It is formed by the sum of the phase contributions, in the kth interval, of the L RVs $\xi_k, \xi_{k-1}, \ldots, \xi_{k-L+1}$. Recalling now the definition (4.142) of a phase state, we can give (4.141) the following expression:

$$\Phi(t; \xi_k, \sigma_k) = \vartheta_k(t) + \varphi_k + \psi_0. \tag{4.149}$$

That is, each phase trajectory in one symbol interval, apart from the initial value ψ_0, is given by the sum of the phase state with the phase branch.

From (4.148) we can conclude that there are M^L different phase branches in each symbol interval and therefore, from (4.149), pM^L different phase trajectories in the same interval. Any sequence of K symbols identifies a phase trajectory over a time span of duration KT. If we draw all the possible phase trajectories starting from an arbitrary initial constant value for the phase, we get the *phase tree* of the modulator.† Notice from (4.149) that the starting phase value in the kth time interval is generally different from the phase state φ_k.

We can conclude this description by saying that the modulation process can be understood by considering both the state trellis of the modulator and the phase tree of the modulated signal. Perhaps the best way to familiarize oneself with these concepts is to reconsider them in the following example.

Example 4.11

Let us assume a binary system ($M = 2$) with modulation index $h = 0.5$. We have $p = 4$ phase states with values $0, \pi/2, \pi, 3\pi/2$. Assume the baseband pulse $g(t)$, and hence $q(t)$, given in Fig. 4.29. We have a corrrelation length $L = 2$ and the total number of

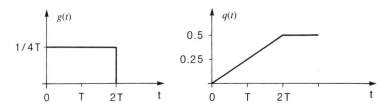

Figure 4.29 Baseband frequency and phase pulses $g(t)$ and $q(t)$ in Example 4.11.

states is, from (4.147), $N_\sigma = 4 \cdot 2^{(2-1)} = 8$. If we choose as an arbitrary initial condition, that the modulator is in the state $\sigma_0 = (0, 1)$, that is, in the phase state $\varphi_0 = 0$ and the correlative state 1, and that it has an initial phase $\psi_0 = -\pi/4$, we get from (4.141) the initial value of any phase trajectory. Recall that $\xi_k = \pm 1$ and $q(T) = \frac{1}{4}$; therefore,

$$\Phi(0; \xi_0, \sigma_0) = \pi\xi_{-1}q(T) + \varphi_0 + \psi_0 = 0.$$

Therefore, each phase trajectory starts from zero phase. Having assumed the stated initial condition, the phase trajectory corresponding to the sequence of symbols $(1, -1, 1, 1, 1, -1)$ is shown in Fig. 4.30 and can be derived directly from (4.141). We can draw the phase tree shown in Fig. 4.31 in an analogous fashion. Notice that the phase transitions are always linear. This is a consequence of the rectangular shape of the pulse $g(t)$. Also note that the slope of each phase transition corresponds to the frequency of the

---

† Notice that the phase trees are usually assumed to start at zero. Notice also that different definitions can be used for the phase states. Therefore, caution must be exercised in reading references on this subject.

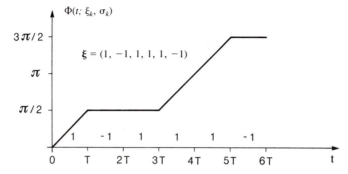

Figure 4.30 Phase trajectory of the sequence used in Example 4.11.

transmitted bandpass signal. Observe, then, that in this case only three frequencies are possible, but due to the constraint introduced by the correlation length of $g(t)$, only two out of these three frequencies are actually used in each node of the phase tree.

Continuing with the description of the phase tree and recalling (4.148), we can draw the possible phase branches that are shown in Fig. 4.32. Due to the choice of the initial phase $\psi_0 = -\pi/4$, it can be verified from (4.149) that the starting phase values at the beginning of each symbol interval coincide with the phase states φ_k. This justifies the choice for ψ_0.

Finally, the state trellis for the modulator is shown in Fig. 4.33. From each one of the eight possible states, two transitions can occur. Which one, depends on the value of

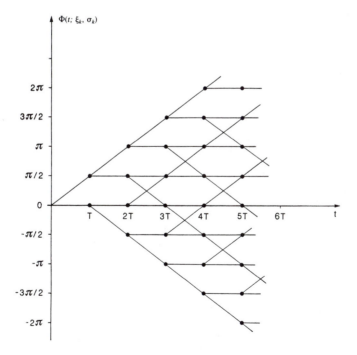

Figure 4.31 Phase tree of the CPM scheme of Example 4.11.

Sec. 4.4 Modulation with Memory and Coherent Demodulation 171

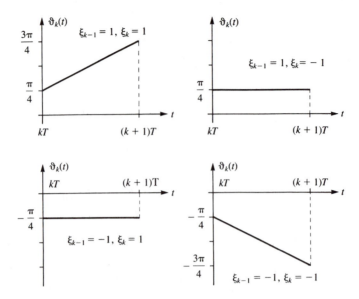

Figure 4.32 Phase branches of the CPM scheme of Example 4.11.

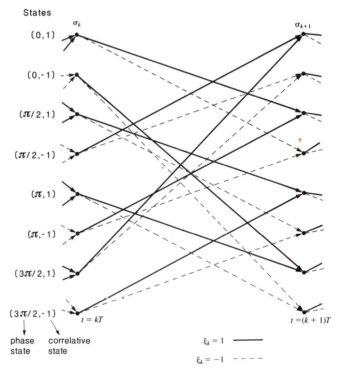

Figure 4.33 State trellis of the CPM scheme of Example 4.11.

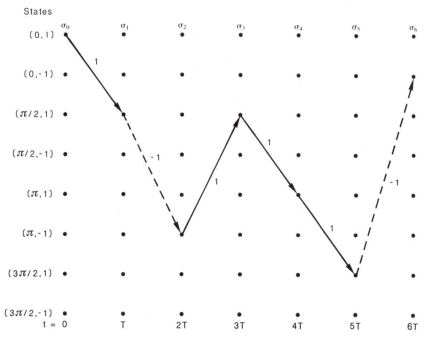

States

| | σ_0 | σ_1 | σ_2 | σ_3 | σ_4 | σ_5 | σ_6 |

$(0,1)$

$(0,-1)$

$(\pi/2,1)$

$(\pi/2,-1)$

$(\pi,1)$

$(\pi,-1)$

$(3\pi/2,1)$

$(3\pi/2,-1)$

t = 0   T   2T   3T   4T   5T   6T

Figure 4.34 Path in the state trellis of the binary sequence whose phase trajectory is shown in Fig. 4.30 (see Example 4.11).

ξ_k. The trellis can be drawn as follows. Assume to be in the state $(\pi/2, -1)$ and that $\xi_k = 1$. Since $\xi_{k-1} = -1$, from (4.145) we get

$$\varphi_{k+1} = \varphi_k + \frac{\pi}{2}\xi_{k-1} = \frac{\pi}{2} - \frac{\pi}{2} = 0.$$

Therefore, the modulator goes into the state $(0, 1)$. Notice that each path into this trellis corresponds to a symbol sequence or equivalently to a state sequence. Figure 4.34 shows the path of the sequence represented in Fig. 4.30. $\square$

4.4.2 Optimum Demodulator Structures

After the description of the main properties of the class of CPM signals, let us now derive the structure of the optimum demodulator, that is, one that implements the ML decision scheme.

A sequence ξ of transmitted symbols corresponds to a unique path into the state trellis of the modulator. The task of the optimum demodulator is that of finding the ML path into the trellis, that is, the one maximizing (4.131). Therefore, the estimated sequence will be

$$\hat{\xi}: l_{\hat{\xi}} = \max_{\xi} l_\xi. \qquad (4.150)$$

At time $t = kT$ the modulator is in state σ_k, that must be one of the possible modulator states

$$S_1, S_2, \ldots, S_{N_\sigma}. \qquad (4.151)$$

All the information sequences of length k that, at time $t = kT$, led into one specific state S_i correspond to a path into the trellis with an associated *metric* $l_{\xi(k)}$ given by (4.133). Since we are looking for the largest metric path through the trellis, for each state S_i at time kT, we keep memory only of the path leading to that state and having the largest value of $l_{\xi(k)}$. All the other paths can be dropped. This is in fact the key point of the Viterbi algorithm. To compute the value of the path metric at the next time $t = (k + 1)T$, the recursion (4.134) is used. Therefore, in each kth interval, the demodulator must compute the metric $l_{\xi k}$ associated with any possible transition from state σ_k to state σ_{k+1}. In conclusion, the optimum demodulator consists of a part devoted to the computation of all the l_{ξ_k}'s followed by a device performing the Viterbi algorithm (*Viterbi processor*). Recalling (4.132) and noticing that the signals have equal energy, we rewrite it disregarding the constant term:

$$l_{\xi k} = \int_{kT}^{(k+1)T} r(t)s(t - kT; \xi_k, \sigma_k)\, dt. \qquad (4.152)$$

The quantities in (4.152) form the sufficient statistics for the received signal $r(t)$:

$$\begin{aligned} r(t) &= \mathfrak{R}\{\tilde{r}(t)e^{j2\pi f_0 t}\} \\ &\triangleq r_P(t) \cos 2\pi f_0 t - r_Q(t) \sin 2\pi f_0 t. \end{aligned} \qquad (4.153)$$

Furthermore, the locally generated replicas of the transmitted signals are

$$s(t - kT; \xi_k, \sigma_k) = \sqrt{\frac{2\mathscr{E}}{T}} \cos\{2\pi f_0 t + \vartheta_k(t) + \varphi_k + \psi_0\}, \quad kT \leqslant t < (k+1)T. \quad (4.154)$$

Remember now that there are M^L possible phase branches $\vartheta_k(t)$, p possible phase states φ_k, and that we can assume $\psi_0 = 0$.

By inserting (4.153) and (4.154) in (4.152) and with the usual assumption that $f_0 T \gg 1$, the RV l_{ξ_k} becomes

$$\begin{aligned} l_i &= \sqrt{\frac{\mathscr{E}}{2T}} \cos \varphi_m \left\{ \int_{kT}^{(k+1)T} r_P(t) \cos \vartheta_j(t)\, dt + \int_{kT}^{(k+1)T} r_Q(t) \sin \vartheta_j(t)\, dt \right\} \\ &\quad + \sqrt{\frac{\mathscr{E}}{2T}} \sin \varphi_m \left\{ \int_{kT}^{(k+1)T} r_Q(t) \cos \vartheta_j(t)\, dt - \int_{kT}^{(k+1)T} r_P(t) \sin \vartheta_j(t)\, dt \right\}, \end{aligned} \qquad (4.155)$$

where

$$m = 1, 2, \ldots, p,$$
$$j = 1, 2, \ldots, M^L,$$

and therefore $i = 1, 2, \ldots, pM^L$. The block diagram of the demodulator based on (4.155) is shown in Fig. 4.35. This demodulator is called a *baseband correlation demodulator*. The term baseband is explained by the fact that correlations are performed over the complex envelope components of the signals instead of directly on the RF signals. It is only a tedious task to interpret each correlation in terms of a corresponding matched filter operation and to get the demodulator block diagram in the form of a baseband matched-filter demodulator.

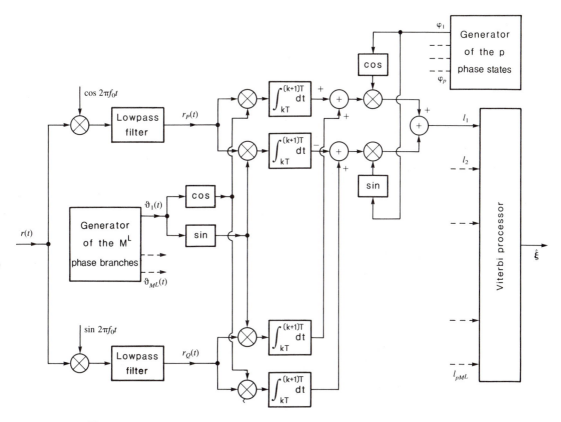

Figure 4.35 Block diagram of the baseband correlation receiver for the ML coherent demodulation of CPM signals.

Let us evaluate the amount of processing required by the demodulator of Fig. 4.35 to supply the Viterbi processor with the pM^L values of l_i in each kth time interval. Apart from the sines and cosines of a small number p of phase states, only the replicas of the sines and cosines of the M^L phase branches $\vartheta(t)$ have to be stored in the demodulator. However, notice from (4.148) that the set of phase branches can be divided in two subsets that are symmetrical around zero. Consequently, the calculations need only be made for one subset because the products with $\cos\vartheta(t)$ are equal for both subsets and the products with $\sin\vartheta(t)$ need only be inverted to obtain the values for the other subset. The computational complexity implied by the Viterbi processor depends on the number N_σ of the modulator states. This subject is discussed in Appendix F.

All the steps presented so far, in a general context, will now be clarified with the aid of the following example.

Example 4.12

Assume a binary system ($M = 2$) with modulation index $h = 0.5$. Recalling Example 4.11, we have $p = 4$ phase states given by 0, $\pi/2$, π, $3\pi/2$. Let us choose the baseband pulse $g(t)$, and hence $q(t)$, given in Fig. 4.36. Since the correlation length is $L = 1$, the

Sec. 4.4   Modulation with Memory and Coherent Demodulation   **175**

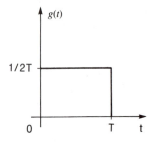
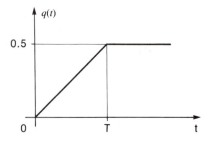

Figure 4.36 Baseband frequency pulse $g(t)$ and phase pulse $q(t)$ of the CPM signals of Example 4.12.

total number of states is $N_\sigma = 4$. Also, they coincide with the phase states, since we do not have correlative states. Using (4.135), (4.136), and (4.141), we get

$$s(t - kT; \xi_k, \sigma_k) = \sqrt{\frac{2\mathscr{E}}{T}} \cos \left\{ 2\pi f_0 t + \frac{\pi}{2T}(t - kT)\xi_k + \varphi_k \right\},$$

(4.156)

$$kT \leq t < (k+1)T.$$

Apart from the slightly different notation, these signals are analogous to those of (4.4) in Example 4.2. The absence of correlative states causes the state trellis to be equal to the phase trellis. That is, the phase tree of Fig. 4.3 reduced mod 2π. This trellis is drawn in Fig. 4.37, where also the sequence $(1, -1, 1, 1, -1)$ is shown. The initial state is the zero state, and solid lines correspond to a $+1$ symbol transition, whereas dashed lines mean a -1 symbol transition.

If we define the two frequencies

$$f_1 \triangleq f_0 + \frac{1}{4T},$$

$$f_2 \triangleq f_0 - \frac{1}{4T},$$

(4.157)

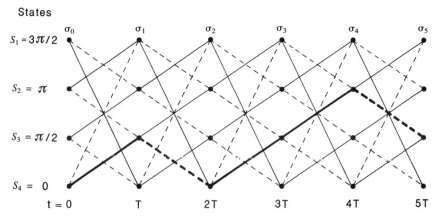

Figure 4.37 Phase trellis of the CPM signals of Example 4.12.

we get from (4.156) that there are only two possible transmitted signals:

$$s_j(t - kT; \varphi_k) = \sqrt{\frac{2\mathscr{E}}{T}} \cos \left\{ 2\pi f_j t + \frac{(-1)^j k\pi}{2} + \varphi_k \right\}, \tag{4.158}$$

$$kT \leq t < (k+1)T,$$

and $j = 1$ for $\xi_k = 1$, $j = 2$ for $\xi_k = -1$.

The phase φ_k at each step represents the (phase) state and guarantees the phase continuity between adjacent time intervals of length T. The transmitted signal corresponding to the sequence shown in Fig. 4.37 is given in Table 4.1. By using (4.152), we can observe that the sufficient statistics is based on the RVs

$$l_i = \int_{kT}^{(k+1)T} r(t) s_j(t - kT; \varphi_m)\, dt, \tag{4.159}$$

where $m = 1, 2, 3, 4$, $j = 1, 2$, and therefore $i = 1, 2, \ldots, 8$. Introducing the signals (4.158) in (4.159), we can get

$$
\begin{aligned}
l_i = {}& \sqrt{\frac{2\mathscr{E}}{T}} \cos \varphi_m \int_{kT}^{(k+1)T} r(t) \cos \left\{ 2\pi f_j t + \frac{(-1)^j k\pi}{2} \right\} dt \\
& - \sqrt{\frac{2\mathscr{E}}{T}} \sin \varphi_m \int_{kT}^{(k+1)T} r(t) \sin \left\{ 2\pi f_j t + \frac{(-1)^j k\pi}{2} \right\} dt.
\end{aligned}
\tag{4.160}
$$

With the definitions

$$P_j \triangleq \sqrt{\frac{2\mathscr{E}}{T}} \int_{kT}^{(k+1)T} r(t) \cos \left\{ 2\pi f_j t + \frac{(-1)^j k\pi}{2} \right\} dt,$$

$$Q_j \triangleq \sqrt{\frac{2\mathscr{E}}{T}} \int_{kT}^{(k+1)T} r(t) \sin \left\{ 2\pi f_j t + \frac{(-1)^j k\pi}{2} \right\} dt,$$

we can rewrite (4.160) as

$$l_i = P_j \cos \varphi_m - Q_j \sin \varphi_m. \tag{4.161}$$

The block diagram of the demodulator is shown in Fig. 4.38.

In the absence of noise, and assuming the transmitted signal given in Table 4.1, we can compute the quantities of (4.160). The result is shown in Table 4.2, where we have set $\mathscr{E} = 1$. Noise, however, is present. Hence, we assume that the values of Table 4.2 are perturbed by the noise terms as shown in Table 4.3. All the branch metrics computed according to (4.161) are shown in Fig. 4.39.

The successive steps performed by the Viterbi processor from time $t = 2T$ up to time $t = 5T$ are shown in Fig. 4.40, where the values of Table 4.3 are used. For each state, at each step, only the largest metric path is drawn. The metric value is also indicated.

To understand how the Viterbi algorithm functions, let us study an example. Assume that we are at time $t = 3T$. We have computed the values of P_j and Q_j (third line of Table 4.3) and we have recorded the situation of the state trellis at time $t = 2T$. We want to make one step forward in the state trellis. Consider, for example, the state $\sigma_3 = S_1$. We can reach this state in two ways:

(a) From state $\sigma_2 = S_2$.

In this case, the branch metric (see Fig. 4.39) is given by $-P_1 = -1.2$ and the metric score at $\sigma_3 = S_1$ would be $1.32 - 1.2 = 0.12$.

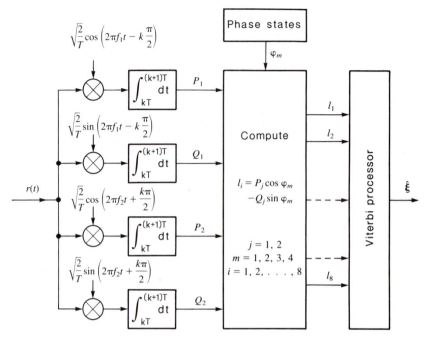

Figure 4.38 Block diagram of the optimum receiver for the CPM signals of Example 4.12.

TABLE 4.1 Transmitted signals corresponding to symbols ξ_k

Time interval	Symbol ξ_k	Waveform
$0 \leq t < T$	$+1$	$\sqrt{\dfrac{2\mathscr{E}}{T}}\cos(2\pi f_1 t)$
$T \leq t < 2T$	-1	$-\sqrt{\dfrac{2\mathscr{E}}{T}}\cos(2\pi f_2 t)$
$2T \leq t < 3T$	$+1$	$-\sqrt{\dfrac{2\mathscr{E}}{T}}\cos(2\pi f_1 t)$
$3T \leq t < 4T$	$+1$	$-\sqrt{\dfrac{2\mathscr{E}}{T}}\cos(2\pi f_1 t)$
$4T \leq t < 5T$	-1	$-\sqrt{\dfrac{2\mathscr{E}}{T}}\cos(2\pi f_2 t)$

(b) From state $\sigma_2 = S_4$.

In this case, the branch metric (see again Fig. 4.39) is given by $P_2 = -0.05$ and the metric score at $\sigma_3 = S_1$ would be $1.92 - 0.05 = 1.87$.

The second case gives, therefore, the surviving path and the first path is discarded. In the same way, we can obtain all the results appearing in Fig. 4.40. At time $t = 5T$ we can order the received estimated sequences in the following way:

Metric score	State sequence	Symbol sequence
5.12	$S_4S_3S_4S_3S_2S_3$	1, −1, 1, 1, −1
4.19	$S_4S_3S_4S_3S_2S_1$	1, −1, 1, 1, 1
2.37	$S_3S_4S_3S_4S_3S_2$	−1, 1, −1, 1, 1
1.47	$S_3S_4S_3S_4S_3S_4$	−1, 1, −1, 1, −1

It is seen that the *ML path* corresponds to the transmitted sequence of Fig. 4.37. The next best path contains only one error located in the last position, since these two paths have diverged only at the previous step. □

TABLE 4.2 Values of P_j and Q_j of (4.161) in the absence of noise (we have set $\mathscr{E} = 1$)

t	P_1	Q_1	P_2	Q_2
T	1.0	0	0	−0.64
$2T$	0.64	0	0	−1.0
$3T$	1.0	0	0	−0.64
$4T$	0	−1.0	−0.64	0
$5T$	0	−0.64	−1.0	0

TABLE 4.3 Same values of Table 4.2 perturbed by noise terms

t	P_1	Q_1	P_2	Q_2
T	1.02	0.1	−0.2	−0.6
$2T$	0.58	−0.3	−0.05	−0.9
$3T$	1.2	0.2	−0.05	−0.59
$4T$	−0.1	−1.1	−0.62	0.05
$5T$	0.03	−0.7	−0.9	0.2

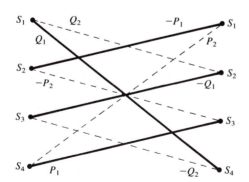

Figure 4.39 Branch metrics for the CPM signals of Example 4.12 and Fig. 4.37.

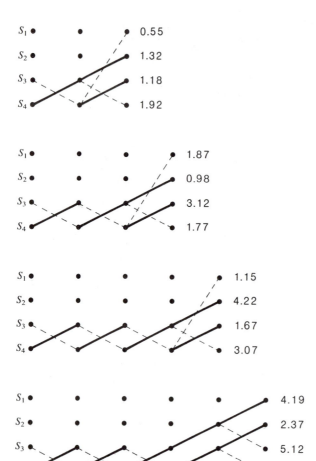

Figure 4.40 Steps of the Viterbi algorithm in Example 4.12.

Values shown on diagram:
- First block: S_1 0.55, S_2 1.32, S_3 1.18, S_4 1.92
- Second block: S_1 1.87, S_2 0.98, S_3 3.12, S_4 1.77
- Third block: S_1 1.15, S_2 4.22, S_3 1.67, S_4 3.07
- Fourth block: S_1 4.19, S_2 2.37, S_3 5.12, S_4 1.47

Time axis: 0, T, 2T, 3T, 4T, 5T

4.4.3 Error Probability Evaluation

The computation of symbol error probability for the signaling schemes described in this section is complex. The most common approaches used to confront this problem resort to bounding techniques and computer simulations. We shall restrict our attention to the class of CPM signals introduced in (4.135).

Recall that the transmitted signal $v_\xi(t)$ is associated with an information sequence $\xi = (\xi_0, \xi_1, \ldots, \xi_{K-1})$ of K RVs, assuming values in the set $\{\pm 1, \pm 3, \ldots, \pm(M-1)\}$. In principle, the Viterbi algorithm allows an ML decision on the whole information sequence at time $t = KT$. However, K may be so large as to make impractical both the path storage and the decision delay implied by the ideal implementation. For this reason, practical Viterbi processors *force* a decision at time $t = NT$, $N \leq K$. This is done, for example, by taking a decision on the first RV ξ_0 based on the observation of the received signal up to $t = NT$. After the decision is taken for ξ_0, the process is repeated for ξ_1 at time $t = (N + 1)T$, and so on. The following analysis will comply with this practical approach.

Since we are dealing with ML decisions taken by a coherent demodulator over a set of transmitted signals, we want to use the minimum-distance concept introduced in Section 4.2 in order to be able to apply the union bound expressed by (4.70). To this end, let us introduce some notation for the forthcoming calculations. First, we define the set of sequences $\{\xi^{(l)}\}$ as the subset of $\{\xi\}$ whose first symbol ξ_0 is equal to l; that is,

$$\xi^{(l)} \triangleq \{\xi \colon \xi_0 = l\}, \tag{4.162}$$

with $l = \pm 1, \pm 3, \ldots, \pm(M - 1)$. The signal associated with the sequence $\xi^{(l)}$ of length N is defined as

$$s(t; \xi^{(l)}) \triangleq \sum_{k=0}^{N-1} s(t - kT; \xi_k^{(l)}, \sigma_k), \tag{4.163}$$

where

$$\xi_0^{(l)} = l, \qquad l = \pm 1, \pm 3, \ldots, \pm(M - 1),$$
$$\xi_k^{(l)} = \xi_k \qquad k > 0.$$

Notice that we have a total of M^N possible signals of the type (4.163) and that M^{N-1} of them have the same value of the index l.

Finally, using (4.26) and the definition (4.27), we can introduce the notation for the squared Euclidean distance between two signals of the type (4.163) as

$$d_N^2(\xi^{(l)}, \xi^{(m)}) \triangleq d^2[s(t; \xi^{(l)}), s(t; \xi^{(m)})]. \tag{4.164}$$

Notice now, that the demodulator must decide on the set of M^N possible signals. But its decision concerns only the first symbol ξ_0 of the sequence, and therefore a correct decision is equivalent to identifying the subset of M^{N-1} signals $s(t; \xi^{(l)})$ that contain as their first symbol the symbol actually transmitted. In other words, if the first transmitted symbol is $\xi_0 = 3$, a correct decision on ξ_0 is achieved whenever any one of the M^{N-1} signals $s(t; \xi^{(3)})$ is chosen.

On the other hand, an error is made if the demodulator chooses one of the $(M - 1)M^{N-1}$ signals that are outside the correct subset. Therefore, applying the union bound (4.70), we obtain an upper bound for the symbol error probability in the form

$$P(e) \leq \frac{(M - 1)M^{N-1}}{2} \, erfc\left(\frac{d_{N,\min}}{2\sqrt{N_0}}\right), \tag{4.165}$$

where

$$d_{N,\min}^2 \triangleq \min_{\substack{l,m \\ l \neq m}} d_N^2(\xi^{(l)}, \xi^{(m)}). \tag{4.166}$$

Therefore, the evaluation of the bound (4.165) is reduced to the computation of $d_{N,\min}$ for the specific choice of parameters of the CPM modulation scheme.

Introducing (4.135) and (4.136) into (4.163) and expanding (4.164), we get the expression

$$d_N^2(\xi^{(l)}, \xi^{(m)}) = 2\mathscr{E}\left\{N - \sum_{k=0}^{N-1} \frac{1}{T} \int_{kT}^{(k+1)T} \cos[\Phi(t; \xi_k^{(l)}, \sigma_k) - \Phi(t; \xi_k^{(m)}, \sigma_k)] \, dt\right\}. \quad (4.167)$$

Let us now define the difference sequence γ as follows:

$$\gamma \triangleq \xi^{(l)} - \xi^{(m)} = (l - m, \xi_1^{(l)} - \xi_1^{(m)}, \xi_2^{(l)} - \xi_2^{(m)}, \ldots), \quad (4.168)$$
$$l, m = \pm 1, \pm 3, \ldots, \pm(M - 1).$$

Using this definition and recalling (4.141), we can rewrite (4.167) as follows:

$$\frac{d_N^2(\xi^{(l)}, \xi^{(m)})}{2\mathscr{E}} = N - \sum_{k=0}^{N-1} \frac{1}{T} \int_{kT}^{(k+1)T} \cos \Phi(t; \gamma_k, \sigma_k) \, dt. \quad (4.169)$$

Therefore, we can study the behavior of the squared distance $d_N^2(\xi^{(l)}, \xi^{(m)})$ by considering the function $\Phi(t; \gamma_k, \sigma_k)$. We have to deal now with the difference sequence γ of (4.168) whose first symbol must be different from zero, since the first pair of data symbols are required to be different. Furthermore, there is a sign symmetry among the difference sequences in (4.169). Hence, due to the cosine function, we can limit ourselves to positive values of γ_0. To compute the values of (4.169) with the aim of finding the minimum (4.166), we have thus to consider the difference sequences γ defined by

$$\gamma_0 = l - m = 2, 4, 6, \ldots, 2(M - 1), \quad (4.170)$$
$$\gamma_k = 0, \pm 2, \pm 4, \ldots, \pm 2(M - 1), \quad k = 1, 2, \ldots, N - 1.$$

Using (4.141) and the sequences (4.170), we can draw the *phase difference tree* that represents all the possible phase difference trajectories to be used in (4.169) to compute all the possible values of $d_N^2(\xi^{(l)}, \xi^{(m)})$. Let us examine these concepts in detail by way of some examples.

Example 4.13

Assume the binary system of Example 4.12. The difference sequences to be used in the phase difference tree are, from (4.170),

$$\gamma_0 = 2, \quad (4.171)$$
$$\gamma_k = 0, \pm 2, \quad k = 1, 2, \ldots, N - 1.$$

The phase difference tree is shown in Fig. 4.41. If, instead, we take the binary system of Example 4.11, we get the phase difference tree of Fig. 4.42. Notice that the phase differences may be reduced mod 2π, so trajectories very far apart may eventually coincide. □

It is intuitive, from the examples just shown, that a brute-force approach for calculating all the allowed values of $d_N^2(\xi^{(l)}, \xi^{(m)})$ in order to find the minimum may require an unrealistic computational effort. There are $(M - 1)(2M - 1)^{N-1}$ different sequences of length N and an exponential growth with N would be inescapable. Efficient recursive algorithms (Saxena, 1983, and Mulligan and Wilson, 1984) have been proposed for this calculation.

It is clear from the definition that $d_{N, \min}^2$ is a nondecreasing function of the observa-

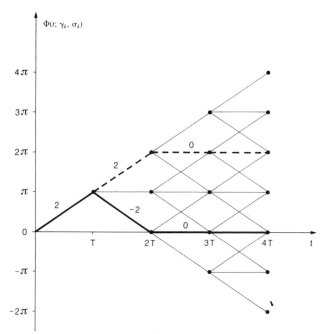

$\Phi(t; \gamma_k, \sigma_k)$

Figure 4.41 Phase difference tree for the CPM signals of Example 4.12. The phase difference is 0 for the solid path and 2π for the dotted path.

tion length N. Therefore, by considering infinitely long sequences, one could define $d^2_{\min}$ as

$$d^2_{\min} \triangleq \lim_{N \to \infty} d^2_{N,\,\min}. \qquad (4.172)$$

The choice of N, in a practical processor, should be such that the value of $d^2_{N,\,\min}$ be close enough to $d^2_{\min}$. Thus, $d^2_{\min}$ is the parameter that dominates the asymptotic ($N_0 \to 0$) behavior of the error probability as in the memoryless case.

Let us examine now the problem of calculating $d^2_{\min}$. Each pair of phase trajectories considered in (4.169) differs in the first symbol and, after some intervals in which symbols may differ or not, the trajectories merge and coincide in the successive intervals. We define this situation as a *merge*. When a merge occurs, the corresponding phase difference trajectory reaches zero (mod 2π) and coincides with the horizontal axis in the successive intervals. These merges can be observed in the examples of Figs. 4.41 and 4.42. In the first case we have a merge after two symbols, in the second after three symbols. In general, it can be seen that, for pulses $g(t)$ of correlation length L, a merge is always possible after $(L + 1)$ intervals from the starting node in the phase difference tree. As a consequence of the definition (4.172), the value of $d^2_{\min}$ cannot exceed the squared distance between two trajectories that merge. We define the upper bound d^2_B to $d^2_{\min}$ as the squared distance measured at the first merge on the phase difference tree.

However, other merges are possible after $(L + 2)$ or more intervals from the starting node, and the corresponding squared distance can be smaller than that found at the first merge. In such a case, the upper bound d^2_B can be tightened. In general, the

Sec. 4.4   Modulation with Memory and Coherent Demodulation   **183**

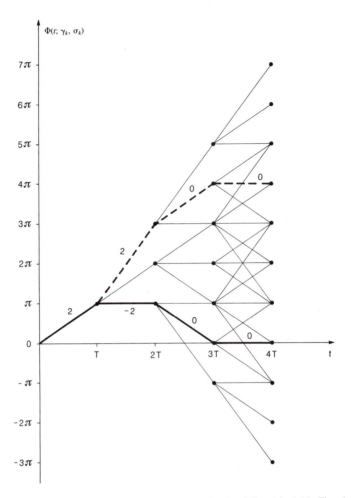

Figure 4.42 Phase difference tree for the CPM signals of Example 4.11. The phase difference is 0 for the solid path and 2π for the dotted path.

exact order of the merge that gives the tightest bound, and hence $d_{\min}^2$, is not known. Only for the case of $L = 1$ is the value of $d_{\min}^2$ found for the first merge.

We can now apply these concepts to the case of pulses of the type of Fig. 4.36 using an arbitrary value for the modulation index h. The phase difference tree is similar to that shown in Fig. 4.41, except that the phase difference values at the nodes are 0, $\pm h\pi$, $\pm 2h\pi$, ... The first merge takes place at $t = 2T$ and is caused by the difference sequence $(2, -2, 0, 0, \ldots)$. Using (4.169),

$$d_B^2 = 2\mathscr{E}\{2 - \frac{1}{T}\int_0^{2T} \cos\{4\pi h[q(t) - q(t-T)]\}\,dt\}$$

$$= 4\mathscr{E}\left(1 - \frac{\sin 2\pi h}{2\pi h}\right).$$

(4.173)

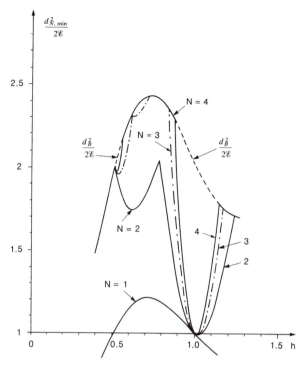

Figure 4.43 Values of normalized $d_{N.\min}^2$ and d_B^2 as functions of h for the case of binary signals and baseband pulses of Fig. 4.36. (Reprinted with permission from Aulin and others, 1982.)

It can be verified that this bound is maximized when $h = 0.715$. The corresponding value of the minimum normalized squared distance is 2.44.

Figure 4.43 shows a plot of this bound as a function of the modulation index h. On the same plot the normalized squared distance with N observation intervals is also shown. Notice that an observation interval of at least three symbols ($N \geq 3$) is required to achieve the best performance for this modulation scheme. Notice also the peculiar behavior at $h = 1$. With this value of the modulation index, we have a merge at $t = T$. This explains the poor behavior in this case. When such situations occur, the corresponding modulation indexes are called *weak modulation indexes*. Clearly, they should be avoided.

If we now take the baseband pulse of Fig. 4.29 and perform similar computations on the phase difference tree (generalized to an arbitrary h) of Fig. 4.42, we can compute the upper bound of the minimum normalized squared distance at $t = 3T$ in the form

$$d_B^2 = 2\mathcal{E}\left\{2\left(1 - \frac{\sin \pi h}{\pi h}\right) + 1 - \cos \pi h\right\}. \tag{4.174}$$

It happens, however, that this bound is not tight for all values of h, since for certain values of h, merges at $4T$ give a tighter bound. A tightest value for the bound is in fact

$$d_B^2 = 2\mathcal{E} \cdot \min\left\{2\left(1 - \frac{\sin \pi h}{\pi h}\right) + 1 - \cos \pi h;\ 4\left(1 - \frac{\sin 2\pi h}{2\pi h}\right)\right\}. \tag{4.175}$$

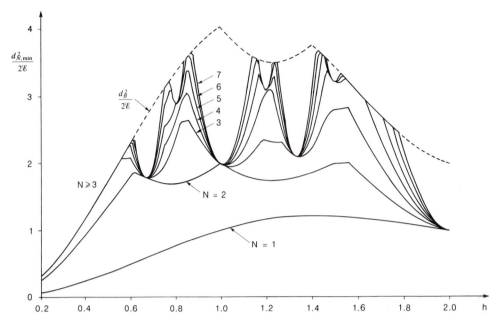

Figure 4.44 Values of normalized $d_{N,\min}^2$ and d_B^2 as functions of h for the case of binary signals and baseband pulses of Fig.4.29. (Reprinted with permission from Aulin and others, 1982.)

This bound is shown in Fig. 4.44. The minimum normalized squared distance for several values of N is also plotted. Considerations similar to those made for Fig. 4.43 are also applicable in this case. Let us now conclude by applying these results to some examples.

Example 4.14

Let us again take the system of Example 4.12 and recall the phase difference tree of Fig. 4.41. By performing a direct calculation or using (4.173) with $h = 0.5$, we get

$$d_{\min}^2 = 4\mathscr{E}. \tag{4.176}$$

Looking at Fig. 4.43, we can in fact verify that the upper bound for $h = 0.5$ is achieved with an observation interval $N = 2$. Finally, inserting (4.176) into (4.165), we get

$$P(e) \leq erfc\left(\sqrt{\frac{\mathscr{E}}{N_0}}\right). \tag{4.177}$$

If we take the system of Example 4.11 and use the phase difference tree of Fig. 4.42, we can get

$$d_{\min}^2 = 2\mathscr{E}\left(3 - \frac{4}{\pi}\right) \cong 3.46\,\mathscr{E}. \tag{4.178}$$

Also in this case, looking at Fig. 4.44, we verify that the upper bound for $h = 0.5$ is achieved with an observation interval $N = 3$. Using (4.178) into (4.165), we get for this case

$$P(e) \leq 2\,erfc\left(\sqrt{0.865\,\frac{\mathscr{E}}{N_0}}\right). \tag{4.179}$$

If we compare these results with the case of binary antipodal signals (4.45), we can conclude that, in the first case, we have the same asymptotic behavior, while in the second there is a loss of 0.62 dB in the ratio $\mathcal{E}/N_0$. This conclusion seems discouraging. Nevertheless, if we use a modulation index $h = 0.7$ and repeat the calculations, we have a gain of 1.61 dB over the binary antipodal case. Then, what is the trade-off for the complexity of these modulation schemes with respect to simple binary antipodal coherent modulation? The first observation is that the parameters that control the performance, that is, the pulse shape $g(t)$, the number of levels M, and the modulation index h, can be chosen in such a way as to improve the error probability behavior with respect to the binary antipodal case. The most important fact, however, is that throughout the chapter we have not considered the bandwidth occupancy of these modulation schemes because a wideband distortionless channel was assumed. It will be seen in Chapter 5 that the spectral occupancy of signals like those considered in this section can be, indeed, lower than that required by memoryless modulation schemes. Consequently, the possible trade-off is a more efficient use of the available bandwidth. ☐

BIBLIOGRAPHICAL NOTES

Much of the material in this chapter is classical in detection and modulation theory and, as such, it can be found in most of the textbooks available on this subject. Only the contents of Section 4.4 represent more recent developments.

The authors are indebted to the excellent books by Wozencraft and Jacobs (1965) and Van Trees (1968). The first in particular, emphasizes the geometric viewpoint. Another pioneering contribution is the book by Kotel'nikov (1959). Among the most recently published textbooks, see Proakis (1983). For the theory of signal spaces, the book by Franks (1969) is a recommended reading. The paper by Arthurs and Dym (1962) contains a clear presentation, in a geometric context, of the problems of coherent and incoherent demodulation of memoryless signals. In particular, our derivation of the upper and lower bound (4.68) and (4.75) follows closely this paper.

A more detailed analysis of the phase coherence of the receiver and of its effects on the demodulated signals can be found in the books by Viterbi (1966) and Lindsey and Simon (1973) and in Viterbi (1965). The problem of incoherent demodulation of two equal-energy signals appears in Helstrom (1958); we have followed closely the general derivation presented in the book by Schwartz, Bennett, and Stein (1966). The material of Example 4.6 is taken from Corazza and Immovilli (1979). All the material related to CPM signals was published in the papers by Aulin and Sundberg (1981) and Aulin, Rydbeck, and Sundberg (1981). Recent results on the computation of the minimum distance are given in Saxena (1983) and Mulligan and Wilson (1984). Further analysis of the error probability evaluation can be found in Lindell, Sundberg, and Svensson (1985).

PROBLEMS†

4.1. Given an orthonormal basis $\{\psi_j(t)\}_{j=1}^N$, show that the RVs defined in (4.23) are independent zero-mean Gaussian RVs with covariance $N_0/2$, where $N_0/2$ is the power spectral density of the Gaussian random process $n(t)$.

4.2. Assume a binary signaling scheme over an AWGN channel with coherent demodulation. Consider three cases $(0 \leq t < T)$:

Case	$s_1(t)$	$s_2(t)$
1	0	$\sqrt{2\mathcal{E}/T}\sin 2\pi f_2 t$
2	$\sqrt{2\mathcal{E}/T}\sin 2\pi f_1 t$	$\sqrt{2\mathcal{E}/T}\sin 2\pi f_2 t$
3	$\sqrt{2\mathcal{E}/T}\sin 2\pi f_1 t$	$-\sqrt{2\mathcal{E}/T}\sin 2\pi f_1 t$

where $f_1 - f_2 = n/T$ for some integer n and $f_1 = m/T$ for some integer m.
(a) Draw the appropriate signal space for each case.
(b) Using (4.44), compute the error probability.
(c) Comment on the relative efficiency of each scheme with regard to the utilization of the average transmitted energy.

4.3. Given a binary scheme as shown in Fig. 4.10b, assume that the two signals are not equally likely. Define $p_1 \triangleq P(\mathbf{s}_1)$, $p_2 \triangleq P(\mathbf{s}_2)$, and show that the optimum demodulator of Fig. 4.7 must have a threshold at the value

$$\frac{N_0}{2d} \ln \frac{p_2}{p_1}.$$

In words, the boundary of the two decision regions is not at zero but is shifted closer to the signal with the lower probability. Find the expression of the resulting error probability.

4.4. Assume a binary antipodal scheme as in Fig. 4.10b.
(a) Evaluate the expression of the error probability of the optimum demodulator of Fig. 4.7 when the decision threshold has an offset A with respect to the optimum zero value.
*(b) For some values of $A/\sqrt{\mathcal{E}}$, obtain error probability curves and compare with Fig. 4.12.

4.5. In this problem we analyze the degradation in performance of a binary demodulation due to the use of a filter different from the optimum matched filter. The system is shown in Fig. P4.1. Assume

$$s_1(t) = 0,$$

$$s_2(t) = \sqrt{\frac{\mathcal{E}}{T}}, \qquad 0 \leq t < T,$$

and $n(t)$ a Gaussian noise with power spectral density $N_0/2$. The optimum demodulator would require

$$h_{\text{opt}}(t) = \sqrt{\frac{1}{T}}, \qquad 0 \leq t < T.$$

† The problems marked with an asterisk should be solved with the aid of a computer.

Figure P4.1

Assume instead a simple RC filter with $h(t) = e^{-At}u_T(t)$.

(a) Compute the error probability for this case.

(b) Find the value of A that minimizes the error probability found in part (a).

(c) Evaluate the increase of transmitted energy required to get the same error probability of the optimum demodulator.

4.6. Two antipodal signals are transmitted over an AWGN channel. The optimum receiver achieves an error probability of 0.1 assuming that $\sqrt{N_0} = 1$.

(a) Compute the capacity of the equivalent binary symmetric channel (see Section 3.3).

***(b)** Modify the receiver by introducing two thresholds of value $\pm A$. Derive the discrete equivalent (binary erasure) channel and compute its capacity as a function of A.

Compare and comment on the two cases.

4.7. Assume a binary scheme with signals

$$s_1(t) = 0, \qquad 0 \le t < T,$$

$$s_2(t) = \sqrt{\frac{2\mathcal{E}}{T}} \sin 2\pi f_2 t, \qquad f_2 = \frac{m}{T}, \ m \text{ integer},$$

and consider a coherent demodulator and an incoherent one.

(a) Obtain for the two cases the equivalent binary discrete channel as a function of $\mathcal{E}/N_0$.

***(b)** Draw a graph of the channel capacity for the two cases as a function of $\mathcal{E}/N_0$.

4.8. Two equally likely signals are transmitted over an AWGN channel with power spectral density $N_0/2$. The two signals are

$$s_1(t) = \sqrt{\frac{2\mathcal{E}}{T}} \cos 2\pi f_1 t,$$

$$s_2(t) = \sqrt{\frac{2\mathcal{E}}{T}} \cos 2\pi (f_1 + \Delta f) t, \qquad 0 \le t < T.$$

Assume $T = 2$ ms and $f_1 = 1$MHz. Consider the two cases of $\Delta f = 500$ Hz and $\Delta f = 1$ kHz.

(a) Compute the error probability for a coherent demodulator.

(b) Compute the error probability for an incoherent demodulator.

4.9. (Wozencraft and Jacobs, 1965) Eight equally likely signals are transmitted over an AWGN channel with noise power spectral density $N_0/2$. The signals are given in Fig. P4.2.

(a) Compute the error probability achieved by an optimum coherent demodulator.

(b) Interpret the result using for the channel a binary symmetric channel model.

4.10. (Lindsey and Simon, 1973) Consider a set of M equally likely signals described in the two-dimensional space as follows:

$$s_i(t) = \sqrt{\frac{2}{T}} a_i \cos 2\pi f_0 t - \sqrt{\frac{2}{T}} b_i \sin 2\pi f_0 t, \qquad 0 \le t < T, \qquad i = 1, 2, \ldots, M.$$

Chap. 4   Problems

189

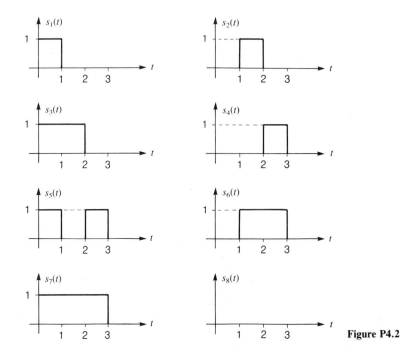

Figure P4.2

The average power of the signal set is

$$\mathcal{P} = \frac{1}{MT} \sum_{i=1}^{M}(a_i^2 + b_i^2).$$

(a) Show that the union bound of (4.68) can be put in the form

$$P(e) \le \frac{1}{2M} \sum_{i=1}^{M} \sum_{\substack{k=1 \\ k \ne i}}^{M} erfc\left\{ \sqrt{\eta}\left(\frac{d_{ik}}{2\sqrt{(1/M)\,\Sigma_{i=1}^{M}|\,\mathbf{s}_i|^2}}\right)\right\},$$

where $\eta = \mathcal{P}T/N_0$.

*(b) Consider the four signal sets of Fig. P4.3 and compare their performance by using the expression found in part (a) and assuming the same value of $\eta = 10$ dB.

4.11. (a) Derive the expression (4.64) for the probability of correct decision in the case of orthogonal signals with coherent demodulation. *Hint:* Observe that

$$P(c\,|\mathbf{s}_1) = P\{r_1 > \max(r_2, \ldots, r_M)|\,\mathbf{s}_1\}.$$

(b) Define a configuration of *biorthogonal signals* as a set of $M = 2N$ signals in an N-dimensional signal space obtained from an original orthogonal signal set by adding to each signal $\mathbf{s}_i$ its opposite $-\mathbf{s}_i$. The case of $N = 2$ is shown in Fig. P4.4. Paralleling the calculations that lead to (4.64), show that for the biorthogonal signals we have

$$P(c\,|\mathbf{s}_i) = \frac{1}{\sqrt{\pi N_0}} \int_0^\infty \exp\left[-\frac{(\alpha - \sqrt{\mathcal{E}})^2}{N_0}\right]\left\{erf\left(\frac{\alpha}{\sqrt{N_0}}\right)\right\}^{(M/2)-1} d\alpha.$$

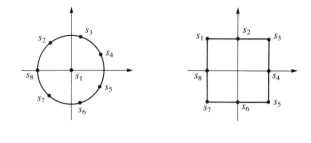

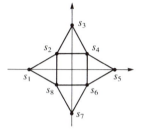
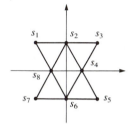

Figure P4.3

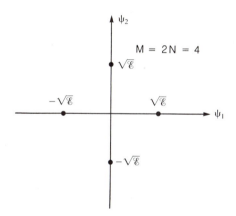

$$M = 2N = 4$$

Figure P4.4

4.12. Given an orthogonal signal set of M equally likely signals of energy $\mathscr{E}$, show that the condition (4.34) for minimum energy requires a translation of the origin given by the vector

$$\mathbf{a} = \frac{1}{M} \sum_{i=1}^{M} \mathbf{s}_i.$$

The resulting set of signals $\{\mathbf{s}_i - \mathbf{a}\}_{i=1}^{M}$ is called a *simplex*. Show that the simplex set has the same error probability as the original one, but requires an average energy $\mathscr{E}[1 - (1/M)]$.

4.13. (Wozencraft and Jacobs, 1965) Assume that a set of M equal-energy signals satisfies the condition

$$\mathbf{s}_i \mathbf{s}_j' = \begin{cases} \mathscr{E}, & i = j, \\ \rho\mathscr{E}, & i \neq j. \end{cases}$$

Chap. 4   Problems

These signals are said to be *equally correlated.*

(a) Prove that

$$-\frac{1}{M-1} \leq \rho \leq 1.$$

Hint: Consider $|\Sigma_{i=1}^{M} \mathbf{s}_i|^2$.

(b) Verify that the minimum value of ρ is achieved by a simplex set (see Problem 4.12).

(c) Prove that, for any ρ, the signal set has the same error probability as the simplex signal set with energy

$$\mathscr{E}_s = \mathscr{E}\left(1 - \frac{1}{M}\right)(1 - \rho).$$

Hint: Consider the set $\{\mathbf{s}_i - \mathbf{a}\}$, with $\mathbf{a} = (1/M) \Sigma_{i=1}^{M} \mathbf{s}_i$.

(d) Verify, as a consequence of part (c), that the signal set has the same error probability as an orthogonal signal set with energy

$$\mathscr{E}_o = \mathscr{E}(1 - \rho).$$

4.14. Using the definition (4.116), prove that the Gaussian RVs n_i satisfy the properties (4.117).

4.15. In this problem we want to derive the result (4.119). Define the two RVs of (4.115) as follows:

$$z_1 = \sqrt{\mathscr{E}} + n_1, \quad z_2 = \rho\sqrt{\mathscr{E}} + n_2$$

and apply the following linear transformation:

$$t_1 = \frac{1}{2}\{z_1(1 + K) + z_2(1 - K)\, e^{-j\phi}\},$$

$$t_2 = \frac{1}{2}\{z_1(1 - K) + z_2(1 + K)\, e^{-j\phi}\},$$

where

$$K = \sqrt{\frac{1 + |\rho|}{1 - |\rho|}}, \quad \cos\phi = \frac{\mathscr{R}(\rho)}{|\rho|}, \quad \sin\phi = \frac{\mathscr{I}(\rho)}{|\rho|}.$$

(a) Show that the two RVs t_1 and t_2 are Gaussian and independent.

(b) Show that $R_1 = |t_1|$ and $R_2 = |t_2|$ are independently distributed Rician variables with pdf given by

$$f_{R_i}(r) = \frac{r}{\sigma_i^2}\exp\left\{-\frac{a_i^2 + r^2}{2\sigma_i^2}\right\} I_0\left(\frac{a_i r}{\sigma_i^2}\right)$$

with $0 \leq r < \infty$, $i = 1, 2$, and

$$a_i \stackrel{\triangle}{=} |\mathrm{E}\{t_i\}|, \quad \sigma_i^2 \stackrel{\triangle}{=} \frac{1}{2}\mathrm{E}\{|t_i - \mathrm{F}\{t_i\}|^2\}.$$

(c) Show that (4.118) can be rewritten as

$$P(e) = P\{|z_2|^2 > |z_1|^2\} = P\{|t_2|^2 > |t_1|^2\} = P\{R_2 > R_1\}.$$

Hint: Write $|t_i|^2 = t_i^* t_i$.

(d) Use the results of Appendix A and get

$$P(e) = Q(\sqrt{a}, \sqrt{b}) - \frac{v^2}{1 + v^2} \exp\left(-\frac{a+b}{2}\right) I_0(\sqrt{ab}),$$

where

$$a \triangleq \frac{a_2^2}{\sigma_1^2 + \sigma_2^2}, \qquad b \triangleq \frac{a_1^2}{\sigma_1^2 + \sigma_2^2}, \qquad v^2 \triangleq \frac{\sigma_1^2}{\sigma_2^2}$$

(e) Use finally the definitions of part (b) and show that in our case

$$a = \frac{\mathcal{E}}{2N_0}(1 - \sqrt{1 - |\rho|^2}), \qquad b = \frac{\mathcal{E}}{2N_0}(1 + \sqrt{1 - |\rho|^2}), \qquad v^2 = 1.$$

4.16. Assume an M-ary modulation scheme with signals given by

$$s_i(t) = A_i \sqrt{\frac{2}{T}} \cos 2\pi f_0 t, \qquad 0 \leq t < T, \qquad i = 1, 2, \ldots, M,$$

and $A_i = (i - 1)d$. Consider an incoherent envelope demodulator that uses the following nonoptimum thresholds

$$b_1 = 0,$$

$$b_i = \left(i - \frac{3}{2}\right)d \sqrt{\frac{2}{N_0}}, \qquad i = 2, 3, \ldots, M$$

$$b_{M+1} = \infty.$$

(a) Extending to this case the analysis that led to (4.101), show that

$$P(c \mid s_i) = Q\left[(i-1)d \sqrt{\frac{2}{N_0}}, b_i\right]$$

$$- Q\left[(i-1)d \sqrt{\frac{2}{N_0}}, b_{i+1}\right], \qquad i = 1, 2, \ldots, M.$$

(b) Show that, since $\mathcal{E}$ is the average energy of the signal set, we have

$$d^2 = \frac{6\mathcal{E}}{(M-1)(2M-1)}.$$

Hint: Use the relation

$$\sum_{i=1}^{n} (i - 1)^2 = \frac{n(n-1)(2n-1)}{6}.$$

*(c) Obtain curves of the error probability

$$P(e) = 1 - \frac{1}{M}\sum_{i=1}^{M} P(c \mid s_i)$$

as a function of the ratio $\mathcal{E}/N_0$.

4.17. Consider a CPM scheme with the following parameters:

number of signals $\qquad M = 2$

modulation index $\qquad h = 0.5$

correlation length $\qquad L = 3$

Chap. 4   Problems   **193**

$$\text{frequency pulse } g(t) = \begin{cases} \dfrac{1}{6T}, & 0 \leqslant t < 3T, \\ 0, & \text{elsewhere.} \end{cases}$$

(a) Paralleling the development of Example 4.11, draw the phase tree, the phase branches, and the state trellis of the modulator.

(b) Derive the phase difference tree and verify that there is a first merge at $t = 4T$. Obtain the value of d_B^2 for this merge.

4.18. Consider a CPM scheme with the following parameters

$$\begin{aligned}
&\text{number of signals} && M = 2 \\
&\text{modulation index} && h \\
&\text{correlation length} && L = 2
\end{aligned}$$

frequency pulse (raised cosine pulse):

$$g(t) = \begin{cases} \dfrac{1}{4T}\left(1 - \cos \dfrac{\pi t}{T}\right), & 0 \leqslant t < 2T, \\ 0, & \text{elsewhere.} \end{cases}$$

(a) Draw the phase tree, the phase branches, and the trellis of the modulator.

(b) Verify that the two sequences $(1, -1, 1)$ and $(-1, 1, 1)$ have a merge at $t = 3T$.

(c) Show that

$$\frac{d_B^2}{2\mathscr{E}} = 3 - 2\frac{1}{T}\int_0^T \cos\left\{ h\pi\frac{t}{T} - h\sin\left(\frac{\pi t}{T}\right)\right\} dt$$

$$-\frac{1}{T}\int_0^T \cos\left\{ 2h\sin\left(\frac{\pi t}{T}\right) + h\pi\right\} dt.$$

(d) Verify that, for $h = 0.5$, $d_B^2 = 1.97$.

CHAPTER 5

Digital Modulation Schemes

This chapter is devoted to the study of certain important classes of digital modulation schemes used in practical applications. The material developed in Chapter 4 will be extensively used to derive the results contained in this chapter. Transmission over an additive white Gaussian noise channel is assumed. The effects of other impairments, besides Gaussian noise, will be considered in Chapter 6.

The aim of this chapter is twofold. First, the digital modulations are considered from the viewpoint of practical applications. Second, we assess how each modulation scheme uses the available resources, that is, power, bandwidth, and complexity, to achieve a preassigned performance objective in terms of error probability. Several constraints and theoretical limitations generate conflicts among the designer's desired goals. Therefore, the whole conceptual framework of the chapter aims at clarifying the trade-offs that are fundamental to the choice of the "best" modulation design.

5.1 INTRODUCTION

As in Chapter 4, we assume transmission over an additive white Gaussian noise (AWGN) channel with a two-sided noise power spectral density $N_0/2$. We denote by $\mathscr{P}$ the average power of the transmitted digital signals, and for simplicity we assume no attenuation on the channel. This means that $\mathscr{P}$ is also the average power of the signal observed at the receiver input. In any case, we must remember that the receiver performance depends on its input power. Therefore, the transmission loss, which is always encountered in practice, must be taken into account when designing a system. It is common engineering practice to define a *signal bandwidth W* and an average *signal-to-noise power ratio* $\mathcal{P}/N_0 W$. Both quantities require some additional considerations.

On the definition of bandwidth

It should not be surprising that there is no unique definition of signal bandwidth. Actually, digital signals strictly time limited to an interval T would have an infinite bandwidth. Conversely, bandlimited signals cannot be constrained into a signaling interval T. The solution to this dilemma will be discussed in Chapter 7. Here we evaluate the power spectral density of the digital signal under consideration and define a width W of this spectrum. To be more precise, let us consider a bandpass linearly modulated digital signal given by

$$v_\xi(t) = \mathcal{R}\left[\sum_{k=-\infty}^{\infty} \xi_k s(t - kT) \, e^{j2\pi f_0 t} \right]. \tag{5.1}$$

Let us assume a stationary sequence (ξ_n) of RVs with zero mean and unit variance, and a rectangular waveform $s(t) = u_T(t)$ of unit amplitude and duration T. Using the results of Example 2.17, we get for the power density spectrum the expression

$$\mathcal{G}_{v_\xi}(f) = \frac{1}{4}[\mathcal{G}(-f-f_0) + \mathcal{G}(f-f_0)], \tag{5.2}$$

where

$$\mathcal{G}(f) = T\left[\frac{\sin \pi f T}{\pi f T}\right]^2.$$

The spectrum $\mathcal{G}(f)$ is shown in Fig. 5.1. It consists of a main lobe and smaller sidelobes, as is generally valid for most digitally modulated signals. It justifies the practical importance of the bandwidth definitions depicted in Fig. 5.1. Recall that the bandwidth of the real signal is defined only for positive frequencies of its spectrum.

(a) *Half-power bandwidth.* It is the interval between the two frequencies at which the power spectrum is 3 dB below its peak value. This definition is a simple and crude indication of the degree of dispersion of the spectrum.

(b) *Equivalent noise bandwidth.* It has been defined in (2.94) for a linear system and represents a more refined measure of the general dispersion of the spectrum. The area of the rectangle in Fig. 5.1 must equal the total signal power.

(c) *Null-to-null bandwidth.* This measure shifts the attention to the spectral sidelobes shape. It represents the width of the main spectral lobe. This definition is probably one of the simplest and most popular measures of the signal bandwidth. Clearly, the implicit assumption is that the spectrum possesses a main lobe containing most of the signal power. Therefore, this mode of defining the bandwidth roughly measures also the spread of the spectrum.

(d) *Fractional power containment bandwidth.* This bandwidth definition states that the occupied bandwidth is the band that leaves outside a given fraction ϵ of the signal power. Thus, this bandwidth contains $(1 - \epsilon)$ of the total signal power.

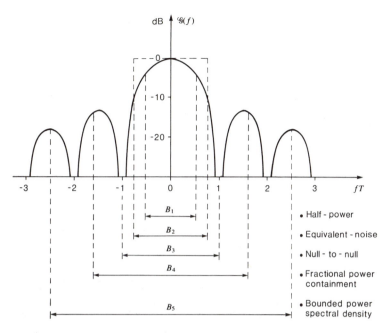

Figure 5.1 Different bandwidth definitions of the power density spectrum of (5.2). B_1 is the half-power bandwidth; B_2 is the equivalent noise bandwidth; B_3 is the null-to-null bandwidth; B_4 is the fractional power containment bandwidth at an arbitrary level; B_5 is the bounded power spectral density at a level of about 18 dB. Notice that the depicted bandwidths are those around the frequency f_0.

(e) *Bounded power spectral density bandwidth.* This is another widely used criterion to specify the signal bandwidth. It states that everywhere outside this bandwidth the power spectral density must have fallen at least to a certain level below its peak. Typical levels might be 35 or 50 dB.

Each bandwidth definition is more appropriate for a given environment and application. For example, the bounded power spectral density bandwidth appears frequently in specifications concerning problems of adjacent channel interference (see Chapter 6). On the other hand, the noise equivalent bandwidth of the receiving filter is necessary to evaluate its output noise power. In this chapter, the comparison among the bandwidth occupancies of different modulation schemes will be based on the null-to-null bandwidth. However, note that the previous discussion cautioned the reader to use, in each case, the most appropriate bandwidth measure.

On the signal-to-noise ratio

Turning our attention to the signal-to-noise power ratio, we must note that this quantity, despite its practical meaning, is not immediately useful when comparing different modulated signals. In fact, we need to emphasize the parameters that define the source independently of the modulation scheme. The source is assumed to emit iid binary digits with rate R_s. The rate, in bits/second, that can be accepted by the modulator is given by

$$R_s = \frac{\log_2 M}{T}, \tag{5.3}$$

where M is the number of signals of duration T available at the modulator, and $1/T$ is the signaling rate. Therefore, the average signal power can be expressed as

$$\mathcal{P} = \frac{\mathcal{E}}{T} = \mathcal{E}_b R_s, \tag{5.4}$$

where $\mathcal{E}$ is the average signal energy and $\mathcal{E}_b = \mathcal{E}/\log_2 M$ is the energy required to transmit 1 bit. As a consequence, we get

$$\frac{\mathcal{P}}{N_0 W} = \frac{\mathcal{E}_b}{N_0} \frac{R_s}{W}. \tag{5.5}$$

This expression shows that the signal-to-noise power ratio is the product of two quantities that are indeed very significant to our purposes. Specifically, the ratio $\mathcal{E}_b/N_0$ is the *energy per transmitted bit divided by twice the noise spectral density*. The results for the symbol error probability $P(e)$, derived in Chapter 4, can be easily expressed in terms of $\mathcal{E}_b/N_0$. The ratio R_s/W represents the *bandwidth efficiency* of a given transmission scheme, since it measures the bits/second that are transmitted over 1 Hz of the bandwidth W. We shall see later in the chapter how significant comparisons among different digital modulations are based on these two quantities.

On the error probability

The performance measure introduced in Chapter 4, the *symbol error probability* $P(e)$, must be slightly refined to allow comparisons among systems with different values of M. Since the user is interested in the received binary sequence, the most useful acceptance criterion is the *bit error probability* $P_b(e)$.

In summary, the evaluation of a given modulation scheme can be based on the three parameters $P_b(e)$, $\mathcal{E}_b/N_0$, and R_s/W. The first is a performance target, the second is a measure of the power expenditure, and the third is a measure of the bandwidth required for a given source rate. According to (5.5), the product of the last two gives the signal-to-noise power ratio for the specified transmission system. As we shall see, the system trade-offs in any digital transmission are compromises between these three parameters. In fact, the goals of a designer are

1. To minimize the bit error probability
2. To minimize the required power
3. To maximize the bandwidth efficiency
4. To minimize the equipment's complexity

It is quite intuitive that these are conflicting goals. This chapter is a description of transmission systems that implement different possible trade-offs among the described parameters.

5.2 PULSE AMPLITUDE MODULATION

When transmitting information over a baseband channel, we can vary the amplitude of a baseband pulse of duration T according to the source symbols. This modulation scheme is referred to as *pulse amplitude modulation (PAM)*. A sequence ξ of K RVs is then represented by the signal

$$v_\xi(t) = \sum_{k=0}^{K-1} \xi_k s(t - kT), \qquad 0 \leq t < KT, \tag{5.6}$$

where the RV ξ_k takes values in the set of amplitudes $\{a_i\}_{i=1}^M$ given by

$$a_i = (2i - 1 - M)\frac{d}{2}, \qquad i = 1, 2, \ldots, M, \tag{5.7}$$

and $M = 2^h$ is the number of possible sequences of h binary digits emitted by the source. The transmitter uses a set of waveforms $\{s_i(t)\}_{i=1}^M = \{a_i s(t)\}_{i=1}^M$ and, if $s(t)$ is assumed to be a unit-energy pulse, we can easily apply the concepts of Section 2.5 to obtain the geometrical representation of the signal set. This is shown in Fig. 5.2 for the cases where $M = 4$, $M = 8$ and $d = 2$.

The transmission scheme just described can also be used over a bandpass channel centered around a frequency f_0. In this case, a sequence ξ of K RVs is associated with the signal

$$
\begin{aligned}
v_\xi(t) &= \mathcal{R}\left[\sum_{k=0}^{K-1} \xi_k \tilde{s}(t - kT)e^{j2\pi f_0 t}\right] \\
&= \sum_{k=0}^{K-1} \xi_k \mathcal{R}[\tilde{s}(t - kT)e^{j2\pi f_0 t}], \qquad 0 \leq t < KT.
\end{aligned}
\tag{5.8}
$$

The usual name for this modulation scheme is *amplitude-shift-keying* (ASK), even if ASK and PAM are in practice synonyms. The geometric representation of the signal sets used in (5.8) is again that of Fig. 5.2. The most important practical case is the coherent demodulation of the signals (5.6) and (5.8). The theory was developed in Section 4.2. The fact that the signal sets use only one basic waveform $s(t)$ allows a

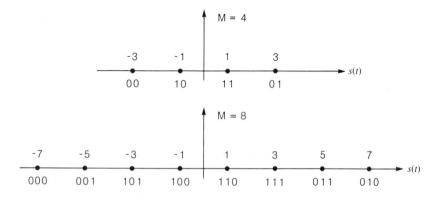

Figure 5.2 Geometrical representation of PAM signal sets.

Sec. 5.2   Pulse Amplitude Modulation   **199**

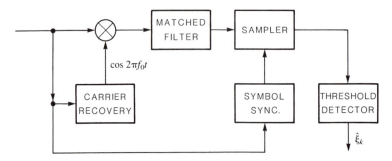

$\cos 2\pi f_0 t$

Figure 5.3 Block diagram of the optimum PAM receiver.

strong simplification of the receiver structure of Figs. 4.4 or 4.5. Only one matched filter or correlator is required, followed by a threshold detector. The block diagram of the optimum ASK receiver is shown in Fig. 5.3. The threshold detector compares the sample of the received signal with the M possible transmitted levels and makes a decision favoring the closest (minimum distance receiver). The diagram of Fig. 5.3 shows some additional blocks requiring comments. The carrier recovery block and the symbol synchronizer are assumed to be ideal, in the sense that they provide a perfect reference carrier and sampling clock. Chapter 6 provides additional insight into these blocks.

The symbol error probability for the PAM signal set with coherent demodulation can be evaluated through the results of Chapter 4 for rectangular signal sets. Using the calculations of Example 4.7 as a guideline, we get

$$P(e) = \frac{M-1}{M} \, erfc\left(\frac{d}{2\sqrt{N_0}}\right). \tag{5.9}$$

To express $P(e)$ as a function of the parameter $\mathcal{E}_b/N_0$, notice that the average energy of the signals is

$$\mathcal{E} = \mathrm{E}\{\xi_k^2\} = \frac{1}{M}\sum_{i=1}^{M}(2i-1-M)^2 \frac{d^2}{4} = \frac{M^2-1}{3}\frac{d^2}{4}. \tag{5.10}$$

Moreover, noticing that $\mathcal{E} = \mathcal{E}_b \log_2 M$, we have, from (5.9),

$$P(e) = \frac{M-1}{M} \, erfc\left(\sqrt{\frac{3\log_2 M}{M^2-1}\frac{\mathcal{E}_b}{N_0}}\right). \tag{5.11}$$

This symbol error probability is plotted in Fig. 5.4. Note that, as M increases, an increase of $\mathcal{E}_b/N_0$ is required to keep the error probability constant. For $M = 4$, the required $\mathcal{E}_b/N_0$ is approximately 4 dB higher than for $M = 2$. For very large M, the increase in $\mathcal{E}_b/N_0$ approaches 6 dB when M is doubled.

As stated previously, we are more interested in the bit error probability $P_b(e)$ than in $P(e)$. The relationship between the two depends on the mapping of the binary sequences onto the set of M transmitted symbols. No simple general result can be given. Instead, simple bounds can be obtained by observing that one symbol error causes at least one and at most $\log_2 M$ binary errors, and therefore

$$\frac{P(e)}{\log_2 M} \le P_b(e) \le P(e). \tag{5.12}$$

200   Digital Modulation Schemes   Chap. 5

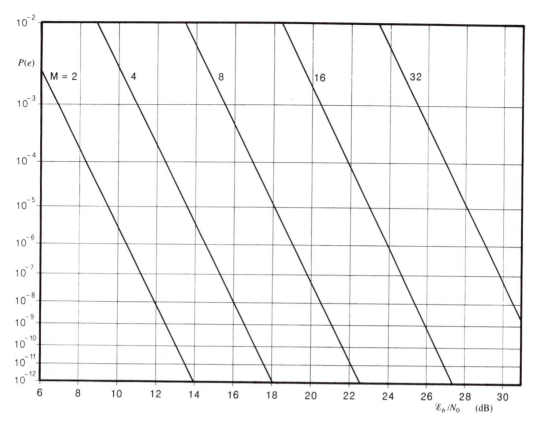

Figure 5.4 Symbol error probability for M-ary PAM (or ASK) signals.

One important practical case refers to the *Gray code* in which adjacent signal points correspond to binary sequences that differ in only one digit. Figure 5.2 shows the application of this code to PAM signals. When a symbol error occurs with this code, and provided that $\mathcal{E}_b/N_0$ is high, it is very likely that instead of the correct signal one of the adjacent signals is chosen. In this case, only one binary digit is in error. Consequently, the bit error probability is well approximated by the lower bound of (5.12).

In PAM modulation the average energy of the transmitted signal is different from the *peak energy* $\mathcal{E}_p$, which is the energy of the maximum amplitude level signal. Sometimes there are design constraints on the peak power of the transmitter. Therefore, it can be useful to relate the average and the peak energy of the transmitted signals. Since $\mathcal{E}_p = (M-1)^2 \dfrac{d^2}{4}$, from (5.10) we can get

$$\frac{\mathcal{E}_p}{\mathcal{E}} = \frac{3(M-1)}{M+1}.\tag{5.13}$$

Using (5.13), the expression (5.11) for $P(e)$ can be simply written in terms of $\mathcal{E}_{bp}/N_0$, the peak energy per bit over the noise spectral density ratio. For example, for $M = 4$

Sec. 5.2   Pulse Amplitude Modulation

201

the same symbol error probability is obtained with a value of $\mathscr{E}_{bp}/N_0$, which is 2.55 dB higher than $\mathscr{E}_b/N_0$.

5.2.1 Bandwidth Requirements of PAM Signals

The bandwidth (whichever definition) of the transmitted signal depends on the choice of the baseband pulse $s(t)$. In fact, the power density spectrum is, from (2.127) and (5.10),

$$\mathscr{G}_{v_\xi}(f) = \frac{\mathscr{E}}{T}|S(f)|^2,$$

where $S(f)$ is the Fourier transform of $s(t)$. Notice that here and in the following, when dealing with power spectral densities, the expression of the modulated signal, as given in (5.6), must have the summation extended from $-\infty$ to $+\infty$ so as to render the signal cyclostationary.

A detailed discussion on the choice of $s(t)$ can be found in Chapter 7, and examples of spectral shapings are illustrated in Section 9.8. Here we are interested in defining the minimum bandwidth required for transmission in an ideal environment. To this purpose, we recall the sampling expansion of bandlimited signals of (2.264) to conclude that the required channel bandwidth W for the case of baseband transmission is $1/2T$.[†] When bandpass transmission is used, the same bandwidth can be maintained either using *single-sideband* (SSB) transmission or *quadrature-amplitude modulation* (QAM), in which the information sequence is split into two parallel subsequences, each transmitted on one of the two quadrature carriers $\cos 2\pi f_0 t$ and $\sin 2\pi f_0 t$. Therefore, the bandwidth efficiency for this modulation system is given by

$$\frac{R_s}{W} = \frac{\log_2 M}{T} \cdot 2T = 2\log_2 M. \tag{5.14}$$

In practice, however, it is difficult to achieve this efficiency due to the practical difficulty of dealing with signals of minimum bandwidth. If a baseband $s(t)$ of a rectangular type is used for transmission, the null-to-null bandwidth is $1/T$ for baseband transmission. This can also be maintained, as said before, for bandpass transmission. In conclusion,

$$\log_2 M \leq \frac{R_s}{W} \leq 2\log_2 M, \tag{5.15}$$

and the values obtained in real systems usually lie between these bounds.

Some conclusions on this type of modulation can be drawn by considering the results expressed in (5.11) and (5.15). The latter shows that the bandwidth required for a fixed source rate R_s decreases with an increase in the number of levels M. However, the former indicates that this improvement in bandwidth efficiency is penalized by the ratio $\mathscr{E}_b/N_0$ required to achieve a specified performance level of error probability.

We conclude this section with the observation that amplitude-modulated digital

---

[†] Notice that in these circumstances the transmitted signal $s(t) = (\sin \pi t/T)/(\pi t/T)$ is not time limited to the interval $(0, T)$. Nevertheless, there is no intersymbol interference because the Nyquist conditions are fulfilled (see Section 7.2).

signals can also be detected by an incoherent demodulator. The reader can find a description of this system in Section 4.3 for the binary case and in Problem 4.16 for the multilevel case.

5.3 PHASE-SHIFT KEYING

When the source-emitted symbols are used to change the phase of a bandpass signal, we have a *digital phase modulation* in which a sequence of K messages is represented by the signal

$$v_\xi(t) = \mathcal{R}\left\{ \sum_{k=0}^{K-1} \xi_k s(t - kT) e^{j2\pi f_0 t} \right\}, \qquad 0 \le t < KT, \tag{5.16}$$

where $\xi_k = e^{j\varphi_k}$, φ_k takes values in the set

$$\left\{ \frac{2\pi}{M}(i - 1) + \Phi \right\}_{i=1}^{M}, \tag{5.17}$$

and Φ is an arbitrary constant phase. When $s(t)$ is a rectangular pulse of amplitude A, that is $s(t) = A u_T(t)$, the modulation is called *phase-shift keying* (PSK), and the signal of (5.16) becomes

$$v_\xi(t) = A \sum_{k=0}^{K-1} u_T(t - kT) \cos(2\pi f_0 t + \varphi_k) = I(t) \cos 2\pi f_0 t - Q(t) \sin 2\pi f_0 t, \tag{5.18}$$

where

$$I(t) \triangleq A \sum_{k=0}^{K-1} \cos \varphi_k u_T(t - kT),$$

$$Q(t) \triangleq A \sum_{k=0}^{K-1} \sin \varphi_k u_T(t - kT).$$

Therefore, the transmitter uses a set of waveforms of duration T and equal energy $\mathscr{E} = A^2 T/2$. The geometric representation of this signal set (see Example 2.24) is shown, for some cases, in Fig. 5.5. The signals lie on a circle. By inspection of this figure it should be intuitive that PSK signals share some general features with PAM. In fact, as M increases ($M > 2$), the dimensionality of the signal space does not change, and this means a bandwidth efficiency increasing with M. On the other hand, the radius of the circle depends on the signal energy. Therefore, when M is increased, the signals get closer if the energy is kept constant. Consequently, the same error performance can be maintained only with an increase of the transmitted energy. These considerations will be verified quantitatively in the forthcoming analysis.

We want now to describe techniques for the generation of the signal sets of Fig. 5.5. First, notice that the source binary symbols are Gray coded into each signal set. As a consequence, adjacent phase signals differ by only one binary digit. This feature will be significant when considering the demodulator performance. The simplest case is the binary PSK (BPSK) modulator. It can be verified immediately that the carrier

Sec. 5.3   Phase-Shift Keying

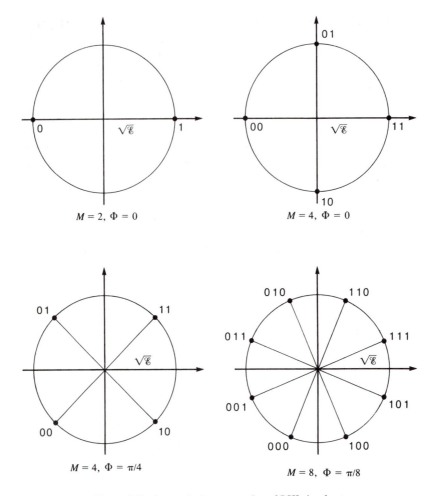

$M = 2, \Phi = 0$

$M = 4, \Phi = 0$

$M = 4, \Phi = \pi/4$

$M = 8, \Phi = \pi/8$

Figure 5.5 Geometrical representation of PSK signal sets.

phase change corresponds to the sign change of its amplitude. Figure 5.6 shows a BPSK modulator. Note that the time-domain multiplication of Fig. 5.6 is equivalent to a double-sideband amplitude modulation (DSB/AM). Therefore, the BPSK signals are equivalent to the binary PAM signals transmitted with DSB modulation.

A quaternary PSK (QPSK) modulator takes two input bits at a time and produces one of the four possible phases of the carrier $A \cos 2\pi f_0 t$. Therefore, a signaling rate reduction of one-half is achieved. The modulated signal consists of the summation of two quadrature sinusoids whose amplitude signs correspond to the transmitted data. The QPSK modulator is shown in Fig. 5.7. This modulator can be considered as made up of two BPSK modulators whose outputs are added together. The two quadrature channels are usually denoted as the *I channel* (in-phase channel) and the *Q channel* (quadrature channel), respectively.

An octonary PSK modulator takes three input bits at a time and produces one of eight possible phases of the carrier $A \cos 2\pi f_0 t$. A signaling rate reduction to one-third

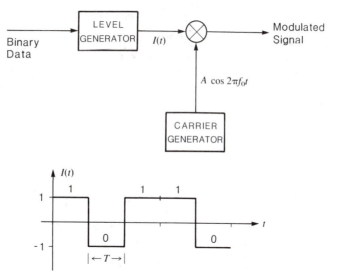

Figure 5.6 Binary PSK modulator.

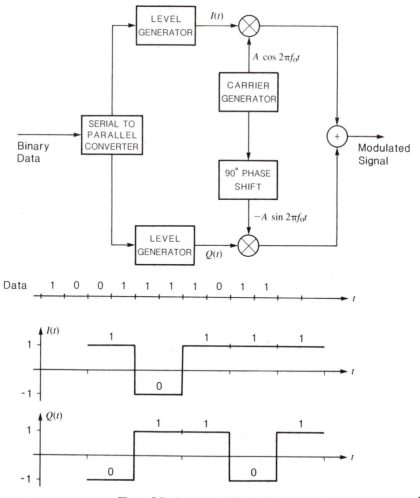

Figure 5.7 Quaternary PSK modulator.

is achieved. The modulator can be viewed as an extension of the QPSK modulator of Fig. 5.7. Two baseband signals with four symmetrical levels are generated for the I and Q channels, respectively. This entails the choice of $\Phi = \pi/8$ for the initial phase in (5.17). Each triplet of source bits is used as follows: the first determines the sign of the level of the I channel, the second the sign of the level of the Q channel, and the third determines for both channels whether the higher or lower level should be generated. The block diagram of the modulator is shown in Fig. 5.8.

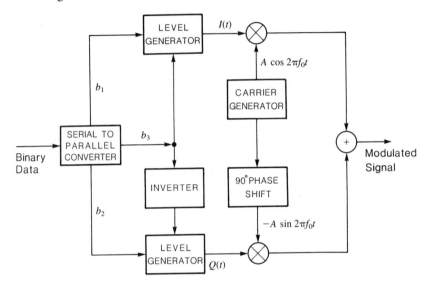

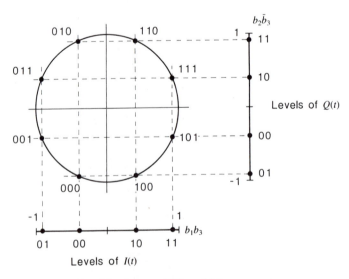

Figure 5.8 Octonary PSK modulator.

Concluding this subject, modern technology is oriented toward the use of completely digital devices. In such an environment, multilevel PSK signals are digitally synthesized and fed to an analog bandpass filter whose output is the desired modulated signal.

5.3.1 Coherent Demodulation of PSK Signals

A coherent demodulation of PSK signals is performed when the receiver has an exact knowledge of the carrier phase. CPSK is the usual abbreviation for such signaling. As already noticed, the binary case is equivalent to a binary PAM tranmission. Therefore, the optimum receiver is that shown in Fig. 5.3, and the symbol error probability is given by (5.11) with $M = 2$. For $M > 2$, the most convenient optimum receiver is obtained by adapting the general block diagram of Fig. 4.6. In fact, PSK signals are always represented in a two-dimensional signal space, and therefore only two matched filters are needed to derive the sufficient statistics. The simplest case is for $M = 4$. The corresponding coherent demodulator is shown in Fig. 5.9. In the general case, the decisions cannot be taken independently on the two recovered baseband signals. The two binary thresholds are replaced by multilevel detectors whose outputs are combined into a logic device that recovers the digital binary stream.

The symbol error probability in the quaternary system is easily computed, as it represents a simple case of rectangular decision regions. Using as a guideline the computations of Example 4.7, we can get

$$P(e) = 1 - (1 - p)^2 = 2p - p^2 \qquad (5.19)$$

with

$$p \triangleq \frac{1}{2} \, erfc \left(\sqrt{\frac{\mathscr{E}_b}{N_0}} \right).$$

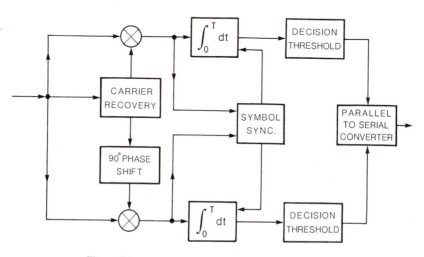

Figure 5.9 Coherent receiver for quaternary PSK signals.

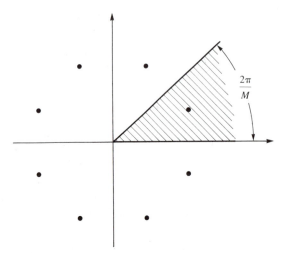

Figure 5.10 Decision region corresponding to an M-ary PSK signal set.

The computation of the symbol error probability for the general multilevel case is a straightforward application of the procedure described in Section 4.2.1. Since the PSK signal sets demonstrate complete symmetry, we can apply (4.38) and (4.35) to get

$$P(e) = 1 - \int_{R_j} f_{\mathbf{n}}(\boldsymbol{\alpha} - \mathbf{s}_j)\, d\boldsymbol{\alpha}, \qquad \text{any } j, \tag{5.20}$$

where R_j is the decision region of the signal vector $\mathbf{s}_j$ shown in Fig. 5.10. Using the minimum-distance concept, this decision region is found to be a sector of width $2\pi/M$ centered on the signal point corresponding to $\mathbf{s}_j$. To evaluate (5.20), we must obtain the pdf $f_{\mathbf{n}}(\boldsymbol{\alpha} - \mathbf{s}_j)$. The two-dimensional signal space basis consists of the two signals $\psi_1(t) = \sqrt{2/T}\cos 2\pi f_0 t$ and $\psi_2(t) = -\sqrt{2/T}\sin 2\pi f_0 t$, $0 \le t < T$. Therefore,

$$f_{\mathbf{n}}(\boldsymbol{\alpha} - \mathbf{s}_j) = \frac{1}{\pi N_0}\exp\left\{ -\frac{1}{N_0}\left[(\alpha_1 - \sqrt{\mathscr{E}}\cos\varphi_j)^2 + (\alpha_2 - \sqrt{\mathscr{E}}\sin\varphi_j)^2 \right] \right\}, \tag{5.21}$$

where α_1 and α_2 are the components of $\boldsymbol{\alpha}$, and φ_j is the phase of $\mathbf{s}_j$. The geometric properties of the problem to be solved (integration in a circular sector of the plane) suggest the transformation of the pdf (5.21) using polar coordinates. Let

$$\alpha_1 = \rho\sqrt{\frac{N_0}{2}}\cos\beta,$$
$$\alpha_2 = \rho\sqrt{\frac{N_0}{2}}\sin\beta. \tag{5.22}$$

Then, introducing (5.22) into (5.21), we get, with a slight abuse of notation,

$$f_{\mathbf{n}}(\rho, \beta - \varphi_j) = \frac{\rho}{2\pi}\exp\left\{ -\frac{1}{2}\left[\rho^2 - 2\rho\sqrt{\frac{2\mathscr{E}}{N_0}}\cos(\beta - \varphi_j) + 2\frac{\mathscr{E}}{N_0} \right] \right\}. \tag{5.23}$$

Let us define the phase $\theta \triangleq \beta - \varphi_j$. This represents the phase displacement of the received signal from the transmitted one. Then, integrating both sides of (5.23) over ρ yields the pdf of θ:

$$f_\theta(x) = \frac{e^{-\mathcal{E}/N_0}}{2\pi}\left\{1 + \sqrt{\frac{\pi\mathcal{E}}{N_0}}\cos x \, e^{(\mathcal{E}/N_0)\cos^2 x}\left[1 + erf\left(\sqrt{\frac{\mathcal{E}}{N_0}}\cos x\right)\right]\right\}, \qquad (5.24)$$

$$-\pi \leqslant x \leqslant \pi.$$

An error is made if, for any j, the noise causes a phase displacement greater than π/M in absolute value, corresponding to a received phase lying outside the jth decision region. Therefore,

$$P(e) = 1 - \int_{-\pi/M}^{\pi/M} f_\theta(x)\, dx. \qquad (5.25)$$

This expression cannot be evaluated in a closed form. For $M \geqslant 4$ and $\mathcal{E}/N_0 \gg 1$, we can resort to some asymptotic approximation of (5.24). In fact, the range of values assumed by $\cos x$ in (5.25) depends on M. More precisely, as M increases, $\cos x$ spans a small interval around 1. Under these hypotheses, we can use the following approximation:

$$erf(x) \sim 1 - \frac{e^{-x^2}}{\sqrt{\pi}x}, \qquad x \gg 1. \qquad (5.26)$$

Using (5.26) in (5.24), we get

$$f_\theta(x) \sim \sqrt{\frac{\mathcal{E}}{\pi N_0}}\cos x \, e^{-(\mathcal{E}/N_0)\sin^2 x}. \qquad (5.27)$$

Finally, introducing (5.27) into (5.25), we arrive at the result

$$P(e) \sim erfc\left(\sqrt{\frac{\mathcal{E}}{N_0}}\sin\frac{\pi}{M}\right) = erfc\left(\sqrt{\frac{\mathcal{E}_b}{N_0}\log_2 M}\sin\frac{\pi}{M}\right). \qquad (5.28)$$

For example, when $M = 4$ we have $P(e) \sim erfc\,(\sqrt{\mathcal{E}_b/N_0})$, which is close to the exact error probability given by (5.19). Notice that the application of the union bound (4.70) gives

$$P(e) \leqslant \frac{M-1}{2}\, erfc\left(\sqrt{\frac{\mathcal{E}}{N_0}}\sin\frac{\pi}{M}\right). \qquad (5.29)$$

Actually, simple and tighter bounds to $P(e)$ are derived by considering Fig. 5.11. The probability that the received vector $\mathbf{r}$ does not belong to the decision region R_j of the transmitted signal $\mathbf{s}_j$ is bounded below and above as follows:

$$P\{\mathbf{r} \in S_1\} \leqslant P(e) < P\{\mathbf{r} \in S_1\} + P\{\mathbf{r} \in S_2\}. \qquad (5.30)$$

But the received vector $\mathbf{r}$ will belong to a half-plane region as S_1 or S_2 if and only if the noise vector component orthogonal to the boundary line is greater than the distance of $\mathbf{s}$ from that boundary (i.e., $\sqrt{\mathcal{E}}\sin \pi/M$). Therefore,

$$P\{\mathbf{r} \in S_1\} = P\{\mathbf{r} \in S_2\} = \frac{1}{2}\, erfc\left(\sqrt{\frac{\mathcal{E}}{N_0}}\sin\frac{\pi}{M}\right). \qquad (5.31)$$

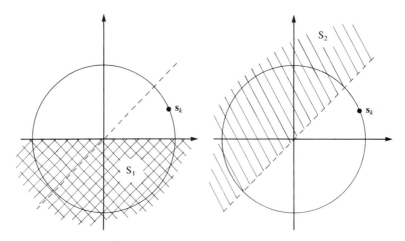

Figure 5.11 Decision regions for bounding $P(e)$ in PSK signals.

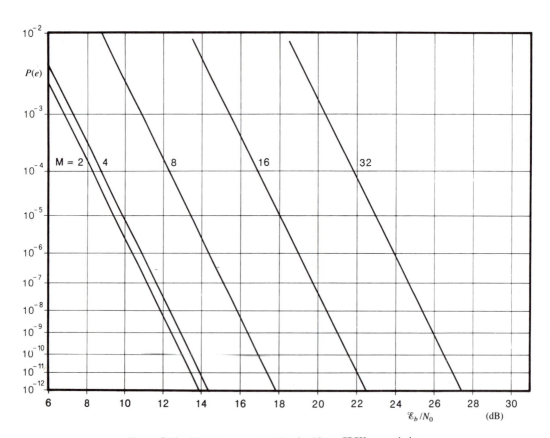

Figure 5.12 Symbol error probability for M-ary CPSK transmission.

Inserting (5.31) into (5.30), we finally obtain

$$\frac{1}{2} erfc \left(\sqrt{\frac{\mathcal{E}}{N_0}} \sin \frac{\pi}{M} \right) \le P(e) < erfc \left(\sqrt{\frac{\mathcal{E}}{N_0}} \sin \frac{\pi}{M} \right). \tag{5.32}$$

Notice that the upper bound in (5.32) is the same as the approximate result (5.28). Recalling the derivation of the latter, we can conclude that the upper bound (5.32) is tight for high signal-to-noise ratios $\mathcal{E}/N_0$.

Curves of the symbol error probability (5.28) are given in Fig. 5.12 for different values of M. The bit error probability $P_b(e)$ for Gray-coded M-ary CPSK signals is well approximated by the lower bound of (5.12).

5.3.2 Differentially Encoded PSK Signals

The coherent demodulation studied in Section 5.3.1 assumes that the receiver provides a perfect carrier reference. However, in practice, the receiver does not achieve this exact phase knowledge. Instead it introduces a certain phase error composed of two parts: one determined by noise and the other by an ambiguity in carrier recovery. The former is a random process $\varphi_e(t)$, while the other is a constant error φ_a (see Chapter 6). Here we are concerned only with the latter. For example, in QPSK the received signal is raised to the fourth power to remove the phase modulation. The frequency of the resulting signal is divided by 4 to provide the reference carrier. This process introduces a phase ambiguity of an integer multiple of $\pi/2$. Notwithstanding this phase ambiguity φ_a, it is possible to transmit information by PSK provided that it is encoded into phase differences between two successive signals. The resulting PSK signals are said to be *differentially encoded*. Let us denote with φ_k the absolute phase of the signal transmitted in the k th interval and by $\hat{\varphi}_k = \varphi_k + \varphi_a$ the phase coherently demodulated in the same interval and in the absence of noise. The set of phase values (5.17) is again used to transmit the source data, but these values are now associated with the differences of the absolute phases in two adjacent intervals, not with the absolute phases themselves. The receiver takes its decision at the end of the k th signaling interval on the basis of the difference $\hat{\varphi}_k - \hat{\varphi}_{k-1} = (\varphi_k + \varphi_a) - (\varphi_{k-1} + \varphi_a) = \varphi_k - \varphi_{k-1}$. Thus the phase ambiguity φ_a can be removed.

To implement this transmission scheme, the modulator first processes the digital binary symbols to achieve differential encoding and then transmits the absolute phases. In other words, the differential encoding is implemented at the digital level. Notice that $(K + 1)$ absolute phases are required to transmit K symbols. The redundancy of one symbol is required to remove the phase ambiguity φ_a. As an example, consider the differential encoder for the binary case shown in Fig. 5.13. The present digit b_k of

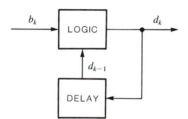

Figure 5.13 Binary differential encoder for BPSK transmission.

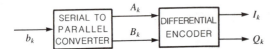

Figure 5.14 Differential encoder for QPSK transmission

the information sequence is compared with the previous digit d_{k-1} of the encoded sequence. If they are equal, then $d_k = 1$; otherwise, $d_k = 0$. The logic relation is

$$d_k = \overline{b_k \oplus d_{k-1}}, \tag{5.33}$$

where $\oplus$ means mod-2 sum and the overbar denotes complementation. Any arbitrary reference digit may be chosen as the initial one of the encoded sequence.

For the quaternary case (Fig. 5.14), the information sequence is split into the two subsequences (A_k) and (B_k), which are encoded into the sequences (I_k) and (Q_k) through the logic relations

$$\begin{aligned}
I_k &= \overline{(A_k \oplus B_k)}(A_k \oplus I_{k-1}) + (A_k \oplus B_k)(B_k \oplus Q_{k-1}), \\
Q_k &= \overline{(A_k \oplus B_k)}(B_k \oplus Q_{k-1}) + (A_k \oplus B_k)(A_k \oplus I_{k-1}).
\end{aligned} \tag{5.34}$$

The two encoded sequences are used to modulate the in-phase and quadrature signals of Fig. 5.7.

Example 5.1

Let us consider a differentially encoded BPSK signal. Equation (5.33) describes the encoder and the transmitted phases are those of the case $M = 2$, $\Phi = 0$ (Fig. 5.5). The following table describes a possible information sequence and the corresponding transmitted phases after encoding.

Information digits	b_k		0	1	1	1	0	1	0	1	1
Encoded digits	d_k	1	0	0	0	0	0	1	1	0	0	0
Transmitted absolute phases	φ_k	0	π	π	π	π	0	0	π	π	π

$\square$

Example 5.2

Consider a differentially encoded QPSK signal. Equations (5.34) describe the encoding operations, while the transmitted phases are those of the case $M = 4$, $\Phi = \pi/4$ (Fig. 5.5). A possible information sequence and the corresponding transmitted phases are shown in the following table.

Information sequence	A_k		1	0	0	1	0	1	0	1
	B_k		1	1	0	0	1	1	1	0
Encoded sequence	I_k	1	0	1	1	0	1	0	0	0
	Q_k	1	0	0	0	0	0	1	0	1
Transmitted absolute phases	φ_k	$\dfrac{\pi}{4}$	$\dfrac{5\pi}{4}$	$\dfrac{7\pi}{4}$	$\dfrac{7\pi}{4}$	$\dfrac{5\pi}{4}$	$\dfrac{7\pi}{4}$	$\dfrac{3\pi}{4}$	$\dfrac{5\pi}{4}$	$\dfrac{3\pi}{4}$

$\square$

5.3.3 Coherent Demodulation of Differentially Encoded PSK Signals

Let us assume that a coherent demodulator is used (as in Section 5.3.1) but that an M-fold phase ambiguity exists in the reference carrier. That is, φ_a takes values in the set

$$\left\{\frac{2\pi}{M} i\right\}_{i=0}^{M-1}. \tag{5.35}$$

The demodulator coherently derives the absolute phase in each interval and removes the phase ambiguity φ_a at a digital level by means of the *differential decoder*.

For the binary case, the estimate $\hat{b}_k$ of the information digit can be recovered from the estimates $\hat{d}_k$ of the encoded digits using the logic relation:

$$\hat{b}_k = \overline{\hat{d}_k \oplus \hat{d}_{k-1}}. \tag{5.36}$$

Similarly, the logic relations for the quaternary case are

$$
\begin{aligned}
\hat{A}_k &= \overline{(\hat{I}_k \oplus \hat{Q}_k)}(\hat{I}_k \oplus \hat{I}_{k-1}) + (\hat{I}_k \oplus \hat{Q}_k)(\hat{Q}_k \oplus \hat{Q}_{k-1}), \\
\hat{B}_k &= \overline{(\hat{I}_k \oplus \hat{Q}_k)}(\hat{Q}_k \oplus \hat{Q}_{k-1}) + (\hat{I}_k \oplus \hat{Q}_k)(\hat{I}_k \oplus \hat{I}_{k-1}).
\end{aligned} \tag{5.37}
$$

The following examples should clarify this point.

Example 5.3

Let us take the sequence of transmitted phases of Example 5.1. Assume a phase ambiguity $\varphi_a = \pi$ and use the decoding relation (5.36).

Transmitted phase	φ_k	0	π	π	π	π	0	0	π	π	π
Estimated phase	$\hat{\varphi}_k$	π	0	0	0	0	π	π	0	0	0
Detected digits	$\hat{d}_k$	0	1	1	1	1	0	0	1	1	1
Decoded information digits	$\hat{b}_k$		0	1	1	1	0	1	0	1	1

It can be verified that the recovered sequence is exactly the transmitted one. $\square$

Example 5.4

For the quaternary case, let us consider the sequence of transmitted phases of Example 5.2. Assume a phase ambiguity $\varphi_a = \pi/2$ and use the decoding relations (5.37).

Transmitted phase	φ_k	$\frac{\pi}{4}$	$\frac{5\pi}{4}$	$\frac{7\pi}{4}$	$\frac{7\pi}{4}$	$\frac{5\pi}{4}$	$\frac{7\pi}{4}$	$\frac{3\pi}{4}$	$\frac{5\pi}{4}$	$\frac{3\pi}{4}$
Estimated phase	$\hat{\varphi}_k$	$\frac{3\pi}{4}$	$\frac{7\pi}{4}$	$\frac{\pi}{4}$	$\frac{\pi}{4}$	$\frac{7\pi}{4}$	$\frac{\pi}{4}$	$\frac{5\pi}{4}$	$\frac{7\pi}{4}$	$\frac{5\pi}{4}$
Detected digits	$\hat{I}_k$	0	1	1	1	1	1	0	1	0
	$\hat{Q}_k$	1	0	1	1	0	1	0	0	0
Detected information digits	$\hat{A}_k$		1	0	0	1	0	1	0	1
	$\hat{B}_k$		1	1	0	0	1	1	1	0

It can be verified that the recovered sequence is exactly the transmitted one. $\square$

Sec. 5.3 Phase-Shift Keying

The modulator for this type of signal must include a differential encoder. For example, the modulator for QPSK is the same as that of Fig. 5.7 except for the serial-to-parallel converter, which includes the encoder. Similarly, the demodulator of Fig. 5.9 remains the same except for the parallel-to-serial converter, which is preceded by a differential decoder.

The evaluation of the symbol error probability for this type of signal is based on the results derived in Section 5.3.1 for the case of coherent demodulation. What is new for differentially encoded symbols is that each decision requires a pair of M-ary decisions on two successive received phases, say β_{k-1} and β_k. Let us introduce the phases ϑ_{k-1} and ϑ_k defined as

$$\vartheta_{k-1} \triangleq \beta_{k-1} - \varphi_a - \varphi_{k-1},$$
$$\vartheta_k \triangleq \beta_k - \varphi_a - \varphi_k. \tag{5.38}$$

They represent the received signal's phase displacement from the transmitted one and also account for the admitted phase ambiguity.

To evaluate the symbol error probability, we consider all possible ways of making a correct decision on the kth transmitted symbol. A correct decision is taken if and only if the two phase displacements ϑ_{k-1} and ϑ_k belong to the same circular sector of the plane. Therefore, there are M mutually exclusive events c_i that exhaust the ways of making a correct decision. These events are defined as

$$c_i \triangleq \left\{ \vartheta_{k-1}, \vartheta_k : \frac{2i-1}{M}\pi \leq \vartheta_{k-1} < \frac{2i+1}{M}\pi, \frac{2i-1}{M}\pi \leq \vartheta_k < \frac{2i+1}{M}\pi \right\}, \tag{5.39}$$

$$i = 0, 1, 2, \ldots, (M-1).$$

To evaluate the probability of these events, notice that the RVs ϑ_{k-1} and ϑ_k are statistically independent and their pdf $f_\vartheta(x)$ is given in (5.24). Therefore, we can conclude that

$$P(c_i) = p_i^2 \tag{5.40}$$

where

$$p_i \triangleq \int_{(2i-1)\pi/M}^{(2i+1)\pi/M} f_\vartheta(x)\, dx. \tag{5.41}$$

The symbol error probability is given finally by

$$P(e) = 1 - \sum_{i=0}^{M-1} p_i^2. \tag{5.42}$$

For the case $M = 4$ a pictorial interpretation of the quantities involved by (5.42) is shown in Fig. 5.15.

It is interesting to relate the symbol error probability (5.42) to that obtained in (5.25), where no differential encoding was present. Rewriting (5.25) with (5.41), we get

$$P(e)|_{CPSK} = 1 - p_0. \tag{5.43}$$

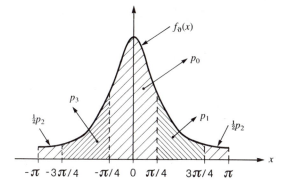

Figure 5.15 Pictorial interpretation of the quantities involved in the derivation of the error probability for DECPSK signals.

Introducing (5.43) into (5.42), we get finally the error probability for differentially encoded CPSK (DECPSK)

$$P(e)|_{\text{DECPSK}} = 2P(e)|_{\text{CPSK}}\left\{1 - \frac{1}{2}P(e)|_{\text{CPSK}} - \frac{1}{2}\frac{\sum_{i=1}^{M-1}p_i^2}{P(e)|_{\text{CPSK}}}\right\}. \tag{5.44}$$

From this result it can be shown that, when error probabilities are very low, the differential encoding simply doubles the symbol error probability. This is the price paid for removing the phase ambiguity. Moreover, it is intuitive that with DECPSK signals errors tend to appear in pairs. In fact, when a demodulated absolute phase is mistaken, it causes an error in two adjacent intervals.

Example 5.5 Binary and Quaternary DECPSK

Special simplified cases of the result (5.44) can be derived for $M = 2$ and $M = 4$. Let us start with the binary case. The only term to be considered in the summation of (5.44) is p_1, which from (5.41) becomes

$$p_1 \triangleq \int_{\pi/2}^{3\pi/2} f_\vartheta(x)\,dx = 2\int_{\pi/2}^{\pi} f_\vartheta(x)\,dx = P(e)|_{\text{CPSK}}.$$

Therefore,

$$P(e)|_{\text{DECPSK}} = 2P(e)|_{\text{CPSK}}\{1 - P(e)|_{\text{CPSK}}\}$$

$$= \left(erfc\ \sqrt{\frac{\mathscr{E}_b}{N_0}}\right)\left(1 - \frac{1}{2}erfc\ \sqrt{\frac{\mathscr{E}_b}{N_0}}\right).$$

For quaternary signals, we can consider that the two in-phase and quadrature channels are independent and therefore we can introduce the results for the binary case into (5.19) and obtain for $M = 4$:

$$P(e)|_{\text{DECPSK}} = 2erfc\ \sqrt{\frac{\mathscr{E}_b}{N_0}} - 2\left(erfc\ \sqrt{\frac{\mathscr{E}_b}{N_0}}\right)^2$$

$$+ \left(erfc\ \sqrt{\frac{\mathscr{E}_b}{N_0}}\right)^3 - \frac{1}{4}\left(erfc\ \sqrt{\frac{\mathscr{E}_b}{N_0}}\right)^4. \quad \square$$

5.3.4 Differentially Coherent Demodulation of Differentially Encoded PSK Signals

The coherent demodulation of PSK signals requires the local generation of a reference carrier. This may be undesirable either because of the amount of circuitry required to recover the carrier or in applications devoid of sufficient time for carrier acquisition. An approach that cleverly avoids the need for a reference carrier consists in accomplishing the demodulation by looking at the phases of the received signal in two successive intervals and estimating their difference. If the information digits have been differentially encoded at the transmitter, then the observed phase difference at the receiver allows the recovery of the information and the removal of the phase ambiguity. The signals are of the type (5.18), but now the information is encoded into the phase differences $\Delta\varphi_k \triangleq \varphi_k - \varphi_{k-1}$, which take values in the set

$$\left\{ \frac{2\pi}{M}(i-1) + \Phi \right\}_{i=1}^{M} \tag{5.45}$$

and Φ is either 0 or π/M.

The demodulator's block diagram, for $M = 4$, is shown in Fig. 5.16. The phase ambiguity φ_a of the received signal is removed, provided only that it remains constant. This demodulator can be proved to be optimum, in the ML sense, for the estimation of the phase differences of the received signals (see Problem 5.3).

The usual approach for the computation of the symbol error probability $P(e)$ of this demodulator is based on the probabilistic description of the received phase differences. However, the result cannot be given in a closed form. We prefer an approach based on the bounding technique described in Fig. 5.11, already used for coherent PSK signals. Notice that in this case we are dealing with phase differences, not with absolute phases. The received signal $r(t)$ in the kth time interval can be written as

$$r(t) = \Re \left\{ \tilde{r}(t)\, e^{j2\pi f_0 t} \right\} \tag{5.46}$$

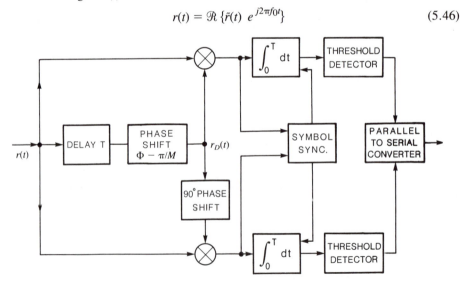

Figure 5.16 Block diagram for the quaternary DCPSK receiver.

where

$$\tilde{r}(t) = Ae^{j\varphi k} + \tilde{n}(t), \qquad kT \leq t < (k + 1)T,$$

and $\tilde{n}(t)$ is the complex envelope of a narrowband Gaussian noise as described in Section 2.4.1 and Fig. 2.20.

The delayed and shifted replica of $r(t)$ can be written as

$$r_D(t) \triangleq \mathcal{R}\{\tilde{r}(t - T)\, e^{j(\Phi - \pi/M)}\, e^{j2\pi f_0 t}\} = \mathcal{R}\{\tilde{r}_D(t)\, e^{j2\pi f_0 t}\}, \tag{5.47}$$

where

$$\tilde{r}_D(t) \triangleq Ae^{j(\varphi k - 1 + \Phi - \pi/M)} + \tilde{n}(t - T)\, e^{j(\Phi - \pi/M)}$$

and in (5.47) we have assumed that $f_0 T$ is an integer. If this were not the case, the phase shifter of Fig. 5.16 could be adjusted to eliminate the unwanted phase shift due to the carrier. The receiver bases its decision on the difference between the phases of the two signals of (5.46) and (5.47); that is,

$$\Delta\beta_k \triangleq \arg\left[\tilde{r}(t)\tilde{r}_D^*(t)\right], \qquad kT \leq t < (k + 1)T. \tag{5.48}$$

When we send a phase difference belonging to the set (5.45), the values assumed by $\Delta\beta_k$, in the absence of noise, belong to the set

$$\left\{\frac{2\pi}{M}\left(i - \frac{1}{2}\right)\right\}_{i=1}^{M}. \tag{5.49}$$

A correct decision is made if the point representing the signal $\tilde{r}(t)\tilde{r}_D^*(t)$ lies inside a sector of width $2\pi/M$ centered around the correct value of $\Delta\beta_k$.

We can notice that the problem presents complete symmetry as for the coherent PSK signals considered in Section 5.3.1. Therefore, we can compute the unconditional error probability considering the case $i = 1$ in (5.49). Using the technique based on Fig. 5.11, reproduced in Fig. 5.17 for the present case, we get for the upper bound

$$P(e) < P\{\gamma_2 < \Delta\beta_k < \gamma_2 + \pi\} + P\{\gamma_1 - \pi < \Delta\beta_k < \gamma_1\}, \tag{5.50}$$

where we have introduced the two phase thresholds

$$\gamma_1 = 0, \qquad \gamma_2 = \frac{2\pi}{M}. \tag{5.51}$$

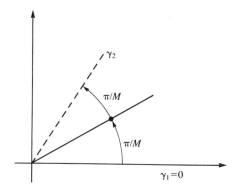

Figure 5.17 Decision regions for bounding $P(e)$ in DCPSK signals.

For computational purposes that will soon become clear, the inequality (5.50) can be given the following form:

$$P(e) < 1 - P\left\{-\frac{3\pi}{2} < \Delta\beta_k - \gamma_2 - \frac{\pi}{2} < -\frac{\pi}{2}\right\}$$

$$+ P\left\{-\frac{3\pi}{2} < \Delta\beta_k - \gamma_1 - \frac{\pi}{2} < -\frac{\pi}{2}\right\}. \tag{5.52}$$

Let us now define, at the k th decision instant $t = t_k$, $kT \le t_k < (k+1)T$, the following RVs:

$$z_1 \triangleq \tilde{r}(t_k) = Ae^{j\varphi k} + \tilde{n}(t_k),$$
$$z_2 \triangleq j\tilde{r}_D(t_k)e^{j\gamma_2} = je^{j(\Phi+\pi/M)}\{Ae^{j\varphi k-1} + \tilde{n}(t_k - T)\}, \tag{5.53}$$
$$z_3 \triangleq j\tilde{r}_D(t_k)e^{j\gamma_1} = je^{j(\Phi-\pi/M)}\{Ae^{j\varphi k-1} + \tilde{n}(t_k - T)\}.$$

Then it is immediately verified that

$$\Delta\beta_k - \frac{2\pi}{M} - \frac{\pi}{2} = \arg\left[z_1 z_2^*\right], \tag{5.54}$$

$$\Delta\beta_k - \frac{\pi}{2} = \arg\left[z_1 z_3^*\right].$$

Using (5.54), we can finally express (5.52) in the form

$$P(e) < 1 - P\{\mathcal{R}(z_1 z_2^*) < 0\} + P\{\mathcal{R}(z_1 z_3^*) < 0\}. \tag{5.55}$$

Using the identity

$$\mathcal{R}(z_i z_j^*) = \left|\frac{z_i + z_j}{2}\right|^2 - \left|\frac{z_i - z_j}{2}\right|^2 \triangleq |\xi_{ij}|^2 - |\eta_{ij}|^2 \tag{5.56}$$

into (5.55), we get

$$P(e) < 1 - P\{|\xi_{12}| < |\eta_{12}|\} + P\{|\xi_{13}| < |\eta_{13}|\}. \tag{5.57}$$

Since the four RVs of (5.57) have a Rician distribution and are independent, the computation is reduced to a classical problem in probability theory. (The same problem was encountered in Section 4.3.1 when dealing with incoherent demodulation of two equal-energy signals.) The computation's detailed steps are suggested in Problem 5.4. The result is

$$P(e) < 1 + Q\left(\sqrt{\frac{\mathcal{E}_b}{N_0}}b_M, \sqrt{\frac{\mathcal{E}_b}{N_0}}a_M\right) - Q\left(\sqrt{\frac{\mathcal{E}_b}{N_0}}a_M, \sqrt{\frac{\mathcal{E}_b}{N_0}}b_M\right), \tag{5.58}$$

where

$$a_M \triangleq \log_2 M\left(1 + \sin\frac{\pi}{M}\right),$$

$$b_M \triangleq \log_2 M \left(1 - \sin\frac{\pi}{M}\right),$$

and $Q(\cdot, \cdot)$ is the Marcum Q function (see Appendix A). For high values of $\mathcal{E}_b/N_0$, the bound (5.58) is very tight. When M also is large, we can use (A.16) and get from (5.58) the approximation

$$P(e) \sim erfc\left\{\sqrt{\frac{\mathcal{E}_b \log_2 M}{2N_0}}\left(\sqrt{1 + \sin\frac{\pi}{M}} - \sqrt{1 - \sin\frac{\pi}{M}}\right)\right\}$$

$$\sim erfc\left(\sqrt{\frac{\mathcal{E}_b \log_2 M}{2N_0}}\sin\frac{\pi}{M}\right). \tag{5.59}$$

Comparing (5.59) with (5.28), we can conclude that there is a 3-dB asymptotic loss with respect to the coherent case.

A special case is the binary case. In fact, for $M = 2$ we can return to the starting point given by (5.50) and observe from Fig. 5.17 that the exact value of $P(e)$ can be written as

$$P(e) = P\{\pi < \Delta\beta_k < 2\pi\}, \tag{5.60}$$

and therefore

$$P(e) = \frac{1}{2}\left[1 - Q\left(\sqrt{\frac{\mathcal{E}_b}{N_0}}a_2, \sqrt{\frac{\mathcal{E}_b}{N_0}}b_2\right) + Q\left(\sqrt{\frac{\mathcal{E}_b}{N_0}}b_2, \sqrt{\frac{\mathcal{E}_b}{N_0}}a_2\right)\right]. \tag{5.61}$$

But we see that $a_2 = 2$ and $b_2 = 0$, and since (see A.11)

$$Q(x, 0) = 1, \qquad Q(0, x) = e^{-x^2/2},$$

we have the final result:

$$P(e) = \frac{1}{2}e^{-\mathcal{E}_b/N_0}. \tag{5.62}$$

Curves of error probability (5.62) and (5.58) are shown in Fig. 5.18. Also, for comparison, those for coherent demodulation are plotted.

We can conclude that the error performance for differentially coherent PSK (DCPSK) is poorer than CPSK. Intuitively, this is the consequence of using a noisy reference signal for demodulation. However, at large signal-to-noise ratios and for $M = 2$, DCPSK is only slightly inferior to CPSK. This explains why DCPSK modulation is often used when the transmission is binary. On the other hand, as $\mathcal{E}_b/N_0 \to \infty$, four-phase DCPSK presents a loss of approximately 2.3 dB with respect to CPSK; this loss can be traded off with a reduced complexity in the implementation of the receiver. The bit error probability $P_b(e)$ for M-ary Gray-coded DCPSK signals is well approximated by the lower bound of (5.12).

The computations that led to the result (5.58) can be extended in a straightforward manner to an interesting practical case, that is, when the noise samples $\tilde{n}(t_k)$ and $\tilde{n}(t_k - T)$ that appear in (5.53) are no longer independent due to filtering effects. This case is proposed to the reader in Problem 5.5.

Sec. 5.3   Phase-Shift Keying

219

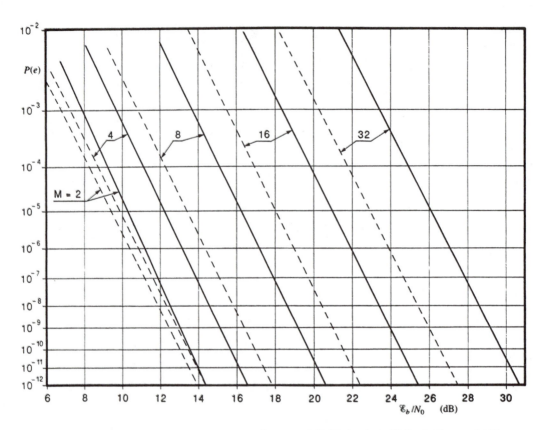

Figure 5.18 Symbol error probability for M-ary DCPSK signals (solid lines). Error probability for CPSK signals (dotted lines) is shown for comparison.

5.3.5 Bandwidth Requirements of PSK Signals

The general expression (5.16) for phase-modulated signals is the same as (5.1). When using rectangular modulating pulses of unit amplitude, the power spectrum is again that of (5.2). The null-to-null bandwidth is $2/T$. However, if minimum bandwidth requirements are sought, as for the case of PAM signals, we see that the minimum RF bandwidth is $1/T$. The maximum bandwidth efficiency for phase-modulated signals is given by

$$\frac{R_s}{W} = \frac{\log_2 M}{T} T = \log_2 M. \tag{5.63}$$

This result shows that the bandwidth efficiency increases by increasing the number of levels M. However, from Fig. 5.18 it is seen that this improvement in bandwidth efficiency is traded off by a penalty in the required $\mathscr{E}_b/N_0$ for a specified error probability $P(e)$.

5.4 COMBINED AMPLITUDE AND PHASE MODULATION

As seen in Sections 5.2 and 5.3, multilevel PAM or PSK can be used to transmit $\log_2 M$ bits of information per symbol or waveform. Both modulations have a bandwidth efficiency increasing as $\log_2 M$. From (5.11) and (5.28) it can also be observed that, for large values of M, a penalty of 6 dB in signal-to-noise ratio must be paid to transmit one additional bit of information at the same symbol error probability. A better result, for the same bandwidth efficiency, can be achieved when the information bits are transmitted using a combination of multiple amplitudes and phases. The signal points should be located on the plane such that the efficiency is increased without decreasing the minimum distance. To this end, the signal points are not constrained to lie on a circle (PSK) or on a line (PAM), but to have a more rational distribution on the plane. The general form of a two-dimensional signal when transmitting a sequence ξ of K RVs is then

$$v_\xi(t) = \mathcal{R}\left\{ \sum_{k=0}^{K-1} \xi_k \tilde{s}(t - kT)\, e^{j2\pi f_0 t}\right\}, \qquad 0 \leq t < KT, \tag{5.64}$$

where the RV ξ_k is defined as

$$\xi_k \triangleq a_{Pk} + ja_{Qk} = A_k e^{j\varphi_k}.$$

When $\tilde{s}(t)$ is a rectangular pulse of unit amplitude, we can rewrite (5.64) as

$$v_\xi(t) = \sum_{k=0}^{K-1} \{a_{Pk} \cos 2\pi f_0 t - a_{Qk} \sin 2\pi f_0 t\}\, u_T(t - kT), \tag{5.65}$$

and the transmitted signal consists of two orthogonal carriers modulated by a set of discrete amplitudes. Such a modulation technique is called *combined amplitude and phase modulation* (AM–PM) and is geometrically represented by a two-dimensional constellation of signal points.

The coherent demodulation of the received signal can be performed with the receiver of Fig. 5.19. According to the theory developed in Section 4.2, the decision is based

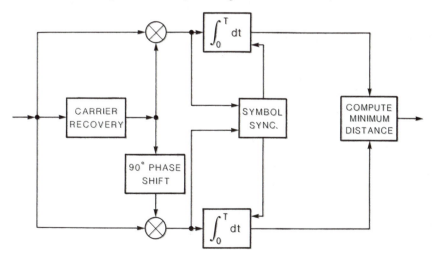

Figure 5.19 Block diagram of the coherent QAM demodulator.

on the minimum distance between signals. In many practical cases the computation of the distances is, as we shall see, an easy task. To obtain some insight into these modulation schemes, let us consider first some simple cases and compare them with the equivalent PSK modulations.

Example 5.6

Consider the case with four signals. The signal points are located in the plane as shown in Fig. 5.20. We have a combined modulation with two amplitudes and four phases. The minimum distance between the signal points is equal to 2, and the average energy is again 2. A four-phase PSK signal set with the same energy has the signal points on a circle of radius $\sqrt{2}$, and the minimum distance of these signals is also 2. Hence, for practical purposes, the error probability of these two signal sets is the same.

Consider now the case of eight signals. A pure eight-phase PSK set with minimum distance 2 has the signal points on a circle of radius $1/\sin(\pi/8)$ and therefore requires an average energy of 6.83. Let us explore combined amplitude and phase signal sets with eight signals and minimum distance 2 and try to obtain a lower average energy. Three of them are shown in Fig. 5.21. The first requires the same energy as the PSK. The second has points on a rectangular grid and gives the same minimum distance with approximately 0.6 dB less energy. The last one is the best, since the improvement with respect to PSK is about 1.6 dB. □

The improvement of AM–PM signal sets with respect to pure PSK may become very significant as M increases. Many efforts have been directed at finding optimum signal constellations. However, if implementation simplicity is considered, the most interesting schemes are the rectangular signal sets shown in Fig. 5.22. They have the advantage of being easily generated as two PAM signals on orthogonal carriers. Moreover, the demodulator of Fig. 5.19 can detect the transmitted symbols by comparing separately the received signal components with a set of thresholds. When $\log_2 M$ is even, this demodulation is optimum; otherwise, it is suboptimum.

For odd numbers of bits/symbol, the signal points form a cross constellation and are taken from the same rectangular grid, except for the $M = 8$ case, which is shown

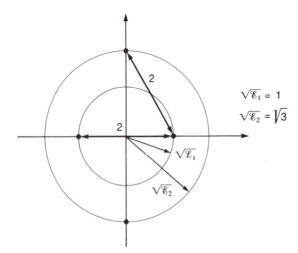

$$\sqrt{\mathcal{E}_1} = 1$$
$$\sqrt{\mathcal{E}_2} = \sqrt{3}$$

Figure 5.20 AM–PM signal constellation with $M = 4$ points.

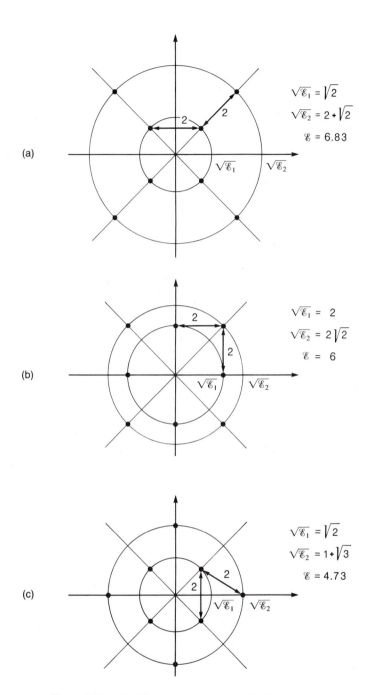

(a)

$\sqrt{\mathcal{E}_1} = \sqrt{2}$

$\sqrt{\mathcal{E}_2} = 2 + \sqrt{2}$

$\mathcal{E} = 6.83$

(b)

$\sqrt{\mathcal{E}_1} = 2$

$\sqrt{\mathcal{E}_2} = 2\sqrt{2}$

$\mathcal{E} = 6$

(c)

$\sqrt{\mathcal{E}_1} = \sqrt{2}$

$\sqrt{\mathcal{E}_2} = 1 + \sqrt{3}$

$\mathcal{E} = 4.73$

Figure 5.21 AM–PM signal constellations with $M = 8$ points.

Sec. 5.4   Combined Amplitude and Phase Modulation   **223**

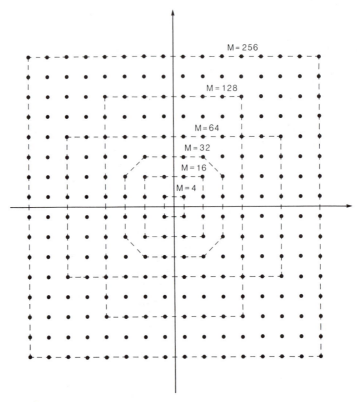

Figure 5.22 Rectangular AM–PM signal constellations. The dotted lines delimit the region of the signal points pertaining to that particular M. Notice the cross constellations for $M = 32 = 2^5$ and $M = 128 = 2^7$.

in Fig. 5.21c. If the signals are scaled so that the minimum distance between any two points is equal to 2, the average signal energy required for transmission is given in Table 5.1. We can see that each additional bit of information requires about 3 dB more energy to keep constant the minimum distance. The modulation schemes consisting

TABLE 5.1 Average signal energy required for the transmission of AM–PM signals. Minimum distance between any two points is 2. The case $M = 8$ refers to Fig. 5.21c.

M	Bits/symbol ($\log_2 M$)	$\mathscr{E}$	$10 \log_{10} \mathscr{E}$
4	2	2	3.0
8	3	4.73	6.75
16	4	10	10.0
32	5	20	13.0
64	6	42	16.2
128	7	82	19.1
256	8	170	22.3

of two independent PAM channels ($\log_2 M$ even) are also called *quadrature-amplitude modulation* (QAM).

The symbol error probability computation is quite straightforward when $\log_2 M$ is even. In fact, it can be obtained from (5.11), considering two independent PAM systems each with $\sqrt{M}$ signals and an average energy one-half that of the QAM system. If p denotes the symbol error probability in each PAM system, we have

$$P(e) = 1 - (1 - p)^2, \tag{5.66}$$

where, from (5.11),

$$p = \left(1 - \frac{1}{\sqrt{M}}\right) erfc\left(\sqrt{\frac{3\log_2 M}{2(M-1)}\frac{\mathscr{E}_b}{N_0}}\right). \tag{5.67}$$

When $\log_2 M$ is odd, there is no equivalently straightforward result, but it is not difficult to compute a useful upper bound (see Problem 5.6) that holds for any value of M:

$$P(e) < 2\, erfc\left(\sqrt{\frac{3\log_2 M}{2(M-1)}\frac{\mathscr{E}_b}{N_0}}\right). \tag{5.68}$$

It is interesting at this point to compare the performance of an M-ary AM–PM system with an M-ary CPSK. To obtain a minimum distance 2 as for the AM–PM of Fig. 5.22, we need a signal energy $1/\sin^2(\pi/M)$. Looking at (5.67), we see that the same result is obtained with AM–PM signals if the average energy increases with M as $2(M-1)/3$. The advantage of AM–PM over CPSK in terms of energy, expressed in decibels, is given in Table 5.2 for different values of M.

As already noted, much effort has been spent to optimize the signal constellations in two dimensions. Few schemes actually outperform the rectangular constellations described in this section, and their advantage over the AWGN channel is always relatively small. The reader is addressed to the Bibliographical Notes and is also invited to solve the proposed problems.

It can also be verified that the power spectra of AM–PM signals are the same as those of PSK. Therefore, the bandwidth efficiency is identical to that of PSK.

TABLE 5.2 Energy saving of AM–PM signals with respect to CPSK

M	Energy saving (dB)
4	0
8	1.6
16	4.14
32	7.01
64	9.95
128	12.89
256	15.90

5.5 FREQUENCY-SHIFT KEYING

When the modulator maps the source bits into frequency shifts of a carrier, we have a *digital frequency modulation* in which a sequence ξ of K RVs is represented by the signal

$$v_\xi(t) = \Re \left\{ \sum_{k=0}^{K-1} \tilde{s}(t - kT) \, e^{j2\pi f_d \xi_k(t-kT)} \, e^{j2\pi f_0 t} \right\}, \qquad 0 \le t < KT, \qquad (5.69)$$

where the RV ξ_k takes values in the set $\{2i - 1 - M\}_{i=1}^{M}$ and $2f_d$ is the separation between adjacent frequencies. If the complex envelope $\tilde{s}(t)$ in (5.69) has constant amplitude A and duration T, the modulation technique is called *frequency-shift keying* (FSK) and the transmitter uses the waveforms

$$s_i(t) = \Re\{\tilde{s}_i(t) \, e^{j2\pi f_0 t}\} = A \cos 2\pi f_i t, \qquad 0 \le t < T,$$
$$f_i = f_0 + (2i - 1 - M)f_d, \qquad\qquad\qquad (5.70)$$
$$i = 1, 2, \ldots, M.$$

They have constant energy $\mathcal{E} = A^2 T/2$ and the modulated signal has constant envelope, as for PSK. Their complex-valued correlation coefficient is

$$\rho = \frac{1}{2\mathcal{E}} \int_0^T \tilde{s}_i(t) \tilde{s}_m^*(t) \, dt$$
$$= \frac{\sin 2\pi f_d T(i - m)}{2\pi f_d T(i - m)} \, e^{j2\pi f_d T(i-m)}, \qquad i, m = 1, 2, \ldots, M. \qquad (5.71)$$

Thus, for a given energy, the distance properties of these waveforms depend only on the choice of their normalized frequency separation $2f_d T$. The generation of these signals may be accomplished with a set of M separate oscillators tuned to the desired frequencies. The major problem with this type of FSK modulation consists in the relatively large spectral sidelobes caused by the phase discontinuities at the switching times. This issue will be discussed in more detail later on in this section.

Another practical possibility is the use of a single oscillator whose frequency is modulated by the source bits. The resulting FSK signal is phase continuous and is called *continuous-phase FSK* (CPFSK). The absence of abrupt phase transitions results in a narrower spectrum. Moreover, this modulation presents memory as a consequence of the phase continuity from one symbol interval to another. Since the information is associated with the frequency and not with the phase, the demodulator can disregard this phase continuity. If it does, there is no difference in performance with respect to the FSK modulation with phase discontinuities. In this section we shall restrict ourselves to the study of certain cases of FSK modulation without phase continuity. A later section will be devoted to the study of CPFSK signals.

5.5.1 Orthogonal FSK with Coherent Demodulation

The FSK modulation technique provides a simple means of generating an orthogonal signal set. In the case of coherent demodulation, the receiver has perfect knowledge of the signal phases, and the orthogonality condition is met when the correlation coefficient of the real signal is zero. That is,

$$\mathcal{R}(\rho) = \frac{\sin 4\pi f_d T(i-m)}{4\pi f_d T(i-m)} = 0. \tag{5.72}$$

This condition is fulfilled when the frequency separation between adjacent signals is such that

$$2f_d T = \frac{m}{2}, \qquad m \text{ any integer.} \tag{5.73}$$

Thus the minimum frequency separation for orthogonality with coherent detection is such that $2f_d T = 0.5$.

The demodulation of these types of signals was discussed in Section 4.2. In particular, the structure of the optimum demodulator is that shown in Fig. 4.4 or 4.5. Notice the complexity of this receiver. The need for a bank of perfectly coherent oscillators renders it rather impractical. The symbol error probability $P(e)$ as a function of $\mathcal{E}_b/N_0$ was derived in Section 4.2 and the result was shown in Fig. 4.19. Those curves show a behavior that is new with respect to the PAM or PSK signals studied so far. In fact, there is an improvement in performance when increasing M, exactly the opposite behavior of the PAM or PSK signals. However, notice that this improvement is obtained at the expense of a larger bandwidth. In fact, increasing M requires more frequencies and therefore more bandwidth.

We now want to relate the symbol error probability to the bit error probability $P_b(e)$. With orthogonal signals, the incorrectly demodulated symbol is equally likely to be any of the remaining $(M-1)$. The number of symbols that contain an incorrect digit in any given position is $M/2$. Therefore, the probability of having an incorrect digit in the $(M-1)$ incorrect symbols is $(M/2)/(M-1)$. As a consequence, we get

$$P_b(e) = \frac{M}{2(M-1)} P(e). \tag{5.74}$$

Finally, note that a tight upper bound to $P_b(e)$ for the case of high $\mathcal{E}_b/N_0$ can be obtained from (4.78) and (5.74) as

$$P_b(e) \leqslant \frac{M}{4} \, erfc \left(\sqrt{\frac{\mathcal{E}_b}{2N_0} \log_2 M} \right). \tag{5.75}$$

5.5.2 Orthogonal FSK with Incoherent Demodulation

Incoherent demodulation of orthogonal FSK signals results in a simpler implementation of the receiver. In such circumstances, the orthogonality condition must be fulfilled independently of the phases of the signals of (5.70). These can be rewritten as

$$s_i(t) = \mathcal{R} \left\{ A e^{j(2\pi f_i t + \vartheta_i)} \right\}, \qquad 0 \leqslant t < T, \tag{5.76}$$

and the reader can verify that the orthogonality condition is satisfied when the correlation coefficient (5.71) is zero. This condition is met when the frequency separation between adjacent signals is

$$2f_d T = m, \qquad m \text{ any integer.} \tag{5.77}$$

Thus minimum frequency separation for orthogonality is such that $2f_d T = 1$. By comparison with (5.73), we can observe that twice as much frequency separation is required for noncoherent demodulation as for coherent demodulation. The demodulation of this type of signals was discussed in Section 4.3 and the structure of the optimum demodulator was given in Fig. 4.21 or 4.22. In essence, the demodulator is composed of a bank of bandpass filters followed by envelope detectors. The matched filters are tuned to each frequency, and the envelope samples are used for comparison in the decision device. The symbol error probability $P(e)$ was given by (4.129) and the curves as a function of $\mathscr{E}_b/N_0$ were plotted in Fig. 4.27. The performance is somewhat inferior to the coherent case, but this is traded off by the easier implementation. All the final observations made for the coherent case also apply to noncoherent demodulation as they are a consequence of the orthogonal structure of the signal set.

5.5.3 Binary FSK with Discrimination Detection

A practical method of detecting FSK signals is *discrimination detection*. The basic structure of the demodulator is shown in Fig. 5.23. It comprises a front-end bandpass filter wide enough to pass all transmitted waveforms without significant distortion. Follow-

Figure 5.23 Block diagram of the FSK demodulator with discrimination detection.

ing the filter is an ideal bandpass limiter, which supplies the frequency discriminator with a constant envelope waveform. The cascade of the limiter and discriminator is assumed to be a perfect instantaneous frequency detector in the sense that its output is the derivative of the instantaneous phase of the input signal. The frequency detector output is sampled and compared with a set of thresholds to decide which frequency was transmitted. We shall examine the performance of this receiver in the case of binary FSK. It is of practical interest for low-speed data transmission because of its low complexity.

The input signal to the frequency detector can be written as

$$r(t) = \mathscr{R}\{\tilde{r}(t)e^{j2\pi f_0 t}\}, \qquad 0 \leq t < T, \tag{5.78}$$

where

$$\tilde{r}(t) = Ae^{j\varphi(t)} + \tilde{n}(t),$$
$$\varphi(t) = \pm 2\pi f_d t, \tag{5.79}$$

and $\tilde{n}(t)$ is the noise complex envelope at the filter output. The two transmitted signals are symmetrically located around the frequency f_0. Adopting the usual terminology, we say that a *mark* is transmitted with the frequency $(f_0 + f_d)$ and a *space* with the frequency $(f_0 - f_d)$. Moreover, we assume that the amplitude of the filter transfer function is symmetric around f_0 and, as a consequence, the noise $\tilde{n}(t)$ has an even power density spectrum.

Recalling (2.178), the instantaneous frequency deviation from f_0 generated by the frequency detector is

$$v(t) = \frac{1}{2\pi} \frac{d}{dt} \arg\left[\tilde{r}(t)\right]. \tag{5.80}$$

The decision circuit may decide by observing the sign of the detector output sampled at $t = t_k$, say $v(t_k)$. If a mark is transmitted, that is, $\varphi(t) = 2\pi f_d t$, then the error probability is

$$P(e|\text{mark}) = P\{v(t_k) < 0\}. \tag{5.81}$$

and similarly for the transmission of a space. The instantaneous frequency in (5.80) can be rewritten as

$$v(t) = \frac{\mathcal{R}\left\{-j\tilde{r}^*(t)\,(d/dt)\,\tilde{r}(t)\right\}}{|\tilde{r}(t)|^2}. \tag{5.82}$$

If we now define the two RVs z_1 and z_2 as

$$z_1 \triangleq -j\frac{d}{dt}\tilde{r}(t)\bigg|_{t=t_k} = A\frac{d\varphi(t)}{dt}e^{j\varphi(t)} - j\frac{d}{dt}\tilde{n}(t)\bigg|_{t=t_k},$$
$$z_2 \triangleq \tilde{r}(t_k) = Ae^{j\varphi(t_k)} + \tilde{n}(t_k), \tag{5.83}$$

the error probability (5.81) becomes

$$P(e|\text{mark}) = P\{\mathcal{R}(z_1 z_2^*) < 0\}. \tag{5.84}$$

The RVs z_1 and z_2 are Gaussian when conditioned on a transmitted symbol and the probability in (5.84) is formally the same as that already encountered in connection with the error probability computation for DCPSK signals [see (5.55)]. The details of the calculation are sketched in Problem 5.11; the result is

$$P(e) = \frac{1}{2}[P(e|\text{mark}) + P(e|\text{space})]$$
$$= \frac{1}{2}[1 - Q(\sqrt{\gamma b}, \sqrt{\gamma a}) + Q(\sqrt{\gamma a}, \sqrt{\gamma b})], \tag{5.85}$$

where

$$a \triangleq \frac{1}{2}\left(1 - \frac{f_d}{f_2}\right)^2, \qquad b \triangleq \frac{1}{2}\left(1 + \frac{f_d}{f_2}\right)^2,$$

$$\gamma \triangleq \frac{A^2}{2P_N}, \qquad f_2^2 \triangleq \frac{\displaystyle\int_{-\infty}^{\infty} f^2 G(f)\,df}{\displaystyle\int_{-\infty}^{\infty} G(f)\,df}.$$

$G(f)$ is the power density spectrum of $\tilde{n}(t)$, and P_N is the noise power at the output of the bandpass filter. Curves of the error probability (5.85) are shown in Fig. 5.24, having f_d/f_2 as a parameter.

Sec. 5.5　Frequency-Shift Keying

229

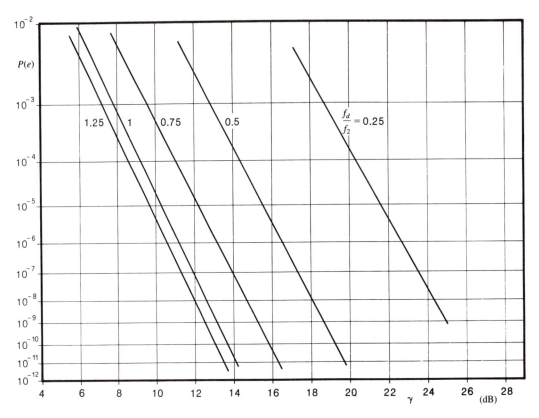

Figure 5.24 Error probability for binary FSK demodulation with discrimination detection. Notice that the definition of signal-to-noise ratio is not consistent with all the other figures of the chapter.

Example 5.7

This example will show how to relate γ to the usual parameter $\mathcal{E}_b/N_0$. The result (5.85) is applied to the ideal case of a rectangular bandpass filter of bandwidth W. First, notice that we have

$$\frac{\mathcal{E}_b}{N_0} = \frac{A^2 T}{2N_0} = \frac{A^2 W T}{2N_0 W} = \gamma W T.$$

Assuming the rather crude approximation that the spectrum of the signal is formed of lobes of width $1/T$ around each of the two frequencies, we have $W \cong (2f_d + 1/T)$. Furthermore, in the case of an ideal rectangular bandpass filter in Fig. 5.23, we have

$$f_2 = \frac{W}{2\sqrt{3}}.$$

Therefore,

$$\frac{f_d}{f_2} = \frac{2\sqrt{3} f_d}{2f_d + (1/T)} = \frac{\sqrt{3}(2f_d T)}{(2f_d T) + 1}.$$

In conclusion, the expression for $P(e)$ given in (5.85) can be computed with the parameters

$$a = \frac{1}{2}\left(1 - \frac{\sqrt{3}(2f_d T)}{(2f_d T) + 1}\right)^2,$$

$$b = \frac{1}{2}\left(1 + \frac{\sqrt{3}(2f_d T)}{(2f_d T) + 1}\right)^2,$$

$$\gamma = \frac{\mathscr{E}_b}{N_0}\frac{1}{(2f_d T) + 1},$$

and the results are shown in Fig. 5.25 for some values of the parameter $2f_d T$. Notice that we now have on the abscissa the same $\mathscr{E}_b/N_0$ as in the previous modulation schemes. The results are therefore directly comparable. The curve (dotted line) for binary incoherent orthogonal FSK is also shown for comparison. It can be seen that the discriminator detection gives roughly the same result as the incoherent detector suggested by optimum decision theory. We have a minimum error probability with discriminator detection when $2f_d T \cong$ 1.25. Incoherent orthogonal FSK, for which $2f_d T = 1$, requires a ratio $\mathscr{E}_b/N_0$ only 0.5 dB smaller to achieve the same error performance. $\square$

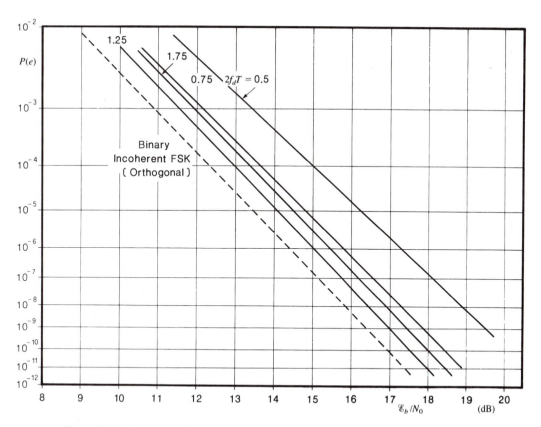

Figure 5.25 Error probability for binary FSK demodulation versus the signal-to-noise ratio $\mathscr{E}_b/N_0$ (discrimination detection). See Example 5.6 for explanation. The dashed curve is for binary incoherent orthogonal FSK (same as Fig. 4.25 with $|\rho| = 0$).

Sec. 5.5   Frequency-Shift Keying

5.5.4 Bandwidth Requirements for FSK Signals

The computation of the power spectra of FSK signals is, in general, a rather heavy task. A crucial problem is the precise definition of the mathematical model for the FSK signals and its relation with their physical generation. In particular, the time behavior of the instantaneous phase of the modulated signal must be precisely defined. The purpose of this section is the detailed analysis of this problem. Notice again that the aspect of phase continuity is irrelevant in the detection process, at least of course for those demodulators that take decisions on a symbol-by-symbol basis. The easiest case is that implied by the signals of (5.70). Statistically independent modulating signals $s_i(t)$ are assumed in each time interval, and no phase dependence among them is assumed. Furthermore, all signals $s_i(t)$ start in each interval with the same initial phase, which, in (5.70), for simplicity has been set to zero. This model fits a situation in which the signal is generated by switching at a rate $1/T$ among M different oscillators that generate signals with always the same initial phase in each time interval. A matching of this model into a physical situation might be the digital generation of the signal waveforms. The situation corresponds to a memoryless nonlinear modulation. From (2.166) and (2.167), we get the following continuous and discrete parts of the spectrum of the complex envelope of the transmitted signal:

$$\mathcal{G}^{(c)}(f) = \frac{1}{MT}\left\{ \sum_{i=1}^{M} |S_i(f)|^2 - \frac{1}{M}\left| \sum_{i=1}^{M} S_i(f) \right|^2 \right\}, \tag{5.86a}$$

$$\mathcal{G}^{(d)}(f) = \frac{1}{(MT)^2}\left| \sum_{i=1}^{M} S_i(f) \right|^2 \sum_{m=-\infty}^{\infty} \delta\left(f - \frac{m}{T}\right), \tag{5.86b}$$

where $S_i(f)$ is the Fourier transform of the signal $s_i(t)$ of (5.70).

Example 5.8

We consider, in this example, binary FSK signals with the model leading to the spectrum (5.86). Using (5.70), the complex envelopes of the two signals are:

$$\tilde{s}_1(t) = Ae^{-j2\pi f_d t},$$
$$\tilde{s}_2(t) = Ae^{j2\pi f_d t}, \qquad 0 \le t < T,$$

and their Fourier transforms are

$$S_1(f) = Ag(f + f_d),$$
$$S_2(f) = Ag(f - f_d),$$

where

$$g(f) = T\frac{\sin \pi f T}{\pi f T} e^{j\pi f T}$$

is the transform of the rectangular pulse $u_T(t)$. Then, using (5.86a), we get for the power spectrum of the complex envelope

$$\frac{4\mathcal{G}^{(c)}(f)}{A^2 T} = \frac{1}{T^2}[|g(f + f_d)|^2 + |g(f - f_d)|^2 - 2\mathcal{R}\{g(f + f_d)g^*(f - f_d)\}], \tag{5.87a}$$

$$\frac{4\mathcal{G}^{(d)}(f)}{A^2} = \frac{1}{T^2} [|g(f+f_d)|^2 + |g(f-f_d)|^2$$

$$+ 2\mathcal{R}\{g(f+f_d)g^*(f-f_d)\}] \sum_{m=-\infty}^{\infty} \delta\left(f - \frac{m}{T}\right) . \tag{5.87b}$$

Finally, the power spectrum of the real signal can be obtained as

$$\mathcal{G}_{v_g}(f) = \frac{1}{4}\{\mathcal{G}^{(c)}(f-f_0) + \mathcal{G}^{(c)}(-f-f_0) + \mathcal{G}^{(d)}(f-f_0) + \mathcal{G}^{(d)}(-f-f_0)\} . \tag{5.88}$$

Some curves of the continuous part of the spectrum (5.87) are shown in Fig. 5.26 for different values of the parameter $2f_dT$ and as a function of fT. The Table 5.3 gives, instead, the coefficients of the line spectrum in the RHS of (5.87b) for the same cases of $2f_dT$ plotted in Fig. 5.26. $\square$

A different situation arises when the modulated signals of the FSK type are generated by switching among M different oscillators, where each of them retains memory of its own phase, in time. Under these conditions, we have no phase discontinuity between two consecutive intervals when the same frequency is transmitted. Moreover, when switching to a given frequency, the signal phase at the beginning of each interval is determined by the evolution of the oscillator phase from its initial value. The difference, with respect to the previous model, is that in Example 5.8 the phase of each oscillator

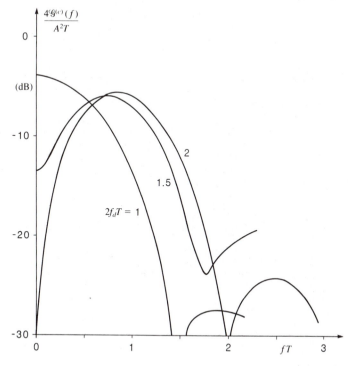

Figure 5.26 Continuous part of the power density spectrum of discontinuous-phase binary FSK. See Example 5.7.

Sec. 5.5   Frequency-Shift Keying

233

TABLE 5.3 Discrete spectrum for the signals of Example 5.8. Coefficients of each line component (in dB).

fT	$2f_dT$		
	1.0	1.5	2.0
0	—	−13.4	—
1	− 7.4	− 6.8	−6.0
2	−15.4	−20.1	—
3	−19.2	−24.6	—
4	−21.8	−27.5	—
5	−23.8	−29.6	—
6	−25.4	−31.3	—
7	−26.8	−32.7	—

was always constricted to keep the same value at the beginning of each transmission interval T. The binary case is analyzed in the following example.

Example 5.9

By considering the two signals to be transmitted, we can account for the phase continuity of each oscillator by rewriting the transmitted signal (5.69) as follows:

$$v_\xi(t) = \frac{A}{2}\Re\left\{ \sum_{k=-\infty}^{\infty} \left[(1 + \xi_k)\, e^{j(2\pi f_1 t + \vartheta_1)} \right.\right. \tag{5.89}$$
$$\left.\left. + (1 - \xi_k)\, e^{j(2\pi f_2 t + \vartheta_2)}]u_T(t - kT) \right\},$$

where ϑ_1 and ϑ_2 are constant phases and (ξ_k) is a sequence of iid RVs taking on values ± 1 with equal probabilities and $f_1 = f_0 - f_d$, $f_2 = f_0 + f_d$. By expanding the products in (5.89), we get the complex envelope

$$\tilde{v}_\xi(t) = \frac{A}{2}[e^{-j(2\pi f_d t - \vartheta_1)} + e^{j(2\pi f_d t + \vartheta_2)}] \sum_{k=-\infty}^{\infty} u_T(t - kT)$$
$$+ \frac{A}{2}[e^{-j(2\pi f_d t - \vartheta_1)} - e^{j(2\pi f_d t + \vartheta_2)}] \sum_{k=-\infty}^{\infty} \xi_k u_T(t - kT). \tag{5.90}$$

By observing that $\sum_{k=-\infty}^{\infty} u_T(t - kT) = 1$ for all T, we recognize that $\tilde{v}_\xi(t)$ is the sum of a pair of sinusoidal signals with frequency f_d and $-f_d$, respectively, and of a zero-mean random signal. These two terms contribute separately to the power spectrum of $\tilde{v}_\xi(t)$ because they are uncorrelated for all t. Some rather tedious but straightforward computations (see Problem 5.12) lead to the following results for the power density spectrum of $\tilde{v}_\xi(t)$:

$$\mathcal{G}^{(c)}(f) = \frac{A^2}{4T}\{|g(f + f_d)|^2 + |g(f - f_d)|^2 - 2\gamma(2f_dT)\Re[e^{j(\vartheta_2 - \vartheta_1)}g(f + f_d)g^*(f - f_d)]\}$$

and

$$\mathcal{G}^{(d)}(f) = \frac{A^2}{4}\delta(f + f_d) + \frac{A^2}{4}\delta(f - f_d),$$

where

$$g(f) = T \frac{\sin \pi fT}{\pi fT} e^{-j\pi fT},$$

$$\gamma(\lambda) = \begin{cases} 1, & \lambda \text{ an integer,} \\ 0, & \text{otherwise.} \end{cases}$$

Notice that the discrete part of the spectrum consists always of only two lines at frequencies $-f_d$ and $+f_d$. This property can be useful for symbol synchronization purposes (see Chapter 6). The reader is invited to verify that this case is reduced to that of Example 5.8 when $f_dT = 1$ and $\vartheta_1 = \vartheta_2 = 0$. □

The final and most interesting case to be considered is that of CPFSK signals. They arise when a single oscillator is used for the generation and, therefore, phase continuity is maintained also when frequency is changed between two successive time intervals of duration T.

To derive the power spectral density of these signals, we can write the signal $v_\xi(t)$ modulated by an infinitely long sequence as follows:

$$v_\xi(t) = \mathscr{R}\{Ae^{j[2\pi f_0 t + 2\pi f_d \int_0^t D(\tau)d\tau + \phi]}\}, \tag{5.91}$$

where ϕ is the initial phase at $t = 0$ and $D(t)$ is the baseband data signal

$$D(t) = \sum_{k=0}^{\infty} \xi_k u_T(t - kT). \tag{5.92}$$

Using (5.92) in (5.91), we can get

$$v_\xi(t) = \mathscr{R}\{\tilde{s}(t)e^{j2\pi f_0 t}\}, \tag{5.93}$$

where

$$\tilde{s}(t) = Ae^{j\phi} \sum_{k=0}^{\infty} q(t - kT; \xi_k, \sigma_k),$$

$$q(t; \xi_k, \sigma_k) \triangleq e^{j2\pi f_d(\sigma_k T + \xi_k t)} u_T(t), \tag{5.94}$$

and σ_k is defined as

$$\sigma_k \triangleq \sum_{i=0}^{k-1} \xi_i, \qquad \sigma_0 \triangleq 0.$$

The general method of Section 2.3.1 can be applied to derive the power spectral density of the signal (5.94). It will represent the spectrum around the carrier frequency f_0. The computational steps are suggested in Problem 5.13 and here only the result is reported, assuming $A = 1$. Define the *characteristic function* $C(p)$ of the RV ξ as

$$C(p) \triangleq E\{e^{j\xi p}\} = |C(p)|e^{j\gamma(p)}$$

and the function

$$X(f) \triangleq \frac{T}{2} e^{-j\pi fT} \frac{\sin \pi fT}{\pi fT}.$$

Then, when $|C(2\pi f_d T)| < 1$, the power density spectrum is continuous and given by

$$\mathscr{G}^{(c)}(f) = \frac{2}{T} E\{|X(f - f_d \xi)|^2\}$$

$$+ \frac{4}{T} \mathscr{R}\left[\frac{e^{-j2\pi f T}}{1 - e^{-j2\pi f T} C(2\pi f_d T)} E\{X(f - f_d \xi)\} E\{X^*(f - f_d \xi) e^{j2\pi f_d T \xi}\} \right]. \tag{5.95}$$

When $|C(2\pi f_d T)| = 1$, the spectrum has a continuous and a discrete part given by

$$\mathscr{G}^{(c)}(f) = \frac{2}{T} E\{|X(f - f_d \xi)|^2\} - \frac{2}{T} |E\{X(f - f_d \xi)\}|^2, \tag{5.96}$$

$$\mathscr{G}^{(d)}(f) = \frac{4\pi}{T} |E\{X(f - f_d \xi)\}|^2 \cos \lambda \sum_{n=-\infty}^{\infty} \delta(\lambda - 2\pi n),$$

where $\lambda \triangleq \gamma (2\pi f_d T) - 2\pi f T$. These results will be used in the following example to get explicit expressions for the power density spectra of CPFSK signals.

Example 5.10

Let us assume an M-ary CPFSK signal. That is, the RV ξ takes equally likely values in the set

$$\{2i - 1 - M\}_{i=1}^{M}.$$

To compute the spectrum (5.95), let us first evaluate the average quantities that appear in the formula.

$$E\{|X(f - f_d \xi)|^2\} = \frac{T^2}{4M} \sum_{i=1}^{M} \left\{ \frac{\sin \pi T \alpha_i}{\pi T \alpha_i} \right\}^2,$$

$$E\{X(f - f_d \xi)\} = \frac{T}{2M} \sum_{i=1}^{M} \frac{\sin \pi T \alpha_i}{\pi T \alpha_i} e^{-j\pi T \alpha_i},$$

$$E\{X^*(f - f_d \xi) e^{j2\pi f_d T}\} = \frac{T}{2M} \sum_{i=1}^{M} \frac{\sin \pi T \alpha_i}{\pi T \alpha_i} e^{-j\pi T [f - (2i + 1 - M)f_d]},$$

where

$$\alpha_i \triangleq f - (2i - 1 - M)f_d.$$

Furthermore,

$$C(2\pi f_d T) = \frac{2}{M} \sum_{i=1}^{M/2} \cos 2\pi (2i - 1 - M) f_d T.$$

Using these quantities and performing the algebraic manipulations in (5.95), we get the final result as

$$\mathscr{G}^{(c)}(f) = \frac{T}{4} \left[\frac{1}{M} \sum_{i=1}^{M} A_i^2(f) + \frac{2}{M^2} \sum_{i=1}^{M} \sum_{j=1}^{M} A_i(f) A_j(f) B_{ij}(f) \right], \tag{5.97}$$

where

$$A_i(f) \triangleq \frac{\sin \pi T \alpha_i}{\pi T \alpha_i},$$

$$B_{ij}(f) \triangleq \frac{\cos(2\pi fT - \beta_{ij}) - C(2\pi f_d T)\cos\beta_{ij}}{1 + C^2(2\pi f_d T) - 2C(2\pi f_d T)\cos 2\pi fT},$$

$$\beta_{ij} = 2\pi(i + j - 1 - M)f_d T.$$

Some plots of (5.97) are shown in Figs. 5.27 and 5.28. It is left as an exercise for the reader (see Problem 5.15) to consider the case of $f_d T = 1$, which gives rise to lines in the spectrum. It can also be verified that the phase continuity causes a smaller bandwidth occupancy than the discontinuous phase. To this purpose, the reader is invited to compare, for a given M and $f_d T$, one case of Examples 5.8, 5.9, and 5.10 on the basis of the fractional power containment bandwidth (see Section 5.1). $\square$

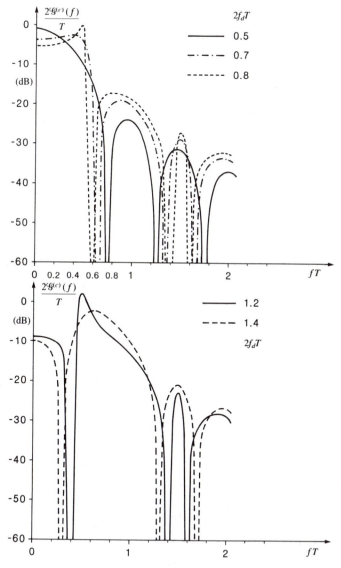

Figure 5.27 Power density spectrum of binary CPFSK signals.

Sec. 5.5   Frequency-Shift Keying

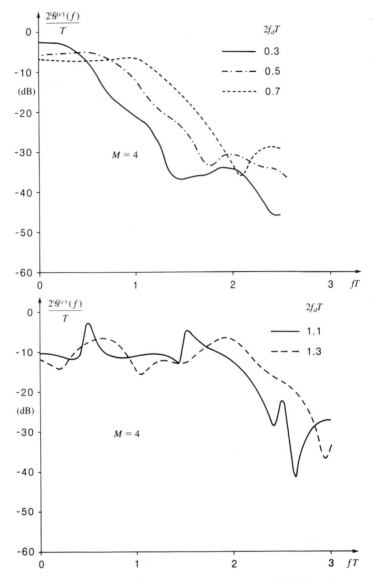

Figure 5.28 Power density spectrum of quaternary CPFSK signals.

The calculations performed on the spectra of FSK signals have shown the relative complexity of the bandwidth evaluation for this case. When only a crude approximation is sufficient, we can use either (5.73) or (5.77) to evaluate the bandwidth efficiency for the cases of, respectively, orthogonal coherent or incoherent FSK signals. Assuming, for the coherent case, that each signal requires a bandwidth $1/2T$, we get

$$\frac{R_s}{W} = \frac{\log_2 M/T}{M/2T} = \frac{2\log_2 M}{M}.$$ (5.98)

Similarly, for orthogonal incoherent FSK signals, we get

$$\frac{R_s}{W} = \frac{\log_2 M/T}{M/T} = \frac{\log_2 M}{M}.$$ (5.99)

From both (5.98) and (5.99), it is evident that the bandwidth efficiency decreases as M increases; actually it goes to zero for $M \to \infty$. Associating this result with the behavior of the previously described error probability with M, we can conclude that FSK signals present properties that are opposite to those of PAM or PSK. In fact, when M is increased, FSK signals require a smaller ratio $\mathcal{E}_b/N_0$ to achieve a specified bit error probability, but this improvement is penalized in terms of bandwidth efficiency.

5.6 OFFSET QPSK AND MINIMUM-SHIFT KEYING MODULATIONS

In the QPSK modulator described in Fig. 5.7, the carrier phase changes only once every $2T_s$ seconds, $1/T_s$ being the source binary rate. When only one of the two quadrature components, either I or Q, changes sign, a phase shift of $\pm 90°$ occurs. A change in both components causes a phase shift of $180°$. These phase jumps, which are ideally instantaneous, are shown in the phasor diagram of Fig 5.29a. Usually, the transmitted QPSK signal is bandlimited by a bandpass filter so as to reduce the out-of-band spectral sidelobes and prevent interference with adjacent channels. A consequence of this filtering is that the bandlimited QPSK signal will no longer present a constant envelope. In fact, the occasional $180°$ phase shifts occur now in a nonzero time and cause the envelope to approach zero, as shown qualitatively in Fig. 5.30. This effect is highly undesirable when the signal undergoes nonlinear amplification, as in satellite repeaters. Actually, a nonlinear amplifier operated at saturation tends to restore the constant envelope of the signal, but at the same time it enhances also the out-of-band spectral sidelobes. Thus, the filtering action at the transmitter is destroyed (a typical example will be given in Section 10.2).

In this section, other modulation techniques are proposed that present a constant envelope and the same performance as QPSK over an AWGN channel, but reduce the described undesired effects.

5.6.1 Offset QPSK Modulation

A reduction of the envelope fluctuations is possible by delaying the Q-channel digits by T_s seconds relative to the I channel, as shown in Fig. 5.31. This modulation is called *offset QPSK* (OQPSK) or sometimes *staggered QPSK* (SQPSK) because the two quadrature components are offset in time by a bit period T_s. This solution eliminates the possibility of $180°$ phase changes. In fact, phase changes of only $\pm 90°$ can occur every T_s seconds. This feature is shown pictorially in the phasor diagram of Fig 5.29b. As a result, *filtered* OQPSK signals present a ratio of roughly $\sqrt{2}$ between the maximum and minimum value of the envelope, compared to infinity for QPSK. Therefore, it may be expected that the undesired features of QPSK are significantly reduced with OQPSK.

Sec. 5.6   Offset QPSK and Minimum-Shift Keying Modulations   **239**

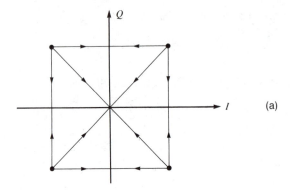

(a)

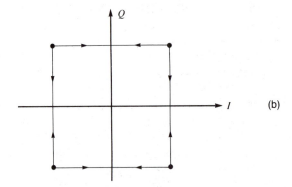

(b)

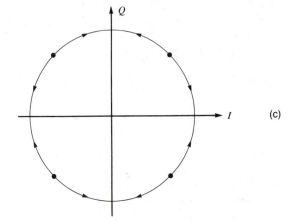

(c)

Figure 5.29 Phasor diagrams of (a) QPSK signals; (b) OQPSK signals; (c) MSK signals.

The transmitted signal, in quadrature form, can be written as

$$v_\xi(t) = I(t) \cos 2\pi f_0 t - Q(t) \sin 2\pi f_0 t, \qquad (5.100)$$

where

$$I(t) = A \sum_k \xi_{2k-1} u_{2T_s}[t - (2k - 1)T_s],$$

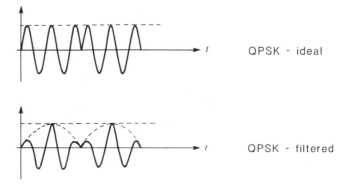

QPSK - ideal

QPSK - filtered

Figure 5.30 Qualitative description of filtering effects over a QPSK signal. The dashed line shows the envelope.

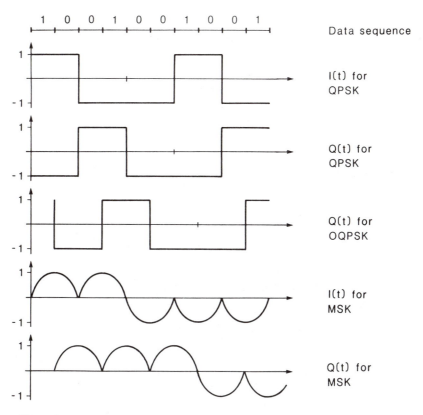

1 0 0 1 0 0 1 0 0 1   Data sequence

I(t) for QPSK

Q(t) for QPSK

Q(t) for OQPSK

I(t) for MSK

Q(t) for MSK

Figure 5.31 In-phase and quadrature baseband components in QPSK, OQPSK, and MSK signals.

$$Q(t) = A \sum_k \xi_{2k} \, u_{2T_s}(t - 2kT_s),$$

and the ξ's are RVs taking values in the set $\{-1, +1\}$. Equation (5.100) implies that the data sequence (ξ_k) be split into odd symbols ξ_{2k-1} and even ones ξ_{2k}. These are used to determine the sign of the shaping waveforms during odd intervals, $(2k - 1)T_s \leq t < (2k + 1)T_s$, and even intervals, $2kT_s \leq t < (2k + 2)T_s$, respectively. The shaping waveform in either channel is a unit square wave of duration $2T_s$. The coherent receiver for OQPSK signals is identical to that shown in Fig. 5.9, with the exception that the Q data stream is delayed by the bit duration T_s. Thus, the error performance and spectral properties of OQPSK are identical to those of QPSK.

5.6.2 Minimum-Shift Keying

This type of modulation can be thought of as a special case of OQPSK in which the rectangular waveform is replaced by a sinusoidal pulse, as shown in Fig. 5.31. Therefore, the $I(t)$ and $Q(t)$ quadrature signals of (5.100) are now

$$I(t) = A \sum_k \xi_{2k-1} u_{2T_s}[t - (2k - 1)T_s] \cos\left(\frac{\pi t}{2T_s}\right),$$

$$Q(t) = -A \sum_k \xi_{2k} u_{2T_s}(t - 2kT_s) \sin\left(\frac{\pi t}{2T_s}\right). \tag{5.101}$$

The shaping waveform in either the I or Q channel is a half-cycle sinusoidal pulse of duration $2T_s$ and alternating sign. This shaping causes the transitions between phases shown in the phasor diagram of Fig. 5.29c. Moreover, these transitions are not abrupt as in QPSK or OQPSK, but instead the phase moves linearly from one value to the other.

This quadrature modulated signal can also be viewed as a particular case of binary FSK with twice the signaling rate. To this purpose, let us define the two frequencies f_1 and f_2 as follows:

$$f_1 \triangleq f_0 + \frac{1}{4T_s}, \qquad f_2 \triangleq f_0 - \frac{1}{4T_s}. \tag{5.102}$$

Then, using standard trigonometric identities, we get from (5.100) and (5.101)

$$v_\xi(t) = \frac{A}{2} \sum_k \xi_{2k-1} u_{2T_s}[t - (2k - 1)T_s]\{\cos 2\pi f_2 t + \cos 2\pi f_1 t\}$$

$$+ \frac{A}{2} \sum_k \xi_{2k} u_{2T_s}(t - 2kT_s)\{\cos 2\pi f_2 t - \cos 2\pi f_1 t\}. \tag{5.103}$$

Inspection of (5.103) reveals that either one of the two frequencies f_1 and f_2 is produced in the interval $(2k - 1)T_s \leq t < 2kT_s$, depending on the combination of the pair of symbols ξ_{2k-1} and ξ_{2k}. Table 5.4 shows the generated signal for each combination of symbols. The table indicates that f_2 is transmitted when $\xi_{2k-1} = \xi_{2k}$, whereas f_1 is transmitted when $\xi_{2k-1} \neq \xi_{2k}$. The polarity of the waveform depends on ξ_{2k-1}. We conclude that the signal of (5.101) is a binary FSK signal with $f_d = 1/4T_s$. Since from

TABLE 5.4 Generated signals in MSK for each combination of information symbols

ξ_{2k-1}	ξ_{2k}	Signal
1	1	$A \cos 2\pi f_2 t$
1	-1	$A \cos 2\pi f_1 t$
-1	1	$-A \cos 2\pi f_1 t$
-1	-1	$-A \cos 2\pi f_2 t$

(5.73) this is the minimum value of f_d that ensures orthogonality, the modulation is usually called *minimum-shift keying* (MSK).

Using the results of Table 5.4 we can resort to a more compact form of (5.103) as follows:

$$v_{\xi}(t) = A \sum_k u_{T_s}(t - kT_s) \cos \left[2\pi f_0 t - (\xi_{k-1}\xi_k)\frac{\pi t}{2T_s} + \psi_k \right], \qquad (5.104)$$

where, if k is odd,

$$\psi_k = \begin{cases} 0, & \text{if } \xi_k = 1, \\ \pi, & \text{if } \xi_k = -1, \end{cases}$$

and, if k is even,

$$\psi_k = \begin{cases} 0, & \text{if } \xi_{k-1} = 1, \\ \pi, & \text{if } \xi_{k-1} = -1. \end{cases}$$

We now demonstrate that these conditions on ψ_k give rise to a binary FSK signal presenting phase continuity at $t = kT_s$. In the kth transmission interval $kT_s \leq t < (k+1)T_s$, the phase variation $\vartheta(t)$ is given by

$$\vartheta_k(t) = -(\xi_{k-1}\xi_k)\frac{\pi t}{2T_s} + \psi_k, \qquad kT_s \leq t < (k+1)T_s.$$

Similarly, we get

$$\vartheta_{k+1}(t) = -(\xi_k\xi_{k+1})\frac{\pi t}{2T_s} + \psi_{k+1}, \qquad (k+1)T_s \leq t < (k+2)T_s.$$

Let us compute the value of these phases at the time $t = (k+1)T_s$. Assuming k even and using the conditions (5.104), we can obtain the values shown in Table 5.5 for all the eight possible cases. These are two important conclusions. First, the phase is continuous at the transition instant between adjacent intervals. Second, this phase can assume only two values, that is, $\pm\pi/2$. The reader is invited to work out the case for k odd and verify that the same conclusions hold true, except for the fact that the two possible phase values at time $t = (k+1)T_s$ are 0 and π. The property of phase continuity shows that MSK is a particular case of binary CPFSK. Additional details

Sec. 5.6   Offset QPSK and Minimum-Shift Keying Modulations   **243**

TABLE 5.5 Values of phases $\vartheta_k[(k + 1)T_s]$ and $\vartheta_{k+1}[(k + 1)T_s]$ for MSK signals. Phase continuity is shown at $t = (k + 1)\,T_s$.

ξ_{k-1}	ξ_k	ξ_{k+1}	ψ_k	ψ_{k+1}	$\vartheta_k[(k + 1)T_s]$	$\vartheta_{k+1}[k + 1)T_s]$
+1	+1	+1	0	0	$-\pi/2$	$-\pi/2$
+1	+1	−1	0	π	$-\pi/2$	$-\pi/2$
+1	−1	+1	0	0	$+\pi/2$	$+\pi/2$
+1	−1	−1	0	π	$+\pi/2$	$+\pi/2$
−1	+1	+1	π	0	$-\pi/2$	$-\pi/2$
−1	+1	−1	π	π	$-\pi/2$	$-\pi/2$
−1	−1	+1	π	0	$+\pi/2$	$+\pi/2$
−1	−1	−1	π	π	$+\pi/2$	$+\pi/2$

can be obtained by solving Problem 5.16. Finally, if we define the RV $b_k \triangleq -(\xi_{k-1}\xi_k)$, which takes values in the set $\{-1, 1\}$, we can rewrite the signal of (5.104) as

$$v_\mathbf{b}(t) = \sum_k s(t - kT_s; b_k, \sigma_k), \tag{5.105}$$

where

$$s(t - kT_s; b_k, \sigma_k) \triangleq A \cos\left\{2\pi f_0 t + \frac{b_k \pi}{2T_s}(t - kT_s) + \varphi_k\right\}, \qquad kT_s \leq t < (k + 1)T_s,$$

and σ_k is the state of the modulator that coincides with the phase φ_k taking values in the set $\left\{0, \dfrac{\pi}{2}, \pi, \dfrac{3\pi}{2}\right\}$. Therefore, the signal in (5.105) is also a CPM signal like those described in Section 4.4. In particular, it is the same signal used in Examples 4.2 and 4.12. Therefore, those examples can be revisited to better understand the MSK signals. The generation of MSK signals can be accomplished by using either (5.105) or (5.101). In the former case, a single oscillator of frequency f_0 is modulated by the stream of binary signals, and the frequency deviation is $f_d = 1/4T_s = R_s/4$. Phase continuity is

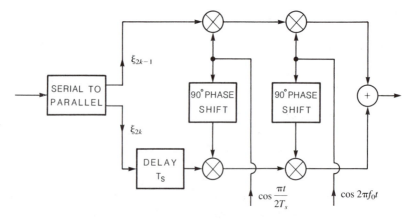

Figure 5.32 Modulator for MSK signals.

assured by the oscillator itself. In the latter case, an OQPSK modulator is used with the sinusoidal shaping of the baseband waveforms, as shown in Fig. 5.32. A one-to-one relationship between input data and transmitted frequency can be achieved by using a differential encoding of type (5.33) on the input binary data. A modification of the scheme of Fig. 5.32 based on (5.103) is proposed in Problem 5.17.

Coherent demodulation of MSK signals

When viewed as a form of OQPSK, MSK signals can be demodulated in quadrature channels as shown in Fig. 5.33. The multipliers and integrators constitute an optimum coherent receiver for the two sinusoidal antipodal signals. Note the integration interval of $2T_s$. This receiver has the same error performance (5.19) as the QPSK receiver. A slight deterioration is caused if differential encoding is used, exactly as for DECPSK in Example 5.5 The same error performance is obtained (see Example 4.12) when considering the phase continuity of the MSK signals. The demodulation is accomplished by using the Viterbi algorithm. This approach exploits the nature of MSK as a CPM signal. On the other hand, MSK can be viewed as a form of orthogonal binary FSK and, as such, it can be coherently demodulated with symbol-by-symbol decisions made every T_s seconds. A degradation of 3 dB with respect to the optimum demodulator will be encountered. Of course, coherent detection requires carrier and clock recovery. An appealing feature of MSK is that both synchronization signals can be easily recovered by manipulating the received signals (De Buda, 1972).

Noncoherent demodulation of MSK signals

Being a type of FSK signals, MSK can be noncoherently demodulated by means of a limiter and frequency discriminator, as discussed in Section 5.5.3. This possibility allows easier and inexpensive demodulation at the price of additional degradation in

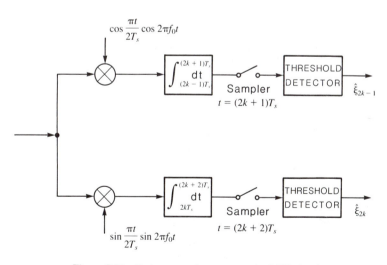

Figure 5.33 Optimum receiver structure for MSK signals.

error performance. This in given in Fig. 5.25 by the curve labeled $2f_dT = 0.5$. A penalty of 3.6 dB is paid with respect to FSK coherent demodulation and of 6.6 dB with respect to the optimum receiver. Due to the memory (phase continuity) of the modulation process, a differential demodulation of MSK is possible (see Problem 5.18) and the resulting performance lies between those for DECPSK and DCPSK.

Bandwidth requirements of MSK signals

Being a particular case of a CPFSK signal, the power spectrum of MSK can be derived from (5.97) by letting $M = 2$ and $2f_dT = 0.5$. We can get

$$\mathscr{G}^{(c)}(f) = \frac{8T_s}{\pi^2} \frac{(1 + \cos 4\pi f T_s)}{(1 - 16f^2 T_s^2)^2}. \tag{5.106}$$

This spectrum is shown in Fig. 5.34 where, for comparison, the spectrum of QPSK (or OQPSK) is also reported. The normalization is with respect to the maximum value of the QPSK spectrum. It is seen that the MSK spectrum falls off at a higher rate than QPSK. In contrast, the main lobe of the MSK spectrum is wider than that of QPSK, the first nulls falling at $f T_s = 0.75$ and $f T_s = 0.5$, respectively. These spectral properties indicate that there is a frequency value beyond which filtering can cause

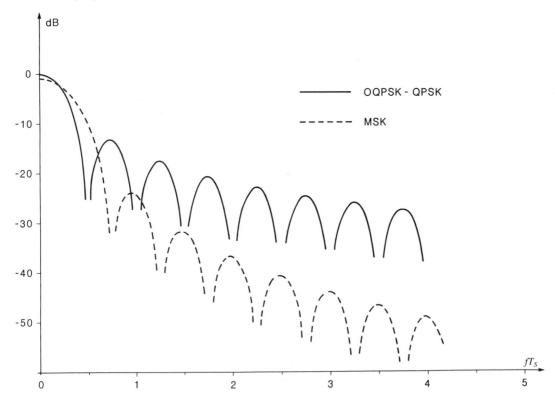

Figure 5.34 Power spectrum of MSK and OQPSK signals, normalized with respect to the maximum value of the QPSK spectrum.

lower distortion on MSK than on OQPSK. However, the precise boundary of the superior performance of the two modulations must be considered in practical situations, as also the channel effects must be included, together with the presence of adjacent signals whose interference is a critical design guideline (see Chapter 6).

Furthermore, to give the same error performance MSK requires less bandwidth than binary PSK for the same source rate and less power than any other binary FSK. For these reasons, MSK is also called fast FSK (FFSK). Its overall performance actually comes close to QPSK and even better in certain practical situations.

5.7 CONTINUOUS PHASE MODULATION

Continuous-phase-modulated (CPM) signals were introduced in Section 4.4. A sequence ξ of K RVs (or symbols) is transmitted using the signal

$$v_\xi(t) = \sum_{k=0}^{K-1} s(t - kT; \xi_k, \sigma_k), \qquad 0 \le t < KT, \tag{5.107}$$

where

$$s(t - kT; \xi_k, \sigma_k) = \sqrt{\frac{2\mathscr{E}}{T}} \cos\left\{2\pi f_0 t + \Phi(t; \xi_k, \sigma_k)\right\}, \qquad kT \le t < (k+1)T. \tag{5.108}$$

The phase function $\Phi(t; \xi_k, \sigma_k)$ is given in (4.137) and decomposed in (4.141) and (4.149). The state σ_k of the modulator is defined in (4.143). Let us rewrite (4.137) for easier reference:

$$\Phi(t; \xi_k, \sigma_k) = 2\pi h \sum_{i=0}^{k} \xi_i q(t - iT) + \psi_0, \qquad kT \le t < (k+1)T, \tag{5.109}$$

where $q(t) \triangleq \int_0^t g(\tau)d\tau$, $g(t)$ being the frequency pulse that satisfies conditions (4.138) and (4.139).

From (5.108) it is seen that CPM signals have a constant envelope, and, looking at (5.105), they can be considered an extension and generalization of MSK. In fact, we can either let the RVs ξ_k assume values in a multilevel set $\{\pm 1, \pm 3, \ldots, \pm(M-1)\}$ or use different pulses $g(t)$ or allow the duration of $g(t)$ to be longer than the symbol interval T. As we shall see, these different choices permit us to achieve different bandwidth occupancies and distances. Therefore, it will be possible to trade bandwidth efficiency for signal-to-noise ratio $\mathscr{E}_b/N_0$. Another important factor coming in with CPM signals is the complexity of both the transmitter and receiver. It is dependent on the number of states given in (4.147). Being a generalization of MSK, the performance of CPM signals will be compared with MSK. A classification of CPM signals, as usual, can derive from the number of bits per symbol transmitted: they can be *binary* or *multilevel*. Another important parameter is the duration of the frequency pulse $g(t)$. When $g(t)$ lasts one symbol period T, we get what is called a *full-response* CPM signal.

On the other hand, *partial-response* CPM signals occur when $g(t)$ is defined over LT periods, $L \ge 2$ being an integer that represents the correlation length of the pulse.

Sec. 5.7   Continuous Phase Modulation   **247**

Of course, different CPM signals are also obtained by appropriate choices of the pulse shape $g(t)$.

5.7.1 Full-Response CPM Signals

As just stated, these signals are characterized by pulse shapes $g(t)$ that are zero outside a symbol period T. A first important class of signals is obtained by using the rectangular pulse of Fig. 5.35a (named concisely 1REC, after Aulin and Sundberg, 1981). They are CPFSK signals, because the frequency is constant over each symbol period and the phase is continuous. The members of this class are identified by the modulation index h and by the number of levels M of the transmitted RVs. It is immediately verified that MSK is a particular member of this class, with $h = 0.5$ and $M = 2$. We want now to analyze the minimum-distance properties and the power-spectrum behavior of CPFSK signals. Starting from expression (4.169) for the normalized squared minimum

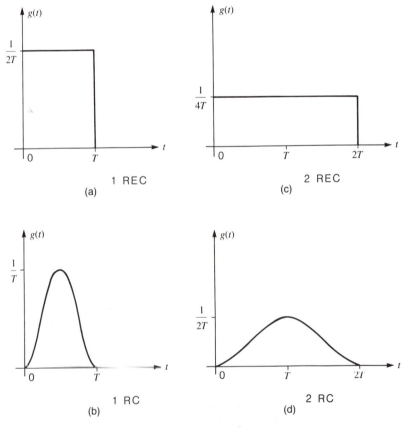

Figure 5.35 Frequency pulses for full-response and partial-response CPM signals.

Euclidean distance and using the same arguments that led to (4.173), we obtain the following result for M-ary CPFSK signals:

$$\frac{d_N^2(\boldsymbol{\xi}^{(l)}, \boldsymbol{\xi}^{(m)})}{2\mathscr{E}_b} \leqslant \frac{d_B^2}{2\mathscr{E}_b}$$

$$= (\log_2 M) \min_{1 \leqslant k \leqslant M-1} \left[2 - \frac{1}{T} \int_0^{2T} \cos\{4\pi hk[q(t) - q(t-T)]\}\, dt \right]. \quad (5.110)$$

Notice that the upper bound $d_B^2/2\mathscr{E}_b$ is normalized with respect to the energy per bit $\mathscr{E}_b$ in order to allow meaningful comparisons. Using in (5.110) the waveform $q(t)$ corresponding to the 1REC pulse, we get

$$\frac{d_B^2}{2\mathscr{E}_b} = (\log_2 M) \min_{1 \leqslant k \leqslant M-1} \left[2\left(1 - \frac{\sin 2\pi hk}{2\pi hk}\right) \right]. \quad (5.111)$$

This result was already derived in (4.173) for the binary case. Figure 4.43 shows a plot of this bound for $M = 2$. Similar plots can be obtained for $M > 2$ (see Problem 5.19). The bound in (5.111) can be optimized with respect to the modulation index h for any given value of M. Some results are shown in Table 5.6. In the table the minimum number N_B of observation intervals required to achieve the bound is also given.

It is interesting to observe that the value $h = 0.715$ that maximizes the distance in the binary case for an observation interval $N \geqslant 3$ also maximizes the distance over a one symbol observation interval (see Example 4.6 and Fig. 4.13). The values given in Table 5.6 for $M > 2$ are only of theoretical interest, since they occur for values of h near 1, which is indeed very critical. In fact, $h = 1$ is a weak modulation index for all M (see Section 4.4.1). Therefore, to closely approach $h = 1$ requires high values of observation intervals to achieve the bound. This causes increasing complexity in the receiver. Moreover, from the spectral properties of CPFSK signals as given in Example 5.10 (notice that $h = 2f_dT$), it can be seen that for h near to 1 the power spectrum tends to exhibit spectral lines.

Nevertheless, Table 5.6 shows, in the last column, substantial possible gains in $\mathscr{E}_b/N_0$ over MSK, for which $d_B^2/2\mathscr{E}_b = 2$. Looking for practical schemes, we have to choose moderate values of M and rational values of h in order to obtain a finite number of states that allows the use of the coherent receiver with the Viterbi processor shown in Fig. 4.35.

TABLE 5.6 CPM full-response signals. Optimum value of the modulation index h for different M's.

M	Optimum h	$d_B^2/2\mathscr{E}_b$	N_B	Gain over MSK (dB)
2	0.715	2.434	3	0.85
4	0.914	4.232	9	3.25
8	0.964	6.141	41	4.87
16	0.983	8.088	178	6.07

TABLE 5.7 Performance parameters for different CPM full-response signals

M	h	$\dfrac{d_B^2}{2\mathscr{E}_b}$	N_B	Gain over MSK (dB)	Bandwidth efficiency (99%)	Number of States
2	1/2	2.00	2	0	0.83	4
4	1/4	1.45	2	−1.38	1.25	8
	2/5	3.04	4	1.82	0.93	5
	4/9	3.60	5	2.56	0.85	9
8	1/8	0.60	2	−5.23	1.85	16
	4/13	3.00	2	1.76	1.00	13
	4/9	5.40	5	4.31	0.72	9

Some results for practical cases are shown in Table 5.7. Notice that we have also given the bandwidth efficiency R_s/W, introduced in (5.5). The bandwidth W is defined here as the 99% fractional power bandwidth. From Table 5.7 we can observe that the quaternary and octonary schemes with $h = \frac{4}{9}$ have approximately the same bandwidth efficiency as MSK, but yield a gain in $\mathscr{E}_b/N_0$ of 2.56 dB and 4.31 dB, respectively. A pictorial comparison of different M-ary CPFSK schemes is also shown in Fig. 5.36. The multilevel signals are superior to binary. Notice also that, for a given M, it is possible to trade bandwidth for signal-to-noise ratio by changing the modulation index h. (In Fig. 5.36 h increases going from left to right.) The complexity of the receiver of Fig. 4.35 depends on the number of states given in (4.147) and shown in Table 5.7.

So far, only CPFSK signals have been considered. A further step is to consider the effects of pulse shaping on the minimum distance and spectrum of full-response

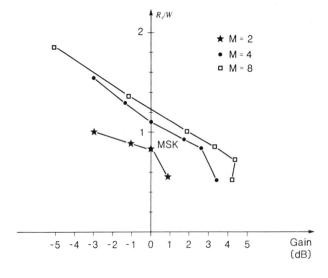

Figure 5.36 Performance of different CPFSK signals. Gain (in decibels) in $\mathscr{E}_b/N_0$ with respect to MSK is shown on the abscissa. Bandwidth efficiency is given on the vertical axis. The bandwidth is that containing 99 percent of the total power.

Digital Modulation Schemes   Chap. 5

CPM signals. We shall devote some discussion to the binary case. Consider frequency pulses $g(t)$ having the symmetry property

$$g(t) = g(T - t), \qquad 0 \leqslant t < T. \tag{5.112}$$

Then it can be shown (see Problem 5.20) that the bound for the minimum distance in the binary case is

$$\frac{(d_B^2)_{sym}}{2\mathscr{E}_b} = 2\left\{1 - \frac{1}{T}\int_0^T \cos[4\pi h q(t)]dt\right\}. \tag{5.113}$$

Since $\cos(\cdot) \geqslant -1$, it can be concluded that $(d_B^2)_{sym}/2\mathscr{E}_b$ can never exceed 4 and, therefore, at most a gain of 3 dB might be attained over MSK by varying the pulse shape. Further insight into the problem of optimizing the pulse shape is gained if (5.113) is developed and written in the new form (see Problem 5.21):

$$\frac{(d_B^2)_{sym}}{2\mathscr{E}_b} = 2\{1 - 2 \cos(\pi h)\frac{1}{T}\int_0^{T/2} \cos[4\pi h q_0(t) - \pi h]dt\}, \tag{5.114}$$

where $q_0(t) \triangleq q(t + T/2)$.

The first important result from (5.114) is that for $h = \frac{1}{2}$ the upper bound is $(d_B^2)_{sym}/2\mathscr{E}_b = 2$ independently of the frequency pulse $g(t)$. This is of particular interest, as this case ($M = 2$, $h = \frac{1}{2}$) has been widely considered due to the possibility of using simple coherent receivers of the offset quadrature type (as that of Fig. 5.32 for MSK). When $h = \frac{1}{2}$, the problem of finding the best pulse $g(t)$ is based on the criterion of the smallest bandwidth in terms of fractional out-of-band power, since the error performance is always the same as MSK.

To reduce the spectral tails, the phase during a symbol interval should change slowly and smoothly. A frequency pulse of interest is a raised-cosine (1RC) function as shown in Fig. 5.35b and given by

$$g(t) = \frac{1}{2T}\left(1 - \cos\frac{2\pi t}{T}\right), \qquad 0 \leqslant t < T. \tag{5.115}$$

This modulation scheme is known as sinusoidal FSK (SFSK). It has a spectrum whose tails decrease faster than MSK (Amoroso, 1976; Ajmone Marsan and Biglieri, 1977).

The problem of finding frequency pulses $g(t)$ that optimize the distance bound d_B^2 is much more difficult for the general M-ary case and no general solution is available. Further understanding of the issue can be obtained by reading the cited references and solving the proposed problems. An interesting generalization of the CPFSK signals discussed in this section is obtained if a set of modulation indexes is used, instead of only one (Multi-h codes). This case is described in Section 9.7.

5.7.2 Partial-Response CPM Signals

If the frequency pulse $g(t)$ is of length greater than T (the symbol interval), we introduce correlation among transmitted waveforms. These signals are called partial-response CPM signals.

With respect to full-response CPM signals, a new parameter, the correlation length L of the pulse $g(t)$, is introduced, and therefore a greater variety of modulation schemes is possible.

Raised cosine (RC) or rectangular (REC) pulses can be used, as shown in Fig. 5.35c and 5.35d for $L = 2$. But other pulses have been considered and their performance evaluated.

The case of a 2REC pulse (Fig. 5.35c) was studied in Example 4.11, and the distance properties of these signals in the binary case are shown in Fig. 4.44, which is a plot of $d_B^2/2\mathscr{E}_b$. Notice the fundamental difference with the binary case and 1REC pulse. Here we have that three (instead of two) instantaneous frequencies are possible in every node of the tree. Due to the correlation introduced by the pulse $g(t)$, only two of these frequencies are used in every node, and the choice is based on the previous symbol. For $h = \frac{1}{2}$, this modulation scheme has been called *duobinary MSK* or *frequency-shift offset quadrature* (FSOQ) modulation. In Problem 5.24 an equivalent and interesting way of looking at this scheme is proposed. Also, the case of a 2RC pulse can be considered.

Some results for these two schemes are shown in Table 5.8. This table gives the answer to the question that was raised in Example 4.14. The duobinary MSK ($h = \frac{1}{2}$, 2REC pulse) presents a loss of 0.62 dB with respect to MSK, but it has a higher bandwidth efficiency. This indeed is progressively better if we consider wider bandwidths. (Compare, in fact, the 99% and 99.9% bandwidth results.)

With the 2RC pulse we can retain the same distance as MSK, while improving the bandwidth efficiency. Better results can be obtained with $h = 0.7$, but at the expense of an increased complexity (number of states).

Generally, the spectral main lobe width is reduced by increasing L, whereas the spectral tails are lowered by choosing a pulse shape $g(t)$ with many continuous derivatives. Concerning the distance d_B^2, the longer the pulses $g(t)$, the larger the distance for large h values is, and the shorter the pulses the larger the distance for low h values. Such behavior gives rise to a large variety of trade-offs between bandwidth and error probability, as shown in Fig. 5.36 for the case of full-response CPM signals.

It is worth mentioning that one of the first methods that achieved a significant improvement on the spectral properties of MSK at the expense of only a 1-dB loss in

TABLE 5.8 Performance parameters for different CPM partial response signals

h	System	$\dfrac{d_B^2}{2\mathscr{E}_b}$	Gain over MSK (dB)	Bandwidth Efficiency (99%)	(99.9%)	Number of States
1/2	MSK	2.00	0	0.83	0.36	4
	2REC	1.73	−0.62	1.11	0.54	8
	2RC	1.97	−0.07	0.94	0.62	8
7/10	2REC	2.90	1.62	0.84	0.47	40
	2RC	2.63	1.19	0.91	0.52	40

$\mathcal{E}_b/N_0$ was the *tamed frequency modulation* (TFM). It can be shown to belong to the class of partial-response CPM signals. This modulation has later been generalized to *correlative phase-shift keying* (CORPSK). The interested reader is referred to the cited references in the Bibliographical Notes.

5.8 DIGITAL MODULATION TRADE-OFFS

Any digital modulation aims at realizing the best possible trade-off in a given situation among the bit error probability $P_b(e)$, the bandwidth efficiency R_s/W, the ratio $\mathcal{E}_b/N_0$, and the complexity of the equipment. This section summarizes the results of the chapter in the framework of a comparison based on these parameters. A concise reference to the various modulation schemes studied in the chapter is given in Table 5.9.

Notice that the indicated bandwidth efficiency is the result of a rather crude evaluation of the bandwidth, as discussed in the previous sections. Should a more refined computation be required, the reader can refer to the various definitions of bandwidth described in Section 5.1. An example of this approach was encountered in Section 5.7 during the discussion of the relative performances of CPM schemes when compared with MSK. These systems are no longer considered here.

For purposes of comparison, it is also interesting to assess what is the ultimate performance of a modulation scheme operating on the AWGN channel. It is described by the channel capacity formula (3.89). Using (5.5) with $R_s = C$, we have

$$\frac{C}{W} = \log_2 \left(1 + \frac{\mathcal{E}_b}{N_0} \frac{C}{W} \right). \tag{5.116}$$

TABLE 5.9 References to the modulation schemes studied in Chapter 5

System	R_s/W	$P(e),\ P_b(e)$
PAM, ASK/SSB	$2 \log_2 M$	(5.11), (5.12), Fig. 5.4
CPSK	$\log_2 M$	(5.25), (5.32), Fig. 5.12
DECPSK	$\log_2 M$	(5.42), (5.44)
DCPSK	$\log_2 M$	(5.58), (5.62), Fig. 5.18
AM–PM	$\log_2 M$	(5.66)
FSK orthogonal (coherent)	$\dfrac{2 \log_2 M}{M}$	(4.35), (4.64), (5.74), Fig. 4.19
FSK orthogonal (incoherent)	$\dfrac{\log_2 M}{M}$	(4.129), (5.74), Fig. 4.27
Binary FSK (discrimination detection)	$\dfrac{1}{1 + 2f_d T}$	(5.85), Fig. 5.25
MSK		(5.106), Fig. 5.34, Section 5.6.2
CPM, full response		Tables 5.6 and 5.7, Fig. 5.36
CPM, partial response		Table 5.8

Hence

$$\frac{\mathscr{E}_b}{N_0} = \frac{2^{C/W} - 1}{C/W}.$$ (5.117)

This expression shows that $\mathscr{E}_b/N_0$ must increase exponentially with C/W. On the other hand, if $C/W \to 0$,

$$\frac{\mathscr{E}_b}{N_0} = \lim_{C/W \to 0} \frac{2^{C/W} - 1}{C/W} = \ln 2,$$ (5.118)

which is -1.6 dB. This result indicates that reliable transmission, with as low a $P_b(e)$ as desired, is possible on the AWGN channel only with a ratio $\mathscr{E}_b/N_0$ of at least this value. Let us apply these concepts to analyze possible trade-offs with CPSK modulation.

Example 5.11

We consider in this example the CPSK signals described in Section 5.3.1. First, the error probability curves of Fig. 5.12 are represented in Fig. 5.37. Instead of the symbol error probability $P(e)$, the bit error probability $P_b(e)$ is plotted by using the assumption of a Gray code and therefore the lower bound in (5.12). Furthermore, each curve is labeled with the parameter R_s/W (bandwidth efficiency), which replaces the number of levels M. Also note that the limit of -1.6 dB on $\mathscr{E}_b/N_0$ is indicated as the Shannon capacity bound.

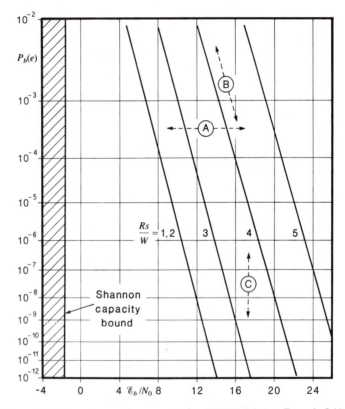

Figure 5.37 Error-plane performance plot for CPSK signals (see Example 5.11).

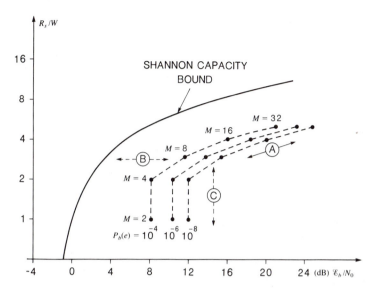

Figure 5.38 Bandwidth-efficiency plane for CPSK signals (see Example 5.11).

We may refer to this figure as the *error probability plane* for this type of modulation. It describes the possible operating points available with this modulation. The allowed trade-offs can be viewed as changes in the operating point on the curves. Such trade-offs are shown by arrows in Figure 5.37.

Moving along the line Ⓐ can be viewed as trading $\mathscr{E}_b/N_0$ versus bandwidth efficiency R_s/W, with $P_b(e)$ fixed. Similarly, the line Ⓑ shows a trade-off of error probability $P_b(e)$ with $\mathscr{E}_b/N_0$, with fixed bandwidth. Finally, movement along line Ⓒ illustrates trading bandwidth efficiency versus $P_b(e)$ performance, with $\mathscr{E}_b/N_0$ fixed. Notice that movement along line Ⓑ is implemented simply by changing the available $\mathscr{E}_b/N_0$, whereas movement along lines Ⓐ or Ⓒ requires a change of the modulation scheme. Therefore, it implies different equipment complexity.

The amount of information contained in Fig. 5.37 can also be presented on the *bandwidth-efficiency plane* of Fig. 5.38. The Shannon capacity bound (5.117) is plotted on this plane. On the error-rate plane we had equibandwidth curves. Hence, here we can plot equi-error-probability "curves," which indeed are meaningful only in the points that represent possible modulation schemes. Moving from curve to curve, $P_b(e)$ can only be inferred while the bandwidth efficiency is explicit, since the ordinate is just R_s/W. This plane is very useful when discussing trade-offs with a fixed $P_b(e)$, this being a usual design constraint. As for the CPSK system, Fig. 5.38 shows that the bandwidth efficiency can be increased at the expense of $\mathscr{E}_b/N_0$ by increasing the number of levels M. Notice in particular that binary PSK and QPSK require the same $\mathscr{E}_b/N_0$, but the latter has twice as much bandwidth efficiency over the former. This feature stems from the fact that QPSK is actually composed of two binary PSK transmitted over orthogonal carriers. □

The bandwidth efficiency plane is a powerful tool for the comparison of different modulation schemes. Such a comparison is illustrated in Fig. 5.39 where a bit error probability $P_b(e) = 10^{-5}$ has been fixed. The Shannon capacity curve shows the bound to reliable transmission of any existing modulation scheme. The graph shows the fact

Sec. 5.8   Digital Modulation Trade-Offs   **255**

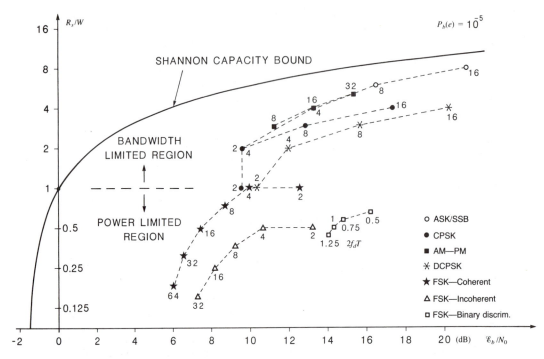

Figure 5.39 Comparison among different modulation schemes on the bandwidth-efficiency plane for a bit error probability $P_b(e) = 10^{-5}$.

that ASK, CPSK, and DCPSK systems are bandwidth-efficient signaling techniques, since they cover the region of the plane where $R_s/W > 1$. In this region, the system bandwidth is limited and it can be traded for power (i.e., $\mathscr{E}_b/N_0$). In fact, for a fixed bandwidth, the bandwidth efficiency can be increased with an increase in the number of levels M. The price paid to achieve the same $P_b(e)$ is an increase in $\mathscr{E}_b/N_0$.

On the other hand, FSK signals make an inefficient use of bandwidth, since they cover the region of the plane where $R_s/W < 1$. But these systems trade bandwidth for a reduction of the $\mathscr{E}_b/N_0$ required to achieve the same $P_b(e)$.

However, notice how the desirable trade-offs associated with each of these regions are achieved. For the bandwidth-limited region, the capacity curve flattens out and ever increasing amounts of $\mathscr{E}_b/N_0$ are required to achieve an improvement in R_s/W. However, in the power-limited region the capacity boundary curve is very steep, and to achieve a small reduction in $\mathscr{E}_b/N_0$ a large increase in bandwidth is required. Nevertheless, it can be shown that in the bandwidth-limited region the ASK/SSB system is asymptotically optimum, in the same way as the FSK orthogonal system with coherent detection is asymptotically optimum in the power-limited region. The former system achieves capacity as $R_s/W \rightarrow 0$ (and consequently $\mathscr{E}_b/N_0 \rightarrow -1.6$ dB). The reader is invited to work out Problems 5.25 and 5.26 to obtain these two results.

In conclusion, some practical considerations can be drawn by inspecting the curves of Fig. 5.39. In the first place, when transmitting data on a voice-grade channel, it should be clear that FSK modulation is indicated for low-speed transmission (up to

1200 bits/s), whereas higher speeds (up to 9600 bits/s) require the use of CPSK, PAM, or AM–PM modulations. Concerning FSK, it can be seen that binary transmission with a discriminator-based receiver has practically the same performance as the optimum incoherent FSK receiver. Trade-offs with equipment complexity are clearly inferred by comparing CPSK with DCPSK or orthogonal FSK and the coherent receiver with the incoherent one. In fact, the latter modulations have inferior performance but simpler circuitry. Notice the position of the AM–PM point ($M = 8$), which achieves a significant improvement over the CPSK modulation. Concerning CPSK, notice that its behavior is similar to ASK/SSB, except for the fact that a larger bandwidth is traded with a lower $\mathscr{E}_b/N_0$. This feature, combined with the constant envelope property of this modulation, justifies why QPSK was the first natural candidate for digital transmission over a satellite link.

Because of their good bandwidth efficiency, multilevel QAM schemes are used on digital radio relay links. A comparison of different schemes for this application can be found in Oetting (1979).

BIBLIOGRAPHICAL NOTES

All books on digital transmission deal with the topics presented in this chapter. The discussion on bandwidth in the Section 5.1 is taken from Amoroso (1980). The material on PAM is standard. For the PSK and FSK modulations, the book by Lindsey and Simon (1973) is a recommended reference; the presentation of some parts of Section 5.3 is based on it. The analysis of DCPSK modulation comes from the papers by Castellani, Lo Presti, and Pent (1974a, 1974b). An interesting reference for this modulation is the book by Schwartz, Bennett, and Stein (1966), which also inspired the discussion of binary FSK with discrimination detection in Section 5.5.3.

The rectangular AM–PM signal constellations were proposed by Campopiano and Glazer (1962); most of the suggestions on AM–PM presented in the problems section are taken from Forney and others (1984). This paper also contains an extensive bibliography on the subject.

Multilevel CPFSK signals with discriminator detection applied to radio links were studied by Corazza, Crippa, and Immovilli (1981).

Results on the spectral properties of FSK signals were first derived by Bennett and Rice (1963), but with a different approach from that presented here.

A tutorial presentation of MSK can be found in Pasupathy (1979). The discussion of optimum demodulation of MSK was first given by De Buda (1972), whereas Gronemeyer and McBride (1976) presented a comparison between MSK and OQPSK. Most of the results on CPM modulation are due to Aulin and Sundberg. The reader can refer to Aulin and Sundberg (1981), Aulin, Rydbeck, and Sundberg (1981), and Aulin, Lindell, and Sundberg (1981). SFSK was introduced by Amoroso (1976). Simon (1976) presented a generalization of MSK-type signals by shaping the symbol pulse. TFM signals were invented by De Jager and Dekker (1978). The schemes known as correlative PSK are described in Muilwijk (1981). An analysis of Duobinary MSK can be found in Elnoubi and Gupta (1981). Discriminator detection of CPM signals was studied by Bellini and

Tartara (1985). The presentation of the final section on digital modulation trade-offs was inspired by the excellent tutorial paper by Sklar (1983).

PROBLEMS†

5.1. A coherent M-ary PAM transmission scheme is used with the constraint that the peak power of each signal be not greater than 1 mW. The noise power spectral density $N_0/2$ is 0.25 μW/Hz. A bit error probability $P_b(e)$ of 10^{-6} is required.

(a) Compute the maximum possible transmission speed in bits/second for $M = 2$, $M = 4$, and $M = 8$.

(b) Which one of the three schemes requires the minimum value of $\mathscr{E}_b$ (energy per bit)?

5.2. Consider an octonary CPSK transmission scheme with the signal space as shown in Fig. 5.5.

(a) Draw a block diagram of the optimum receiver with a logic device that takes decisions based only on the sign of the received signal components.

*(b) Compute the capacity of the equivalent discrete transmission channel by using for $P(e)$ the upper bound of (5.32). Assume that $\mathscr{E}/N_0$ is so high that only errors between adjacent signals in the signal space can be taken into account. Compare this capacity with that of binary and quaternary CPSK as a function of $\mathscr{E}_b/N_0$.

(c) Compute the exact (closed-form) expression of the bit error probability $P_b(e)$. Consider a Gray code as in Fig. 5.5.

5.3. (Arthurs and Dym, 1962) This problem is proposed in order to show that the demodulator of Fig. 5.16 is an optimum ML receiver for the detection of phase differences.

(a) First, consider (2.324) and notice that the ML receiver looks for the maximum conditional pdf $f_{r|s j}(\mathbf{r}|\mathbf{s}_j)$, where $\mathbf{r}$ is the vector of the received signal that contains the sufficient statistics.

(b) The couple of received signals used in the detection of the transmitted signal $\mathbf{s}_j$ is the following:

$$r_1(t) = \sqrt{\frac{2\mathscr{E}}{T}} \cos(2\pi f_0 t + \varphi_a) + n(t),$$

$$r_2(t) = \sqrt{\frac{2\mathscr{E}}{T}} \cos\left\{2\pi f_0 t + \varphi_a + \frac{2\pi}{M}(j-1) + \Phi\right\} + n(t - T),$$

where $n(t)$ is a white Gaussian noise. Notice that the hypothesis of signal $\mathbf{s}_j$ means that the phase difference is

$$\Delta\varphi_j = \frac{2\pi}{M}(j-1) + \Phi.$$

(c) Show that the received signal points coordinates in the plane

$$\psi_1(t) = \sqrt{\frac{2}{T}} \cos 2\pi f_0 t, \qquad \psi_2(t) = -\sqrt{\frac{2}{T}} \sin 2\pi f_0 t$$

† The problems marked with an asterisk should be solved with the aid of a computer.

have the form

$$r_{11} = \sqrt{\mathcal{E}} \cos \varphi_a + n_{11},$$
$$r_{12} = \sqrt{\mathcal{E}} \sin \varphi_a + n_{12},$$
$$r_{21} = \sqrt{\mathcal{E}} \cos \left[\varphi_a + \frac{2\pi}{M} (j-1) + \Phi \right] + n_{21},$$
$$r_{22} = \sqrt{\mathcal{E}} \sin \left[\varphi_a + \frac{2\pi}{M} (j-1) + \Phi \right] + n_{22}.$$

(d) Define the vector $\mathbf{r} \triangleq [r_{11}, r_{12}, r_{21}, r_{22}]$ of the received signals and show that the conditional pdf is given by

$$f_{\mathbf{r}|s_j,\varphi_a}(\mathbf{r}|s_j, \varphi_a) = A \exp \{C \cos \varphi_a + D \sin \varphi_a\},$$

where

$$A \triangleq \left(\frac{1}{\pi N_0} \right)^2 \exp \left\{ -\frac{1}{N_0} (|\mathbf{r}|^2 + 2\mathcal{E}) \right\},$$

$$C \triangleq \frac{2\sqrt{\mathcal{E}}}{N_0} \{r_{11} + r_{21} \cos \Delta\varphi_j + r_{22} \sin \Delta\varphi_j\},$$

$$D \triangleq \frac{2\sqrt{\mathcal{E}}}{N_0} \{r_{12} - r_{21} \sin \Delta\varphi_j + r_{22} \cos \Delta\varphi_j\},$$

(e) Assume that the phase ambiguity φ_a has a uniform distribution and average it out. You will get

$$f_{\mathbf{r}|s_j}(\mathbf{r}|s_j) = AI_0(\sqrt{C^2 + D^2}),$$

where $I_0(\cdot)$ is the modified Bessel function of order zero. Choosing the maximum pdf is equivalent to choose the maximum of $(C^2 + D^2)$, since A does not depend on $\Delta\varphi_j$ and $I_0(\cdot)$ is a monotone increasing function of its argument.

(f) Switch to polar coordinates and show that

$$C^2 + D^2 = \frac{8\mathcal{E}\rho}{N_0} [\rho + \cos(\Delta\beta - \Delta\varphi_j)],$$

where

$$\rho = |\mathbf{r}_1| = |\mathbf{r}_2|,$$
$$\Delta\beta = \arg(\mathbf{r}_2) - \arg(\mathbf{r}_1),$$

and

$$\mathbf{r}_i = [r_{i1}, r_{i2}].$$

(g) Decide on the optimality of the receiver.

5.4. In this problem we outline the computations that led to the result (5.58) for DCPSK signals. Let us start from (5.55).

(a) Compute first $P\{\mathcal{R}(z_1 z_2^*) < 0\}$. Define the two Gaussian RVs

$$\xi_{12} = \frac{z_1 + z_2}{2}, \qquad \eta_{12} = \frac{z_1 - z_2}{2}$$

Chap. 5   Problems

259

and show that they are independent under the assumption that

$$E\{\tilde{n}^*(t_k)\tilde{n}(t_k - T)\} = 0.$$

(b) The RVs $R_1 \triangleq |\xi_{12}|$ and $R_2 \triangleq |\eta_{12}|$ are independent Rician variables with pdf given by

$$f_{R_i}(x) = \frac{x}{\sigma_i^2} \exp - \frac{a_i^2 + x^2}{2\sigma_i^2} I_0\left(\frac{a_i x}{\sigma_i^2}\right),$$

with $0 \le x < \infty$, $i = 1, 2$, and

$$a_1^2 = |E\{\xi_{12}\}|^2 = \frac{A^2}{2}\left(1 - \sin\frac{\pi}{M}\right),$$

$$a_2^2 = |E\{\eta_{12}\}|^2 = \frac{A^2}{2}\left(1 + \sin\frac{\pi}{M}\right),$$

$$\sigma_1^2 = \frac{1}{2}E\{|\xi_{12} - E\{\xi_{12}\}|^2\} = \frac{N_0}{2T},$$

$$\sigma_2^2 = \frac{1}{2}E\{|\eta_{12} - E\{\eta_{12}\}|^2\} = \frac{N_0}{2T} = \sigma_1^2.$$

(c) Use formulas of Appendix A and get

$$P\{\Re(z_1 z_2^*) < 0\} = P\{R_1 < R_2\} = \frac{1}{2}[1 - Q(\sqrt{b}, \sqrt{a}) + Q(\sqrt{a}, \sqrt{b})],$$

where

$$a \triangleq \frac{a_2^2}{\sigma_1^2 + \sigma_2^2}, \qquad b \triangleq \frac{a_1^2}{\sigma_1^2 + \sigma_2^2}.$$

(d) Following the same procedure, show that

$$P\{\Re(z_1 z_3^*) < 0\} = \frac{1}{2}[1 - Q(\sqrt{a}, \sqrt{b}) + Q(\sqrt{b}, \sqrt{a})].$$

(e) Conclude and obtain (5.58).

5.5. Removing the assumption of Problem 5.4, assume that the noise samples of (5.53) are correlated and define

$$\frac{T}{2N_0} E\{\tilde{n}^*(t_k)\tilde{n}(t_k - T)\} \exp\left[j\left(\gamma_m + \Phi - \frac{\pi}{M} + \frac{\pi}{2}\right)\right] \triangleq \rho_{cm} + j\rho_{sm}, \qquad m = 1, 2,$$

and

$$\rho_D \triangleq \frac{T}{2N_0} E\{\tilde{n}^*(t_k)\tilde{n}(t_k - T)\}.$$

Therefore,

$$\rho_{cm} = -\rho_D \sin\left(\gamma_m + \Phi - \frac{\pi}{M}\right) = -\rho_D \sin\psi_m,$$

$$\rho_{sm} = \rho_D \cos\left(\gamma_m + \Phi - \frac{\pi}{M}\right) = \rho_D \cos\psi_m, \qquad m = 1, 2.$$

(a) Consider first the case of the computation of $P\{\mathcal{R}(z_1 z_2^*) < 0\}$. Define two RVs with the following transformation:

$$t_1 = \frac{(z_1 + z_2)(1 + K) + (z_1 - z_2)(1 - K)\,\exp(-j\alpha)}{4},$$

$$t_2 = \frac{(z_1 + z_2)(1 - K) + (z_1 - z_2)(1 + K)\,\exp(-j\alpha)}{4},$$

where

$$K = \sqrt{\frac{1 + \rho_{s2}}{1 - \rho_{s2}}}, \qquad \alpha = -\frac{\pi}{2}.$$

Show that the two RVs t_1 and t_2 are independent.

(b) Show that

$$P\{\mathcal{R}(z_1 z_2^*) < 0\} = P\{|t_1|^2 < |t_2|^2\}.$$

Therefore, we are reduced again to the case of two Rician independent variables.

(c) Repeat steps (a) and (b) with the two RVs z_1 and z_3 (using ρ_{s1} for the definition of K).

(d) Using the formulas of Appendix A, show that the conditional error probability can be put in the form

$$P(e\,|\,\Delta\varphi_i) < 1 + \frac{1}{2}\{Q(\sqrt{b_2},\ \sqrt{a_2}) - Q(\sqrt{a_2},\ \sqrt{b_2})$$

$$+ \frac{c_2}{2}\exp\left(-\frac{a_2 + b_2}{2}\right)I_0(\sqrt{a_2 b_2}) - Q(\sqrt{b_1},\ \sqrt{a_1})$$

$$+ Q(\sqrt{a_1},\ \sqrt{b_1}) - \frac{c_1}{2}\exp\left(-\frac{a_1 + b_1}{2}\right)I_0(\sqrt{a_1 b_1})\}, \qquad i = 1, 2, \ldots, M,$$

where

$$\begin{Bmatrix} a_m \\ b_m \end{Bmatrix} \triangleq \frac{T}{4N_0(1 - \rho_D^2 \cos^2\psi_m)}\left[\,|x|^2 + |y_m|^2\right.$$

$$\left. + 2\rho_D \cos\psi_m\,\mathcal{I}(xy_m^*) \mp 2\sqrt{1 - \rho_D^2 \cos^2\psi_m}\,\mathcal{R}(xy_m^*)\right],$$

$$c_m \triangleq -\frac{\rho_D \sin\psi_m}{\sqrt{1 - \rho_D^2\cos^2\psi_m}}, \qquad m = 1, 2,$$

and

$$x \triangleq E\{z_1\}, \qquad y_1 \triangleq E\{z_2\}, \qquad y_2 \triangleq E\{z_3\}.$$

5.6. Consider an AM–PM signal set of the type shown in Fig. 5.22. By assuming as decision regions squares of side $2d$, derive the upper bound (5.68).

5.7. Consider the signal constellation of 64 points of Fig. P5.1. Notice that the four outer points of the square constellation of Fig. 5.22 are moved to the axes, as was done for the eight-point constellation of Fig. 5.21c. The resulting constellation is "more circular" and should be better than the corresponding square with 64 points. Evaluate the amount of energy improvement (in decibels) with respect to the square constellation.

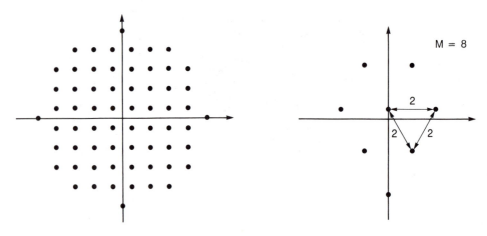

Figure P5.1   **Figure P5.2**

5.8. Consider the cases of Figs. P5.2, P5.3, and P5.4, where the minimum distance is always 2 and the center of gravity is at the origin. Compute the average energy of each signal set. Verify, by comparing with Table 5.1, the amount of energy gain for each case. Notice that the structure with $M = 16$ signals is the best 16-point constellation known.

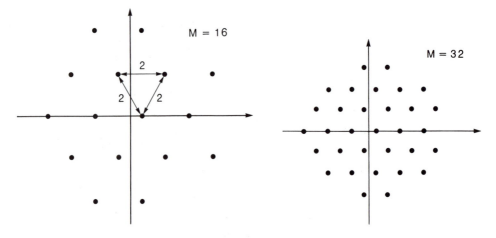

Figure P5.3   **Figure P5.4**

5.9. Consider the suboptimal ''double diamond'' eight-point constellation of Fig. P5.5 and compare its average energy with the other eight-point constellations of Figs. 5.21 and P5.2.

5.10. Assume that we want to transmit $(h + \frac{1}{2})$ bits per symbol. To this purpose we use a signal constellation with 2^h inner points taken from a regular grid and 2^{h-1} outer points drawn from the same grid with the goal of symmetry and minimum energy. Two examples are shown in Figs. P5.6 and P5.7 for $h = 4$ and $h = 5$ (respectively, 24 and 48 signal points). The transmission goes as follows.

1. Group the incoming bits into blocks of $2h + 1$ bits to be sent with two waveforms $(h + \frac{1}{2}$ bits per symbol).

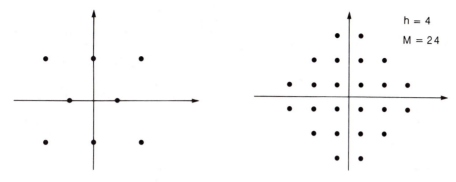

Figure P5.5   **Figure P5.6**

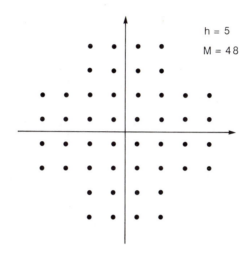

Figure P5.7

2. The first bit determines whether or not any outer point is to be used.

3. If not, the remaining $2h$ bits are used to select two inner points for transmission in two successive periods.

4. If yes, one additional bit selects which of the two signals should be an outer point and the remaining $h - 1$ and h bits select, respectively, the outer and the inner point to be sent.

On the average, one outer point is transmitted every other four transmissions. Compute the average energy of the signal sets of the figure and verify that the transmission of the additional $\frac{1}{2}$ bit per symbol requires about 1.5 dB more energy. Generalize the scheme to transmit $h + 2^{-t}$ bits/symbol, with t an integer greater than 1.

5.11. In this problem we sketch the steps of the computation to obtain the result (5.85) for FSK with discrimination detection.

(a) Assume that a mark is transmitted and show that

$$\sigma_{z_1}^2 = \frac{1}{2} \mathrm{E}\{(z_1 - \mathrm{E}\{z_1\})^*(z_1 - \mathrm{E}\{z_1\})\} = P_N (2\pi f_2)^2,$$

$$\sigma_{z_2}^2 = \frac{1}{2} E\{(z_2 - E\{z_2\})^*(z_2 - E\{z_2\})\} = P_N,$$

$$\sigma_{z_1}\sigma_{z_2}(\rho_c + j\rho_s) = \frac{1}{2} E\{(z_1 - E\{z_1\})^*(z_2 - E\{z_2\})\} = 0,$$

where P_N is the noise power at the output of the front-end filter and

$$f_2^2 \triangleq \frac{\displaystyle\int_{-\infty}^{\infty} f^2 G(f)\, df}{\displaystyle\int_{-\infty}^{\infty} G(f)\, df},$$

with $G(f)$ being the power density spectrum of $\tilde{n}(t)$.

(b) Apply the following linear transformation:

$$t_1 = \frac{z_1 + Kz_2}{2},$$

$$t_2 = \frac{z_1 - Kz_2}{2},$$

$$K = \frac{\sigma_{z_1}}{\sigma_{z_2}} = 2\pi f_2,$$

and show that t_1 and t_2 are independent nonzero-mean complex Gaussian RVs.

(c) Show that

$$P\{\Re(z_1 z_2^*) < 0\} = P(|t_1|^2 < |t_2|^2).$$

(d) The RVs $R_1 \triangleq |t_1|$ and $R_2 \triangleq |t_2|$ are independently distributed Rician variables with pdf given by

$$f_{R_i}(x) = \frac{x}{\sigma_i^2} \exp\left(-\frac{a_i + x^2}{2\sigma_i^2}\right) I_0\left(\frac{a_i x}{\sigma_i^2}\right),$$

with $0 \leq x < \infty$, $\quad i = 1, 2$, and

$$a_1^2 = |E\{t_1\}|^2 = \frac{A^2}{4} [2\pi f_d + K]^2,$$

$$a_2^2 = |E\{t_2\}|^2 = \frac{A^2}{4} [2\pi f_d - K]^2,$$

$$\sigma_1^2 = \frac{1}{2} E\{(t_1 - E\{t_1\})^*(t_1 - E\{t_1\})\} = \frac{P_N K^2}{2},$$

$$\sigma_2^2 = \sigma_1^2.$$

(e) Use formulas of Appendix A and get

$$P(e \mid \text{mark}) = P\{\Re(z_1 z_2^*) < 0\} = P\{R_1 < R_2\}$$

$$= \frac{1}{2} [1 - Q(\sqrt{b}, \sqrt{a}) + Q(\sqrt{a}, \sqrt{b})],$$

with

$$a \triangleq \frac{a_2^2}{\sigma_1^2 + \sigma_2^2}, \qquad b \triangleq \frac{a_1^2}{\sigma_1^2 + \sigma_2^2}.$$

(f) Show that

$$P(e \mid \text{space}) = P(e \mid \text{mark})$$

and therefore

$$P(e) = P(e \mid \text{mark}).$$

5.12. Compute the power density spectrum of the signal (5.89) and derive the results of Example 5.9. *Hint*: As the signal may not be cyclostationary, use (2.99) and (2.100).

5.13. Compute the power density spectrum of signal (5.94), and derive the results (5.95) and (5.96). *Hint*: Show first that, if $V(f)$ denotes the Fourier transform of $v_\xi(t)$,

$$\Gamma_{v_\xi}(f_1, f_2) \triangleq E[V(f_1)V^*(f_2)] = 4 \sum_{n=0}^{\infty} \sum_{m=0}^{\infty} e^{-j2\pi(nf_1 - mf_2)T}$$

$$E\{e^{j2\pi f_d T(\sigma_n - \sigma_m)} X(f_1 - f_d \xi_n)X^*(f_2 - f_d \xi_m)\}.$$

The expectation in the RHS of this equation can be evaluated by considering separately the terms for $n = m$, $n > m$, and $n < m$. The following formula (Jones, 1966, p. 137)

$$\sum_{n=1}^{\infty} e^{jnxy} = \frac{\pi}{x} \sum_{n=-\infty}^{\infty} \delta\left(y - n\frac{2\pi}{x}\right) + \frac{j}{2} \operatorname{ctg} \frac{xy}{2} - \frac{1}{2}$$

is useful.

5.14. Consider the digital signal

$$v_\xi(t) = \sum_{n=-\infty}^{\infty} q(t - nT; \xi_n),$$

$$q(t; \xi_n) = \xi_n' g'(t) + \xi_n'' g''\left(t - \frac{T}{2}\right)$$

and ξ_n', ξ_n'' are iid random variables taking equally likely values $\pm A$, and

$$g'(t) = g''(t) = 0, \qquad \text{for } |t| > \frac{T}{2}.$$

(a) Show that when

$$g'(t) = g''(t) = \begin{cases} 1, & |t| \leq \frac{T}{2}, \\ 0, & \text{elsewhere,} \end{cases}$$

$v_\xi(t)$ is an offset QPSK signal. Also, when

$$g'(t) = g''(t) = \begin{cases} \cos \frac{\pi}{T} t, & |t| \leq \frac{T}{2}, \\ 0, & \text{elsewhere,} \end{cases}$$

$v_\xi(t)$ is an MSK signal.

(b) Show that the power density spectrum of $v_\xi(t)$ is given by

$$\mathcal{G}_{v_\xi}(f) = \frac{A^2}{T}\{|G'(f)|^2 + |G''(f)|^2\}$$

Chap. 5   Problems

where $G'(f)$, $G''(f)$, are the Fourier transforms of $g'(t)$, $g''(t)$, respectively. Use this result to derive the power density spectra of offset PSK and MSK.

(c) Derive the power density spectrum of $v_\xi(t)$ when

$$g'(t) = g''(t) = \begin{cases} \cos\left(\dfrac{\pi}{T}t - \dfrac{1}{4}\sin\dfrac{4\pi}{T}t\right), & |t| \le \dfrac{T}{2}, \\ 0, & \text{elsewhere.} \end{cases}$$

The resulting modulation scheme is called SFSK defined through (5.115).

(d) Compare the asymptotic behavior, for $f \to \infty$, of the power spectral densities of offset PSK, MSK, and SFSK. Verify in particular that the spectrum of offset PSK is $O(f^{-2})$, that of MSK is $O(f^{-4})$ and that of SFSK is $O(f^{-6})$. How can this asymptotic behavior of the spectrum be derived from the expression of $g'(t)$, $g''(t)$?

*5.15. Consider a set of CPFSK signals as given by (5.91). Assume $M = 2$ and $f_d T = 1$, and compute the power spectrum using the computations of Example 5.10. Verify that the result is the same as in Examples 5.8 and 5.9 for the same values of M and $f_d T$. (Assume $\vartheta_2 = \vartheta_1 = 0$ in Example 5.9. Moreover, observe that as the model (5.91) implies a causal $v_\xi(t)$, the spectrum obtained should be multiplied by a factor of 2.)

5.16. Consider the signal (5.104) and, after defining the RV b_k as

$$b_k \triangleq -(\xi_{k-1}\xi_k),$$

show that it can be written in the quadrature form (5.100) where now

$$I(t) = A \sum_k u_{2T_s}[t - (2k - 1)T_s] \cos \psi_k \cos\left(\frac{\pi t}{2T_s}\right),$$

$$Q(t) = A \sum_k u_{2T_s}[t - 2kT_s](b_k \cos \psi_k) \sin\left(\frac{\pi t}{2T_s}\right).$$

It appears again as a quadrature signal in which the data terms are $\cos \psi_k$ for the I channel and $b_k \cos \psi_k$ for the Q channel, respectively. Since the RVs ξ_k can change every T_s seconds, it might appear that also $\cos \psi_k$ and $b_k \cos \psi_k$ can change every T_s seconds. Show that as a result of the phase continuity constraints on ψ_k in (5.104) the term $\cos \psi_k$ can change value only at the zero crossings of $\cos(\pi t/2T_s)$, and the term $b_k \cos \psi_k$ can change value only at the zero crossings of $\sin(\pi t/2T_s)$. Conclude that this result is equivalent to (5.101).

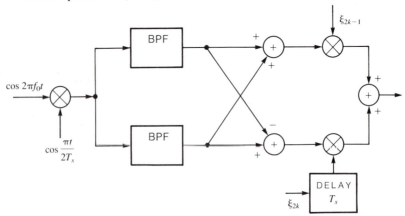

Figure P5.8

Digital Modulation Schemes   Chap. 5

5.17. Using (5.103), show that the block diagram of Fig. P5.8 generates an MSK signal. The two bandpass filters are centered at f_2 and f_1, respectively.

5.18. Consider an MSK signal whose instantaneous phase is given through (5.105) as

$$\Phi(t) = \frac{\pi b_k}{2T_s}(t - kT_s) + \varphi_k \qquad kT_s \le t < (k + 1)T_s,$$

where b_k is a RV taking values in the set $\{-1, +1\}$ depending on the binary source data 0 and 1. Consider the phase $\Phi(t - T_s)$ and show that the information digits can be recovered by using the decision function

$$\sin\{\Phi(t) - \Phi(t - T_s)\}.$$

Draw a block diagram of this differential MSK demodulator (Masamura and others, 1979).

*5.19. Use (5.111) to derive $d_B^2/2\mathcal{E}_b$ for full-response CPFSK signals with $M = 4$. With the help of (4.169), check the distance for various observation lengths N and verify that the bound $d_B^2/2\mathcal{E}_b$ is achieved with $N = 9$ and $h = 0.914$ (see Table 5.6).

5.20. Use the symmetry property (5.112), corresponding to

$$q(t) + q(T - t) = \frac{1}{2}, \qquad 0 \le t < T,$$

to elaborate (5.110) and show that

$$\frac{(d_B^2)_{\text{sym}}}{2\mathcal{E}_b} = (\log_2 M) \min_{1 \le k \le M-1} \left\{ 2\left(1 - \frac{1}{T} \int_0^T \cos[4\pi hkq(t)]\, dt\right) \right\}.$$

For $M = 2$, this gives (5.113).

5.21. Define

$$q_0(t) \triangleq q\left(t + \frac{T}{2}\right)$$

and use this definition in the result of Problem 5.20 to show equation (5.114).

5.22. Use expression (5.113) to get the bound $(d_B^2)_{\text{sym}}/2\mathcal{E}_b$ for the binary signal with the $g(t)$ of (5.115). Compare with the 1REC case.

5.23. Use (5.114) to show that, for any binary scheme, $(d_B^2)_{\text{sym}}/2\mathcal{E}_b$ is bounded by

$$2(1 - \cos 2\pi h) \le \frac{(d_B^2)_{\text{sym}}}{2\mathcal{E}_b} \le 1 - \cos 2\pi h, \qquad 0 \le h < 0.5,$$

$$(1 - \cos 2\pi h) \le \frac{(d_B^2)_{\text{sym}}}{2\mathcal{E}_b} \le 2(1 - \cos \pi h), \qquad 0.5 \le h < 1.$$

Fit into these bounds the values of $(d_B^2)_{\text{sym}}/2\mathcal{E}_b$ obtained for the pulses $g(t)$ of the type 1REC and 1RC (Fig. 5.35a and 5.35b).

5.24. Show that the CPM binary signal that uses the frequence pulse $g(t)$ of Fig. 5.35d (Duobinary MSK, FSOQ) can be obtained with the block diagram of Fig. P5.9. The binary symbols a_k belong to the set $\{0, 1\}$ and the level generator gives

$$u_k = a_k - \frac{1}{2}.$$

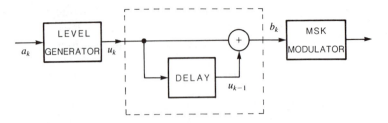

Figure P5.9

The duobinary encoder generates the transmitted ternary symbols b_k as

$$b_k = u_k + u_{k-1}.$$

Prove that it is possible to remove correlation between symbols using a differential encoder. Derive the phase trellis and the Viterbi processor for the demodulator that operates on this CPM signal.

5.25. In this problem we want to show that the ASK/SSB modulation achieves capacity as $R_s/W \to \infty$. Use first the expression (3.89) for channel capacity and show that if the transmitted power is increased by a factor 4^n (i.e., $P_s' = 4^n P_s$) then

$$\left(\frac{C}{W}\right)' \cong \frac{C}{W} + 2n.$$

That is, the bandwidth efficiency is incremented by $2n$ bits/s per 1 Hz. Take now the error probability (5.11) and show that the same increase of power corresponds to

$$\left(\frac{R_s}{W}\right)' \cong \frac{R_s}{W} + 2n.$$

Reach a conclusion for $n \to \infty$.

5.26. In this problem we want to show that the FSK orthogonal modulation achieves capacity as $R_s/W \to 0$. Start from (4.64) and write the symbol error probability $P(e)$ in the form

$$P(e) = \frac{1}{\sqrt{2\pi}} \int_{-\infty}^{\infty} \left\{ 1 - \left[1 - \frac{1}{2} erfc\left(\frac{x}{\sqrt{2}}\right) \right]^{M-1} \right\} e^{-(x-\sqrt{2\eta})^2/2} \, dx,$$

where $\eta \triangleq \mathcal{E}/N_0$. Use the following two bounds:

$$\left\{ 1 - \left[1 - \frac{1}{2} erfc\left(\frac{x}{\sqrt{2}}\right) \right]^{M-1} \right\} \leq \frac{M-1}{2} erfc\left(\frac{x}{\sqrt{2}}\right) < M e^{-x^2/2}, \quad \text{for } x \text{ large,}$$

$$\left\{ 1 - \left[1 - \frac{1}{2} erfc\left(\frac{x}{\sqrt{2}}\right) \right]^{M-1} \right\} \leq 1, \quad \text{for } x \text{ small.}$$

Therefore,

$$P(e) < \frac{1}{2} \int_{-\infty}^{x_0} e^{-(x-\sqrt{2\eta})^2/2} \, dx + \frac{M}{\sqrt{2\pi}} \int_{x_0}^{\infty} e^{-x^2/2} e^{-(x-\sqrt{2\eta})^2/2} \, dx.$$

268   Digital Modulation Schemes   Chap. 5

Optimize x_0 and show that $x_0 = \sqrt{2 \log_2 M \ln 2}$. Using simple exponential bounds for the two integrals, show that

$$P(e) < \begin{cases} 2e^{-\log_2 M(\eta_b - 2\ln 2)/2}, & \dfrac{\eta}{4} \geq \ln M, \\ 2e^{-\log_2 M(\sqrt{\eta_b} - \sqrt{\ln 2})/2}, & \dfrac{\eta}{4} < \ln M \leq \eta, \end{cases}$$

where $\eta = (\log_2 M)\eta_b$. Notice that $P(e) \to 0$ as $M \to \infty$ provided that $\eta_b > \ln 2$ ("Shannon bound").

CHAPTER 6

Digital Transmission over Real Channels

From Chapter 5 we know the theoretical limits of a digital modulation system's performance and how the most common of them compare with respect to those limits for an AWGN channel. In many practical situations this channel is not a realistic model. However, in most cases the performance of the different modulation schemes on such a channel can be considered as an upper bound of the actual performance. Moreover, in a Gaussian noise environment the probability of error depends only on one parameter, the signal-to-noise ratio. Thus, meaningful comparisons among the different modulation schemes can be obtained with only a moderate computational effort. Finally, it is hoped (and often true) that the hierarchy between different systems obtained on an AWGN is maintained in real channels, although the absolute performance may change.

In this chapter we shall consider a more realistic model of the system in which additional impairments that contribute to the degradation of the overall performance, besides Gaussian noise, will be assumed. Emphasis will be placed on the *intersymbol interference* (ISI) caused by the linear distortion introduced by the finite bandwidth and the nonideal characteristics of the devices used in the system, such as filters and amplifiers. In addition to ISI, other factors affecting the system performance will be given some consideration:

(i) *Cochannel* and *interchannel interferences*, which arise in systems sharing a common medium (e.g., in frequency-division multiplexing, FDM).

(ii) *Phase* and *timing jitter* due to imperfect synchronization in synchronous and coherent modulation schemes.

(iii) *Selective fading* typically present in terrestrial radio-relay links.

270

We shall focus on coherent modulation schemes whose representative signal points lie in a one- or two-dimensional space. This choice allows us to unify the treatment of different modulation methods by exploiting the concept of *analytic signal* as introduced in Section 2.4. Note that it encompasses a wide range of practical cases. The first section of the chapter is devoted to this unifying analysis of coherent digital systems.

Besides its effects on performance, ISI highly complicates the performance computation. The second section of the chapter deals with this subject. It presents some methods that allow a reasonably fast and accurate evaluation of the error probability; the analytical details can be found in Appendix E.

We have already mentioned the problem of imperfect synchronization in digital transmission systems. Synchronization, in its main aspects of carrier and timing acquisition, is a crucial point in all digital receivers. It is the subject of entire books and has been treated by many authors in an impressive number of papers. We devote to it only a few pages in the third section of this chapter. Some useful references can be found in the Bibliographical Notes at the end of the chapter.

A particular class of digital transmission systems, the terrestrial microwave radio links, are the subject of the last section of the chapter. In particular, we analyze the problem of modeling how the frequency selective fading affects the performance and the possible countermeasures.

In this section, the system aspects may prevail relative to the analytical details, as the major part of the results in this field have been obtained through extensive field trials, laboratory tests, and computer simulations. Thus, we have employed descriptive modeling techniques using real data, rather than the classic analytical approach typically applied to the modeling of fading channels.

6.1 ANALYSIS OF COHERENT DIGITAL SYSTEMS

In this section we analyze the coherent modulation schemes whose signal points lie in one- or two-dimensional spaces, like PAM, CPSK, and AM–PM. Their performance on the AWGN channels, as well as the modulators and demodulators block diagrams, were described in Chapter 5.

A block diagram of the transmission system we consider here is shown in Fig. 6.1. The bit stream at the output of the source is first sent to a serial-to-parallel converter that groups the binary digits in blocks of length h. Then the signal enters the modulator, which performs a memoryless mapping between the $M = 2^h$ input sequences and its alphabet of M waveforms. As we know from Chapter 4, each waveform can be represented in this case as a point in a one- or two-dimensional Euclidean space, characterized by two real coordinates or, equivalently, by a complex number. The modulated signal is transmitted over the channel, in which Gaussian noise is added. The bandpass filter in Fig. 6.1 represents, without loss of generality, the cascade of the transmitter filter, the channel filter, and the receiving filter. The received signal is fed to the carrier recovery device, which supplies the reference carrier to the coherent demodulator. The main features of the carrier recovery devices will be studied later. At this point, we assume that the recovered carrier is affected by a *phase jitter* $\vartheta(t)$, whose variations with time are so slow that it can be considered as a RV with known pdf. Then the demodulated

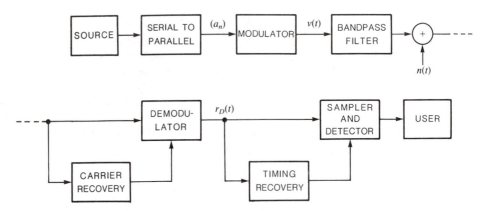

Figure 6.1 Transmission system with a coherent receiver.

signal is sampled at the symbol rate $1/T$, in correspondence with the sampling instants provided by the timing recovery device. Finally, the detector makes a decision on which signal was transmitted based on the samples thus obtained and according to decision regions that depend on the particular modulation scheme.

A general representation of the modulated signal $v(t)$ is as follows:

$$v(t) = v_P(t) \cos 2\pi f_0 t - v_Q(t) \sin 2\pi f_0 t, \tag{6.1}$$

where f_0 is the carrier frequency and†

$$v_P(t) = \sum_n a_{Pn} s_P(t - nT) - \sum_n a_{Qn} s_Q(t - nT), \tag{6.2}$$

$$v_Q(t) = \sum_n a_{Qn} s_P(t - nT) + \sum_n a_{Pn} s_Q(t - nT). \tag{6.3}$$

In (6.2) and (6.3), a_{Pn} and a_{Qn} are the coordinates of the signal point in the nth signaling interval $(nT, (n + 1)T)$ and can take M values in the sets A_P and A_Q. The waveforms $s_P(t)$ and $s_Q(t)$ are suitable baseband shaping functions. The representation (6.1) includes the case of a baseband signal, for which $f_0 = 0$.

Example 6.1

In the case of M-ary PAM modulation, we have $A_P = \{(2k - M - 1)d/2\}_{k=1}^M$ and $s_Q(t) \equiv 0$, and, for instance, $s_p(t) = u_T(t)$. □

The sets A_P and A_Q and $s_P(t)$ and $s_Q(t)$ for the different modulation schemes are presented in Table 6.1, where A_k, ϕ_k represent the amplitude and phase of the signal points in the case of two-dimensional modulation schemes. In the case of PAM–SSB, $s_Q(t)$ is obtained as the Hilbert transform of $s_P(t)$. According to the theory of the complex envelope representation of bandpass signals, developed in Section 2.4, we can represent $v(t)$ by its complex envelope $\tilde{v}(t)$:

$$\tilde{v}(t) = v_P(t) + j v_Q(t). \tag{6.4}$$

---

† From here on the symbol $\sum_n$ will mean summation over all integers n from $-\infty$ to $+\infty$.

TABLE 6.1 Coordinates of signal points and shaping functions for coherent modulation schemes

Modulation Scheme	A_P	A_Q	$s_P(t)$	$s_Q(t)$
PAM-DSB	$\{(2k - M - 1)d/2\}_{k=1}^{M}$	0	$s(t)$	0
PAM-SSB	$\{(2k - M - 1)d/2\}_{k=1}^{M}$	0	$s(t)$	$\hat{s}(t)$
CPSK	$A \cos \phi_k$	$A \sin \phi_k$	$s(t)$	0
AM-PM	$A_k \cos \phi_k$	$A_k \sin \phi_k$	$s(t)$	0

Moreover, the bandpass filtering operated by the channel on $v(t)$ can be represented by the filtering operated by the low-pass equivalent channel on $\tilde{v}(t)$. Thus, defining

$$\tilde{g}(t) = g_P(t) + jg_Q(t), \tag{6.5}$$

where $g(t) = \mathcal{R}\{\tilde{g}(t)e^{j2\pi f_0 t}\}$ is the impulse response of the bandpass filter in Fig. 6.1, we can write the complex envelope of the received signal $r(t)$ as

$$\tilde{r}(t) = \tfrac{1}{2}\,\tilde{v}(t) * \tilde{g}(t) + \tilde{n}(t), \tag{6.6}$$

where $\tilde{n}(t) = n_P(t) + jn_Q(t)$ is the complex envelope of the bandpass Gaussian noise process, and $n_P(t)$, $n_Q(t)$ are baseband Gaussian processes whose samples are Gaussian RVs with zero mean and variance σ_n^2. This variance σ_n^2 is obtained as $N_0 B_{eq}$, with $N_0/2$ being the two-sided power spectral density of the channel white noise and B_{eq} the equivalent noise bandwidth of the receiving filter (see Fig. 2.20). Using now (6.2), (6.3), and (6.4), we can write

$$\tilde{v}(t) = \sum_n a_n \tilde{s}(t - nT), \tag{6.7}$$

having defined

$$\begin{aligned} a_n &\triangleq a_{Pn} + ja_{Qn} \\ \tilde{s}(t) &\triangleq s_P(t) + js_Q(t). \end{aligned} \tag{6.8}$$

Thus, finally, the received signal $\tilde{r}(t)$ can be given the following expression:

$$\tilde{r}(t) = \sum_n a_n \tilde{h}(t - nT) + \tilde{n}(t), \tag{6.9}$$

where

$$\tilde{h}(t) \triangleq \tfrac{1}{2}\,\tilde{s}(t) * \tilde{g}(t) = h_P(t) + jh_Q(t). \tag{6.10}$$

From here on, the (a_n) is assumed to be a sequence of iid RVs. Recalling (6.5) and (6.8), the convolution in (6.10) gives rise to

$$h_P(t) = \tfrac{1}{2}\,[s_P(t) * g_P(t) - s_Q(t) * g_Q(t)], \tag{6.11}$$

$$h_Q(t) = \tfrac{1}{2}\,[s_P(t) * g_Q(t) + s_Q(t) * g_P(t)]. \tag{6.12}$$

Example 6.2

Consider a PAM transmission and a bandpass filter whose transfer function $G(f)$ satisfies the following symmetry conditions for every f:

$$G_R^+(f_0 + f) = G_R^+(f_0 - f),$$
$$G_I^+(f_0 + f) = -G_I^+(f_0 - f),$$

where G_R^+ and G_I^+ are the real and imaginary parts of the transfer function

$$G^+(f) \triangleq \begin{cases} G(f), & f \geq 0, \\ 0, & f < 0. \end{cases}$$

To compute $\tilde{h}(t)$ according to (6.10), we need $\tilde{g}(t)$, that is, the inverse Fourier transform of $G^+(f + f_0)$. But $G^+(f + f_0)$ exhibits the symmetries of $G^+(f)$ around the origin $f = 0$, and this makes $\tilde{g}(t) \equiv g_P(t)$ real. Thus we have

$$\tilde{h}(t) = \tfrac{1}{2} s(t) * g_P(t), \quad \text{(real)}, \quad \text{for PAM–DSB},$$

and

$$\tilde{h}(t) = \tfrac{1}{2} [s(t) + j\hat{s}(t)] * g_P(t), \quad \text{for PAM–SSB.} \ \square$$

The actual signal that enters the demodulator of Fig. 6.1 can be obtained simply as $r(t) = \Re\{\tilde{r}(t)e^{j2\pi f_0 t}\}$. The task performed by the demodulator is represented by the block diagram of Fig. 6.2. In fact, this is the general form of the demodulator, which simplifies and reduces to the upper branch only when a one-dimensional modulation scheme like PAM is used. It can be seen (see Problem 6.1) that the two outputs $r_{DP}(t)$ and $r_{DQ}(t)$ from the branches of the demodulator of Fig. 6.2 are the same as the outputs of the system shown in Fig. 6.3. At its input the complex envelope of $r(t)$ is presented. The presence of nonideal low-pass filters in the demodulator could also be easily accounted for by including their transfer functions in the overall low-pass equivalent filter represented by $\tilde{h}(t)$ in (6.10). Then we are immediately able to write the expressions of the demodulated signals. For the sake of clarity, let us consider separately the one- and two-dimensional cases.

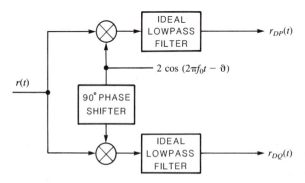

Figure 6.2 Demodulation of a two-dimensional modulation scheme.

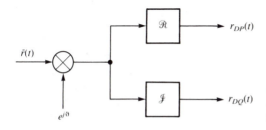

Figure 6.3 Demodulator equivalent to that of Fig. 6.2 using complex envelope representation.

PAM modulation

The demodulated signal is given by

$$r_D(t) \triangleq r_{DP}(t) = \mathcal{R}\{\tilde{r}(t)\, e^{j\vartheta}\}$$

$$= \cos \vartheta \sum_n a_n h_P(t - nT) - \sin \vartheta \sum_n a_n h_Q(t - nT) + v_P(t). \qquad (6.13)$$

In (6.13), the baseband random process $v_P(t) = n_P(t) \cos \vartheta - n_Q(t) \sin \vartheta$ is conditionally Gaussian, with zero mean and variance σ_n^2, like $n_P(t)$ and $n_Q(t)$ (see Problem 6.2). Equation (6.13) makes evident the different sensitivities of DSB and SSB to the phase jitter. In fact, suppose that the channel transfer function $G(f)$ satisfies the symmetry conditions of Example 6.2. Then $h_Q(t)$ is equal to zero for the PAM–DSB modulation. In this case, the presence of the phase jitter reduces to an attenuation of the received signal by $\cos \vartheta$. However, for SSB systems, $h_Q(t)$ is not zero. Thus, the second summation in the RHS of (6.13) contributes to the performance degradation.

CPSK and AM–PM modulations

The two demodulated signals are given by

$$r_{DP}(t) \triangleq \mathcal{R}\{\tilde{r}(t) e^{j\vartheta}\} = \sum_n \mathcal{R}\{a_n \tilde{h}(t - nT) e^{j\vartheta}\} + v_P(t) \qquad (6.14)$$

$$r_{DQ}(t) \triangleq \mathcal{I}\{\tilde{r}(t) e^{j\vartheta}\} = \sum_n \mathcal{I}\{a_n \tilde{h}(t - nT) e^{j\vartheta}\} + v_Q(t), \qquad (6.15)$$

where $v_P(t)$ is the same as before, and $v_Q(t) = n_P(t) \sin \vartheta + n_Q(t) \cos \vartheta$ is a conditionally Gaussian baseband process with zero mean and variance σ_n^2. Moreover, samples of $v_P(t)$ and $v_Q(t)$ taken at the same time instant are conditionally independent RVs (see Problem 6.2).

Example 6.3

Consider an AM–PM system without phase jitter ($\vartheta = 0$) and a bandpass transfer function $G(f)$ exhibiting the symmetries of Example 6.2. Using (6.11) and (6.12), we have

$$h_P(t) = \tfrac{1}{2} s(t) * g_P(t),$$

$$h_Q(t) = 0,$$

and, from (6.14) and (6.15),

$$r_{DP}(t) = \sum_n a_{Pn} h_P(t - nT) + n_P(t),$$

$$r_{DQ}(t) = \sum_n a_{Qn} h_P(t - nT) + n_Q(t). \quad \square$$

In the detector, decisions on the transmitted a_n are taken by comparing sampled values of $r_{DP}(t)$ and $r_{DQ}(t)$ [or only $r_{DP}(t)$ in the PAM case] with suitable thresholds. In other words, the receiver is the same as for the Gaussian channel described in Chapter 5. The sampling times form a sequence $(t_0 + iT)_{i=-\infty}^{\infty}$, where $0 \leq t_0 < T$ is the optimum (in some sense) timing instant depending on the impulse response $\tilde{h}(t)$. Assuming that the sequence (a_n) is stationary, the processes $r_{DP}(t)$ and $r_{DQ}(t)$ are cyclostationary random processes with period T (see Section 2.2.2). Thus the performance of the system does not depend on the particular signaling interval. We shall consider the sampling instant t_0.

The following shorthand notation will be used in this chapter for all the time functions:

$$y_n \triangleq y(t_0 - nT), \qquad \text{for all integers } n.$$

The sampled demodulated signals are then given by the following expressions:

PAM modulation†

$$
\begin{aligned}
r_{D0} = a_0(h_{P0} \cos \vartheta - h_{Q0} \sin \vartheta) \\
+ \sum_n{}' a_n(h_{Pn} \cos \vartheta - h_{Qn} \sin \vartheta) + v_{P0}.
\end{aligned}
\tag{6.16}
$$

CPSK and AM–PM modulations

$$r_{DP0} = r_{DP}(a_0) + \sum_n{}' r_{DP}(a_n) + v_{P0}, \tag{6.17}$$

$$r_{DQ0} = r_{DQ}(a_0) + \sum_n{}' r_{DQ}(a_n) + v_{Q0}, \tag{6.18}$$

where

$$r_{DP}(a_n) = (a_{Pn}h_{Pn} - a_{Qn}h_{Qn}) \cos \vartheta - (a_{Pn}h_{Qn} + a_{Qn}h_{Pn}) \sin \vartheta, \tag{6.19}$$
$$r_{DQ}(a_n) = (a_{Pn}h_{Qn} + a_{Qn}h_{Pn}) \cos \vartheta + (a_{Pn}h_{Pn} - a_{Qn}h_{Qn}) \sin \vartheta. \tag{6.20}$$

In (6.16), (6.17), and (6.18) the term with $n = 0$ has been given special consideration as it contains the required information about the symbol a_0 on which we are deciding. The summations on the RHS represent the contribution (unwanted!) to the sample taken at $t = t_0$ of the past and future symbols in the sequence (a_n). These terms are called *intersymbol interference* (ISI) and may represent a major cause of system performance inpairment.

Looking at (6.17) to (6.20), an important fact can be observed. Even in the absence of phase jitter, we have an interaction between the in-phase and quadrature channels whenever $h_Q(t)$ is not zero at the sampling instants. This happens when the transfer function of the channel $G(f)$ does not satisfy the symmetry conditions of Example 6.2.

Example 6.4

Consider a binary PAM system, with $d = 2$ and $s_P(t) = u_T(t)$, transmitted over a channel with $\tilde{g}(t) = 2Te^{-t/T}$, $t \geq 0$. Using (6.10), we have (see Fig. 6.4)

$$\tilde{h}(t) = \begin{cases} T^2(1 - e^{-t/T}), & 0 \leq t \leq T, \\ T^2(e - 1)e^{-t/T}, & t > T. \end{cases}$$

† Hereafter the symbol Σ' will mean summation over all integers $n \neq 0$ from $-\infty$ to $+\infty$.

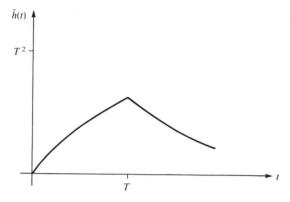

Figure 6.4 Example of low-pass equivalent impulse response.

Using now (6.13) and assuming as the transmitted sequence (a_n) the sequence $+1$, -1, -1, $+1$, and $\vartheta = 0$, we obtain in the absence of noise the received signal $r_D(t)$ shown in Fig. 6.5b obtained by summing the various contributions of Fig. 6.5a. $\square$

An effective way of displaying the qualitative effects of ISI is the construction of the *eye pattern*. It consists in slicing the demodulated signal (in the absence of noise) in segments of T seconds duration and superimposing the various slices in the interval $(0, T)$ as in Fig. 6.6, which refers to Example 6.4. The eye pattern is obtained by observing the data signal through an oscilloscope whose time axis is synchronized at the symbol rate. In the binary antipodal case, the typical aspect of the eye pattern is as in Fig. 6.7, where the sampling instant is shown to correspond with the maximum eye opening, yielding the greatest protection against the noise. In Fig. 6.7 the *amplitude peak distortion* is also indicated. It is obtained as the maximum value assumed by the

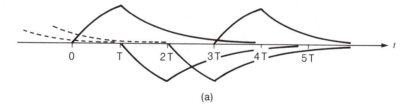

(a)

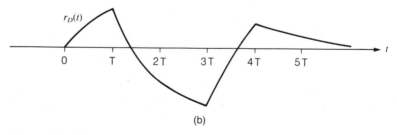

(b)

Figure 6.5 (a) Successive component waveforms of the received signal; (b) received signal for the impulse response of Fig. 6.4 corresponding to the binary data sequence $+1$, -1, -1, $+1$ in the absence of noise.

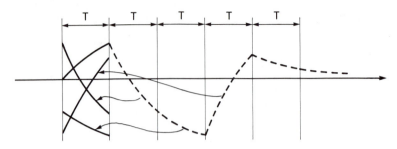

Figure 6.6 Construction of the eye pattern for the signal of Fig. 6.5.

ISI over all the possible transmitted sequences (a_n). Using (6.16), with $\vartheta = 0$, it is given by

$$D_P \triangleq \max_{a_n} \sum_n{}' a_n h_{Pn} = \sum{}' |h_{Pn}|. \tag{6.21}$$

The concept of eye pattern and peak distortion can be generalized to the multilevel PAM and two-dimensional modulation systems.

The general form of the overall low-pass equivalent impulse response $\tilde{h}(t)$ is shown in Table 6.2, together with the expressions of r_{DP0} and r_{DQ0}. We have proved that the system shown in Fig. 6.8 permits us to obtain the real and imaginary parts of the demodulated signal $r_D(t)$ of Fig. 6.1. Note that in Fig. 6.8 the modulating and demodulating carriers have disappeared. Besides its great simplicity and conciseness, this result proves to be very useful in the computer simulation of bandpass data-transmission systems. In fact, using the model of Fig. 6.8, the frequency at which signals must be sampled before being processed by the computer is seen to be related to the bandwidth of the *modulating signal*, not to the *carrier frequency*, which is usually much larger.

To conclude this part, let us summarize step by step how the signal analysis we have just described can be done. This analysis is the preliminary step in the computation of the error probability, as we shall see in the next sections.

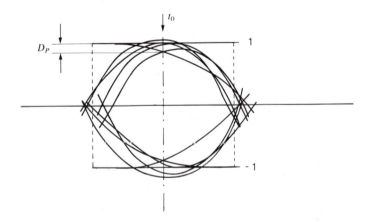

Figure 6.7 Example of eye pattern.

TABLE 6.2 Low-pass equivalent impulse responses and in-phase and quadrature samples of the received signal for coherent modulation schemes

Modulation scheme	h_P	h_Q	r_{DP0}	r_{DQ0}
PAM-DSB	$s * g_P$	$s * g_Q$	$\sum_n a_n (h_{Pn} \cos \vartheta - h_{Qn} \sin \vartheta) + v_{P0}$	0
PAM-SSB	$s * g_P - \hat{s} * g_Q$	$s * g_Q + \hat{s} * g_P$	$\sum_n a_n (h_{Pn} \cos \vartheta - h_{Qn} \sin \vartheta) + v_{P0}$	0
CPSK, AM-PM	$s * g_P$	$s * g_Q$	$\sum_n (a_{Pn} h_{Pn} - a_{Qn} h_{Qn}) \cos \vartheta$ $- \sum_n (a_{Pn} h_{Qn} + a_{Qn} h_{Pn}) \sin \vartheta$ $+ v_{P0}$	$\sum_n (a_{Pn} h_{Qn} + a_{Qn} h_{Pn}) \cos \vartheta$ $+ \sum_n (a_{Pn} h_{Pn} - a_{Qn} h_{Qn}) \sin \vartheta$ $+ v_{Q0}$

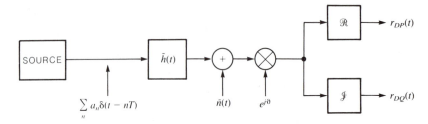

Figure 6.8 Equivalent block diagram of a linear digital transmission system using complex envelope representation.

Step 1. Given the modulation scheme and the shaping filter $s(t)$, use Table 6.1 to obtain the transmitted signal $\tilde{v}(t)$ of (6.2) to (6.4).

Step 2. Cascade the bandpass transmitter, channel, and receiver filters to obtain the transfer function $G(f)$.

Step 3. Compute the low-pass equivalent impulse response $\tilde{g}(t)$ by taking the inverse Fourier transform of $G^+(f + f_0)$ and use it in the convolutions (6.11) and (6.12) to obtain the real and imaginary parts of $\tilde{h}(t)$.

Step 4. Find the expressions of r_{DP0} and r_{DQ0} in Table 6.2 as a function of a_{Pn}, a_{Qn} (Table 6.1), and h_P, h_Q computed in Step 3.

The computational tools normally used in a digital computer to evaluate the convolutions in Step 3 are the fast Fourier transform in the frequency domain and the state variable technique or the bilinear z-transform in the time domain. The interested reader is invited to consult the Bibliographical Notes at the end of this chapter for relevant references.

6.2 EVALUATION OF THE ERROR PROBABILITY

The received signal, after demodulation and sampling, enters the decision device, which locates it in one of the decision regions and chooses the corresponding point in the signal space as the transmitted one. In the practice, the decision regions are optimum under the criterion of minimum distance (i.e., optimum for the AWGN channel). Thus, computing the error probability for a given transmitted signal point entails evaluating the probability that the point (r_{DP0}, r_{DQ0}) lies in a two-dimensional region, depending on the particular modulation scheme adopted. The computation is usually performed in two steps:

Step 1. Compute the probability that a point lies in a two-dimensional (or one-dimensional in the PAM case) region, taking advantage of the fact that the RVs r_{DP0} and r_{DQ0} are independent Gaussian. This probability is precisely the error probability conditioned on ISI and ϑ.

Step 2. Compute the expectation of the result obtained in step 1 with respect to (a) the two RVs (one for PAM) representing the ISI affecting the in-phase and quadrature components of the received signal and (b) the random phase ϑ.

As we shall see, step 1 can be achieved analytically in an exact or approximate manner for almost all coherent modulation schemes. Indeed, there is no difference from the AWGN channel discussed in detail in Chapter 5. What really complicates the situation is step 2. In fact, although in most cases all the values assumed by the ISI RVs could be exhaustively enumerated and the conditional error probabilities computed, such a procedure would take an extremely long time. Thus, it is often impractical. We shall verify this conclusion with reference to the PAM transmission system.

6.2.1 PAM Modulation Scheme

The sampled received signal's expression was given by (6.16), where the symbols forming the sequence (a_n) can take the values shown in Table 6.1 with equal probabilities $1/M$. With the following shorthand notations,

$$h_n(\vartheta) \triangleq h_{Pn} \cos \vartheta - h_{Qn} \sin \vartheta, \tag{6.22}$$

$$X_n(\vartheta) \triangleq a_n h_n(\vartheta), \tag{6.23}$$

$$X(\vartheta) \triangleq \sum_n{}' X_n(\vartheta), \tag{6.24}$$

the received signal r_{D0} becomes

$$r_{D0} = a_0 h_0(\vartheta) + X(\vartheta) + v_{P0}.$$

Assuming M even and $h_0(\vartheta) > 0$, the decisions at the receiver are taken by comparing r_{D0} with the following thresholds:

$$-\left(\frac{M}{2} - 1\right) d h_0(\vartheta), \ldots, -d h_0(\vartheta), 0, d h_0(\vartheta), \ldots, \left(\frac{M}{2} - 1\right) d h_0(\vartheta).$$

Thus, following the analysis of Section 5.2, the error probability is easily computed:

$$P(e) = \mathrm{E}_\vartheta\{P(e|\vartheta)\} = \frac{M-1}{M} \mathrm{E}_\vartheta \mathrm{E}_X \left\{ erfc \left[\frac{(d/2)\, h_0(\vartheta) - X(\vartheta)}{\sqrt{2}\, \sigma_n} \right] \right\}, \tag{6.25}$$

where E_ϑ denotes averages over the RV ϑ. We shall discuss later how to perform this average.

The problem, then, is the computation of the expectation with respect to $X(\vartheta)$, that is, of the integral

$$I \triangleq \int_{\mathscr{X}} erfc \left[\frac{(d/2)\, h_0 - x}{\sqrt{2}\, \sigma_n} \right] f_X(x)\, dx. \tag{6.26}$$

For simplicity, in (6.26) we have dropped the coefficient $(M-1)/M$ and the dependence on ϑ. In the integral (6.26), $\mathscr{X}$ and $f_X(x)$ represent, respectively, the range and the pdf of the RV X.

Some facts about the RV X

Looking at (6.24) and (6.23), the RV X is seen to be the sum of a number N of RVs X_n. The number N depends on the duration of the impulse response $h(t)$ through its samples h_n. In principle, N may be infinite. However, in practice, only a finite number of samples significantly contribute to performance degradation. The interested reader wanting a deeper insight into the structure of X for N infinite is referred to Hill and Blanco (1973). The situation is such that one is tempted to invoke the central limit theorem and assume that X is a Gaussian RV. Unfortunately, the central limit theorem cannot be applied as, in practice, the range of X is almost always a bounded interval, and its variance is limited (see Loève, 1963, p. 277). In fact, the largest value taken by X cannot exceed

$$x_{\sup} \triangleq (M - 1) \frac{d}{2} \sum_{n}' |h_n|. \tag{6.27}$$

The value $x_{\sup}$ is assumed by X with our assumption that (a_n) is a sequence of independent RVs. When N is infinite, $x_{\sup}$ is still bounded if the asymptotic decay of the impulse response $h(t)$ is faster than $1/t$. In the practice this is *always* the case. What happens when we try to apply the central limit theorem to this case is shown in the following example.

Example 6.5

Consider a PAM transmission system for which $x_{\sup} < h_0 d/2$. This means that the eye pattern of the system is open, so we transmit with zero error probability in the absence of noise. Applying the central limit theorem (which leads to the *Gaussian assumption* for X), the RV X is treated like a Gaussian RV, with zero mean and variance $\sigma_X^2 = EX^2$, independent of the noise. Thus, the sum $X + v_{P0} \sim \mathcal{N}(0, \sigma_X^2 + \sigma_n^2)$, and the integral in (6.26) becomes

$$I_G = erfc \left(\frac{dh_0}{2\sqrt{2} \sqrt{\sigma_X^2 + \sigma_n^2}} \right).$$

Now, increasing the signal-to-noise ratio in the channel by letting $\sigma_n \to 0$, we get

$$I_G\big|_{\sigma_n=0} = erfc \left(\frac{dh_0}{2\sqrt{2}\,\sigma_X} \right),$$

which leads to an asymptotic error probability value different from zero. This clearly contrasts with the hypothesis $x_{\sup} < h_0 d/2$. However, when ISI is small, this asymptotic value may be so low that in the region of interest the curve for the Gaussian assumption gives a reasonable approximation of the error probability. $\square$

Exact value of I

Henceforth, we shall suppose that N is finite. Although this is not always true, in practice it is possible to find an N large enough to make immaterial the error due to the truncation of $\bar{h}(t)$. In Prabhu (1971), the problem of bounding the error due to the impulse response truncation was examined. X is then a discrete RV, assuming values $\{x_i\}_{i=1}^{L}$, with probabilities $\{p_i\}_{i=1}^{L}$, and its pdf $f_X(x)$ can be written as

$$f_X(x) = \sum_{i=1}^{L} p_i \delta(x - x_i). \tag{6.28}$$

Inserting (6.28) into (6.26), we immediately get

$$I = \sum_{i=1}^{L} p_i\, erfc \frac{(d/2)h_0 - x_i}{\sqrt{2}\sigma_n} \tag{6.29}$$

and the problem is solved. The ease in obtaining the *true value* of I should make the reader suspicious. In fact, what often renders (6.29) useless is the number L, as it can be extremely large. Suppose, for example, that we have an octonary PAM with a channel memory $N = 30$. Then L is given by

$$L = M^N = 8^{30} \cong 1.24 \cdot 10^{27}.$$

If we could use a computer able to compute 10^6 complementary error functions in 1 second, it would take only slightly more than $3 \cdot 10^{13}$ years to compute I's value. Not to mention the money! That alone seems a good motivation for the research done in this area. Many methods have been proposed in the literature, during a decade of research, to obtain approximations of I in (6.26) with different trade-offs between accuracy and computer time.

Since this book's aim is not to emphasize the historical development of the topics, but, instead, to provide modern and helpful tools to the reader, we only reference some of the earlier methods and outline in the problem section some of their applications. Here we propose the simplest upper bound, known as the *worst-case bound*, and use the *Gauss quadrature rules* (*GQR*) method, described in Appendix E, as it has emerged as one of the most efficient in approximating integrals like I in (6.26).

Worst-case bound

The worst-case bound is an upper bound to I in (6.26) computed through the substitution of the RV X with the constant x_{sup} defined in (6.27). Thus, we have

$$I \leqslant erfc \, \frac{(d/2)h_0 - x_{\text{sup}}}{\sqrt{2}\, \sigma_n} \, . \tag{6.30}$$

Since $erfc\,(\cdot)$ is a monotonic decreasing function, the RHS of (6.30) is clearly an upper bound to the RHS of (6.26). The term $(d/2)h_0 - x_{\text{sup}}$ is precisely the semi-opening of the eye pattern at the sampling instant. The worst-case bound is very easily computed. The approximation involved is reasonable when one interfering sample is dominant with respect to the others. Otherwise, the bound is too loose. A better upper bound, based on Chernoff bound, is described in Saltzberg (1968) (see also Problem 6.7) and will be used later in the examples.

Method of Gauss quadrature rules

The method of GQR is described in detail in Appendix E. It seems to be one of the best compromises between accuracy and computer time. Essentially, it allows one to compute an approximation of I in (6.26) in the form

$$I \cong \sum_{j=1}^{J} w_j \, erfc \, \frac{(d/2)h_0 - x_j}{\sqrt{2}\, \sigma_n} \, . \tag{6.31}$$

The $\{x_j\}_{j=1}^{J}$ and $\{w_j\}_{j=1}^{J}$ are called, respectively, the *abscissas* and the *weights* of the quadrature rule. They can be obtained through a numerical algorithm based on the knowledge of the first $2J$ moments of RV X. Comparing (6.31) with the I's exact value (6.29), one immediately realizes the similarity. The great difference lies in the value of J in (6.31), which is usually much less than the value of L in (6.29). The tightness of the bounds depends on J (i.e., on the number of known moments $K = 2J$). Computational experience shows that a value of K between 10 and 20 leads to good approximations of the true value of I. If, for some reasons, the number of known moments is too low to get a good approximation, upper and lower bounds can be obtained, as shown in Appendix E. The same method can be used to evaluate the average with respect to the RV ϑ in (6.25) once the moments of ϑ are known. Thus, our attention is now focused on the computation of the moments μ_k of the RV X:

$$\mu_k \triangleq E\{X^k\}, \qquad k = 1, 2, \ldots, 2J, \qquad (6.32)$$

without resorting to the pdf of X.

Computation of the moments of X

X is the sum of N independent RVs X_n. A recursive method to compute the moments of X particularly suited for computer implementation is described step by step in the following, where the X_n's have been renumbered from X_1 to X_N.

Step 1. Compute the moments of the individual RVs X_n defined in (6.23):

$$\gamma_k^{(n)} \triangleq E\{X_n^k\} = h_n^k \, E\{a_n^k\}$$

$$= h_n^k \frac{d^k}{2^k M} \sum_{i=1}^{M} (2i - M - 1)^k, \qquad n = 1, \ldots, N, \qquad k = 0, \ldots, K.$$

Step 2. Define the partial sums

$$Y_m \triangleq \sum_{n=1}^{m} X_n$$

with $Y_N \equiv X$ and compute recursively, for each k, the moments

$$\lambda_k^{(m)} \triangleq E\{Y_m^k\}$$

through

$$\lambda_k^{(m+1)} = E\{Y_{m+1}^k\} = E\{(Y_m + X_{m+1})^k\}$$

$$= \sum_{j=0}^{k} \binom{k}{j} E\{Y_m^j\} E\{X_{m+1}^{k-j}\} = \sum_{j=0}^{k} \binom{k}{j} \lambda_j^{(m)} \gamma_{k-j}^{(m+1)},$$

where the independence between Y_m and X_{m+1} has been exploited.

When $m = N$ stop the procedure, since

$$\lambda_k^{(N)} = E\{Y_N^k\} = E\{X^k\} = \mu_k.$$

The method for computing the moments, although described with reference to PAM, is fairly general, since it only requires X to be written as a sum of independent

RVs. When this is not possible, as for instance in systems employing line codes (see Section 9.8) that correlate the symbols in (a_n), a more elaborate approach must be used to compute the error probability (see Cariolaro and Pupolin, 1975).

Example 6.6

In this example the methods described to compute the error probability in the presence of ISI will be applied, for the sake of comparison, to the case of binary PAM transmission, with $\vartheta = 0$ and

$$h_P(t) = \frac{\sin (\pi t/T)}{\pi t/T}. \tag{6.33}$$

The impulse response of (6.33) is that of an ideal low-pass filter with cutoff frequency $1/2T$. The transfer function of the filter satisfies the Nyquist criterion (see Section 7.2) and, thus, it does not give rise to ISI when properly sampled at $t = 0, \pm T, \pm 2T, \ldots$. We will suppose that the timing recovery circuit is not ideal, so the sampling instants will be $t_n = t_0 + nT$, $n = -\infty, \ldots, +\infty$, with $t_0 \neq 0$, and we define the normalized sampling time deviation $\Delta \triangleq t_0/T$.

The methods discussed for computing the error probability are the worst-case bound (curves labeled ① in the figures), the Chernoff bound (curves labeled ②), the series expansion described in Appendix E (curves labeled ③), and the GQR (curves labeled ④). In Fig. 6.9 the error probability is plotted as a function of Δ for a signal-to-noise ratio at the nominal sampling instant ($t_0 = 0$) SNR $\triangleq 1/2\sigma_n^2$ of 15 dB. The impulse response has been truncated to $N = 50$. The curve ③, relative to the series expansion method, stops at $\Delta = 0.15$, since the summation of the series exhibits numerical instability for greater values of Δ. This is visualized in Fig. 6.10, where the exact error probability, computed through (6.29) for $N = 10$, and the error probability estimated either with the series expansion or with the GQR method are reported for $\Delta = 0.2$ as a function of J, the number of terms used in the series or the GQR. The curve giving the results of the series expansion method ends with $J = 8$, since the successive nine-terms approximation gives a negative value for $P(e)$. The processing time required for the computation on a medium-sized computer is less than a few seconds for all the methods described. It is practically constant with N for the worst-case bound, whereas with the other methods it grows linearly. $\square$

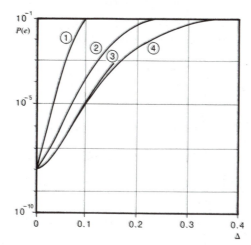

Figure 6.9 Error probability for binary PAM as a function of the normalized sampling time deviation Δ. The impulse response is that of an ideal low-pass filter with cutoff frequency $1/2T$. The labels of the curves are as follows: ① worst-case bound, ② Chernoff bound, ③ series expansion method, ④ GQR method, SNR = 15 dB.

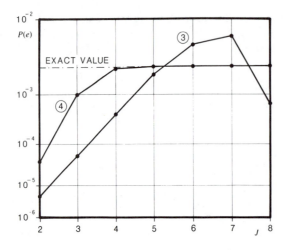

Figure 6.10 Error probability of the system of Fig. 6.9 with $\Delta = 0.2$ as a function of the number of terms used in the series expansion or in the GQR. Labels of curves as in Fig. 6.9. The exact value is also given.

6.2.2 Two-Dimensional Modulation Schemes

Expressions of the sampled in-phase and quadrature received signals were given in (6.17) and (6.18). The error probability will involve in general, as a final step, the average with respect to the RV ϑ, as for PAM. For simplicity, let us assume $\vartheta = 0$. With shorthand notations,

$$X_P \triangleq \sum_n{}' r_{DP}(a_n) = \sum_n{}' (a_{Pn}h_{Pn} - a_{Qn}h_{Qn}), \tag{6.34}$$

$$X_Q \triangleq \sum_n{}' r_{DQ}(a_n) = \sum_n{}' (a_{Pn}h_{Qn} + a_{Qn}h_{Pn}), \tag{6.35}$$

the in-phase and quadrature received signals become

$$r_{DP0} = r_{DP}(a_0) + X_P + v_{P0}, \tag{6.36}$$

$$r_{DQ0} = r_{DQ}(a_0) + X_Q + v_{Q0}. \tag{6.37}$$

The decisions at the receiver are taken through a rule that partitions the two-dimensional space of the received signal points into M regions R_k. The error probability is thus given by (4.32), which is here rewritten as

$$P(e) = 1 - \frac{1}{M} \sum_{\alpha_k \in A} P\{\mathbf{r}_{D0} \in R_k | a_0 = \alpha_k\}, \tag{6.38}$$

where $\mathbf{r}_{D0}$ is the received vector with components r_{DP0}, r_{DQ0}, and $A = \{\alpha_k\}_{k=1}^{M}$ is the set of values assumed by a_0.

The probabilities in the RHS of (6.38) can be computed in two steps

$$P\{\mathbf{r}_{D0} \in R_k | a_0 = \alpha_k\} = E_{X_P, X_Q} P\{\mathbf{r}_{D0} \in R_k | a_0 = \alpha_k, X_P, X_Q\}. \tag{6.39}$$

The first step consists in evaluating the conditional probability in RHS of (6.39). The received vector $\mathbf{r}_{D0}$, conditioned on α_k, X_P, and X_Q, is a Gaussian vector with independent components r_{DP0} and r_{DQ0}. Thus, the evaluation of (6.39) involves integration of a bivariate Gaussian RV with independent components within the region R_k. This problem has been discussed in Chapter 5 for the most important two-dimensional coherent modulation schemes. If we define

$$D_k(\alpha_k, X_P, X_Q) \triangleq P\{\mathbf{r}_{D0} \in R_k | a_0 = \alpha_k, X_P, X_Q\},$$

the second step to get the probability in LHS of (6.39) becomes the evaluation of the integral

$$I_k(\alpha_k) \triangleq \iint_{\mathcal{X}} D_k(\alpha_k, x_P, x_Q) f_{X_P X_Q}(x_P, x_Q) \, dx_P dx_Q, \tag{6.40}$$

where $\mathcal{X}$ and $f_{X_P X_Q}(x_P, x_Q)$ represent the joint range and pdf of X_P and X_Q, respectively.

In Appendix E the method of *cubature rules* is outlined to approximate integrals like (6.40) on the basis of the knowledge of a certain number of joint moments of the RVs X_P and X_Q. These moments can be computed using an extension of the recursive algorithm already explained for the one-dimensional case (see Problem 6.6). In some cases, owing to the symmetry of the modulation scheme, the two-dimensional problem can be reduced to the product of two one-dimensional problems, or even to a single one-dimensional problem. An example is the CPSK.

CPSK modulation

The complete symmetry of the signal set allows us to simplify the error probability (6.38) as

$$P(e) = 1 - P\{\mathbf{r}_{D0} \in R_1 | a_0 = A\}$$

$$= 1 - P\left\{-\frac{\pi}{M} < \phi_{D0} \leq \frac{\pi}{M}\right\}. \tag{6.41}$$

In (6.41) we have assumed that the phase zero has been transmitted (see Table 6.1), and we have defined the phase of the received vector $\mathbf{r}_{D0}$ as

$$\phi_{D0} \triangleq \tan^{-1} \frac{r_{DQ0}}{r_{DP0}}. \tag{6.42}$$

A straightforward extension of the bounding technique that led to (5.32) for the AWGN channel results in the following bounds for the error probability:

$$\max(I_1, I_2) \leq P(e) < I_1 + I_2, \tag{6.43}$$

where

$$I_1 = \frac{1}{2} \int_{\mathcal{L}} erfc\left(\frac{\lambda_0^+ + \lambda}{\sqrt{2}\,\sigma_n}\right) f_\Lambda(\lambda) \, d\lambda, \tag{6.44}$$

$$I_2 = \frac{1}{2} \int_{\mathcal{L}} erfc\left(\frac{\lambda_0^- - \lambda}{\sqrt{2}\,\sigma_n}\right) f_\Lambda(\lambda) \, d\lambda, \tag{6.45}$$

and

$$\Lambda \triangleq A \sum_n{}' \left[h_{Pn} \sin\left(\frac{\pi}{M} + \phi_n\right) + h_{Qn} \cos\left(\frac{\pi}{M} + \phi_n\right) \right], \tag{6.46}$$

$$\lambda_0^{\pm} \triangleq A \left(h_{P0} \sin\frac{\pi}{M} \pm h_{Q0} \cos\frac{\pi}{M} \right). \tag{6.47}$$

Looking at (6.44) to (6.47), we can see that the evaluation of the bounds of the error probability for CPSK modulation has been reduced to the computation of two one-dimensional integrals like (6.26). Thus, all the methods introduced in the PAM case directly apply; in particular, we can apply the GQR method.

Example 6.7

Consider a CPSK modulation scheme that uses a channel modeled as a third-order Butterworth filter with 3-dB bandwidth B_0. In Fig. 6.11 the error probability for binary PSK computed by GQR is plotted as a function of the number of points J of the quadrature formula or of the terms retained in the series expansion. The dashed line is the exact value of $P(e)$ obtained by means of (6.29). The number of interfering samples chosen equals 20. It can be seen that even with a small value of J the GQR offers a high accuracy.

The difference in the computer times needed to obtain the two curves of Fig. 6.11 is enormous and such as to prevent the use of the direct enumeration when the number of phases increase. In Fig. 6.12 the signal-to-noise ratio $\mathcal{E}_b/N_0$ necessary to obtain an error probability of 10^{-6} for quaternary CPSK is plotted as a function of the normalized bandwidth $2B_0T$. The two curves refer to the Chernoff bound and to the GQR methods. The asymptotic value represents the case of no ISI. It can be seen that the Chernoff bound is rather loose, leading to an asymptotic difference of about 1 dB in signal-to-noise ratio. $\square$

Example 6.8

In this example we want to show how the computational techniques that have been presented in this chapter can be applied to the analysis and design of a digital transmission system operating in an FDM multichannel environment. The system model employing CPSK modulation is presented in Fig. 6.13. The attention is focused on one particular channel (the "useful" channel), disturbed by two adjacent channels, working at the same signaling rate, giving rise to interchannel interference, and by one channel at the same frequency (separated through cross-polarization), originating cochannel interference. The transmitter

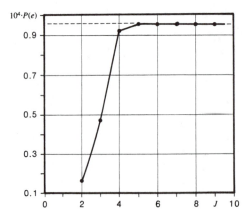

Figure 6.11 Error probability as a function of the number of points of the quadrature rule for a binary CPSK system with third-order Butterworth filter; the dashed line represents the exact value.

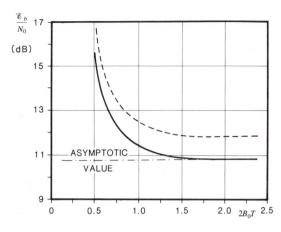

Figure 6.12 Signal-to-noise ratio $\mathcal{E}_b/N_0$ necessary to obtain a symbol error probability $P(e) = 10^{-6}$ for quaternary CPSK system with third-order Butterworth filter as a function of the normalized 3-dB bandwidth $2B_0T$.

filters are assumed to have the same transfer function, except for the frequency location. In other words, let

$$G_i^+(f) \triangleq G_0^+(f + if_d), \qquad i = -1, 0, 1, \tag{6.48}$$

be the transfer function of the ith channel transmitter filter for positive frequencies, where f_d is the frequency spacing between two adjacent channels. For simplicity, we shall assume that $G_i(f)$ satisfies the symmetry conditions of Example 6.2 with respect to its center frequency $f_0 + if_d$, f_0 being the carrier frequency of the useful signal.

If

$$\tilde{v}(t) = \sum_n \exp(j\phi_n)s(t - nT) \tag{6.49}$$

is the complex envelope of the useful signal at the modulator output, the ith interfering signal can be written as

$$\tilde{v}_i(t) = V_i \sum_{m_i} \exp[j\phi_{m_i} + j(2\pi i f_d t + \vartheta_i)]s(t - \tau_i - m_iT), \tag{6.50}$$

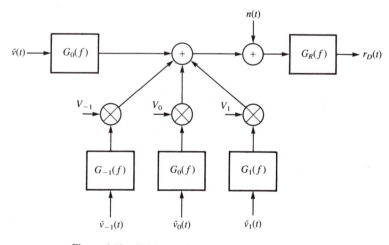

Figure 6.13 CPSK system model with interfering channels.

where the meaning of the symbols is as follows:

τ_i accounts for the possible misalignment of signaling intervals in different channels. It may be modeled as a uniformly distributed RV in the interval $(0, T)$.

ϑ_i is a RV uniformly distributed in the interval $(0, 2\pi)$ and accounts for the lack of coherence among the different carriers.

(ϕ_{m_i}) is the sequence of information phases pertaining to the i th channel.

The bounding technique described for the case of CPSK with ISI can be applied here for estimating the error probability. Moreover, the GQR method can also handle this situation, provided that the error probability conditioned on given values of τ_i and ϑ_i is first computed and the averages over τ_i and ϑ_i are performed later using standard quadrature rules.

From the system engineer's viewpoint, the main design parameters are the frequency spacing f_d between adjacent channels, the cross-polarization attenuation of a carrier at the same frequency as the useful one, the choice of the transmitter and receiver filters (types and bandwidths), and the signal-to-noise ratio required to get a desired value of the error probability. The choice of these parameters is usually accomplished through a cut-and-try approach, which requires repeated analyses of the system and, hence, the availability of a flexible tool to evaluate system performance in a fast and accurate way.

As usual in PSK, we shall assume that the shaping function $s(t)$ is rectangular. Both the transmitter and receiver filters are assumed to be Butterworth. Consider now the following parameters defining the system:

n_T, n_R: the order of transmitter and receiver filters, respectively.

$(B_{eq}T)_T$, $(B_{eq}T)_R$: equivalent noise bandwidths of the transmitter and receiver filters normalized to the symbol period T.

$D = f_d T$: frequency spacing between two adjacent channels normalized to the symbol period T.

$\mathcal{E}_b/N_0$: signal-to-noise ratio, being $N_0/2$ the two-sided power spectral density of the noise.

The results that follow have been obtained by choosing as the sampling instant t_0 the time value corresponding to the maximum of the response of the overall system without interchannel and cochannel interferences.

Interchannel interference

Two symmetrically located interfering channels are present at the same power level as the one interfered with. The modulation is assumed to be quaternary CPSK. The first parameter considered for optimization is the normalized bandwidth of the receiver filter. In Fig. 6.14 the signal-to-noise ratio necessary to obtain an error probability equal to 10^{-6} is plotted as a function of the normalized receiver filter bandwidth. The symbol intervals in the three channels are first assumed to be time aligned (i.e., $\tau_i = 0$ for both the interfering channels). It can be seen that a value of the normalized bandwidth around 1.1 is optimum. In the remaining curves of this example, the normalized receiver bandwidth will be assumed equal to 1.1. Let us now consider the choice of the channel spacing. In Fig. 6.15 the signal-to-noise ratio necessary to obtain an error probability of 10^{-6} is plotted as a function of the normalized channel spacing D. The three curves refer to different values of the transmitter filter bandwidths (the value ∞ means absence of the transmitter filter). It is seen that the presence of a transmitter filter with bandwidth equal to 2.4 significantly improves the performance of the system.

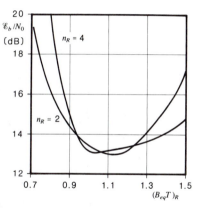

Figure 6.14 Quaternary CPSK system: signal-to-noise ratio $\mathcal{E}_b/N_0$ necessary to obtain a $P(e) = 10^{-6}$ as a function of the normalized equivalent noise bandwidth of the receiving filter; $n_T = 6$, $(B_{eq}T)_T = 2.5$, and $D = 1.6$.

This result is confirmed by Fig. 6.16, where the only difference is represented by the fact that there is a random misalignment among the modulating bit streams. Thus the error probability is evaluated through an average over the RV τ_i.

Cochannel interference

Finally, in Fig. 6.17, the presence of one interfering channel at the same frequency as the useful one is considered. The modulating bit stream on the interfering channel is supposed to have a random misalignment. The curves plot the signal-to-noise ratio necessary to obtain an error rate of 10^{-6} as a function of the attenuation of the interfering channel. It is seen that the attenuation has to be of the order of 14, 16, or 20 dB for the cases of binary, quaternary, and octonary PSK, respectively, to ensure a negligible degradation of performance, as compared with the case of no interference. $\square$

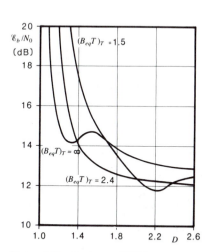

Figure 6.15 Quaternary CPSK system: signal-to-noise ratio $\mathcal{E}_b/N_0$ necessary to obtain a $P(e) = 10^{-6}$ as a function of the normalized frequency displacement of two symmetrically located interfering channels modulated by time-aligned bit streams; $n_T = 6$, $n_R = 2$, $(B_{eq}T)_R = 1.1$.

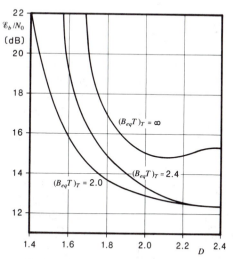

Figure 6.16 Quaternary CPSK system: same situation as in Fig. 6.15 except for the random time misalignment between the bit streams of the interfering and interfered channels.

Sec. 6.2 Evaluation of the Error Probability

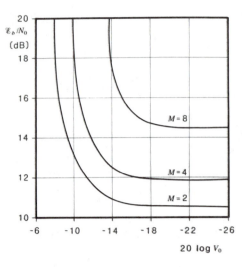

Figure 6.17 Cochannel interference effects in binary, quaternary, and octonary CPSK systems. The signal-to-noise ratio $\mathscr{E}_b/N_0$ necessary to obtain a $P(e) = 10^{-6}$ is plotted as a function of the attenuation of the interfering channel. The modulating bit streams are assumed to be randomly misaligned; $n_R = 2$, $(B_{eq}T)_T = \infty$, $(B_{eq}T)_R = 1.1$.

6.3 CARRIER AND CLOCK SYNCHRONIZATION

So far, in computing the performance of a digital communication system, we assumed implicitly that the same clock controlled both the transmitter and the receiver operations. This means that corresponding events in the transmitter and receiver are synchronous (i.e., they occur at the same time instants, or at time instants that differ by a fixed and constant delay).

In Chapter 5 we saw that the most efficient modulation schemes are coherent; they make use of the phase information of the carrier. Optimum demodulation requires a local carrier at the receiver side whose frequency and phase are in perfect agreement with that of the transmitted signal. In principle, two pairs of ideal identical oscillators at the transmitter and receiver sides could ensure the synchronization and coherence required for a proper operation of the system. However, the signals emitted by a pair of oscillators with the same nominal frequency f_0 can be written as

$$z_1(t) = A_1 \cos\left[2\pi f_0 t + \vartheta_1(t)\right], \tag{6.51}$$

$$z_2(t) = A_2 \cos\left[2\pi f_0 t + \vartheta_2(t)\right], \tag{6.52}$$

where each $\vartheta_i(t)$ can be modeled as a Wiener random process with $\vartheta_i(0) = 0$, zero mean, and variance equal to t/τ_i, $i = 1, 2$. This random process is a nonstationary Gaussian process defined in the interval $(0, \infty)$. Thus, the variance of the random process representing the phase difference between the two oscillators is given by

$$E[\vartheta_1(t) - \vartheta_2(t)]^2 - \frac{t}{\tau_1} + \frac{t}{\tau_2} = t\frac{\tau_1 + \tau_2}{\tau_1\tau_2} \triangleq \frac{t}{\tau_{12}}, \tag{6.53}$$

where we have defined the *joint coherence time* τ_{12} of the two oscillators. Since the variance (6.53) increases with time, we can conclude that a pair of independent oscillators cannot maintain the synchronization indefinitely. They need to mutually exchange certain informations, that is, to be in some way *locked*.

Acquisition and tracking

So far we have supposed $\vartheta_1(0) - \vartheta_2(0) = 0$. This is certainly not true when we switch on the modems to start the transmission. The two oscillators are completely incoherent and we need an initial period of time, before the transmission of data can be started, to synchronize the oscillators. This is usually known as *acquisition time* or *acquisition phase*. At the end of the acquisition phase, the two oscillators are locked and data transmission starts. During the data transmission, we also need to keep the phase difference between the two oscillators within certain specified bounds. This operation is known as the *tracking phase*. It is needed only when the transmission time is significantly larger than the joint coherence time of the oscillators. When this is not the case, as in the transmission of characters from a terminal keyboard, we have an asynchronous transmission.

Different levels of synchronization are often required in the system. As an example, consider a TDM–PCM system using binary CPSK modulation. We need the four levels of synchronization shown in Fig. 6.18: the frame, word, symbol, and carrier synchronization. Here we will only be concerned with the last two, carrier and symbol synchronization.

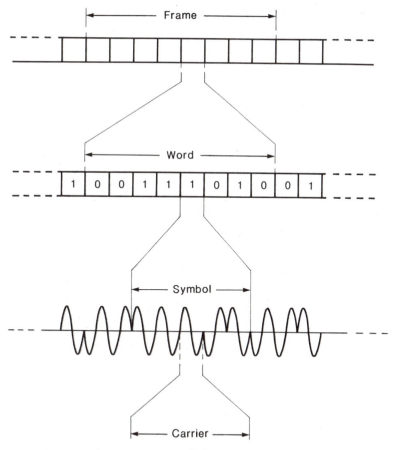

Figure 6.18 Different synchronization levels in digital transmission.

Sec. 6.3   Carrier and Clock Synchronization   **293**

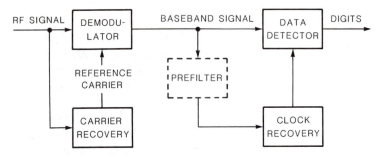

Figure 6.19 Receiver illustrating locations of synchronizers.

The location of carrier and clock synchronizers within a typical receiver is shown in Fig. 6.19.

We have seen from the spectral analysis of modulated signals in Chapter 5 that the most efficient digital modulation techniques suppress the carrier completely; all transmitted power resides in the sidebands and none is "wasted" on a spectral line at the carrier frequency. Also, efficient line codes (see Chapter 9) give rise to data streams not containing spectral lines at the clock frequency. Thus, any carrier or clock synchronizer will be composed of two conceptually distinct parts: (1) a suitable nonlinear circuit that regenerates a carrier or clock from the data signal that contains neither and (2) a narrowband device (typically a tuned filter or a *phase-locked loop*, PLL) that separates the regenerated carrier or clock from background disturbances.

To give a theoretically sound justification of the structure of the PLL, which is the most widely used circuit for synchronization purposes, let us consider a simplified situation in which the regenerated carrier is only affected by the addition of Gaussian noise; that is,

$$z(t, \vartheta) = A \sin (2\pi f_0 t + \vartheta) + n(t). \tag{6.54}$$

We want to obtain the "best" estimate of the unknown phase ϑ based on the observation of $z(t)$ in an interval of length nT_0, with $T_0 = 1/f_0$. A straightforward application of detection theory (Section 2.6.2) to the continuous case leads to the following expression for the log-likelihood function:

$$\lambda(\Theta) = \int_0^{nT_0} [z(t, \vartheta) - A \sin (2\pi f_0 t + \Theta)]^2 \, dt. \tag{6.55}$$

The optimum unbiased estimate $\hat{\vartheta}$ of ϑ under the ML criterion is the one minimizing the RHS of (6.55) or, equivalently, satisfying the equation

$$\int_0^{nT_0} z(t, \vartheta) \cos (2\pi f_0 t + \hat{\vartheta}) \, dt = 0. \tag{6.56}$$

A block diagram of the ML estimator of ϑ is presented in Fig. 6.20.

To obtain an approximate expression of the variance of the estimate $\hat{\vartheta}$, let us assume that the noise power is low so that we can write

$$\cos (2\pi f_0 t + \hat{\vartheta}) \cong \cos (2\pi f_0 t + \vartheta) + (\hat{\vartheta} - \vartheta) \frac{\partial}{\partial \vartheta} \cos (2\pi f_0 t + \vartheta). \tag{6.57}$$

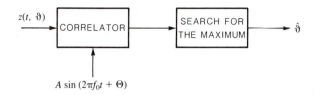

$z(t, \vartheta)$ → CORRELATOR → SEARCH FOR THE MAXIMUM → $\hat{\vartheta}$

$A \sin (2\pi f_0 t + \Theta)$

Figure 6.20 ML estimator for the carrier phase.

Substitution of (6.54) and (6.57) in (6.56) leads to

$$\hat{\vartheta} - \vartheta \cong \frac{\displaystyle\int_0^{nT_0} n(t) \cos (2\pi f_0 t + \vartheta) \, dt}{A \displaystyle\int_0^{nT_0} \sin^2 (2\pi f_0 t + \vartheta) \, dt}. \tag{6.58}$$

Now, accounting for the fact that $n(t)$ is a white Gaussian noise with power spectral density $N_0/2$, we can easily obtain for the variance of the estimate the following expression:

$$E (\hat{\vartheta} - \vartheta)^2 \cong \frac{N_0}{A^2 nT_0}. \tag{6.59}$$

From examination of (6.59), one can easily conclude that the variance of the estimate is inversely proportional to the signal-to-noise ratio and to the length of the observation interval. Figure 6.20 also shows that the optimum estimator has an open-loop structure. However, practical considerations render the solution of Fig. 6.20 unacceptable. In fact, the estimate of the unknown phase is available only at the end of the observation interval. Since the phase estimate has to be used for coherent demodulation with the final goal of deciding on the transmitted data, these data should be stored for a time equal to nT_0 in order to postpone any decision about them. This procedure should also be repeated periodically in order to follow the slow variations of the carrier phase during the tracking period. On the whole, this solution is impractical.

A possible way of overcoming these difficulties consists in obtaining the desired estimate using an iterative procedure. Suppose that at the kth carrier period we have the estimate $\hat{\vartheta}_k$ of the unknown phase, and that we want to modify this estimate on the basis of the observation of the received signal in the following carrier period. Consider the average of the quantity (6.56) in an observation interval of length T_0 conditioned on the value $\hat{\vartheta}_k$ of ϑ_k obtained in the previous period:

$$E \left\{ \frac{d\lambda (\Theta)}{d \Theta} \Big| \Theta = \hat{\vartheta}_k \right\} = E \left\{ \int_{kT_0}^{(k+1)T_0} z(t, \vartheta) \cos (2\pi f_0 t + \hat{\vartheta}_k) \, dt \right\}$$
$$= \frac{AT_0}{2} \sin (\hat{\vartheta}_k - \vartheta). \tag{6.60}$$

In Fig. 6.21 the behavior of this average is shown as a function of $\hat{\vartheta}_k$. If we have a value of $\hat{\vartheta}_k$ close to ϑ (which is reasonable at the end of the acquisition phase), the average (6.60) is a good error indicator for the estimate at hand. In fact, whenever it is positive, we know that $\hat{\vartheta}_k$ is greater than ϑ; on the other hand, when it is negative, we know that $\hat{\vartheta}_k$ is smaller than ϑ. Moreover, its magnitude tells us how far from ϑ our past estimate is.

Sec. 6.3   Carrier and Clock Synchronization

295

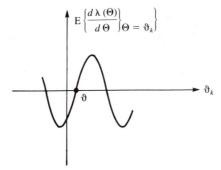

Figure 6.21 Error indicator function for recursive phase estimation.

A reasonable recursive algorithm to update our estimate is thus the following:

$$\hat{\vartheta}_{k+1} = \hat{\vartheta}_k - \alpha_k E\{e(\hat{\vartheta}_k)\} \qquad (6.61)$$

where $e(\hat{\vartheta}_k)$ is defined as

$$e(\hat{\vartheta}_k) \triangleq \left. \frac{d\lambda(\Theta)}{d\Theta} \right|_{\Theta=\hat{\vartheta}_k.} \qquad (6.62)$$

In (6.61) the similarities with the gradient algorithms described in Chapter 8 are evident. Since we are not able to compute the average in RHS of (6.61), it seems appropriate to modify the recursive algorithm (6.61) as

$$\hat{\vartheta}_{k+1} = \hat{\vartheta}_k - \frac{1}{k}\sum_{i=0}^{k}\alpha_i e(\hat{\vartheta}_i), \qquad (6.63)$$

where we have extended the influence of the past estimates to the whole time interval $(0, kT_0)$.

6.3.1 Phase-Locked Loop

A practical implementation of the recursive algorithm (6.63) is provided by the circuit of Fig. 6.22, which is called a *phase-locked loop* (PLL). In it, the received signal $z(t, \vartheta)$ is multiplied by the output of a *voltage-controlled oscillator* (VCO). This device generates a carrier whose frequency varies linearly with the amplitude of a control signal. The product is low-pass filtered and input to the VCO, whose instantaneous angular frequency is changed according to

$$\dot{\hat{\vartheta}}(t) \triangleq \frac{d\hat{\vartheta}(t)}{dt} = k_2 e(t). \qquad (6.64)$$

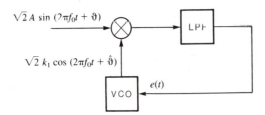

Figure 6.22 Block diagram of PLL.

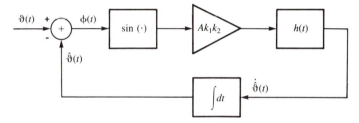

Figure 6.23 Equivalent block diagram of a PLL for the phases.

Under reasonable simplifications, it can be shown (see Problem 6.14) that the PLL implements a relationship like (6.63) between successive estimates of ϑ.

Let us now analyze in some detail the behavior of the PLL, which is the heart of many synchronization circuits. Suppose for the moment that the received signal is noiseless. Let us denote by $h(t)$ the impulse response of the low-pass filter in Fig. 6.22 and suppose that it filters out the high-frequency component at the output of the multiplier. Thus, having defined the phase error

$$\phi(t) \triangleq \vartheta(t) - \hat{\vartheta}(t), \tag{6.65}$$

we obtain the following nonlinear equation governing the behavior of the PLL:

$$\dot{\phi}(t) = \dot{\vartheta}(t) - Ak_1k_2 \int_0^t h(t-\tau) \sin\phi(\tau)\, d\tau. \tag{6.66}$$

In Fig. 6.23 a block diagram that functionally represents equation (6.66) is shown. While in general its behavioral analysis is difficult, it simplifies drastically in the case of a phase error $\phi(t)$ small enough to justify the approximation $\sin\phi \cong \phi$, which linearizes (6.66) and leads to the circuit of Fig. 6.24. Its analysis is straightforward using the Laplace transforms. We obtain the transfer function of interest:

$$H_{eq}(s) \triangleq \frac{\hat{\Theta}(s)}{\Theta(s)} = \frac{Ak_1k_2H(s)}{s + Ak_1k_2H(s)}, \tag{6.67}$$

where $\Theta(s)$ is the Laplace transform of $\vartheta(t)$.

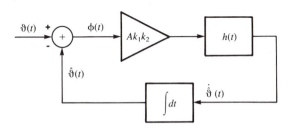

Figure 6.24 Linearized version of the block diagram of Fig. 6.23.

Order of the PLL

The order of the PLL is defined according to the degree of the denominator of $H_{eq}(s)$, which in turn depends on the loop filter transfer function $H(s)$. Thus we have

First-order PLL

$$H(s) = 1 \quad \Rightarrow \quad H_{eq}(s) = \frac{Ak_1k_2}{s + Ak_1k_2}. \tag{6.68a}$$

Second-order PLL with active filter

$$H(s) = \frac{1 + s\tau_2}{s\tau_1} \quad \Rightarrow \quad H_{eq}(s) = \frac{4\zeta\pi f_n s + (2\pi f_n)^2}{s^2 + 4\zeta\pi f_n s + (2\pi f_n)^2}, \tag{6.68b}$$

where the two parameters f_n and ζ, called respectively the *natural frequency* and the *damping factor* of the loop, are given by

$$f_n = \frac{1}{2\pi} \sqrt{\frac{Ak_1k_2}{\tau_1}}, \tag{6.69}$$

$$\zeta = \frac{\tau_2}{2} \sqrt{\frac{Ak_1k_2}{\tau_1}}. \tag{6.70}$$

We observe that the order of the PLL corresponds to the number of perfect integrators within the loop. For a first-order PLL, we can control only one parameter, Ak_1k_2, which is the 3-dB bandwidth of $H_{eq}(s)$, whereas the second-order loop gives us the two parameters f_n and ζ.

The magnitude of the frequency response $H_{eq}(j2\pi f)$ of a second-order loop for several values of ζ is plotted in Fig. 6.25. It can be seen that the loop performs a low-pass filtering operation on phase inputs. Using root-locus plot characteristics, it can be shown that first- and second-order loops are always stable, whereas third- and higher-order loops can be stable under certain conditions. Besides the stability considerations, it is important to know the steady-state behavior of the PLL (i.e., the steady-state phase and frequency errors in the presence of particular inputs). We shall examine two different cases of input phases:

$$\vartheta_A(t) = 2\pi t \, \Delta f + \Delta\vartheta, \tag{6.71}$$

$$\vartheta_B(t) = \pi t^2 \, \dot{\Delta f} + 2\pi t \, \Delta f + \Delta\vartheta. \tag{6.72}$$

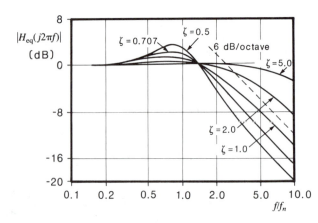

Figure 6.25 Frequency response of a second-order loop with active loop filter.

TABLE 6.3 Steady-state phase and frequency errors for PLL of various order

Errors	1st order PLL	2nd order PLL	3rd order PLL
$\Delta\phi_A$	$\dfrac{2\pi\Delta f}{Ak_1k_2}$	0	0
$\Delta\phi_B$	∞	$\dfrac{2\pi\Delta f\tau_1}{Ak_1k_2}$	0
Δf_A	0	0	0
Δf_B	$\dfrac{\Delta f}{Ak_1k_2}$	0	0

The first case is the most important in data transmission, since it refers to a situation in which the received carrier presents a frequency displacement Δf (e.g., due to an FDM mu-demultiplexing) and an initial phase shift $\Delta\vartheta$. The second case can happen when there is an accelerated motion between transmitter and receiver, as in a mobile radio communication system. Using the final-value theorem of the Laplace transform (see Problem 6.13), we obtain the steady-state errors of Table 6.3, where $\Delta f_{A,B}$ and $\Delta\phi_{A,B}$ represent the frequency and phase errors, respectively.

All the preceding results were based on the assumption that the phase error is sufficiently small, thus allowing the loop to be considered linear in its operation. This assumption becomes progressively less useful as error increases until, finally, the loop drops out of lock and the assumption becomes worthless. Through the analysis of the nonlinear model of the PLL, one can identify important parameters like, for example, the *hold-in range* (i.e., the input frequency range over which the loop will hold lock) or the *acquisition time* (i.e., the time required by the loop to reduce the phase error under a given threshold). A detailed analysis of the behavior of the PLL without the linear assumption can be found in Viterbi (1966) and Lindsey (1972).

When a Gaussian noise is present in additive form at the input of the PLL, an approximate linear analysis is still possible when the signal-to-noise ratio is sufficiently high. This leads (see Viterbi, 1966) to the functional block diagram of Fig. 6.26, where $n'(t)$ is a Gaussian noise process independent of $n(t)$, with the same spectral properties as the input noise. To see the effect of the noise, we assume that the input phase $\vartheta(t)$ is constant so that the fluctuations in the phase of the VCO signal can be entirely attributed to the noise. From Fig. 6.26 we can obtain the transfer function between the noise $n'(t)$ and the recovered phase $\hat{\vartheta}(t)$:

$$H_n(s) \triangleq \frac{\hat{\Theta}(s)}{N'(s)} = \frac{k_1k_2H(s)}{s + Ak_1k_2H(s)} = \frac{1}{A}H_{eq}(s). \tag{6.73}$$

The noise power contribution to $\vartheta(t)$ is then

$$\sigma_\vartheta^2 = \frac{1}{A^2}\int_{-\infty}^{\infty} G_{n'}(f)|H_{eq}(j2\pi f)|^2\,df = \frac{N_0}{A^2}B_{eq}, \tag{6.74}$$

Sec. 6.3   Carrier and Clock Synchronization

299

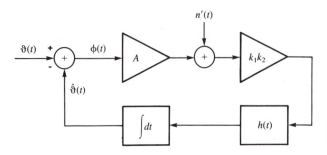

Figure 6.26 Equivalent block diagram of a linearized PLL including noise.

where we have defined the equivalent noise bandwidth of the PLL. For first- and second-order PLL, it is given by the following expressions:

$$B_{eq} = \frac{Ak_1k_2}{4} \quad \text{(first order)}, \tag{6.75}$$

$$B_{eq} = \pi f_n \sqrt{\zeta + \frac{1}{4\zeta}} \quad \text{(second order)}. \tag{6.76}$$

It can be seen from (6.75) and Table 6.3 that, for a first-order PLL, the needs for a small steady-state phase error and a small noise bandwidth are in conflict. For a second-order PLL, a good compromise is achieved with the value $\zeta = 0.707$.

When the linear analysis is valid, the VCO phase jitter is a Gaussian random process. In general, this is not true. Nonlinear analysis of a PLL has been concerned with deriving the pdf of the phase error ϕ. This pdf (see Viterbi, 1966) is found as the steady-state solution of a nonlinear stochastic partial-differential equation known as the Fokker–Planck equation. Without going into the details, the resulting Tikhonov pdf is

$$f_\phi(\varphi) = \frac{\exp(\rho \cos \varphi)}{2\pi I_0(\rho)}, \quad |\varphi| \le \pi, \tag{6.77}$$

where ρ is the signal-to-noise ratio of the loop and $I_0(\rho)$ is the modified Bessel function of the first kind and order zero. The pdf (6.77) approaches a Gaussian one for large ρ.

In its essence, the PLL acts as a narrowband filter whose central frequency tracks the frequency of the received signal (6.54) within a reasonable range without affecting its noise bandwidth. As already stated, it requires a spectral line at the frequency to be tracked. Thus, beyond the PLL, suitable nonlinear circuits are integral portions of a synchronizer. In the following, we shall consider some of the most common carrier and clock synchronizers. Only a qualitative description of their behavior will be presented. The reader interested in the detailed performance analysis may refer to the Bibliographical Notes at the end of the chapter.

6.3.2 Carrier Synchronizers

To simplify, let us consider initially a binary CPSK signal written in its bandpass form:

$$v(t) = v_P(t) \cos(2\pi f_0 t + \vartheta), \tag{6.78}$$

where

300   Digital Transmission over Real Channels   Chap. 6

$$v_P(t) = \sum_n a_n s(t - nT), \tag{6.79}$$

$$s(t) = u_T(t), \tag{6.80}$$

and the information symbols a_n take the values ± 1.

There are three main types of carrier synchronizers, the *squaring loop*, the *remodulator*, and the *Costas loop*. They differ in the position of the nonlinearity, which is entirely separated from the PLL in the squaring loop, whereas it is included in the phase detector for the remodulator and the Costas loop. The block diagram of the squaring loop is shown in Fig. 6.27.

Its nonlinearity is a square-law device, so its output

$$z(t) = v^2(t) = \tfrac{1}{2}[1 + \cos(4\pi f_0 t + 2\vartheta)] \tag{6.81}$$

contains a spectral line at frequency $2f_0$ that can be tracked by a conventional PLL. The VCO output is divided by 2 to provide the desired carrier. In the divide-by-2 operation, there is a phase indeterminacy, which makes it impossible to decide whether the current symbol is 1 or -1. This phase ambiguity is inherent in all phase-shift modulation techniques. It can be resolved by special encoding, like the differential encoding described in Section 5.3.2.

A remodulator synchronizer is shown in Fig. 6.28. The received signal is demodulated and the message $v_P(t)$ is recovered. It is used to remodulate the received signal so as to remove the modulation. If the baseband waveforms are time aligned, the output of the balanced modulator has a pure carrier component that can be tracked by the PLL. The delay τ in Fig. 6.28 is required to compensate for the delay of the low-pass filter following the demodulator. In the figure, the relationships explaining the behavior of the synchronizer in the absence of noise are also given.

A block diagram of the Costas loop is shown in Fig. 6.29. Its behavior should be explained by the relationships indicated in the figure, which are valid in the absence of noise.

The carrier recovery circuits described before can be generalized to the situation in which the digital information is transmitted via M-ary CPSK modulation. An Mth power synchronizer is shown in Fig. 6.30. Its operation is easily understood by simple extension of the squaring loop. Because of their wide applications, block diagrams of

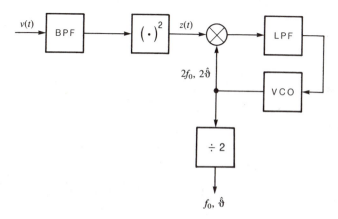

Figure 6.27 Block diagram of the squaring loop for carrier recovery.

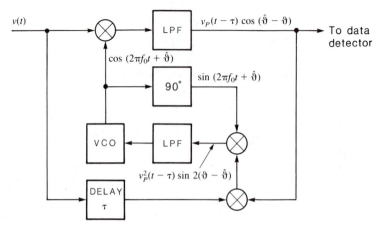

Figure 6.28 Block diagram of the remodulator for carrier recovery in the case of binary CPSK.

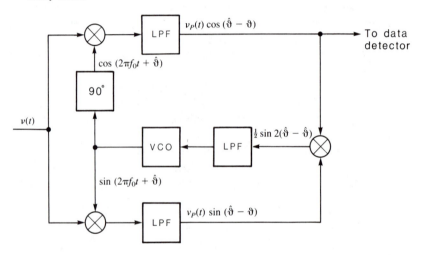

Figure 6.29 Block diagram of the Costas loop for carrier recovery in the case of binary CPSK.

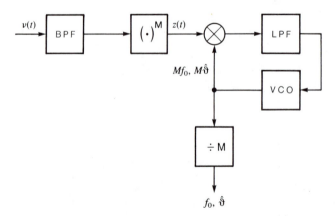

Figure 6.30 Block diagram of the Mth power synchronizer for carrier recovery.

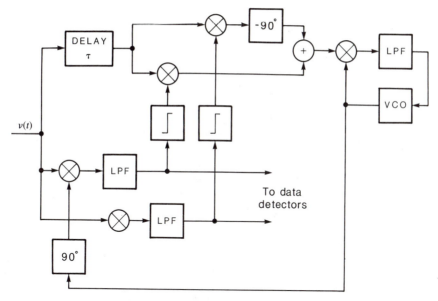

Figure 6.31 Block diagram of the remodulator for QPSK.

the QPSK remodulator and the Costas loop are shown in Figs. 6.31 and 6.32. A stable lock can be achieved at any of four different phases. There is an inherent fourfold ambiguity that must be resolved by other means. Both the remodulator and the Costas loop perform a multiplication by the demodulated message in analog form to remove modulation. Better noise rejection would be possible if the detected digital message were used for modulation removal. This is done in *decision-directed* synchronizers. Unfortunately, a decision-directed synchronizer cannot acquire the carrier until the clock has been obtained. Thus, it is not suited for applications requiring fast acquisition.

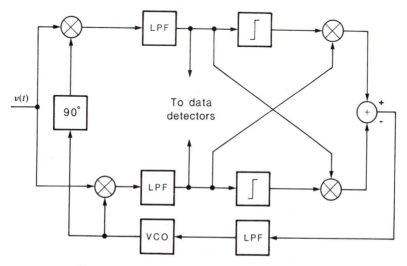

Figure 6.32 Block diagram of the Costas loop for QPSK.

Sec. 6.3   Carrier and Clock Synchronization

303

In general, the performance analysis of a synchronizer is very difficult because of the presence of the nonlinear regenerator. Without going into details, we can write the noise-caused VCO phase jitter variance for an M-phase synchronizer as

$$\sigma_\vartheta^2 = M^2 \left(\frac{N_0 B_{eq}}{A^2} \right) L_M(\rho_i), \qquad (6.82)$$

where M^2 shows the M-fold phase magnification, $L_M(\rho_i)$ is the loss caused by noise intermodulation in the nonlinearity, and ρ_i is the signal-to-noise ratio in the input filter. The quantity within the parentheses represents the jitter variance of an ordinary PLL [see (6.74)]. For the special case of an Mth-power nonlinearity (Butman and Lesh, 1977), some typical losses are given in Table 6.4.

6.3.3 Clock Synchronizers

We assume that carrier and clock recovery is made in two distinct steps: first, the phase ϑ of the carrier is estimated; then the timing wave is extracted from the demodulated baseband signal. In other words, we shall not consider the approach of the simultaneous estimate of the carrier and clock. This omission does not imply a merit judgment, since in some cases superior performance can be obtained with joint estimation methods. Some comments can be found in the Bibliographical Notes at the end of this chapter.

Consider now the baseband signal obtained from the demodulator:

$$r_D(t) = \sum_n a_n h(t - nT), \qquad (6.83)$$

where (a_n) is the message sequence, which is assumed to be a zero-mean stationary discrete random process formed by iid RVs. The objective of the timing synchronization circuit is to extract from $r_D(t)$ a periodic wave with period T and a proper phase indicating the sampling instant within each period. Clock synchronizers can be categorized according to the bandwidth of the communication system as wideband or narrowband. We are interested in situations where bandwidth occupancy approaches the Nyquist limit of $1/2T$ Hz at baseband. More precisely, we assume that the bandwidth does not exceed $1/T$ Hz. In this case, the pulse $h(t)$ spreads over many symbol intervals, giving rise to

TABLE 6.4
Intermodulation losses of
Mth-law regenerators

M	$L_M(\rho_i)$
1	1
2	$1 + \dfrac{1}{2\rho_i}$
3	$1 + \dfrac{2}{\rho_i} + \dfrac{2}{3\rho_i^2}$
4	$1 + \dfrac{3}{\rho_i} + \dfrac{6}{\rho_i^2} + \dfrac{3}{\rho_i^3}$

ISI. As we shall see in Chapter 7, to avoid ISI the pulses are given a Nyquist shaping. This allows us to eliminate ISI at nominal sampling instants, but it is not sufficient to eliminate the effects of ISI on the clock synchronizer. In general, the recovered clock is affected by a jitter component, called *self-noise* or *data noise*, caused by ISI. In many applications, this self-noise is predominant with respect to the Gaussian noise. For this reason, we have not included the additive Gaussian noise in the RHS of (6.83).

As for carrier acquisition, the available signal $r_D(t)$ has no spectral lines at frequency $1/T$. In fact, $r_D(t)$ is easily recognized as a cyclostationary random process (see Section 2.2.2) with period T, zero mean, and mean-square value

$$E\{r_D^2(t)\} = E\{a^2\} \sum_n h^2(t - nT). \tag{6.84}$$

Equation (6.84) shows that the square of $r_D(t)$ does possess a periodic mean value.

Using the Poisson sum formula (see Problem 6.15), we can express (6.84) in the more convenient form of a Fourier series whose coefficients are obtained from $H(f)$, the Fourier transform of $h(t)$:

$$E\{r_D^2(t)\} = \frac{E\{a^2\}}{T} \sum_l \mu_l \exp\left(j \frac{2\pi l}{T} t\right), \tag{6.85}$$

where

$$\mu_l = \int_{-\infty}^{\infty} H^*\left(f - \frac{l}{T}\right) H(f)\, df. \tag{6.86}$$

Due to the hypothesis of bandwidth limitation for $H(f)$, only the three terms with $l = 0, \pm 1$ in the summation of (6.85) are different from zero. The first corresponds to a dc component, whereas the other two give a sinusoidal signal with frequency $1/T$ and amplitude

$$\mu_1 = \int_{-\infty}^{\infty} H^*\left(f - \frac{1}{T}\right) H(f)\, df. \tag{6.87}$$

Note that the sinusoidal component at the clock frequency vanishes when $H(f)$ is strictly bandlimited in the interval $(-1/2T, 1/2T)$. Thus, also in this case, for signals exhibiting some extra bandwidth beyond the Nyquist frequency we can use a nonlinearity (e.g., a square-law rectifier) to restore the desired spectral line followed by a tuned filter or a PLL that tracks the restored timing wave. Alternate zero-crossings of the reference waveform $w(t)$ are used by the pulse generator of Fig. 6.33 as indications of the correct sampling instants.

Remembering the spectral analysis of cyclostationary processes of Section 2.3, we realize that the spectrum of $y(t) = r_D^2(t)$ presents a continuous part, besides the discrete one giving rise to the desired spectral line. Thus, even in the absence of additive Gaussian noise, we have a self-noise entering the tuned filter (or PLL) of Fig. 6.33

Figure 6.33 Block diagram of a clock synchronizer.

Sec. 6.3 Carrier and Clock Synchronization **305**

and causing a fluctuation of the zero crossings of $w(t)$ around the nominal sampling instants, the *timing jitter*. To understand better, consider the complex envelope $\tilde{h}(t)$ of $h(t)$ defined with respect to the frequency $1/2T$, so that

$$h(t) = \mathfrak{R}\left\{\tilde{h}(t) \exp\left(j\frac{\pi}{T}t\right)\right\},$$

and write $r_D^2(t)$ in terms of $\tilde{h}(t)$:

$$r_D^2(t) = y(t) = \frac{1}{2}\mathfrak{R}\left\{\left[\sum_n (-1)^n a_n \tilde{h}^2(t - nT)\right] \exp\left(j\frac{2\pi}{T}t\right)\right\}$$

$$+ \frac{1}{2}\left|\sum_n (-1)^n a_n \tilde{h}(t - nT)\right|^2. \qquad (6.88)$$

The second term in (6.88) can be disregarded as it is not passed through the tuned filter (or PLL). The first term can be rewritten as

$$A \cos\frac{2\pi}{T}t + b_P(t)\cos\frac{2\pi}{T}t + b_Q(t)\sin\frac{2\pi}{T}t. \qquad (6.89)$$

In (6.89) we can recognize the desired periodic component as well as two in-phase and quadrature disturbances. It is precisely the quadrature component $b_Q(t)$ of the complex envelope that produces timing jitter. Using (6.88), this component is

$$b_Q(t) = \sum_k \sum_m a_k a_m (-1)^{k+m} h_P(t - kT) h_Q(t - mT), \qquad (6.90)$$

where $h_P(t)$ and $h_Q(t)$ are the real and imaginary parts of $\tilde{h}(t)$. It is evident from (6.90) that the timing jitter is strongly dependent on the shape of the date pulse $h(t)$ at the input of the nonlinearity. For this reason, some authors have suggested the insertion of a suitable prefilter before the nonlinearity of Fig. 6.33 in order to eliminate or greatly reduce the data noise (Franks and Bubrouski, 1974; Mengali, 1983). In particular, it has been shown that using a tuned filter (or a PLL) with a transfer function exhibiting a conjugate symmetry around the symbol rate $1/T$ and a transfer function $H(f)$ limited in bandwidth to the interval $(1/4T, 3/4T)$ with a conjugate symmetry around $1/2T$, one can completely eliminate the data noise if the nonlinearity is a square-law rectifier.

6.3.4 Effect of Phase and Timing Jitter

At the beginning of the chapter, we saw how to compute the symbol error probability conditioned on a given value of the phase jitter considered as a RV with a known pdf. Later, we introduced the Tikhonov pdf [see (6.77)], which describes the statistical behavior of the phase error at the output of a PLL. We mentioned that it approaches a Gaussian pdf for large values of the loop signal-to-noise ratios. To give an idea of the effect of the phase error on the average error probability, we shall consider a binary CPSK system without ISI affected by a phase error in the recovered carrier with Gaussian pdf. Figure 6.34 shows the behavior of the average error probability as a function of the signal-to-noise ratio SNR. Different curves are labeled according to a value of the standard deviation of the phase jitter σ_ϑ in radians. The curves have been obtained by calculating the conditional error probability and, then, by averaging with respect to the

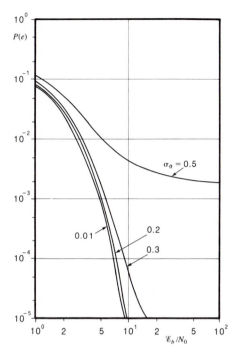

Figure 6.34 Binary CPSK symbol error probability with imperfect carrier synchronization as a function of the signal-to-noise ratio $\mathcal{E}_b/N_0$. (J. J. Stiffler, *Theory of Synchronous Communications*, © 1971, p. 270. Reprinted by permission of Prentice-Hall, Inc., Englewood Cliffs, N.J.)

pdf of the phase jitter. We see that there is a rather strong threshold effect on σ_ϑ at the higher values of SNR.

The effect of a symbol synchronization jitter on the error probability is shown in Fig. 6.35. Also here we assume a simplified situation in which the elementary pulse is

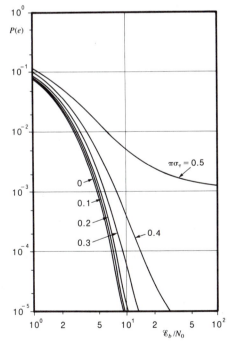

Figure 6.35 Binary CPSK symbol error probability with imperfect symbol synchronization as a function of the signal-to-noise ratio $\mathcal{E}_b/N_0$. (J. J. Stiffler, *Theory of Synchronous Communications*, © 1971, p. 271. Reprinted by permission of Prentice-Hall, Inc., Englewood Cliffs, N.J.)

rectangular, so the only effect of the timing error is to reduce the signal-to-noise ratio at the output of the correlator that implements the optimum receiver for binary CPSK. The timing error is supposed to be a Gaussian RV. The parameter σ_τ in the curves of Fig. 6.35 is the standard deviation of the relative error $\tau = (t_0 - \hat{t}_0)/T$ in the receiver symbol clock.

When we consider a multilevel signaling scheme employing two quadrature carriers, the effect of the carrier jitter is enhanced, because the receiver must choose from four states instead of two. Also, with a phase error we introduce, as seen at the beginning of this chapter, a cochannel interference besides the simple attentuation of the binary case. Thus, the accuracy requirements of the carrier recovery circuits become more stringent.

6.4 DIGITAL TRANSMISSION OVER FADING MULTIPATH CHANNELS

In this section, we consider the effects of radio channels having randomly time variant impulse responses. Among the various radio channels that suffer from fading due to multiple propagation paths, we shall consider in particular the terrestrial microwave radio systems employing digital modulation techniques. After their introduction for use in the 1930s in France, digital microwave radio systems had a fairly long maturing period before being extensively applied after about 1970. Since then the use of high-speed digital radio systems working below 12 GHz has grown rapidly in many countries. High-capacity digital radio is an economically attractive candidate to satisfy the growing requirements for interconnecting digital switching machines and metropolitan digital carrier concentrations. This application calls for a system capable of adding or dropping voice circuits at points separated by a few tens of kilometers. Relative to analog facilities, the lower costs of digital multiplexers and channel banks offset the higher digital radio line costs.

To obtain competitive transmission with large-capacity analog FM systems, highly spectrally efficient modulation techniques (e.g., 64 QAM) capable of sending $4 \div 5$ bits per hertz are necessary. A major source of impairment in high-capacity digital radio systems is selective fading due to multipath propagation.

6.4.1 Multipath Fading

In this section we discuss first the single-frequency fading laws. They are derived from measurements at the carrier frequency and are valid for narrowband analog FM systems. Then we extend the description to frequency selective fading, as the single-frequency characterization of multipath fading does not satisfactorily explain experimental results for digital radio systems.

The radio-frequency (RF) power received after transmission over a microwave radio hop is never absolutely constant; even at noon, when the atmosphere has "stabilized," there can be fractional decibel excursions several times per second, as well as slower excursions of one or two decibels. In propagation experiments, the normal value

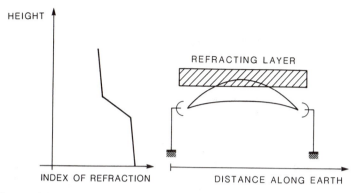

HEIGHT

REFRACTING LAYER

INDEX OF REFRACTION   DISTANCE ALONG EARTH

Figure 6.36   Refraction from a single layer. (Reprinted with permission from the *Bell System Technical Journal*. Copyright 1971, AT&T.)

of the received signal is determined from the peak in a signal-level histogram obtained at least one-half hour at or near noon. This *free-space* value of the received signal is used as a reference to identify periods during which enhanced or depressed signals have resulted from the relatively steady atmospheric focusing or defocusing. Unusual atmospheric conditions may support microwave propagation over two or more distinct paths between two line-of-sight radio antennas. The various signal paths will typically differ in their propagation delay, and therefore the signals traveling over the separate paths will undergo unequal phase shifts. If the received signal, which is the vector sum of the signals received from those paths, is reduced substantially below its free-space value by this mechanism, multipath fading is said to occur. Most severe fading occurs during clear summer nights, when temperature inversions and associated meteorological effects produce negative gradients in the index of refraction of the atmosphere. Figure 6.36 illustrates a simple profile of the refraction index that can produce two or three transmission paths between transmitter and receiver. Two transmission paths are shown.

The time during which a signal envelope is below a certain level is called *duration of fade* of that level. An example is shown in Fig. 6.37 (Vigants, 1975). A convenient expression found experimentally for determining a fade's average duration $\bar{t}$ is (Vigants, 1975)

$$\bar{t} = 410\,L \text{ seconds}, \qquad L < 0.1, \tag{6.91}$$

where L is the level relative to normal.

To characterize the fading, the probability distribution function of the received RF envelope v must be known. It can be obtained from plots of the sum T of the durations of all fades of a particular depth L, called *time below level L*, against fade depth in decibels. As an example, values of T for a 42.8-km path and average terrain are shown in Fig. 6.38 (Vigants, 1975) for a heavy fading month. The curves show a slope of a decade of time per 10 dB, typical of multipath fading. Also, they permit us to obtain an approximate expression for the distribution function of v in the form

$$P\{v \leqslant L\} = rL^2, \tag{6.92}$$

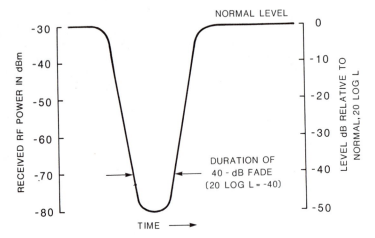

Figure 6.37 Definitions of L and fade duration (-30 dBm assumed as normal as an example). (Reprinted with permission from the *Bell System Technical Journal*. Copyright 1971, AT&T.)

where r is called the *multipath occurrence factor* and can be approximated for heavy fading months as

$$r = 6 \cdot c \cdot f_0 \cdot D^3 \cdot 10^{-7}, \tag{6.93}$$

where f_0 is the carrier frequency in gigahertz, D is the path length in kilometers, and c is a parameter depending on the environment:

$$c = \begin{cases} 1, & \text{average terrain,} \\ 4, & \text{water,} \\ 0.25, & \text{mountains and dry climate.} \end{cases}$$

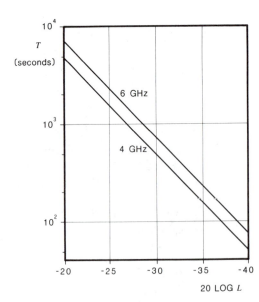

Figure 6.38 Time below level in a heavy fading month ($D = 42.8$ km, $c = 1$, $T = 31$ days or $2.68 \cdot 10^6$ seconds). (Reprinted with permission from the *Bell System Technical Journal*. Copyright 1971, AT&T.)

Note that, when $r = 1$, (6.92) approximates the Rayleigh distribution for the envelope

$$P\{v \leqslant L\} = 1 - e^{-L^2} \cong L^2, \qquad \text{for } L < 0.1, \tag{6.94}$$

which is a widely used approximation in many situations different from line-of-sight microwave radio, like tropospheric and mobile radio propagation. As an example, with $c = 1, f = 6$ GHz, and $D = 50$ km, we get $r = 0.45$ from (6.93).

Deep multipath fading can cause service interruptions or *outages*. The outage can be defined as a situation in which the bit error probability $P_b(e)$ exceeds a preassigned value, for example 10^{-3}. The reliability objectives for a digital radio link are usually specified in terms of a maximum allowable time of outage attributable to all causes, expressed as a percentage of total service time during a given period over a given route length.

Example 6.9

United States telephone companies' reliability objectives for short-haul systems (total length less than 402 km) limit the two-way service failure time to 0.02% annually. Usually, one-half of this is allocated to causes associated with equipment, maintenance, and plant errors, and the rest, 0.01%, is allocated to fading. If we consider a typical hop of 42.8 km, its one-way annual fading allocation equals 5.2×10^{-4} percent, corresponding to 165 seconds per year. ☐

The estimated annual time below level is obtained from (6.92) as

$$T = rT_0 L^2, \tag{6.95}$$

with T_0 describing the duration in seconds of the fading season, assumed to be equal to three months as a geographic average. The curves of Fig. 6.39 (Vigants, 1975) are then obtained, in which the performance objective is also indicated. For a 4-GHz link, it corresponds to roughly 40 dB. Thus, to meet the reliability objectives, the system must be capable of working with a $P_b(e)$ less than 10^{-3} in the presence of a received signal faded by up to 40 dB. This is obtained by increasing by the necessary amount (40 dB in our case) the transmitted level that guarantees a $P_b(e)$ of 10^{-3} in normal conditions. This amount is called the *fade margin*.

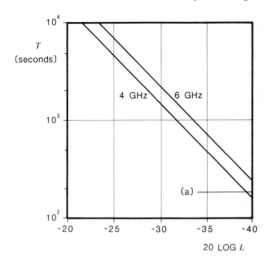

Figure 6.39 Annual time below level (average case). Objective: (a) short haul = 165 seconds per year. (Reprinted with permission from the *Bell System Technical Journal*. Copyright 1971, AT&T.)

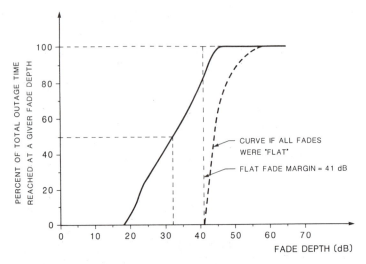

Figure 6.40 Outage distribution for $P_b(e) > 10^{-3}$. (Reprinted with permission. Copyright © 1981 IEEE.)

In Example 6.9, the fade margin has been evaluated under the hypothesis that the only effect of multipath fading over the received signal is a constant attenuation in the signal bandwidth (flat fading). However, the history of microwave digital radio systems has taught that outages often occur long before the attenuation reaches the fade margin. Figure 6.40 (Giger and Barnett, 1981) shows field data taken over a period of about 2.5 months on a radio hop carrying a 45 Mbit/s, 8-CPSK signal in a 20-MHz bandwidth centered around 4 GHz. The continuous curve represents the percent of total outage time reached at a given fade depth versus the fade depth in decibels. It can be seen that 50 percent of the total outage time is caused by fade depths of less than 32 dB. A design making the erroneous flat-fade assumption would predict no outage for such a fade depth, leading to a behavior as shown by the dashed curve in the figure. We are in a situation in which the limiting factor is ISI instead of attenuation. The ISI is produced by channel distortion (i.e., the amount of fading introduced by multipath propagation is varying within the signal bandwidth). The fading becomes a frequency selective phenomenon. This explains how increasing the transmitted level does not improve the performance. That is, it also enhances the ISI level. Remedies other than increasing the fade margin have to be employed to meet the outage objectives. However, before seeking them, we need to characterize the disease.

6.4.2 Models of the Frequency Selective Fading

The calculation and prediction of multipath fading outage for microwave digital radio systems requires a statistical model of the fading on the path. As seen, experimental results showed that the single frequency characterization (flat-fading assumption) gives poor $P_b(e)$ performance indications. We need a channel model in the form of a transfer function whose parameters are statistically described. The model's choice (besides the phenomenon's physical aspects) is dictated by simplicity considerations, analytical handiness, and the ease of fitting the parameters' statistics to the experimental data. In the literature, two models have been proposed and extensively used for performance evaluations. Both are based, for what concerns the parameters statistics, on propagation data obtained in the summer of 1977 from a 42.8-km hop near Atlanta, Georgia. In the following, we describe the *delay network* or *three-path* model, due to Rummler (1979). It is preferred by most researchers in the area. A description of the second model, known as the *first-degree polynomial model*, can be found in Greenstein and Czekaj (1981).

Delay-network model

The delay-network model is based on a three discrete rays model of the multipath propagation. It allows the prediction of behaviors that agree very well with measured results. Generally, the amplitude of the signal on each of these three paths is different. Also, the second and third paths are delayed relative to the first, respectively, by τ_1 and τ_2 seconds, where $\tau_2 > \tau_1$. We define the *simple* three-path model. It requires the delay τ between the first two paths to be sufficiently small, that is,

$$(f_2 - f_1)\tau \ll 1, \tag{6.96}$$

where f_2 and f_1 are the highest and lowest frequencies in the signal bandwidth. This assumption allows one to obtain the low-pass equivalent transfer function of the channel with respect to its center frequency as

$$H(f) = a[1 - be^{j(2\pi f \tau + \phi)}], \tag{6.97}$$

where a, b, and ϕ are the quasi-static fitting parameters and τ is a fixed delay parameter of value 6.3 ns. The function in (6.97) can be interpreted as a channel response that provides a direct transmission path with amplitude a and a second path with a relative amplitude b, a delay τ, and a phase ϕ (independently controllable) at the channel's center frequency. Fitting the function in (6.97) to experimental amplitude data produces an inherent twofold ambiguity. The two possible solutions for $H(f)$ consist of one with $b < 1$ (*minimum phase function*) and one with $b > 1$ (*nonminimum phase function*).

A typical plot of the attenuation produced by this model is shown in Fig. 6.41 (Lundgren and Rummler, 1979). The parameters a and b control the depth and shape of the channel transfer function. The parameter $f_N \triangleq \phi/2\pi\tau$ determines the position of the minimum (or *notch*). If the position of the minimum is within the channel bandwidth, the model can generate notches with a wide range of levels and notch widths. With the minimum out of band, it can generate a wide range of level combinations, slopes, and curvatures within the channel bandwidth. To completely specify the model transfer function (6.97), we need to know the parameters' statistics. In the rest of this section, all numerical quantities in braces $\{\cdot\}$ are derived from experimental data and may vary with path length, antennas, sites, year, and so on. It is hoped, however, that the underlying model structure (i.e., the form of the transfer function and the statistical models for the parameters) is generally applicable.

The notch depth's b distribution is best described in terms of the relative notch depth in decibels (see also Fig. 6.41):

$$B \triangleq -20 \log (1 - b).$$

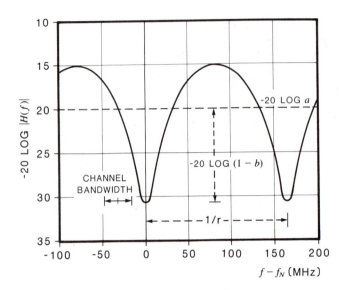

Figure 6.41 Attenuation of the delay-network model for $a = 0.1$ and $b = 0.7$. (Reprinted with permission from the *Bell System Technical Journal*. Copyright 1971, AT&T.)

The resulting distribution function is approximately exponential. A good approximation for the experimental case considered at 6 GHz was

$$P\{B \leq x\} = 1 - e^{-x\{3.8\}}. \tag{6.98}$$

Although the distribution of the parameter $A = -20 \log a$ is dependent on the value of B, the dependence is limited and may often be ignored. This is always true for current digital radio systems, which will tolerate flat fades of 40 dB or so. The distribution of A is well approximated by a normal distribution with a standard deviation equal to 5 and mean value 24.6; thus

$$P\{A \leq y\} = \frac{1}{2}\left[1 - erf\left(\frac{y - \{24.6\}}{\sqrt{2}\,\{5\}}\right)\right]. \tag{6.99}$$

The distribution of ϕ was found to be independent of A and B. Its pdf can be described as being uniform at two levels, with values less than $90°$ being five times more likely than values greater than $90°$:

$$f_\phi(z) = \begin{cases} \dfrac{1}{216}, & |z| < 90°, \\[2mm] \dfrac{1}{1080}, & 90° \leq |z| \leq 180°. \end{cases} \tag{6.100}$$

To be useful for outage calculations, the probability distributions (6.98) and (6.99) must be scaled to seconds of occurrence in a heavy fading month. This was achieved according to the procedure that led to (6.95) and (6.93), obtaining

$$T = \{0.0175\}\, c \cdot f_0 \cdot D^3, \tag{6.101}$$

where the meaning of symbols is the same as in (6.93).

6.4.3 Performance of Digital Radio Systems Under Multipath Fading

The scope of accurately modeling the radio channel under selective multipath fading is to develop an analytic tool for evaluating the performance of different modulation schemes and for identifying possible remedies against fading, like space diversity and/or adaptive equalization. The performance evaluation can be undertaken, in a simplified environment, analytically or by simulation. Another important application of the model (Lundgren and Rummler, 1979) is to build a multipath fading simulator to be used in laboratory tests prior to investing in the equipment's extensive field testing.

The best measure of the performance of microwave digital radio systems is, undoubtedly, the outage probability P_o, defined as the probability that $P_b(e)$ exceeds some specified threshold value P_t, given that multipath fading is occurring. The threshold is typically 10^{-3} for voice circuit applications. In principle, the evaluation of P_o requires the calculation of the expression of the error probability as a function of the random parameters characterizing the channel model and then the integration of the joint pdf of the parameters in the region of the space where the error probability is greater than the threshold P_t. In formulas, referring to the delay network model, we should find the *outage domain* $\mathcal{D}(a, b, \phi)$ defined as

$$\mathcal{D}(a, b, \phi) \triangleq \{(a, b, \phi): P_b(e|a, b, \phi) \geq P_t\} \qquad (6.102)$$

and then compute P_o as

$$P_o = \iiint\limits_{\mathcal{D}(a, b, \phi)} f_{a, b, \phi}(x, y, z)dx\, dy\, dz, \qquad (6.103)$$

where $f_{a, b, \phi}(x, y, z)$ is the joint pdf of the RVs a, b, and ϕ.

The evaluation of the error probability $P_b(e)$ conditioned on a set of the channel parameter's values is not different from that already studied in this chapter, since it reflects a situation of digital transmission with ordinary ISI. Unfortunately, the phenomenon is not stationary and a change in the parameter's value would change the channel transfer function and require a new computation of the error probability. The whole procedure of finding the outage region in the parameters' space is therefore a difficult if not prohibitive task. In the following, we present some interesting results obtained by partially simplifying these difficulties. The aim is to offer both an insight into the multipath fading effects and some methodological hints to system engineers on how to handle the analysis of digital radio systems.

A very useful method for comparing different systems in the presence of multipath fading is that of measuring the *signature* of an equipment (Giger and Barnett, 1981). This method is based on the delay-network model. It consists of a laboratory test: a network with transfer function (6.97) is inserted between the transmitter and the receiver, and the set of parameters A, B, and ϕ is measured which give a constant error rate. An example is shown in Fig. 6.42 for a 78 Mbit/s, 8-CPSK system operating at 6 GHz with 30-MHz channel spacing and a $P_b(e)$ of 10^{-3}. In the figure, plots of the parameter B versus $f_N - f_0$ are presented with A as a parameter. The curve labelled $A = 0$ dB is called the *signature* of the system. For each value of $f_N - f_0$, it gives the

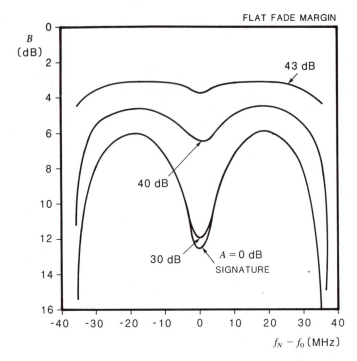

Figure 6.42 Signature for a 6-GHz, 78-Mbit/s, 8-CPSK digital radio system with 30-MHz channel spacing. (Reprinted with permission. Copyright © 1981 IEEE.)

largest value of B that the system can tolerate, no matter what the signal-to-noise ratio is. The system is limited only by the selectivity regardless of signal attenuation. Notice also that the curves vary little over a range of A from 0 to 20 dB. The curve corresponding to the flat-fade margin $A = 46$ dB does not tolerate any dispersive variations across the band and is therefore coincident with the line $B = 0$ dB. The dip in the center of the curves indicates a lesser susceptibility to band-centered fades. The reason is that off-centered fades produce a transfer function that lacks conjugate symmetry around the carrier frequency, thus giving rise to crosstalk between in-phase and quadrature components (see also Section 6.1).

Signatures of systems can be used to estimate P_o. According to Lundgren and Rummler (1979), a very good approximation for the outage due to selective fading can be obtained by considering only the critical value of B, say $B_c(\phi)$. This stems from the fact that most of the outage is caused by selectivity, that is, fades characterized by points (B, f_N) lying below the signature of the system. Thus we can write

$$P_o \doteq \int_{-180}^{180} f_\phi(z)dz \int_{B_c(\phi)}^{\infty} f_B(y)\, dy, \qquad (6.104)$$

where $f_B(y)$ is the pdf of B.

Examples of signatures for three commercial digital radio systems operating at 6 GHz are shown in Fig. 6.43 (Giger and Barnett, 1981). Curve I is identical to B_c in Fig. 6.42. Curve II is for a 90-Mbit/s, 8-CPSK system operating in a 30-MHz channel

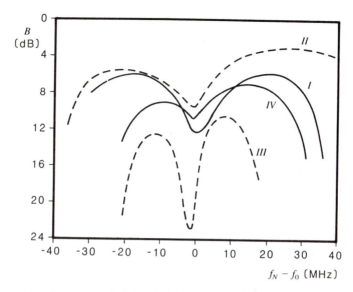

Figure 6.43 Signatures for 6-GHz digital radio systems with 30-MHz channel spacing. (Reprinted with permission. Copyright © 1981 IEEE.)

spacing. The better bandwidth utilization is paid for by system II in terms of a worse signature. The insertion of an equalizer improves the signature (curve III). Finally, curve IV refers to a 16-QAM system operating at 90 Mbit/s in a 30-MHz channel spacing.

We have seen how, in general, the exact computation of the outage probability P_o is an impossible achievement. To simplify the analysis, one can compute the probability P_c of eye pattern closure, given that multipath fading is occurring. This probability can be used as a relative measure for the comparison of different modulation schemes. This approach is justified by noting that most outages are caused by signal distortion alone (i.e., they are due to a closed eye condition). Because, during multipath fading, P_c is the expected fraction of time that a system will be in this vulnerable state, it provides a useful basis for comparisons. Two steps are needed to evaluate P_c:

Step 1. The identification of the region of the parameters characterizing the model over which the eye pattern is closed.

Step 2. The integration of the joint pdf of the parameters over that region.

The two steps are conceptually the same as those for the computation of P_o. Here the identification of the region is simpler.

This approach has been followed by Andrisano (1982) with the delay-network model for QPSK modulation and by Greenstein and Czekaj (1982) with the polynomial model for various modulation schemes, like QPSK, 8-CPSK, and 16-QAM, taking into account carrier and timing recovery and including an equalizer into the receiver. Step 1 is performed using a computer to obtain the samples of the received signals, and search algorithms to find the parameters region where eye closure takes place. Step 2 is done through the numerical integration of the joint pdf of the model parameters.

Sec. 6.4   Digital Transmission over Fading Multipath Channels   **317**

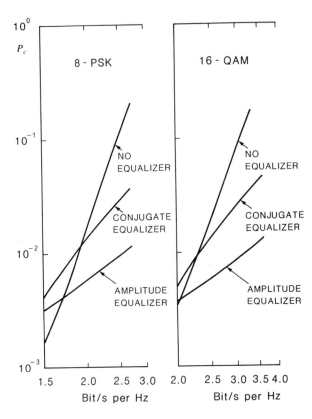

8 - PSK

16 - QAM

NO EQUALIZER

NO EQUALIZER

CONJUGATE EQUALIZER

CONJUGATE EQUALIZER

AMPLITUDE EQUALIZER

AMPLITUDE EQUALIZER

Bit/s per Hz

Bit/s per Hz

Figure 6.44 Closed-eye probability versus spectral efficiency for 8-CPSK and 16-QAM under several equalization conditions (channel spacing of 30 MHz, raised-cosine shaping). (Reprinted with permission. Copyright © 1982 IEEE.)

Two possible equalization approaches are suggested, the conjugate equalizer and the amplitude equalizer. The first leads to a composite response (cascade of channel and equalizer) with conjugate symmetry around the center frequency, thus avoiding interferences between in-phase and quadrature components. The second equalizer presents a composite response that eliminates ISI when the channel's transfer function has minimum phase.

Results are presented in Fig. 6.44 (Greenstein and Czekaj, 1982) for 8-CPSK and 16-QAM in terms of P_c versus the bandwidth utilization. The curves reveal the notable equalization benefits at high bandwidth utilization values when the system is degraded by strong interchannel interference. From the many ways in which digital radio systems under multipath fading conditions have been observed by field experiments or laboratory tests and analyzed through computer simulations, we can conclude generally that system performance cannot be improved so as to meet reliability objectives by simply increasing the transmitted power. Other remedies have to be tried. The final part of this section is devoted to a brief digression about this point.

6.4.4 Countermeasures for Multipath Fading

As seen, multipath fading is an interference phenomenon that is both frequency and space selective. It is thus natural that the best known and used techniques to attenuate its effects are multiple-receiver methods that combine techniques categorized as *frequency* or *space diversity*. The first consists in switching from a faded unserviceable radio channel to a protection channel operating at a different radio frequency. It is the most frequently used technique for the protection of FDM–FM microwave systems and their performance improvement. However, in high fading activity areas, frequency diversity alone cannot provide the desired transmission availability for digital radio systems.

Space diversity is an increasingly used alternative for protection from the effects of multipath fading. Diversity results from vertically separating the two receiver antennas. Its effectiveness is based on the fact that deep fades of signals received on two vertically separated receiving antennas rarely overlap in time. As an example, in Fig. 6.45 (Vigants, 1975) the curve giving the annual simultaneous time T_s below level for two antennas separated vertically by 9 meters has been added to the curves already presented in Fig. 6.39. It can be seen that the simultaneous time below level is less than the short-haul objective of 165 seconds per year for fade margins larger than about 30 dB.

The improvement presented in Fig. 6.45 is based on the performance of an idealized comparator where the diversity signal is chosen at every instant as the stronger of the two received signals. Space diversity switching, however, is unsuitable for a digital radio system because of the excessive number of interruptions that such an arrangement would produce during fading. The most frequently adopted way of using the two diversity signals is intermediate frequency (IF) in-phase combining. The two diversity signals are added coherently, so the combiner yields a better signal-to-noise ratio than switched space diversity.

So far we have seen the benefits of space diversity in terms of an increase of the flat-fade margin. On the other hand, we know that digital radio systems are particularly sensitive to in-band distortions due to the fading's frequency selectivity. In this context,

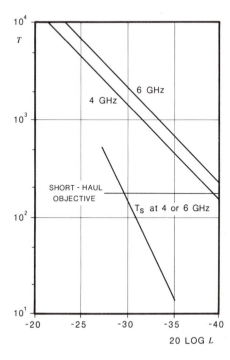

Figure 6.45 Annual simultaneous time below level for 9-meter separated antennas. (Reprinted with permission from the *Bell System Technical Journal*. Copyright 1971, AT&T.)

in-phase combining may lead to noticeable amplitude dispersion, particularly in the case where one flat and another notch spectra are received at the space diversity antennas. The explanation is that the in-phase combiner would add one side of the notch in phase, whereas the other side would suffer cancellation due to the phase discontinuity at the notch frequency. A different way of combining the diversity signals has been proposed (Komaki, Okamoto, and Tajima, 1980), aiming at minimizing the in-band dispersion, instead of maximizing the received level.

A general conclusion drawn as a consequence of the intensive field trials is that space diversity helps during selective multipath fading activity but, in itself, it is not a solution to the problem. It has to be accompanied by *adaptive equalization* of the residual in-band distortion. Adaptive equalization can be done either at IF or at baseband. From the originally proposed single-frequency-domain adaptive equalizers, attention has been focused on extending the time-domain adaptive equalizers from voiceband applications to digital microwave radio (Chapter 8 is devoted entirely to the subject of adaptive equalization). Due to the high bit rate involved, only tapped-delay-line equalizers with a few taps (less than 10) can be realized. In Fig. 6.46 (Fenderson and others, 1984), experimental signatures of a 16-QAM digital radio system with two receivers are shown. The first receiver was equipped with an adaptive slope equalizer, whereas the second was equipped with both the adaptive IF slope equalizer and a five-tap baseband adaptive transversal equalizer. Both minimum and nonminimum phase fading were considered. While IF equalizers can be designed to condition a channel properly for minimum phase fading, they double the delay distortion during periods of nonminimum phase fading. This effect naturally affects the outage of those digital radio systems that rely solely on amplitude correction. Comparing both sets of curves, we observe significant improvement in equipment signature performance that can be ascribed to the transversal equalizer alone.

Finally, it has been observed that the combination of space diversity and adaptive

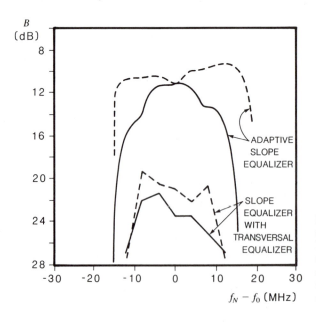

Figure 6.46 Measured equipment signatures for 16-QAM digital radio. Performance with adaptive slope equalization and adaptive slope and transversal equalization. (Reprinted with permission from the *Bell System Technical Journal*. Copyright 1984, AT&T.)

equalization provides a greater improvement than that which may be expected from the product of the two improvements taken separately. The synergy of the two techniques seems to come from the combiner's ability to replace in-band notches (which the equalizer is not able to compensate for without great noise enhancement) with simple slopes (which the equalizer can easily correct).

BIBLIOGRAPHICAL NOTES

The state-variable method to simulate linear filtering using a digital computer can be found in Smith (1977). The z transform and the FFT methods to simulate linear filtering in the time and frequency domains, respectively, are described in Chapters 4 and 6 of Rabiner and Gold (1975).

The problem of digital transmission systems performance evaluation in the presence of additive Gaussian noise and ISI has received considerable research attention since the late 1960s. The first approach was to find an upper bound to the error probability using the Chernoff inequality (Saltzberg, 1968; Lugannani, 1969). Other authors computed the error probability using a Hermite polynomials series expansion (Ho and Yeh, 1970, 1971) or a Gram–Charlier expansion (Shimbo and Celebiler, 1971). A different bounding technique based on the first moments of the RV representing the ISI has been described by Glave (1972) and refined by Matthews (1973). The Gauss quadrature rule approach to the evaluation of the error probability was first proposed by Benedetto, De Vincentiis, and Luvison (1973). New classes of upper and lower bounds based on the moments and related to the Gauss quadrature rules approach have been proposed by Yao and Biglieri (1980) (see also Appendix E). Algorithms for the recursive computation of the moments of the ISI RV are described in (Prabhu, 1971) for the case of independent data sequences and in Cariolaro and Pupolin (1975) for the case of correlated data sequences.

Although tailored for PAM modulation, almost all the aforementioned methods have been applied to the evaluation of the symbol error probability of coherent and noncoherent modulation schemes in the presence of ISI and adjacent channel interferences. A useful reference for these applications can be found in the second part of the IEEE reprints collection edited by Stavroulakis (1980).

A huge literature exists in the field of synchronization of digital communication systems. The following books are focused on the PLL theory and applications: Viterbi (1966), Lindsey (1972), Blanchard (1976), and Gardner (1979). In particular, Viterbi's exact analysis of the first-order PLL solving the Fokker–Planck equation has provided much insight for understanding nonlinear operations, whereas Gardner's book is very useful for practicing engineers. The problems related to the design and analysis of digital PLL have not been considered in this book, although they are becoming ever more important with the fast improvement in integrated digital circuitry. A good starting point for the interested reader is the tutorial papers by Gupta (1975) and Lindsey and Chie (1981). A survey of the peculiar methodology used in the analysis of digital PLL without noise can be found in D'Andrea and Russo (1983).

The general problem of carrier and symbol synchronizers is faced by Stiffler (1971), Lindsey and Simon (1973), and Franks (1983). A comprehensive tutorial paper has been written by Franks (1980). The joint recovery of carrier phase and symbol timing has been analyzed by Mengali (1977), Mancianti, Mengali, and Reggianini (1979), and Meyers and Franks (1980). The effect of imperfect synchronization on system performance is the subject of certain papers in Stavroulakis (1980). The treatment of this subject in this book is attributable to Franks (1980), Gardner (1979), and Mengali (1979).

A multipath-fading physical model for line-of-sight microwave radio systems was proposed in (Ruthroff, 1971). The spectral characteristics of nondiversity and space diversity narrowband radio channels subject to multipath fading were estimated from measured data by Babler (1973). The empirical probability distribution (6.92) was established by Barnett (1972). The delay-network model for selective fading was proposed by Rummler (1979); a polynomial model for it was described by Greenstein and Czekaj (1981). The concept of equipment signature to characterize the performance of a digital radio system was introduced by Giger and Barnett (1981). The performance of digital microwave radio links under selective fading conditions has been evaluated by Greenstein and Czekaj (1982) and Andrisano (1982). Comparisons between calculated and observed performance of digital radio systems under selective fading have been presented in Lundgren and Rummler (1979) and Rummler (1982). A different method of combining space diversity signals was proposed by Komaki, Okamoto, and Tajima (1980). Results showing the benefits of adaptive equalization to the performance of digital radio systems under selective fading can be found in Wong and Greenstein (1984), Moreno and Salerno (1983), Fenderson and others, (1984), and Leclert and Vandamme (1985).

PROBLEMS†

6.1. Show that the outputs of the block diagram of Fig. 6.2 are the same as those of the block diagram of Fig. 6.3.

6.2. Given two independent Gaussian random processes $n_P(t)$ and $n_Q(t)$ with zero mean and equal variance σ_n^2, find the first-order pdf of the processes $v_P(t) = n_P(t) \cos \vartheta - n_Q(t) \sin \vartheta$ and $v_Q(t) = n_P(t) \sin \vartheta + n_Q(t) \cos \vartheta$, where ϑ is a constant. Prove that samples of $v_P(t)$ and $v_Q(t)$ taken at the same time instant are statistically independent.

*6.3. Write a computer program implementing the recursive algorithm described in Section 6.2.1 in the case of multilevel PAM transmission.

*6.4. Write a computer program implementing the algorithm described in Golub and Welsch (1969) (see also Appendix E) to construct a Gauss quadrature rule starting from the first $2J$ moments of a RV.

*6.5. Use the programs available from Problems 6.3 and 6.4 to evaluate the error probability for an octonary PAM system with ISI due to a raised cosine impulse response $h(t)$ with a roll-off factor $\alpha = 0.5$ [for the impulse response of raised cosine type, see (7.22)] in the presence of a normalized sampling time deviation of 0.05, 0.1, 0.15, and 0.2. Assume an SNR that gives an error probability of 10^{-6} at the nominal sampling instants and use a 15-sample approximation for the impulse response.

† Problems marked with an asterisk should be solved with the aid of a computer.

6.6. Find a recursive algorithm for the computation of the joint moments of the RVs X_P and X_Q defined in (6.34) and (6.35) assuming that the a_n's are iid discrete RVs with known moments. *Hint*: Generalize the procedure described in Section 6.2.1 for one RV.

*6.7. For the same case of Problem 6.5, compute the Chernoff bound to the error probability extending the method described in Saltzberg (1968) to the multilevel case.

*6.8. For the same case of Problem 6.5, compute the error probability using the series expansion method described in Ho and Yeh (1970) and in Appendix E.

6.9. Particularize the result (6.39) to the case of 16-QAM modulation with $\vartheta = 0$.

6.10. Generalize the results (6.43) to (6.47) to the case of a phase offset $\vartheta \neq 0$.

*6.11. Extend the program developed in Problem 6.3 to the case of an M-ary CPSK modulation scheme.

*6.12. Using the program of Problem 6.11 and the results of Problems 6.4 and 6.10, compute the error rate for a QPSK modulation using a second-order Butterworth filter impulse response $h(t)$ (with a normalized 3-dB bandwidth of 1.1) as a function of the phase offset ϑ. Assume the signal-to-noise ratio that gives an error probability of 10^{-6} in ideal conditions (no ISI) and truncate the impulse response to 10 samples.

6.13. Using the final-value theorem of the Laplace transform, verify the steady-state frequency and phase errors of Table 6.3.

6.14. Show that the PLL of Fig. 6.22 implements a relationship like (6.63) between successive estimates of ϑ. *Hint*: Start from the differential equation (6.66) and suppose that the variations of $\hat{\vartheta}(t)$ are so slow that it is possible to write

$$\dot{\hat{\vartheta}}(t) \cong \frac{\hat{\vartheta}(t) - \hat{\vartheta}(t - T)}{T}.$$

6.15. **(a)** Show that (Poisson sum formula)

$$\sum_k h(t - kT) = \frac{1}{T} \sum_m H\left(\frac{m}{T}\right) e^{j2\pi mt/T}.$$

Hint: Find first the Fourier series expansion of the periodic factor

$$\sum_k h(t - kT).$$

(b) Using the result in part (a), verify (6.85).

6.16. Using the result of (6.87), evaluate the magnitude of the discrete component at frequency $1/T$ at the output of a square-law device for a raised-cosine impulse response $h(t)$ [see (7.22) for the expression of $h(t)$] as a function of the roll-off factor α.

6.17. Consider the transmission of binary antipodal PAM signals over a linear channel perturbed by additive Gaussian noise and an intersymbol interference $X = \sum_{i=1}^{N} a_i h_i$. (Assume $d = 2$.) Denote the resulting error probability by $P_N(e)$, and the error probability without intersymbol interference by $P_0(e)$.

(a) Prove that, if the eye pattern is open, i.e., $h_0 > \sum_{i=1}^{N} |h_i|$, then we have $P_0(e) \leqslant P_N(e)$, i.e., intersymbol interference increases the error probability.

(b) Generalize the result of (a) by deriving an inequality involving $P_{N'}(e)$, $N' < N$ (i.e., the error probability obtained by retaining only N' out of N interfering samples).

(c) Show, through an example, that if the eye pattern is *not* open we may have $P_0(e) > P_N(e)$.

CHAPTER 7

System Design for Channels with Intersymbol Interference

In this chapter we shall consider some design problems related to the transmission of linearly modulated signals over a time-dispersive linear channel, that is, a channel perturbed by intersymbol interference (ISI) as well as additive noise. Whereas in Chapter 6 we examined the effect of ISI on system performance, here we show how the system designer can cope with it. The first problem we shall take into consideration is the design of a system in which the receiver has the form of a linear filter followed by a sampler and a detector that takes decisions on a sample-by-sample basis. Two design criteria will be considered under this constraint. The first is the elimination of ISI from the sequence of samples to be processed by the detector. The second is the minimization of the joint effects of ISI and noise on the same sequence.

If the receiver structure is not constrained, an optimum receiver can be designed performing maximum-likelihood (ML) estimation of the sequence of data transmitted. A receiver based on this design criterion could not produce final decisions before the entire transmitted data sequence has been received, and its complexity would be prohibitive. Hence, a practically implementable, approximate version of this receiver will be derived based on an application of the Viterbi algorithm.

7.1 INTRODUCTION

The transmission model assumed in this chapter is presented in Fig. 7.1. The source and the modulator are modeled assuming that the data to be transmitted form a stationary random sequence (a_l) of independent, identically distributed (iid) complex random variables (RV) with mean zero and

324

$$\sigma_a^2 \triangleq E|a_l|^2. \tag{7.1}$$

The data sequence (a_l) is sent to a linear modulator. For mathematical convenience this is modeled as the cascade of a modulator having the ideal impulse $\delta(t)$ as its basic waveform and of a shaping filter with an impulse response $s(t)$ and a frequency response $S(f)$. The number of symbols to be transmitted per second (i.e., the signaling rate) is denoted by $1/T$. Thus, the modulated signal is $\sum_{l=-\infty}^{\infty} a_l \delta(t - lT)$, and the signal sent to the channel is $\sum_{l=-\infty}^{\infty} a_l s(t - lT)$.

The channel section is represented by a time-invariant linear system having known frequency response $C(f)$ and impulse response $c(t)$ and a generator of additive noise. The noise process $w(t)$ is assumed to be Gaussian, independent of the data sequence, to have zero mean, finite power, and a known power density spectrum $G_w(f)$. Thus, the signal observed at the output of the channel section can be written as

$$r(t) = \sum_{l=-\infty}^{\infty} a_l p(t - lT) + w(t), \tag{7.2}$$

where $p(t)$ is the response of the noiseless part of the channel to the waveform $s(t)$ or, equivalently, the convolution

$$p(t) = s(t) * c(t). \tag{7.3}$$

The first problem that we shall consider in this chapter is the design of a receiver (see Fig. 7.2) having the form of a linear filter (hereafter referred to as the *receiving filter*) followed by a sampler. After linear filtering, the received signal is sampled every T seconds and the resulting sequence (x_l) is sent to a detector. The detector takes decisions, on a sample-by-sample basis, according to a defined rule. We consider the minimum-distance rule described in Chapter 4.

The first criterion considered in the design of the receiver shown in Fig. 7.2 concerns the elimination of the effects of ISI from the sampled sequence (x_l). Such a criterion, the Nyquist criterion, will define the constraints on the overall channel impulse response, say $S(f)C(f)U(f)$. A method to cope with ISI is to put all the burden of eliminating ISI on the receiving filter $U(f)$. Another possibility is to choose both $S(f)$

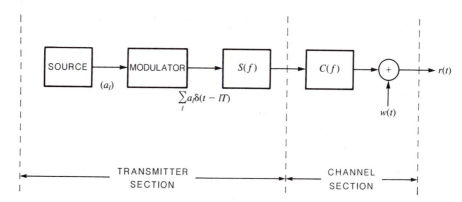

Figure 7.1 Model for the transmission of linearly modulated data over a time-dispersive channel.

Sec. 7.1   Introduction

325

Figure 7.2 Sampling receiver for the transmission system of Fig. 7.1.

and $U(f)$ so as to meet the specified needs for their product: in this case, the choice of $S(f)$ and $U(f)$ can be made so as to minimize the effects of additive noise at the detector input, and hence to minimize the probability of error for the transmission system under the constraint of no ISI.

The second design criterion is the technique of mean-square error optimization. This takes into account the joint effects of intersymbol interference and additive noise. It is based on the minimization of the mean-square difference between the samples x_l and the transmitted data symbols a_l. This can be achieved by choosing the receiving filter $U(f)$ and the shaping filter $S(f)$.

Assume finally that the constraint of a linear receiver is relaxed. In other words, consider the problem of designing an optimum receiver without the restriction regarding symbol-by-symbol decisions and the receiver structure. This problem will be solved by using the ML criterion for the detection of the whole sequence of data (a_l) and by devising a "best" estimate selection strategy (the Viterbi algorithm).

7.2 ELIMINATING INTERSYMBOL INTERFERENCE: THE NYQUIST CRITERION

Consider the transmission system shown in Fig. 7.1 and the sampling receiver of Fig. 7.2. Denote by $q(t)$ the convolution

$$q(t) = p(t) * u(t)$$
$$= s(t) * c(t) * u(t) \tag{7.4}$$

and by $n(t)$ the convolution

$$n(t) = w(t) * u(t), \tag{7.5}$$

where $u(t)$ is the impulse response of the receiving filter. Thus, at the sampler input we have

$$x(t) = \sum_{k=-\infty}^{\infty} a_k q(t - kT) + n(t) \tag{7.6}$$

and hence

$$x_l = \sum_{k=-\infty}^{\infty} a_k q_{l-k} + n_l, \tag{7.7}$$

where the signal samples are defined by

$$x_l \triangleq x(t_0 + lT), \tag{7.8}$$

$$q_l \triangleq q(t_0 + lT), \tag{7.9}$$

$$n_l \triangleq n(t_0 + lT), \tag{7.10}$$

and $t_0 + lT$, $-\infty < l < \infty$, are the sampling instants. In what follows, for the sake of clarity, we simplify the exposition by assuming $t_0 = 0$.

For error-free transmission, allowing for a delay of D symbol intervals between transmission and reception of a given symbol, we have to satisfy the condition that x_l is equal to a_{l-D}. However, from (7.7) we obtain

$$x_l = q_D a_{l-D} + \sum_{k \neq l-D} a_k q_{l-k} + n_l. \tag{7.11}$$

The factor q_D of (7.11) is a complex number representing a constant change of scale and possibly a phase shift if the channel is bandpass (see Chapter 6). Under the hypothesis of a known channel, it can be easily compensated for. Thus, we assume that $q_D = 1$. The second term of (7.11) represents the contribution of ISI. It is seen that it depends on the entire transmitted sequence (a_k), as weighted by the samples q_{l-k} of the impulse response of the overall channel. This is the effect of the tails and precursors of the waveforms overlapping the one carrying the information symbol a_{l-D}. The third term in (7.11) represents the effect of the additive noise.

Once the sample sequence (x_l) is obtained, it has to be processed to get an estimate $(\hat{a}_l)$ of the transmitted symbol. Of course, a reasonable way to do this is to perform symbol-by-symbol decisions (i.e., to use only x_l to get an estimate of a_{l-D}, $-\infty < l < \infty$). This procedure is the simplest, but suboptimum as the samples x_l given by (7.11) are correlated due to the effect of intersymbol interference. Hence, for an optimum decision the whole sequence (x_l) should be processed.

In the framework proposed, what seems at first a reasonable approach to the problem of optimizing the transmission system is to attempt to eliminate the ISI term in (7.11). If this is undertaken, the problem is reduced to a situation in which only additive Gaussian noise is present. Hence, a symbol-by-symbol decision rule based on minimum distance is optimum. We shall first examine this solution.

To avoid the appearance of the ISI term in (7.11) the overall channel impulse response sample sequence (q_l) should satisfy the condition

$$q_l = \begin{cases} 0, & l \neq D, \\ 1, & l = D. \end{cases} \tag{7.12}$$

This condition can also be expressed by observing that, with $\Delta_T(t)$ denoting a periodic train of delta functions spaced T seconds apart, that is,

$$\Delta_T(t) = \sum_{k=-\infty}^{\infty} \delta(t - kT), \tag{7.13}$$

(7.12) is equivalent to

$$q(t) \cdot \Delta_T(t) = \delta(t - DT) \tag{7.14}$$

(see Fig. 7.3). Taking the Fourier transform of both sides of (7.14), with the definition

$$Q(f) \triangleq \mathscr{F}[q(t)] = S(f)C(f)U(f), \tag{7.15}$$

we get

$$\frac{1}{T} Q(f) * \Delta_{1/T}(f) = \exp(-j2\pi f DT). \tag{7.16}$$

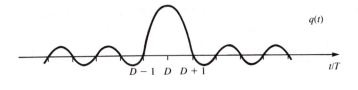

Figure 7.3 Eliminating ISI from the samples of the channel output.

The effect of convolving $Q(f)$ with the train $\Delta_{1/T}(f)$ of spectral lines spaced at $1/T$ Hz is to obtain a train of replicas of $Q(f)$ spaced $1/T$ Hz apart (Fig. 7.4). Denoting this convolution by $Q_{eq}(f)$,

$$Q_{eq}(f) \triangleq \sum_{k=-\infty}^{\infty} Q\left(f + \frac{k}{T}\right), \qquad (7.17)$$

Eq. (7.16) requires that $Q_{eq}(f)$ have a constant magnitude and a linear phase. It is easily seen that, for any $Q(f)$, $Q_{eq}(f)$ is a periodic function of f with period $1/T$.

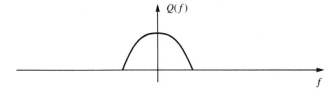

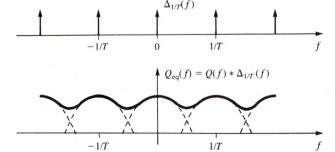

Figure 7.4 Convolution of $Q(f)$ with a train of spectral lines.

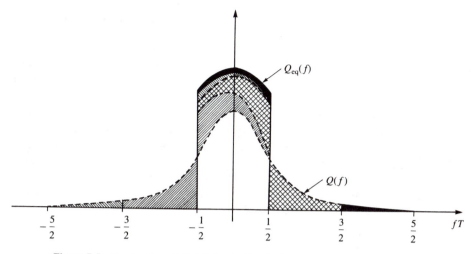

Figure 7.5 Construction of $Q_{eq}(f)$ in the Nyquist interval $(-1/2T, 1/2T)$. $Q_{eq}(f)$ is assumed to be real.

Thus, without loss of generality, we can confine our consideration of this function to the fundamental interval $(-1/2T, 1/2T)$, and express condition (7.16) in the form

$$Q_{eq}(f) = T \cdot \exp(-j2\pi fDt), \qquad |f| < \frac{1}{2T}. \tag{7.18}$$

Condition (7.18) for the removal of ISI is called the (first) *Nyquist criterion* and the interval $(-1/2T, 1/2T)$ the *Nyquist interval*. This criterion says that, if the frequency response $Q(f)$ of the overall channel is cut in slices of width $1/T$ and these are piled up in the Nyquist interval with the proper phases (see Fig. 7.5), ISI is eliminated from the sample sequence (x_l) when the resulting *equivalent spectrum* $Q_{eq}(f)$ has a constant magnitude and a linear phase.

7.2.1 Raised Cosine Spectrum

If $Q(f)$ is nonzero outside the Nyquist interval, many classes of responses satisfy (7.18). Thus, the Nyquist criterion does not uniquely specify the shape of the frequency response $Q(f)$. On the contrary, if $Q(f)$ is limited to an interval smaller than Nyquist's, it is impossible for (7.18) to hold. Thus, the ISI cannot be removed from the received signal. If $Q(f)$ is exactly bandlimited in the Nyquist interval, (7.18) requires that

$$Q(f) = \begin{cases} Q_{eq}(f) = T \cdot \exp(-j2\pi fDT), & |f| < \dfrac{1}{2T} \\ 0, & \text{elsewhere.} \end{cases} \tag{7.19}$$

That is, $Q(f)$ is the "brickwall" frequency response of the ideal low-pass filter with delay DT.

With $Q(f)$ as in (7.19), the overall channel impulse response $q(t)$ becomes

$$q(t) = \frac{\sin \pi(t/T - D)}{\pi(t/T - D)}, \tag{7.20}$$

Sec. 7.2   Eliminating Intersymbol Interference: The Nyquist Criterion   **329** a noncausal function (for any finite D) that decays for large t as $1/t$. Even if such a response were physically realizable, this type of pulse shape would be impractical, because every real-world system will exhibit channel perturbations resulting in erroneous sampling times. These perturbations would cause the eye pattern to close simply because the series $\sum_{k=-\infty}^{\infty} q(\tau + kT)$ is not absolutely summable for $\tau \neq 0$ when $q(t)$ is as in (7.20) (see Section 6.2.1). For this reason, it appears sensible to trade a wider bandwidth for a reduced sensitivity to inaccuracies in sampling times (and possibly for an easier implementation). Since it is recognized that the problem with the response (7.20) is due to its slow rate of decay, and since the rate of decay of a pulse is intimately related to the discontinuities of its Fourier transform, it is reasonable to investigate classes of responses that satisfy the Nyquist criterion with a minimum of discontinuities, (considering also the discontinuities in the derivatives). This can be obtained, for example, as shown in Fig. 7.6. Let α, $0 \leq \alpha \leq 1$, be the allowed relative amount of bandwidth in excess of Nyquist's; that is, let $Q(f)$ be strictly bandlimited to the interval $|f| \leq (1 + \alpha)/2T$. Letting $D = 0$ for simplicity, choose

(a) $Q(f) = T$,   for $|f| \leq (1 - \alpha)/2T$;

(b) $Q(f)$ real, decaying from T to zero for $(1 - \alpha)/2T \leq f \leq (1 + \alpha)/2T$, and exhibiting symmetry with respect to the points with abscissas $\pm 1/2T$ and ordinate $T/2$. This *roll-off spectrum* must be chosen in such a way that it presents a minimum of discontinuities at $|f| = (1 + \alpha)/2T$, the band edges.

Choice of a sinusoidal form for the roll-off spectrum leads to the *raised cosine response* defined as follows:

$$Q(f) = \begin{cases} T, & |f| \leq \dfrac{1-\alpha}{2T}, \\[2ex] \dfrac{T}{2}\left\{1 - \sin\left[\dfrac{\pi T}{\alpha}\left(|f| - \dfrac{1}{2T}\right)\right]\right\}, & \dfrac{1-\alpha}{2T} \leq |f| \leq \dfrac{1+\alpha}{2T}, \\[2ex] 0, & |f| \geq \dfrac{1+\alpha}{2T}. \end{cases} \qquad (7.21)$$

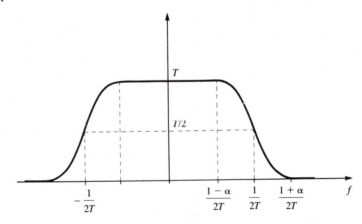

Figure 7.6 Example of a real $Q(f)$ satisfying the Nyquist criterion.

The impulse response corresponding to a raised cosine spectrum is

$$q(t) = \frac{\sin \pi t/T}{\pi t/T} \cdot \frac{\cos \alpha \pi t/T}{1 - (2\alpha t/T)^2} \tag{7.22}$$

and decays asymptotically as $1/t^3$ for $t \to \infty$.

Figure 7.7 shows the raised cosine spectra and the corresponding impulse responses

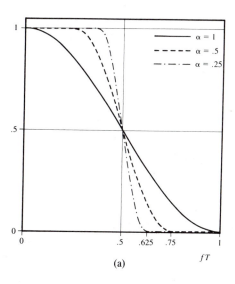

(a)

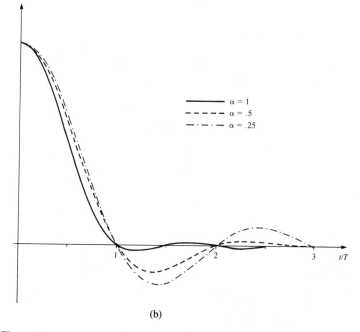

(b)

Figure 7.7 (a) Raised cosine spectra; (b) impulse response of raised cosine filters.

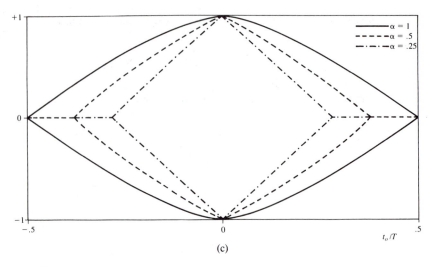

(c)

Figure 7.7 (cont.) (c) Inner envelopes of eye patterns resulting from antipodal binary transmission over a channel with a raised-cosine transfer function.

for $\alpha = 0.25$, 0.5, and 1, together with the inner envelopes of the corresponding eye patterns for binary transmission with symbols ± 1. It is seen from Fig. 7.7c that the immunity to erroneous sampling times augments with increasing α. In particular, with a 100 percent roll-off open-eye transmission is possible even with a sampling time error approaching $0.5T$ in absolute value. With smaller values of α, the margin against erroneous sampling decreases and is zero when $\alpha = 0$ (corresponding to the brickwall frequency response).

Notice also that $q(t)$ in (7.22) is not causal and hence not physically realizable. However, approximate realizations can be obtained by considering a delay D so large that a causal approximation to $q(t - D)$ gives a performance satisfactorily close to that predicted by the theory. Raised cosine spectra are often considered for practical applications.

7.2.2 Optimum Design of Shaping and Receiving Filter

Assume now that $Q(f)$ has been chosen so as to satisfy the Nyquist criterion. Thus, freedom from ISI is assured by taking the shaping filter and the receiving filter such that

$$S(f)C(f)U(f) = Q(f). \tag{7.23}$$

Thus, for a given $C(f)$ the actual design of $S(f)$ and $U(f)$ still leaves a degree of freedom, as only their product is specified. This freedom can be taken advantage of by imposing one further condition, that is, the minimization of the effect of the noise at the sampler input or, equivalently, the minimization of the error probability (in fact, in the absence of ISI, errors are caused only by the additive noise).

The average noise power at the receiving filter output is, from (7.5),

$$\sigma_n^2 = \int_{-\infty}^{\infty} G_w(f)|U(f)|^2 \, df. \tag{7.24}$$

Minimization of σ_n^2 without constraints would lead to the trivial solution $|U(f)| \equiv 0$, which is not compatible with (7.23). To avoid this situation, we constrain the signal power at the channel input to a finite value. This results in a constraint on $S(f)$. This, in turn, prevents [through (7.23)] $U(f)$ from assuming too small values.

Since the overall channel frequency response is the fixed function $Q(f)$, the signal power spectral density at the shaping filter output is, from (2.127),

$$\frac{\sigma_a^2}{T}|S(f)|^2 = \frac{\sigma_a^2}{T}\frac{|Q(f)|^2}{|C(f)U(f)|^2}, \tag{7.25}$$

and the corresponding signal power is

$$\mathcal{P} = \frac{\sigma_a^2}{T}\int_{-\infty}^{\infty}\frac{|Q(f)|^2}{|C(f)U(f)|^2} \, df. \tag{7.26}$$

Minimization of σ_n^2 under the constraint (7.26) can be performed using Lagrange multiplier and variational techniques (see Appendix C). Omitting an unessential factor, the minimizing $U(f)$ is given by the equation

$$|U(f)| = \frac{|Q(f)|^{1/2}}{G_w^{1/4}(f)|C(f)|^{1/2}}, \tag{7.27}$$

and the corresponding shaping filter characteristic is obtained through

$$S(f) = \frac{Q(f)}{C(f)U(f)} \tag{7.28}$$

[in (7.27) and (7.28) it is assumed that $Q(f)$ is zero at those frequencies for which the denominators are zero]. Notice that the phase characteristic of $U(f)$ is not specified and is therefore arbitrary [of course, $S(f)$ in (7.28) is such that $Q(f)$ has a linear phase, as required by the Nyquist criterion]. In the special case of white noise and $C(f)$ = constant, it is seen from (7.27) and (7.28) that $U(f)$ and $S(f)$ can be identical apart from an irrelevant scale factor, so only one design has to be implemented for both filters.

7.3 MEAN-SQUARE ERROR OPTIMIZATION

In the last section we saw how a system free of ISI can be designed. After choosing the overall channel transfer function, the optimum design of shaping and receiving filters was achieved by minimizing the noise power at the sampler's input. Although this procedure sounds reasonable, it is by no means certain that it achieves minimum error probability. In fact, it might happen that, by trading a small ISI for a lower additive noise power, a better error performance is obtained. On the other hand, system optimization under the criterion of a minimum error probability is a rather complex task. This suggests that we look for a criterion leading to a more manageable problem.

Thus, in this section we shall consider the mean-square error (MSE) criterion for system optimization; this choice allows ISI and noise to be taken jointly into account and in most practical situations leads to values of error probability very close to the minimum.

Consider again the system model shown in Figs. 7.1 and 7.2. Instead of constraining the noiseless samples to be equal to the transmitted symbols, we can take into account the presence of additive noise and try to minimize the mean-squared difference between the sequence of transmitted symbols (a_l) and the sampler outputs (x_l). By allowing for a channel delay of D symbol intervals, we shall determine the shaping filter $S(f)$ and the receiving filter $U(f)$ so that the mean-square value of

$$\epsilon_l \triangleq x_l - a_{l-D} \tag{7.29}$$

is minimized. This will result in a system that, although not specifically designed for optimum error performance, should provide a satisfactory performance even in terms of error probability.

We shall begin by deriving an expression for the MSE at the detector input, defined as

$$\mathcal{E} \triangleq E|\epsilon_l|^2 = E|x_l - a_{l-D}|^2. \tag{7.30}$$

From (7.11), ϵ_l can be given the form

$$\epsilon_l = a_{l-D}(q_D - 1) + \sum_{k \neq l-D} a_k q_{l-k} + n_l, \tag{7.31}$$

so that, due to the independence of the terms summed up in the RHS, we obtain

$$\mathcal{E} = \sigma_a^2 |q_D - 1|^2 + \sigma_a^2 \sum_{k \neq D} |q_k|^2 + \sigma_n^2 \tag{7.32}$$

$$= \sigma_a^2 [1 - 2\,\mathcal{R}(q_D)] + \sigma_a^2 \sum_{k=-\infty}^{\infty} |q_k|^2 + \sigma_n^2.$$

Now we want to express $\mathcal{E}$ by using frequency-domain quantities. By assuming as usual that t_0 is equal to zero, we get

$$q_k = \int_{-\infty}^{\infty} Q(f) e^{j2\pi fkT}\, df, \tag{7.33}$$

and consequently, by direct calculation,

$$\sum_{k=-\infty}^{\infty} |q_k|^2 = \frac{1}{T} \sum_{k=-\infty}^{\infty} \int_{-\infty}^{\infty} Q^*(f) Q\left(f + \frac{k}{T}\right) df. \tag{7.34}$$

Thus, (7.32) can be rewritten, using also (7.24), in the form

$$\mathcal{E} = \sigma_a^2 \left[1 - 2\,\mathcal{R} \int_{-\infty}^{\infty} Q(f) e^{j2\pi fDT}\, df + \frac{1}{T} \int_{-\infty}^{\infty} Q^*(f) \sum_{k=-\infty}^{\infty} Q\left(f + \frac{k}{T}\right) df \right] \tag{7.35}$$

$$+ \int_{-\infty}^{\infty} G_w(f) |U(f)|^2\, df.$$

Upon inspection of (7.35), we immediately observe that the MSE is the sum of two terms. The first (enclosed in square brackets) represents the overall channel ISI, while the second represents the contribution of the additive noise. These terms are not independent, as any change in $U(f)$ will also affect $Q(f)$. Qualitatively, it can be said that, if the bandwidth of the receiving filter $U(f)$ is reduced in order to decrease the value of the noise term, this will result in a corresponding increase of the overall channel ISI.

Example 7.1

Consider a baseband transmission system with white Gaussian noise having power spectral density $N_0/2$, data with $\sigma_a^2 = 1$,

$$s(t) = u_T(t), \tag{7.36}$$

a channel modeled through a fourth-order low-pass Butterworth filter with 3-dB frequency B_C Hz, and a second-order low-pass Butterworth receiving filter with 3-dB frequency B_U Hz. In Fig. 7.8 the dashed lines represent the contribution of the noise (which augments with increasing B_U and N_0) and the continuous lines the contribution of the overall channel MSE (which augments with decreasing B_U and B_C). The total MSE $\mathscr{E}$ is obtained by summing up the two contributions, which results in a minimum for an optimum value of B_U. Notice

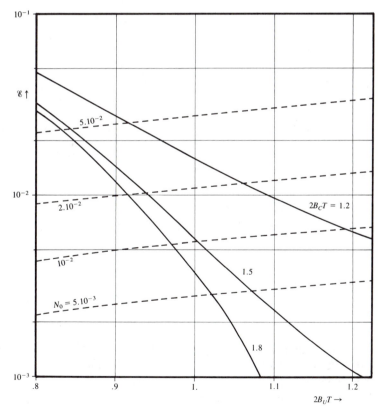

Figure 7.8 Contributions to mean-square error of intersymbol interference (continuous line) and additive noise (dashed line) in the situation of Example 7.1.

Sec. 7.3   Mean-Square Error Optimization

that these results do not depend on the particular modulation scheme used for the channel (provided of course that $\sigma_a^2 = 1$). $\square$

7.3.1 Optimizing the Receiving Filter U(f)

We shall now consider the selection of a transfer function $U(f)$ that gives a minimum for $\mathscr{E}$ when $S(f)$, as well as $C(f)$, are given. By using (7.35) and applying standard variational techniques (see Appendix C), it can be proved that a necessary and sufficient condition for $U(f)$ to minimize $\mathscr{E}$ is that

$$\frac{1}{T} S^*(f)C^*(f) \sum_{k=-\infty}^{\infty} S\left(f+\frac{k}{T}\right)C\left(f+\frac{k}{T}\right)U\left(f+\frac{k}{T}\right)$$
$$+ \frac{1}{\sigma_a^2}G_w(f)U(f) = S^*(f)C^*(f)e^{-j2\pi fDT} \tag{7.37}$$

be satisfied. In spite of its formidable appearance, (7.37) is amenable to a closed-form solution, which in turn admits an interesting interpretation. To see this, let us first show that the optimum receiving filter, say $U_{\mathrm{opt}}(f)$, has the following expression:

$$U_{\mathrm{opt}}(f) = \frac{P^*(f)}{G_w(f)} \Gamma(f), \tag{7.38}$$

where $\Gamma(f)$ is a periodic function with period $1/T$, and

$$P(f) \triangleq S(f)C(f) \tag{7.39}$$

is the transfer function of the cascade of shaping filter and channel (we assume here for simplicity that $G_w(f)$ is nonzero everywhere). By substituting (7.38) into (7.37) and observing that $\Gamma(f + k/T) = \Gamma(f)$ for all k due to the periodicity assumption, we get

$$\frac{1}{T}P^*(f)\Gamma(f) \sum_{k=-\infty}^{\infty} \frac{|P(f+k/T)|^2}{G_w(f+k/T)} + \frac{1}{\sigma_a^2}P^*(f)\Gamma(f) = P^*(f)e^{-j2\pi fDT}. \tag{7.40}$$

For all the frequencies at which $P(f)$ vanishes, (7.37) shows that $U_{\mathrm{opt}}(f)$ must also be zero, so that (7.38) is true. For $P(f) \neq 0$, (7.40) gives

$$\Gamma(f) = \frac{\sigma_a^2 e^{-j2\pi fDT}}{1 + \sigma_a^2 L(f)}, \tag{7.41}$$

where

$$L(f) \triangleq \frac{1}{T} \sum_{k=-\infty}^{\infty} \frac{|P(f+k/T)|^2}{G_w(f+k/T)} \tag{7.42}$$

is periodic with period $1/T$, as required. This shows that the solution to (7.37) has the form (7.38). Insight into the behavior of the optimum receiving filter can be gained by considering the special case of a channel bandlimited to the Nyquist interval $(-1/2T, 1/2T)$. In this case, (7.42) specializes to

$$L(f) = \frac{1}{T} \frac{|P(f)|^2}{G_w(f)}, \tag{7.43}$$

and from (7.38) and (7.41) we get

$$U_{opt}(f) = \frac{\sigma_a^2 P^*(f)}{G_w(f) + (\sigma_a^2/T)|P(f)|^2} e^{-j2\pi fDT}. \tag{7.44}$$

Equation (7.44) shows that, in the absence of noise, the optimum receiving filter is simply the inverse of $P(f)$, an obvious result, since in this situation ISI is the only contribution to the MSE and this can be reduced to zero by forcing the overall channel to a flat frequency response in the Nyquist band. However, when $G_w(f) \neq 0$, elimination of ISI does not provide the best solution. To the contrary, for spectral regions where the denominator of the RHS of (7.44) is dominated by $G_w(f)$, $U_{opt}(f)$ (apart from a scale factor and a delay term) approaches the matched filter characteristics $P^*(f)/G_w(f)$.

More generally, for a channel not constrained to have a zero transfer function outside the Nyquist interval, (7.38) can be interpreted by observing that $P^*(f)/G_w(f)$ is the transfer function of a filter matched to the impulse response $p(t)$ of the cascade of the shaping filter and the channel. Also, $\Gamma(f)$, being a periodical transfer function with period $1/T$, can be thought of as the transfer function of a transversal filter whose taps are spaced T seconds apart. Thus, we can affirm that the optimum receiving filter is the cascade of a matched filter and a transversal filter. The former reduces the noise effects and provides the principal correction factor when the signal-to-noise ratio is small. The latter reduces ISI and in the situation of high signal-to-noise ratio attempts to suppress it.

7.3.2 Performance of the Optimum Receiving Filter

Let us now evaluate the MSE of a system in which $S(f)$ and $C(f)$ are given and $U(f)$ has been optimized. Substituting (7.38) for $U(f)$ in (7.35) and using (7.43) and (7.44), we get, after algebraic manipulations,

$$\mathcal{E} = \sigma_a^2 \left[1 - \int_{-\infty}^{\infty} \frac{|P(f)|^2}{G_w(f)} \frac{\sigma_a^2}{1 + \sigma_a^2 L(f)} df \right]. \tag{7.45}$$

For a more compact form of the error expansion, the integral appearing in (7.45) is rewritten as follows:

$$\int_{-\infty}^{\infty} \frac{|P(f)|^2}{G_w(f)} \frac{\sigma_a^2}{1 + \sigma_a^2 L(f)} df$$

$$= \sum_{k=-\infty}^{\infty} \int_{(2k-1)/2T}^{(2k+1)/2T} \frac{|P(f)|^2}{G_w(f)} \frac{\sigma_a^2}{1 + \sigma_a^2 L(f)} df$$

$$= \int_{-1/2T}^{1/2T} \left(\sum_{k=-\infty}^{\infty} \frac{|P(f + k/T)|^2}{G_w(f + k/T)} \right) \frac{\sigma_a^2}{1 + \sigma_a^2 L(f)} df.$$

Also, using (7.42), we can express (7.45) in the form

$$\mathcal{E} = \sigma_a^2 \left[1 - T \int_{-1/2T}^{1/2T} \frac{\sigma_a^2 L(f)}{1 + \sigma_a^2 L(f)} df \right], \tag{7.46}$$

Sec. 7.3   Mean-Square Error Optimization   **337** and, finally

$$\mathscr{E} = T \int_{-1/2T}^{1/2T} \frac{\sigma_a^2}{1 + \sigma_a^2 L(f)} \, df, \tag{7.47}$$

which, in conjunction with (7.42), is the expression of the MSE achievable by optimizing the receiving filter $U(f)$ for a given channel and a given shaping filter.

Example 7.2

Let us consider again Example 7.1, in which the goal is to optimize the receiving filter. We assume here $B_c T = 0.6$. The MSE for such a system is depicted in Fig. 7.9. The dotted line refers to a second-order Butterworth receiving filter whose bandwidth has been chosen so as to minimize $\mathscr{E}$, while the dashed line refers to the optimum receiving filter given by (7.38). It is observed that the effectiveness of the optimization increases as N_0 decreases (i.e., the system performance is limited by ISI rather than by additive noise). $\square$

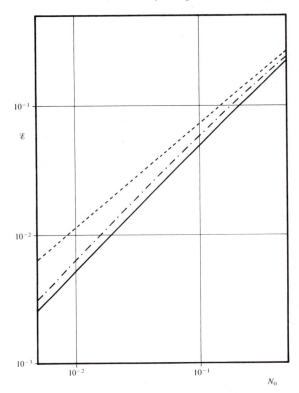

Figure 7.9 Performance of the transmission system of Example 7.1 to 7.3. Dotted line: the receiving filter is second-order Butterworth whose 3-dB bandwidth is chosen so as to minimize $\mathscr{E}$. Dashed line: the receiving filter has the optimum transfer function given by Eq. (7.38). Continuous line: both shaping and receiving filters are optimum in the MSE sense.

7.3.3 Optimizing the Shaping Filter S(f)

A further step toward system optimization can now be taken by looking for the optimum shaping filter $S(f)$. To do this, $\mathscr{E}$ in (7.35) should be minimized with respect to $S(f)$ subject to the power constraint at the shaping filter output:

$$\mathscr{P} = \frac{\sigma_a^2}{T} \int_{-\infty}^{\infty} |S(f)|^2 \, df, \tag{7.48}$$

which is the same as in (7.26).

The resulting equation, expressing a necessary and sufficient condition for $S(f)$ to minimize $\mathscr{E}$, does not seem amenable to a simple closed-form solution. Hence, we shall omit the details of the derivation. Here we shall restrict ourselves to a general description of the solution and an example of its application.

The optimum shaping filter transfer function can be obtained as follows:

(1) For every f in the Nyquist interval $(-1/2T, 1/2T)$, determine the integer k_f such that $|C(f + k/T)|^2/G_w(f + k/T)$ takes on its maximum value with respect to k. We shall define F as the set of frequencies that can be written in the form $f + k_f/T$, $f \in (-1/2T, 1/2T)$.

(2) Choose $\lambda > 0$ and define the subset F_λ of F such that, for $f \in F_\lambda$,

$$\frac{|C(f)|^2}{G_w(f)} > \frac{1}{\lambda}. \tag{7.49}$$

(3) Take

$$|S_{\text{opt}}(f)|^2 = \begin{cases} -\dfrac{TG_w(f)}{\sigma_a^2|C(f)|^2} + \dfrac{T}{\sigma_a^2|C(f)|}\sqrt{\lambda G_w(f)}, & f \in F_\lambda, \\ 0, & \text{elsewhere.} \end{cases} \tag{7.50}$$

Then compute $U(f)$ according to (7.38) and (7.39) and choose the phases of $S_{\text{opt}}(f)$, $U(f)$ so that $Q(f) = S_{\text{opt}}(f)C(f)U(f)$ is real and positive. Inspection of (7.35) demonstrates that the MSE depends on the phase of $S(f)U(f)$, but is independent of the way it is distributed between $S(f)$ and $U(f)$.

(4) Evaluate the resulting average channel input power by substituting for $|S(f)|^2$ in (7.48) the expression obtained from (7.50). The value computed will generally be different from the constraint value $\mathscr{P}$, so steps (2) to (4) should be repeated for different values of λ until the average channel input power is equal to $\mathscr{P}$.

From this procedure, it is seen that the optimum shaping filter, and hence the whole channel transfer function $Q(f)$, is generally bandlimited to the frequency set F, which has measure $1/T$ (this set is usually referred to as a *generalized Nyquist set*). The pulses transmitted through the channel have their energy confined in this set, whose frequencies are chosen, according to step (1), in order to afford the largest possible contribution to $L(f)$ in (7.42), and hence by rendering $L(f)$ as large as possible to minimize the RHS of (7.47). This simply shows that the pulse energy must be allocated at those frequencies where the channel performs better in the sense that $|C(f)|$ is large and/or $G_w(f)$ is small. In step (2), the set F is further reduced in order to preserve only those frequencies at which the ratio $|C(f)|^2/G_w(f)$ lies above a certain level depending on the value of λ. Actually, λ, a Lagrange multiplier in the constrained optimization problem, turns out to be proportional to the signal-to-noise ratio, defined as the ratio of the average transmitted power to the average noise power at the output of the receiving filter.

Example 7.3

Let us regard again the situations described in Examples 7.1 and 7.2 and consider the optimization of the shaping filter. The noise is white, and $|C(f)|^2$ is assumed to be a monotonically decreasing function of $|f|$ (in fact, the function has the form $|C(f)|^2 = [1 + (f/B_C)^8]^{-1}$). Thus, it follows that k_f, as already defined in step (1), is always zero, and hence $F = (-1/2T, 1/2T)$. Furthermore, $F_\lambda = (-1/2T', 1/2T')$, where for high λ values (i.e., high signal-to-noise ratios) $T' = T$, while for low λ values [i.e., in the situation that (7.49) does not hold for all $f \in F$] $T' > T$. Figure 7.9 shows the MSE obtained after optimizing both the shaping and the receiving filter in the situation dealt with here (continuous line). $\square$

7.4 OPTIMIZATION UNDER DIFFERENT CRITERIA

One conclusion arrived at in the last section was that, under the minimum mean-square error criterion, the optimum receiving filter $U(f)$ in the system of Figs. 7.1 and 7.2 has the form of a matched filter cascaded to a transversal filter with taps spaced T seconds apart. Another conclusion was that the optimum shaping and receiving filters are strictly bandlimited, in the sense that both have a zero transfer function outside a generalized Nyquist set of measure $1/T$.

Now, obviously, the MSE is not the only performance index, and other sensible optimization criteria are possible. The most natural of them seems to be the minimization of the error probability due to the combined effects of ISI and noise (Aaron and Tufts, 1966; Yao, 1972). Another approach leading to far simpler calculations is the maximization of the signal-to-noise ratio at the sampling instants (see Problem 7.2). The solutions to these problems share some common features. In fact, all of them result in the ubiquitous structure matched filter–transversal filter, and shaping and receiving filters are strictly bandlimited to a generalized Nyquist set if they are jointly optimized. Actually, it can be proved that *any* reasonable optimization criterion will lead to solutions exhibiting those features. In this section, we shall demonstrate this for the structure of the receiving filter. For the proof of the bandlimitedness of the optimum shaping filter, the reader is referred to Problem 7.8.

First, consider the scheme of Fig. 7.10, obtained by retaining the features that are relevant to the subject of this section from Figs. 7.1 and 7.2 [including definition (7.39)]. Due to the sampling performed at rate $1/T$, the effects of filters $P(f)$ and $U(f)$ with respect to any input are completely described by the equivalent transfer function (7.17). Here this is rewritten in the form

$$Q_{eq}(f) = \sum_{k=-\infty}^{\infty} P\left(f + \frac{k}{T}\right) U\left(f + \frac{k}{T}\right). \tag{7.51}$$

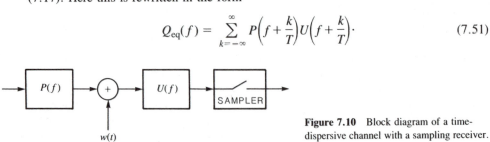

Figure 7.10 Block diagram of a time-dispersive channel with a sampling receiver.

Similarly, since the noise is Gaussian, its effect with respect to the system output is completely described by the power spectral density $G_w(f)|U(f)|^2$. Also, because of the sampler, the relevant quantity is the *equivalent power spectral density*

$$G_{eq}(f) \triangleq \sum_{k=-\infty}^{\infty} G_w\left(f + \frac{k}{T}\right)\left|U\left(f + \frac{k}{T}\right)\right|^2. \tag{7.52}$$

Given the preceding, let us now consider the choice of the optimization criterion. We shall restrict its choice to the realm of reasonable criteria. By "reasonable" we mean that the performance is assumed to be impaired by any introduction of additional noise in the system. Equivalently, we assume that any increase of the spectral density (7.52) will result in a deterioration. Let $U_{opt}(f)$ be the receiving filter, as optimized under the given criterion. We shall prove, assuming for $U_{opt}(f)$ the form (7.38), with $\Gamma(f)$ periodic with period $1/T$, that $Q_{eq}(f)$ will not be affected and that $G_{eq}(f)$ will not increase.

To see this, take for $\Gamma(f)$ the expression

$$\Gamma(f) = \left[\sum_{k=-\infty}^{\infty} P\left(f + \frac{k}{T}\right) U_{opt}\left(f + \frac{k}{T}\right)\right]\left[\sum_{k=-\infty}^{\infty} \frac{|P(f + k/T)|^2}{G_w(f + k/T)}\right]^{-1}, \tag{7.53}$$

and substitute it in (7.38). Simple calculations show that $Q_{eq}(f)$ in (7.51) is left unchanged. This is equivalent to saying that the input signal is still processed optimally. Consider then the equivalent power spectral density (7.52). This is changed to

$$\begin{aligned} G'_{eq}(f) &= |\Gamma(f)|^2 \sum_{k=-\infty}^{\infty} \frac{|P(f + k/T)|^2}{G_w(f + k/T)} \\ &= \left|\sum_{k=-\infty}^{\infty} P\left(f + \frac{k}{T}\right) U_{opt}\left(f + \frac{k}{T}\right)\right|^2 \cdot \left[\sum_{k=-\infty}^{\infty} \frac{|P(f + k/T)|^2}{G_w(f + k/T)}\right]^{-1}. \end{aligned} \tag{7.54}$$

Application of Schwarz's inequality shows that

$$\left|\sum_{k=-\infty}^{\infty} P\left(f + \frac{k}{T}\right) U_{opt}\left(f + \frac{k}{T}\right)\right|^2 \\ \leqslant \left[\sum_{k=-\infty}^{\infty} \frac{|P(f + k/T)|^2}{G_w(f + k/T)}\right] \cdot \left[\sum_{k=-\infty}^{\infty} G_w\left(f + \frac{k}{T}\right)\left|U_{opt}\left(f + \frac{k}{T}\right)\right|^2\right], \tag{7.55}$$

which, in conjunction with (7.54), yields

$$G'_{eq}(f) \leqslant \sum_{k=-\infty}^{\infty} G_w\left(f + \frac{k}{T}\right)\left|U_{opt}\left(f + \frac{k}{T}\right)\right|^2. \tag{7.56}$$

The comparison of (7.56) and (7.52) shows that $G'_{eq}(f)$ is not larger than the equivalent noise power spectral density achieved by using $U_{opt}(f)$ as the receiving filter. This, under our assumptions, means that no loss of performance is incurred if $U_{opt}(f)$ is chosen according to (7.38).

Sec. 7.4   Optimization Under Different Criteria   **341**

7.5 MAXIMUM-LIKELIHOOD SEQUENCE RECEIVER

In this section an entirely different approach will be adopted in the design of the optimum receiver for the system shown in Fig. 7.1. In particular, we shall apply the theory of ML reception outlined in Chapter 2 to a channel with ISI. We shall demonstrate that this approach provides a conceptually simple (but not always practical) solution to the optimization problem. Our assumptions are that the noise $w(t)$ is white and that the filters $S(f)$, $C(f)$ have a finite-length impulse response. A consequence of the latter assumption is that, before the addition of the noise, the waveforms at the channel output, as considered in any finite time interval, can only take a *finite* number of shapes (this number can be very large, but conceptually this is not a hindrance). Thus, the ML reception of a finite-length message is equivalent to the detection of one out of a finite set of waveforms in AWGN, so the theory developed in Chapters 2 and 4 is valid. In particular, the optimum receiver consists of a bank of matched filters, one for each waveform. Their outputs are sampled at the end of the transmission, and the largest sample is used to select the most likely symbol sequence.

However, in practice, this solution would be unacceptable due to its excessive complexity. In fact, for a message of K M-ary symbols, M^K matched filters might be necessary, with about M^K comparisons to be performed to select their largest sampled output. Thus, to provide a practical solution to the problem of ML reception, we must overcome several difficulties that appear intrinsic to it. First is the complexity induced by the large number of matched filters needed. Second is that induced by the large number of comparisons necessary to make a decision. Third we have the size of the memory required to store all the possible transmitted sequences and the delay involved in the detection process. As we shall see, satisfactory solutions can be found for these difficulties. In fact, only *one* matched filter is sufficient, due to the channel linearity. Furthermore, we can specify an algorithm whereby the number of computations necessary for the selection of the most likely symbol sequence and the memory size grow only *linearly* with respect to the message length K. Also, a suboptimum version of this algorithm can be adopted that allows decisions to be taken about the first transmitted symbols before the whole sequence has been received.

7.5.1 Maximum-Likelihood Sequence Detection Using the Viterbi Algorithm

The key to the ML receiver design is the expression of the log-likelihood ratio for the detection of the finite sequence of symbols $\mathbf{a} \triangleq (a_0, a_1, \ldots, a_{K-1})$ based on the observation of the waveform

$$r(t) \triangleq \sum_{l=0}^{K-1} a_l p(t - lT) + w(t), \qquad t \in \mathrm{I}, \tag{7.57}$$

where I is a time interval long enough to assure that $p(t)$, $p(t - T)$, $\ldots$, $p(t - (K - 1)T)$ are identically zero outside it. Definition (7.57) is derived from (7.2) by considering a finite symbol sequence instead of an infinite one. Moreover, we assume for simplicity that we are dealing with real signals only. The extension to the complex case is straightforward and requires only some minor changes of notation (see Section

2.6.3). In (7.57), $w(t)$ denotes white Gaussian noise with power spectral density $N_0/2$. We also assume that the sequence length K is large enough to disregard certain end effects. This concept is made more precise when the need arises. The log-likelihood ratio for $\mathbf{a}$ is then

$$\lambda_{\mathbf{a}} = \frac{2}{N_0} \int_I v_{\mathbf{a}}(t)r(t)\,dt - \frac{1}{N_0} \int_I [v_{\mathbf{a}}(t)]^2\,dt, \qquad (7.58)$$

where $v_{\mathbf{a}}(t)$ is the noiseless waveform corresponding to the symbol sequence $\mathbf{a}$:

$$v_{\mathbf{a}}(t) \triangleq \sum_{l=0}^{K-1} a_l p(t - lT). \qquad (7.59)$$

Using (7.59), we can rewrite (7.58) in the form

$$\lambda_{\mathbf{a}} = \frac{2}{N_0}\left\{\sum_{l=0}^{K-1} a_l \int_I p(t - lT)r(t)\,dt\right\} - \frac{1}{N_0}\sum_{l=0}^{K-1}\sum_{m=0}^{K-1} a_l a_m$$
$$\cdot \int_I p(t - lT)p(t - mT)\,dt. \qquad (7.60)$$

For notational simplicity, it is convenient to define the following quantities:

$$Z_l \triangleq \int_I p(t - lT)r(t)\,dt \qquad (7.61)$$

and

$$s_{l-m} \triangleq \int_I p(t - lT)p(t - mT)\,dt. \qquad (7.62)$$

With the assumption of a finite duration for the waveform $p(t)$, say

$$p(t) = 0, \qquad t < 0,\, t > (L + 1)T \qquad (7.63)$$

(the value L, $L \ll K$, will be referred to hereafter as the *memory* of the channel), the sequence Z_l, $l = 0, 1, \ldots, K - 1$, can be obtained by sampling at times $(L + l + 1)T$ the output of a filter matched to the waveform $p(t)$ when its input is the received signal $r(t)$. Notice also that (7.63) implies that $I = (0, (K + L)T)$. Strictly speaking, the RHS of (7.62) depends on l and m separately. However, we assume that the choice of K makes the interval I long enough for it to depend on $l - m$ only. Moreover, due to the assumption of a finite-memory channel, s_{l-m} can be nonzero only for a finite set of values of $l - m$. In fact, we have

$$s_k = 0, \qquad |k| \geq L + 1. \qquad (7.64)$$

Finally, observe that, under the hypothesis of a known function $p(t)$, the values of s_k are also known. Use now (7.61) and (7.62) in (7.60). Upon multiplication by the constant factor N_0, it is seen that the ML sequence $\hat{\mathbf{a}}$ is the one that minimizes the quantity

$$l_{\mathbf{a}} \triangleq -2\sum_{l=0}^{K-1} a_l Z_l + \sum_{l=0}^{K-1}\sum_{m=0}^{K-1} a_l a_m s_{l-m}. \qquad (7.65)$$

Now we observe that one of the results anticipated at the beginning of this section can be proved. In fact, all we need in order to compute $l_{\mathbf{a}}$ for every vector $\mathbf{a}$ is the sample sequence $(Z_l)_{l=0}^{K-1}$ obtained at the output of a *single matched filter*. Precisely, this set of samples provides a sufficient statistics for $r(t)$. This means that all we need to know about the received signal is contained in these samples.

The ML decision requires $l_\mathbf{a}$ to be minimized over the whole set of possible sequences $\mathbf{a}$. Thus, the matched filter must be followed by a processor, the *ML sequence detector*, determining as the most likely transmitted data sequence, say $\hat{\mathbf{a}}$, the one minimizing $l_\mathbf{a}$. The direct computation of $l_\mathbf{a}$ for all possible $\mathbf{a}$ to find the minimum is impractical due to the sheer number of computations involved. However, a sequential algorithm is available that performs such a selection in a computationally efficient manner. This is the celebrated *Viterbi algorithm* described in Appendix F. It performs the minimization of a function of several variables and is applicable to minimization problems that can be formulated as the search for the minimum-length path in a finite trellis. The significance of the Viterbi algorithm is that the number of computations required for the ML detection of a sequence of length K *grows only linearly with K*.

We shall now show how the Viterbi algorithm can be applied to our problem. (From now on we shall assume that the reader is familiar with Appendix F.) Essentially, our task is to show that $l_\mathbf{a}$ can be reduced to a sum of terms where each corresponds to the length of a branch in a suitable trellis diagram.

To do this, the first step is to rewrite $l_\mathbf{a}$, as defined in (7.65), in the form

$$l_\mathbf{a} = \left\{ -2\left(\sum_{l=0}^{K-2} a_l Z_l\right) + \sum_{l=0}^{K-2} \sum_{m=0}^{K-2} a_l a_m s_{l-m} \right\}$$
$$+ \left\{ -2(a_{K-1} Z_{K-1}) + 2\left(a_{K-1} \sum_{m=K-L-1}^{K-2} a_m s_{K-1-m}\right) + a_{K-1}^2 s_0 \right\}, \tag{7.66}$$

where (7.64) and the property $s_{-l} = s_l$ have been used. In (7.66) we have decomposed $l_\mathbf{a}$ into the sum of two bracketed terms. The first is similar to the RHS of (7.65) (the only change is the upper summation limit), and the second is a function only of the $L + 1$ symbols $a_{K-L-1}, a_{K-L}, \ldots, a_{K-1}$, not of the entire vector $\mathbf{a}$. Our decomposition of $l_\mathbf{a}$ into a sum of functions suitable for the application of the Viterbi algorithm will be based on repeated application of such decompositions. At this point it is convenient to define the variables

$$\sigma_l \triangleq (a_{l-1}, a_{l-2}, \ldots, a_{l-L}), \qquad l = L, \ldots, K, \tag{7.67}$$

and the quantities

$$U_{k+1}(\sigma_L, \ldots, \sigma_{k+1}) \triangleq -2\left(\sum_{l=0}^{k} a_l Z_l\right) + \sum_{l=0}^{k} \sum_{m=0}^{k} a_l a_m s_{l-m}, \tag{7.68}$$
$$k = L - 1, \ldots, K - 1,$$

and

$$V_{k+1}(\sigma_k, \sigma_{k+1}) \triangleq -2(a_k Z_k) + 2\left(a_k \sum_{m=k-L}^{k-1} a_m s_{k-m}\right) + a_k^2 s_0, \tag{7.69}$$
$$k = L, \ldots, K - 1.$$

Now, observing that

$$l_\mathbf{a} = U_K(\sigma_L, \ldots, \sigma_K), \tag{7.70}$$

we can rewrite (7.66) in the form

$$U_K(\sigma_L, \ldots, \sigma_K) = U_{K-1}(\sigma_L, \ldots, \sigma_{K-1}) + V_K(\sigma_{K-1}, \sigma_K),$$

and generalize it to show that

$$U_{k+1}(\sigma_L, \ldots, \sigma_{k+1}) = U_k(\sigma_L, \ldots, \sigma_k) + V_{k+1}(\sigma_k, \sigma_{k+1}),$$
$$k = L, \ldots, K - 1. \tag{7.71}$$

Repeated application of (7.71) yields

$$l_{\mathbf{a}} = U_L(\sigma_L) + V_{L+1}(\sigma_L, \sigma_{L+1}) + V_{L+2}(\sigma_{L+1}, \sigma_{L+2})$$
$$+ \cdots + V_K(\sigma_{K-1}, \sigma_K), \tag{7.72}$$

which is the required decomposition.

Our next step will be to exhibit a trellis such that we can associate to its branches the values taken on by the functions $V_{k+1}(\sigma_k, \sigma_{k+1})$, $k = L, \ldots, K - 1$. This task is simplified by a proper interpretation of the meaning of the variables σ_k defined in (7.67).

Recall that we have assumed the channel to have a finite memory L. This assumption is expressed mathematically by (7.63) and can be interpreted in the following manner. At any given time t, the received signal $r(t)$ defined in (7.57) depends on a set of $L + 1$ consecutive symbols, say $a_l, a_{l-1}, \ldots, a_{l-L}$. The last L of these symbols has been defined to form σ_l.

This is then called the *state of the channel* at time t. The transmission of the symbol a_l when the channel state is σ_l will then bring the channel to the succeeding state $\sigma_{l+1} = (a_l, a_{l-1}, \ldots, a_{l-L+1})$, and so forth, for symbols $a_{l+1}, a_{l+2}, \ldots$. Thus, we have set a one-to-one correspondence between the sequence of transmitted symbols $a_0, a_1, \ldots, a_{K-1}$ and the sequence of states $\sigma_L, \ldots, \sigma_K$. Therefore, the problem of selecting the most likely symbol sequence is equivalent to that of selecting the most likely sequence of states. This can also be seen directly from (7.72).

We are now able to define the trellis structure needed for the application of the Viterbi algorithm. For each value of the index l, $l = L, L + 1, \ldots, K$, associate a set of M^L nodes where each corresponds to a value of σ_l. Each node has M branches stemming from it, one for each value taken by a_l. Also, the branches represent the transition from the state σ_l to the next state σ_{l+1} as shown in Fig. 7.11. An example will help clarify these procedures.

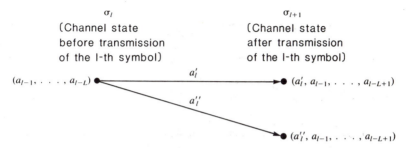

Figure 7.11 Transition from one state to the next and construction of the trellis diagram for application of the Viterbi algorithm. a_l' and a_l'' are two possible values taken on by the lth data symbol.

Example 7.4

Assume a binary baseband modulation with symbols ± 1 and a channel with a finite memory $L = 2$. The trellis for this situation consists of the states $\sigma_l = (a_{l-1}, a_{l-2})$. Each state can assume four values for each l (see Fig. 7.12), and the branches joining adjacent states represent the structure of the state vectors. For instance, the two branches stemming from the state $(a_l, a_{l-1}) = (1, -1)$ connect it to the allowable successor states (a_{l+1}, a_l) $= (-1, 1)$ or $(1, 1)$, corresponding to the symbols $a_{l+1} = -1$ and 1, respectively. Conversely, given the state $(a_{l+1}, a_l) = (1, 1)$, there are two allowable predecessor states $(a_l, a_{l-1}) = (1, 1)$ and $(1, -1)$, corresponding to the symbols $a_{l-1} = 1$ and -1, respectively. $\square$

The ML detection problem has now been reduced to the selection of a path through the trellis just described once the branches joining states σ_l and σ_{l+1} have been assigned the values taken by the function $V_{l+1}(\sigma_l, \sigma_{l+1})$, usually referred to as the *metric*. The minimum-metric path corresponds to the most likely sequence of states, and hence to the most likely sequence of symbols. The Viterbi algorithm is then applicable as follows:

Step 1. Observe the values of Z_0 at time $(L + 1)T$, Z_1 at time $(L + 2)T$, . . . , Z_{L-1} at time $2LT$. Let $l = L$ and use (7.68) to compute $U_L(\sigma_L)$ for each value of σ_L. Store the values of $U_L(\sigma_L)$.

Step 2. Let $l \rightarrow l + 1$. Observe the value of Z_l at time $(L + l + 1)T$, and use (7.69) to compute $V_{l+1}(\sigma_l, \sigma_{l+1})$ for each pair of states σ_l, σ_{l+1} such that the transition from σ_l to σ_{l+1} is allowed by the trellis structure.

Step 3. For each state σ_{l+1}, compute

$$u_{l+1}(\sigma_{l+1}) \triangleq \min_{\sigma_l} [u_l(\sigma_l) + V_{l+1}(\sigma_l, \sigma_{l+1})], \qquad (7.73)$$

where the minimum is taken over the values of σ_l compatible with σ_{l+1}, and $u_L(\cdot)$ $\triangleq U_L(\cdot)$. The quantity $u_{l+1}(\sigma_{l+1})$ is the minimum length of the paths leading to σ_{l+1}; store this quantity and this minimum-length path for each value of σ_{l+1}. If $l = K$, go to step 5.

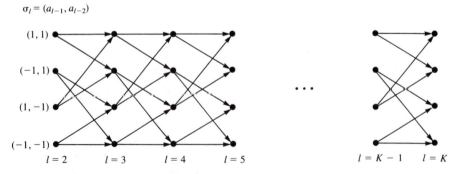

Figure 7.12 Trellis diagram for the situation of Example 7.4.

Step 4. Go to step 2.

Step 5. Compute $\min_{\sigma_K} u_K(\sigma_K)$; this is the minimum length of the paths through the trellis. The minimum-length path corresponds to the most likely sequence of states.

Example 7.5

Consider the situation of Example 7.4 and assume $s_0 = 1$, $s_1 = s_{-1} = 0.4$, $s_2 = s_{-2} = -0.2$, $s_k = 0$, $|k| > 2$. Assume also that $K = 8$, and $Z_0^2 = 1.0$, $Z_1 = -1.2$, $Z_2 = 0.5$, $Z_3 = -1.5$, $Z_4 = -0.2$, $Z_5 = 1.0$, $Z_6 = 0.8$, and $Z_7 = 0.9$. Upon reception of the matched filter outputs Z_0 and Z_1, $U_2(\sigma_2)$ can be computed for the four values of σ_2; we get

$$U_2(-1, -1) = -1.6, \qquad U_2(1, -1) = 1.6, \qquad U_2(-1, 1) = 0.8, \qquad U_2(1, 1) = 7.2.$$

Then, after receiving each value of Z_l, $l = 2, \ldots, 7$, $V_{l+1}(\sigma_l, \sigma_{l+1})$ can be computed. The corresponding values are shown in Fig. 7.13 together with those of $u_l(\sigma_l)$, $l = 2, \ldots, 8$. The minimum-length paths stored in step 3 of the algorithm are shown by the solid lines.

Application of the Viterbi algorithm shows that the ML path joins the states $(-1, -1)$, $(1, -1)$, $(-1, 1)$, $(1, -1)$, $(-1, 1)$, $(1, -1)$, and $(-1, 1)$, corresponding to the data sequence $-1, -1, 1, -1, 1, -1, 1, -1$. □

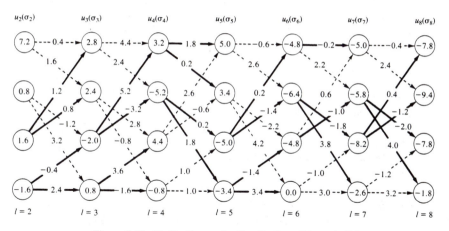

Figure 7.13 Trellis diagram for the situation of Example 7.5.

7.5.2 Error Probability for the Maximum-Likelihood Sequence Receiver

In the following, the performance of the ML sequence receiver will be evaluated by computing upper and lower bounds to the probability of a symbol error

$$P(e) \triangleq P\{\hat{a}_l \neq a_l\}, \tag{7.74}$$

where a_l denotes the lth transmitted symbol and $\hat{a}_l$ its estimate. Strictly speaking, this probability is a function of index l as our model is not stationary due to the consideration of a finite symbol sequence $a_0, a_1, \ldots, a_{K-1}$. However, under the usual assumption that K is large enough, we shall disregard this difficulty and assume that the RHS in

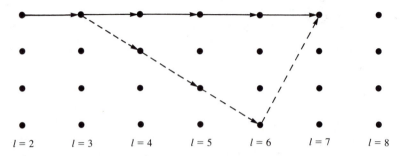

| $l=2$ | $l=3$ | $l=4$ | $l=5$ | $l=6$ | $l=7$ | $l=8$ |

Figure 7.14 An error event of length 3. The continuous line is the path corresponding to the transmitted sequence of states. The dashed line corresponds to the estimated sequence of states.

(7.74) does not depend on l. Since the ML sequence detection can be viewed as the choice of a path in the state trellis, for errors to occur it is necessary that the ML path *diverge* for a certain index, say l_1, from the path representing the transmitted symbol sequence, and *remerge* later, say for index $l_1 + H$. When this happens, we say that an *error event of length $H - 1$* has taken place (Fig. 7.14). The concept of error event specifies mathematically the fact that, when a sequence is estimated, consecutive symbol errors do not occur independently, but in finite clumps.

If we define

$$e_l \triangleq \hat{a}_l - a_l, \qquad l = 0, 1, \ldots, K-1, \tag{7.75}$$

and recall definition (7.67), it is seen that an error event starting at index l_1 and extending up to index $l_1 + H$, say

$$\{\hat{\sigma}_l = \sigma_l, l = l_1, l_1 + H; \hat{\sigma}_l \neq \sigma_l, l_1 < l < l_1 + H\}$$

corresponds to a sequence of symbol errors

$$\mathbf{e} \triangleq (e_{l_1}, e_{l_1+1}, \ldots, e_{l_1+H-L}), \qquad e_{l_1} \neq 0, e_{l_1+H-L} \neq 0.$$

(Incidentally, this shows that $H \geq L$; i.e., the error events are always at least as long as the channel memory.) If we let U be the set of all nonzero error events, $w(\mathbf{e})$ the number of decision errors entailed by the error event $\mathbf{e}$ (i.e., the number of nonzero entries in $\mathbf{e}$), and $P\{\mathbf{e}\}$ the probability of $\mathbf{e}$ to occur, we have

$$P(e) = \sum_{\mathbf{e} \in U} w(\mathbf{e}) P\{\mathbf{e}\}. \tag{7.76}$$

Since the exact computation of error probability using (7.76) does not seem feasible, we shall resort to evaluation of upper and lower bounds to $P(e)$.

An upper bound to $P(e)$

Computation of an upper bound to $P(e)$ will be based on the approximate evaluation of $P\{\mathbf{e}\}$ in (7.76). Let $A(\mathbf{e})$ be the event that the transmitted sequence $\mathbf{a}$ of data symbols (with the same length of $\mathbf{e}$) is compatible with the occurrence of $\mathbf{e}$. Then, for $\mathbf{e}$ to occur, it is necessary that $A(\mathbf{e})$ occur and that $\mathbf{a} + \mathbf{e}$ have a likelihood greater than any other possible sequence of source symbols, including $\mathbf{a}$. Since this latter event is included in the event $\{\lambda_{\mathbf{a}+\mathbf{e}} > \lambda_{\mathbf{a}}\}$, the probability of $\mathbf{e}$ can be bounded above as follows:

$$P\{\mathbf{e}\} \leq P\{\lambda_{\mathbf{a}+\mathbf{e}} > \lambda_{\mathbf{a}} | A(\mathbf{e})\} \cdot P\{A(\mathbf{e})\}. \tag{7.77}$$

We shall now proceed to evaluate separately the two factors in the RHS of (7.77). For a stationary sequence of independent source symbols, we have, for an error event of length H,

$$P\{A(\mathbf{e})\} = \prod_{l=0}^{H-L} P\{A(e_l)\}. \tag{7.78}$$

Example 7.6

For example, if a_l takes on values ± 1 with equal probabilities, only $a_l = +1$ is compatible with $e_l = -2$, only $a_l = -1$ is compatible with $e_l = +2$, and both values $a_l = \pm 1$ are compatible with $e_l = 0$. Thus, $P\{A(2)\} = P\{A(-2)\} = \frac{1}{2}$ and $P\{A(0)\} = 1$, which yields

$$P\{A(\mathbf{e})\} = \prod_{l=0}^{H-L} \left(1 - \frac{|e_l|}{4}\right). \tag{7.79}$$

More generally, it can be easily proved that, if a_l can take on the M values $-M + 1, -M + 3, \ldots, M - 1$ with equal probabilities, we have

$$P\{A(\mathbf{e})\} = \prod_{l=0}^{H-L} \left(1 - \frac{|e_l|}{2M}\right). \quad \square \tag{7.80}$$

Consider then the event $\{\lambda_{\mathbf{a}+\mathbf{e}} > \lambda_{\mathbf{a}} | A(\mathbf{e})\}$. By recalling (7.60) and using (7.61) and (7.62), the inequality $\lambda_{\mathbf{a}+\mathbf{e}} > \lambda_{\mathbf{a}}$ can be rewritten, after some algebraic manipulations, in the form

$$d^2(\mathbf{e}) \leq 2\left(\sum_{l=0}^{H-L} e_l v_l\right), \tag{7.81}$$

where

$$d^2(\mathbf{e}) \triangleq \sum_{l=0}^{H-L} \sum_{m=0}^{H-L} e_l e_m s_{l-m} \tag{7.82}$$

is called the *distance of the error event*, and

$$v_l \triangleq \int_I p(t - lT) w(t) \, dt, \qquad l = 0, 1, \ldots, H - L, \tag{7.83}$$

are Gaussian random variables with zero mean and covariance

$$E\{v_l v_k\} = \frac{N_0}{2} s_{l-k}. \tag{7.84}$$

The RHS of (7.81) turns out to be a Gaussian RV with zero mean and variance $(N_0/2)d^2(\mathbf{e})$. Therefore,

$$P\{\lambda_{\mathbf{a}+\mathbf{e}} > \lambda_{\mathbf{a}} | A(\mathbf{e})\} = \frac{1}{2} erfc\left(\frac{d(\mathbf{e})}{2\sqrt{N_0}}\right). \tag{7.85}$$

Sec. 7.5   Maximum-Likelihood Sequence Receiver   **349**

Using finally (7.76), (7.77), and (7.85), we get

$$P(e) \leq \sum_{\mathbf{e} \in U} w(\mathbf{e}) \cdot \frac{1}{2} \, erfc \left(\frac{d(\mathbf{e})}{2\sqrt{N_0}} \right) P\{A(\mathbf{e})\}, \tag{7.86}$$

where $P\{A(\mathbf{e})\}$ can be computed through (7.78). Equation (7.86), in spite of the considerable effort spent to derive it, is still not in a usable form, mainly because of the difficulty involved in enumerating the elements of U. Thus, we shall resort to an approximation of the RHS of (7.86), valid for small values of N_0 and based on the steep decrease of the function erfc ($\cdot$). It suffices to observe that the terms in the summation (7.86) will be dominated, as $N_0 \to 0$, by the terms involving the smallest value of $d(\mathbf{e})$, which we shall denote by $d_{\min}$:

$$d_{\min} \triangleq \min_{\mathbf{e} \in U} d(\mathbf{e}). \tag{7.87}$$

Hence we get the approximate upper bound

$$P(e) \stackrel{\sim}{\leq} \frac{1}{2} \psi'(d_{\min}) erfc \left(\frac{d_{\min}}{2\sqrt{N_0}} \right), \tag{7.88}$$

where

$$\psi'(d_{\min}) \triangleq \sum_{\mathbf{e} \in U(d_{\min})} w(\mathbf{e}) P\{A(\mathbf{e})\}, \tag{7.89}$$

and $U(d_{\min})$ is the subset of U including the error events with distance $d_{\min}$. Notice, in particular, that $\psi'(d_{\min})$ is a constant independent of N_0.

Example 7.7

Consider a binary transmission system with equally likely symbols ± 1, and assume that $U(d_{\min})$ includes the two error sequences $(+2, -2)$ and $(-2, +2)$. The set $A(+2, -2)$ includes only the data sequence $(-1, +1)$, and $A(-2, +2)$ includes only $(+1, -1)$. Thus,

$$\psi'(d_{\min}) = 2 \cdot \tfrac{1}{4} + 2 \cdot \tfrac{1}{4} = 1.$$

In the same conditions, if $U(d_{\min}) = \{(-2), (+2)\}$, we have $A(-2) = \{(+1)\}$ and $A(+2) = \{(-1)\}$, so

$$\psi'(d_{\min}) = \tfrac{1}{2} + \tfrac{1}{2} = 1. \quad \square$$

A lower bound to $P(e)$

We shall now proceed to evaluate a lower bound for the symbol error probability. To do this, we consider the ideal situation in which the detection process is aided by a genie supplying to the receiver some side information on the transmitted symbols. If the receiver takes its decisions by exploiting optimally the genie information, it is clear that it cannot be outperformed by any receiver working without the genie's aid. Thus, if $P_G(e)$ denotes the symbol error probability achieved by the genie-aided receiver, we have, for every real-life receiver (and hence for the ML sequence detector),

$$P(e) \geq P_G(e). \tag{7.90}$$

Assume now that the genie operates as follows: when the sequence $\mathbf{a} = (a_0, a_1, \ldots, a_{K-1})$ is transmitted, he chooses at random another sequence $\mathbf{a}' = \mathbf{a} + \mathbf{e}$, which has an error on the lth symbol a_l (and possibly others). Then he tells the receiver that either $\mathbf{a}$ or $\mathbf{a}'$ was transmitted. In this situation, the task of the receiver is to choose one out of two known signals perturbed by white Gaussian noise. This can be achieved optimally with a probability of error

$$\frac{1}{2} erfc\left(\frac{d(\mathbf{e})}{2\sqrt{N_0}}\right).$$

Thus, the probability that the genie-aided receiver makes an incorrect decision on a_l is

$$P_G(e) = \sum_{\mathbf{e}\in U} \frac{1}{2} erfc\left(\frac{d(\mathbf{e})}{2\sqrt{N_0}}\right) P\{A(\mathbf{e})\}, \tag{7.91}$$

where $A(\mathbf{e})$ can now be interpreted as the event that the data sequence chosen by the genie is compatible with $\mathbf{e}$. Equivalently, $P\{A(\mathbf{e})\}$ is the ratio between the number of sequences $\mathbf{a}'$ such that $\mathbf{a}' = \mathbf{a} + \mathbf{e}$ for some $\mathbf{a}$ and the total number of data sequences with length K.

By combining (7.90) and (7.91) and discarding from the summation of (7.91) all those sequences $\mathbf{e}$ for which $d(\mathbf{e}) > d_{min}$, we have

$$P(e) \geqslant \frac{1}{2} \psi''(d_{min}) erfc\left(\frac{d_{min}}{2\sqrt{N_0}}\right), \tag{7.92}$$

where

$$\psi''(d_{min}) \triangleq \sum_{\mathbf{e}\in U(d_{min})} P\{A(\mathbf{e})\} \tag{7.93}$$

is the probability that a data sequence chosen at random has a sequence $\mathbf{e}$ compatible with it and such that $d(\mathbf{e}) = d_{min}$. The result (7.92) generalizes the lower bound (4.77) obtained for symbol-by-symbol detection.

Example 7.8

Under the same conditions as for Example 7.7, if $U(d_{min}) = \{(-2, +2), (+2, -2)\}$, we get $\psi''(d_{min}) = \frac{1}{4} + \frac{1}{4} = \frac{1}{2}$, and if $U(d_{min}) = \{(-2), (+2)\}$, we get $\psi''(d_{min}) = \frac{1}{2} + \frac{1}{2} = 1$. $\square$

7.5.3 Significance of d_{min} and its Computation

The results obtained so far in this section, and in particular the upper and lower bounds (7.88) and (7.92) of error probability, show that the key parameter for the performance evaluation of the ML sequence detector is the minimum distance d_{min} defined by (7.87) and (7.82) or, equivalently, by

$$d^2_{min} \triangleq \min_{\mathbf{e}\neq\mathbf{0}} \sum_l \sum_m e_l e_m s_{l-m} \tag{7.94}$$

or

$$d^2_{min} \triangleq \min_{\mathbf{a}'\neq\mathbf{a}} \int_I [v_\mathbf{a}(t) - v_{\mathbf{a}'}(t)]^2 \, dt, \tag{7.95}$$

Sec. 7.5   Maximum-Likelihood Sequence Receiver   **351**

where (7.59) has been used. In words, d^2_{min} can be viewed from (7.94) as arising from minimization over error patterns, or from (7.95) as the square of the smallest possible distance between distinct signals at the output of the deterministic part of the channel. It is easily seen that, in the special case of transmission of independent symbols over the AWGN channel without intersymbol interference, (7.95) reduces to the minimum distance between signal points, as defined in Chapter 4. Also, it is interesting to observe that inequality (7.92) provides a bound to the symbol error probability of *any* real-life receiver that can be conceived to detect a sequence of data transmitted on a channel with intersymbol interference. Thus, computation of d_{min} provides an important parameter for judging the quality of the channel itself.

Unfortunately, direct computation of d_{min} involves a minimization problem that may be hard to solve, as the number of relevant error patterns $\mathbf{e}$ or, equivalently, of symbol sequence pairs $(\mathbf{a}, \mathbf{a}')$ that have to be tested can be prohibitively large. Tree-search algorithms for the determination of d_{min} have been proposed (see Fredricsson, 1974; Messerschmitt, 1973b; Anderson and Foschini, 1975). For channels with a short memory, a convenient procedure to find d_{min} has been proposed by Anderson and Foschini (1975). This procedure, stemming from an approach combining functional analysis and computer search, is based on the selection, from the full set of error sequences $\mathbf{e}$, of a small subset of crucial sequences such that at least one element of the subset attains the minimum distance. As a result, d_{min} can be obtained from (7.94) by evaluating $\sum_l \sum_m e_l e_m s_{l-m}$ for every element of this subset and choosing the smallest value found.

Tables 7.1 and 7.2 contain some examples of these "sufficient" sets of error patterns obtained for a baseband data transmission with $M = 2$ (symbols ± 1), $M = 3$ (symbols $0, \pm 2$), and $M = 4$ (symbols $\pm 1, \pm 3$) over channels with memory $L = 2$ or 3. As an example, if the channel has memory 2, only the error sequences $\mathbf{e} = (2), (-2), (2, -2)$, and $(2, 2)$ must be tested for a binary system. If $M = 3$, the sequences to be tested are the preceding plus $(2, -4, 4, -2)$ and $(2, 4, 4, 2)$. An efficient algorithm for the computation of d_{min} based on dynamic programming will be described in Section 10.5.2 in a more general context.

TABLE 7.1 Sufficient error sequences for the computation of d_{min} in a channel with memory 2

		$L = 2$					
$M = 2$		2					
		-2					
		2	-2				
		2	2				
$M = 3$		2	-4	4	-2		
		2	4	4	2		
$M = 4$		2	-4	6	-6	4	-2
		2	4	6	6	4	2

TABLE 7.2 Same as in Table 7.1 for a channel with memory 3

			$L = 3$									
$M = 2$	2											
	−2											
	2	−2										
	2	2										
	2	−2	2									
	2	2	2									
	2	−2	−2	2								
	2	2	−2	−2								
$M = 3$	2	−4	4	−2								
	2	4	4	2								
	2	−4	4	−4	2							
	2	4	4	4	2							
	2	−4	4	0	−4	4	−2					
	2	4	4	0	−4	−4	−2					
	2	−2	4	−2	2							
	2	2	4	2	2							
	2	−2	4	−4	4	−2	2					
	2	2	4	4	4	2	2					
$M = 4$	2	−4	6	−6	4							
	2	4	6	6	4							
	2	−4	6	−6	6	−4	2					
	2	4	6	6	6	4	2					
	2	−2	4	−4	6	−4	4	−2	2			
	2	2	4	4	6	4	4	2	2			
	2	−2	4	−4	6	−6	6	−6	4	−2	2	
	2	2	4	4	6	6	6	6	4	2	2	

Example 7.9

Consider a channel bandlimited to the Nyquist interval $(-\frac{1}{2}, \frac{1}{2})$. If the samples of its response are $p_0 = 1$, $p_1 = p_2 = \epsilon$, $0 \le \epsilon \le 1$, from (7.62) we get $s_0 = 1 + 2\epsilon^2$, $s_1 = s_{-1} = \epsilon + \epsilon^2$, $s_2 = s_{-2} = \epsilon$, and $s_l = 0$, $|l| > 2$. For a binary baseband transmission with symbols ± 1, Table 7.1 shows that $d^2_{\min}$ can be found by looking for the smallest between the quantities $4(1 + 2\epsilon^2)$, corresponding to the error sequences (± 2); $8(1 - \epsilon + \epsilon^2)$, corresponding to the error sequence $(+2, -2)$; and $8(1 + \epsilon + 3\epsilon^2)$, corresponding to the error sequence $(+2, +2)$. Hence, we get

$$d^2_{\min} = \min \{4(1 + 2\epsilon^2), 8(1 - \epsilon + \epsilon^2)\}.$$

It can be seen that, for $0 \le \epsilon < \frac{1}{2}$, we have $d^2_{\min} = 4(1 + 2\epsilon^2)$, which entails that the most probable error event includes a single error. Similarly, for $\frac{1}{2} < \epsilon \le 1$, we have $d^2_{\min} = 8(1 - \epsilon + \epsilon^2)$, and the double-error event $(+2, -2)$ is more probable. For $\epsilon = \frac{1}{2}$, single- and double-error events are equally likely to occur.

Sec. 7.5   Maximum-Likelihood Sequence Receiver

It is interesting to observe that for $0 \leq \epsilon \leq \frac{1}{2}$ (i.e., when the channel eye pattern is either open or just closed at the sampling instant), we get $d^2_{min} = p^2_0 + p^2_1 + p^2_2$ (i.e, d^2_{min} is equal to the energy of the channel response). Thus, apart from a constant independent of N_0, the ML sequence detector provides the same error rate that we would get on a channel without ISI by using the same pulse energy. In other words, no degradation is suffered because of ISI (provided of course that the ML sequence detector is used). The situation changes as ϵ grows larger, for in this case $d^2_{min} < p^2_0 + p^2_1 + p^2_2$, and ISI causes a degradation that cannot be compensated for, not even by the optimum receiver. $\square$

Example 7.10

Consider an M-ary baseband transmission with symbols $-M + 1, -M + 3, \ldots, M - 1$ and a channel bandlimited to $(-1/2T, 1/2T)$ with $p_0 = 1, p_1 = -1$. Thus, we have $s_0 = 2T$, $s_1 = s_{-1} = -T$, and $s_l = 0$, $|l| > 1$. In this case (see Problem 7.9), $d^2_{min} = 8T$, which is achieved by the error events $(\pm 2, \pm 2, \ldots, \pm 2)$, $m = 1, 2, \ldots$

Hence, using (7.80) and (7.89), we obtain

$$\psi'(d_{min}) = 2 \sum_{m=1}^{\infty} m \cdot \left(1 - \frac{1}{M}\right)^m = 2M(M - 1).$$

Similarly, (7.93) yields

$$\psi''(d_{min}) = 2 \sum_{m=1}^{\infty} \left(1 - \frac{1}{M}\right)^m = 2(M - 1).$$

The symbol error probability is then bounded as follows:

$$(M - 1) \, erfc \left(\sqrt{\frac{2T}{N_0}}\right) \leq P(e) \leq M(M - 1) \, erfc \left(\sqrt{\frac{2T}{N_0}}\right).$$

For a channel without ISI, M-ary transmission using pulses with energy $2T$ would result in an error probability of

$$P(e) = \frac{M - 1}{M} \, erfc \left(\sqrt{\frac{2T}{N_0}}\right). \quad \square$$

7.5.4 Implementation of Maximum-Likelihood Sequence Detectors

Even with present-day technology, the implementation of an ML sequence detector can be difficult in high-speed data transmission due to the processing requirements of the Viterbi algorithm. In fact, the number of values taken by the state variables σ_l, and hence the number of quantities to be stored and processed per received symbol, grows as M^L, and the demand on the processor speed increases with the symbol rate for a given value of M^L. For binary symbols and a very short channel memory (say, $L = 1$ to 3), there may not be any problem with this complexity at low-speed data transmission. However, for many real-life channels M^L can be so large as to make implementation of a Viterbi receiver unfeasible even at low data rates. As an example, for typical voiceband telephone channels with efficient use of the bandwidth, the complexity of the receiver becomes prohibitive due to values of L lying in the range $10 \div 100$.

Also, a truly optimum receiver delays its decision on the symbol sequence until it has been received in its entirety. In certain cases, a decision can be made *before* the entire sequence $(Z_l)_{l=0}^{K-1}$ has been observed and processed; this occurs when during the computations it is seen that all the M^L trellis paths that have been stored leading to the nodes corresponding to state σ_{l_2} (say) pass through a single node corresponding to state σ_{l_1}, $l_1 < l_2$. In this situation, it is said that a *merge* has occurred for $l = l_1$, and a decision can be taken on the first states from σ_L to σ_{l_1}. For example, in Fig. 7.13 a merge occurs for $l = 4$ in the state $(-1, 1)$; this is detected for $l = 6$, and a decision can be taken on the states σ_2, σ_3, σ_4.

In general, merges occur at random, and in certain unfortunate cases they may never occur during the transmission of a finite sequence. Thus, in practice, it is necessary to *force* decisions about the first transmitted symbols when the area allocated for the paths storage is liable to be exceeded. Qureshi (1973a) has shown by analysis and computer simulation that in most practical situations the probability of additional errors due to premature decisions becomes irrelevant if the decisions are taken after a reasonable delay. In many cases, it will be sufficient to choose a delay just beyond twice the channel memory L, provided of course that the decisions are made by selecting the sequence that has the greatest likelihood at the moment of the decisions.

To limit the receiver complexity due to the channel memory length, an approach that has often been adopted is to use a linear filter preceding the optimum receiver in order to reduce the channel memory to a small value. With this prefilter taking care of limiting L, the Viterbi algorithm can be implemented with a tolerable complexity. However, any linear filtering of the received signal will also affect the noise. Thus, any attempt to compensate for the nulls or near nulls in the channel equivalent transfer function results in prefilter characteristics that, by trying to invert the channel transfer function, will increase the noise power at the receiver input. Thus, linear prefiltering designed to condition optimally the channel impulse response should also take into account the output noise variance. To do this, the desired frequency response of the combination channel–prefilter should be close to the channel's in those frequency intervals where the channel cannot be equalized without excessive noise enhancement.

Several solutions to the problem of optimum prefilter design have been proposed. Qureshi and Newhall (1973) use a mean-square criterion to force the overall response of the channel plus the prefilter to approximate a truncated version of the original channel impulse response. Falconer and Magee (1973) show how the desired response can be chosen to minimize the noise variance. The approach of Messerschmitt (1974) is to minimize the noise variance while retaining the first nonzero sample of the desired response fixed. Beare's (1978) design method results in a transfer function for the cascade of the channel and the prefilter that is as close as possible to that of the original channel under the constraint of the memory length. Notice that the process of truncating the impulse response of the channel will never be perfect, so that the receiver will ignore some of the input ISI. The performance of this "mismatched" receiver has been considered by Divsalar (1978). McLane (1980) has derived an upper bound to the bit error probability due to this residual neglected channel memory. Other approaches to the reduction of complexity of the optimum receiver have been taken. Vermeulen and Hellman (1974) and Foschini (1977) consider the choice of a reduced-state trellis in order to simplify the Viterbi algorithm; Lee and Hill (1977) embed a decision-feedback equalizer (see Chapter 8) into the receiver structure.

BIBLIOGRAPHICAL NOTES

The paper by Nyquist (1928), a classic in the field of data transmission, and the subsequent paper by Gibby and Smith (1965), include the formulation of what has been named the Nyquist criterion. The generalization of Nyquist criterion to a situation in which the transmission is assumed to be affected by both ISI and crosstalk interference was considered by Shnidman (1967) and Smith (1968a).

The design of signal pulses subject to criteria other than the elimination of ISI only has been the subject of several studies. Chalk (1950) finds the time-limited pulse shape that minimizes adjacent-channel interference. Spaulding (1969) considers the design of networks whose response simultaneously minimizes ISI and bandwidth occupancy; his procedure generates better results than the approximation of raised-cosine responses. Mueller (1973) designs a transversal filter whose impulse response is constrained to give zero ISI and has minimum out-of-band energy. Mueller's theory has been generalized by Boutin, Morissette, and Porlier (1982). Franks (1968) selects pulses that minimize the effect of the ISI resulting from a small deviation from the proper timing instants $t_0 + lT$, $-\infty < l < \infty$.

For a receiver with the structure shown in Fig. 7.2, the most natural approach to the optimization of the filter $U(f)$ is to choose the error probability as a performance criterion. This was done by Aaron and Tufts (1966), whereas Yao (1972) provided a more efficient computational technique. A simpler approach is to constrain the ISI to be zero and then minimize the error probability, as described in Section 7.2; this was considered by Lucky, Salz, and Weldon (1968). Yet another approach is to maximize the signal-to-noise ratio at the sampling instants (George, 1965).

Joint optimization of shaping and receiving filters under a minimum MSE criterion was considered by Smith (1968a) and Berger and Tufts (1967) (our handling of the issue follows closely the latter paper). A different derivation of Berger and Tufts's results was obtained by Hänsler (1971). Ericson (1971 and 1973) proved that for every reasonable optimization criterion the optimum shaping filter is bandlimited, and that the optimum receiving filter can be realized as the cascade of a matched filter and a tapped-delay line. Our treatment in Section 7.3 is based on Ericson's papers.

Nonlinear receivers have also been studied. Since maximum a posteriori or ML detection seems at first to lead to a receiver complexity that grows exponentially with the length K of the sequence to be detected, sequential algorithms were investigated in order to reduce this complexity. Chang and Hancock (1966) developed a sequential algorithm for a maximum a posteriori detection of sequences of length $L + 1$, leading to a receiver structure whose complexity grows only linearly with K. Abend and Fritchman (1970) obtained a similar algorithm for symbol-by-symbol detection. The idea of using the Viterbi algorithm for ML detection of data sequences for baseband transmission channels was developed, independently and almost simultaneously, by Forney (1972b), Kobayashi (1971a), and Omura (1971). The case of complex symbols (i.e., carrier-modulated signals) was considered by Ungerboeck (1974), whereas Foschini (1975) provided a mathematically rigorous derivation of error probability bounds. Our treatment follows those of Forney (1972b), Ungerboeck (1974), Foschini (1975), and Hayes (1975).

PROBLEMS†

7.1. Consider a raised-cosine transfer function $Q(f)$ with roll-off α and its powers $Q^\gamma(f)$, $0 < \gamma < 1$.
   (a) Compute the equivalent noise bandwidth of $Q^\gamma(f)$ for several values of α and γ [see (2.94)].
   (b) Compute the error probability in a binary baseband PAM system modeled as in Figs. 7.1 and 7.2 with $S(f) = \beta \, Q^\gamma(f)$, $U(f) = (1/\beta) \, Q^{1-\gamma}(f)$, symbols $a_n = \pm 1$, rate 1200 bits/s, and a white Gaussian noise with power spectral density $G_w(f) = 10^{-5}$ W/Hz. The constant β is chosen so as to have a unit power at the output of the shaping filter.

7.2. In the system of Figs. 7.1 and 7.2, for given transfer functions $S(f)$ and $C(f)$, choose the receiving filter $U(f)$ so as to maximize at its output, for a given sampling instant, the ratio between the instantaneous signal power and the average power of ISI plus noise. Show that this filter can be implemented in the form of a matched filter cascaded to a transversal filter.

7.3. Consider a binary PAM data-transmission system. Data must be transmitted at a rate of 9600 bits/s with a bit error probability lower than 10^{-5}. The channel transfer function is given by

$$C(f) = \begin{cases} 1, & |f| < 6000 \text{ Hz}, \\ 0, & \text{elsewhere}. \end{cases}$$

The noise is white Gaussian with a power spectral density $G_w(f) = 10^{-6}$ W/Hz. Choose the shaping filter $S(f)$ and the receiving filter $U(f)$ so as to minimize the average transmitted power while getting rid of the intersymbol interference at the sampling instants. Compute the signal power at the output of the shaping filter.

*7.4. In a binary baseband PAM system modeled as in Figs. 7.1 and 7.2, the cascade of $S(f)$, $C(f)$, and $U(f)$ has a raised-cosine response with roll-off α, $0 < \alpha < 1$. The sampling instants are affected by a constant offset of 5 percent with respect to the nominal values, so ISI is present. Assuming that the transmitted symbols are ± 1 and that the noise is white Gaussian with a power spectral density $G_w = 10^{-6}$ W/Hz, compute the bit error probability of the system as a function of α using one of the techniques described in Section 6.2.

*7.5. Consider a bandpass transmission system operating at a signaling rate of $1/T$ on a channel with a flat transfer function $C(f)$. The shaping filter is a fourth-order Butterworth with a 3-dB bandwidth B_S, and the receiving filter is a second-order Butterworth with 3-dB bandwidth B_U. Assuming a white Gaussian noise, determine, for $1.2 < B_S T < 2$, the values of $B_U T$ that give the minimum bit error probability for an M-ary coherent PSK modulation ($M = 2$, 4, and 8). For every situation, choose the signal-to-noise ratio $\mathcal{E}_b/N_0$ so that this minimum probability is 10^{-6}.

7.6. Consider a binary baseband PAM data-transmission system operating at a rate of 4800 bits/s and modeled as in Figs. 7.1 and 7.2. Assume $s(t) = u_T(t)$,

$$C(f) = \frac{1}{1 + jf/f_c}, \qquad f_c = 2400 \text{ Hz},$$

and a white Gaussian noise with power spectral density $G_w(f) = 10^{-7}$ W/Hz.

† The problems marked with an asterisk should be solved with the aid of a computer.

(a) Determine the shape of the receiving filter that minimizes the bit error probability while removing ISI at the sampling instants.

(b) Determine the shape of the receiving filter that minimizes the mean-square error at the sampler's output.

*(c) Compare the error probabilities obtained with the systems designed in parts (a) and (b).

7.7. Consider a filter with impulse response

$$s(t) = \sum_{l=-Nm}^{Nm} s_l b\left(t - \frac{lT}{N}\right)$$

(N an integer > 1). This can be modeled as a linear transversal filter cascaded to a linear system with impulse response $b(t)$. Define the $(2Nm + 1)$-dimensional vectors

$$\mathbf{s} \triangleq [s_{-Nm}, \ldots, s_0, \ldots, s_{Nm}]'$$

and

$$\mathbf{z} \triangleq [z^{-Nm}, \ldots, z^0, \ldots, z^{Nm}]',$$

with $z \triangleq \exp(j2\pi fT/N)$; assume that $b(t)$ has energy $\mathscr{E}$ and a duration $\leq T/N$.

(a) Show that if $s(t)$ is the impulse response of the shaping filter of a data-transmission system with independent symbols (a_n) and a signaling rate $1/T$, the power density spectrum of its output signal can be written in the form

$$\frac{\sigma_a^2}{T}|B(f)|^2 \mathbf{s}^\dagger \mathbf{z}\, \mathbf{z}^\dagger \mathbf{s},$$

where $B(f)$ denotes the Fourier transform of $b(t)$.

(b) Let $\mathscr{P}_F$ denote the power of the signal at the output of the shaping filter in the frequency interval $(-F, F)$. Show that the shaping filter coefficients vector $\mathbf{s}$ that maximizes the ratio $\mathscr{P}_F/\mathscr{P}_\infty$ [relative power in the frequency interval $(-F, F)$] is the eigenvector of a symmetric matrix $\mathbf{R}$ corresponding to the maximum eigenvalue λ_{max}. Determine the entries of $\mathbf{R}$, and show that λ_{max} is the maximum value of $\mathscr{P}_F/\mathscr{P}_\infty$.

(c) Assume that $s(t)$ must satisfy the Nyquist criterion for intersymbol interference-free transmission; that is,

$$s_{\pm lN} = 0, \qquad \text{for} \qquad l = \pm 1, \pm 2, \ldots, \pm m.$$

How can this constraint be included in part (b)?

*(d) Assuming $b(t) = \delta(t)$, derive the shaping filter that gives a Nyquist-type response with maximum relative power in the frequency interval $|f| < 2400$ Hz for $m = 8$ (impulse response limited in duration to $|t| < 8T$) and $N = 4$ (four samples per signaling interval T) (Mueller, 1973).

7.8. The joint optimization of the shaping filter $S(f)$ and the receiving filter $U(f)$ under the power constraint (7.48) leads to filters that are strictly bandlimited to a generalized Nyquist set of measure $1/T$.

(a) Prove this statement for the minimum-mean-square error criterion of Section 7.2.

(b) Prove this statement for a different criterion that is "reasonable" in the sense of Section 7.3.

Hint: (a) Using variational arguments, find a necessary and sufficient condition for $S(f)$ to minimize $\mathscr{E}$ for a fixed receiving filter $U(f)$, subject to the power constraint (7.48). Use this condition and (7.37) to prove that for an optimal system

$$|S_{opt}(f)|^2 = KG_w(f)|U_{opt}(f)|^2,$$

where K is a constant. Use then (7.38) to prove that at every frequency f a jointly optimum system must satisfy either

$$S_{opt}(f) = U_{opt}(f) = 0$$

or

$$\frac{K'}{1 + \sigma_a^2 L(f)} = \frac{G_w(f)}{|C(f)|^2}, \tag{A}$$

(where K' is another constant) and observe that the LHS of (A) is periodic, whereas the RHS is usually aperiodic. Thus, if (A) is satisfied at some frequency f_0, it cannot be satisfied at any other frequency of the form $f_0 + k/T$, $k \neq 0$.

(b) Let $S(f)$ be any shaping filter satisfying the given power constraint. Define another shaping filter $S'(f)$ such that

$$|S(f)|^2 = \begin{cases} \dfrac{G_w(f)}{|C(f)|^2} \displaystyle\sum_k \dfrac{|P(f + k/T)|^2}{G_w(f + k/T)}, & f \in F, \\ 0, & f \notin F, \end{cases}$$

where F is a set of frequencies such that for $f \in F$

$$\frac{|C(f)|^2}{G_w(f)} \geq \frac{|C(f + k/T)|^2}{G_w(f + k/T)}, \qquad k = \pm 1, \pm 2, \dots.$$

Show that if $S'(f)$ is used in the system in lieu of $S(f)$, the system performance is not impaired and the power constraint is still satisfied (Berger and Tufts, 1967)

7.9. Assume that the channel transfer function $P(f)$ is bandlimited in the Nyquist interval $(-\frac{1}{2}, \frac{1}{2})$, and denote by p_l, $-\infty < l < \infty$, the samples, taken every second, of its response.

(a) Derive an expression of the minimum distance for this channel in terms of the discrete convolution between the sequence (p_l) and the sequence (e_l) of symbol errors.

(b) Using the result obtained in part (a), derive the minimum distance for a channel with memory $L = 1$ when the data symbols take on the values $0, 1, \dots, M - 1$.

(c) Consider a channel with impulse response samples $p_l = 1/\sqrt{n}$, $l = 0, 1, \dots, n - 1$, and binary symbols ± 1. Derive the minimum distance for this channel, and verify that $d_{min} \to 0$ as $n \to \infty$.

7.10. Show that, with the notations of Section 7.4, the inequality

$$\sum_{l \neq 0} |s_l| < s_0 \tag{B}$$

is a sufficient condition for the nonexistence of error events whose distance is smaller than that achieved by a single error. *Hint*: If δ_0 denotes the minimum nonzero value of $|e_l|$, show that

$$\left| \sum_{k=0}^{H} e_{l+k} e_l \right| \leq \sum_{k=0}^{H} e_k^2 - \delta_0^2, \qquad l \neq 0,$$

and use this result to prove that, if (B) holds, d_{min}^2 cannot be smaller than $\delta_0^2 s_0$, the minimum distance achieved by a single error.

Chap. 7   Problems

CHAPTER 8

Adaptive Receivers and Channel Equalization

The theory developed in Chapter 7 concerning the optimum design of the receiver was based on the assumption of an exact knowledge of the channel characteristics (i.e., of its transfer function or impulse response). However, often this assumption is far from realistic. In fact, whereas it is generally true that the designer knows the basic features of the channel, this knowledge may not be sufficiently accurate to allow system optimization. This occurs, for example, because the channel, although time invariant, has been selected randomly from an ensemble. This is typical of dial-up telephone lines. Another possibility is that the channel varies randomly with time. This is typical of radio-link channels affected by fading (see Section 6.4). Consequently, it is often required that the receiver, designed to cope with the unwanted effects of intersymbol interference (ISI) and additive noise, be self-optimizing or *adaptive* (i.e., have parameters that are automatically adjusted to an optimum operating point, possibly keeping track of the changing conditions).

Two philosophies can be used to design an adaptive receiver. The first is based on the scheme of Fig. 8.1. It assumes that the relevant channel parameters are first estimated and then fed to an optimum (or suboptimum) receiver, which makes use of them to adjust itself to the transmission environment. The receiver can be, for example, a Viterbi receiver in which the channel impulse response samples must be known for ideal operation. This leads to the problem of *adaptive channel identification*, which is addressed in Section 8.1. Another approach is depicted in Fig. 8.2. Here a device, called an *equalizer*, compensates for the channel unwanted features and presents the receiver with a sequence of samples that have been in a sense ''cleaned-up'' from the effects of ISI and noise. This is by far the most popular approach, one of the reasons being that the equalizer can be realized using a transversal filter, that is, a *tapped*

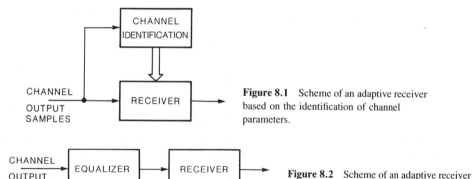

Figure 8.1 Scheme of an adaptive receiver based on the identification of channel parameters.

Figure 8.2 Scheme of an adaptive receiver based on channel equalization.

delay line (TDL) with adjustable weights. The availability of simple algorithms for adaptation of weights has made the TDL equalizer the most widely used solution to the problem discussed. Later in the chapter, we shall analyze in some depth the behavior of this equalizer and some modifications that have been suggested to improve its performance.

8.1 ADAPTIVE CHANNEL IDENTIFICATION

We shall assume that the source symbols form a stationary sequence (a_l) of identically distributed, zero-mean complex random variables (RV), not necessarily independent. The modulation scheme and the channel are both linear (see Chapter 10 for an analysis of the nonlinear case). For mathematical convenience, it is also assumed that the discrete overall channel [i.e., the discrete-time system that accepts as input the sequence (a_l) and that outputs the sequence (x_l); see Figs. 7.1 and 7.2], has a finite memory L. Thus, the relationship between the sequences (a_l) and (x_l) is given by

$$x_l = \sum_{i=0}^{L} q_i a_{l-i} + n_l, \tag{8.1}$$

where $(q_l)_{l=0}^{L}$ is the impulse response of the overall channel and (n_l) is the sequence of noise samples, resulting from a white sequence independent of the source symbols passed through the receiving filter. The notations are the same as in Section 7.2.

Clearly, the effect of the channel on any input sequence (a_l) is fully described by the $L + 1$ complex numbers $q_0, \ldots, q_L$. Hence, the channel's identification involves the estimation of these numbers.

Equation (8.1) can be rewritten in a compact vector form by introducing the vector

$$\mathbf{q} = \begin{bmatrix} q_0 \\ q_1 \\ \cdot \\ \cdot \\ \cdot \\ q_L \end{bmatrix} \quad L + 1 \tag{8.2}$$

Sec. 8.1   Adaptive Channel Identification

361

of the channel impulse response, and the vector

$$\mathbf{a}_l = \begin{bmatrix} a_l \\ a_{l-1} \\ \cdot \\ \cdot \\ \cdot \\ a_{l-L} \end{bmatrix} \quad\updownarrow\; L+1 \tag{8.3}$$

of the source symbols that affect the output x_l. With these notations, (8.1) takes the form

$$x_l = \mathbf{q}'\, \mathbf{a}_l + n_l, \tag{8.4}$$

so that the problem of channel identification can be handled as follows: derive an estimate of the vector $\mathbf{q}$ based on the observation of the received sequence (x_l) and on the knowledge of the transmitted symbol sequence (a_l). If $\hat{\mathbf{q}}$ denotes an estimate of $\mathbf{q}$, we can construct a sequence (x_l) approximating the received sequence as follows:

$$\hat{x}_l = \hat{\mathbf{q}}'\, \mathbf{a}_l \tag{8.5}$$

and measure the accuracy of the channel estimate by comparing $(\hat{x}_l)$ with the actual sequence (x_l). If $\mathbf{B}$ denotes the $(L+1) \times (L+1)$ matrix

$$\mathbf{B} \triangleq \mathrm{E}[\mathbf{a}_l^*\, \mathbf{a}_l'] \tag{8.6}$$

and

$$\sigma_n^2 \triangleq \mathrm{E}|n_l|^2, \tag{8.7}$$

the mean-square error $\mathrm{E}|x_l - \hat{x}_l|^2$ can be given the form

$$\mathrm{E}|x_l - \hat{x}_l|^2 = (\mathbf{q} - \hat{\mathbf{q}})^\dagger \mathbf{B}(\mathbf{q} - \hat{\mathbf{q}}) + \sigma_n^2. \tag{8.8}$$

When $\mathbf{B}$ is positive definite, that is, the source symbols are linearly independent (see Example B.2 of Appendix B), the minimum achievable mean-square error (MSE) is σ_n^2, which corresponds to the situation $\hat{\mathbf{q}} = \mathbf{q}$ (i.e., perfect identification). In the following we shall make this assumption about $\mathbf{B}$ without further comments.

Consider now an algorithm for automatic identification. The condition $\hat{\mathbf{q}} = \mathbf{q}$ occurs when the MSE (8.8) is minimum. Thus a sensible identification algorithm should be based on the search for this minimum. Several minimization algorithms are available from numerical analysis. However, not all of them suit our requirements. First, we want to use an iterative scheme; in fact, iterative algorithms are usually simpler to implement, and they may be faster than matrix inversion or the direct solution of a set of equations. Moreover, it is necessary to average out the effects of noise and round-off errors due to the digital (i.e., finite accuracy) implementation of the algorithm. This can be achieved using an iterative scheme operating on the time span of a large number of received symbols. Also, certain algorithms that achieve a very fast convergence but need highly accurate computations must be excluded. For example, the conjugate gradient method of Hestenes and Stiefel (1952) achieves a very fast convergence when operated in the absence of noise (see Kobayashi, 1971c). Unfortunately, it can be highly divergent when computations are perturbed by noise and round-off errors due to the digital implementation of the algorithm (Schonfeld and Schwartz, 1971b). The most widely used minimization algorithm for identification and for equalization is the steepest-descent or *gradient algorithm*. To understand its behavior, observe that the MSE, a quadratic function of the estimated vector $\hat{\mathbf{q}}$, can be viewed geometrically as a bowl-shaped surface in the space of the complex vectors with $L + 1$ components. As the minimum of MSE corresponds to the bowl's bottom, minimization algorithms seek this bottom. In the steepest-descent method, one starts by choosing arbitrarily an "initial" vector $\hat{\mathbf{q}}^{(0)}$, which corresponds to a point of the surface. The $(L + 1)$-component gradient vector of the MSE with respect to $\hat{\mathbf{q}}$ is then computed at this point. As the negative of the gradient is parallel to the direction of steepest descent, a step is performed on the surface in the direction of the negative of the gradient. If the step is short enough, the new point $\hat{\mathbf{q}}^{(1)}$ will be closer to the bottom of the surface. Hence, the gradient is computed at $\hat{\mathbf{q}}^{(1)}$ and the procedure repeated. Under certain conditions (which we shall explore later), this process will eventually converge to the bottom of the bowl (where the gradient is zero) regardless of the choice of the initial point.

Consider now (8.8). The gradient of the MSE with respect to $\hat{\mathbf{q}}$ is $-2\mathbf{B}(\mathbf{q} - \hat{\mathbf{q}})$ (see Section B.5), a linear function of the overall channel's estimated impulse response (this is due to the fact that the MSE is a quadratic function). Besides providing simpler implementation of the minimization algorithm, it also implies the absence of relative minima. The gradient algorithm will then take the form

$$\hat{\mathbf{q}}^{(n+1)} = \hat{\mathbf{q}}^{(n)} + \alpha\mathbf{B}(\mathbf{q} - \hat{\mathbf{q}}^{(n)}), \qquad n = 0, 1, \ldots, \tag{8.9}$$

where $\hat{\mathbf{q}}^{(n)}$ denotes the value assumed by the estimated impulse response at the nth iteration step, and α is a positive constant small enough to ensure convergence of the iterative procedure. This subject will be discussed shortly.

Equation (8.9) does not appear to be in a form useful for the algorithm implementation. In fact, it involves the vector $\mathbf{q}$, which is obviously not available. Using (8.5) and (8.6), we can change (8.9) to

$$\hat{\mathbf{q}}^{(n+1)} = \hat{\mathbf{q}}^{(n)} + \alpha\mathrm{E}[(x_n - \hat{x}_n)\mathbf{a}_n^*], \tag{8.10}$$

which is expressed in terms of the observable quantities x_n, $\hat{x}_n$ and of the vector $\mathbf{a}_n$, assumed to be known. The difficulty now is that exact evaluation of the expectation in the RHS of (8.10) is not practically achievable. In fact, explicit computation of this expectation requires knowledge of the channel impulse response, which is not available. If suitable ergodicity assumptions are made, the expectation can be evaluated as a time average performed over a sufficiently long time interval (theoretically infinite). But this would prevent real-time operation. Thus, we must resort to an approximation of the RHS of (8.10).

Before discussing this approximation problem, we shall analyze the performance of algorithm (8.9) or (8.10). This relatively simple analysis will provide some insight into the behavior of the implementable algorithm described afterward.

8.1.1 Gradient Algorithm for Identification

Consider again (8.9) rewritten in the form

$$\hat{\mathbf{q}}^{(n+1)} = (\mathbf{I} - \alpha\mathbf{B})\,\hat{\mathbf{q}}^{(n)} + \alpha\mathbf{B}\,\mathbf{q}, \tag{8.11}$$

where $\mathbf{I}$ denotes the $(L + 1) \times (L + 1)$ identity matrix. By subtracting $\mathbf{q}$ from both sides and defining the *estimation error* $\boldsymbol{\epsilon}^{(n)}$ at the nth iteration as

$$\boldsymbol{\epsilon}^{(n)} = \hat{\mathbf{q}}^{(n)} - \mathbf{q}, \tag{8.12}$$

we get a first-order homogeneous recursion describing the evolution of $\boldsymbol{\epsilon}^{(n)}$:

$$\boldsymbol{\epsilon}^{(n+1)} = (\mathbf{I} - \alpha\mathbf{B})\,\boldsymbol{\epsilon}^{(n)}, \qquad n = 0, 1, \ldots. \tag{8.13}$$

From (8.13) it follows that

$$\boldsymbol{\epsilon}^{(n)} = (\mathbf{I} - \alpha\mathbf{B})^n \boldsymbol{\epsilon}^{(0)}, \tag{8.14}$$

so that, taking the norms of both sides (see Section B.2),

$$\|\boldsymbol{\epsilon}^{(n)}\| \leq \|\mathbf{I} - \alpha\mathbf{B}\|^n \|\boldsymbol{\epsilon}^{(0)}\| = \rho^n \|\boldsymbol{\epsilon}^{(0)}\| \tag{8.15}$$

where

$$\rho \triangleq \|\mathbf{I} - \alpha\mathbf{B}\| \tag{8.16}$$

denotes the spectral radius of the matrix $\mathbf{I} - \alpha\mathbf{B}$. From (8.15) is follows that $\|\boldsymbol{\epsilon}^{(n)}\| \rightarrow 0$ as $n \rightarrow \infty$ (that is, the algorithm converges for any $\|\boldsymbol{\epsilon}^{(0)}\|$), if $\rho^n \rightarrow 0$ as $n \rightarrow \infty$. The latter condition means that the maximum eigenvalue of $\mathbf{I} - \alpha\mathbf{B}$ must have a magnitude smaller than 1. As the eigenvalues of $\mathbf{I} - \alpha\mathbf{B}$ are given by $1 - \alpha\lambda_i$, $i = 1, 2, \ldots,$ $L + 1$, where the eigenvalues λ_i of $\mathbf{B}$ are real and positive, we obtain the following sufficient condition for convergence:

$$\max_i |1 - \alpha\lambda_i| < 1, \tag{8.17}$$

which is equivalent to (see Fig. 8.3)

$$0 < \alpha < 2/\lambda_{\max}, \tag{8.18}$$

where $\lambda_{\max}$ denotes the largest eigenvalue of $\mathbf{B}$. Equation (8.15) shows, in particular, that the norm of the estimation error is reduced at each iteration by at least a factor of ρ.

It may be useful to bound the allowable values of α without actually evaluating the eigenvalues of $\mathbf{B}$. To this end, observe that

$$\lambda_{\max} \leq \sum_{i=1}^{L+1} \lambda_i = \operatorname{tr}\mathbf{B} = (L + 1)\sigma_a^2, \tag{8.19}$$

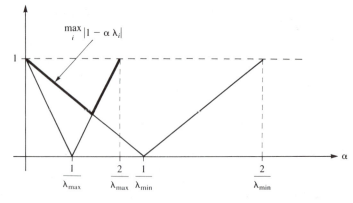

Figure 8.3 Set of α values that assure convergence of algorithm (8.11).

where σ_a^2 denotes the absolute second moment of the source symbols

$$\sigma_a^2 \triangleq E|a_n|^2, \tag{8.20}$$

and the last equality in (8.19) occurs as all entries in the main diagonal of $\mathbf{B}$ are equal to σ_a^2. We get the sufficient condition for convergence:

$$0 < \alpha < \frac{2}{(L+1)\sigma_a^2}. \tag{8.21}$$

Condition (8.21) is tighter than (8.18). Consider in fact the case of independent source symbols. This implies $\mathbf{B} = \sigma_a^2\mathbf{I}$, and hence $\lambda_1 = \lambda_2 = \cdots = \lambda_{L+1} = \sigma_a^2$. Condition (8.21) results in an upper bound on α that is $(L+1)$ times tighter than (8.18).

8.1.2 Stochastic-Gradient Algorithm for Identification

Let us now consider an approximation of (8.9) in a form useful for real-time implementation. The simplest such approximation, and by far the most widely used, is obtained by disregarding the expectation operator in (8.10). We obtain the new algorithm

$$\hat{\mathbf{q}}^{(n+1)} = \hat{\mathbf{q}}^{(n)} + \alpha(x_n - \hat{x}_n)\,\mathbf{a}_n^*. \tag{8.22}$$

The second term in the RHS of (8.22) has obviously the same expected value as the corresponding term in the gradient algorithm. Hence it can be viewed as an unbiased estimate of that quantity.

Implementation of (8.22) is shown in Fig. 8.4 for real data symbols. Each iteration is performed every T seconds. When a new source symbol a_n is fed into the TDL, the

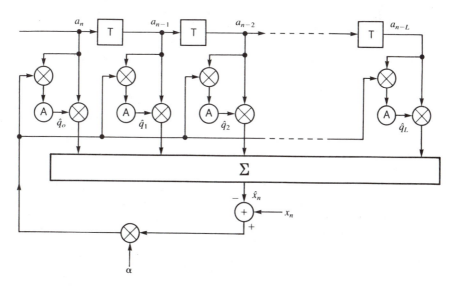

Figure 8.4 Implementation of the stochastic-gradient algorithm for channel identification (real data symbols a_n are assumed). The blocks labeled T denote delay elements; the blocks labeled A denote accumulators.

channel output estimate $\hat{x}_n$ is obtained by combining linearly $a_n, \ldots, a_{n-L}$ according to (8.5). The error signal $x_n - \hat{x}_n$ is then formed, and after having been multiplied by the scaling factor α, this value is also multiplied by the content of the TDL (the vector $\mathbf{a}_n$). The resulting values are used finally to update the accumulators containing the actual estimates $\hat{q}_0, \ldots, \hat{q}_L$.

The algorithm just described is usually denoted as the *stochastic-gradient* (or *estimated-gradient*) algorithm because the directions along which the algorithm moves on the MSE surface are *random*.

8.1.3 Convergence of the Stochastic-Gradient Algorithm for Identification

Consider now the convergence properties of this algorithm. Their study is far more difficult than with the "true" gradient algorithm (8.9). In fact, we have from (8.22)

$$\boldsymbol{\epsilon}^{(n+1)} = (\mathbf{I} - \alpha\, \mathbf{a}_n^* \mathbf{a}_n')\, \boldsymbol{\epsilon}^{(n)}, \tag{8.23}$$

a version of (8.13) with the matrix $\mathbf{B}$ changed to $\mathbf{a}_n^* \mathbf{a}_n'$. The complication here arises from the fact that $\mathbf{a}_n^* \mathbf{a}_n'$ is a random matrix, which in turn is not statistically independent of $\boldsymbol{\epsilon}^{(n)}$. To proceed further, we need an approximation proposed and widely employed to make mathematically tractable the analysis of the stochastic-gradient algorithm. This approximation, referred to hereafter as the *independence assumption*, implies assuming that $(\mathbf{a}_n)$ is a sequence of iid zero-mean vectors. This assumption is obviously rather crude. Nevertheless, it has been widely verified that analyses of the convergence of the stochastic-gradient algorithm based on the independence assumption are in close agreement with the results of experiments and simulation, provided that the step size α is sufficiently small (see, e.g., Ungerboeck, 1972; Widrow and others, 1976; Gitlin and Weinstein, 1979). This fact has also been verified analytically (see, e.g., Mazo, 1979; Jones, Cavin, and Reed, 1982).

Since in (8.23) $\boldsymbol{\epsilon}^{(n)}$ depends only on the sequence $\mathbf{a}_0, \mathbf{a}_1, \ldots, \mathbf{a}_{n-1}$, the independence assumption entails that $\boldsymbol{\epsilon}^{(n)}$ be independent of $\mathbf{a}_n$. Thus, if we take the expectation of both sides of (8.23), we get

$$E[\boldsymbol{\epsilon}^{(n+1)}] = (\mathbf{I} - \alpha\mathbf{B})E[\boldsymbol{\epsilon}^{(n)}]. \tag{8.24}$$

As no random quantity appears in (8.24), we can repeat the convergence analysis carried out for the true gradient algorithm. This enables us to conclude that the average error vector $E[\boldsymbol{\epsilon}^{(n)}]$ tends to the null vector, as $n \to \infty$, if the condition (8.18) holds. Nonetheless, this result is incomplete, as the behavior of $E[\boldsymbol{\epsilon}^{(n)}]$ does not give us a complete picture of the convergence of the algorithm. In fact, nothing prevents this vector from being very close to the null vector, while $\boldsymbol{\epsilon}^{(n)}$ itself exhibits large deviations around its average. A deeper analysis of the behavior of $\boldsymbol{\epsilon}^{(n)}$ is therefore necessary. This can be obtained by studying the quadratic error $E\|\boldsymbol{\epsilon}^{(n)}\|^2$.

If we define

$$\mathcal{E}^{(n)} \triangleq E\|\boldsymbol{\epsilon}^{(n)}\|^2 = E[\boldsymbol{\epsilon}^{(n)\dagger}\boldsymbol{\epsilon}^{(n)}], \tag{8.25}$$

we get from (8.23) the following recursive equation:

$$\mathcal{E}^{(n+1)} = E[\boldsymbol{\epsilon}^{(n)\dagger}(\mathbf{I} - \alpha\, \mathbf{a}_n^* \mathbf{a}_n')^2 \boldsymbol{\epsilon}^{(n)}] = E[\boldsymbol{\epsilon}^{(n)\dagger}\mathbf{C}\, \boldsymbol{\epsilon}^{(n)}], \tag{8.26}$$

where

$$\mathbf{C} \triangleq \mathbf{I} - 2\alpha\mathbf{B} + \alpha^2 E[\mathbf{a}_n^* \mathbf{a}_n']^2. \tag{8.27}$$

To proceed further, we introduce the additional simplifying assumption that the symbols a_n are zero-mean independent random variables. Thus $\mathbf{B} = \sigma_a^2 \mathbf{I}$ and $E[\mathbf{a}_n^* \mathbf{a}_n']^2 = \sigma_a^4 (k_a + L) \mathbf{I}$, where

$$k_a \triangleq \frac{E\{|a_n|^4\}}{\sigma_a^4}. \tag{8.28}$$

With this assumption, repeated application of (8.26) yields

$$\mathscr{E}^{(n)} = [1 - 2\sigma_a^2 + \alpha^2 \sigma_a^4 (k_a + L)]^n \mathscr{E}^{(0)}, \tag{8.29}$$

so the quadratic error's convergence is assured provided that the quantity in brackets is less than 1.

Example 8.1

A simple special case occurs when the random variables a_n take on values ± 1 with equal probabilities and, hence, $\sigma_a^2 = k_a = 1$. We have

$$\mathscr{E}^{(n)} = [1 - 2\alpha - \alpha^2(L + 1)]^n \mathscr{E}^{(0)}. \tag{8.30}$$

The quantity in brackets can be minimized with respect to the choice of α, giving

$$\mathscr{E}^{(n)} = \left(\frac{L}{L+1}\right)^n \mathscr{E}^{(0)}. \tag{8.31}$$

Notice how convergence is slowed down as L increases. $\square$

8.1.4 Some Further Problems

Several other features of automatic channel identification might be worth analyzing here. However, many of them are shared with adaptive equalization. Thus, to avoid unnecessary duplications, we shall only deal with them in the framework of equalization, leaving to the interested reader the task of extrapolating their analysis to identification. Here, we shall focus our attention on only three problems, the first two because of their relevance, and the third because it is typical of adaptive channel identification, as handled in this section.

The first issue concerns the behavior of the stochastic-gradient algorithm when the channel is not stationary (i.e., its impulse response changes with time). This is the problem of *adaptive* identification. If changes occur slowly enough with respect to the signaling rate, we can expect that the algorithm will allow the channel estimate to track continually the channel characteristics. Another problem relates to the assumption that the source data sequence (a_n) is known to the receiver, a situation not met in the practice. It can be solved by first sending through the channel a known sequence, which will provide a reasonably good channel estimate. Afterward the receiver should provide a sufficiently good performance to assume that most of its estimates of transmitted symbols are correct. In this situation, the assumption that (a_n) is known is reasonable.

Finally, consider the effect on automatic identification of inaccurate knowledge of the true channel memory span L. Clearly, if the TDL of Fig. 8.4 has a number $\hat{L}$ of delay elements larger than L, then at the end of the identification process $\hat{L} - L$ tap weights will assume zero values in identification. If instead $\hat{L} < L$ (i.e., the number of delay elements in the TDL is smaller than the channel memory), it can be shown (see Problem 8.1) that $\hat{L} + 1$ of the channel impulse response samples can still be identified correctly, provided that the data symbols are uncorrelated and have zero mean.

8.2 TAPPED DELAY LINE EQUALIZER

Until the early 1960s, equalization of voice-grade channels employed for data transmission was generally undertaken by using a fixed or manually adjustable receiver filter to compensate for the nonideal channel characteristics. When the demand for increased data-transmission speed exceeded a certain limit, it became apparent that these equalizers were no longer suitable. In particular, they could not remove the distortions introduced by variations between connections in a switched service. In fact, in this case the assigned channel may differ considerably from the average because of differences either in the characteristics of the links to be connected or in their number. Compensation of the average channel characteristics with a compromise equalizer may be adequate for medium-speed voiceband modems (from 1200 to 2400 bits/s). However, for higher speeds this solution must be abandoned, and modems must incorporate an equalizer able to adjust itself to compensate for a broad range of channels.

The first comprehensive solution to this problem was proposed by Lucky (1965); see also the Bibliographical Notes at the end of the chapter. He studied an equalizer based on an algorithm for automatically adjusting the coefficients of a TDL. The choice of a TDL structure, besides being justified by theoretical considerations (we have seen in Chapter 7 that the optimum linear receiver includes such a TDL), makes the equalizer relatively simple to implement and to analyze, mainly because of the linear relation between the tap weights and the equalizer output. In the following, we shall assume a transmission scheme like that depicted in Figs. 7.1 and 7.2, with (a_n) an iid sequence and $E\{a_n\} = 0$.

The TDL equalizer operating on the samples (x_n) of the received signal has the structure depicted in Fig. 8.5. An N-tap TDL stores N samples that are linearly combined to produce the equalizer output

$$y_n = \mathbf{c}' \, \mathbf{x}_n, \tag{8.32}$$

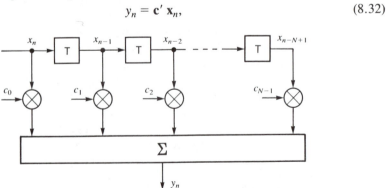

Figure 8.5 The TDL equalizer.

where $\mathbf{c}$ is the tap-weight vector

$$\mathbf{c} = \begin{bmatrix} c_0 \\ c_1 \\ \cdot \\ \cdot \\ \cdot \\ c_{N-1} \end{bmatrix} \Bigg\} N \tag{8.33}$$

and $\mathbf{x}_n$ denotes the TDL's content at discrete time n:

$$\mathbf{x}_n = \begin{bmatrix} x_n \\ x_{n-1} \\ \cdot \\ \cdot \\ \cdot \\ x_{n-N+1} \end{bmatrix} \Bigg\} N. \tag{8.34}$$

Ideally, we would like the sequence (y_n) at the output of the equalizer to reproduce the sequence (a_n) of transmitted data symbols, except perhaps for a finite delay D. As we cannot expect to achieve this, even in the absence of noise, with a finite-length TDL, a reasonable goal would be to find $\mathbf{c}$ so as to minimize a suitable distortion measure, with the constraints of the equalizer length N and the delay D. If we choose as a distortion measure the mean-square error between y_n and a_{n-D}, D a fixed integer, as in Gersho (1969a) and Proakis and Miller (1969), we have to minimize the MSE:

$$\mathcal{E}(\mathbf{c}) \triangleq \mathrm{E}|y_n - a_{n-D}|^2 = \mathbf{c}^\dagger \mathbf{X}\,\mathbf{c} - 2\mathcal{R}[\mathbf{c}^\dagger \mathbf{v}] + \sigma_a^2, \tag{8.35}$$

where σ_a^2 is as in (8.20); $\mathbf{v}$ is the correlation vector defined as

$$\mathbf{v} \triangleq \mathrm{E}[a_{n-D}\,\mathbf{x}_n^*], \tag{8.36}$$

and $\mathbf{X}$ is the $N \times N$ matrix

$$\mathbf{X} \triangleq \mathrm{E}[\mathbf{x}_n^*\mathbf{x}_n']. \tag{8.37}$$

This matrix is positive definite. In fact, for every complex N vector $\mathbf{a}$ we have

$$\mathbf{a}^\dagger \mathbf{X}\mathbf{a} = \mathrm{E}|\mathbf{a}'\,\mathbf{x}_n|^2. \tag{8.38}$$

The RHS of (8.38) can be viewed as the average output power from an equalizer with a tap-weight vector $\mathbf{a}$. This power cannot be zero because of the random noise added to the samples at the channel's output. In the absence of noise, (8.38) can be zero only if the samples x_n are linearly dependent. With this exception (which we want to discard), $\mathbf{a}^\dagger \mathbf{X}\mathbf{a} > 0$ (i.e., $\mathbf{X}$ is positive definite). To define the value of the tap-weight vector $\mathbf{c}$ that minimizes (8.35), we must find the value of $\mathbf{c}$ that gives a null value of the gradient (see Examples B.5 and B.6):

$$\nabla\mathcal{E}(\mathbf{c}) = 2(\mathbf{X}\,\mathbf{c} - \mathbf{v}). \tag{8.39}$$

As the gradient has a unique zero for

$$\mathbf{c} = \mathbf{c}_{\mathrm{opt}} \triangleq \mathbf{X}^{-1}\mathbf{v}, \tag{8.40}$$

we obtain the minimum value of the MSE:

$$\mathscr{E}_{min} = \mathscr{E}(\mathbf{c}_{opt}) = \sigma_a^2 - \mathbf{v}^\dagger \mathbf{X}^{-1} \mathbf{v}. \tag{8.41}$$

An alternative form for $\mathscr{E}(\mathbf{c})$ in (8.35) is then

$$\mathscr{E}(\mathbf{c}) = \mathscr{E}_{min} + (\mathbf{c} - \mathbf{c}_{opt})^\dagger \mathbf{X} (\mathbf{c} - \mathbf{c}_{opt}). \tag{8.42}$$

This explicitly shows the quadratic nature of the functional $\mathscr{E}(\mathbf{c})$ and separates the contributions of the minimum achievable MSE, $\mathscr{E}_{min}$, from a term depending on nonoptimum weight setting.

8.2.1 Performance of the Infinitely Long Equalizer

Our next task is to analyze in greater detail the optimum equalizer's performance. We shall do so by relating the achievable MSE to the transmission channel characteristics. Consider a situation in which the number N of TDL taps is large enough to justify the substitution of the Toeplitz matrix $\mathbf{X}$ with a circulant matrix whose rows are cyclic shifts of one of them (see Section B.3 for the relevant definitions). This approximation, asymptotically valid as $N \to \infty$, entails neglecting that the true $\mathbf{X}$ still differs from a circulant matrix in the lower-left and the upper-right corners. As eigenvalues and eigenvectors of a circulant matrix have simple expressions, our approximation yields simple, closed-form expressions for the quantities of interest.

Consider the diagonal decomposition of $\mathbf{X}$

$$\mathbf{X} = \mathbf{U} \mathbf{\Lambda} \mathbf{U}^{-1}, \tag{8.43}$$

where $\mathbf{\Lambda}$ is the diagonal matrix of the eigenvalues $\mu_0, \ldots, \mu_{N-1}$, of $\mathbf{X}$, and $\mathbf{U}$ is the $N \times N$ unitary matrix of its eigenvectors (see Section B.4). As $\mathbf{X}$ is circulant, from Example B.3 the entries of $\mathbf{U}$ are

$$u_{ik} = \frac{1}{\sqrt{N}} w^{ik}, \qquad i = 0, 1, \ldots, N-1, \qquad k = 0, 1, \ldots, N-1, \tag{8.44}$$

where

$$w \triangleq e^{j2\pi/N}. \tag{8.45}$$

Consider then the eigenvalues μ_i, $i = 0, \ldots, N-1$, of the matrix $\mathbf{X}$. From (B.39) and the definition of $\mathbf{X}$, we get, as $N \to \infty$,

$$\mu_i = \sum_{l=-N/2}^{N/2} E[x_{n+l} x_n^*] \, e^{j2\pi li/N}. \tag{8.46}$$

Define now the signal $x_{eq}(t)$ as the time function, bandlimited in the interval $(-1/2T, 1/2T)$, whose sample sequence is (x_n). From the sampling expansion we obtain its Fourier transform:

$$X_{eq}(f) = \sum_{n=-\infty}^{\infty} x_n e^{-j2\pi fnT}, \qquad 0 \leq f < \frac{1}{T}. \tag{8.47}$$

If we apply the techniques introduced in Section 2.3.1, we can derive the power spectrum of $x_{eq}(t)$:

$$\mathcal{G}_{eq}(f) = \frac{1}{T} \sum_{l=-\infty}^{\infty} E[x_{n+l} x_n^*] \, e^{-j2\pi f l T}, \qquad 0 \leqslant f < \frac{1}{T}. \tag{8.48}$$

The comparison of (8.46) with (8.48) shows that, as $N \to \infty$,

$$\mu_i = T \mathcal{G}_{eq}\left(\frac{i}{NT}\right), \qquad i = 0, 1, \ldots, N-1; \tag{8.49}$$

that is, the eigenvalues of $\mathbf{X}$ are the values that the power spectrum of the bandlimited signal $x_{eq}(t)$ takes on at equally spaced frequencies i/NT. Let us now derive an expression for $\mathcal{G}_{eq}(f)$ depending on channel parameters. Using (7.6) and standard spectral computations techniques, we obtain

$$\mu_i = \frac{\sigma_a^2}{T^2} \left| Q_{eq}\left(\frac{i}{NT}\right) \right|^2 + \frac{1}{T} G_n(f), \tag{8.50}$$

where $Q_{eq}(f)$ is defined as in (7.17) and $G_n(f)$ is the noise power spectral density at the receiver filter output. For simplicity, we assume that the noise is bandlimited in $(-1/2T, 1/2T)$.

We are now ready to compute the performance of the infinite-length equalizer. Using (8.40) and (8.43), we get for the optimum tap-weight vector

$$\mathbf{c}_{opt} = \mathbf{U} \, \boldsymbol{\Lambda}^{-1} \mathbf{U}^\dagger \mathbf{v}. \tag{8.51}$$

Due to (8.44) and (8.45), premultiplication of a vector by $\mathbf{U}^\dagger$ gives the Fourier transform of its components for $f = i/N$, $i = 0, 1, \ldots, N-1$. If we denote by $C_{opt}(f)$ and $V(f)$ the Fourier transforms of the components of $\mathbf{c}_{opt}$ and $\mathbf{v}$, respectively, in the form

$$C_{opt}(f) \triangleq \sum_{i=0}^{N-1} [\mathbf{c}_{opt}]_i \, e^{-j2\pi f i T}, \tag{8.52a}$$

$$V(f) \triangleq \sum_{i=0}^{N-1} [\mathbf{v}]_i \, e^{-j2\pi f i T}, \tag{8.52b}$$

we derive from (8.51)

$$C_{opt}\left(\frac{i}{NT}\right) = \frac{1}{\mu_i} V\left(\frac{i}{NT}\right), \qquad i = 0, 1, \ldots, N-1. \tag{8.53}$$

Since in (8.50) the μ_i have already been expressed in terms of the channel parameters, we need an analogous expression for $V(\cdot)$. If we write x_l explicitly as in (8.1), from definition (8.36) we obtain

$$\mathbf{v} = \sigma_a^2 \begin{bmatrix} q_D^* \\ q_{D-1}^* \\ \cdot \\ \cdot \\ \cdot \\ q_{D-N+1}^* \end{bmatrix}, \tag{8.54}$$

where the q_l are the samples of the impulse response of the overall channel preceding the sampler. Then

$$V(f) = \sigma_a^2 \sum_{i=0}^{N-1} q_{D-i}^* e^{-j2\pi f i T}$$

$$= \sigma_a^2 \sum_{i=0}^{N-1} \int_{-\infty}^{\infty} Q^*(f') e^{-j2\pi f' DT} e^{-j2\pi(f-f')iT} df'$$

and, as $N \to \infty$, using the equality (2.115),

$$V(f) = \frac{\sigma_a^2}{T} \sum_{m=-\infty}^{\infty} \int_{-\infty}^{\infty} Q^*(f') e^{-j2\pi f' DT} \delta\left(f - f' - \frac{m}{T}\right) df'$$

$$= \frac{\sigma_a^2}{T} \sum_{m=-\infty}^{\infty} Q^*\left(f - \frac{m}{T}\right) e^{-j2\pi f DT} \tag{8.55}$$

$$= \frac{\sigma_a^2}{T} Q_{\text{eq}}^*(f) e^{-j2\pi f DT}.$$

By combining (8.49), (8.53), and (8.55), we get finally

$$C_{\text{opt}}\left(\frac{i}{NT}\right) = \frac{\sigma_a^2 Q_{\text{eq}}^*(i/NT) e^{-j2\pi i D/N}}{G_n(i/NT) + (\sigma_a^2/T)|Q_{\text{eq}}(i/NT)|^2}, \qquad i = 0, 1, \ldots, N-1. \tag{8.56}$$

As $N \to \infty$, we can assume that the transfer function of the optimum infinitely long equalizer is given by

$$C_{\text{opt}}(f) = \frac{\sigma_a^2 Q_{\text{eq}}^*(f)}{G_n(f) + (\sigma_a^2/T)|Q_{\text{eq}}(f)|^2} e^{-j2\pi f DT}, \qquad 0 \leqslant f < \frac{1}{T}. \tag{8.57}$$

By comparing this result with (7.44), it can be seen that, under the minimum MSE criterion, $C_{\text{opt}}(f)$ is indeed the optimum receiving filter, provided that $Q_{\text{eq}}(f) = Q(f)$; that is the channel is bandlimited in the interval $(-1/2T, 1/2T)$. If the latter condition is not fulfilled, the TDL equalizer fails to be the optimum filter. We shall return to this point later when discussing the fractionally spaced equalizer.

Next we evaluate the infinite-length equalizer performance. The quadratic form appearing in (8.41) is computed using (8.43):

$$\mathbf{v}^\dagger \mathbf{X}^{-1} \mathbf{v} = \mathbf{v}^\dagger \mathbf{U} \boldsymbol{\Lambda}^{-1} \mathbf{U}^\dagger \mathbf{v}$$

$$= \frac{1}{N} \sum_{i=0}^{N-1} \frac{|V(i/NT)|^2}{\mu_i} \tag{8.58}$$

$$= \frac{1}{N} \sum_{i=0}^{N-1} \frac{\sigma_a^4 |Q_{\text{eq}}(i/NT)|^2}{TG_n(i/NT) + \sigma_a^2 |Q_{\text{eq}}(i/NT)|^2}.$$

By taking the limit of (8.58), which can be done by using the Toeplitz distribution theorem (B.36), we have, as $N \to \infty$,

$$\mathbf{v}^\dagger \mathbf{X}^{-1} \mathbf{v} = T \int_{-1/2T}^{1/2T} \frac{\sigma_a^4 |Q_{\text{eq}}(f)|^2}{TG_n(f) + \sigma_a^2 |Q_{\text{eq}}(f)|^2} df, \tag{8.59}$$

and finally

$$\mathscr{E}_{\min} = T \int_{-1/2T}^{1/2T} \frac{\sigma_a^2 G_n(f)}{G_n(f) + (\sigma_a^2/T)|Q_{eq}(f)|^2} \, df. \tag{8.60}$$

This is the minimum MSE achievable using an infinitely long TDL equalizer. Equation (8.60) shows how it is related to both the noise power spectral density and the channel transfer function.

Several interesting conclusions can be drawn from (8.57) and (8.60), as follows:

(a) In the absence of noise, (8.57) reduces to

$$C_{opt}(f) = TQ_{eq}^{-1}(f)e^{-j2\pi f DT}, \tag{8.61}$$

which shows that the equalizer has a transfer function proportional to the inverse of $Q_{eq}(f)$. A delay of DT seconds is also introduced.

(b) From (8.60) we see that $\mathscr{E}_{\min}$ attains its maximum value σ_a^2 when $|Q_{eq}(f)| = 0$. Qualitatively, it can be observed that deep nulls in the function $|Q_{eq}(f)|$ will contribute to the increase of the integral value. This suggests that in the presence of noise the linear equalizer performance is limited by large depressions in the frequency response of the channel. This fact can also be understood by observing that the equalizer will try to compensate for a deep null by synthesizing a large gain at the corresponding frequencies. But this large gain will enhance the effect of the noise at the same frequency, thus preventing a perfect compensation or even leading to a serious performance degradation.

(c) Telephone channels do not usually exhibit deep spectral nulls. However, while $Q(f)$ does not, its "aliased" version $Q_{eq}(f)$ may. It happens that when the channel bandwidth exceeds $(-1/2T, 1/2T)$ the choice of the sampling instant, which does not affect $|Q(f)|$, does indeed affect $|Q_{eq}(f)|$ (this was shown in Section 2.5.1). Thus, for a channel whose frequency response extends beyond the Nyquist interval, inappropriate choice of the sampling instants can produce nulls in the equivalent channel response. Other channels (e.g., the multipath radio channel) exhibit nulls in their frequency response. Hence, a linear TDL equalizer may be inadequate to compensate for ISI.

(d) The minimum MSE of an infinitely long equalizer [see (8.60)] does not show dependence on the allowed delay D. However, the performance of a finite equalizer does depend on the choice of D. This delay is introduced because of the following consideration. Assume for simplicity that a single impulse is transmitted and that there is no noise on the channel. The TDL will contain, at each time instant, N samples of the impulse response of the channel. In particular, at a certain time the sample with the largest magnitude will appear in the TDL; NT seconds later, it will disappear. Between these two instants, at a specific time, the TDL will contain the samples that include most of the energy of the channel impulse response, and, in a sense, at this moment maximum information about the channel is available to the equalizer. Thus, it appears reasonable to assume as DT the difference between this time and the time at which the impulse has been sent. In practice, however, besides the difficulty of implementing an algorithm based on this principle, a criterion for choosing D as just described may not lead to an optimum. In fact, although for a symmetric impulse response the optimum D is given by $(L + N)/2$, where L is the channel memory span, the conjecture that this value should be decreased when the impulse response of the channel has a short precursor and a long tail (Niessen and Willim, 1970) has been proved untrue (Qureshi, 1973b). We shall return to this problem in Section 8.4, where a practical procedure for the selection of D will be described.

8.2.2 Gradient Algorithm for Equalization

We shall now describe a gradient algorithm for the automatic adjustment of the tap-weight values vector $\mathbf{c}$ to its optimum value. As the gradient of the MSE is given by (8.39), the gradient algorithm for equalization is described by the recursion

$$\mathbf{c}^{(n+1)} = \mathbf{c}^{(n)} - \alpha(\mathbf{X}\mathbf{c}^{(n)} - \mathbf{v}) = \mathbf{c}^{(n)} - \alpha E[(y_n - a_{n-D})\mathbf{x}_n^*], \qquad (8.62)$$

where $\mathbf{c}^{(n)}$ denotes the tap-weight vector at the nth iteration step, and α is a positive constant small enough to ensure convergence. The difficulty encountered in the implementation of (8.62) is that the average cannot be computed in real time [see the discussion following Eq. (8.10)]. Thus, as for automatic channel identification, we must resort to a stochastic gradient algorithm. However, before doing that it is expedient to analyze the convergence properties of the ''true gradient'' algorithm (8.62). By duplicating arguments used in connection with the convergence analysis of the channel identification algorithm (8.10), it can be proved that the tap-weight error at the nth iteration,

$$\boldsymbol{\varepsilon}^{(n)} \triangleq \mathbf{c}^{(n)} - \mathbf{c}_{\text{opt}} \qquad (8.63)$$

satisfies the recursion

$$\boldsymbol{\varepsilon}^{(n+1)} = (\mathbf{I} - \alpha\mathbf{X})\boldsymbol{\varepsilon}^{(n)}, \qquad n = 0, 1, \ldots , \qquad (8.64)$$

so that

$$\boldsymbol{\varepsilon}^{(n)} = (\mathbf{I} - \alpha\mathbf{X})^n \boldsymbol{\varepsilon}^{(0)}, \qquad (8.65)$$

and convergence of the tap-weight error is assured for any $\boldsymbol{\varepsilon}^{(0)}$ provided that

$$0 < \alpha < 2/\mu_{\text{max}}, \qquad (8.66)$$

where μ_{max} denotes the largest eigenvalue of the matrix $\mathbf{X}$.

Define now the excess MSE at the nth iteration step, that is, when $\mathbf{c} = \mathbf{c}^{(n)}$. We have, using (8.42),

$$\Delta^{(n)} \triangleq \mathscr{E}(\mathbf{c}^{(n)}) - \mathscr{E}_{\text{min}} = \boldsymbol{\varepsilon}^{(n)\dagger} \mathbf{X} \boldsymbol{\varepsilon}^{(n)}. \qquad (8.67)$$

Substitution of (8.65) in (8.67) yields

$$\Delta^{(n)} = \boldsymbol{\varepsilon}^{(0)\dagger}(\mathbf{I} - \alpha\mathbf{X})^n \mathbf{X}(\mathbf{I} - \alpha\mathbf{X})^n \boldsymbol{\varepsilon}^{(0)}, \qquad (8.68)$$

or, observing that $\mathbf{X}$ commutes with $\mathbf{I} - \alpha\mathbf{X}$, and hence with any of its powers,

$$\Delta^{(n)} = \boldsymbol{\varepsilon}^{(0)\dagger}(\mathbf{I} - \alpha\mathbf{X})^{2n} \mathbf{X} \boldsymbol{\varepsilon}^{(0)}. \qquad (8.69)$$

Our next step in the analysis of the convergence of $\Delta^{(n)}$ is based on the diagonal decomposition of $\mathbf{X}$, in the form [see (B.45)]

$$\mathbf{X} = \mathbf{U}\,\mathbf{\Lambda}\,\mathbf{U}^{-1}. \tag{8.70}$$

This yields

$$(\mathbf{I} - \alpha\mathbf{X})^{2n}\mathbf{X} = \mathbf{U}(\mathbf{I} - \alpha\mathbf{\Lambda})^{2n}\,\mathbf{\Lambda}\,\mathbf{U}^{-1}, \tag{8.71}$$

so that we get finally

$$\Delta^{(n)} = \sum_{i=0}^{N-1} \beta_i^{(0)}\mu_i(1 - \alpha\mu_i)^{2n}, \tag{8.72}$$

where $\beta_i^{(0)}$, $i = 0, \ldots, N - 1$, is the square magnitude of the ith element of the vector $\mathbf{U}^\dagger\boldsymbol{\varepsilon}^{(0)}$. From (8.72) it is seen that $\Delta^{(n)}$ can be decomposed as the sum of N exponentials, all of them decaying to zero if (8.66) holds. For fast convergence, the step size α can be chosen so as to minimize the quantity

$$r(\alpha) \triangleq \max_i |1 - \alpha\mu_i|. \tag{8.73}$$

This choice will speed up the convergence of the slowest decaying exponential term in (8.72). This "optimum" α satisfies the condition

$$1 - \alpha\mu_{min} = -(1 - \alpha\mu_{max}) \tag{8.74}$$

(see Fig. 8.6), where μ_{min} is the smallest eigenvalue of $\mathbf{X}$. Thus

$$\alpha_{opt} = \frac{2}{\mu_{max} + \mu_{min}}, \tag{8.75}$$

and

$$r(\alpha_{opt}) = \frac{\mu_{max}/\mu_{min} - 1}{\mu_{max}/\mu_{min} + 1}. \tag{8.76}$$

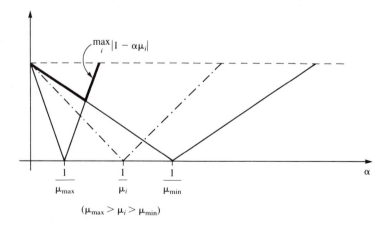

$(\mu_{max} > \mu_i > \mu_{min})$

Figure 8.6 Choice of the "optimum" value of α.

Sec. 8.2 Tapped Delay Line Equalizer

375

Thus, we have proved that the maximum convergence speed is, in a sense, dominated by the eigenvalue spread μ_{max}/μ_{min}. In fact, the smaller this value is, the faster the convergence of the true-gradient algorithm that can be achieved by a suitable choice of the step size α.

8.2.3 Stochastic-Gradient Algorithm for Equalization

We shall now consider the stochastic-gradient version of algorithm (8.62):

$$\mathbf{c}^{(n+1)} = \mathbf{c}^{(n)} - \alpha(y_n - a_{n-D})\mathbf{x}_n^*. \tag{8.77}$$

Figure 8.7 shows how this algorithm can actually be implemented. We shall assume, for the moment, that the source symbol sequence (a_n) is known at the receiver and disregard the dashed box (we shall consider its operation later). Each iteration is performed every T seconds. When a new sample x_n enters the TDL, a value y_n is computed by combining linearly the N samples contained in the TDL according to (8.32). After subtraction of a_{n-D}, the stochastic gradient is formed by multiplying this "error signal" by the samples x_{n-i}, $i = 0, 1, \ldots, N - 1$. The values obtained, after rescaling by a factor $-\alpha$, are added to the values of the tap weights stored in their accumulators so as to provide their updated version.

To analyze the convergence properties of the stochastic-gradient algorithm for equalization, we must resort, as for channel identification, to the simplifications allowed by the independence assumption. In our situation this consists in assuming that $(\mathbf{x}_n)$ is a sequence of iid zero-mean vectors. Again, this simplification does not make much sense mathematically, but offers results validated experimentally.

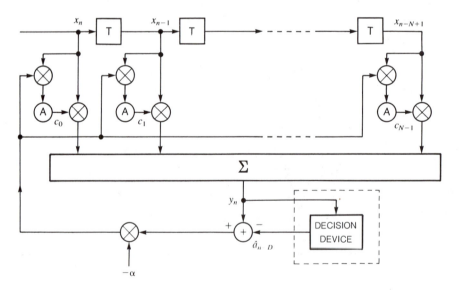

Figure 8.7 Implementation of the stochastic-gradient algorithm for channel equalization (real signals are assumed).

Consider first the tap-weight vector $\mathbf{c}^{(n)}$. From (8.77), it depends on $\mathbf{x}_0, \mathbf{x}_1, \ldots,$ $\mathbf{x}_{n-1}$. Given the independence assumption, $\mathbf{c}^{(n)}$ is independent of $\mathbf{x}_n$; thus, averaging both sides of (8.77) and using (8.63), we get

$$E[\boldsymbol{\varepsilon}^{(n+1)}] = (\mathbf{I} - \alpha \mathbf{X})E[\boldsymbol{\varepsilon}^{(n)}]. \tag{8.78}$$

This recursion shows that the average tap-weight error $E[\boldsymbol{\varepsilon}^{(n)}]$ converges to zero subject to condition (8.66). Thus, if (8.66) is satisfied, the stochastic-gradient algorithm is stable on the average (i.e., the average tap-weight vector will converge to its optimum value irrespective of the initial tap-weight setting).

Consider then the evolution of the mean-square error. If we define the excess MSE $\delta^{(n)}$ as

$$\delta^{(n)} \triangleq E|y_n - a_{n-D}|^2 - E|y_{\text{opt}}^{(n)} - a_{n-D}|^2, \tag{8.79}$$

where

$$y_{\text{opt}}^{(n)} \triangleq \mathbf{c}'_{\text{opt}}\mathbf{x}_n, \tag{8.80}$$

we have, after some manipulations,

$$\delta^{(n)} = E|y_n - y_{\text{opt}}^{(n)}|^2 + 2\,\mathcal{R}E[(\mathbf{c}^{(n)} - \mathbf{c}_{\text{opt}})^\dagger \mathbf{x}_n^*(y_{\text{opt}}^{(n)} - a_{n-D})]. \tag{8.81}$$

The second term in the RHS of (8.81) is zero. To show this, use (8.77) and the independence assumption to verify that $\mathbf{c}^{(n)}$ depends on $\mathbf{x}_m$ and a_{m-D} only for $m < n$, and hence that

$$E[(\mathbf{c}^{(n)} - \mathbf{c}_{\text{opt}})^\dagger \mathbf{x}_n^*(y_{\text{opt}}^{(n)} - a_{n-D})] = E[(\mathbf{c}^{(n)} - \mathbf{c}_{\text{opt}})^\dagger]E[\mathbf{x}_n^*(y_{\text{opt}}^{(n)} - a_{n-D})]. \tag{8.82}$$

Moreover,

$$\begin{aligned} E[\mathbf{x}_n^*(y_{\text{opt}}^{(n)} - a_{n-D})] &= E[\mathbf{x}_n^* \mathbf{x}_n']\mathbf{c}_{\text{opt}} - \mathbf{v} \\ &= \mathbf{X}\,\mathbf{c}_{\text{opt}} - \mathbf{v} = \mathbf{0}. \end{aligned} \tag{8.83}$$

In conclusion, we have

$$\begin{aligned} \delta^{(n)} &= E|y_n - y_{\text{opt}}^{(n)}|^2 \\ &= E[\boldsymbol{\varepsilon}^{(n)\dagger}\mathbf{x}_n^* \mathbf{x}_n'\,\boldsymbol{\varepsilon}^{(n)}] \\ &= E[\boldsymbol{\varepsilon}^{(n)\dagger}\mathbf{X}\,\boldsymbol{\varepsilon}^{(n)}], \end{aligned} \tag{8.84}$$

where the independence assumption has again been invoked in the last equality, and $\boldsymbol{\varepsilon}^{(n)}$ is defined as in (8.63). It is worthwhile emphasizing that (8.84) is, in general, invalid if both $(\mathbf{x}_n)$ and (a_n) are not independent sequences.

To proceed further, we first rewrite the stochastic-gradient recursion (8.77) in the form

$$\boldsymbol{\varepsilon}^{(n+1)} = \mathbf{A}_n\boldsymbol{\varepsilon}^{(n)} + \mathbf{b}_n, \tag{8.85}$$

where

$$\mathbf{A}_n \triangleq \mathbf{I} - \alpha\mathbf{x}_n^* \mathbf{x}_n' \tag{8.86}$$

and

$$\mathbf{b}_n \triangleq -\alpha(y_{\text{opt}} - a_{n-D})\mathbf{x}_n^*. \tag{8.87}$$

Thus, from (8.84) we get

$$\begin{aligned}
\delta^{(n+1)} = &\, \mathrm{E}\{\boldsymbol{\varepsilon}^{(n)\dagger}\mathrm{E}[\mathbf{A}_n\,\mathbf{X}\,\mathbf{A}_n]\boldsymbol{\varepsilon}^{(n)}\} \\
&+ 2\mathrm{E}[\boldsymbol{\varepsilon}^{(n)\dagger}]\mathrm{E}[\mathbf{A}_n\,\mathbf{X}\,\mathbf{b}_n] \\
&+ \mathrm{E}[\mathbf{b}_n^{\dagger}\,\mathbf{X}\,\mathbf{b}_n].
\end{aligned} \tag{8.88}$$

To simplify the analysis, we shall limit ourselves to the case in which the elements of $\mathbf{x}_n$ are independent, and $y_{\text{opt}} - a_{n-D}$ is also independent of $\mathbf{x}_n$. Thus, recalling that $\mathrm{E}[a_n] = 0$ (which implies $\mathrm{E}[\mathbf{x}_n] = \mathbf{0}$), we have $\mathbf{X} = \sigma_x^2\,\mathbf{I}$, where $\sigma_x^2 \triangleq \mathrm{E}\{|x_n|^2\}$, and the middle term in (8.88) vanishes. In fact, we have

$$\mathrm{E}[\mathbf{A}_n\,\mathbf{X}\,\mathbf{b}_n] = \mathbf{0}. \tag{8.89}$$

Furthermore, the last term in (8.88) reduces to

$$\mathrm{E}[\mathbf{b}_n^{\dagger}\,\mathbf{X}\,\mathbf{b}_n] = \alpha^2\,\mathrm{E}|y_{\text{opt}}^{(n)} - a_{n-D}|^2\,\mathrm{E}[\mathbf{x}_n'\,\mathbf{X}\,\mathbf{x}_n^*] = \alpha^2 N \mathscr{E}_{\min}\sigma_x^4. \tag{8.90}$$

Finally, the matrix in the first term of the RHS of (8.88) reduces to

$$\mathrm{E}[\mathbf{A}_n\,\mathbf{X}\,\mathbf{A}_n] = [(1 - \alpha\sigma_x^2)^2 + \alpha^2\rho]\sigma_x^2\,\mathbf{I}, \tag{8.91}$$

where

$$\rho\mathbf{I} \triangleq \mathrm{E}[\mathbf{x}_n^*\,\mathbf{x}_n']^2 - \sigma_x^4\mathbf{I} = [\mathrm{E}|x_n|^4 + (N - 2)\sigma_x^4]\,\mathbf{I}. \tag{8.92}$$

From the substitution of (8.89) to (8.92) into (8.88) and the observation that under our assumptions $\mathrm{E}[\boldsymbol{\varepsilon}^{(n)\dagger}\boldsymbol{\varepsilon}^{(n)}] = \sigma_x^{-2}\,\delta^{(n)}$, we finally obtain

$$\delta^{(n+1)} = \gamma\delta_n + \alpha^2 N \mathscr{E}_{\min}\sigma_x^4, \tag{8.93}$$

where

$$\gamma = (1 - \alpha\sigma_x^2)^2 + \alpha^2[\mathrm{E}|x_n|^4 + (N - 2)\sigma_x^4]. \tag{8.94}$$

Thus, $\gamma < 1$ turns out to be a necessary and sufficient condition for the convergence of the excess MSE $\delta_n^{(n)}$. We can see, for example, that the convergence is adversely affected by the number N of tap weights in the equalizer, as well as by the fourth absolute moment of the received samples $\mathrm{E}\{|x_n|^4\}$. Also, if $\gamma < 1$ and $n \to \infty$, the excess MSE tends to the residual value

$$\delta^{(\infty)} \triangleq \frac{\alpha^2 N \mathscr{E}_{\min}\sigma_x^4}{1 - \gamma}. \tag{8.95}$$

Notice in particular from (8.95) that the excess mean-square error does not tend to zero, while it does in the true-gradient algorithm, but to a value approximately proportional to α^2, at least for small γ values. This shows, for example, that the choice of the step size α in the stochastic-gradient algorithm entails a trade-off between fast convergence and small residual MSE.

We conclude this section with the observation that by using an iterative algorithm the equalizer can work adaptively by tracking and compensating for channel changes, provided that they are sufficiently slow with respect to the settling time of the equalizer.

8.3 FRACTIONALLY SPACED EQUALIZER

We have assumed so far that the signal $x(t)$ received at the channel's output is first filtered and then sampled every T seconds before being sent to the TDL with adjustable weights and elementary delays T. As the decision device operates on samples taken at the equalizer output every T seconds, the choice to process the received signal sampled at rate $1/T$ appears quite natural. However, it is far from being innocuous. We have seen (Section 2.5) that the process of sampling a signal at rate $1/T$ superimposes its spectral components spaced $1/T$ hertz apart (the "aliasing" effect). As seen at the end of Section 8.2.1, a result is that the quality of equalization can be adversely affected by the appearance of deep "nulls" in the equivalent channel transfer function.

Another way of seeing this is by observing that the transfer function of the T-spaced equalizer is periodic with period $1/T$ hertz; thus, spectral components of the incoming signal lying at frequencies spaced $1/T$ hertz apart cannot be processed independently by adjusting the tap weights. Moreover, this periodicity does not allow the noise-frequency components lying outside the interval $(-1/2T, 1/2T)$ to be suppressed. This task is assigned to the receiving filter preceding the equalizer. Ideally, this filter should approach the performance of a matched filter (indeed, it was found that the combination of a matched filter with a TDL is optimum; see Sections 7.3 and 7.4).

Instead, assume that $x(t)$ is sampled every $T' < T$ seconds, and consequently the TDL elementary delay is T'. In particular, if α denotes the excess bandwidth, that is if the received signal $x(t)$ is confined to the frequency interval $[-(1 + \alpha)/2T, (1 + \alpha)/2T)$, we can choose $T' \leq (1 + \alpha)^{-1}T$. With this choice, the equalizer transfer function becomes sufficiently large to accommodate for the whole signal spectrum. Moreover, $Q_{eq}(f) = Q(f)$; hence the sampling instant becomes irrelevant and the appearance of deep nulls because of a badly chosen sampling instant is avoided. Finally, from (8.57) we see that the equalizer provides the optimum (MSE) receiving filter, thus avoiding the need for a separate filter to suppress the noise.

It must be kept in mind that the signal at the output of the equalizer is still sampled at a rate of $1/T$. But, since its input is sampled at rate $1/T'$, the equalizer acts on the received signal before aliasing its frequency components. An equalizer based on this principle is called a *fractionally spaced equalizer*.

A convergence analysis, similar to that of Section 8.2, can be carried out for fractionally spaced equalizers. Simulation of equalizers with $T' = T/2$, QAM modulation and data transmission over typical voice-grade circuits seems to confirm the improvement predicted by the theory over nonfractionally spaced equalization. In particular, (1) the $T/2$ equalizer performs almost as well or even better than a T equalizer with the same number of TDL segments (i.e., half the time span), (2) a receiving filter preceding the equalizer [$U(f)$ in Fig. 7.2] is not required with a $T/2$ equalizer, and (3) for channels with severe band-edge delay distortion, the $T/2$ equalizer outperforms the T equalizer regardless of the choice of the sampling instant (Qureshi, 1982).

8.4 TRAINING THE EQUALIZER

In Section 8.2, our analysis of the TDL equalizer performance and convergence properties assumed that the data sequence (a_n) needed to evaluate the error gradient was known at the receiver's front end. A widely used method to render this assumption realistic in practice is now described. In an initial (training or start-up) period, a particular data sequence, known and available in a proper time alignment at the receiver, is transmitted. Once the equalized channel quality has become so good that decisions on transmitted symbols can be made with small enough error probability, the gradient is computed by replacing the estimated data symbol sequence $(\hat{a}_n)$ for the transmitted one (a_n) (see the dashed box in Fig. 8.7). Simulations and experimental studies indicate that for reasonable error rates this replacement does not alter the convergence of the equalizer.

In some cases, the error probability before equalization is so small that the training period can be avoided. The equalizer is then said to work in a "bootstrap" mode, this name being derived from the old saying about pulling oneself up by one's own bootstraps. However, in most situations the equalizer must be trained before it can be switched to a decision-directed mode of operation.

Cyclic equalization

Concerning the selection of the training sequence, a good choice is a periodic sequence whose period equals N, the number of TDL taps. This choice, which gives rise to *cyclic equalization*, enables us to solve a problem arising in the start-up procedure and concerning the best choice of the delay D to use in the definition (8.35) of the mean-square error to be minimized. In fact, when aligning in time the training sequence generated locally with that sent by the source, D should be chosen in order to best compensate for the delay introduced by the channel. In principle, we can make this choice by a trial-and-error technique, computing the minimum MSE for a number of possible D values and choosing the least error value. But obviously this is impractical. A technique for adaptively adjusting D to minimize the MSE has been proposed by Qureshi (1973b). However, its convergence cannot be guaranteed. Cyclic equalization provides a rule to choose D such that the minimum MSE may not be achieved, but a relatively simple implementation is obtained, coupled with adequate performance.

Consider a training data sequence with period N. Assume for the moment that the channel is noiseless and distortionless. Thus, the received samples are just a delayed version of the transmitted symbol sequence. We also assume that $N > D$. After convergence of the equalizer, only one of the tap weights will have a nonzero value. The corresponding tap position informs us about the time shift between the received sequence and the one generated locally; in particular, any unit time shift in the sequence generated locally will cause the unique nonzero-weight tap to move by one position in the TDL. Let us now return to a channel affected by linear distortion, but without noise. The received sampled sequence is once again periodic with period N. One full period is stored in the TDL. After convergence, the start-up procedure will finish with a set of tap-weight values that needs to be only *cyclically* shifted for proper time alignment. As any cyclic shift between the received sample sequence and the data sequence generated locally causes a cyclic shift of the set of tap-weight values, it is not necessary to achieve time alignment before start-up. This can be done *after* start-up by cycling the tap-weight values such that the largest absolute value is found in a reference position (e.g., the center tap).

An equalizer scheme based on this principle is shown in Fig. 8.8. The periodic sequence generator outputs the training word. All tap weights are preset to identical values to reflect that the location of the largest weight is not known a priori. To begin the start-up procedure, the switch at the bottom is set to position ① . After training, the tap weight values are aligned by cyclic shifts, as just indicated. The equalizer can now be operated in a decision-directed mode by moving the switch to position ② .

It remains now to analyze the effects that can be expected from the use of a periodic sequence like that employed in the start-up phase. As without noise, the sequence (x_n) of received samples is periodic. It has a periodic correlation, which makes the matrix $\mathbf{X}$ circulant. Thus, the analysis carried out in Section 8.2.1 based on decomposition (8.43) becomes exact. In particular, from (8.56) we have

$$C_{\text{opt}}\left(\frac{i}{NT}\right) = TQ_{\text{eq}}^{-1}\left(\frac{i}{NT}\right) e^{-j2\pi iD/N}, \qquad i = 0, 1, \ldots, N - 1. \qquad (8.96)$$

This shows that equalization of the channel is achieved at a set of N equally spaced frequencies in a frequency interval of width $1/T$ hertz. In other words, the inverse channel response is approximated through interpolation at equally spaced points. Thus, equalization after start-up, although nonoptimum in the MSE sense for random transmitted data, can be expected to be reasonably close to the optimum when N, the number of taps in the TDL, is sufficiently large.

Concerning the choice of the periodic symbol sequence to be employed in the

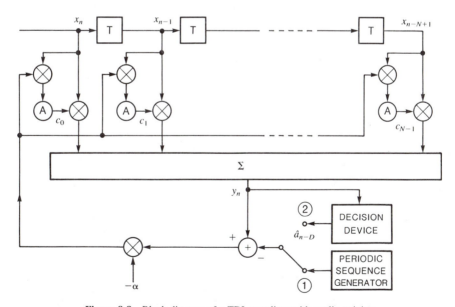

Figure 8.8 Block diagram of a TDL equalizer with cyclic training.

start-up phase, it has been proved (see Godard, 1981) that the best sequences in the presence of noise are those whose periodic autocorrelation

$$R_k \triangleq \sum_{n=0}^{N-1} a_{(n+k)\bmod N}\, a_n^*$$

is exactly zero for $k \neq iN$, $i = 0, \pm 1, \pm 2, \ldots$. As sequences with this property can be generated for any period N using constant-amplitude (i.e., purely phase modulated) signals, they have been called *constant-amplitude, zero-autocorrelation* (CAZAC) sequences.

Example 8.2

For data transmission at 2400 bit/s in the duplex mode on telephone lines, CCITT recommends channel separation by frequency division (with carrier frequencies of 1200 and 2400 Hz), and transmission of 600 symbols per second with 16-QAM modulation. The combined response of the shaping filter and the receiving filter is a raised cosine with roll-off 0.75. In the training mode of the equalizer, a sequence with period 8 can be used, obtained by transmitting a subset of the 16 signals available from the modulator (those with the largest energy). A CAZAC sequence for this situation is obtained by transmitting sequentially the phases $\pi/4$, $7\pi/4$, $3\pi/4$, $3\pi/4$, $\pi/4$, $3\pi/4$, $3\pi/4$, $7\pi/4$. Figure 8.9 shows the convergence behavior of a stochastic-gradient equalizer with eight taps in the training mode for transmission over a channel modeled after a low-quality telephone line. The MSE is obtained by time averaging the instantaneous squared error $|y_n - a_{n-D}|^2$ over 10 symbol intervals. It is seen, in particular, how the choice of the iteration step α affects

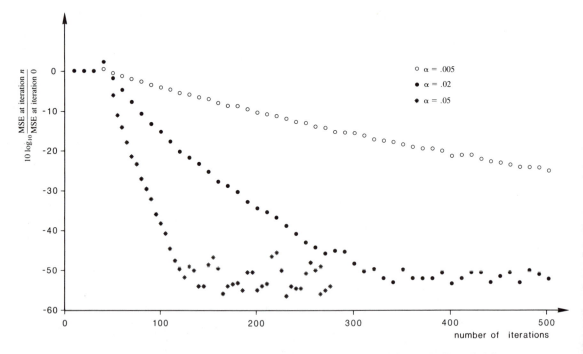

Figure 8.9 Convergence of an eight-tap equalizer in the training mode (Example 8.2).

Adaptive Receivers and Channel Equalization   Chap. 8

the convergence: a smaller step size provides a slower, but smoother convergence, while a large step size gives a faster convergence, but a larger variance of the MSE fluctuations. $\square$

8.5 VARIATIONS ON THE GRADIENT'S THEME

As briefly mentioned in previous sections, the basic gradient algorithm (8.62) and its stochastic version (8.77) are not the only algorithms that can be used to adjust the tap weights to their optimum values. Indeed, many variations have been proposed; several among them seem to have only a theoretical interest, while some have found practical applications and even been incorporated into commercial modems. These variations of the basic algorithms can be grouped, according to the goal that they are expected to achieve, into three basic classes. To be more specific, modified algorithms have been proposed:

1. To simplify the implementation of the iterative algorithm.
2. To speed up the convergence of the tap weights to their optimum values.
3. To improve the approximation of the steady-state tap weights to their ideal values.

Consider first class (1), the algorithms aimed at simplifying the equalizer implementation. The simplest algorithms that can be derived from (8.77) use only the sign information contained in the error signal $y_n - a_{n-D}$ and/or in the components of $\mathbf{x}_n$. Thus, we have the following algorithms:

$$\mathbf{c}^{(n+1)} = \mathbf{c}^{(n)} - \alpha \, \text{sgn}[y_n - a_{n-D}]\mathbf{x}_n^*, \qquad (8.97)$$

$$\mathbf{c}^{(n+1)} = \mathbf{c}^{(n)} - \alpha[y_n - a_{n-D}] \, \text{sgn}[\mathbf{x}_n^*], \qquad (8.98)$$

$$\mathbf{c}^{(n+1)} = \mathbf{c}^{(n)} - \alpha \, \text{sgn}[y_n - a_{n-D}] \, \text{sgn}[\mathbf{x}_n^*], \qquad (8.99)$$

where $\text{sgn}[\mathbf{x}_n^*]$ is the vector whose components are the signs of the components of $\mathbf{x}_n^*$, and the sign of a complex quantity $a + jb$ is defined as

$$\text{sgn}[a + jb] = \text{sgn}[a] + j \, \text{sgn}[b]. \qquad (8.100)$$

The algorithms (8.97) to (8.99) may be viewed as extreme cases of quantization of the received sample values and of the error values to a single bit. The price one should expect to pay in exchange for increased simplicity is the decreased accuracy of the steady-state tap-weight setting and slower convergence (see Duttweiler, 1982).

Consider next the class 2 algorithms. A relatively easy, yet conceptually satisfying, way of modifying the gradient algorithm is to employ a variable step size in (8.77). That is, change α to $\alpha^{(n)}$ and derive a suitable sequence $(\alpha^{(n)})_{n=0}^{\infty}$ in order to improve the convergence speed. The step-size influence on the convergence of the true-gradient algorithm can be analyzed by inspection of (8.72). We have already observed that fast convergence can be achieved by choosing $\alpha = \alpha_{\text{opt}}$ according to (8.75). This value of the step size maximizes the convergence rate of the slowest among the N decaying exponentials in (8.72). However, this slowest term has the smallest multiplicative factor, μ_{min}. Thus, in initial iterations when the slowest exponential is not yet dominant, a value of α larger than α_{opt} will cause $\Delta^{(n)}$ to decrease faster. All this suggests a "gear-shifting'' adaptation strategy based on a progressive reduction of the step size [notice also that a small final value for α would result in a small residual value of the excess MSE in the stochastic gradient algorithm; see (8.95)]. Simulation results (Ungerboeck, 1972) show that the choice $\alpha = 1/\text{tr}[\mathbf{X}]$ during the training mode, with a reduction to $\alpha = 1/5\ \text{tr}[\mathbf{X}]$ when $\Delta^{(n)}$ is small enough for the equalizer to be switched to the decision-directed mode, gives excellent convergence properties.

For a more refined solution to the optimum step-size sequence problem in the true-gradient algorithm, use (8.64) and (8.67) to get

$$\Delta^{(n+1)} = \boldsymbol{\varepsilon}^{(n)\dagger}(\mathbf{I} - \alpha\mathbf{X})^2 \mathbf{X}\ \boldsymbol{\varepsilon}^{(n)}. \tag{8.101}$$

Take then the derivative of $\Delta^{(n+1)}$ with respect to α and set it equal to zero so as to obtain

$$\alpha = \alpha^{(n)} = \frac{\boldsymbol{\varepsilon}^{(n)\dagger}\mathbf{X}^2\boldsymbol{\varepsilon}^{(n)}}{\boldsymbol{\varepsilon}^{(n)\dagger}\mathbf{X}^3\boldsymbol{\varepsilon}^{(n)}}. \tag{8.102}$$

Since $\boldsymbol{\varepsilon}^{(n)}$ depends on $\boldsymbol{\alpha}^{(n)}$, this is an implicit equation that does not seem to be amenable to a simple solution. However, simple explicit bounds on $\alpha^{(n)}$ can be obtained (Gardner, 1984; see also Problem 8.4):

$$\frac{1}{\mu_{\max}} \le \alpha^{(n)} \le \frac{1}{\mu_{\min}}, \tag{8.103}$$

where $\mu_{\max}$ and $\mu_{\min}$ are, respectively, the largest and the smallest eigenvalue of $\mathbf{X}$. Schonfeld and Schwartz (1971a) consider the true-gradient algorithm and a step-size sequence designed so as to minimize the tap-weight error $\boldsymbol{\varepsilon}^{(n)\dagger}\boldsymbol{\varepsilon}^{(n)}$ after a specified number of iterations, while disregarding its values for intermediate iterations. More complex algorithms, based on higher-order gradient methods, were studied by Devieux and Pickholtz (1969) and Schonfeld and Schwartz (1971b). The iteration scheme is

$$\mathbf{c}^{(n+1)} = \mathbf{c}^{(n)} - \alpha_0^{(n)}\mathbf{g}^{(n)} - \alpha_1^{(n)}\ \mathbf{g}^{(n-1)} - \cdots - \alpha_K^{(n)}\mathbf{g}^{(n-K)},$$

where $\mathbf{g}^{(l)}$ is the gradient vector (or an estimate thereof) evaluated at the lth iteration.

Other techniques aimed at improving the convergence speed reduce the eigenvalue spread $\mu_{\max}/\mu_{\min}$ by finding a sequence of matrices $(\mathbf{D}^{(n)})$, converging to $\mathbf{X}^{-1}$, that are used instead of the scalar α to premultiply the stochastic gradient. The adjustment algorithm is then

$$\mathbf{c}^{(n+1)} = \mathbf{c}^{(n)} - \mathbf{D}^{(n)}(y_n - a_{n-D})\mathbf{x}_n^*.$$

Several techniques are available for the computation of $\mathbf{D}^{(n)}$ (see Gitlin and Magee, 1977), but the speed improvement achieved does not seem to compensate for the increase in complexity.

Consider finally class 3. Inspection of (8.95) shows that the excess MSE converges, as the number of iterations grows to infinity, to a value that is roughly proportional to α^2. This shows that α must be as small as possible to improve the accuracy of the tap-weight adjustment. But a small value of α constrasts with the need for a large α to achieve a fast convergence. Again, a gear-shifting strategy, leading to a decreasing sequence $(\alpha^{(n)})$ of step sizes, seems convenient. A special case of such a sequence is amenable to exact analysis by using stochastic approximation theory. Any sequence of step sizes such that

$$\sum_{n=0}^{\infty} \alpha^{(n)} = \infty, \qquad \sum_{n=0}^{\infty} [\alpha^{(n)}]^2 < \infty,$$

leads to a stochastic-gradient algorithm in which $c^{(n)} \to c_{opt}$ with probability 1 as $n \to \infty$ (Gersho, 1969a).

Another approach stems from the observation that the constant term in (8.93), which prevents the excess MSE from converging to a zero value as $n \to \infty$, is due to approximation of the true gradient with a crude estimate, as obtained by removing the expectation operator from (8.62). Thus, it can be reasonably expected that any improvement in the gradient's estimate will reduce the steady-state value of $\delta^{(n)}$. This is actually the case with the algorithm

$$c^{(n+1)} = c^{(n)} - \alpha \frac{1}{K} \sum_{l=0}^{K-1} (y_{nK+l} - a_{nK+l-D}) \, x_{nK+l}^*, \qquad (8.104)$$

which is obtained from (8.77) by replacing the expectation with a time average performed over K symbols. This includes the stochastic-gradient algorithm as a special case in which $K = 1$. As averaging takes K symbol intervals, successive adjustments of tap weights are K times less frequent than with a stochastic gradient. Hence, convergence is slowed down. However, as shown by Gardner (1984), the constant term in (8.93) is reduced by a factor of K, so by choosing a large enough K the steady-state excess MSE can be made as small as desired.

8.6 NON-MSE CRITERIA FOR EQUALIZATION

The previous sections were devoted to the analysis of equalizers based on a given structure, the TDL, and on a given optimization criterion, the minimum MSE. Although this combination has proved most fruitful for application purposes, it is by no means the only one, and considerable effort has been spent to devise and analyze different equalization criteria and/or structures. Hereafter, we shall review some of the more significant solutions provided in this framework. In particular, in this section we shall describe three non-MSE criteria for TDL equalization, and defer to later sections the description of other equalizer structures.

As mentioned before, minimization of the MSE $E\{|y_n - a_{n-D}|^2\}$ between equalized samples and transmitted symbols provides a simple and useful criterion for optimization of the TDL weights. However, it is apparent that a more meaningful criterion would be the minimization of the error probability $P\{\hat{a}_n \neq a_n\}$. Unfortunately, error probability is a highly complicated function of the tap weights. Thus, equalization algorithms based on its minimization do not appear practical. On the other hand, it can be expected that a MSE-optimized TDL equalizer is very close to providing minimum error probability.

8.6.1 Zero-Forcing Equalizer

The first approach to automatic equalization assumed a peak-distortion performance criterion. Peak distortion can be derived from the eye pattern of the received signal and is closely related to the worst-case bound to error probability (see Section 6.2).

Consider for simplicity that a binary (± 1) stream of independent symbols is transmitted. We define the normalized peak distortion as

$$\mathscr{D}(\mathbf{c}) \triangleq \frac{1}{h_0} \sum_k{}' |h_k|, \tag{8.105}$$

where (h_k) is the impulse response of the cascade of the discrete-time channel with a TDL with tap-weight vector $\mathbf{c}$. It is assumed that $h_0 = \max_k |h_k|$. In words, $\mathscr{D}(\mathbf{c}) \cdot h_0$ represents the maximum value of the ISI affecting the equalized signal. $\mathscr{D}(\mathbf{c}) = 0$ means that there is no ISI, whereas $\mathscr{D}(\mathbf{c}) \geq 1$ denotes that the eye pattern is completely closed (hence reliable transmission is impossible, irrespective of the noise power level).

If the tap-weight vector $\mathbf{c}$ is chosen to minimize $\mathscr{D}(\mathbf{c})$, it can be adjusted by using a gradient algorithm (Lucky, 1965, 1966). This gives the *zero-forcing* algorithm. The reason for its name lies in the fact that, if the normalized peak distortion is smaller than 1 (i.e., the unequalized eye pattern is open), $\mathscr{D}(\mathbf{c})$ is minimized by constraining $q_k = 0, k \neq 0$, for the samples within the TDL range. This algorithm is easily implemented. However, it has the fundamental drawback that convergence to a minimum of $\mathscr{D}(\mathbf{c})$ is not assured when the unequalized eye is not open. Moreover, the effects of noise are not included in the performance criterion. Nevertheless, due to their simplicity, zero-forcing equalizers were the first incorporated into commercial modems.

8.6.2 Least-Squares Algorithms

More recently, the expansion of data-transmission systems requiring quick setup and response has created the requirement for equalizers in which a short training time is a premium. This occurs, for example, in multipoint networks, where the tributary terminals may transmit only when polled by the control modem. The messages from the tributary to a control station are often short, and the control modem must adjust its equalizer whenever a message is received. A quickly convergent equalization algorithm has been sought either by modifying the basic gradient algorithm under a MSE criterion (we discussed this approach in the last section) or by devising other performance criteria. The latter approach can be pursued by introducing a *least-squares* (LS) criterion, that is, the sequence of cost functions

$$\mathscr{L}(\mathbf{c}^{(n)}) \triangleq \sum_{k=1}^{n} |\mathbf{c}^{(n)\prime}\mathbf{x}_k - a_{k-D}|^2, \qquad 1 \leq n < \infty, \tag{8.106}$$

to be minimized over the tap-weight vector $\mathbf{c}^{(n)}$. In words, a $\mathbf{c}^{(n)}$ is sought that minimizes the sum of the squared errors that would be obtained if $\mathbf{c}^{(n)}$ were used with all the past received signal samples. Algorithms matched to the cost function (8.106), called *LS algorithms*, have been proved to provide fast convergence as required.

By taking the gradient of (8.106) and setting it equal to the null vector, the following equation for the optimum tap-weight vector is obtained:

$$\mathbf{X}^{(n)} \mathbf{c}_{\text{opt}}^{(n)} = \mathbf{v}^{(n)}, \tag{8.107}$$

where

$$\mathbf{X}^{(n)} \triangleq \sum_{k=1}^{n} \mathbf{x}_k^* \mathbf{x}_k' \tag{8.108}$$

and

$$\mathbf{v}^{(n)} \triangleq \sum_{k=1}^{n} a_{k-D}\, \mathbf{x}_k^*. \qquad (8.109)$$

It should be observed that, apart from a constant factor $1/n$, $\mathbf{X}^{(n)}$ and $\mathbf{v}^{(n)}$ are the time-average counterparts of $\mathbf{X}$ and $\mathbf{v}$, as defined respectively in (8.37) and (8.36). Thus, $\mathbf{X}^{(n)}$ and $\mathbf{v}^{(n)}$ can be viewed as estimates of $\mathbf{X}, \mathbf{v}$.

Solution of (8.107) by matrix inversion can be complicated by the fact that $\mathbf{X}^{(n)}$, being only an estimate of $\mathbf{X}$, need not be positive definite. Hence, its inverse may not exist. This problem can be circumvented by simply adding to $\mathbf{X}^{(n)}$, as defined in (8.108), a scalar matrix $\delta\mathbf{I}$, where δ is a positive constant included to ensure that $\mathbf{X}^{(n)}$ is nonsingular for all n. Also, the cost function (8.106) can be slightly modified to include a desirable feature if the channel is time varying: by introducing a geometric attenuation factor $0 < \lambda < 1$, that is, by considering the new cost function

$$\mathscr{L}_\lambda(\mathbf{c}^{(n)}) \triangleq \sum_{k=1}^{n} \lambda^{n-k} |\mathbf{c}^{(n)\prime}\mathbf{x}_k - a_{k-D}|^2, \qquad (8.110)$$

the present influences the tap-weight update more than the past. Thus, slow channel variations with time can be tracked.

For the update of $\mathbf{c}^{(n)}$ several algorithms have been proposed. The *Kalman algorithm* (Godard, 1974) assures rapid start-up, but requires matrix operations; therefore, the number of calculations required is proportional to N^2, where, as usual, N is the number of taps in the TDL. A similar algorithm, usually referred to as the *fast Kalman algorithm* (Falconer and Ljung, 1978), improves the Kalman algorithm as it achieves a lower complexity (linear growth with N) without performance degradation because it is mathematically equivalent to the latter. These algorithms have been compared by simulation over several channels by Lim and Mueller (1980). Their convergence properties have been proved to be very similar. They require roughly one-third as many iterations as the stochastic-gradient algorithm. The price for this increase in speed is complexity: the fast Kalman algorithm, which has the lowest complexity, requires about 10 times as many multiplications as the stochastic gradient. Notice also that the fast Kalman algorithm may be unstable when implemented digitally with insufficient accuracy (Lim and Mueller, 1980).

8.6.3 Self-Recovering Equalization

The third non-MSE criterion considered in this section was devised to avoid the need for an initial training period for the equalizer. In fact, we shall see that, by a proper choice of the distortion function to be minimized, a tap-weight-setting algorithm can be found that does not depend on the transmitted data. This makes it possible to work in a bootstrap mode. These equalizers that do not require an initial training period are sometimes referred to as *blind equalizers*.

The following assumptions on the source symbols are made: the RV a_n are complex and iid. Moreover,

$$\mathrm{E}[a_n^2] = 0 \qquad (8.111)$$

Sec. 8.6   Non-MSE Criteria for Equalization   **387**

and

$$2m_2^2 > m_4, \tag{8.112}$$

where

$$m_2 \triangleq E\{|a_n|^2\} \tag{8.113}$$

and

$$m_4 \triangleq E\{|a_n|^4\}. \tag{8.114}$$

Conditions (8.111) and (8.112) imply certain symmetries in the signal constellation used by the digital modulator. For example, (8.111) excludes binary PSK and one-dimensional modulation schemes. The distortion criterion is then

$$Q(\mathbf{c}) \triangleq E\left[|y_n|^2 - \frac{m_4}{m_2}\right]^2. \tag{8.115}$$

If we denote by (h_n) the impulse response of the discrete channel extending from the source to the equalizer output, so that

$$y_n = \sum_{k=-\infty}^{\infty} a_{n-k} h_k, \tag{8.116}$$

we have, after some algebra,

$$Q(\mathbf{c}) = (m_4 - m_2^2) \sum_{k=-\infty}^{\infty} |h_k|^2 \tag{8.117}$$

$$+ 2m_2^2 \left[\sum_{k=-\infty}^{\infty} |h_k|^2\right]^2 - 2m_4 \sum_{k=-\infty}^{\infty} |h_k|^2 + \frac{m_4^2}{m_2^2}.$$

Now, if $\sum_k'$ denotes summation for k from $-\infty$ to $+\infty$ with deletion of the term with $k = 0$, (8.117) can be rewritten in the form

$$Q(\mathbf{c}) = m_4(1 - |h_0|^2)^2 + m_4 \sum_k' |h_k|^4$$

$$+ 2m_2^2 \left[\left(\sum_k' |h_k|^2\right)^2 - \sum_k' |h_k|^4\right] \tag{8.118}$$

$$+ [4m_2^2|h_0|^2 - 2m_4] \sum_k' |h_k|^2 + \frac{m_4^2}{m_2^2} - m_4.$$

We have derived (8.118) to make clear the choice of (8.115) for the distortion function. To show that this choice makes sense, we shall prove that it leads approximately to the same results of the distortion:

$$Q'(\mathbf{c}) = E[|y_n|^2 - |a_n|^2]^2, \tag{8.119}$$

which appears at a first glance to be a sensible one, but which does not solve our problem because it incorporates a_n. Expanding (8.119), we get

$$Q'(\mathbf{c}) = m_4(1 - |h_0|^2)^2 + m_4 \sum_k{}' |h_k|^4$$
$$+ 2m_2^2 \left[\left(\sum_k{}' |h_k|^2 \right)^2 - \sum_k{}' |h_k|^4 \right] \qquad (8.120)$$
$$+ [4m_2^2 |h_0|^2 - 2m_2^2] \sum_k{}' |h_k|^2.$$

Now (8.120) has a minimum when $|h_0|^2$ is close to unity and the ISI samples h_k, $k \neq 0$, are small in magnitude. Moreover, comparison of (8.118) and (8.120) shows that

$$Q(\mathbf{c}) - Q'(\mathbf{c}) = \frac{m_4}{m_2} - m_4 - 2[m_2^2 - m_4] \sum_k{}' |h_k|^2. \qquad (8.121)$$

The difference (8.121) is almost independent of (h_n), and hence of $\mathbf{c}$, when the distortion term is small enough. So it can be expected that minimizing $Q(\mathbf{c})$ will also provide a minimum for $Q'(\mathbf{c})$. The condition for this to be true is that in (8.118) the term $[4m_2^2|h_0|^2 - 2m_4]$ does not become negative near the minimum; but this is assured, because near the minimum $|h_0|^2 \cong 1$, and (8.112) is assumed.

A gradient equalization algorithm can now be exhibited. Since $y_n = \mathbf{c}'\mathbf{x}_n$ as usual, the gradient of $Q(\mathbf{c})$ taken with respect to the tap-weight vector $\mathbf{c}$ is

$$\nabla Q(\mathbf{c}) = 4\mathrm{E} \left[y_n \left(|y_n|^2 - \frac{m_4}{m_2} \right) \mathbf{x}_n^* \right], \qquad (8.122)$$

so that the following stochastic-gradient algorithm can be used:

$$\mathbf{c}^{(n+1)} = \mathbf{c}^{(n)} - \alpha y_n \left(|y_n|^2 - \frac{m_4}{m_2} \right) \mathbf{x}_n^*. \qquad (8.123)$$

Comparison of (8.123) with (8.77) shows that the stochastic-gradient term does not depend on the symbols a_n, as required. Also, it can be verified that when $\mathbf{c}'\mathbf{x}_n = a_n$ (i.e., perfect equalization is achieved) the gradient (8.122) is zero. Incidentally, this result would not hold if in lieu of m_4/m_2 another constant were selected in the brackets of (8.115) (see Godard, 1980, for further details). As a result, but in exchange for a lower convergence speed, (8.123) offers a significant advantage over (8.77). Specifically, a start-up period involving a known training sequence is not required.

8.7 NON-TDL EQUALIZER STRUCTURES

In this section we shall briefly deal with linear equalizer structures that are not transversal filters. In the next, an important nonlinear structure will be introduced and analyzed. It should be observed that no linear equalizer structure can outperform a TDL of an appropriate length. In fact, we have seen that the latter structure can be made to approach the optimum linear filter as its length grows to infinity. Also, making a non-TDL structure adaptive is generally more complicated. However, the study of these different structures may show better convergence properties or better performance with comparable complexity.

One of these alternative structures is obtained by using the *Kalman filter* as an equalizer (Lawrence and Kaufman, 1971; Benedetto and Biglieri, 1974). The Kalman filter, a version of the minimum-MSE linear receiver, has a recursive structure. Comprehensive simulation results (Benedetto and Biglieri, 1974) have shown that the performance of this linear filter is not significantly better than that of a TDL equalizer of comparable complexity.

Another equalizer structure, which seems far more promising than the latter and which has attracted considerable attention in the last few years, is based on *lattice filters*. A lattice filter is obtained by cascading a number of stages like the one shown in Fig. 8.10. The operation of the mth stage is described by the recursions

$$b_{n,1} = f_{n,1},$$
$$f_{n,m+1} = f_{n,m} - k'_m b_{n-1,m}, \qquad\qquad\qquad (8.124)$$
$$b_{n,m+1} = -k''_m f_{n,m} + b_{n-1,m}, \qquad m = 1, 2, \ldots, N,$$

where we have introduced the double-index sequences of the *forward* and *backward* error residuals $f_{n,m}$, $b_{n,m}$, and the *reflection*, or *partial correlation*, coefficients k'_m, k''_m. The index m refers to the stage of the lattice filter, the index n to discrete time. A fundamental property of the lattice filter is that if the quantities $E\{|f_{n,m}|^2\}$ and $E\{|b_{n,m}|^2\}$ are minimized for all m, which can be obtained by a proper choice of the coefficients k'_m, k''_m, then the backward coefficients $b_{n,m}$ are uncorrelated; that is, assuming them to have zero mean, $E[b_{n,l} b^*_{n,m}] = 0$ for $l \neq m$. Thus, if the sequence (x_n) of samples to be equalized is sent into a lattice filter whose reflection coefficients take on proper values, we derive from it a sequence $b_{n,1}, \ldots, b_{n,N}$ of samples that can be used, in lieu of the "plain" delayed samples $x_{n-1}, \ldots, x_{n-N+1}$, in a transversal equalizer (see Fig. 8.11). This situation is equivalent to having a TDL equalizer in which the correlation matrix $\mathbf{X}$ is *diagonal*, and hence the convergence of the gradient algorithm can be expected to be very fast. In practice, this is indeed the case.

Other properties of lattice equalizers, besides that of fast convergence, make them worthy of special attention as implementation candidates. Among them, we should mention the high insensitivity of lattice filters to round-off noise deriving from finite precision digital implementation (see Lim and Mueller, 1980). Also, as optimal reflection coefficients are independent of the number of stages in the filter, lattice stages can be added or deleted without the need of recalculating already existing coefficients. This is in

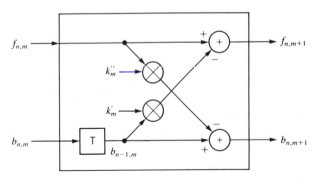

Figure 8.10 Basic stage of a lattice filter.

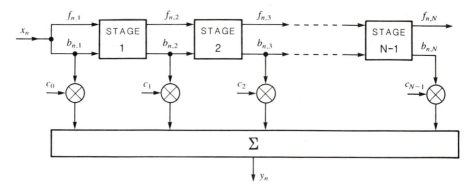

Figure 8.11 Structure of a lattice equalizer.

marked contrast to the behavior of the optimal tap weights in a TDL, whose values are interdependent.

Satorius and Pack (1981) compare the convergence properties of lattice equalizers based on the minimization of MS or LS error with those of an MSE TDL equalizer. By simulation, the LS lattice equalizer is shown to converge in 40 to 50 iterations in a situation where the MSE lattice equalizer needs about 120 iterations. These figures are independent of the eigenvalue spread of the matrix $\mathbf{X}$. On the other hand, an MSE TDL working with the stochastic gradient algorithm requires about 600 iterations for its convergence when the eigenvalue spread $\mu_{max}/\mu_{min} = 11$, and about 1000 when $\mu_{max}/\mu_{min} = 21$. Thus, not only does the lattice equalizer converge faster, but its convergence properties do not depend, to a certain extent, on the channel. The price paid for this improved performance is of course increased complexity. In fact, the LS lattice equalizer must perform $12N$ multiplications, $11N$ additions, and $3N$ divisions at each step, while the TDL equalizer needs only $2N$ multiplications and $2N$ additions (Schichor, 1982). The LS lattice equalizer needs even more operations than the fast Kalman algorithm mentioned in Section 8.6 (which requires $10N$ multiplications, $9N$ additions, and 2 divisions.) On the other hand, the fast Kalman algorithm performs much the same as the LS lattice in terms of convergence speed (see Lim and Mueller, 1980). The reader interested in delving deeper into the topic of adaptive lattice equalizers is referred to the Bibliographical Notes at the end of this chapter.

8.8 DATA-AIDED EQUALIZATION

In this section, we shall examine a class of nonlinear equalizers that appear particularly useful for channels with severe distortion. The basic idea of this class is that of *data-aided equalization*. We explain this before giving a detailed description of two equalizer structures based on it. Let us recall that, due to propagation delay in the transmission channel, the received signal sample x_l is used to make a decision on the symbol that was emitted by the source D discrete-time instants before, say a_{l-D}. The impulse response (q_l) of the discrete channel whose input is the sequence (a_l) and whose output is the sequence (x_l) is sketched qualitatively in Fig. 8.12. The samples q_l, $l < D$, are called

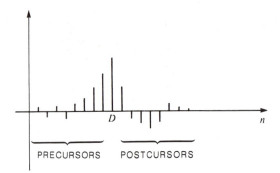

D

n

PRECURSORS POSTCURSORS

Figure 8.12 Qualitative impulse response of a discrete channel to be equalized.

the *precursors*, while the samples q_l, $l > D$, are called the *tails*, or *postcursors*, of the impulse response. Assume for a moment that the impulse response and the source symbol a_{l-D} are perfectly known. Since

$$x_l = \sum_{m=-\infty}^{\infty} a_{l-m} q_m + n_l, \tag{8.125}$$

we can subtract the known quantities $a_{l-D} q_{k+D}$ from the samples x_{l+k}, $k \neq 0$, thus eliminating all the ISI due to symbol a_{l-D}. This is the essence of data-aided equalization: if a number of source symbols are correctly detected and the channel impulse response is known, then the ISI can be reconstructed and therefore canceled from the received signal. By implementing this idea when the channel suffers a large amount of amplitude distortion, we can expect a performance improvement with respect to standard equalizers. In this situation, an ordinary linear equalizer would considerably enhance the noise, while the data-aided equalizer would not play any role in determining the noise power of the equalizer output. In fact, it will just provide a weighted sum of noise-free symbols to be subtracted from the received symbols. The reader is warned, however, that the assumption of a known transmitted sequence makes the preceding statements only approximately true. Actually, in a real setting there is no hope of canceling completely the ISI without enhancing noise. This is because the minimum distance $d_{\min}$ (see Section 7.5.3), which depends on the source symbol structure and on the channel, imposes a limit to the error performance of *any conceivable receiver*, and hence of any receiver based on data-aided equalization.

The block diagram of an ideal data-aided equalizer is shown in Fig. 8.13. The ISI canceler is a transversal filter with tap weights $\{c_k\}$. Denoting by S the index set $\{0, 1, \ldots, D-1, D+1, \ldots, N\}$, the equalized signal takes the form

$$y_l = \sum_{m=-\infty}^{\infty} a_{l-m} q_m - \sum_{m \in S} a_{l-m} c_m + n_l. \tag{8.126}$$

Notice that S does not include the index D. This is because we want to restrict the role of the canceler to remove the ISI without altering the useful signal $a_{l-D} q_D$. It can be proved that the canceler weights $\{c_k\}$ that minimize the MSE $\mathscr{E} \triangleq E\{|y_l - a_{l-D}|^2\}$ are

$$c_k = q_k, \qquad k \in S. \tag{8.127}$$

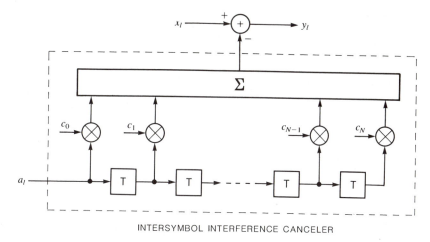

INTERSYMBOL INTERFERENCE CANCELER

Figure 8.13 Block diagram of an ideal data-aided equalizer.

8.8.1 Decision-Feedback Equalizer

Consider now the practical implementation of data-aided equalization. Because the source symbol sequence (a_l) appearing in Fig. 8.13 is not available at the receiver, it must be estimated from the received samples. Thus, it must be assumed that, after a suitable training period, the equalizer provides an error rate so low that perfect detection is not an unrealistic assumption. Also, as the channel is now known in advance (and can vary with time), the canceler will have adaptively varying coefficients.

The most studied scheme of data-aided equalization is the *decision-feedback equalizer* (DFE). In it, the set S includes only the integers $D + 1, \ldots, N$, so the canceler operates only on the postcursors of the channel impulse response. With this choice for S, we see from (8.126) that at time l the source symbols needed for cancellation are $a_{l-D-1}, a_{l-D-2}, \ldots, a_{l-D-N}$. Since at time l a decision is taken on a_{l-D}, the symbols needed can be obtained from the previous decisions.

In conclusion, a DFE is based on a canceler that takes care of the postcursors of the channel impulse response. As such, it cannot be self-sufficient, because the precursors also have to be accounted for. Thus, a preliminary equalizer should precede the canceler; its task can be viewed as the elimination of the precursors. Hence, in a sense, the preliminary equalizer and the canceler perform complementary functions.

The DFE scheme is presented in Fig. 8.14. In it, as is customary, the canceler is referred to as the feedback filter. Notice that the inclusion of the decision device into a loop renders this equalizer intrinsically nonlinear (as generally in all the data-aided structures).

To analyze in detail the DFE behavior, assume that the preliminary equalizer and the feedback filter are TDL with length N and N' and weight vectors $\mathbf{c}$ and $\mathbf{f}$, respectively. Thus, the signal sample at the input of the decision device can be written in the form

$$y_l = \mathbf{c}'\mathbf{z}_l - \mathbf{f}'\hat{\mathbf{a}}_l, \tag{8.128}$$

Sec. 8.8   Data-Aided Equalization   **393**

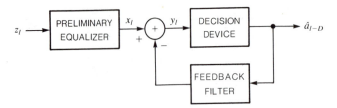

Figure 8.14 Block diagram of a decision-feedback equalizer.

where $\mathbf{z}_l \triangleq [z_l, z_{l-1}, \ldots, z_{l-N+1}]'$ and $\hat{\mathbf{a}}_l \triangleq [\hat{a}_l, \hat{a}_{l-1}, \ldots, \hat{a}_{l-N'+1}]'$. By defining the two vectors

$$\mathbf{b} \triangleq \begin{bmatrix} \mathbf{c} \\ \mathbf{f} \end{bmatrix} \updownarrow N + N' \tag{8.129}$$

and

$$\mathbf{u}_l \triangleq \begin{bmatrix} \mathbf{z}_l \\ -\hat{\mathbf{a}}_l \end{bmatrix} \updownarrow N + N', \tag{8.130}$$

Eq. (8.128) can be rewritten in a more compact form:

$$y_l = \mathbf{b}' \mathbf{u}_l. \tag{8.131}$$

Since (8.131) bears a close resemblance to (8.32), it is not difficult to duplicate the arguments of Section 8.2 to find the $\mathbf{b}$, and hence the $\mathbf{c}$ and the $\mathbf{f}$, that minimize the MSE and to devise a gradient algorithm for minimization. (Minimization of MSE is not the only design criterion; for example, a zero-forcing criterion is also applicable; see Price, 1972). By assuming $\hat{a}_l = a_l$ for all l, we get

$$E\{|y_l - a_{l-D}|^2\} = \mathbf{b}^\dagger \mathbf{U} \mathbf{b} - 2\mathcal{R}[\mathbf{b}^\dagger \mathbf{w}] + \sigma_a^2, \tag{8.132}$$

where

$$\mathbf{U} \triangleq E[u_l^* u_l'] \tag{8.133}$$

and

$$\mathbf{w} \triangleq E[a_{l-D} u_l^*]. \tag{8.134}$$

By taking the gradient of (8.132) with respect to $\mathbf{b}$ and setting it equal to the null vector, we obtain the following equation for the optimum tap-weight vector $\mathbf{b}_{\text{opt}}$:

$$\mathbf{U} \mathbf{b}_{\text{opt}} = \mathbf{w}. \tag{8.135}$$

Since $\mathbf{U}$ is not necessarily positive definite, a form for $\mathbf{b}_{\text{opt}}$ similar to (8.40) may not be available. However, a gradient algorithm can be displayed:

$$\mathbf{b}^{(l+1)} = \mathbf{b}^{(l)} - \alpha E[(y_l - a_{l-D}) \mathbf{u}_l^*], \tag{8.136}$$

which converges to a tap-weight vector achieving the minimum MSE.

By taking the limit as N and N' both tend to infinity, an expression can be obtained for the minimum achievable MSE (Salz, 1973):

$$[\mathscr{E}_{\min}]_{\text{DFE}} = \exp\left\{ T \int_{-1/2T}^{1/2T} \ln \frac{\sigma_a^2 G_n(f)}{G_n(f) + (\sigma_a^2/T)|Q_{\text{eq}}(f)|^2} \, df \right\}. \tag{8.137}$$

By comparing (8.137) with (8.60), the analogous expression for the infinite-length linear TDL equalizer, one can prove that $[\mathscr{E}_{min}]_{DFE}$ is always less than or equal to the value given by (8.60). This shows that in the absence of decision errors the DFE performs asymptotically better than the linear equalizer. Unfortunately, however, there is no definite answer to the question of whether a finite-length DFE achieves a lower MSE than a linear equalizer with the same overall number of taps. In fact, the relative performance of the two equalizers depends on the actual channel characteristics, on the number of taps, and on the choice of the delay D (Qureshi, 1982). Simulation results (Salz, 1973; Proakis, 1975) confirm what is intuitively expected: the DFE is markedly superior to the linear equalizer with the same finite length when operating on channels with spectral nulls in the Nyquist interval. This is the situation where the linear equalizer suffers a considerable noise enhancement. In addition, the DFE performance is less sensitive to the sampling time (Qureshi, 1982).

Finally, it must be observed that our computations were based on the assumption that $\hat{a}_l = a_l$ for all l (i.e., the decision process is error free). Now it is reasonable to ask to what extent the DFE performance is degraded by decision errors. Notice that decision errors tend to propagate because they produce wrong cancellation of tails, and hence a reduced noise margin for future decisions. In turn, this entails a higher probability of future incorrect decisions, and so on. Unfortunately, no complete analysis of the performance of a DFE in the presence of decision errors is available. Simulation results show that the error propagation is not catastrophic. In fact, on typical channels errors tend to cluster in bursts short enough to only slightly degrade performance.

8.8.2 A Data-Aided Equalizer

Derivation of the decision-feedback equalizer was based on the choice to retain in the tap-weight index set S only integers greater than the channel delay D. With this choice, only previously detected data symbols are needed by the canceler. Assume now that this restriction is removed; hence $S = \{0, 1, \ldots, D - 1, D + 1, \ldots, N\}$. In this situation, even not yet detected symbols should be made available at the canceler input. This seems to prevent its real-time implementation. This difficulty can be overcome, as proposed by Gersho and Lim (1981), by generating a sequence $(\bar{a}_l)$ of preliminary decisions on the source symbols, and using it as an input to the canceler. After cancellation of ISI, final decisions can be taken. A scheme of the data-aided equalizer proposed by Gersho and Lim is shown in Fig 8.15. Here a preliminary fractionally spaced equalizer operates on the received sample sequence in order to ensure the small error rate needed for good preliminary estimates. A $T/2$-spaced "matching filter" is used as a receiving filter. To make the canceling procedure causal, this filter should be preceded by a block introducing a delay $(D + \frac{1}{2})T$. For an adaptive structure, the error $\bar{y}_n - \bar{a}_{n-D}$ of preliminary decisions will be used to adaptively adjust the preliminary equalizer's tap weights, and the error $y_n - \hat{a}_{n-D}$ of final decisions will simultaneously adjust those of the matching filter and the canceler.

We shall now discuss the performance of data-aided equalizers for the three channel models as shown in Fig. 8.16. Channel 1 has a flat amplitude response and a small delay distortion over the entire voiceband frequency, except near the lower band edge. Channel 2 has moderate amplitude and delay distortion. It just meets the basic conditions

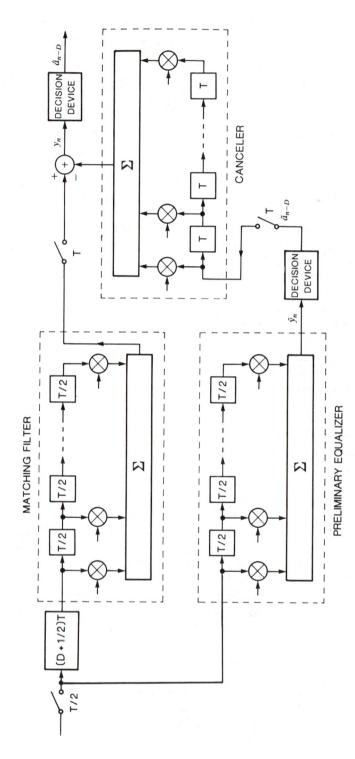

Figure 8.15 Implementation of a data-aided equalizer.

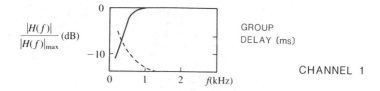

CHANNEL 1

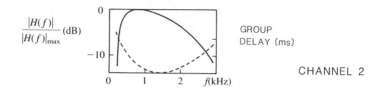

CHANNEL 2

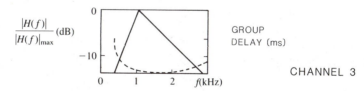

CHANNEL 3

Figure 8.16 Channel characteristics for evaluating the performance of data-aided equalizers. Solid lines: amplitude characteristics. Dotted lines: delay characteristics. (Redrawn with permission from Gersho and Lim, 1981; *Bell System Technical Journal*, © 1981, AT&T.)

for private lines. Channel 3 has severe amplitude distortion. The modulation format is QAM, with data symbols a_n whose real and imaginary parts independently take on values in the set ± 1, ± 3. The data rate is 9.6 kbits/s. In the scheme of Fig. 8.14, it is assumed that the matching filter has 62 $T/2$ taps, and the canceler 31; the preliminary equalizer has 64 $T/2$ taps.

Table 8.1 summarizes simulation results obtained in this situation (Gersho and Lim, 1981), using as a performance measure the output SNR, defined as

$$\text{Output SNR} \triangleq \frac{\text{E}\{|a_n|^2\}}{\text{E}\{|y_n - a_{n-D}|^2\}}.$$

The input SNR is defined as the ratio of the average data symbol power $\text{E}\{|a_n|^2\}$ to the Gaussian noise power. Results for channel 1 show that with the linear equalizer, the decision-feedback equalizer, or the data-aided equalizer, ISI does not cause any significant degradation (i.e., the output SNR is approximately equal to the input SNR). For channel 2, the data-aided equalizer has a gain of approximately 2 dB relative to the linear equalizer and approximately 1 dB relative to the decision-feedback equalizer. Similar results are obtained for channel 3. It can be observed that, due to the considerable amplitude distortion, the linear equalizer performs significantly worse on this channel.

Sec. 8.8   Data-Aided Equalization

397

TABLE 8.1 Comparison of linear equalizer, decision-feedback equalizer, and data-aided equalizer performances. (Reprinted with permission from Gersho and Lim (1981), *The Bell System Technical Journal*, Copyright 1981, AT&T).

		Input SNR (dB)					
		Channel 2			Channel 3		
		20	24	30	20	24	30
Output SNR (dB)	Linear equalizer	18.0	22.1	26.8	15.4	20.4	25.5
	Decision-feedback equalizer	19.1	23.0	27.9	17.3	22.0	27.1
	Data-aided equalizer	19.7	23.8	29.2	18.5	23.1	28.7

The data-aided equalizer outperforms the linear equalizer by about 3 dB and the decision-feedback equalizer by 1 dB.

8.9 MORE ON COMPLEX EQUALIZERS

Throughout this chapter, extensive use has been made of complex notations to denote samples of bandpass signals and channel responses. As a special case, baseband signals and channels can be dealt with by obvious changes in the formulas. In this section, we shall expand briefly on certain features of two-dimensional system equalization.

Hitherto we have implicitly assumed that the carrier phase for demodulation has been properly estimated. This estimate can be performed in a decision-directed mode with the receiver arrangement of Fig. 8.17. In it, equalization is performed after coherent demodulation and inside the loop for decision-directed phase compensation (Matyas and McLane, 1974). With this arrangement, as the equalizer itself introduces a many-symbol-interval delay between input and output, the estimated phase sequence $(\hat{\vartheta}_n)$ is a delayed version of the true phase sequence (ϑ_n). This delay prevents the receiver from correctly compensating any time-varying phase shift introduced by the channel. To avoid this impairment source, two different receiver structures can be used (Falconer, 1976). The first one (Fig. 8.18) places phase compensation after the equalizer, while demodulation is performed using a free-running oscillator before the equalizer. The second one (Fig. 8.19) places both demodulation and phase compensation after the equalizer. The latter now operates on bandpass samples. Hence, it is referred to as a *bandpass equalizer*. If equal phase compensation is assumed for both schemes, mathematically these structures are exactly equivalent.

A feature of these two structures worth observing here is the *tap-rotation property* (Gitlin, Ho, and Mazo 1973). Assume that the samples at the input of the equalizer are rotated by an angle φ. Then the vector $\mathbf{x}_n$ defined in (8.34) is changed into $\mathbf{x}_n e^{j\varphi}$. Consider the effect of this phase rotation on the operation of a linear equalizer. From

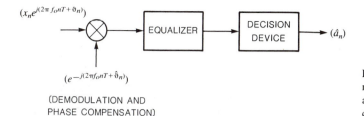

$(x_n e^{j(2\pi f_o nT + \vartheta_n)})$

EQUALIZER → DECISION DEVICE → $(\hat{a}_n)$

$(e^{-j(2\pi f_o nT + \hat{\vartheta}_n)})$

(DEMODULATION AND PHASE COMPENSATION)

Figure 8.17 Discrete-time model of a receiver in which equalization is performed after demodulation and carrier-phase compensation.

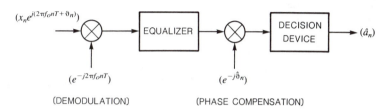

$(x_n e^{j(2\pi f_o nT + \vartheta_n)})$

EQUALIZER → DECISION DEVICE → $(\hat{a}_n)$

$(e^{-j2\pi f_o nT})$ (DEMODULATION)

$(e^{-j\hat{\vartheta}_n})$ (PHASE COMPENSATION)

Figure 8.18 Discrete-time model of a receiver in which equalization is performed after demodulation and before carrier-phase compensation.

definitions (8.36) and (8.37), it can be easily seen that $\mathbf{X}$ is left unchanged, whereas $\mathbf{v}$ is changed into $\mathbf{v}e^{-j\varphi}$. In turn, this implies that the optimum tap-weight vector (8.40) is changed into $\mathbf{c}_{opt}e^{-j\varphi}$ (i.e., the equalizer taps are rotated by $-\varphi$), the net result being that when the equalizer has settled at its optimum the equalized output samples will not be affected by the phase rotation.

Furthermore, consider the effect on the operation of these two equalizer structures of two transmission impairments typical of telephone lines, phase jitter and frequency offset. Phase jitter acts as a real random sequence (φ_n) affecting the phase angle of the channel output samples. When its time constant is much larger than the equalizer settling time, it can be assumed to cause a constant phase shift. Thus, due to the tap-rotation property, it can be compensated for by the equalizer. Similar considerations hold for a frequency offset (i.e., the perturbation of the carrier frequency f_0 by a small amount Δf). In conclusion, the equalizer, due to its tap-rotation property, can track small amounts of phase jitter and frequency offset so that the phase compensation loop implicit in Figs. 8.18 and 8.19 is not required. When phase jitter or frequency offset are significant, this phase compensation loop is needed.

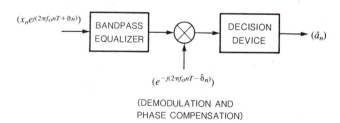

$(x_n e^{j(2\pi f_o nT + \vartheta_n)})$

BANDPASS EQUALIZER → DECISION DEVICE → $(\hat{a}_n)$

$(e^{-j(2\pi f_o nT - \hat{\vartheta}_n)})$

(DEMODULATION AND PHASE COMPENSATION)

Figure 8.19 Discrete-time model of a receiver in which equalization is performed before demodulation and carrier-phase compensation.

Sec. 8.9 More on Complex Equalizers

BIBLIOGRAPHICAL NOTES

Prior to the mid-1960s, considerable research effort was directed to the specification of digital data transmission receivers for channels affected by ISI (some of the resulting papers were referenced in Chapter 7). From this work, the structure of an adaptive receiver based on a TDL and an iterative optimization technique for adjusting the tap weights eventually emerged. The history of the TDL filter as an equalizer dates back to Nyquist (1928). Accounts of it can be found in the papers by Rudin (1967) and Di Toro (1968). The fundamental ideas on which automatic equalization is based were not unknown in the mid-1960s (see, e.g., Goldenberg and Klovsky, 1959, a paper that one of the authors claims to be the first to describe time-domain adaptive equalization: Klovsky and Nikolaev, 1978, p. 40; or the papers by Kettel (1961, 1964)). However, there is no doubt that it was R. W. Lucky's early work (Lucky, 1965, 1966) that provided the major breakthrough in the problem of equalization for intersymbol interference. The TDL equalizer based on an MSE criterion was first analyzed by Proakis and Miller (1969) and Gersho (1969a). Its convergence properties in the training mode were studied by, among others, Ungerboeck (1972), Mazo (1979), and Gardner (1984) using the "independence assumption." Our treatment of the subject is based on the latter paper, which contains the most comprehensive convergence analysis published so far. Convergence analysis of the MSE TDL equalizer working in the tracking mode with the stochastic-gradient algorithm is even more complicated. For details, the reader is referred to Macchi and Eweda (1984) and to the references therein.

Recently, interest has centered on developing adaptive equalizers with fast convergence. The approach of transforming the equalizer input in order to have uncorrelated samples was taken by Chang (1971) and Gitlin and Magee (1977). Godard (1974) obtained an adaptive algorithm that has the structure of a Kalman filter and shows a particularly fast convergence. In Gitlin and Magee (1977) it is shown that Godard's algorithm owes its convergence properties to its capability of decorrelating the equalizer inputs. The "fast Kalman algorithm," having the same convergence properties as Godard's, but requiring a lower complexity, was proposed by Falconer and Ljung (1978). Later, attention was attracted by lattice filters. As lattice filters perform a Gram–Schmidt orthogonalization procedure on the samples entering it, their application to the fast converging algorithm problem is natural. An overview of the properties and the applications of lattice filters can be found in Makhoul (1978), Friedlander (1982), and in the book by Honig and Messerschmitt (1984). Application of lattice filters to adaptive equalization was first suggested by Satorius and Alexander (1979) and Makhoul (1978). A problem with adaptive lattice filters is that their outputs are uncorrelated only after the adaptation algorithm has reached a steady state. Thus, the equalizer convergence may not be as fast as with Godard's algorithm. An adaptive lattice algorithm with uncorrelated outputs at any time was discovered by Morf (1977) (see also Morf, Vieira, and Lee, 1977) and applied to equalization by Schichor (1982) and Satorius and Pack (1981). A simplified derivation of the algorithm can be found in McWhirter and Shepherd (1983). Complex adaptive lattice structures are considered in Symons (1979). The Kalman, fast Kalman, and adaptive lattice algorithms are extended to complex fractionally spaced equalizers by Mueller (1981).

The idea of fractionally spaced TDL equalizers dates back to 1969 (Gersho, 1969b) and was rediscovered a few years later (see Guidoux, 1975; Macchi and Guidoux, 1975). Ungerboeck (1976), Qureshi and Forney (1977), and Gitlin and Weinstein (1981) analyze their performance and convergence properties.

The idea of using past decisions to cancel intersymbol interference, and hence the concept of decision-feedback equalization, was introduced by Austin (1967). An overview of the work done in this area before 1978 is contained in Belfiore and Park (1979). A unified theory of data-aided equalization is provided by the paper of Mueller and Salz (1981). Our treatment of the subject is based on it.

The first blind equalizer was designed by Sato (1975). Recent work on blind equalizers is reported in Benveniste and Goursat (1984) and Foschini (1985). Our treatment is based on Godard (1980).

Because, in practice, adaptive equalizers are implemented digitally, their parameters, as well as the signal samples, are quantized to a finite number of levels. The effects of digital implementation of the TDL equalizer are considered in Gitlin, Mazo, and Taylor (1973) and Gitlin and Weinstein (1979). If the tap-weight values are constrained to be power of 2 (or sums thereof), the multipliers in the TDL structure are substituted for by shifters (or shifters and adders), as this simplifies implementation (see, e.g., Pirani and Zingarelli, 1984).

The scheme of Fig. 8.1 was analyzed by Magee and Proakis (1973) and Proakis (1974).

PROBLEMS†

8.1. Consider the identification problem of Section 8.1. Assume that the TDL used for channel identification has $\hat{L}$ delay elements, while the channel memory span is $L > \hat{L}$. This situation can be dealt with by writing, in lieu of (8.4) and (8.5),

$$x_l = \mathbf{q}' \, \mathbf{a}_l + n_l$$

and

$$\hat{x}_l = \hat{\mathbf{q}}_0' \, \mathbf{a}_l,$$

where

$$\mathbf{q} \triangleq \begin{bmatrix} \mathbf{q}_a \\ \mathbf{q}_b \end{bmatrix}, \qquad \hat{\mathbf{q}}_0 = \begin{bmatrix} \hat{\mathbf{q}} \\ \mathbf{0} \end{bmatrix},$$

and $\hat{\mathbf{q}}$, $\mathbf{q}_a$ have $\hat{L} + 1$ components, and $\mathbf{0}$ is the null vector with $L - \hat{L}$ entries. With these assumptions, derive an expression for the minimum mean-square identification error. Also, when the source symbols (a_n) are uncorrelated and have zero mean, show that the minimum MSE is achieved for $\hat{\mathbf{q}} = \mathbf{q}_a$.

*8.2. Consider a binary PAM transmission with source symbols ± 1 using a channel bandlimited in the Nyquist interval $(-1/2T, 1/2T)$. The impulse response samples, taken every T seconds, are 0.833, 1.0, and 0.583. The noise is additive Gaussian, and the receiving filter is an

---

† The problems marked with an asterisk should be solved with the aid of a computer.

ideal low-pass filter with cutoff frequency $1/2T$. Compute the bit error probability versus $\mathcal{E}_b/N_0$ in the following situations:

(a) Unequalized channel.

(b) Channel equalized using a 5-tap minimum-MSE TDL equalizer (choose the optimum value of the delay D).

(c) Same as in part (b), with a 7-tap TDL.

(d) Same as in part (b), with a 15-tap TDL.

Compare the results obtained in parts (a) to (d) with the error probability that would be achieved if the ISI were completely removed (and hence the only disturbance were AWGN).

8.3. Consider a linearly modulated signal transmitted over a dispersive channel. Assume that an infinitely long zero-forcing equalizer is used that completely eliminates ISI.

(a) Derive the transfer function of this equalizer.

(b) Derive an expression for the bit error probability of this transmission system that takes into account the noise enhancement introduced by the equalizer.

8.4. Derive the bounds (8.103). *Hint*: Use the following result from matrix theory. If $\mathbf{A}$ is a Hermitian matrix, then

$$\lambda_{min} \leq \frac{\mathbf{x}^\dagger \mathbf{A} \mathbf{x}}{\mathbf{x}^\dagger \mathbf{x}} \leq \lambda_{max},$$

where λ_{max}, λ_{min} are the largest and the smallest eigenvalues of $\mathbf{A}$, respectively.

*8.5. Consider a binary PAM transmission with source symbols ± 1 using a channel bandlimited in the Nyquist interval $(-1/2T, 1/2T)$. The channel has a sampled overall impulse response (h_n), the noise is additive Gaussian, and the receiving filter is an ideal low-pass filter with cutoff frequency $1/2T$. Assume that the channel is equalized using a three-tap minimum-MSE TDL equalizer, and compute the bit error probability versus $\mathcal{E}_b/N_0$, with the delay D as a parameter, for these two situations:

(a) $(h_n) = (0.5, 1.0, 0.5)$.

(b) $(h_n) = (1.0, 0.67, 0.45, 0.3, 0.2, 0.135)$.

*8.6. Consider the transmission of a four-level PAM signal with source symbols -3, -1, $+1$, $+3$ at 4800 bits/s over channel 3 of Fig. 8.15. Choosing an appropriate signal shape, compare the minimum MSE achievable using a T-spaced and a $T/2$-spaced minimum-MSE TDL equalizer with the same number of taps (say, 31).

CHAPTER 9

Reliable Transmission with Encoded Waveforms

The effect of channel impairments such as noise, distortions, and fading is that the binary stream delivered to the user contains errors. The error probability over this received binary sequence is one of the most important design parameters in the communication link. Usually it must be kept smaller than a given preassigned value, depending on the application. Techniques to control the error probability are based on the addition of *redundancy* to the information sequence. The most trivial example is the repetition of the same message on the transmission channel. It is intuitive that redundancy, and therefore reliability, must be traded against transmission efficiency.

Let us first assume that redundancy is added to the source binary sequence. In this case, a source sequence of k binary digits is represented by a longer sequence of n ($n > k$) binary digits. Therefore, the 2^k possible source sequences are mapped into a set of code words of length n. The *channel encoder* is the device that performs such mapping. The set of 2^k code words is called a *binary channel code*. When the encoding process undertaken by the channel encoder is memoryless, the resulting code is a *block code*. On the other hand, when it has memory, the resulting code is a *tree code*. The codes are also classified as *error-detecting codes* or *error-correcting codes* for the following reason. When the *channel decoder* observes the received sequence, it attempts to detect possible errors and to correct them. The first part of the chapter deals with these codes, their properties, implementation, and expected performance.

Instead of applying the technique of using longer binary sequences, redundancy can be added directly to the signal set. In fact, the 2^k possible source sequences can be mapped into a set of 2^n signals ($n > k$) with an encoding process with memory. These codes are called *signal-space codes*. They permit good error performance capabilities

403

combined with satisfactory spectral properties of the transmitted signal. The second part of the chapter will present some examples of this type of codes.

Another approach toward the goal of better reliability is that of using redundancy *to prevent* the effects of channel impairments other than noise. An example could be the control of the spectral shape of the transmitted signal. Codes that achieve this goal are called *line codes*. They will be discussed in the final part of the chapter.

9.1 INTRODUCTION

As already stated, in principle, channel encoding for error detection and correction is achieved by adding to the source binary sequence a controlled amount of redundancy. This helps protect the information against possible errors. The encoding rule adopted will heavily affect the complexity of the *decoding algorithm*, that is, the processing to be performed on the received sequence to recover the transmitted information.

Two different strategies can be used in the channel decoder. Conceptually, these strategies can be related to Fano's inequality (3.73). The first strategy is called *error detection*. The decoder observes the received sequence and detects whether or not errors have occurred. A certain measure of uncertainty is eliminated, which corresponds to the term $H(e)$ in (3.73). Error detection is used to implement one of two possible schemes: *error monitoring* or *automatic repeat request* (ARQ). In the case of error monitoring, the decoder supplies to the user a continuous indication regarding the quality of the received sequence. In the case of ARQ, the transmitter is asked to repeat unsuccessful transmissions. To this end, a feedback channel from the receiver to the transmitter must be available.

The second strategy is called *forward error correction* (FEC). The decoder attempts to restore the correct transmitted sequence whenever errors are detected in the received sequence. In this case, an additional measure of uncertainty must be removed corresponding to the term $P(e) \log (N - 1)$ of (3.73). It is intuitive that this strategy requires more redundancy and more complex decoding algorithms. The choice between the two strategies depends on the particular application and on the complexity of the transmission system considered. For example, the ARQ scheme is usually applied in intercomputer communication, since a two-way transmission channel is available together with large buffers that store information while performing the retransmission procedure. On the other hand, FEC is adopted when information must be protected on a one-way channel.

To analyze the benefits to be gained by using channel encoding in comparison with the uncoded schemes of Chapter 5, let us consider the model of Fig. 9.1. We restrict our attention to binary sources and binary encoders.

The source emits binary digits at a rate of R_s bits/s and the encoder represents each sequence of k source digits using $n = k/R_c$ binary digits. R_c is the encoder rate defined in (3.76). The transmission speed on the channel must be increased to the value R_s/R_c binary symbols per second, and thus the required bandwidth must also be increased. As a result, the use of channel encoding decreases the bandwidth efficiency with respect to the uncoded transmission.

The binary symbols produced by the channel encoder are presented to the modulator and converted into a sequence of waveforms using one of the modulation schemes

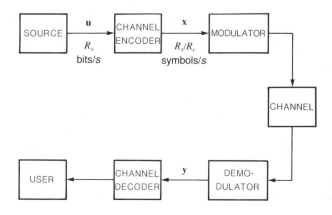

Figure 9.1 Model of the transmission system.

described in Chapter 5. Without loss of generality and for the purposes of this preliminary discussion, we assume that the modulator uses an antipodal binary modulation over an AWGN channel (see Example 4.4). With this modulation scheme, each binary encoded symbol is associated on the channel with a binary waveform of duration $T = T_c = R_c/R_s$ seconds. This duration is shorter than that used in the uncoded case. If we denote with $\mathscr{E}_b$ the energy per information bit and assume that the transmitted power is kept constant, we can conclude that coding decreases the energy per channel symbol to the value $\mathscr{E} = \mathscr{E}_b R_c$. As a result, on the average, more channel symbols will be in error than with uncoded transmission.

These observations about coding seem rather discouraging. In fact, bandwidth efficiency is decreased and more errors in the received sequence are to be expected. Nevertheless, in a well-designed coded system the larger number of channel symbol errors will be compensated for by the error-correcting capabilities of the decoder. Therefore, a coded transmission should trade off the bandwidth efficiency for a better overall error performance, using the same transmission power.

Let us describe the processing that must be performed at the channel output by the demodulator and by the channel decoder to achieve such result. Consider first the case in which the demodulator is used to take decisions on whether each binary waveform represents a 0 or a 1. To this purpose, the demodulator output is quantized to two levels denoted by 0 and 1 and is said to take *hard decisions*. The sequence of binary digits from the modulator is fed into the decoder. This attempts to recover the information sequence by using the redundancy for either detecting or correcting the errors that are present at the demodulator output. Such a decoding process is called *hard-decision decoding*. In this model, the combination of modulator, channel, and demodulator is equivalent to a *binary symmetric channel* (BSC). Its transition probability is the error probability (4.45) of a binary antipodal modulation scheme. The overall error performance of the coded scheme depends on the implementation of efficient algorithms for error detection and correction.

At the other extreme, consider the case in which the unquantized output of the demodulator is fed to the decoder. This stores the n outputs corresponding to each sequence of n binary waveforms and builds 2^k decision variables. The decoding process consists of choosing one out of the 2^k possible transmitted sequences as the ML one. Thus, one information sequence is recovered. Such a decoding process is called *unquan-*

Sec. 9.1   Introduction

405

tized soft-decision decoding. In this model, the combination of modulator, channel, and demodulator is equivalent to a binary-input, continuous-output channel. It is intuitive that this approach presents higher reliability than that achieved with the hard-decision scheme. That is, the decoder can take advantage of the additional information contained in the unquantized samples that represent each individual binary transmitted waveform.

An intermediate case, called *soft-decision decoding*, is represented by a demodulator whose output is quantized to Q levels, with $Q > 2$. In this case, the combination of modulator, channel, and demodulator is equivalent to a binary input, Q-ary output discrete channel. The decoding process is based on algorithms that manipulate Q-ary symbols from the channel output to recover the binary information sequence associated with the transmitted sequence of waveforms. The advantage over the analog (unquantized) case is that all the processing can be accomplished with digital circuitry. Therefore, it represents an approximation of the unquantized soft-decision decoding.

The advantage of a coded transmission scheme in relation to an uncoded one is measured by its *coding gain*. This is defined as the difference (in decibels) in the required value of $\mathcal{E}_b/N_0$ to achieve a given bit error probability with respect to an ideal binary antipodal uncoded transmission. This concept is represented in Fig. 9.2. The curve of the ideal binary antipodal scheme is reproduced and the channel capacity limit of -1.6

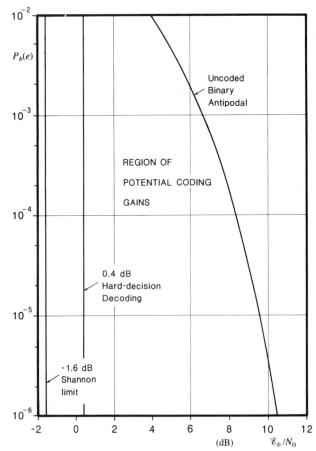

Figure 9.2 Potential coding gains of coded transmission with respect to binary uncoded antipodal transmission.

dB is also shown. For a bit error probability of 10^{-5}, a potential coding gain of 11.2 dB is theoretically available in the case of unquantized soft-decision decoding. The other limit of 0.4 dB regards the case of hard-decision decoding and entails a loss of about 2 dB with respect to the Shannon limit. This limit will be further discussed when analyzing in detail the performance of coded transmission.

We conclude this introduction with the observation that coded transmissions can be useful in the power-limited region of Fig. 5.39. In fact, in this region the bandwidth efficiency is rather poor, but the channel capacity can be approached for small values of E_b/N_0.

We shall now describe the more important properties and performance characteristics of those channel-encoding schemes that attempt to realize in practice the potential coding gains promised by theory. Two classes of codes are available to this end. The block codes require no memory. The tree codes include memory in the encoding operation. Only a particular class of the latter codes will be considered, the *convolutional codes*.

9.2 BLOCK CODES

The basic feature of block codes is that the block of n digits (code word) generated by the encoder depends only on the corresponding block of k digits generated by the source. Therefore, the encoder is memoryless. A great deal of block code theory is an extension of the notion of *parity check*. Take a sequence of k binary digits. Transform it into a sequence of length $n = k + 1$ digits by simply adding in the last position a new binary digit, following the rule that the number of ones in the new sequence must be even. This redundant digit is called a parity-check digit. In this way, any error event that changes the parity of the sequence from even to odd can be detected by the decoder.

Parity-check codes are a particular class of block codes in which the digits of the code word are a set of n parity checks performed on the k information digits. The code is usually referred to as an (n, k) code. A code is called *systematic* when the first k digits are a replica of the information digits and the remaining $(n - k)$ ones are parity checks on the k information digits. Parity checks in binary sequences are formally dealt with using modulo-2 arithmetic in which the rules of ordinary arithmetic hold true except that the sum $(1 + 1)$ is 0, not 2. Throughout this chapter, modulo-2 arithmetic will be used unless otherwise specified.

A functional block diagram of the encoder is shown in Fig. 9.3. It consists of a k-stage input shift register, n modulo-2 adders, and an n-stage output shift register.

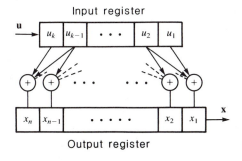

Figure 9.3 Encoder block diagram for a parity-check code.

Each adder is connected to a subset of stages of the input register in order to perform the desired parity check. The vector $\mathbf{u} = [u_1, u_2, \ldots, u_k]$ of k information digits is fed into the input register. When this register is loaded, the content of each adder is fed in parallel into the corresponding stage of the output register, which shifts out the code word $\mathbf{x} = [x_1, x_2, \ldots, x_n]$. While shifting out one code word, the input register is reloaded and the whole operation is repeated. The clocks for the two registers are different and the higher output rate is self-evident. The following simple examples will clarify these concepts.

Example 9.1 Repetition code (3, 1)

In this code, each code word of length $n = 3$ is defined by the relations:

$$x_1 = u_1, \qquad x_2 = u_1, \qquad x_3 = u_1. \tag{9.1}$$

The encoder is sketched in Fig. 9.4. Obviously, the adders are omitted in this case. The resulting coding rule is

$$\begin{matrix} 0 \\ 1 \end{matrix} \Leftrightarrow \begin{matrix} 000 \\ 111 \end{matrix}$$

Notice that only two of the eight sequences of length 3 are used for transmission. □

Example 9.2 Parity-check code (3, 2)

This is a systematic code in which the third digit is a parity check on the first two digits. The code word is defined by the relations

$$x_1 = u_1, \qquad x_2 = u_2, \qquad x_3 = u_1 + u_2. \tag{9.2}$$

The encoder is shown in Fig. 9.5. The resulting coding rule is

$$\begin{matrix} 00 \\ 01 \\ 10 \\ 11 \end{matrix} \Leftrightarrow \begin{matrix} 000 \\ 011 \\ 101 \\ 110 \end{matrix}$$

Notice that only four of the eight sequences of length 3 are used for transmission. □

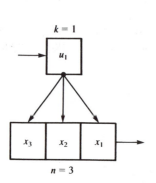

Figure 9.4 Encoder for a (3, 1) repetition code.

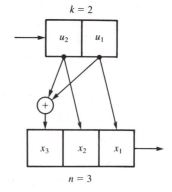

Figure 9.5 Encoder for a (3, 2) parity-check code.

Example 9.3   Hamming code (7, 4)

Again, this is a systematic code defined by the relations

$$x_i = u_i, \quad i = 1, 2, 3, 4,$$
$$x_5 = u_1 + u_2 + u_3,$$
$$x_6 = u_2 + u_3 + u_4, \tag{9.3}$$
$$x_7 = u_1 + u_2 + u_4.$$

The corresponding encoder is shown in Fig. 9.6. The resulting coding rule is

0000	0000 000
0001	0001 011
0010	0010 110
0011	0011 101
0100	0100 111
0101	0101 100
0110	0110 001
0111	0111 010
$\Leftrightarrow$	
1000	1000 101
1001	1001 110
1010	1010 011
1011	1011 000
1100	1100 010
1101	1101 001
1110	1110 100
1111	1111 111

Notice that only 16 of 128 sequences of length 7 are used for transmission. $\square$

These examples show that all the information required to specify the encoder operation is contained in relations of the type of (9.1), (9.2), and (9.3). With reference

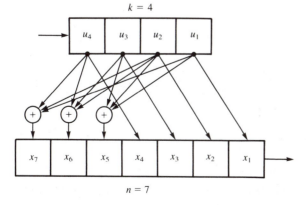

Figure 9.6   Encoder for a (7, 4) Hamming code.

Sec. 9.2   Block Codes

409

to Figs. 9.4, 9.5, and 9.6, these relations specify the connections between the input register stages and the adders. In fact, if the code is systematic, only the $(n - k)$ parity-check equations of the redundant digits must be assigned.

The information that specifies the encoding rule can be represented concisely in matrix form. Define a $k \times n$ binary matrix $\mathbf{G}$ in which the (i, j) entry is 1 if the ith stage of the input register is connected to the jth adder. Otherwise, it is 0. The matrix $\mathbf{G}$ is called the *generator matrix* of the block code. Using a vector notation for the source sequences and the code words, the encoding rule can be described through the equation

$$\mathbf{x} = \mathbf{u}\,\mathbf{G}. \tag{9.4}$$

It can be seen that obtaining a code word $\mathbf{x}$ through (9.4) is equivalent to summing the rows of the matrix $\mathbf{G}$ corresponding to the ones contained in the information sequence $\mathbf{u}$. For example, if the information sequence digits u_i, u_j, and u_l are ones and the remaining digits are zeros, then the corresponding code word is obtained by summing the ith, jth, and lth rows of the generator matrix.

Example 9.4

For the (7, 4) code of Example 9.3, the generator matrix $\mathbf{G}$ can be obtained by inspection of Fig. 9.6 as follows:

$$\mathbf{G} = \begin{array}{cc} \begin{array}{cccc} x_1 & x_2 & x_3 & x_4 \end{array} \quad \begin{array}{ccc} x_5 & x_6 & x_7 \end{array} & \\ \left[\begin{array}{cccc:ccc} 1 & 0 & 0 & 0 & 1 & 0 & 1 \\ 0 & 1 & 0 & 0 & 1 & 1 & 1 \\ 0 & 0 & 1 & 0 & 1 & 1 & 0 \\ 0 & 0 & 0 & 1 & 0 & 1 & 1 \end{array} \right] & \begin{array}{c} u_1 \\ u_2. \\ u_3 \\ u_4 \end{array} \end{array} \tag{9.5}$$

If we want the code word corresponding to the information sequence $\mathbf{u} = (1100)$, we must sum the first two rows of $\mathbf{G}$. We have

$$\begin{array}{l} 1\ 0\ 0\ 0\ 1\ 0\ 1 \\ \underline{0\ 1\ 0\ 0\ 1\ 1\ 1} \\ 1\ 1\ 0\ 0\ 0\ 1\ 0 \end{array}$$

and the result checks with the table of the code given in Example 9.3. □

When the code is systematic, the matrix $\mathbf{G}$ has the form

$$\mathbf{G} = [\mathbf{I}_k \mid \mathbf{P}], \tag{9.6}$$

where $\mathbf{I}_k$ is the $k \times k$ identity matrix and $\mathbf{P}$ is a $k \times (n - k)$ matrix containing the information regarding the parity checks. The knowledge of $\mathbf{P}$ completely defines the encoding rule for a systematic code.

The following important properties of parity check codes can be proved:

Property 1. Each code word is a sum of rows of the generator matrix.

Property 2. The block code consists of all the possible sums of the rows of the generator matrix.

Property 3. The sum of two code words is still a code word.

Property 4. The sequence of all zeros is always a code word of a parity check code.

Because of these properties, parity-check codes are also called *linear codes*. Linear block codes can also be interpreted as being a subspace of the vector space containing all the 2^n binary n-tuples. From this algebraic viewpoint, the rows of the generator matrix $\mathbf{G}$ are a basis of the subspace and contain k linearly independent code words. In fact all their 2^k linear combinations generate the entire subspace, that is, the code. For these reasons it is straightforward to conclude that any generator matrix of an (n, k) block code can be reduced, by way of row operations and column permutations, to the systematic form (9.6). Thus every (n, k) block code is equivalent to a systematic (n, k) block code (see Problem 9.5). Therefore, we can consider only systematic codes without loss of generality. An important parameter of a code word is its *Hamming weight*, that is, the number of ones that it contains. The set of all distinct weights in a code together with the number of code words of that weight is the *weight distribution* of the code.

Given two code words $\mathbf{x}_i$ and $\mathbf{x}_j$, it is useful to define a quantity to measure their difference, that is, the number of positions in which they differ. This quantity is the *Hamming distance* d_{ij} between the two code words, and clearly d_{ij} satisfies the condition $0 < d_{ij} \le n$. The smallest of these distances is called the *minimum distance* $d_{\min}$ of the code. The following property allows an easy computation of $d_{\min}$ for linear codes.

Property 5. The minimum distance of a linear block code is the minimum weight of its nonzero code words.

In fact, the distance between two binary sequences is equal to the weight of their modulo-2 sum and the sum of two code words is still a code word (Property 3).

Example 9.5

Consider again the (7, 4) Hamming code of Example 9.3. From the code table, we get the following weight distribution

Weight	Number of code words
0	1
3	7
4	7
7	1

Using Property 5, we get $d_{\min} = 3$. □

Error-detecting and error-correcting capabilities of a block code

Assume that the demodulator takes hard decisions so that the discrete channel between the channel encoder and decoder can be modeled as a BSC. Each transmitted code word $\mathbf{x}$ is received at the decoder input as a sequence $\mathbf{y}$ of n binary digits (Fig. 9.1). The code is systematic. Therefore, the first k digits of $\mathbf{y}$ are the received information digits, while the remaining $(n - k)$ ones are the received check digits. The sequence $\mathbf{y}$ can contain independent random errors caused by channel noise. Let us define a binary vector $\mathbf{e}$ called an *error vector*:

$$\mathbf{e} \triangleq [e_1, e_2, \ldots, e_n]. \tag{9.7}$$

Each component e_i is 1 if the channel has changed the ith transmitted digit; otherwise, it is 0. The received vector is then

$$\mathbf{y} = \mathbf{x} + \mathbf{e}, \tag{9.8}$$

where $\mathbf{x}$ is the transmitted code word.

The decoder task is to compare each of the $(n - k)$ received parity-check digits with the linear combination of the digits that formed the corresponding parity-check digit at the transmitter. If they match, the received sequence is a code word. Otherwise, an error has been detected. Therefore, at least for error detection, the decoding rule is very simple: an error pattern is detected whenever at least one of the $(n - k)$ controls on parity checks fails.

Let us define a vector $\mathbf{s}$ that contains the pattern of failures in the parity checks. Its $(n - k)$ binary digits are zeros for all parity checks that are satisfied and ones for those that are not. The vector $\mathbf{s}$ is called the *syndrome* of the received vector $\mathbf{y}$. Recalling the definition (9.6) of the generator matrix $\mathbf{G}$ of a systematic code, it can be verified that the syndrome can be obtained with the equation

$$\mathbf{s} = \mathbf{y}\,\mathbf{H}', \tag{9.9}$$

where we have introduced the *parity-check matrix* $\mathbf{H}$ defined as

$$\mathbf{H} \triangleq [\mathbf{P}' \mid \mathbf{I}_{n-k}]. \tag{9.10}$$

It is an $(n - k) \times n$ matrix whose rows represent the parity-check controls performed by the decoder. It follows that

$$\mathbf{G}\,\mathbf{H}' = \mathbf{0}, \tag{9.11}$$

where $\mathbf{0}$ is a $k \times (n - k)$ matrix with all-zero elements.

Example 9.6

Consider again the (7, 4) Hamming code of Example 9.3. The three parity-check controls performed by the decoder on the received sequence $\mathbf{y}$ can be written by inspection of (9.3) as follows:

$$\begin{aligned}
s_1 &= (y_1 + y_2 + y_3) + y_5, \\
s_2 &= (y_2 + y_3 + y_4) + y_6, \\
s_3 &= (y_1 + y_2 + y_4) + y_7.
\end{aligned} \tag{9.12}$$

The parity check matrix is therefore

$$\mathbf{H} = \begin{bmatrix} 1 & 1 & 1 & 0 & 1 & 0 & 0 \\ 0 & 1 & 1 & 1 & 0 & 1 & 0 \\ 1 & 1 & 0 & 1 & 0 & 0 & 1 \end{bmatrix}. \tag{9.13}$$

It can be verified that (9.13) is also obtained from (9.5) using the definition (9.10). The property (9.11) can also be verified. $\square$

From the definition of the syndrome associated with a received sequence $\mathbf{y}$, the following two properties can be verified:

Property 6. The syndrome associated with a sequence $\mathbf{y}$ is a zero vector if and only if $\mathbf{y}$ is a code word.

Property 7. The decoder can detect all the channel errors represented by vectors $\mathbf{e}$ that are not code words.

Since the channel can introduce 2^n different error vectors, only 2^k of them are not detected by the decoder, that is, those corresponding to the set of code words. Finally, since no code word exists with a weight less than d_{min} (except, of course, the all-zero code word), the following theorem can be proved.

THEOREM 9.1 A linear block code (n, k) with minimum distance d_{min} can detect all the error vectors of weight not greater than $(d_{min} - 1)$. ■

Until now, we have only explored the error-detection capabilities of a hard-decision decoder. The problem of error correction is more complicated since the syndrome does not contain sufficient information to locate the errors. Using (9.8), the expression (9.9) for the syndrome can be rewritten as

$$\mathbf{s} = \mathbf{y}\,\mathbf{H}' = (\mathbf{x} + \mathbf{e})\,\mathbf{H}', \tag{9.14}$$

where $\mathbf{x}$ is a code word. Since $\mathbf{x}\mathbf{H}' = \mathbf{0}$ (Property 6), there are 2^k different sequences $\mathbf{y}$ that give the same syndrome. They are obtained by summing to a given error vector $\mathbf{e}$ the 2^k code words. Therefore, given a transmitted code word $\mathbf{x}$, there are 2^k error vectors that give the same syndrome. Which one actually occurred is an uncertainty that cannot be removed using only the syndrome.

A suitable decoding algorithm must be elaborated. Assume that ML decisions are taken by the decoder. This means that it achieves minimum error probability on the received code words when they are equally likely. If p is the transition probability of the equivalent BSC implied by hard decisions, we have

$$P(\mathbf{y}|\mathbf{x}_i) = p^{d_i}(1 - p)^{n - d_i}, \tag{9.15}$$

where n is the block length and d_i is the Hamming distance between the received sequence $\mathbf{y}$ and the transmitted code word $\mathbf{x}_i$. Since $p < \frac{1}{2}$, the probability $P(\mathbf{y}|\mathbf{x}_i)$ is a monotonic decreasing function of d_i. Therefore, ML decoding is accomplished with *minimum distance decisions*. The "best" decoding algorithm decides for the code word $\mathbf{x}_i$, which is closer to $\mathbf{y}$. Recalling the discussion regarding (9.14), we can conclude that the minimum-distance decoding algorithm assumes that the error vector $\mathbf{e}$ that actually occurred is the minimum-weight error vector in the set of the 2^k error vectors that give the syndrome associated with the received sequence $\mathbf{y}$. Before considering this decoding rule in detail, using the minimum-distance decoding algorithm, let us relate the error-correcting capabilities of a block code to the code parameter d_{min}.

THEOREM 9.2 A linear block code (n, k), with minimum distance d_{min}, can correct all the error vectors containing no more than $t = \lfloor(d_{min} - 1)/2\rfloor$ errors, where $\lfloor a \rfloor$ denotes the largest integer contained in a. This code is denoted as a t-error-correcting code.

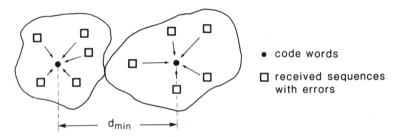

• code words

□ received sequences with errors

Figure 9.7 Qualitative representation of the decision regions assigned to the code words.

Proof. The decoding algorithm is implemented by assigning to each code word a decision region containing the subset of all the received sequences that are closer to it than to any other (see Fig. 9.7). An error vector with no more than $[(d_{min} - 1)/2]$ errors produces a received sequence lying inside the correct decision region. Error correction is therefore possible. ■

The results of Theorems 9.1 and 9.2 are summarized in Table 9.1.

Therefore, a design goal for a block-code (n, k) is to use the redundancy to achieve the largest possible minimum distance d_{min}. So far, no generalized solution to this problem is known. Instead, upper and lower bounds to d_{min} are used. Some of them will be described in Section 9.5.1.

TABLE 9.1 Error-correction and error-detection capabilities of a code with a given d_{min}

d_{min}	Errors detected	Errors corrected
2	1	0
3	2	1
4	3	1
5	4	2
6	5	2
7	6	3

Decoding table and standard array of a block code

Using the minimum-distance decoding algorithm just described, the decoding operation can be performed by looking for the code word nearest to the received sequence. This approach requires the storage of 2^k code words and repeated comparisons with the received sequence. The total storage requirement is of the order of $n2^k$ bits. Hence, the approach becomes rapidly impractical even for moderate-sized codes. Also, the comparison process is unacceptably long when n and k are large.

A more efficient approach is to evaluate the syndrome associated with the received sequence **y** assuming that the error vector **e** that actually occurred is the minimum-weight vector in the set of the 2^k ones that generate the same syndrome. With this approach we can build a decoding table by associating with each possible syndrome the error vector of minimum weight that generated it. The positions of the ones in the error vector indicate the digits that must be corrected in the received sequence $\mathbf{y}$. This approach is better clarified by the following example.

Example 9.7

The (7, 4) Hamming code has minimum distance 3. Thus, it is expected to correct all single errors. There are, of course, 128 possible received sequences and only 8 different syndromes. All these sequences are included in Table 9.2. They are grouped in rows containing all sequences that share the same syndrome. This is shown as the first entry in each row. The first column of the table contains all the error vectors of minimum weight. It can be verified by inspection that each error vector containing only one error has a different syndrome, and hence that it can be corrected.

Therefore, the decoding table for this code is the following:

Syndrome	Error vector	Digit in error
000	0000000	None
001	0000001	7
010	0000010	6
011	0001000	4
100	0000100	5
101	1000000	1
110	0010000	3
111	0100000	2

Obviously, the syndrome 000 corresponds to the set of the 16 code words. The syndrome 111 locates an error in the second position of the received sequence, and so on. Table 9.2 can also be interpreted as follows. Assume that the sequence 1101010 is received. The corresponding syndrome is 011. Therefore, an error in position 4 is assumed and corrected. The code word obtained, which is 1100010, appears at the top of the column containing the received sequence. □

A table such as Table 9.2, containing all the 2^n possible sequences of length n organized in that order, is called the *standard array* of the code. It has 2^k columns and 2^{n-k} rows. The rows are called *cosets*. The first sequence in each row is nominated a *coset leader*. The top sequence in a column is a code word, and each coset leader is the minimum-weight sequence that generates the syndrome that is common to all the sequences of that coset.

The decoding table is built by simply associating with each syndrome the coset leaders of the standard array. The coset leaders are therefore the correctable error patterns; if the error pattern is not a coset leader, then an incorrect decoding will be performed. To minimize the average error probability, the coset leaders must be the error patterns that are the most likely to occur. For a BSC, the coset leaders are the minimum-weight sequences associated with a given syndrome. Therefore, the decoding algorithm works as follows:

1. Compute the syndrome for the received sequence.
2. Find the correctable error pattern (coset leader) in the decoding table.

Sec. 9.2   Block Codes   **415**

TABLE 9.2 Standard array of the (7, 4) Hamming code

Syn-dromes	Coset leaders	Error patterns														
								Other errors								
000	0000000	1000101	0100111	0010110	0001011	1100010	0110001	1001110	0110010	0101100	0011101	0111010	1011000	1101001	1110100	1111111
001	0000001	1000100	0100110	0010111	0001010	1100011	0110000	1001111	0110010	0101101	0011100	0111011	1011001	1101000	1110101	1111110
010	0000010	1000111	0100101	0010100	0001001	1100000	1010001	1001100	0110011	0101111	0011111	0111000	1011010	1101011	1110110	1111101
011	0001000	1001101	0101111	0011110	0000011	1101010	1011001	1000110	0111001	0100100	0010101	0110010	1010000	1100001	1111100	1110111
100	0000100	1000001	0100011	0010010	0001111	1100110	1010101	1001010	0110101	0101000	0011001	0111110	1011100	1101101	1110000	1111011
101	1000000	0000101	1100111	1010110	1001011	0100010	0010001	0001110	1110001	1101110	1011101	1111010	0011000	0101000	0110100	0111111
110	0010000	1010101	0110111	0000110	0011011	1110010	1000001	1011110	0100001	0111100	0001101	0101010	1001000	1111001	1100100	1101111
111	0100000	1100101	0000111	0110110	0101011	1000010	1110001	1101110	0010001	0001100	0111101	0011010	1111000	1001001	1010100	1011111

416

3. Get the estimated code word by adding the correctable error pattern to the received sequence.

The decoding table requires a storage of 2^{n-k} syndromes of length $(n - k)$ and of 2^{n-k} error patterns of length n: a total of $2^{n-k} \times (2n - k)$ bits. For high rate codes $(k \cong n)$, the storage requirement is close to $n2^{n-k}$, considerably less as compared to the $n2^k$ bits required by an exhaustive search. In spite of this interesting result, the decoding table becomes impractical when n and k are large numbers. In that case a much more elaborate algebraic structure must be assigned to the code in order that decoding strategies based on computational algorithms, rather than on table look-ups, can be employed.

9.2.1 Hamming Codes

Equation (9.14) can be rewritten in the form

$$\mathbf{s} = \mathbf{e} \, \mathbf{H}'. \tag{9.16}$$

Therefore, the syndrome of a given sequence is the sum of the columns of $\mathbf{H}$ corresponding to the position of the ones in the error vector. Consequently, if a column of $\mathbf{H}$ is zero, an error in that position cannot be detected. Furthermore, if two columns of $\mathbf{H}$ are equal, a single error in one of those two positions cannot be corrected since the two syndromes are not distinct. We can conclude that a block code can correct all single errors if and only if the columns of the parity-check matrix $\mathbf{H}$ are nonzero and all different.

Hamming codes are characterized by a matrix $\mathbf{H}$ whose columns are all the possible sequences of $(n - k)$ binary digits except the zero sequence. For every $l = 2, 3, 4, \ldots$, there is a Hamming code $(2^l - 1, 2^l - 1 - l)$. These codes have $d_{\min} = 3$ and are thus capable of correcting all single errors. Their rate increases with increasing l. Some of these codes are listed in Table 9.3.

TABLE 9.3
Some Hamming codes

l	(n, k)
2	(3, 1)
3	(7, 4)
4	(15, 11)
5	(31, 26)
6	(63, 57)

Example 9.8

The parity-check matrix of the Hamming code (15, 11) is the following:

$$\mathbf{H} = \begin{bmatrix} 0 & 0 & 0 & 0 & 0 & 0 & 0 & 1 & 1 & 1 & 1 & 1 & 1 & 1 & 1 \\ 0 & 0 & 0 & 1 & 1 & 1 & 1 & 0 & 0 & 0 & 0 & 1 & 1 & 1 & 1 \\ 0 & 1 & 1 & 0 & 0 & 1 & 1 & 0 & 0 & 1 & 1 & 0 & 0 & 1 & 1 \\ 1 & 0 & 1 & 0 & 1 & 0 & 1 & 0 & 1 & 0 & 1 & 0 & 1 & 0 & 1 \end{bmatrix}. \tag{9.17}$$

Notice that $\mathbf{H}$ is not written in the systematic form of (9.10). It can be reduced to systematic form by a simple rearrangement of columns. The interesting property of (9.17) is that each column is a binary number that identifies the column position. Therefore, an error vector with a single error will generate a syndrome that gives, in binary form, the position of the error in the received sequence. □

The Hamming codes have an interesting property that can be verified by inspection of the standard array [see Table 9.2 for the (7, 4) code]. All the possible received sequences have distance 1 from one of the code words. Codes of this type are called *perfect codes*. Another property of the Hamming codes is that they are one of the few classes of codes for which the complete weight distribution is known. If we define the *weight-enumerating function* of the code as the polynomial in the indeterminate z,

$$A(z) = \sum_{i=0}^{n} A_i z^i, \tag{9.18}$$

where A_i is the number of code words in the code with weight i, then for the Hamming codes we have

$$A(z) = \frac{1}{n+1}[(1+z)^n + n(1+z)^{(n-1)/2}(1-z)^{(n+1)/2}]. \tag{9.19}$$

The result of Example 9.5 can be checked with (9.19).

Each Hamming code can be converted to a new code by appending one additional parity symbol that checks all the previous n symbols of the code word. This results in a class of $(2^l, 2^l - 1 - l)$ block codes called *extended Hamming codes*. Their parity-check matrix $\mathbf{H}_{ext}$ is obtained by adding a new row to the Hamming parity-check matrix $\mathbf{H}$ as follows:

$$\mathbf{H}_{ext} = \left[\begin{array}{c|c} & \begin{matrix} 0 \\ 0 \\ \cdot \\ \cdot \\ \cdot \\ 0 \end{matrix} \\ \mathbf{H} & \\ \hline 1\ 1\ .\ .\ .\ 1 & 1 \end{array} \right]. \tag{9.20}$$

The last row represents the overall parity-check digit. Since, with an overall parity check, the weight of every code word must be even, the extended Hamming codes have $d_{min} = 4$. The particular structure of these codes makes it possible to detect all double errors while *simultaneously* correcting all single errors (as in the original Hamming codes). In fact, the syndromes for double errors form a subset distinct from that of the syndromes for single errors. The decoding algorithm works as follows:

1. If the last digit of the syndrome is 1, then the number of errors must be odd. Using the minimum-distance algorithm, correction can be performed as for the Hamming codes.

2. If the last digit of the syndrome is 0, but the syndrome is not all-zero, then no correction is possible since at least two errors must have occurred. Double errors are therefore detected.

This property of extending a code by the addition of an overall parity check can be applied to any linear block code other than the Hamming codes. In particular, any linear (n, k) block code with an odd minimum distance can be converted into an extended $(n + 1, k)$ block code with a minimum distance increased by one.

9.2.2 Dual Codes

The generator matrix $\mathbf{G}$ and the parity-check matrix $\mathbf{H}$ of a linear (n, k) block code are related by (9.11). This relation can be rewritten as

$$\mathbf{H\,G'} = \mathbf{0}. \tag{9.21}$$

Thus the two matrices can be interchanged and the $\mathbf{H}$ matrix can be the generator matrix of a new $(n, n - k)$ block code. Codes that are so related are said to be *dual codes*. There is a very interesting relationship between the weight distributions of two dual codes. Let $A(z)$ be the weight enumerator of the (n, k) block code and $B(z)$ the weight enumerator of its $(n, n - k)$ dual code. Then the two weight enumerators are related by the identity (MacWilliams and Sloane, 1977)

$$B(z) = 2^{-k}(1 + z)^n A\left(\frac{1 - z}{1 + z}\right). \tag{9.22}$$

This relationship has proved to be very useful in determining the weight structure of many block codes through a computer exhaustive search performed on the dual codes.

9.2.3 Maximal-Length Codes

The *maximal-length codes* are the duals of the Hamming codes. Therefore, for every $l = 2, 3, 4, \ldots$ there is a maximal-length code $(2^l - 1, l)$. Its generator matrix is the parity-check matrix of the corresponding Hamming code. One property of these codes is that their weight distribution can be easily determined by introducing (9.19) into (9.22). The weight enumerator $B(z)$ for the maximal-length codes is thus found to be

$$B(z) = 1 + (2^l - 1)z^{2^{l-1}}. \tag{9.23}$$

Hence, all the nonzero code words have identical weight 2^{l-1}. Also, this is the minimum distance of the code. These codes are also called *equidistant* or *simplex* codes. Additional insight into the properties of these codes will be obtained later when discussing cyclic codes.

9.2.4 Reed–Muller Codes

The *Reed–Muller codes* are a class of linear block codes covering a wide range of rates and minimum distances. They offer very interesting properties. However, a serious disadvantage is that they have low code rates. These codes are defined in the following fashion. For any m and $r < m$, there is a Reed–Muller code with parameters given by

$$n = 2^m, \qquad k = \sum_{i=0}^{r} \binom{m}{i}, \qquad d_{\min} = 2^{m-r}. \tag{9.24}$$

The generator matrix $\mathbf{G}$ of the rth-order Reed–Muller code is defined by assigning a set of vectors as follows. Let $\mathbf{v}_0$ be a vector whose 2^m components are all ones and let $\mathbf{v}_1, \mathbf{v}_2, \ldots, \mathbf{v}_m$ be the rows of a matrix with all possible m-tuples as columns. The rows of the rth-order generator matrix are the vectors $\mathbf{v}_0, \mathbf{v}_1, \ldots, \mathbf{v}_m$ and all their products two at a time, three at a time, up to r at a time. Here the product vector $\mathbf{v}_i\mathbf{v}_j$ has components given by the products of the corresponding components of $\mathbf{v}_i$ and $\mathbf{v}_j$.

Example 9.9

In this example, we demonstrate the generation of Reed–Muller codes when $m = 3$. There are two codes. They have the following parameters:

r	n	k	$d_{\min}$
1	8	4	4
2	8	7	2

The vectors used for building the generator matrices are given in Table 9.4. The first-order code ($r = 1$) is generated by using the vectors $\mathbf{v}_0, \mathbf{v}_1, \mathbf{v}_2, \mathbf{v}_3$ as rows of the generator matrix. The second-order code ($r = 2$) is generated by augmenting this matrix with the additional three rows of Table 9.4. □

TABLE 9.4 Vectors for deriving the generator matrices of Reed–Muller codes with $m = 3$

$\mathbf{v}_0$	1	1	1	1	1	1	1	1
$\mathbf{v}_1$	0	0	0	0	1	1	1	1
$\mathbf{v}_2$	0	0	1	1	0	0	1	1
$\mathbf{v}_3$	0	1	0	1	0	1	0	1
$\mathbf{v}_1\mathbf{v}_2$	0	0	0	0	0	0	1	1
$\mathbf{v}_1\mathbf{v}_3$	0	0	0	0	0	1	0	1
$\mathbf{v}_2\mathbf{v}_3$	0	0	0	1	0	0	0	1

The first-order Reed–Muller codes are closely related to the maximal-length codes. If a maximal-length code is extended by adding an overall parity check, we obtain an *orthogonal code*. This code has 2^m code words. Each has a weight 2^{m-1}, except for the all-zero code word. Therefore, every code word agrees in 2^{m-1} positions and disagrees in 2^{m-1} positions with every other code word. If this code is transmitted using an antipodal signaling scheme, each code word is represented by one out of 2^m orthogonal signals. This explains the name "orthogonal" code. For the case $m = 3$, the code generator matrix consists of the three rows $\mathbf{v}_1, \mathbf{v}_2,$ and $\mathbf{v}_3$ of Table 9.4. In fact, the first column represents the overall parity-check digit, whereas the other columns are all the seven possible triples of binary digits. The first-order Reed–Muller code is obtained from this code (the orthogonal code) by adding to the generator matrix the all-ones vector (vector $\mathbf{v}_0$). In terms of the transmitted signals, this operation adds to the original orthogonal signal set the opposite of each signal. For this reason, the code is also called a *biorthogonal code*. Finally, notice that the rth-order Reed–Muller code is the dual of the code of order $(m - r - 1)$.

9.2.5 Cyclic Codes

The *cyclic codes* are a special class of parity-check codes that present a large amount of mathematical structure. These codes, of course, share all the properties previously described for parity-check codes, but in addition they have peculiar properties that allow easy encoding operations and usually simple decoding algorithms. Cyclic codes are, for this reason, of great practical interest. An (n, k) linear block code is a cyclic code *if and only if any cyclic shift of a code word produces another code word.*

Example 9.10

It can be verified that the (7, 4) Hamming code of Example 9.3 is a cyclic code. Take, for instance, the code word 0111010. There are six different cyclic shifts of this code word.

$$1110100 \quad 1101001 \quad 1010011 \quad 0100111 \quad 1001110 \quad 0011101.$$

They all belong to the code word set. The same is true for all the code words. $\square$

In dealing with cyclic codes it is useful to represent a binary sequence with the coefficients of a polynomial in the indeterminate D. The binary digits of a code word will be numbered in decreasing order from $(n - 1)$ to 0 so that each index matches the exponent of D. Indeed, the algebraic structure of cyclic codes is related to the algebraic properties of polynomials. A code word $\mathbf{x} = (x_{n-1}, x_{n-2}, \ldots, x_0)$ of a cyclic code is represented by the code polynomial $x(D)$ as follows:

$$x(D) = x_{n-1}D^{n-1} + x_{n-2}D^{n-2} + \cdots + x_1 D + x_0. \tag{9.25}$$

The coefficients of this polynomial are binary. They will be manipulated with the rules of modulo-2 arithmetic. In this new notation the code words of an (n, k) linear block code are in a one-to-one correspondence with code polynomials of degree not greater than $(n - 1)$. If $x(D)$ is a code polynomial of a cyclic code, then a cyclic shift of the code word (say to the left) of i positions generates another code polynomial that we denote $x^{(i)}(D)$. Theorem 9.3 defines this major property of cyclic codes.

THEOREM 9.3 The code polynomial $x^{(i)}(D)$ is the remainder resulting from dividing $D^i x(D)$ by $(D^n + 1)$; that is,

$$D^i x(D) = q(D)(D^n + 1) + x^{(i)}(D), \tag{9.26}$$

where $q(D)$ is the quotient polynomial of degree not greater than $(i - 1)$.

Proof

$$D^i x(D) = x_{n-1}D^{n-1+i} + x_{n-2}D^{n-2+i} + \cdots + x_{n-i}D^n + \cdots + x_0 D^i.$$

Sum to this expression twice the terms $x_{n-1}D^{i-1}, x_{n-2}D^{i-2}, \ldots, x_{n-i}$; this is possible because $(x_{n-i} + x_{n-i}) = 0$. We get

$$D^i x(D) = x_{n-1}D^{i-1}(D^n + 1) + x_{n-2}D^{i-2}(D^n + 1) + \cdots + x_{n-i}(D^n + 1)$$
$$+ x_{n-i-1}D^{n-1} + \cdots + x_1 D^{i+1} + x_0 D^i + x_{n-1}D^{i-1} + \cdots + x_{n-i},$$

and finally,

$$D^i x(D) = (x_{n-1}D^{i-1} + x_{n-2}D^{i-2} + \cdots + x_{n-i})(D^n + 1)$$
$$+ x_{n-i-1}D^{n-1} + \cdots + x_0 D^i + \cdots + x_{n-i}$$
$$= q(D)(D^n + 1) + x^{(i)}(D). \blacksquare$$

As we are interested in binary sequences of length n, in the following every polynomial of degree greater than n must be reduced mod $(D^n + 1)$.

Example 9.11

Let us take again the code word 0111010 of Example 9.10. The corresponding code polynomial is

$$x(D) = D^5 + D^4 + D^3 + D.$$

A shift of four positions to the left generates the code polynomial $x^{(4)}(D)$, which is obtained by dividing $D^4 x(D)$ by $(D^7 + 1)$ as follows:

$$
\begin{array}{r|l|l}
D^7 + 1 & D^9 + D^8 + D^7 + D^5 & D^2 + D + 1 \\
\cline{1-1}
& D^9 + D^2 & \text{quotient} \\
\cline{2-2}
& D^8 + D^7 + D^5 + D^2 & \\
& D^8 + D & \\
\cline{2-2}
& D^7 + D^5 + D^2 + D & \\
& D^7 + 1 & \\
\cline{2-2}
& D^5 + D^2 + D + 1 & \text{remainder}
\end{array}
$$

The remainder is $D^5 + D^2 + D + 1$, that is, the sequence 0100111. This sequence is obtained from the original one with a four-position shift to the left. $\square$

Using the polynomial description of cyclic codes, we now want to exploit the algebraic properties of their generator matrices. Let us first proceed through an example that will also enable us to define an important theorem.

Example 9.12

The (7, 4) Hamming code of Example 9.3 was already claimed to be cyclic in Example 9.10. Let us rewrite its generator matrix (9.5) in polynomial form as follows:

$$\mathbf{G}(D) = \begin{bmatrix} D^6 + & & & D^2 + & 1 \\ & D^5 + & & D^2 + D + 1 \\ & & D^4 + & D^2 + D \\ & & & D^3 + & D + 1 \end{bmatrix}. \tag{9.27}$$

Consider the last row of this generator matrix, that is, the polynomial

$$g(D) = D^3 + D + 1. \tag{9.28}$$

This code polynomial must have a 1 in the last position (coefficient of D^0); otherwise, six cyclic shifts to the left would generate a code word with $k = 4$ information digits equal to 0 and a parity-check section containing $n - k = 3$ ones, which is impossible. Furthermore, this is the only polynomial of degree $n - k = 3$ in the code. In fact, if there were another, it could be added to $g(D)$ to generate another code word presenting again an all-zero information section with a nonzero parity section. As a conclusion, there is a unique code polynomial $g(D)$ of degree $n - k = 3$, and this polynomial has always the form

$$g(D) = D^3 + \cdots + 1. \qquad (9.29)$$

Let us now derive the remaining rows of the generator matrix (9.27). The third row can be obtained with one cyclic shift to the left of the last row. Should a 1 appear in the fourth position, the last row can be added to cancel it. Each row of $\mathbf{G}(D)$ can be obtained in a similar way from the row below. The result is

$$\mathbf{G}(D) = \begin{bmatrix} (D^3 + D + 1)\, g(D) \\ (D^2 + 1) \qquad g(D) \\ D \qquad g(D) \\ g(D) \end{bmatrix}. \qquad (9.30)$$

All the rows of $\mathbf{G}(D)$ are multiples of the polynomial $g(D)$. But, since all code words in the code are linear combinations of the rows of $\mathbf{G}(D)$, we can conclude that all the code polynomials are multiples of the polynomial $g(D)$. $\square$

The important result obtained in the previous example is generalized in Theorem 9.4.

THEOREM 9.4   Given an (n, k) cyclic code, there is a unique code polynomial of degree $(n - k)$ that has the form

$$g(D) = D^{n-k} + \cdots + 1. \qquad (9.31)$$

All the other $2^k - 1$ code polynomials are multiples of $g(D)$, and every polynomial of degree $(n - 1)$ or less that is divisible by $g(D)$ must be a code polynomial.

Proof.   Generalize the development of Example 9.12. ∎

The polynomial $g(D)$ defined by Theorem 9.4 is called the *generator polynomial* of the cyclic code. Any cyclic code is completely defined by its generator polynomial. The natural question now is whether there exists an (n, k) cyclic code for any n and k and which is the corresponding generator polynomial. Theorem 9.5 is the answer.

THEOREM 9.5   The generator polynomial $g(D)$ of an (n, k) cyclic code is a divisor of $(D^n + 1)$. Conversely, every divisor of $(D^n + 1)$ of degree $(n - k)$ generates an (n, k) cyclic code.

Proof. Consider the polynomial $D^k g(D)$ of degree n and divide it by $(D^n + 1)$. We get

$$D^k g(D) = (D^n + 1) + g^{(k)}(D), \qquad (9.32)$$

where $g^{(k)}(D)$ is a polynomial of degree not greater than $(n - 1)$. Using (9.26), we can conclude that $g^{(k)}(D)$ is a code polynomial obtained with k cyclic shifts to the left of $g(D)$. Therefore, it is also a multiple of $g(D)$, say $m(D)g(D)$. From (9.32), we get

$$D^n + 1 = \{D^k + m(D)\}g(D) = h(D)g(D) \qquad (9.33)$$

and the direct part of the theorem is proved.

Conversely, let $g(D)$ be a divisor of $(D^n + 1)$ of degree $(n - k)$ and consider the k polynomials $g(D)$, $Dg(D)$, . . . , $D^{k-1}g(D)$ of degree $(n - k)$, $(n - k + 1)$, . . . , $(n - 1)$. There are 2^k linear combinations of these k polynomials. Each of them is a multiple of $g(D)$ and together they form an (n, k) linear code. Let $x(D)$ be one of these code polynomials and consider

$$D^i x(D) = q(D)(D^n + 1) + x^{(i)}(D), \qquad (9.34)$$

where $x^{(i)}(D)$ is a cyclic shift of $x(D)$. Since both $x(D)$ and $(D^n + 1)$ are multiples of $g(D)$, then $x^{(i)}(D)$ is also a multiple of $g(D)$. Furthermore, it can be expressed as a linear combination of the aforementioned k polynomials. It follows that $x^{(i)}(D)$ is a code polynomial and that the linear code is cyclic. ∎

Finally, notice that if $g(D)$ divides $(D^m + 1)$ as well as $(D^n + 1)$, with $m < n$, then $(D^m + 1)$ is a code word in the cyclic code (n, k) whose minimum distance is therefore 2. To avoid this drawback, n must be taken as the smallest integer, such as $(D^n + 1)$ is a multiple of $g(D)$.

Considerable algebraic results are available regarding the properties of the polynomials $(D^n + 1)$. In particular, tables of divisors for different values of n can be found. One of these is reproduced in Table 9.5. These tables are very useful because the design of a cyclic code with preassigned properties reduces to an appropriate selection of divisors for the generator $g(D)$.

Notice that the Table 9.5 considers only odd values of n, because in binary algebra we have $D^{2m} + 1 = (D^m + 1)^2$. Furthermore, the values of $n = 3, 5, 11, 13, 19, 29, 37, 53, 59, 61$ are omitted from the table. In fact, for these values the factorization is simply $(D^n + 1) = (D + 1)(D^{n-1} + D^{n-2} + \cdots + D + 1)$. Therefore, Table 9.5 gives the factors of $(D^n + 1)$ for $n \leq 63$ and $n = 127$. The factors are given in octal notation, with the lowest-degree terms on the left. As an example, the second line of the table means that the coefficients of the factors are, respectively, 110, 111, and 100100100, and we obtain

$$1 + D^9 = (1 + D)(1 + D + D^2)(1 + D^3 + D^6).$$

Encoding algorithms for cyclic codes

Given the generator polynomial $g(D)$ of a cyclic code (n, k), the code polynomial $x(D)$ corresponding to an information sequence $u(D)$ can be obtained from Theorem 9.4 as

$$x(D) = u(D)g(D). \qquad (9.35)$$

This simple algorithm does not actually generate a systematic code, as is verified in the following example.

TABLE 9.5 Factors of the polynomial ($D^n + 1$). Each polynomial factor is given in octal notation with the lowest-degree terms on the left. (Reprinted with permission from MacWilliams and Sloane, 1977.)

n	Factors
7	6.54.64.
9	6.7.444.
15	6.7.46.62.76.
17	6.471.727.
21	6.7.54.64.534.724.
23	6.5343.6165.
25	6.76.4102041.
27	6.7.444.4004004.
31	6.45.51.57.67.73.75.
33	6.7.4522.6106.7776.
35	6.54.64.76.57134.72364.
39	6.7.57074.74364.77774.
41	6.5747175.6647133.
43	6.47771.52225.64213.
45	6.7.46.62.76.444.40044.44004.
47	6.43073357.75667061.
49	6.54.64.40001004.40200004.
51	6.7.433.471.637.661.727.763.
55	6.76.7776.5551347.7164555.
57	6.7.5604164.7565674.7777774.
63	6.7.54.64.414.444.534.554.604.634.664.714.724.
127	6.406.422.436.442.472.516.526.562.576.602.626.646.652. 712.736.742.756.772.

Example 9.13

The (7, 4) Hamming code described in Example 9.12 has the generator $g(D) = D^3 + D + 1$. Let us find the code word corresponding to the information sequence 1101. Since

$$u(D) = D^3 + D^2 + 1,$$

we get

$$x(D) = (D^3 + D^2 + 1)(D^3 + D + 1) = D^6 + D^5 + D^4 + D^3 + D^2 + D + 1.$$

The code word is then 1111111 and the code is not systematic. All the other code words can be obtained in the same way. $\square$

The algorithm based on (9.35) can be modified to get the code words of the systematic cyclic code (n, k). Given the information sequence $u(D)$, let us multiply it by D^{n-k} and divide by the generator polynomial $g(D)$. We have

$$D^{n-k}u(D) = q(D)g(D) + r(D), \tag{9.36}$$

where $q(D)$ and $r(D)$ are, respectively, the quotient and the remainder of the division. Notice that $r(D)$ must have a degree $(n - k - 1)$ or less, since the degree of $g(D)$ is $(n - k)$. Rearranging (9.36), we get

$$D^{n-k}u(D) + r(D) = q(D)g(D). \tag{9.37}$$

This is the key for the desired encoding algorithm. In fact, the LHS of this equation is a polynomial of degree $(n - 1)$ or less, a multiple of $g(D)$, and hence a code polynomial in the cyclic code generated by $g(D)$. Let us write it explicitly:

$$u_{k-1}D^{n-1} + u_{k-2}D^{n-2} + \cdots + u_0 D^{n-k} + r_{n-k-1}D^{n-k-1} + \cdots + r_0. \quad (9.38)$$

Therefore, the code word consists of the k information digits followed by the $(n - k)$ parity check digits and the code is systematic. In conclusion, *the parity check section of each code word in a systematic cyclic code (n, k) is obtained as the remainder of the division of $D^{n-k} u(D)$ by the generator polynomial $g(D)$.*

Example 9.14

Given the (7, 4) cyclic code with the generator polynomial $g(D) = D^3 + D + 1$, let us find the code word for the information sequence 1101 using the encoding algorithm just described. We have

$$D^3 u(D) = D^6 + D^5 + D^3.$$

Therefore,

$$
\begin{array}{r|l|l}
D^3 + D + 1 & D^6 + D^5 + \qquad D^3 & D^3 + D^2 + D + 1 \\
\hline
 & D^6 \qquad + D^4 + D^3 & \text{quotient} \\
\hline
 & D^5 + D^4 \\
 & D^5 \qquad + D^3 + D^2 \\
\hline
 & D^4 + D^3 + D^2 \\
 & D^4 \qquad + D^2 + D \\
\hline
 & \qquad D^3 \qquad + D \\
 & \qquad D^3 \qquad + D + 1 \\
\hline
 & \qquad\qquad\qquad 1 \\
\end{array}
$$

The code word is $x(D) = D^6 + D^5 + D^3 + 1$, corresponding to the binary form 1101001. The code is systematic. □

The encoding algorithm based on (9.37) requires the division of $D^{n-k}u(D)$ by the generator polynomial $g(D)$ to get the remainder $r(D)$. This is the parity-check section of the code polynomial corresponding to $u(D)$. Therefore, the implementation of the algorithm requires a circuit that performs a division. This task can be accomplished by a shift register having $(n - k)$ stages, the degree of the divisor, and suitable feedback connections that correspond to the coefficients of the divisor. The circuit shown in Fig. 9.8 performs the division described in Example 9.14. At each clock pulse, the digits of the dividend are fed in leftwise starting with the most significant digit, and the quotient is shifted out rightward. The remainder is what remains in the register when all seven digits of the dividend have been fed in. Notice in particular that the feedback connections correspond to the structure of the divisor.

A first possible version of the encoder for the (7, 4) code of Example 9.14 is

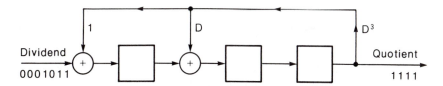

Figure 9.8 Circuit to divide by $g(D) = D^3 + D + 1$. When all seven digits of the dividend have been fed in, the contents $r_2 r_1 r_0$ of the register are the remainder of the division. For the example shown, the remainder will be 001.

shown in Fig. 9.9. The switches have three position: First, at Ⓐ for four clock pulses, during which the four information digits are fed into the register and sent into the channel; second, at Ⓑ for three clock pulses, while three zeros enter the register; third, at Ⓒ for three clock pulses, while the remainder of the division is sent into the channel. The disadvantage of this implementation is that the channel remains idle while the switches are at Ⓑ. To overcome this, the message digits can be fed into the right end of the shift register. This is equivalent to multiplying the symbols by D^3 as they come in. Hence, the divisor circuit of Fig. 9.10 is used, instead of that of Fig. 9.8. The remainder of the division is now available in the register as soon as the last message digit has been fed in. The implementation of the encoder based on this concept is

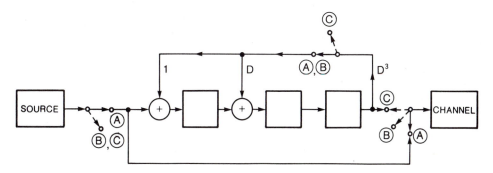

Figure 9.9 First implementation of the encoder for the (7, 4) code with generator polynomial $g(D) = D^3 + D + 1$. The switches are at position Ⓐ for 4 clock pulses, at Ⓑ for 3 clock pulses, and at Ⓒ for 3 clock pulses.

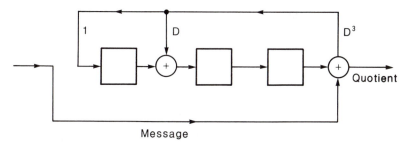

Figure 9.10 Circuit to divide by $g(D) = D^3 + D + 1$. Only the four message digits are fed to the shift register at its right end.

shown in Fig. 9.11. The switches are at Ⓐ for four clock pulses and at Ⓑ for three clock pulses. The operation of this encoder is described in detail in the following example.

Example 9.15

We reproduce here the situation described in Example 9.14. The code word for the sequence 1101 is obtained by shifting in the circuit of Fig. 9.11 the information sequence and shifting out the remainder of the division. The contents of the shift register, at each step, are

Input digit	Register contents		
u_i	r_0	r_1	r_2
—	0	0	0
1	1	1	0
1	1	0	1
0	1	0	0
1	1	0	0

As soon as the four information digits are entered into the register and delivered to the channel, the register contains the sequence 001. This corresponds to the remainder $r(D) = 1$. □

An encoder similar to that of Fig. 9.11 will work for any cyclic code. It requires $(n - k)$ delay elements in the shift register, and the generator polynomial is reflected in the feedback connections structure. For codes with $k < (n - k)$, a simpler circuit with a k-stage shift register can be implemented. To this purpose, let $x(D) = q(D)g(D)$ be a code polynomial, and define the *parity-check polynomial* as

$$h(D) \triangleq \frac{D^n + 1}{g(D)}. \tag{9.39}$$

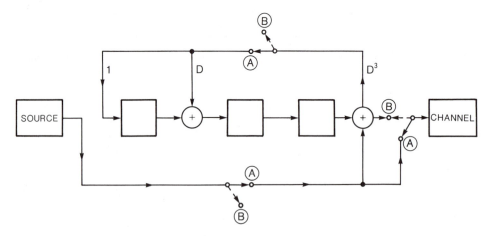

Figure 9.11 Second implementation of the encoder for the (7, 4) code with generator polynomial $g(D) = D^3 + D + 1$. The switches are at Ⓐ for 4 clock pulses and at Ⓑ for 3 clock pulses.

Its degree is k. Using (9.39), we have

$$x(D)h(D) = q(D)(D^n + 1). \tag{9.40}$$

Since the degree of $q(D)$ is $(k - 1)$ or less, we can conclude that the polynomial $x(D) h(D)$ does not contain the powers from D^k to D^{n-1}. This property will be used to implement another encoding circuit. If the code is systematic, the first k coefficients of $x(D)$, that is, $x_{n-1}, x_{n-2}, \ldots, x_{n-k}$, correspond to the k information digits u_{k-1}, $u_{k-2}, \ldots, u_0$, while $x_{n-k-1}, \ldots, x_1, x_0$ are the $(n - k)$ parity-check digits. Let h_i be the coefficients of $h(D)$. Notice that $h_0 = 1$ and $h_k = 1$ as a consequence of the properties of $g(D)$. The property stated for the coefficients of (9.40) can now be written explicitly as

$$\sum_{i=0}^{k} h_i x_{n-i-j} = 0, \qquad 1 \le j \le (n - k), \tag{9.41}$$

and therefore

$$x_{n-k-j} = \sum_{i=0}^{k-1} h_i x_{n-i-j}, \qquad 1 \le j \le (n - k). \tag{9.42}$$

This equation permits us to derive all the $(n - k)$ parity-check digits of each code word from the k information digits. For example, letting $j = 1$, we get from (9.42) the first parity-check digit as

$$\begin{aligned}
x_{n-k-1} &= h_0 x_{n-1} + h_1 x_{n-2} + \cdots + h_{k-1} x_{n-k} \\
&= u_{k-1} + h_1 u_{k-2} + \cdots + h_{k-1} u_0.
\end{aligned} \tag{9.43}$$

Thus, the (n, k) cyclic code generated by $g(D)$ is also completely specified by the parity-check polynomial $h(D)$. The corresponding encoder for the $(7, 4)$ code generated by $g(D) = D^3 + D + 1$ is shown in Fig. 9.12. Its operation is examined in detail in the following example.

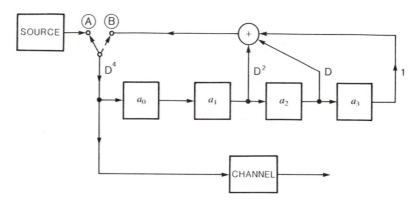

Figure 9.12 Encoder for the $(7, 4)$ code based on the parity-check polynomial $h(D)$ $= D^4 + D^2 + D + 1$. The switch is at Ⓐ for 4 clock pulses and at Ⓑ for 3 clock pulses.

Example 9.16

The (7, 4) Hamming code with generator polynomial $g(D) = D^3 + D + 1$ has a parity-check polynomial given by

$$h(D) = \frac{D^7 + 1}{g(D)} = D^4 + D^2 + D + 1. \tag{9.44}$$

Since $h_4 = 1$, $h_3 = 0$, $h_2 = 1$, $h_1 = 1$, and $h_0 = 1$, we get from (9.42) the following three equations:

$$x_2 = x_6 + x_5 + x_4 = u_3 + u_2 + u_1,$$
$$x_1 = x_5 + x_4 + x_3 = u_2 + u_1 + u_0, \tag{9.45}$$
$$x_0 = x_4 + x_3 + x_2 = u_3 + u_2 + u_0.$$

Taking into account the different notation for the symbols x and u, it can be verified that the parity check equations (9.45) are those indicated in the rows of the parity-check matrix **H** given for the same code by (9.13). The three parity-check symbols of (9.45) are derivd from the circuit of Fig. 9.12. The switch is at Ⓐ for four clock cycles and at Ⓑ for three clock cycles. In the first time interval, the four message digits are fed into the channel and into the shift register. In the second interval, the circuit computes successively the three check digits according to the following table (message sequence is 1101):

	Register contents				
Clock cycle	a_0	a_1	a_2	a_3	Output digit
4	1	0	1	1	—
5	0	1	0	1	0
6	0	0	1	0	0
7	1	0	0	1	1

The generated code word is therefore again 1101001. ◻

The encoder of Fig. 9.12 requires k delay elements in the shift register, and the connections to the adder reflect the structure of the polynomial $h(D)$. This encoder will work for any cyclic code and will be more convenient than the circuit of Fig. 9.11 only when $k < (n - k)$.

Finally, notice that the polynomial $h(D)$ contains the information represented by the parity check matrix **H**. Therefore, $h(D)$ can be used as the generator polynomial of the dual of the code generated by $g(D)$. The two polynomials are related by (9.39).

Error detection and error correction with cyclic codes

Assume that a code polynomial is transmitted over a noisy channel. Analogously with (9.8), the received sequence can be written in polynomial form as

$$y(D) = x(D) + e(D), \tag{9.46}$$

where $x(D)$ is the code polynomial and $e(D)$ is the error polynomial. Let us now divide $y(D)$ by the generator polynomial of the code. We get

$$y(D) = m(D)g(D) + s(D), \tag{9.47}$$

where $m(D)$ is the quotient and $s(D)$ the remainder of the division. Since only the code polynomials are multiples of the generator polynomial, $y(D)$ will be a code word if and only if the polynomial $s(D)$ is zero. This polynomial of degree not greater than $(n - k - 1)$ is the syndrome of $y(D)$. Notice that, since $x(D) = q(D) g(D)$, we can compare (9.46) with (9.47) and obtain

$$e(D) = [m(D) + q(D)]g(D) + s(D). \tag{9.48}$$

This equation shows that $s(D)$ is also the syndrome of $e(D)$. In conclusion, error detection can be accomplished by simply checking the remainder of the division of the received polynomial $y(D)$ by the generator $g(D)$. The detection circuit can be implemented with a circuitry similar to that shown in Fig. 9.8. The register is first set to zero and the received sequence is shifted in. The contents of the register will represent the syndrome $s(D)$ as soon as the last digit of $y(D)$ is entered.

One additional property of the syndrome is stated in Theorem 9.6.

THEOREM 9.6   If $s(D)$ is the syndrome of an error polynomial $e(D)$, the syndrome of $D^i e(D)$, that is, of $e(D)$ shifted cyclically i places to the left, is obtained by shifting $s(D)$ i places inside the division circuit.

Proof. First notice that we have

$$e(D) = q(D)g(D) + s(D) \tag{9.49}$$

and therefore

$$D^i e(D) = D^i q(D)g(D) + D^i s(D). \tag{9.50}$$

This shows that the remainder of the division of $D^i e(D)$ by $g(D)$ is simply the remainder of the division of $D^i s(D)$ by $g(D)$. Remembering the operation of the circuit of Fig. 9.8, this remainder is obtained by shifting the syndrome i times into the division circuit. ∎

The syndrome $s(D)$ of a received sequence $y(D)$ can be obtained using the encoder circuit of the type of Fig. 9.9. Also the circuit of Fig. 9.10 (or of Fig. 9.11) can be used to the same end. With this circuit, however, the calculated syndrome is that of the sequence $D^{n-k}y(D)$.

The properties of the syndrome generating circuits are best understood through an example.

Example 9.17

Let us use again the (7, 4) cyclic code with generator polynomial $g(D) = D^3 + D + 1$ and assume a received sequence $y(D) = D^6 + D^5 + 1$. First, let us derive the syndrome $s(D)$ from (9.47).

$$
\begin{array}{r|l|l}
D^3 + D + 1 & D^6 + D^5 \qquad\qquad\qquad + 1 & D^3 + D^2 + D \\
\hline
 & D^6 \qquad + D^4 + D^3 & \text{quotient} \\
\hline
 & D^5 + D^4 + D^3 \qquad + 1 & \\
 & D^5 \qquad + D^3 + D^2 & \\
\hline
 & D^4 \qquad + D^2 + 1 & \\
 & D^4 \qquad + D^2 + D & \\
\hline
 & D + 1 & \text{remainder}
\end{array}
$$

This syndrome corresponds to the vector 011. This can also be obtained by using the parity-check matrix (9.13). The circuit of Fig. 9.13 can be used to derive the syndrome $s(D) = D + 1$. The contents of the register at the successive steps are given in the following table:

Received digit	Register contents		
	s_0	s_1	s_2
—	0	0	0
1	1	0	0
1	1	1	0
0	0	1	1
0	1	1	1
0	1	0	1
0	1	0	0
1	1	1	0

If, instead, the circuit of Fig. 9.14 is used as syndrome generator, we get in a similar manner

Received digit	Register contents		
	s_0	s_1	s_2
—	0	0	0
1	1	1	0
1	1	0	1
0	1	0	0
0	0	1	0
0	0	0	1
0	1	1	0
1	1	0	1

This syndrome, that is, $s(D) = D^2 + 1$, is that of the sequence $D^3y(D) = D^3 + D^2 + D$ mod $(D^7 + 1)$. Using Theorem 9.6, the same syndrome can be obtained by shifting three steps the syndrome $s(D) = D + 1$ in the register of Fig. 9.13. The result is given in the following table:

Shifts	Register contents		
	s_0	s_1	s_2
—	1	1	0
1	0	1	1
2	1	1	1
3	1	0	1

$\square$

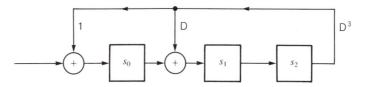

Figure 9.13 First implementation of the syndrome generator for the (7, 4) code with generator polynomial $g(D) = D^3 + D + 1$.

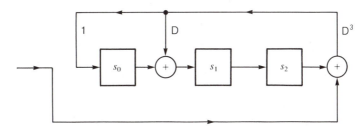

Figure 9.14 Second implementation of the syndrome generator for the (7, 4) code with generator polynomial $g(D) = D^3 + D + 1$.

Also for cyclic codes, error correction is more complicated than error detection. In principle, the decoder must find a correctable error pattern $e(D)$ from the syndrome $s(D)$. The transmitted code word $x(D)$ is then obtained by adding $e(D)$ to the received sequence $y(D)$. Whether or not this is a practical operation depends on the complexity of the decoder that computes the correctable error pattern $e(D)$. Special classes of codes have been developed that lead to practical algorithms. Some of them will be mentioned later. But the thorough description of error-correcting schemes is beyond the scope of this book. However, one simple technique that is applicable in the cases of single-error correction will be described hereafter. It is based on the following general theorem.

THEOREM 9.7 If the errors of $e(D)$ are confined to the $(n - k)$ parity-check positions of $y(D)$, the syndrome $s(D)$ is identical to $e(D)$.

Proof. The assumption is equivalent to saying that $e(D)$ is a polynomial of degree not greater than $(n - k - 1)$. Therefore, the division by $g(D)$, which has degree $(n - k)$, gives as a remainder just $e(D)$. ∎

Under the conditions of Theorem 9.7, error correction is accomplished by simply adding the syndrome to the $(n - k)$ received parity-check digits. Should the errors be confined to $(n - k)$ consecutive digits different from the parity-check section, then the use of Theorem 9.6 allows again for error correction. In fact, the errors can also in this case be confined in the parity-check section by shifting cyclically the received sequence to the left by i places. The syndrome of this new sequence is that of $D^i y(D)$. Let us apply these concepts to an example.

Example 9.18

Let us use once again the (7, 4) code with generator $g(D) = D^3 + D + 1$. Assume that

$$x(D) = D^6 + D^5 + D^3 + 1,$$
$$e(D) = D^3.$$

Therefore, as in Example 9.17, the received sequence is $y(D) = D^6 + D^5 + 1$. Since $y(D)$ is shifted into the syndrome generator from the rightmost stage, it corresponds to a preshifted sequence $D^{n-k}y(D) = D^3y(D)$. Using the syndrome generator of Fig. 9.14, we get from Example 9.17 that $s(D) = D^2 + 1$. Therefore, an error located in position D^j in $y(D)$ corresponds to an error in position $D^{n-k+j} = D^{3+j}$ in the preshifted sequence. When $j = n - 1 = 6$, an error occurs in the first position of $y(D)$ and appears in position $(n - k + j) = 9$ of the preshifted sequence. Taking into account the end-around shift, this position is the highest of the parity-check section (in fact, $9 - 7 = 2$). Due to Theorem 9.7, the syndrome corresponding to this situation is $s(D) = D^2$ (sequence 100). We can now apply Theorem 9.6.

The syndrome $D^2 + 1$ is shifted inside the division circuit. When the syndrome D^2 is identified, this means that the single error is in the first position of the cyclically shifted received sequence.

These concepts are applied to the error-correcting circuit of Fig. 9.15. It consists of the syndrome generator of Fig. 9.14, a buffer, and an AND gate with $(n - k) = 3$ inputs. The received sequence is shifted into both the buffer and the syndrome generator. While it is read out from the buffer, the syndrome is simultaneously shifted into the register to identify the register contents corresponding to $s(D) = D^2$. When this is the case, the error

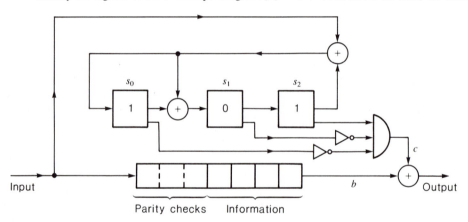

Syndrome			Shift	b	c	Output
s_0	s_1	s_2				
1	0	1	–	1	0	1
1	0	0	1	1	0	1
0	1	0	2	0	0	0
0	0	1	3	0	1	1
1	1	0	4	0	0	0
0	1	1	5	0	0	0
1	1	1	6	1	0	1

Figure 9.15 Error-correction circuit for the (7, 4) cyclic code generated by $g(D) = D^3 + D + 1$. The table shows the correction operation that is accomplished while the received sequence is shifted out from the buffer and the initial syndrome 101 is shifted cyclically inside the register.

is identified and the digit that will be shifted out will be corrected. The reader is invited to work out the details using Fig. 9.15. □

The example has shown that one particular syndrome, that is, $s(D) = D^2$ corresponding to 100, allows the location and capture of the single error while the received sequence is shifted out from the buffer. This technique for error correction is called *error-trapping decoding*. It can be extended to cases other than single-error correction (see Section 9.2.8).

9.2.6 Special Classes of Cyclic Codes

Cyclic Hamming codes

The Hamming codes described in Section 9.2.1 can also be shown to be a special case of cyclic codes. To this purpose, let us define an *irreducible polynomial* as a polynomial of degree l that is not divisible by any polynomial of degree less than l and greater than zero. Furthermore, an irreducible polynomial $g(D)$ of degree l is called *primitive* when the smallest integer n, such that $(D^n + 1)$ is a multiple of $g(D)$, is $2^l - 1$. Therefore, invoking Theorem 9.5, this primitive polynomial can generate a $(2^l - 1, 2^l - l - 1)$ cyclic code that is a Hamming code. It has been shown that this code can correct all single errors. Let us reconsider the proof using polynomial notation. Let two single-error patterns be $e_i(D) \triangleq D^i$ and $e_j(D) \triangleq D^j$, where $0 \leq i < j < n$. It will be sufficient to show that the two corresponding syndromes, say $s_i(D)$ and $s_j(D)$, are different. Using (9.49), we have

$$e_i(D) \triangleq D^i = q_i(D)g(D) + s_i(D), \tag{9.51}$$
$$e_j(D) \triangleq D^j = q_j(D)g(D) + s_j(D),$$

Summing these two equations, we get

$$D^i(D^{j-i} + 1) = [q_i(D) + q_j(D)]g(D) + s_i(D) + s_j(D). \tag{9.52}$$

Since $j - i < n = 2^l - 1$ and $g(D)$ is primitive, then $g(D)$ cannot divide $(D^{i-j} + 1)$ and, consequently, $s_i(D) \neq s_j(D)$ for $i \neq j$. Cyclic Hamming codes can be decoded by using the error-trapping algorithm as in Example 9.18.

A list of primitive polynomials that generate Hamming codes is given in Table 9.6 for different values of l. The polynomials are given in octal notation, with the lowest-degree terms on the left. For example, the first line of the table means

$$64 \Rightarrow 110100 \Rightarrow g(D) = 1 + D + D^3.$$

Golay codes

In searching for perfect codes, Golay discovered a (23, 12) code that is a cyclic code with generator polynomial

$$g(D) = D^{11} + D^9 + D^7 + D^6 + D^5 + D + 1 \tag{9.53}$$

and with minimum distance $d_{\min} = 7$. Therefore, triple error correction is possible. The important point is that this code is the only possible nontrivial linear binary perfect

TABLE 9.6 Primitive polynomials of degree l. Each polynomial is given in octal notation with the lowest-degree terms on the left.

l	Primitive polynomial	l	Primitive polynomial
3	64	14	60421
4	62	15	600004
5	51	16	640042
6	604	17	440001
7	442	18	4020004
8	561	19	7100002
9	4204	20	4400001
10	4402	21	50000004
11	5001	22	60000002
12	62404	23	41000001
13	66002	24	702000004

code with multiple error-correcting capabilities. Besides the Hamming single-error-correcting codes, the repetition codes (with n odd), and the Golay codes, no other linear binary perfect codes exist (see MacWilliams and Sloane, 1977, Chapter 6).

Bose–Chaudhuri–Hocquenghem (BCH) codes

This class of cyclic codes is one of the most useful for correcting random errors mainly because decoding algorithms can be implemented with an acceptable amount of complexity. For any pair of positive integers m and t, there is a binary BCH code with the following parameters:

$$n = 2^m - 1, \qquad n - k \leq mt, \qquad d_{\min} \leq 2t + 1.$$

This code can correct all combinations of t or fewer errors. Therefore, it is called a t-error-correcting code. The generator polynomial for this code can be constructed from factors of $(D^{2^m - 1} + 1)$. Unfortunately, this procedure is not straightforward and is beyond the scope of this book. A list of generator polynomials for BCH codes of different parameters is given in Table 9.7. The polynomials are represented in octal notation, with the highest-degree terms on the left.† Thus, the third line of the table means

$$721 \Rightarrow 111010001 \Rightarrow g(D) = D^8 + D^7 + D^6 + D^4 + 1.$$

Notice that this polynomial can be factored as $g(D) = (D^4 + D + 1)(D^4 + D^3 + D^2 + D + 1)$; it can be verified from Table 9.5 that these two factors are factors of $(D^{15} + 1)$.

The BCH codes provide a large class of codes. They are useful not only because of flexibility in the choice of parameters (block length and code rate), but also because

---

† Be careful with this table, because the octal notation must be decoded differently from Tables 9.5 and 9.6.

TABLE 9.7 List of generator polynomials for BCH codes. Each polynomial is represented in octal notation with the highest-degree terms on the left. (Copyright © 1964 IEEE; reprinted with permission from Stenbit, 1964.)

n	k	t	$g(D)$
7	4	1	13
15	11	1	23
	7	2	721
	5	3	2467
31	26	1	45
	21	2	3551
	16	3	107657
	11	5	5423325
	6	7	313365047
63	57	1	103
	51	2	12471
	45	3	1701317
	39	4	166623567
	36	5	1033500423
	30	6	157464165547
	24	7	17323260404441
	18	10	1363026512351725
	16	11	6331141367235453
	10	13	472622305527250155
	7	15	5231045543503271737
127	120	1	211
	113	2	41567
	106	3	11554743
	99	4	3447023271
	92	5	624730022327
	85	6	130704476322273
	78	7	26230002166130115
	71	9	6255010713253127753
	64	10	1206534025570773100045
	57	11	335265252505705053517721
	50	13	54446512523314012421501421
	43	14	17721772213651227521220574343
	36	15	31460746665220750447645747217 35
	29	21	403114461367670603667530141176155
	22	23	123376070404722522435445626637647043
	15	27	22057042445604554770523013762217604353
	8	31	7047264052751030651476224271567733130217
255	247	1	435
	239	2	267543
	231	3	156720665
	223	4	75626641375
	215	5	23157564726421
	207	6	16176560567636227
	199	7	7633031270420722341
	191	8	2663470176115333714567
	187	9	52755313540001322236351
	179	10	22624710717340432416300455
	171	11	15416214212342356077061630637
	163	12	7500415510075602551574724514601

TABLE 9.7 (*Continued*)

n	k	t	$g(D)$
	155	13	37575130054076650157225064646777633
	147	14	164213017353716552530416530544101711
	139	15	461401732060175561570722730247453567445
	131	18	215713331471510151261250277442142024165471
	123	19	12061405224206600371721032651614122262 72506267
	115	21	60526665572100247263636404600276352556 313472737
	107	22	22205772322066256312417300235347420176 574750154441
	99	23	10656667253473174222741416201574332252 411076432303431
	91	25	67502650303274441727236317247325110755 50762720724344561
	87	26	11013676341474323643523163430717204620 6722545273311721317
	79	27	66700035637657500020270344207366174621 01532671176654134235
	71	29	24024710520644321515554172112331163205 44425036255764322170603
	63	30	10754475055163544325315217357707003666 11172645526761365670254330
	55	31	73154252035011001330152753060320543254 14326755010557044426035473617
	47	42	25335420170626465630330413774062331751 2333414544604500506602455254317
	45	43	15202056055234161131101346376423701563 67002447076237303320215702505154
	37	45	51363302550670074141774472454375304207 3570617432343234764435473740304400
	29	47	30257155366730714655270640123613771153 42242324201174114060254757410403565037
	21	55	12562152570603326560017731536076121032 27341405653074542521153121614466513473725
	13	59	46417320050525645444265737142500660043 30677445476561403174677213570261344 60500547
	9	63	15726025217472463201031043255355134614 162367212044074545112766115547705561677516057

at block lengths of a few hundred or less many of these codes are among the best-known codes of the same length and rate. For the decoding algorithms, see Berlekamp (1968, Chapter 7), Clark and Cain (1981, Chapter 5), and Blahut (1983, Chapters 7 and 9).

Reed–Solomon codes

These codes are a subclass of BCH codes generalized to the nonbinary case, that is, to a symbol set of size $q = 2^m$. Thus, each symbol can still be represented as a binary m-tuple, and the code can be considered as a special type of binary code (see Blahut, 1983, Chapter 7). The parameters of a Reed–Solomon code are the following:

Symbol	m binary digits
Block length	$n = (2^m - 1)$ symbols
	$= m(2^m - 1)$ binary digits
Parity checks	$(n - k) = 2t$ symbols
	$= 2mt$ binary digits

These codes are capable of correcting all combinations of t or fewer symbol errors. Alternatively, interpreted as binary codes, they are well suited for correction of bursts of errors (see Section 9.2.8). However, the most important application of these codes is in the concatenated coding scheme described in Section 9.4.

Shortened cyclic codes

Since the generator polynomial of a cyclic code must be a divisor of $(D^n + 1)$, it often happens that its possible degree $(n - k)$ does not cover all combinations of n and k that satisfy practical needs. To avoid this difficulty, cyclic codes are often used in a shortened form. To this purpose, the first i information digits are assumed to be always zero and are not transmitted. In this way a new $(n - i, k - i)$ code is derived whose code words are a subset of the code words of the original code. The code is called *shortened cyclic code* although it is not cyclic. The new code has at least the same minimum distance as the code from which it is derived. The encoding and syndrome calculation can be accomplished by the same circuits employed in the original code, since the leading string of zeros does not affect the parity-check computations. Error correction can be accomplished by prefixing to each received vector a string of i zeros (or by modifying accordingly the related circuitry). Therefore, these codes share all the implementation advantages of cyclic codes and are also of practical interest.

Majority logic decodable codes

Most of these codes are cyclic codes and their interest relies on a decoding scheme that can be implemented using simple circuits based on layers of majority gates. These codes are slightly inferior to BCH codes in terms of error-correcting capabilities and efficiency, but decoding is implemented in an easier way.

The Reed–Muller codes described in Section 9.2.4 belong to this class. They are extended cyclic codes, that is, cyclic codes with an overall parity check on each code word. Reed–Muller codes are one of the oldest and best understood families of codes and will be used in the following example to illustrate the idea of majority logic decoding.

Example 9.19

Consider the first-order ($r = 1$) Reed–Muller code (8, 4) whose generator matrix (see Example 9.9) is

$$\mathbf{G} = \begin{bmatrix} 11111111 \\ 00001111 \\ 00110011 \\ 01010101 \end{bmatrix} \begin{matrix} \mathbf{v}_0 \\ \mathbf{v}_1 \\ \mathbf{v}_2 \\ \mathbf{v}_3 \end{matrix}. \tag{9.54}$$

This code has $d_{min} = 4$. Therefore, it can correct all single errors. The information sequence $\mathbf{u} = (u_0, u_1, u_2, u_3)$ is coded into the code word $\mathbf{x} = (x_0, x_1, \ldots, x_7)$ with the product $\mathbf{x} = \mathbf{uG}$. For instance, the sequence $\mathbf{u} = (0101)$ gives the code word $\mathbf{x} = (01011010)$. It can be verified that the three symbols u_1, u_2, and u_3 satisfy the following equations:

$$\begin{aligned}
u_1 &= x_0 + x_1, & u_2 &= x_0 + x_2, & u_3 &= x_0 + x_4, \\
u_1 &= x_2 + x_3, & u_2 &= x_1 + x_3, & u_3 &= x_1 + x_5, \\
u_1 &= x_4 + x_5, & u_2 &= x_4 + x_6, & u_3 &= x_2 + x_6, \\
u_1 &= x_6 + x_7, & u_2 &= x_5 + x_7, & u_3 &= x_3 + x_7.
\end{aligned} \tag{9.55}$$

Therefore, we have four "votes" for each of the three symbols. If only one error occurs, the majority "vote" is still correct and the symbols u_1, u_2, and u_3 can be correctly decoded. Assume, for instance, that the received sequence $\mathbf{y}$ is 11011010; that is, there is an error in the first position. Equations (9.55) give

	"Votes" for:		
	u_1	u_2	u_3
	0	1	0
	1	0	1
	1	0	1
	1	0	1
Majority decision	1	0	1

Therefore, a majority decision recovers correctly these three symbols. It remains to determine u_0. To this end, form the vector

$$\mathbf{x}' = \mathbf{y} + \sum_{i=1}^{3} u_i \mathbf{v}_i.$$

We obtain

11011010	$\mathbf{y}$
00001111	$u_1 \mathbf{v}_1$
00000000	$u_2 \mathbf{v}_2$
01010101	$u_3 \mathbf{v}_3$
10000000	$\mathbf{x}'$

Again, a majority "vote" gives the correct decision, that is, $u_0 = 0$. This idea of majority decisions can be generalized to obtain a decoding algorithm for all Reed–Muller codes. □

9.2.7 Maximal-Length (Pseudonoise) Sequences

The code words of the cyclic $(2^l - 1, l)$ simplex (or maximal-length) code of Section 9.2.3 resemble random sequences of zeros and ones. In fact, we shall see that any nonzero code word of this code has many of the properties that we would expect from a binary sequence obtained by tossing a coin $2^l - 1$ times.

Maximal-length codes are the duals of the Hamming codes. Remember that a Hamming code of length $2^l - 1$ is generated by a primitive polynomial $g(D)$ of degree l. The dual code of the same length can be obtained by letting the same $g(D)$ to be its parity-check polynomial. The dual code can therefore be generated by using an l-stage encoder of the type of Fig. 9.12 with feedback connections reflecting the structure of $g(D)$. For purposes of clarification, we use the following example.

Example 9.20

The dual code of the (7, 4) Hamming code generated by $g(D) = D^3 + D + 1$ is a (7, 3) cyclic code with $g(D)$ as the parity-check polynomial. A three-stage encoder for the dual code is shown in Fig. 9.16. This scheme is a slight modification of the encoder type shown in Fig. 9.12. The register is first loaded (from left to right) with the information sequence. Then the register content is shifted out (seven steps) from the right. In Fig. 9.16 the generation of the code word, corresponding to the sequence 100, is shown. The successive states of the register are also shown in Fig. 9.16. The last column of the table is the desired code word. In the dual code, all the code words, with the exception of that which is all zero, are different cyclic shifts of a single code word. This property is understood by considering the evolution of the states of the shift register of the encoder of Fig. 9.16. When the register is initially loaded and shifted $2^3 - 1$ times, it cycles through each possible $2^3 - 1$ states. Then it returns to the original one. The output sequence, when indefinitely shifted out, is periodic with period $2^3 - 1$. Since there are only $2^3 - 1$ possible states, this period corresponds to the largest possible one in this register. This explains the name of maximal-length sequence and why the $2^3 - 1$ code words of the cyclic code are different cyclic shifts of one code word. □

The example can be generalized to show that the encoder of a maximal-length code can be used to generate maximal-length sequences of period $2^l - 1$. Primitive

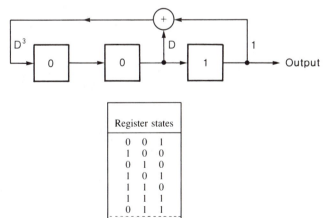

Figure 9.16 Shift-register circuit for encoding the dual code of the Hamming (7, 4) code with generator $g(D) = D^3 + D + 1$. The circuit generates also a PN sequence of length $2^3 - 1 = 7$.

polynomials (see Table 9.6) are suitable for the generation of these sequences. As already stated, these sequences are also called *pseudonoise (PN) sequences*. They present the following pseudorandomness properties:

Property 1. In any segment of length $2^l - 1$ of the sequence, there are exactly 2^{l-1} ones and $2^{l-1} - 1$ zeros. That is, the number of ones and the number of zeros are nearly equal. This property is an immediate consequence of the fact that the considered binary sequence is a code word of the simplex code, whose weight is constant and always equal to 2^{l-1} [see (9.23)].

Property 2. If we define a *run* to be the maximal string of consecutive identical symbols, then in any segment of the PN sequence of length $2^l - 1$ one-half of the runs have length 1, one-quarter have length 2, one-eighth have length 3, and so on. In each case, the number of runs of zeros is equal to the number of runs of ones.

Property 3. The most relevant property is related to the autocorrelation function of the PN sequence. Let us define the autocorrelation function of an infinite real sequence (a_i) of period n as

$$r_m \triangleq \frac{1}{n} \sum_{i=0}^{n-1} a_i a_{i+m}, \qquad m = 0, \pm 1, \pm 2, \ldots \qquad (9.56)$$

Notice that r_m is periodic, of period n. If the sequence is binary, let us replace 1's by -1's and 0's by $+1$'s. Thus, from (9.56), we get

$$r_m = \frac{1}{n} \sum_{i=0}^{n-1} (-1)^{a_i + a_{i+m}} = \frac{A - D}{n}, \qquad (9.57)$$

where A and D are the number of places where the sequence $(a_0 a_1 a_2 a_3 \ldots a_{n-1})$ and its cyclic shift $(a_m a_{m+1} \ldots a_{m+n-1})$ agree and disagree, respectively (so $A + D = n$). Therefore, for a sequence of period $n = 2^l - 1$, we have

$$r_0 = 1$$

$$r_m = -\frac{1}{n}, \qquad \text{for } l \le m \le 2^l - 2. \qquad (9.58)$$

In the sense of minimizing the magnitude of r_m, for $m \ne 0$, this is the "best" possible autocorrelation function of any binary sequence of period n.

PN sequences are very useful in practice when it is desired to obtain sequences with random properties. To this purpose, the same PN sequence is added modulo-2 to the sequence at hand both at the transmitter and receiver side (Fig. 9.17). This is

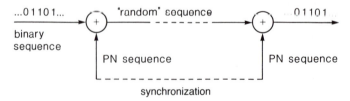

Figure 9.17 Scrambling and descrambling a binary sequence.

possible as the PN sequence is deterministic. The only requirement is that in the two additions the two PN sequences be synchronized. The operation is called *scrambling*.

9.2.8 Codes for Burst-Error Detection and Correction

In this section, we abandon the model of BSC and assume a channel model in which errors tend to be clustered in *bursts*. This is a typical situation in certain communication media such as magnetic tapes, magnetic disks, and magnetic memories. In other cases, the channel is basically an AWGN that is occasionally disturbed by large bursts of noise or radio-frequency interference. In general, when burst errors dominate, codes designed for correcting random errors are very inefficient. Nevertheless, again cyclic codes are very useful in this situation.

Let us define a burst of length b as an error pattern in which the errors are confined to b consecutive positions. Therefore, a burst-error pattern of length b can be represented by the polynomial:

$$e(D) = D^i e_b(D), \tag{9.59}$$

where D^i locates the burst into the error sequence of length n and $e_b(D)$ is a polynomial of the type $e_b(D) = D^{b-1} + \cdots + 1$.

THEOREM 9.8   Any cyclic code (n, k) can detect all bursts whose length is not greater than $(n - k)$.

Proof. The syndrome of such bursts is the remainder of the division of $D^i e_b(D)$ by the generator polynomial $g(D)$. But this syndrome is always different from zero since neither D^i nor $e_b(D)$ are multiples of $g(D)$, provided that $b \leq (n - k)$. ∎

When error correction is required, Theorem 9.9 provides a lower bound on the degree of the generator polynomial of the code.

THEOREM 9.9   A burst-error correcting code can correct all bursts of length b or less provided that the number of check digits satisfies the inequality (*Reiger bound*)

$$n - k \geq 2b. \tag{9.60}$$

Proof. Consider that to correct all bursts of length b the bursts of length $2b$ (or less) must be different from each code word. In fact, if a code word is a burst of length $2b$ (or less), it can be expressed as the sum of two bursts of length b (or less). Consider the standard array of the code. If one of the two bursts (the correctable one) is a coset leader, the other, as a consequence of the assumption made on the code word, must be in the same coset. Therefore, the second burst can not be corrected. In conclusion, no burst of length $2b$ (or less) can be allowed to be a code word in order to correct all bursts of length b. When this condition is met, the number of check digits is at least $2b$. In fact, consider the sequences whose nonzero components are confined to the first $2b$ positions. There are 2^{2b} such sequences. These sequences must be in different cosets of the standard array. Otherwise, their sum would be a code word corresponding to a burst of length $2b$ (or less). Since the cosets are 2^{n-k}, then the inequality (9.60) follows. ∎

As a consequence of Theorem 9.9, the ratio

$$z \triangleq \frac{2b}{n-k} \qquad (9.61)$$

can be assumed as a measure of the burst-correcting efficiency of the code. Some decoding algorithms for burst-error correction are based on error-trapping techniques (see Peterson and Weldon, 1972, Chapters 8 and 11).

A list of efficient cyclic codes and shortened cyclic codes for correcting short bursts is given in Table 9.8. The polynomials are again represented in octal notation, with the highest-degree terms on the left, as in Table 9.7.

TABLE 9.8 Efficient cyclic codes and shortened cyclic codes for burst correction. The generator polynomial is represented in octal notation with the highest-degree terms on the left. (Reprinted with permission from Lin, 1970.)

$n-k-2b$	Code (n, k)	Burst-correcting ability, b	Generator polynomial
0	(7,3)	2	35
	(15,9)	3	171
	(19,11)	4	1151
	(27,17)	5	2671
	(34,22)	6	15173
	(38,24)	7	114361
	(50,34)	8	224531
	(56,38)	9	1505773
	(59,39)	10	4003351
1	(15,10)	2	65
	(27,20)	3	311
	(38,29)	4	1151
	(48,37)	5	4501
	(67,54)	6	36365
	(103,88)	7	114361
	(96,79)	8	501001
2	(31,25)	2	161
	(63,55)	3	711
	(85,75)	4	2651
	(131,119)	5	15163
	(169,155)	6	55725
3	(63,56)	2	355
	(121,112)	3	1411
	(164,153)	4	6255
	(290,277)	5	24711
4	(511,499)	4	10451
5	(1023,1010)	4	22365

Fire codes

These codes are a versatile and systematic class of cyclic codes designed for correcting or detecting a single burst of length b in a block of n digits. Let $p(D)$ be an irreducible polynomial of degree $m \geq b$, and let e be the smallest positive integer such that $p(D)$ divides $(D^e + 1)$. Furthermore, assume that e and $(2b - 1)$ are relatively prime integers. Then the polynomial

$$g(D) = (D^{2b-1} + 1)p(D) \tag{9.62}$$

is the generator of a b burst-error correcting Fire code of length $n = \text{LCM}(e, 2b - 1)$, where LCM means least common multiple. Notice that the number of parity-check digits in these codes is $m + 2b - 1$. For the limit case of $m = b$, we get a burst-correcting efficiency that cannot exceed $\frac{2}{3}$. Under the same conditions as before, given two integers b and d, we can generate a Fire code capable of correcting any burst of length b (or less) and simultaneously detecting any burst of length up to $d \geq b$ by using the generator polynomial

$$g(D) = (D^c + 1)p(D), \tag{9.63}$$

where $c \geq b + d - 1$ (see Peterson and Weldon, 1972, Chapter 11).

Example 9.21

We want to design a Fire code to correct all bursts of length $b = 7$ (or less) and to detect all bursts of lengths 8 up to 10. We get $c \geq 16$ and $m = 7$. If we choose the primitive polynomial in Table 9.6, we get the following generator:

$$g(D) = (D^{16} + 1)(D^7 + D^3 + 1).$$

Since $p(D)$ is primitive, we have $e = 2^7 - 1 = 127$ and the length of the code is $n = 16 \times 127 = 2032$. Thus, the code is a (2032, 2009) Fire cyclic code with a high rate ($R_c = 0.99$) and a burst-correcting efficiency $z = 0.6$. Notice that the low value of $(n - k)$ makes it easy to implement the code. On the other hand, these codes usually have a high length, even for a modest burst-correcting capability. This is a disadvantage, since only one burst per each block length is correctable or detectable. Therefore, a very high guard space between successive bursts is required. $\square$

Interleaved codes

A practical technique to cope with burst errors is that of using random-error-correcting codes in connection with a suitable interleaver/deinterleaver pair. An *interleaver* is a device that rearranges the ordering of a sequence of symbols in a deterministic manner. The *deinterleaver* applies the inverse operation to restore the sequence to its original ordering. Given an (n, k) cyclic code, an (in, ik) *interleaved code* can be obtained by arranging i code words of the original code into i rows of a rectangular array that will be transmitted by columns. The parameter i is called the *interleaving degree* of the code. If the original code corrects up to t random errors, the interleaved code will have the same random-error-correction capability, but in addition it will be able to correct all bursts of length it (or less). The use of this technique is shown in Fig. 9.18 and explained in the following example.

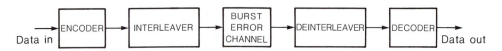

Data in → ENCODER → INTERLEAVER → BURST ERROR CHANNEL → DEINTERLEAVER → DECODER → Data out

Figure 9.18 Block diagram for the application of the interleaver–deinterleaver pair.

Example 9.22

Consider a BCH (15, 5) code, whose generator polynomial is, from Table 9.7, $g(D) = D^{10} + D^8 + D^5 + D^4 + D^2 + D + 1$. This code corrects all random errors with $t = 3$ (or less) errors in the sequences of length $n = 15$. Taking $i = 5$, we can derive a (75, 25) interleaved code. The arrangement of the code words is shown in Fig. 9.19. An information sequence of 25 digits is divided into five 5-digit message blocks and five code words of length 15 are generated using $g(D)$. These code words are arranged as five rows of the 5 $\times$ 15 matrix shown. The columns of the matrix are transmitted, in the indicated order, as a code word of length 75. Each burst of length 15 (or less) produces no more than three errors in each row of the matrix. A burst from position 18 to position 32 is shown by dashed squares in the figure. Therefore, the decoder can correct the errors by operating on each row. The interleaving process has, in fact, diffused the burst into isolated errors, and all error patterns containing three errors or less in each row of the matrix are correctable. □

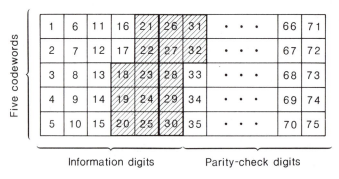

Figure 9.19 Scheme for the interpretation of a (75, 25) interleaved code derived from a (15, 5) BCH code. A burst of length $b = 15$ is spread into $t = 3$ error patterns in each of the five code words of the interleaved code.

9.2.9 Performance of Block Codes with Hard Decisions

In Chapter 5, different modulation schemes were compared on the basis of the bit error probability $P_b(e)$. The scope of this section is to provide useful tools for extending those comparisons to coded transmission. When block codes are used, the most natural performance measure is the *word error probability* $P_w(e)$, that is, the probability of delivering to the user a block of digits different from that transmitted. Let us start by discussing properly this performance measure.

We indicate with $P(\mathbf{y}_j \mid \mathbf{x}_i)$ the probability of receiving the sequence $\mathbf{y}_j$ when the code word $\mathbf{x}_i$ of the (n, k) code is transmitted over a BSC with transition probability p. Then the word error probability is given by

$$P_w(e) = 1 - \frac{1}{M} \left\{ \sum_{i=1}^{M} \sum_{j \in S_i} P(\mathbf{y}_j|\mathbf{x}_i) \right\}, \tag{9.64}$$

where $M = 2^k$ indicates the number of code words assumed to be equally likely and S_i the set of subscripts j of received sequences $\mathbf{y}_j$, which are decoded into the code word $\mathbf{x}_i$ according to Theorem 9.2. Notice that j runs from 1 to 2^n and that S_i is a subset of these integers. Unfortunately, the evaluation of this apparently simple expression is very hard (Elia, 1983) unless one resorts to exhaustive computations. Therefore, upper bounds to (9.64) have been sought.

If a t-error-correcting code is used, one simple assumption is that the error patterns with t (or less) errors are corrected, whereas all error patterns with more than t errors produce an erroneous decision. We get

$$P_w(e) \leq \sum_{m=t+1}^{n} P(m, n), \tag{9.65}$$

where $P(m, n)$ indicates the probability of an error pattern of weight m in a sequence of length n. On a BSC, all error patterns with the same weight are equally likely with probability

$$P(m, n) = \binom{n}{m} p^m (1 - p)^{n-m}. \tag{9.66}$$

Notice that the equal sign in (9.65) holds only for perfect codes.
When $np \ll 1$, (9.65) can be approximated by

$$P_w(e) \gtrsim P(t + 1, n) = \binom{n}{t + 1} p^{t+1}(1 - p)^{n-t-1}. \tag{9.67}$$

A general upper bound on $P_w(e)$ is obtained by following a derivation similar to that used to obtain the union bound (4.68). Let us define the error event e_{il} as follows. Whenever the transmitted code word is $\mathbf{x}_i$, the received sequence $\mathbf{y}_j$ lies closer to at least one code word $\mathbf{x}_l$ different from $\mathbf{x}_i$. Therefore,

$$e_{il} \triangleq \{\mathbf{y}_j: P(\mathbf{y}_j|\mathbf{x}_l) > P(\mathbf{y}_j|\mathbf{x}_i)|\mathbf{x}_i\}. \tag{9.68}$$

Denoting by S_{il} the set of subscripts j for which e_{il} occurs, we have

$$P(e_{il}) = \sum_{j \in S_{il}} P(\mathbf{y}_j|\mathbf{x}_i) \tag{9.69}$$

and the union bound (4.66) gives

$$P_w(e|\mathbf{x}_i) \leq \sum_{\substack{l=1 \\ l \neq i}}^{M} P(e_{il}). \tag{9.70}$$

To proceed further, we invoke the *uniform error property* of linear codes used on the BSC (see Problem 9.23). That is, the error probability for the ith code word is the same for all i. Therefore, we can assume that the all-zero code word $\mathbf{x}_1$ was transmitted and get

Sec. 9.2   Block Codes   **447**

$$P_w(e) = P_w(e|\mathbf{x}_1) \leq \sum_{l=2}^{M} P(e_{1l}). \tag{9.71}$$

Now we derive an upper bound for $P(e_{il})$ of (9.69). This result, introduced in (9.71), will give us the final answer.

Defining the function $f_l(\mathbf{y}_j)$ as

$$f_l(\mathbf{y}_j) \triangleq \begin{cases} 1, & \text{for } j \in S_{il}, \\ 0, & \text{for } j \notin S_{il}, \end{cases} \tag{9.72}$$

we can rewrite (9.69) as

$$P(e_{il}) = \sum_{j=1}^{2^n} f_l(\mathbf{y}_j)P(\mathbf{y}_j|\mathbf{x}_i), \tag{9.73}$$

where the summation has been extended over the whole set of received sequences $\mathbf{y}_j$. Now, we may easily bound $f_l(\mathbf{y}_j)$ by

$$f_l(\mathbf{y}_j) \leq \sqrt{\frac{P(\mathbf{y}_j|\mathbf{x}_l)}{P(\mathbf{y}_j|\mathbf{x}_i)}}. \tag{9.74}$$

Because of (9.68), this bound is verified for $j \in S_{il}$, whereas for $j \notin S_{il}$ it is trivial. Introducing (9.74) into (9.73), we get

$$P(e_{il}) \leq \sum_{j=1}^{2^n} \sqrt{P(\mathbf{y}_j|\mathbf{x}_l)P(\mathbf{y}_j|\mathbf{x}_i)}. \tag{9.75}$$

This expression is called the *Bhattacharyya bound*. Using the memoryless property of the BSC, (9.75) can be shown to lead to the result (see Problem 9.24):

$$P(e_{1l}) \leq \left[\sum_{y \in Y} \sqrt{P(y|x = 1)P(y|x = 0)} \right]^{w_l}, \tag{9.76}$$

where $Y = \{0, 1\}$, and w_l is the weight of the code word $\mathbf{x}_l$. Using the transition probability p of the BSC and introducing (9.76) into (9.71), we get

$$P_w(e) \leq \sum_{l=2}^{M} [\sqrt{4p(1 - p)}]^{w_l}. \tag{9.77}$$

A simpler, but weaker, bound is obtained if in (9.77) we employ the minimum distance d_{min} in place of the weight distribution of the code. That is,

$$P_w(e) \leq (M - 1)[\sqrt{4p(1 - p)}]^{d_{min}}. \tag{9.78}$$

We want now to relate $P_w(e)$ to the average bit error probability $P_b(e)$ on the message information digits. When at least $(t + 1)$ errors occur, the decoder delivers to the user an erroneous code word containing at least $d_{min} = 2t + 1$ errors in the code word of n digits. Therefore, $k(2t + 1)/n$ is the average number of erroneous message digits. Considering that in one code word there are k message digits, we have

$$kP_b(e) \cong (2t + 1)\frac{k}{n}P_w(e) \tag{9.79}$$

and therefore

$$P_b(e) \cong \frac{2t + 1}{n} P_w(e). \tag{9.80}$$

When only error detection is employed, it is of great interest to consider the probability $P_w(d)$ of error detection. If $P_w(c)$ is the probability of correctly decoding a code word, then we have

$$P_w(c) + P_w(d) + P_w(e) = 1. \tag{9.81}$$

In this case, the word error probability is equal to the probability that the error pattern is one of the $2^k - 1$ nonzero code words. Therefore,

$$P_w(e) = \sum_{m=d_{\min}}^{n} \frac{w(m)}{\binom{n}{m}} P(m, n), \tag{9.82}$$

where $w(m)$ is the number of code words of weight m. Hence, the computation of (9.82) requires knowledge of the weight distribution of the code. A rule of thumb that works for many codes (MacWilliams and Sloane, 1977) is that the weights $w(m)$ have an approximately binomial distribution. That is,

$$(9.83) \qquad\qquad w(m) \cong \frac{\binom{n}{m}}{2^{n-k}}.$$

Using this approximation in (9.82), we get

$$P_w(e) \cong \frac{1}{2^{n-k}} \sum_{m=d_{\min}}^{n} P(m, n). \tag{9.84}$$

If we introduce (9.84) in (9.81), noticing that $P_w(c) = P(0, n)$, we finally obtain

$$P_w(d) \cong \sum_{m=1}^{n} P(m, n) - \frac{1}{2^{n-k}} \sum_{m=d_{\min}}^{n} P(m, n). \tag{9.85}$$

In conclusion, consider that error detection is usually employed with transmission schemes involving retransmission. Therefore, the transmitter has to repeat the message whenever an error is detected. With these schemes, an important performance measure is the *throughput rate*, that is, the fraction of the source rate R_s at which information is delivered to the user. Of course, a low throughput rate entails an inefficient use of the channel even if the word error probability $P_w(e)$ is very low. All the results developed in this section will now be used to explore some examples.

Example 9.23

We consider here the performance of some of the BCH codes whose generator polynomials were given in Table 9.7. To this purpose, combining (9.80) with (9.67), we get for the bit error probability the following approximation:

$$P_b(e) \cong \frac{2t + 1}{n} \binom{n}{t + 1} p^{t+1}(1 - p)^{n-t-1}. \tag{9.86}$$

Remember that the block length n of the code and its error-correcting capability t are numbers that can be taken from Table 9.7. To calculate the error probability p of the BSC, we assume that a binary antipodal modulation is used. Therefore,

$$p = \frac{1}{2} erfc \left(\sqrt{R_c \frac{\mathcal{E}_b}{N_0}} \right). \tag{9.87}$$

Using (9.87) in (9.86), we can obtain $P_b(e)$ as a function of $\mathcal{E}_b/N_0$ for a given (n, k) BCH code. Some curves are plotted in Fig. 9.20. All the codes considered have a rate R_c of about 0.5. Thus, the required bandwidth expansion is about 2. Comparing Fig. 9.20 with Fig. 9.2, the reader can gain an immediate perception of how much these codes fill the region of potential coding gains. Indeed, it can be verified that substantial coding gains can be obtained, for example, at $P_b(e) = 10^{-5}$. With a code length $n = 511$, about a 4-dB gain is achievable, which is almost half that promised in Fig. 9.2. Increasing this coding gain is not easy. Indeed, Fig. 9.20 shows that the curves tend to cluster as n increases. It was shown in Wozencraft and Jacobs (1965) that a rather broad maximum of coding gain versus code rate R_c occurs for each block length of BCH codes. This maximum lies in the range from one-third to three-quarters for R_c. $\square$

Example 9.24

We now consider the (7, 4) Hamming code that has been used in the examples throughout the chapter. The aim of this example is to assess the bounds introduced in this section and to clarify the meaning of the various error probabilities introduced. We assume again a binary antipodal transmission on the channel. Therefore, the symbol error probability is given by (9.87). Let us first consider the word error probability $P_w(e)$. The numerical results are plotted in Fig. 9.21. The exact result for $P_w(e)$ is obtained from (9.65). This gives

$$P_w(e) = \sum_{m=2}^{7} P(m, 7) = \sum_{m=2}^{7} \binom{7}{m} p^m (1-p)^{7-m}.$$

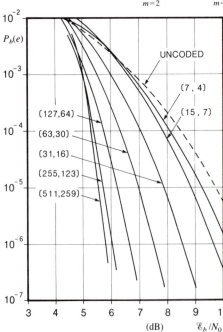

Figure 9.20 Bit error probability for some of the BCH codes of Table 9.7. The curves are obtained using the approximate expression (9.86). Binary antipodal transmission is assumed. (Reprinted with permission from Clark and Cain, 1981).

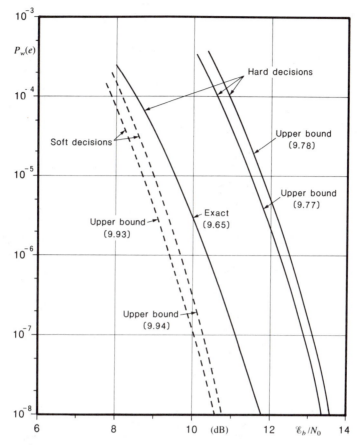

Figure 9.21 Word error probability for the (7, 4) Hamming code. Hard-decision and soft-decision decoding curves. Binary antipodal transmission.

The two upper bounds (9.77) and (9.78) can be derived using the weight distribution of the code (see Example 9.5) and remembering that $d_{min} = 3$. From (9.77) we get the tighter bound as

$$P_w(e) \leq 7z^3 + 7z^4 + z^7,$$

where $z \triangleq \sqrt{4p(1-p)}$. Similarly, from (9.78) we get

$$P_w(e) \leq 15z^3.$$

The two bounds are plotted in Fig. 9.21. It can be verified that the tighter of the two bounds differs by about 1.75 dB from the exact error probability at $P_w(e) = 10^{-6}$. This difference decreases as the error probability decreases. For longer codes, the reader can verify that this bound is tighter than in this example. The results for the bit error probability $P_b(e)$ are shown in Fig. 9.22. These are all exact.

Curve ① gives the error probability on the binary digits transmitted on the channel when the code is used. This curve plots (9.87) with $R_c = \frac{4}{7} \cong 0.571$. Curve ② gives the bit error probability for uncoded transmission. Again (9.87) is used, but now $R_c = 1$. Comparing these two curves, we better understand the terms of the discussion presented

Sec. 9.2   Block Codes

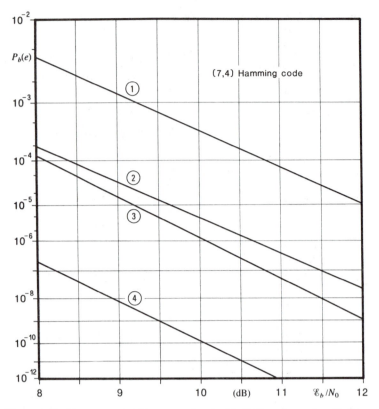

Figure 9.22 Bit error probability curves for the (7, 4) Hamming code with binary antipodal transmission: ① Symbol error probability on the channel with coding; ② symbol error probability on the channel without coding; ③ bit error probability with coding; ④ bit error probability when only error detection and retransmission are used.

in the introduction to this chapter. In fact, the comparison shows that, when the code is used, the error probability on the binary transmitted symbols is higher than with uncoded transmission. The two curves differ by 10 log $R_c \cong 2.4$ dB. It is expected that the error-correcting capabilities of the code are able to eliminate this disadvantage and, it is hoped, to achieve a coding gain. The actual result for the Hamming code is shown by curve ③. This curve plots (9.80); that is,

$$P_b(e) \cong \frac{3}{7} \sum_{i=2}^{7} \binom{7}{i} p^i (1 - p)^{7-i}.$$

For $P_b(e) = 10^{-6}$, a coding gain of 0.6 dB is present.

Finally, curve ④ represents the bit error probability when only error detection is used. This curve plots (9.80) with (9.82) for the definition of $P_w(e)$. The apparent dramatic improvement cannot be directly compared with curve ③, because in this case we also have a retransmission strategy. Therefore, at least the throughput rate should be taken into account. □

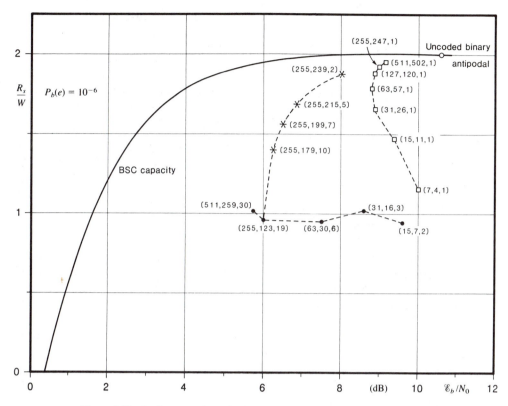

Figure 9.23 Performance chart of different BCH and Hamming codes (see Example 9.25). Each code is identified with three numbers: the block length n, the number of information bits k, and the corrected errors t.

Example 9.25

We show here in a more complete way the performance of a group of BCH codes and of the Hamming codes. Each code is identified by the triplet (n, k, t) where t denotes its correcting capability. The BCH codes are those of Table 9.7. The results were obtained using the formulas and approximations of the two previous examples. All the details of the computations are omitted. The results are presented in the graph of Fig. 9.23. Notice that this presentation is identical to that used for the comparison of different modulation schemes (Chapter 5). The two parameters in the figure are the bandwidth efficiency R_s/W and the ratio $\mathscr{E}_b/N_0$. All the results are given for a fixed bit error probability $P_b(e) = 10^{-6}$. Notice, first, the performance point for uncoded binary antipodal transmission. Comparing with Fig. 5.39, it can be seen that the codes fill the region of potential gains in the power-limited region. The bandwidth expansion of the codes is also evident. Let us briefly comment on the results. To this purpose, three groups of codes are identified by different symbols (stars, black dots, and squares) and connected by dashed lines. These have no other meaning than that of identifying the group.

The black dots refer to the BCH codes of Fig. 9.20. Actually, the points are obtained by sectioning the curves of Fig. 9.20 at $P_b(e) = 10^{-6}$. The rate of these codes is practically constant and roughly 0.5. These black dots give pictorial evidence to the essence of the channel-coding theorem [see (3.83)]. We can improve performance (in this case we save signal power) by coding with increasing block length n.

A second group of codes, the squares, represents Hamming codes. They are single-error-correcting codes and their coding gain is rather poor. Increasing the block length, we improve both the coding gain and the bandwidth efficiency. The best code at $P_b(e) = 10^{-6}$ is the (63, 57) code. If we continue to increase the block length, the bandwidth efficiency continues to increase, but the coding gain becomes poorer.

The last group of codes, the stars, are BCH codes of length 255 and different rates, starting from the code (255, 239) down to the code (255, 123). These stars clearly show the trade-off between coding gain and bandwidth efficiency when the block length is kept constant. $\square$

9.2.10 Performance of Block Codes with Unquantized Soft Decisions

As described in Section 9.1, unquantized soft-decision decoding entails no quantization of the channel output. Therefore, each binary waveform is demodulated by the optimum demodulator (a matched filter followed by a sampler) and a code word is represented by a sequence of n random variables. To make our ideas more precise, let us refer to Example 4.3. Let $\mathscr{E}$ denote the energy of one of these n binary waveforms representing one code word. Then, dropping an irrelevant constant, each binary decision variable can be written as

$$z_i = \begin{cases} \sqrt{\mathscr{E}} + v_i, & \text{if the } i\text{th digit is 1,} \\ -\sqrt{\mathscr{E}} + v_i, & \text{if the } i\text{th digit is 0,} \end{cases} \tag{9.88}$$

and $i = 1, 2, \ldots, n$. The variables $\{v_i\}$ are samples of the Gaussian noise with zero mean and variance $N_0/2$.

From knowledge of the $M = 2^k$ code words and upon reception of the sequence $z_1, z_2, \ldots, z_n$ from the demodulator, the decoder forms M decision variables as follows:

$$L_i = \sum_{j=1}^{n} (2x_{ij} - 1)z_j, \qquad i = 1, 2, \ldots, M, \tag{9.89}$$

where x_{ij} denotes the digit in the jth position of the ith code word. In this way, the decision variable corresponding to the actual transmitted code word will have a mean value $n\sqrt{\mathscr{E}}$, while the other $(M - 1)$ ones will have smaller mean values. Correct decoding is achieved by selecting the largest among the L_i's of (9.89).

Although the computations involved in this decoding process are very simple, it may soon become impractical to implement this algorithm because of the exponential growth of M with k. Several different types of soft-decision decoding algorithms have been invented to circumvent this difficulty. Some reference is given to them in the Bibliographical Notes.

The derivation of the error probability in the decoding process is not straightforward, as it is complicated by the correlations between the decision variables. Therefore, we resort to a union bound. Recalling the uniform error property, we can assume that the all-zero code word is transmitted. Let us define the probability $P(e_{1m})$ as

$$P(e_{1m}) \triangleq P\{L_m > L_1 | \mathbf{x}_1\}, \tag{9.90}$$

which is the probability that the decoder chooses the mth code word when the all-zero code word is transmitted. The union bound (4.66) gives

$$P_w(e) \leq \sum_{m=2}^{M} P(e_{1m}). \tag{9.91}$$

But the error probability $P(e_{1m})$ depends only on the Euclidean distance d_{1m}^2 between the two code words. If the mth code word has weight w_m, then it differs from the all-zero code word in w_m positions. Therefore,

$$d_{1m}^2 = 4w_m \mathscr{E} = 4w_m R_c \mathscr{E}_b. \tag{9.92}$$

Introducing this value into (4.41), we get $P(e_{1m})$ and from (9.91) we conclude with

$$P_w(e) \leq \frac{1}{2} \sum_{m=2}^{M} erfc \left(\sqrt{R_c w_m \frac{\mathscr{E}_b}{N_0}} \right). \tag{9.93}$$

A looser bound is obtained using $d_{\min}$ in (9.93). We get

$$P_w(e) \leq \frac{M-1}{2} erfc \left(\sqrt{R_c d_{\min} \frac{\mathscr{E}_b}{N_0}} \right). \tag{9.94}$$

We can compare this result with (4.45) for binary antipodal transmission. Using in both cases the bound $erfc \, (\sqrt{z}) < e^{-z}$, we can affirm that the coding gain is approximately given by

$$10 \log_{10} \left(R_c d_{\min} - \frac{k \ln 2}{\mathscr{E}_b / N_0} \right).$$

Notice that it depends on both the code parameters and the signal-to-noise ratio. The asymptotic value is $10 \log_{10}(R_c \, d_{\min})$.

Example 9.26

Let us take the (7, 4) Hamming code and specify for it the two bounds (9.93) and (9.94). We get

$$P_w(e) \leq \frac{7}{2} erfc \left(\sqrt{\frac{12}{7} \frac{\mathscr{E}_b}{N_0}} \right) + \frac{7}{2} erfc \left(\sqrt{\frac{16}{7} \frac{\mathscr{E}_b}{N_0}} \right) + \frac{1}{2} erfc \left(\sqrt{4 \frac{\mathscr{E}_b}{N_0}} \right)$$

$$\leq \frac{15}{2} erfc \left(\sqrt{\frac{12}{7} \frac{\mathscr{E}_b}{N_0}} \right).$$

These two bounds are plotted for comparison in Fig. 9.21. The tighter one, at $P_w(e) = 10^{-8}$, shows an improvement of roughly 1.5 dB with respect to hard-decision decoding. For lower values of $P_w(e)$, this advantage reaches the 2 dB expected from Fig. 9.2. $\square$

9.3 CONVOLUTIONAL CODES

A binary convolutional encoder is a finite-memory system that generates n_0 binary digits for every k_0 message digits presented at its input. Again the code rate is defined as $R_c = k_0/n_0$. In contrast with block codes, k_0 and n_0 are usually small numbers. A scheme

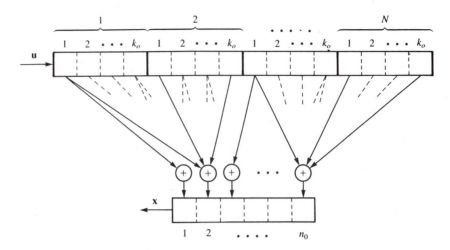

Figure 9.24 General block diagram of a convolutional encoder for an (n_0, k_0) code with constraint length N.

for a convolutional encoder is shown in Fig. 9.24. The message digits are introduced k_0 at a time into the input shift register, which has Nk_0 positions. As a block of k_0 digits enters the register, the n_0 modulo-2 adders feed the output register with the n_0 digits and these are shifted out. Then the input register is fed with a new block of k_0 digits, and the old blocks are shifted to the right, the oldest one being lost. And so on. We can conclude that in a convolutional code the n_0 digits generated by the encoder depend not only on the corresponding k_0 message digits, but also on the previous $(N - 1)k_0$ ones. Such a code is called an (n_0, k_0) convolutional code of *constraint length N*. This encoding process makes the encoded sequence depend on the whole message sequence. To understand the difference, a block code can be considered to be the limiting case of a convolutional code, with constraint length $N = 1$.

If we define **u** to be the semi-infinite message vector and **x** the corresponding encoded vector, we want now to describe how to get **x** from **u**. As for the block codes, to describe the code we only need to know the connections between the input and output registers of Fig. 9.24. This approach enables us to show both the analogies and the differences with respect to block codes. But, if pursued further, it leads to complicated notations and tends to emphasize the algebraic structure of convolutional codes. This is less interesting for decoding purposes. Therefore, we shall only present this approach briefly. Later the description of the code will be restated from a completely different viewpoint.

To describe the encoder of Fig. 9.24, we can use N submatrices $\mathbf{G}_1, \mathbf{G}_2, \ldots,$ $\mathbf{G}_N$ containing k_0 rows and n_0 columns. The submatrix $\mathbf{G}_i$ describes the connections of the ith segment of k_0 cells of the input register with the n_0 cells of the output register. The n_0 entries of the first row of $\mathbf{G}_i$ describe the connections of the first cell of the ith input register segment with the n_0 cells of the output register. A 1 means a connection, while a 0 means no connection. We can now define the *generator matrix* of the convolutional code as

$$\mathbf{G}_\infty \triangleq \begin{bmatrix} \mathbf{G}_1\,\mathbf{G}_2\ldots\mathbf{G}_N & & & \\ & \mathbf{G}_1\,\mathbf{G}_2\ldots\mathbf{G}_N & & \\ & & \mathbf{G}_1\,\mathbf{G}_2\ldots\mathbf{G}_N & \\ & & & \mathbf{G}_1\,\mathbf{G}_2\ldots\mathbf{G}_N \\ & & \ldots\ldots\ldots \end{bmatrix}. \tag{9.95}$$

All other entries in $\mathbf{G}_\infty$ are zero. This matrix has the same properties as for block codes, except that it is semi-infinite (it extends arbitrarily downward and to the right). Therefore, given a semi-infinite message vector $\mathbf{u}$, the corresponding coded vector is

$$\mathbf{x} = \mathbf{u}\,\mathbf{G}_\infty. \tag{9.96}$$

This equation is formally identical to (9.4). A convolutional code is said to be *systematic* if in each segment of n_0 digits generated by the encoder the first k_0 are a replica of the corresponding message digits. It can be verified that this condition is equivalent to have the following submatrices:

$$\mathbf{G}_1 = \left.\begin{bmatrix} 1 & 0 & 0 & \ldots & 0 & \vdots & \\ 0 & 1 & 0 & \ldots & 0 & \vdots & \mathbf{P}_1 \\ \cdot & \cdot & \cdot & \cdot & \cdot & \vdots & \\ 0 & & \cdots & & 1 & \vdots & \end{bmatrix}\right\} k_0,$$

$$\mathbf{G}_i = \left.\begin{bmatrix} 0 & 0 & 0 & \ldots & 0 & \vdots & \\ 0 & 0 & 0 & \ldots & & \vdots & \mathbf{P}_i \\ \cdot & \cdot & \cdot & \cdot & \cdot & \vdots & \\ 0 & 0 & 0 & \ldots & 0 & \vdots & \end{bmatrix}\right\} k_0, \qquad i = 2, 3, \ldots, N. \tag{9.97}$$

All these concepts are better clarified with two examples.

Example 9.27

Consider a $(3, 1)$ convolutional code with constraint length $N = 3$. The encoder is shown in Fig. 9.25a. The output register is replaced by a commutator that reads sequentially the

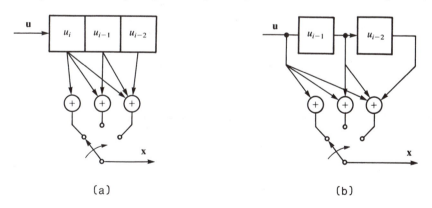

(a)   (b)

Figure 9.25 Convolutional encoder for the $(3, 1)$ code of Example 9.27. The constraint length is $N = 3$.

outputs of the three adders. A more convenient scheme for the same encoder is shown in Fig. 9.25b. The code is defined by the following three submatrices (actually, three row vectors, since $k_0 = 1$):

$$\mathbf{G}_1 = [1 \quad 1 \quad 1],$$
$$\mathbf{G}_2 = [0 \quad 1 \quad 1],$$
$$\mathbf{G}_3 = [0 \quad 0 \quad 1].$$

The generator matrix, from (9.95), becomes

$$\mathbf{G}_\infty = \begin{bmatrix} 111 & 011 & 001 & 000 & \cdots & \cdots & \cdots \\ 000 & 111 & 011 & 001 & 000 & \cdots & \cdots \\ 000 & 000 & 111 & 011 & 001 & 000 & \cdots \\ \cdots & \cdots & \cdots & \cdots & \cdots & \cdots & \cdots \end{bmatrix}.$$

It can be verified, from (9.96), that the information sequence $\mathbf{u} = (11011 \ldots)$ is encoded into the sequence $\mathbf{x} = (111100010110100 \ldots)$. The code is systematic. $\square$

Example 9.28

Consider a (3, 2) code with constraint length $N = 2$. The encoder is shown in Fig. 9.26. The code is now defined by the two submatrices

$$\mathbf{G}_1 = \begin{bmatrix} 1 & 0 & 1 \\ 0 & 1 & 0 \end{bmatrix}, \qquad \mathbf{G}_2 = \begin{bmatrix} 0 & 0 & 1 \\ 0 & 0 & 1 \end{bmatrix}.$$

The code is systematic, since (9.97) are satisfied. The generator matrix is now given by

$$\mathbf{G}_\infty = \begin{bmatrix} 101 & 001 & 000 & \cdots & \cdots \\ 010 & 001 & 000 & \cdots & \cdots \\ 000 & 101 & 001 & 000 & \cdots \\ 000 & 010 & 001 & 000 & \cdots \\ 000 & 000 & 101 & 001 & 000 \\ 000 & 000 & 010 & 001 & 000 \\ \cdots & \cdots & \cdots & \cdots & \cdots \end{bmatrix}.$$

The information sequence $\mathbf{u} = (11011011 \ldots)$ is now encoded into the sequence $\mathbf{x} = (111010100110 \ldots)$. The encoder of Fig. 9.26 requires a serial input. The $k_0 = 2$ input digits can also be presented in parallel, and the corresponding encoder is given in Fig. 9.27. The output digits are also presented in parallel form. $\square$

From the first example, it can be verified that, in general, the operation of the

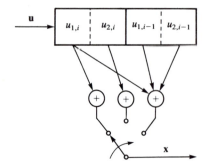

Figure 9.26 Convolutional encoder for the (3, 2) code of Example 9.28. The constraint length is $N = 2$.

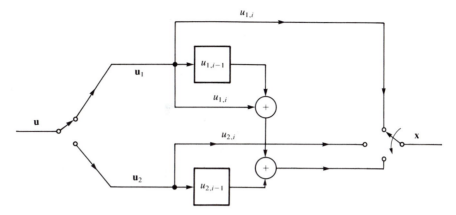

Figure 9.27 Parallel implementation of the same convolutional encoder of Fig. 9.26.

encoder for an $(n_0, 1)$ code is to generate n_0 digits of the sequence $\mathbf{x}$ for each digit u_i according to the following expression:

$$(x_{i1}, x_{i2}, x_{i3}, \ldots, x_{in_0}) = u_i\mathbf{G}_1 + u_{i-1}\mathbf{G}_2 + \cdots + u_{i-N+1}\mathbf{G}_N. \qquad (9.98)$$

This expression is recognized as the *discrete convolution* of the vectors $\mathbf{G}_1$, $\mathbf{G}_2$, . . . $\mathbf{G}_N$ and the N-digit input sequence u_i, u_{i-1}, . . . , u_{i-N+1}. The term convolutional code comes from this observation. Notice that, in general, the number of modulo-2 adders in the encoder is smaller than the constraint length of the code. In fact, code rates of $\frac{1}{2}$ or $\frac{1}{3}$ are widely used and in these cases we have only two or three adders, respectively. For this reason, instead of describing the code with the N submatrices $\mathbf{G}$ previously defined, it is more convenient to describe the encoder connections by using the *generators* $\mathbf{g}_i$, defined as

$$\mathbf{g}_i = (g_{i1}, g_{i2}, \ldots, g_{iN}), \qquad i = 1, 2, \ldots, n_0, \qquad k_0 = 1. \qquad (9.99)$$

These vectors describe the connections among the ith adder of Fig. 9.24 and the N stages of the encoder input shift register.

Example 9.29

Let us reconsider the code of Example 9.27. This code has $k_0 = 1$ and $n_0 = 3$. Therefore, it can be described with the following three generators:

$$\mathbf{g}_1 = (100),$$
$$\mathbf{g}_2 = (110),$$
$$\mathbf{g}_3 = (111).$$

The advantage of this is not immediately apparent. As in Example 9.27, we have three vectors. But, for example, in the case of a $(3, 1)$ code with constraint length $N = 10$, this second representation always requires three generators of length 10, whereas in the other case we need 10 vectors (the submatrices $\mathbf{G}_i$) of length 3. No doubt the first description is more practical. For codes with $k_0 > 1$, which description is better depends on the specific case. □

State diagram representation of convolutional codes

As already noted, there is a powerful and practical alternative to the algebraic description of convolutional codes. This alternative is based on the observation that the convolutional encoder is a finite-memory system whose output sequence depends on the input sequence and on the state of the device. The description we are looking for is called the *state diagram* of the convolutional code. We shall illustrate the concepts involved in this description by taking as an example the encoder of Fig. 9.25b. This encoder refers to the (3, 1) code with constraint length $N = 3$ described in Example 9.27. Notice that each output triplet of digits depends on the input digit and on the content of the shift register that stores the oldest two input digits. We say that the encoder has memory $L = N - 1 = 2$. Let us define the state σ_l of the encoder at time l as the content of its memory at the same time. That is,

$$\sigma_l \triangleq (u_{l-1}, u_{l-2}). \tag{9.100}$$

There are $2^L = 4$ possible states. That is, 00, 01, 10, and 11. Looking at Fig. 9.25(b), assume, for example, that the encoder is in state 10. When the input digit is 1, the encoder produces the output digits 100 and moves to the state 11. This type of behavior is completely described by the state diagram of Fig. 9.28. Each of the four states is represented in a circle. A solid edge represents a transition between two states when the input digit is 0, whereas a dashed edge represents a transition when the input digit is 1. The label on each edge represents the output digits corresponding to that transition. Using the state diagram of Fig. 9.28, the computation of the encoded sequence is quite straightforward. Starting from the initial state 00, we jump from one state to the next one following a solid edge when the input is 0 or a dashed edge when the input is 1. If we define the states to be $S_1 \triangleq (00)$, $S_2 \triangleq (01)$, $S_3 \triangleq (10)$, and $S_4 \triangleq (11)$, we can easily check that the input sequence $\mathbf{u} = (11011 \ldots)$, already considered in Example 9.27, corresponds to the *path* $S_1S_3S_4S_2S_3S_4$ through the state diagram, and the output sequence is $\mathbf{x} = (111\ 100\ 010\ 110\ 100 \ldots)$, as found by writing down the sequence of edge labels. The concept of state diagram can be applied to any (n_0, k_0) code of constraint length N and memory $L = N - 1$. The number of states is $2^{k_0 L}$.

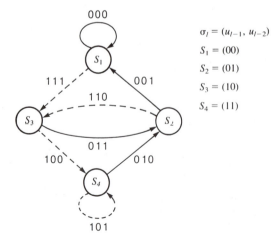

$\sigma_l = (u_{l-1}, u_{l-2})$
$S_1 = (00)$
$S_2 = (01)$
$S_3 = (10)$
$S_4 = (11)$

Figure 9.28 State diagram for the (3, 1) convolutional code of Example 9.27.

There are 2^{k_0} edges entering each state and 2^{k_0} edges leaving each state. The labels on each edge are sequences of length n_0. As L increases, the size of the state diagram grows exponentially and becomes very hard to handle. Usually, only for $L \leq 10$ can the state diagram be a practical working tool. As we are "walking inside" the state diagram following the guidance of the input sequence, it soon becomes difficult to keep track of the past path, because we travel along the same edges many times. Therefore, it is desirable to modify the concept of state diagram by introducing time explicitly. This result is achieved if we have a replica of the states at each time step, as shown in the diagram of Fig. 9.29. This is called a *trellis diagram*. It refers to the state diagram of Fig. 9.28. In this trellis, the four nodes on the same vertical represent the four states at the same discrete time l, which is called the *depth* into the trellis. Dashed and solid edges have the same meaning as in the state diagram. The input sequence is now represented by the path $\sigma_0 = S_1$, $\sigma_1 = S_3$, $\sigma_2 = S_4$, . . . , and so on. Any encoder output sequence can be found by walking through the appropriate path into the trellis.

Finally, a different representation of the code can be given by expanding the trellis diagram of Fig. 9.29 into the *tree diagram* of Fig. 9.30. In this diagram, the encoding process can be conceived as a walk through a *binary tree*. Each encoded sequence is represented by one particular path into the tree. The encoding process is guided by binary decisions (the input digit) at each *node* of the tree. This tree has an exponential growth. At the depth l, there will be 2^l possible paths representing all the possible encoded sequences of that length. The path corresponding to the input sequence 11011 is shown as an example in Fig. 9.30. The nodes of the tree are labeled with reference to the states of the state diagram shown in Fig. 9.28.

Distance properties and transfer functions of convolutional codes

As for block codes, the error-detection and error-correction capabilities of a convolutional code are directly related to the distance properties of the encoded sequences. Due to the uniform error property of linear codes, we assume, without loss of generality, that the all-zero sequence is transmitted in order to determine the performance of the convolutional code.

Let us start with some definitions. Consider a pair of encoded sequences up to

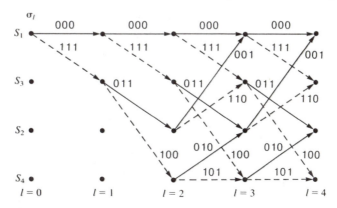

Figure 9.29 Trellis diagram for the (3, 1) convolutional code of Example 9.27.

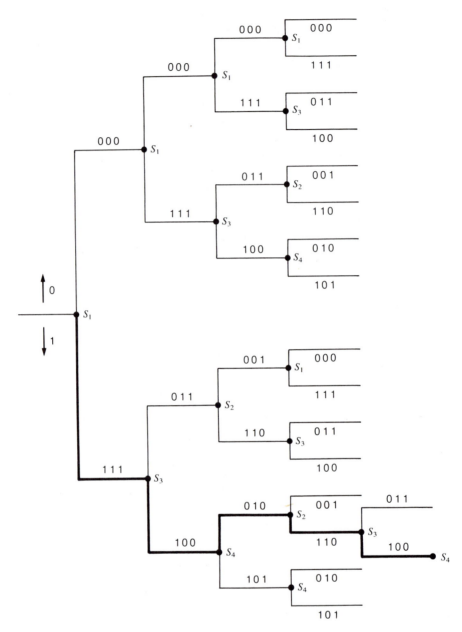

Figure 9.30 Tree diagram for the (3, 1) convolutional code of Example 9.27. The solid path corresponds to the input sequence 11011.

the depth l into the code trellis and assume that they disagree at the first branch. We define the lth-order *column distance* $d_c(l)$ as the minimum Hamming distance between all pairs of such sequences. For the computation of $d_c(l)$, one of the sequences of the pair can be the all-zero sequence. Therefore, we have to consider all sequences, up to the depth l in the code trellis, such that they disagree at the first branch from the all-zero sequence. The column distance $d_c(l)$ is the minimum weight of this set of code sequences. The column distance $d_c(l)$ is a nondecreasing function of the depth l. By specifying the value of l, we get two important particular cases.

The first is the *minimum distance* d_{min} of the convolutional code, obtained when $l = N$, N being the code constraint length. Therefore,

$$d_{min} \triangleq d_c(N). \qquad (9.101)$$

The second case is the *free distance* d_f of the convolutional code, defined as

$$d_f = \lim_{l \to \infty} d_c(l). \qquad (9.102)$$

From (9.102), we see that the free distance of the code is the minimum Hamming distance between infinitely long encoded sequences. It can be found on the code trellis by looking for those sequences (paths) that, after diverging from the all-zero sequence, merge again into it. The free distance is the minimum weight of this set of encoded sequences.

A simple algorithm to compute d_f is based on the following steps:

1. Compute $d_c(l)$ for $l = 1, 2, \ldots$.
2. If the sequence giving $d_c(l)$ merges into the all-zero sequence, keep its weight as d_f.

Example 9.30

We want to reconsider the (3, 1) convolutional code, whose trellis is given in Fig. 9.29, to find the distances just defined. Let us consider Fig. 9.31. Part of the trellis is reproduced here, with the following features. Only the all-zero sequence and the sequences diverging from it at the first branch are reproduced. Furthermore, each transition edge is now labeled with the weight of the encoded sequence. The column distance of the code can be found by inspection of Fig. 9.31. We get

l	$d_c(l)$
1	3
2	4
3	$5 \leftarrow d_{min}$
4	$6 \leftarrow d_f$
5	6

Since the constraint length of this code is $N = 3$, we have also $d_{min} = 5$. Furthermore, notice that for $l = 3$ we also have the first merge of one sequence into the all-zero sequence. However, the merging sequence does not give $d_c(3)$ and we must keep looking for d_f. For $l = 4$ we have a merging sequence giving $d_c(4)$. Its weight is 6 and therefore $d_f = 6$. $\square$

The computation of d_f, although straightforward, may require consideration of exceedingly long sequences. In practice, the problem is amenable to an algorithmic solution based on the state diagram of the code. We take again the case of Fig. 9.28 as a guiding example. The state diagram is redrawn in Fig. 9.32 with certain modifications

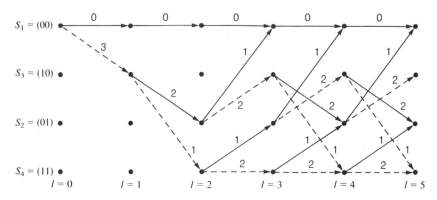

Figure 9.31 Part of the trellis diagram for the (3, 1) code of Example 9.30 for the computation of the code distance properties. The trellis is the same as Fig. 9.29, but the labels represent now the weight of the output sequence associated with each transition.

made in view of our goal. First, the edges are labeled with an indeterminate D raised to an exponent that denotes the weight of the encoded sequence corresponding to that transition. Furthermore, the self-loop at state S_1 is eliminated, since its contribution to the weight of a sequence is nothing. Finally, the state S_1 is split into two states, one of which represents the input and the other the output of the state diagram.

Let us now define the label of a path as the product of the labels of all its edges. Therefore, among all the infinitely many paths starting in S_1 and merging again into S_1, we are looking for the path whose label D is raised to the minimum possible power. This label is indeed D^{df}. By inspection of Fig. 9.32, we can verify that the path $S_1 S_3 S_2 S_1$ (see Example 9.30) has label D^6, and indeed this code has $d_f = 6$.

We can define a *generating function* $T(D)$ of the output sequence weights as a series that gives all the information about the weights of the paths starting from S_1 and merging again into S_1. This generating function can be regarded as the *transfer function* of the signal-flow graph of Fig. 9.32. Using standard techniques for the study of directed graphs (see Appendix D), the transfer function for the graph of Fig. 9.32 is given by

$$T(D) = \frac{2D^6 - D^8}{1 - (D^2 + 2D^4 - D^6)} = 2D^6 + D^8 + 5D^{10} + \cdots = \sum_{d=6}^{\infty} a(d)D^d. \qquad (9.103)$$

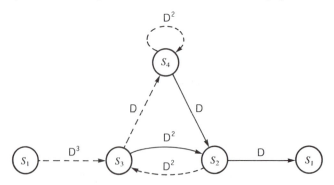

Figure 9.32 State diagram for the (3, 1) code of Fig. 9.28. The labels allow computation of the generating function $T(D)$.

This result means that there are two paths of weight 6, one path of weight 8, five paths of weight 10, and so on. This transfer function can be used to provide additional information on the code properties. This is done by considering the modified graph of Fig. 9.33. Each edge has a label L that counts the length of a path. Therefore, the exponent of L for a given path gives the number of path's branches. Finally, a label J is included into an edge if the corresponding transition is caused by an input digit 1 (the dashed edges in the diagram). This augmented transfer function comes out as

$$T(D, L, J) = \frac{D^6 L^3 J(1 + LJ - D^2 LJ)}{1 - D^2 LJ(1 + D^2 L + D^2 L^2 J - D^4 L^2 J)}$$

$$= D^6 L^3 J(1 + LJ) + D^8 L^5 J^3 + \cdots$$

(9.104)

Of course, $T(D)$ can be obtained from $T(D, L, J)$ by setting $L = J = 1$. Notice that (9.104) says that the two paths of weight 6 have lengths 3 and 4, respectively, and that the ones in the input sentences are 1 and 2. The path of weight 8 has length 5, and the corresponding input sequence has weight 3. And so on. These numbers can be checked immediately in Example 9.30.

Thus, we have fully determined the properties of all code paths of this simple convolutional code. The same techniques can be applied to any code of arbitrary rate and constraint length. We shall see in the next sections how the generating function of the code can be used to bound the error probability of the convolutional codes.

9.3.1 Best Known Short-Constraint-Length Convolutional Codes

When considering the generating function $T(D)$ of a convolutional code, it was implicitly assumed that $T(D)$ converges. Otherwise, the expansions of (9.103) and (9.104) are not valid. This convergence is not always true. When this occurs, the code is called *catastrophic*. An example is given in Problem 9.28. This code is a (2, 1) code with constraint length $N = 3$. The state diagram of this code shows that the self-loop at state S_4 does not increase the distance from the all-zero sequence. Therefore, the path

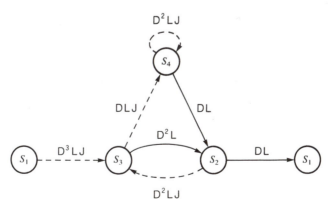

Figure 9.33 State diagram for the (3, 1) code of Fig. 9.28. The labels allow computation of the generating function $T(D, L, J)$.

$S_1S_3S_4 \ldots S_4S_2S_1$ will be at distance 6 from the all-zero path no matter how many times it circulates in the self-loop at state S_4. Thus it is possible to have an arbitrarily large number of decoding errors even for a fixed finite number of channel errors. This explains the name given to these codes.

Conditions can be established on the code generators of (9.99) to avoid catastrophic codes. These conditions can be found in Massey and Sain (1968). Also Forney (1970) and Rosenberg (1971) explore the question of the relative fraction of catastrophic codes in the ensemble of all convolutional codes of a given rate and constraint length.

An important feature of this problem is that systematic convolutional codes cannot be catastrophic. However, unfortunately, the free distances that are achievable with systematic codes realized with the schemes of Fig. 9.25 are usually lower than for nonsystematic codes of the same constraint length N, at least for high values of N. Table 9.9 shows the maximum free distances achievable with systematic codes and nonsystematic noncatastrophic codes of rates $\frac{1}{2}$ and $\frac{1}{3}$.

Finally, it must be noticed that one interesting difference between linear block codes and convolutional codes is that the choice of a linear encoder for the latter is a crucial issue. This problem and the complexity of convolutional encoders are examined in Forney (1970).

TABLE 9.9 Maximum free distances achievable with systematic codes and nonsystematic noncatastrophic codes with constraint length N and rates $\frac{1}{2}$ and $\frac{1}{3}$

N	d_f(Rate $\frac{1}{2}$)		d_f(Rate $\frac{1}{3}$)	
	Systematic	Nonsystematic	Systematic	Nonsystematic
2	3	3	5	5
3	4	5	6	8
4	4	6	8	10
5	5	7	9	12
6	6	8	10	13
7	6	10	12	15
8	7	10	12	16

Computer search methods have been used to find convolutional codes optimum in the sense that, for a given rate and a given constraint length, they have the largest possible free distance. These results were obtained by Odenwalder (1970), Larsen (1973), Paaske (1974), and Daut, Modestino, and Wismer (1982). They are reproduced in part in Tables 9.10 through 9.15.

The generators of the code are those defined in (9.99). They are given in octal notation. The tables also give simple upper bounds on d_f derived in Heller (1968) for codes of rate $1/n_0$ and extended to codes of rate k_0/n_0 by Daut, Modestino, and Wismer (1982).

Example 9.31

The rate $\frac{1}{2}$ convolutional code of constraint length $N = 4$ of Table 9.10 has generators 15 and 17, which means

$$\mathbf{g}_1 = (1101),$$

$$\mathbf{g}_2 = (1111).$$

The block diagram of the encoder is shown in Fig. 9.34. For the rate $\frac{2}{3}$ code of constraint length $N = 2$, the three generators are

$$\mathbf{g}_1 = (1111),$$

$$\mathbf{g}_2 = (0110),$$

$$\mathbf{g}_3 = (1101),$$

and the block diagram of the encoder is shown in Fig. 9.35. □

TABLE 9.10 Maximum free distance convolutional codes of rate $\frac{1}{2}$ and constraint length N. (Copyright © 1973 IEEE, reprinted with permission from Larsen, 1973).

Constraint length N	Generators in octal		d_f	Upper bound on d_f
3	5	7	5	5
4	15	17	6	6
5	23	35	7	8
6	53	75	8	8
7	133	171	10	10
8	247	371	10	11
9	561	753	12	12
10	1167	1545	12	13
11	2335	3661	14	14
12	4335	5723	15	15
13	10533	17661	16	16
14	21675	27123	16	17

TABLE 9.11 Maximum free distance convolutional codes of rate $\frac{1}{3}$ and constraint length N. (Copyright © 1973 IEEE, reprinted with permission from Larsen, 1973.)

Constraint length N	Generators in octal			d_f	Upper bound on d_f
3	5	7	7	8	8
4	13	15	17	10	10
5	25	33	37	12	12
6	47	53	75	13	13
7	133	145	175	15	15
8	225	331	367	16	16
9	557	663	711	18	18
10	1117	1365	1633	20	20
11	2353	2671	3175	22	22
12	4767	5723	6265	24	24
13	10533	10675	17661	24	24
14	21645	35661	37133	26	26

TABLE 9.12 Maximum free distance convolutional codes of rate $\frac{1}{4}$ and constraint length N. (Copyright © 1973 IEEE, reprinted with permission from Larsen, 1973.)

Constraint length N	Generators in octal				d_f	Upper bound on d_f
3	5	7	7	7	10	10
4	13	15	15	17	13	15
5	25	27	33	37	16	16
6	53	67	71	75	18	18
7	135	135	147	163	20	20
8	235	275	313	357	22	22
9	463	535	733	745	24	24
10	1117	1365	1633	1653	27	27
11	2387	2353	2671	3175	29	29
12	4767	5723	6265	7455	32	32
13	11145	12477	15537	16727	33	33
14	21113	23175	35527	35537	36	36

TABLE 9.13 Maximum free distance convolutional codes of rate $\frac{1}{5}$ and constraint length N. (Copyright © 1982 IEEE, reprinted with permission from Daut, Modestino, and Wismer, 1982.)

Constraint length N	Generators in octal					d_f	Upper bound on d_f
3	7	7	7	5	5	13	13
4	17	17	13	15	15	16	16
5	37	27	33	25	35	20	20
6	75	71	73	65	57	22	22
7	175	131	135	135	147	25	25
8	257	233	323	271	357	28	28

TABLE 9.14 Maximum free distance convolutional codes of rate $\frac{2}{3}$ and constraint length N. (Copyright © 1982 IEEE, reprinted with permission from Daut, Modestino, and Wismer, 1982.)

Constraint length N	Generators in octal			d_f	Upper bound on d_f
2	17	06	15	3	4
3	27	75	72	5	6
4	236	155	337	7	7

TABLE 9.15 Maximum free distance convolutional codes of rates $\frac{3}{4}$ and $\frac{3}{8}$ and constraint length $N = 2$. (Copyright © 1982 IEEE, reprinted with permission from Daut, Modestino, and Wismer, 1982.)

Rate	Constraint length N	Generators in octal				d_f	Upper bound on d_f
$\frac{3}{4}$	2	13	25	61	47	4	4
$\frac{3}{8}$	2	15	42	23	61	8	8
		51	36	75	47		

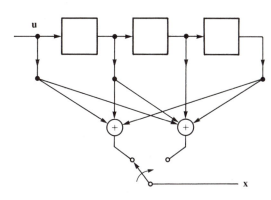

Figure 9.34 Encoder for a (2, 1) convolutional code of constraint length $N = 4$ and generators $\mathbf{g}_1 = (15)$ and $\mathbf{g}_2 = (17)$.

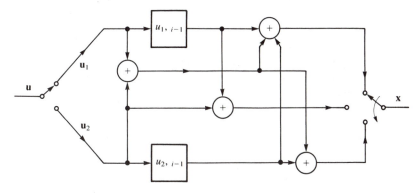

Figure 9.35 Encoder for a (3, 2) convolutional code of constraint length $N = 2$ and generators $\mathbf{g}_1 = (17)$, $\mathbf{g}_2 = (06)$, and $\mathbf{g}_3 = (15)$.

9.3.2 Maximum-Likelihood Decoder for Convolutional Codes: The Viterbi Algorithm

We have already seen that ML decoding of block codes is achieved when the decoder selects the code word whose distance from the received sequence is the minimum. In the case of hard decoding, the distance considered is the Hamming distance, while for soft decoding it is the Euclidean distance. Unlike a block code, a convolutional code has no fixed block length. But it is intuitive that the same principle works also for convolutional codes. In fact, each possible encoded sequence is a path into the code trellis. Therefore, the optimum decoder must choose that path into the trellis that is closest to the received sequence. Also in this case, the distance measure will be the Hamming distance for hard decoding and the Euclidean distance for unquantized soft decoding.

Let us start with the case of hard decoding. We assume binary antipodal modulation, and consequently the equivalent discrete channel is a BSC with error probability p. If we denote with $\mathbf{y}_l$ the sequence of n_0 binary digits supplied to the decoder by the demodulator between discrete times l and $(l + 1)$ and with $\mathbf{x}_l^{(r)}$ the n_0-tuple of digits of the rth path in the code trellis between states σ_l and σ_{l+1}, then the optimum decoder must choose the path r of the trellis for which the conditional probability $P(\mathbf{y}_l|\mathbf{x}_l^{(r)})$ is maximum. We can take the logarithm of this probability as well. Therefore, the ML decoder must find the path for which

$$U(\sigma_{K-1}) \triangleq \max_r U^{(r)}(\sigma_{K-1}) \triangleq \max_r \left\{ \ln \prod_{l=0}^{K-1} P(\mathbf{y}_l|\mathbf{x}_l^{(r)}) \right\} \quad (9.105)$$

$$= \max_r \left\{ \sum_{l=0}^{K-1} \ln P(\mathbf{y}_l|\mathbf{x}_l^{(r)}) \right\}$$

is a maximum. In (9.105), K indicates the length of the path into the trellis or, equivalently, Kn_0 is the length of the binary received sequence.

The maximization of (9.105) is already formulated in terms suitable for the application of the Viterbi algorithm and, henceforth, it is assumed that the reader is familiar with the contents of Appendix F. The metric for each branch of the code trellis is defined as

$$V_l^{(r)}(\sigma_{l-1}, \sigma_l) \triangleq \ln P(\mathbf{y}_l|\mathbf{x}_l^{(r)}) \quad (9.106)$$

and therefore

$$U(\sigma_{K-1}) = \max_r \sum_{l=0}^{K-1} V_l^{(r)}(\sigma_{l-1}, \sigma_l). \quad (9.107)$$

If we denote with $d_l^{(r)}$ the Hamming distance between the two sequences $\mathbf{y}_l$ and $\mathbf{x}_l^{(r)}$ and use (9.15), we can rewrite (9.106) as

$$V_l^{(r)}(\sigma_{l-1}, \sigma_l) = (-d_l^{(r)}) \ln \frac{1-p}{p} + n_0 \ln(1 - p) = -\alpha d_l^{(r)} - \beta, \quad (9.108)$$

with α and β positive constants (if $p < 0.5$).

Using (9.108) into (9.107) and dropping unessential constants, the problem is reduced to finding the minimum of

$$U'(\sigma_{K-1}) \triangleq \min_r \sum_{l=0}^{K-1} d_l^{(r)}. \quad (9.109)$$

As expected intuitively, (9.109) states that ML decoding requires the minimization of the Hamming distance between the received sequence and the chosen path into the code trellis. The form of (9.109) is such that the minimization can be accomplished with the Viterbi algorithm, the metric on each branch being the Hamming distance between binary sequences.

Example 9.32

We apply the Viterbi decoding algorithm to the code whose trellis is shown in Fig. 9.29, corresponding to the state diagram of Fig. 9.28. We know already (see Example 9.30) that this code has $d_f = 6$. Assume the information transmitted sequence is 01000000 . . . , whose corresponding encoded sequence is 000 111 011 001 000 000 000 000 . . . . Furthermore, assume the received sequence is 110 111 011 001 000 000 000 000 . . . . It contains two errors in the first triplet of digits and therefore it does not correspond to any path into the trellis. To apply the Viterbi algorithm, it is more useful to refer to a trellis similar to Fig. 9.31. The successive steps of the algorithm are shown in Fig. 9.36. The algorithm, at each step l into the trellis, stores for each state the *surviving path* (the minimum distance path from the starting state $\sigma_0 = S_1$) and the corresponding *accumulated metric*. Consider, for example, the situation at step $l = 4$. We have

State σ_4	Surviving path	Metric
S_1	$S_1 S_1 S_3 S_2 S_1$	2
S_3	$S_1 S_3 S_2 S_1 S_3$	5
S_2	$S_1 S_3 S_4 S_4 S_2$	7
S_4	$S_1 S_3 S_4 S_4 S_4$	6

Therefore, at step $l = 4$, the ML path (the one with the smallest distance) is $S_1 S_1 S_3 S_2 S_1$, and the corresponding information sequence is 0100 (follow the dashed and solid edges on the trellis). Consequently, in spite of the two initial channel errors, already at this step the correct information sequence is identified.

Let us see in detail how the algorithm proceeds one step farther to compute the situation at $l = 5$. Consider first the state $\sigma_5 = S_1$. From the trellis diagram of Fig. 9.29, we can verify that the state S_1 can be reached either from the state S_1 with a transition corresponding to an encoded triplet 000 or from the state S_2 with a transition corresponding to an encoded triplet 001. The received triplet during this transition is 000. Therefore, from (9.108) and (9.107) we have

$$U'(\sigma_5)|_{\sigma_5 = S_1} = U'(\sigma_4)|_{\sigma_4 = S_1} + V_5^{(1)}(S_1, S_1) = 2 + 0 = 2,$$

$$U'(\sigma_5)|_{\sigma_5 = S_2} = U'(\sigma_4)|_{\sigma_4 = S_2} + V_5^{(2)}(S_1, S_2) = 7 + 1 = 8.$$

Thus, the minimum-distance path leading to S_1 at $l = 5$ comes from S_1, and the transition from S_2 is dropped. The metric $U'(\sigma_5)$ at S_1 will be 2 and the surviving path will be that of $\sigma_4 = S_1$ (i.e., 0100) with a new 0 added (i.e., 01000). The interesting feature is that at $l = 6$ all surviving paths *merge* at state $\sigma_6 = S_1$. This means that already at this step the first four digits are uniquely decoded in the correct way and the two channel errors are corrected. □

For a general (n_0, k_0) convolutional code with constraint length N, there are $2^{k_0(N-1)}$ states at each step in the trellis. Consequently, the Viterbi decoding algorithm requires the storage of $2^{k_0(N-1)}$ surviving paths and $2^{k_0(N-1)}$ metrics. At each step, there are 2^{k_0} paths reaching each state and therefore 2^{k_0} metrics must be computed for each state.

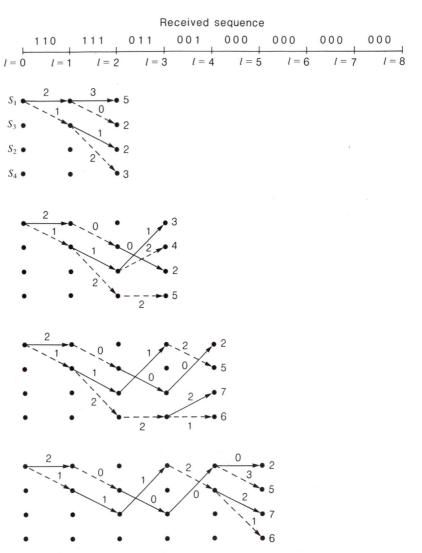

Figure 9.36 Viterbi decoding algorithm applied to the (3, 1) convolutional code of Fig. 9.28. The decoded sequence is 01000000.

Only one of the 2^{k_0} paths reaching each state does survive, and this is the minimum-distance path from the received sequence up to that transition. Thus, the complexity of the Viterbi decoder grows exponentially with k_0 and N. For this reason, practical applications are confined to the cases for which $k_0 N$ is of the order of 10. However, the Viterbi algorithm is basically simple, and practical hardware implementations can be produced using the information provided in this simple description. Actually, Viterbi decoding has been widely applied and is presently one of the most practical techniques for providing large coding gains.

The trellis structure of the decoding process has the following consequence. If at some point an incorrect path is chosen, it is highly probable that it will merge with the

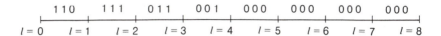

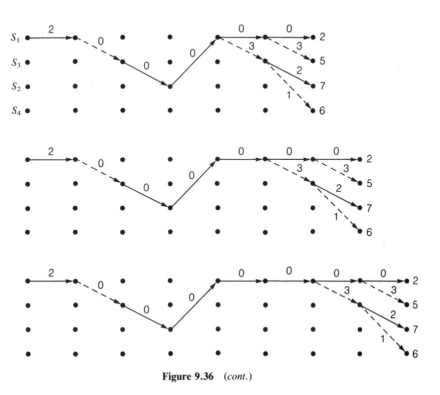

Figure 9.36 (*cont.*)

correct path at a later time. Therefore, the typical error sequences of convolutional codes, when decoded by a Viterbi decoder, result in bursts of errors due to the incorrect path diverging from the correct one and soon merging again into it. One final consideration concerns the technique used to output the decoded digits. The optimum procedure would be to decode the sequence only at the end of the whole receiving process. However, this would cause too long decoding delays and excessive memory storage for the surviving sequences. But we have seen in the example that all surviving paths tend to merge into one single path when proceeding deeply enough into the trellis. A solution to this problem is to modify the Viterbi algorithm in such a way as to force the decision on the oldest symbol of the minimum distance path after a fixed and sufficiently long delay. Computer simulations have shown that a delay of the order of $5N$ results in a negligible degradation with respect to the optimum performance.

An important feature of the Viterbi algorithm is that soft-decision decoding, unlike block codes, requires a trivial modification of the procedure discussed previously. In fact, it suffices to replace the Hamming metric with the Euclidean metric and all the other decoding operations remain the same. Therefore, the implementation complexity for soft-decision decoding is not significantly different from the hard-decision case.

Let us now derive an expression for the branch metric (9.106) in the case of unquantized soft decisions. If $\{y_{jl}\}_{j=1}^{n_0}$ is the set of the received demodulator outputs in the case of binary antipodal modulation and lth branch, we have for y_{jl} the expression

$$y_{jl} = \sqrt{\mathscr{E}}(2x_{jl}^{(r)} - 1) + v_j, \tag{9.110}$$

which is obtained from (9.89). Here $x_{jl}^{(r)}$ is a binary digit and v_j is a Gaussian RV with zero mean and variance $N_0/2$. Therefore, from (9.110) we get

$$P(\mathbf{y}_l|\mathbf{x}_l^{(r)}) = \prod_{j=1}^{n_0} P(y_{jl}|x_{jl}^{(r)}) \tag{9.111}$$

$$= \prod_{j=1}^{n_0} \frac{1}{\sqrt{\pi N_0}} \exp\left\{ -\frac{[y_{jl} - \sqrt{\mathscr{E}}(2x_{jl}^{(r)} - 1)]^2}{N_0} \right\}.$$

Inserting (9.111) into (9.106) and neglecting all terms that are common to all branch metrics, we get

$$V_l^{(r)}(\sigma_{l-1}, \sigma_l) = \sum_{j=1}^{n_0} y_{jl}(2x_{jl}^{(r)} - 1). \tag{9.112}$$

This is the branch metric to be used by the soft-decision Viterbi decoder.

9.3.3 Other Decoding Techniques for Convolutional Codes

The computational effort and the storage size required to implement the Viterbi algorithm confine its application to convolutional codes with a small constraint length ($N \leqslant 10$). Other decoding techniques have been discovered for convolutional codes. These techniques preceded the Viterbi algorithm historically. They are quite useful in certain applications since they can use longer code constraint lengths than those allowed by the Viterbi algorithm.

Sequential decoding techniques

As already pointed out, the operation of a convolutional encoder can be described as the choice of a path through a binary tree in which each path represents an encoded sequence. The sequential decoding techniques share with the Viterbi algorithm the idea of a probabilistic search of the correct path, but, unlike that case, the search does not extend to all paths that can potentially be the best. Only some subsets of possible paths are extended. These are the paths that appear to be the most probable. For this reason, sequential decoding is not an optimum algorithm as Viterbi's. Nevertheless, sequential decoding is one of the most powerful tools for decoding convolutional codes of long constraint length ($N > 10$). Its error performance is comparable to that of Viterbi decoding.

The decoding approach can be conceived as a trial-and-error technique for searching out the correct path into the tree. Let us consider a qualitative example by looking at the code tree of Fig. 9.30. In the absence of noise, the code sequences of length $n_0 =$

3 are received without errors. Consequently, the receiver can start its walk into the tree from the root and then follow a path by simply replicating at each node the encoding process and taking its binary decision after comparing the locally generated sequence with the received one. The transmitted message will be recovered directly from the path followed into the tree.

The presence of noise introduces errors, and hence the decoder can find itself in a situation in which the decision entails risk. This happens when the received sequence is different from all the possible alternatives that are locally generated by the receiver. Assume, for instance, the received sequence 111 100 110 . . . and consider the code tree of Fig. 9.30. Starting from the root, the first two choices are not ambiguous. But, when reaching the second-order node, the decoder must choose between the upward path (sequence 010) and the downward path (sequence 101), having received the sequence 110. The choice that sounds more reasonable is to go upward in the tree, but the decoder would at this point proceed on a wrong path in the tree and the continuation would be in error. If, for instance, the branch metric is the Hamming distance, the decoder can track the cumulative Hamming distance between the received sequence and the path followed into the tree and can eventually notice that this distance grows higher than expected. In this case, the decoder can decide to go back to the node at which an apparent error was made and try the other choice. This process of going forward and backward into the tree is the main concept of sequential decoding. This movement can be guided by modifying the metric of the Viterbi algorithm (the Hamming distance for hard decisions) with the addition of a negative constant at each branch. The value of this constant is selected such that the metric for the correct path increases on the average, while that for any incorrect path decreases. By comparing the accumulated metric with a moving threshold, the decoder can detect and discard the incorrect paths. Besides their nonoptimality, the sequential algorithms trade with Viterbi's a larger decoding delay with a smaller storage need.

Sequential decoding was first introduced by Wozencraft (1957) and subsequently modified by Fano (1963). An interesting class of sequential decoding algorithms are the *stack algorithms*, which were proposed independently by Jelinek (1969) and Zigangirov (1966). These algorithms are computationally simpler than Fano's but require more storage. An analysis of the computational problems implied by sequential decoding can be found in Wozencraft and Jacobs (1965), Savage (1966), and Forney (1974).

Syndrome decoding techniques

Unlike sequential decoding, these techniques are deterministic and rely on the algebraic properties of the code. Typically, a syndrome sequence is calculated (as for block codes). It provides a set of linear equations that can be solved for determining the minimum-weight error sequence. The two most widely used of such techniques are *feedback decoding* (Heller, 1975) and *threshold decoding* (Massey, 1963). They have the advantage of simple circuitry and small decoding delays, thus allowing high-speed decoding. However, since the allowable codes presenting the required algebraic properties are rather poor, only moderate coding gain values are achievable with these techniques.

Sec. 9.3   Convolutional Codes   **475**

9.3.4 Performance of Convolutional Codes with Hard Decisions and Viterbi Decoding

Before discussing techniques for bounding the bit error probability $P_b(e)$, it is first necessary to analyze in some detail what is an *error event* in the Viterbi decoding. Since we consider only linear convolutional codes and the uniform error property holds for them, we can assume that the all-zero sequence is transmitted and evaluate error probabilities under this hypothesis. We denote as the *correct path* the horizontal path at the top of the trellis diagram (Fig. 9.37). The decoder bases its decisions on the received noisy version of the transmitted sequence and can choose a path different from the correct one on the basis of the accumulated metric.

An error event is defined as an incorrect path segment that, after diverging from the correct path at a certain node, merges again into it at a successive node. Figure 9.37 shows three error events corresponding to the sequences x_2, x_3, and x_4. Notice also that the dotted path, corresponding to the sequence x_2', may have a higher metric than the correct path and yet not be selected, because its accumulated metric is smaller than that of the solid path corresponding to the sequence x_2. We may conclude that a necessary condition for an error event to occur at a certain node is that the metric of an incorrect path, diverging from the correct one at that node, accumulates higher metric increments than the correct path over the unmerged path segment. If we denote by x_d a sequence of weight d corresponding to one of these incorrect paths, an error event is the union of all the events for which the path metric corresponding to x_d is larger than that of the correct path (sequence x_1). Therefore, the probability $P(e)$ of an error event can be upper bounded, using the union bound, in the form

$$P(e) \leq \sum_{d=d_f}^{\infty} a(d)P(e_{1d}), \qquad (9.113)$$

where $a(d)$ is the number of paths of weight d given by $T(D)$ and $P(e_{1d})$ is the pairwise error probability between the all-zero path and that of weight d. Using (9.77), we have

$$P(e_{1d}) < [\sqrt{4p(1-p)}]^d. \qquad (9.114)$$

Introducing (9.114) into (9.113) and recalling the definition (9.103) for $T(D)$, we finish with

$$P(e) < T(D)\big|_{D=\sqrt{4p(1-p)}}. \qquad (9.115)$$

This result emphasizes the role of the transfer function $T(D)$ for the computation of the probability $P(e)$ of an error event.

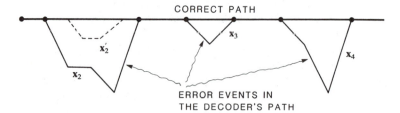

Figure 9.37 Trellis paths showing possible error events of a Viterbi decoder.

Turning now to the bit error probability $P_b(e)$, notice that the expected number of bit errors in the decoded sequence can be bounded by weighting each term in the union bound (9.113) with the number of bit errors that occurred on that incorrect path. If $k_0 = 1$, this number corresponds to the number of ones in the decoded sequence over the incorrect path that is selected. If $a(d, i)$ is the number of paths of weight d diverging from the correct one and containing i ones in the decoded sequence over the unmerged path segment, then we have

$$P_b(e) \le \sum_{i=1}^{\infty} \sum_{d=d_f}^{\infty} ia(d,i)P(e_{1d}). \tag{9.116}$$

But, going back to the definition (9.104) of the augmented transfer function $T(D, L, J)$, we can write, for $L = 1$,

$$T(D, J) = \sum_{i=1}^{\infty} \sum_{d=d_f}^{\infty} a(d, i)D^d J^i. \tag{9.117}$$

Taking the derivative of $T(D, J)$ with respect to J and letting $J = 1$, we obtain

$$\left.\frac{\partial T(D, J)}{\partial J}\right|_{J=1} = \sum_{i=1}^{\infty} \sum_{d=d_f}^{\infty} ia(d, i)D^d. \tag{9.118}$$

Therefore, comparing (9.118) with (9.116) and using (9.114), we have finally

$$P_b(e) < \left.\frac{\partial T(D, J)}{\partial J}\right|_{J=1, D=\sqrt{4p(1-p)}}. \tag{9.119}$$

When $k_0 > 1$, one branch of the trellis corresponds to k_0 information bits, and therefore the result given in (9.119) should be divided by k_0.

When p is very small, the general result (9.119) can be approximated by taking only the smallest power term in the summation. Thus we get

$$P_b(e) \le A[4p(1 - p)]^{d_f/2} \cong A2^{d_f}p^{d_f/2}, \tag{9.120}$$

where A is a constant depending on the code. The importance of the free distance d_f is thus fully evident. A somewhat tighter upper bound can be obtained in (9.116) by using for $P(e_{1d})$ an expression different from the Bhattacharyya bound (see Problem 9.29).

9.3.5 Performance of Convolutional Codes with Unquantized Soft Decisions and Viterbi Decoding

The case of soft decoding is the same as the hard-decision one, except that the metric is the Euclidean distance. Define the metrics of the two paths merging at node B, the all-zero path, and the path with Euclidean distance d from it. From (9.107) and (9.112), we get

$$U^{(r)}(\sigma_B) = \sum_{l=0}^{B} \sum_{j=1}^{n} y_{jl}(2x_{jl}^{(r)} - 1). \tag{9.121}$$

If $r = 1$, the metric (9.121) identifies the all-zero path and with $r = d$ the path at Hamming distance d from it. Then the pairwise error probability between these two paths is

$$P(e_{1d}) = \frac{1}{2} erfc \left(\sqrt{dR_c \frac{\mathcal{E}_b}{N_0}} \right).$$ (9.122)

And finally, from (9.114), we get the union bound:

$$P(e) \leq \sum_{d=d_f}^{\infty} \frac{1}{2} a(d) erfc \left(\sqrt{dR_c \frac{\mathcal{E}_b}{N_0}} \right).$$ (9.123)

With the usual bound $erfc (\sqrt{z}) < e^{-z}$, we can express (9.123) as

$$P(e) < \frac{1}{2} T(D) \Big|_{D=e^{-R_c(\mathcal{E}_b/N_0)}}.$$ (9.124)

Also, in analogy with the derivation of (9.119), we have, for unquantized soft-decision decoding, the following bit error probability bound:

$$P_b(e) < \frac{1}{2} \frac{\partial T(D, J)}{\partial J} \Big|_{J=1, D=e^{-R_c(\mathcal{E}_b/N_0)}}.$$ (9.125)

For $\mathcal{E}_b/N_0 \to \infty$, this general result can be given the approximate form

$$P_b(e) \leq \frac{A}{2} e^{-d_f R_c (\mathcal{E}_b/N_0)},$$ (9.126)

where A is a constant depending on the code [the same as (9.120)]. We can conclude that the behavior of $P_b(e)$ for convolutional codes is dominated in all cases by the free distance d_f, which plays the same role as $d_{\min}$ for the block codes. The bound of (9.125) can be slightly refined and made tighter (see Problem 9.30).

Finally, comparing (9.126) with the bit error probability for uncoded transmission (4.45), we can verify that an upper bound for the asymptotic coding gain with soft decision is given by $10 \log_{10} d_f R_c$.

Example 9.33

In this example we want to derive the bit error probability $P_b(e)$ for the $(3, 1)$ convolutional code of Fig. 9.28. This code has constraint length $N = 3$, rate $R_c = \frac{1}{3}$, and $d_f = 6$ (see Example 9.30). Recalling (9.104), we also have

$$T(D, J) = D^6(J + J^2) + D^8 J^3 + \cdots$$

To apply the bounds (9.119) and (9.125), we have

$$\frac{\partial T(D, J)}{\partial J} \Big|_{J=1} = 3D^6 + 3D^8 + \cdots$$

From (9.119), the case of hard-decision Viterbi decoding, we get the following approximate bound:

$$P_b(e) \leq 3[\sqrt{4p(1 - p)}]^6 \cong 192[p(1 - p)]^3.$$

Using instead the approach proposed in Problem 9.29, we can derive a tighter upper bound in the form

$$P_b(e) \leq \frac{3}{2} \binom{6}{3} p^3 (1 - p)^3 \cong 30[p(1 - p)]^3.$$

These two bounds are shown in Fig. 9.38. From (9.125), the case of soft-decision Viterbi decoding, we get the following approximate bound:

$$P_b(e) \leq \frac{3}{2} \exp\left(-2\frac{\mathcal{E}_b}{N_0}\right).$$

The corresponding curve is also shown in Fig. 9.38. Soft-decision decoding presents a coding gain of about 2 dB at $P_b(e) = 10^{-6}$. At this $P_b(e)$, hard-decision decoding causes a loss of almost 2 dB in coding gain, but this loss increases as $P_b(e)$ decreases. Notice that the asymptotic coding gain for soft-decision decoding is for this code 10 log 6 $\frac{1}{3}$ = 3 dB. $\square$

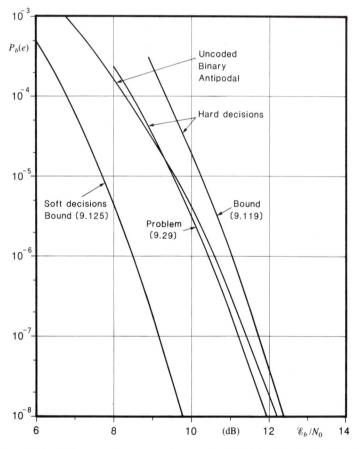

Figure 9.38 Performance bounds with hard- and soft-decision Viterbi decoding for the (3, 1) convolutional code of Fig. 9.28 (see Example 9.33).

Sec. 9.3   Convolutional Codes   479

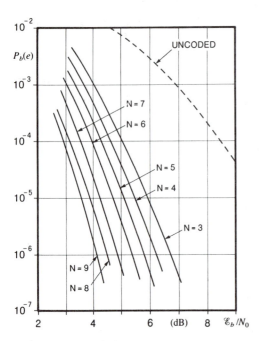

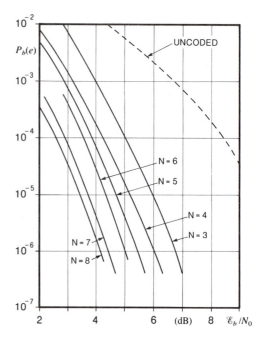

Figure 9.39. Performance bounds of different convolutional codes of rate $\frac{1}{2}$ with unquantized soft-decision Viterbi decoding. The codes are those listed in Table 9.10 (Reprinted with permission from Clark and Cain, 1981.)

Figure 9.40 Performance bounds of different convolutional codes of rate $\frac{1}{3}$ with unquantized soft-decision Viterbi decoding. The codes are those listed in Table 9.11 except that of constraint length 7, whose generators in octal are 133, 165, 171 and whose d_f is 14. (Reprinted with permission from Clark and Cain 1981.)

Example 9.34

To get a more complete idea of the coding gains achievable with convolutional codes and soft-decision Viterbi decoding, some of the codes of rate $\frac{1}{2}$ (Table 9.10) and rate $\frac{1}{3}$ (Table 9.11) and constraint length N have been considered. Their $P_b(e)$ has been evaluated using the bound (9.125). The results are shown in Figs. 9.39 and 9.40. Notice that the potential coding gains are quite significant and they increase as $P_b(e)$ decreases. As the constraint length N is increased for the codes of the same rate, the coding gain increases by about

TABLE 9.16 Achievable coding gains with some convolutional codes of rate R_c and constraint length N. Eight-level quantization soft-decision Viterbi decoding is used. The last line gives the upper bound $10 \log_{10} d_f R_c$. (Copyright © 1974 IEEE, reprinted with permission from Jacobs, 1974.)

$\mathscr{E}_b/N_0$ for uncoded transmission (dB)	$P_b(e)$	$R_c = \frac{1}{3}$		$R_c = \frac{1}{2}$			$R_c = \frac{2}{3}$		$R_c = \frac{3}{4}$	
		N		N			N		N	
		7	8	5	6	7	6	8	6	9
6.8	10^{-3}	4.2	4.4	3.3	3.5	3.8	2.9	3.1	2.6	2.6
9.6	10^{-5}	5.7	5.9	4.3	4.6	5.1	4.2	4.6	3.6	4.2
11.3	10^{-7}	6.2	6.5	4.9	5.3	5.8	4.7	5.2	3.9	4.8
$\to \infty$	$\to 0$	7.0	7.3	5.4	6.0	7.0	5.2	6.7	4.8	5.7

0.3 to 0.4 dB for each increase of one unity in N. On the other hand, increasing the code rate for the same N entails a loss in coding gain of about 0.4 dB.

The curves shown assume unquantized soft decisions. However, the predicted coding gains are quite real. In fact, if an eight-level quantization is used rather than infinite quantization, the observed degradation in performance is of the order of 0.25 dB. Achievable coding gains with eight-level quantization soft-decision Viterbi decoding are given in Table 9.16. The upper bound is $10 \log_{10} d_f R_c$, and is also indicated. Notice finally that for hard decisions the degradation is of the order of 2.2 dB. $\square$

9.4 CONCATENATED CODES

This coding technique is a practical tool for implementing very long codes with large error-correcting capabilities. The basic concept of *concatenated codes* is illustrated in Fig. 9.41. One code, called the *inner code*, is a binary (n, k) code; the other code, called the *outer code*, is a nonbinary (N, K) code, usually a Reed–Solomon code (see Section 9.2.6). Encoding is performed as follows. A block of Kk information digits is divided into K bytes of k binary symbols each. Each byte of k symbols is encoded first by the outer encoder into an N-byte code word. Then each of the N bytes of k binary digits is encoded by the inner encoder into a binary code word of n symbols. The set of N code words of the inner code is a code word in the concatenated (Nn, Kk) code of rate $R_c = Kk/Nn$. In spite of the possibly large overall length, the complexity of the decoder is significantly reduced by the structure of the concatenation that breaks decoding into two steps. In fact, decoding is first done on the inner code, and the N inner code words are reduced to N bytes of k symbols. Then these N bytes are decoded by the outer decoder into K bytes of k binary symbols each.

Concatenated codes are effective against a mixture of both random and burst-error patterns. In fact, the inner code can correct random errors, while bursts can be left to the outer code.

As said, the outer code is usually a Reed–Solomon code. Different choices are possible for the inner code. Two cases are presented and discussed hereafter.

9.4.1 Reed–Solomon and Convolutional Codes

A convolutional code can be used as an inner code. The convolutional codes that are usually chosen are short constraint length codes with soft-decision Viterbi decoding. Interleaving could also be used in connection with the outer Reed-Solomon code in order to contrast the tendency of the Viterbi decoder to correlate errors. Results for this case are shown in Fig. 9.42. The inner code is a rate $\frac{1}{2}$, constraint length 6 convolutional code. The Reed–Solomon code uses symbols of from 6 up to 9 bits. For each value of N, the code selected is that with the best performance. Comparing these results with

Figure 9.41 Block diagram for concatenated codes.

Fig. 9.2, we see that coding gains of up to 7 dB are available at $P_b(e) = 10^{-5}$. Also notice that the curves are so steep that a moderate increase in $\mathscr{E}_b/N_0$ achieves very low bit error probabilities.

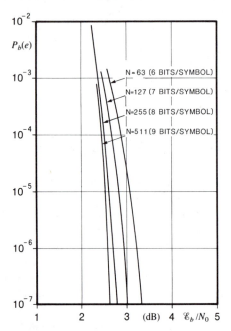

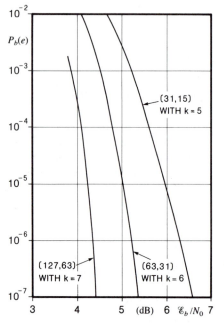

Figure 9.42 Bit error probability for a concatenated code formed by an inner convolutional code of rate $\frac{1}{2}$ and constraint length 6 and an outer Reed–Solomon code. (Reprinted with permission from Clark and Cain, 1981.)

Figure 9.43 Bit error probability for a concatenated code. The inner code is a set of 2^k orthogonal signals with noncoherent demodulation. The outer code is an (N, K) Reed–Solomon code. (Reprinted with permission from Clark and Cain, 1981.)

9.4.2 Reed–Solomon Codes and Orthogonal Modulation

An interesting choice of the inner code is that of representing each k-digits symbol of the Reed–Solomon code with one of the 2^k signals of an orthogonal signal set. The decoder complexity grows with k, since a separate correlator is required for each of the 2^k signals of the inner code. Therefore, only small values of k are of practical interest. Furthermore, a large bandwidth is required. These codes are used in deep-space communication systems. Figure 9.43 shows typical results concerning the use of orthogonal signals with noncoherent demodulation. There is an available coding gain of the order of 9 to 10 dB [at $P_b(e) = 10^{-5}$] with respect to binary orthogonal signaling and noncoherent demodulation.

9.5 CODING BOUNDS

The first of these bounds concerns the minimum distance for block codes or the free distance for convolutional codes. When attempting to find new codes, it would be very useful to know how close we are to the best possible code. A meaningful parameter to assess the goodness of a code is indeed its minimum (or free) distance. Another type of bounds concerns the error performance of a code. These bounds are all derived with random coding arguments and, therefore, provide indications regarding the average performance of a class of codes. For this reason, these bounds may be useful in assessing the performance of a channel when coding is used. Some significant results are presented hereafter.

9.5.1 Bounds on Minimum Distance

We start with block codes. The design goal when dealing with an (n, k) block code is that of achieving the largest possible d_{min} with the highest possible code rate k/n. This means that the best use has been made of the available redundancy. Here we shall discuss some results obtained in this area. Their derivation usually requires long calculations that are beyond the scope of this book and hence are omitted. The interested reader can find the details in MacWilliams and Sloane (1977).

The first bound is an upper bound known as the *Hamming*, or *sphere-packing*, *bound*. The maximum achievable d_{min} is given implicitly by the expression

$$\sum_{i=0}^{t} \binom{n}{i} \leqslant 2^{n-k}, \tag{9.127}$$

where t is the maximum number of correctable errors and $d_{min} = 2t + 1$. The equality sign in (9.127) holds only for perfect codes. This bound is tight for high-rate codes.

A tight upper bound for low-rate codes is the *Plotkin bound*, given by

$$2(d_{min} - 1) - \log_2 d_{min} \leqslant n - k. \tag{9.128}$$

An interesting question can be raised at this point. Given that an (n, k) code with an error-correcting ability better than that obtained from (9.127) or (9.128) cannot exist, what can actually be achieved? A partial answer to this question is given by the *Varshamov–Gilbert bound*, which states that it is always possible to find an (n, k) code with minimum distance *at least* given by

$$\sum_{i=1}^{d_{min}-2} \binom{n-1}{i} < 2^{n-k}. \tag{9.129}$$

This result represents a lower bound to the achievable d_{min}.

Example 9.35

We want to find a block code of length $n = 127$ and error-correcting capability $t = 5$ (i.e., $d_{min} = 11$). Its rate should be the largest possible. The Hamming bound (9.127) gives

$$\sum_{i=0}^{5} \binom{127}{i} \leqslant 2^{127-k}.$$

From this we get $k \leqslant 99$. Instead, the Plotkin bound (9.128) gives

$$2(11 - 1) - \log_2 11 \leqslant 127 - k$$

from which we get $k \leqslant 110$. On the other hand, the Varshamov–Gilbert bound (9.129) gives

$$\sum_{i=0}^{9} \binom{126}{i} < 2^{127-k}$$

from which we get that codes exist with $k > 82$. Therefore, the maximum value of k cannot exceed 99 and should be greater than 82. From Table 9.7 we can observe that a (127, 92) BCH code exists that provides a satisfactory answer to our problem. $\square$

For convolutional codes of rate $1/n_0$, a simple upper bound that was already used to present results in Tables 9.10 through 9.13 is the *Heller bound*, given by

$$d_f \leqslant \min_{r>0} \left[\frac{2^{r-1}}{2^r - 1} (N + r - 1)n_0 \right], \qquad (9.130)$$

where N is the code constraint length and r is an integer. This bound has been extended to codes of any rate by Daut, Modestino, and Wismer (1982).

9.5.2 Bounds on Performance

These bounds are obtained using random coding techniques, that is, evaluating the average performance of an ensemble of codes. This implies the existence of specific codes that behave better than the average. The most important result of this approach was already mentioned in Chapter 3 as the *channel-coding theorem*. This theorem states that the word error probability of a code can be driven to zero by simply increasing its length n, provided only that the code rate does not exceed the channel capacity C.

Given the ensemble of binary block codes of length n and rate R_c, the minimum attainable error probability over any discrete memoryless channel is bounded by

$$P_w(e) \leqslant 2^{-nE(R_c)}, \qquad R_c \leqslant C. \qquad (9.131)$$

The achievable performance is determined by the *reliability function* $E(R_c)$, whose typical behavior for a discrete memoryless channel is given in Fig. 9.44. Notice that R_{cr}, called the *critical rate* of the channel, is the rate at which the tangent to $E(R_c)$ has slope -1. This tangent intercepts the horizontal axis at a value of R_c that we call the *cutoff rate* R_0 of the channel. Therefore, we can write a simpler upper bound in the form

$$P_w(e) < 2^{-n(R_0 - R_c)}, \qquad R_c \leqslant R_0. \qquad (9.132)$$

The parameter R_0 plays an important role in modulation and coding theory. Since it is a characteristic of the channel, it allows comparisons of different channels with respect to an ensemble of codes. We shall now derive the expression of R_0 for the ensemble of binary block codes using a binary antipodal modulation over an unquantized soft-decision AWGN.

Let us consider the ensemble C of all binary block codes c of length n and rate

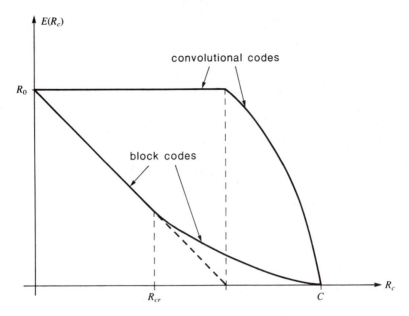

Figure 9.44 Typical behavior of the reliability function $E(R_c)$ on the AWGN channel.

R_c. Each code c has $M = 2^{nR_c}$ code words, and there is a total of 2^{nM} possible codes. This ensemble also includes some very bad codes, such as those having all equal code words. Nevertheless, the bounding technique gives useful results. If we select at random one code c, the error probability over the code ensemble is given by

$$P_w(e) = 2^{-nM} \sum_{c \in C} P_w(e|c). \tag{9.133}$$

Assume now that the code word $\mathbf{x}_i$ of c is transmitted. Then

$$P_w(e|c) = \sum_{i=1}^{M} P_w(e|c, \mathbf{x}_i) P(\mathbf{x}_i). \tag{9.134}$$

The conditional error probability $P_w(e|c, \mathbf{x}_i)$ can be upper bounded by using the union bound (9.71). We get

$$P_w(e|c, \mathbf{x}_i) \le \sum_{\substack{j=1 \\ j \ne i}}^{M} P_w(e_{ij}|c), \tag{9.135}$$

where $P_w(e_{ij}|c)$ denotes the pairwise error probability between the two code words $\mathbf{x}_i$ and $\mathbf{x}_j$ of code c.

Introducing (9.135) into (9.134) and going back to (9.133), we obtain

$$P_w(e) \le 2^{-nM} \sum_{c \in C} \sum_{i=1}^{M} P(\mathbf{x}_i) \sum_{\substack{j=1 \\ j \ne i}}^{M} P_w(e_{ij}|c). \tag{9.136}$$

The crucial step in this derivation is the interchange of the summations order in (9.136) to get

$$P_w(e) \leq \sum_{i=1}^{M} P(\mathbf{x}_i) \sum_{\substack{j=1 \\ j \neq i}}^{M} \left\{ 2^{-nM} \sum_{c \in C} P_w(e_{ij}|c) \right\}. \tag{9.137}$$

The quantity in braces is the average of the pairwise error probability $P_w(e_{ij}|c)$ over the ensemble of codes and is quite straightforward to compute. Since the code c, and hence the code words, are chosen at random, we perform the average of $P_w(e_{ij}|c)$ by considering the pairs $\mathbf{x}_i$, $\mathbf{x}_j$ of randomly chosen code words that differ in h symbols (i.e., whose Hamming distance is $d_{ij} = h$). We have then

$$2^{-nM} \sum_{c \in C} P_w(e_{ij}|c) = \sum_{h=0}^{n} P(d_{ij} = h) P_w(e_{ij}|d_{ij} = h). \tag{9.138}$$

The probability that two code words of length n selected at random differ in h symbols is

$$P(d_{ij} = h) = \binom{n}{h} 2^{-n}. \tag{9.139}$$

Furthermore, from (9.92) and (4.41), we get

$$P_w(e_{ij}|d_{ij} = h) = \frac{1}{2} \, erfc\left(\sqrt{\frac{h R_c \mathcal{E}_b}{N_0}} \right) \leq e^{-hR_c(\mathcal{E}_b/N_0)}. \tag{9.140}$$

Substituting (9.140) and (9.139) into (9.138) and using the binomial expansion, we can obtain

$$2^{-nM} \sum_{c \in C} P_w(e_{ij}|c) \leq 2^{-n}(1 + e^{-R_c(\mathcal{E}_b/N_0)})^n. \tag{9.141}$$

Introducing (9.141) into (9.137) and observing that the RHS of (9.141) is independent of i and j, we get

$$P_w(e) \leq (M-1)2^{-n}(1 + e^{-R_c(\mathcal{E}_b/N_0)})^n < M2^{-n}(1 + e^{-R_c(\mathcal{E}_b/N_0)})^n. \tag{9.142}$$

From (9.142) we can obtain the bound (9.132) by letting

$$R_0 = 1 - \log_2(1 + e^{-R_c(\mathcal{E}_b/N_0)}). \tag{9.143}$$

Let us emphasize again that (9.143) represents the cutoff rate of an unquantized AWGN channel with binary antipodal modulation. Similar computations can be extended to derive the cutoff rate for different types of modulation (see Problem 9.31) on the same channel. When hard-decision decoding is used, we have the general model of a discrete memoryless channel with N_X input symbols and N_Y output symbols. The cutoff rate of such a channel, when the input symbols are equally likely, was shown to be (Gallager, 1965)

$$R_0 = -\log_2 \left\{ \sum_{j=1}^{N_Y} \left[\frac{1}{N_X} \sum_{i=1}^{N_X} \sqrt{P(y_j|x_i)} \right]^2 \right\}. \tag{9.144}$$

For the BSC, (9.144) specializes to

$$R_0 = 1 - \log_2\{1 + 2\sqrt{p(1-p)}\}. \tag{9.145}$$

An important result is that the cutoff rate R_0 of a channel plays a similar role also in determining the average asymptotic behavior of a special class of convolutional codes. This class is that of time-varying convolutional codes for which the encoder connections vary with time (see Chapter 5 of Viterbi and Omura, 1979). The probability $P(e)$ of an error event is bounded by an expression similar to (9.131); that is,

$$P(e) \leq A(R_c)\, 2^{-n_0 NE(R_c)}, \qquad R_c < C, \tag{9.146}$$

where $A(R_c)$ is a small constant depending only on R_c. The reliability function $E(R_c)$ for convolutional codes is sketched as well in Fig. 9.44. Therefore, also in this case, a simpler upper bound can be written in the form

$$P(e) < A(R_c)2^{-n_0 NR_0}, \qquad R_c \leq R_0. \tag{9.147}$$

The first significant conclusion can be drawn from these results in comparing the relative performance of block and convolutional codes. If we take (9.132) and (9.147) with $n = n_0 N$, we compare block and convolutional encoders in a homogeneous way. Choosing n as a reasonable indicator of the decoder complexity for both block and convolutional codes, we may achieve an exponent in (9.147) for convolutional codes that is much greater than the one in (9.132) for block codes (Fig. 9.44).†

Another important conclusion concerns the role of the cutoff rate R_0. This unique parameter indicates for the two types of codes *both* a region of rates in which arbitrarily small error probability is achievable *and* an exponent in the bounding expression of the error probability. In fact, the cutoff rate R_0 of a channel was claimed to be the most sensible parameter for comparing different modulation schemes when coded transmission is used (Massey, 1974). When soft decisions are used, R_0 is the only meaningful parameter. In these cases, the symbol error probability at the demodulator output is a very poor (practically useless) indication of the quality of the system in the presence of coding. Actually, the demodulator symbol error probability is a straight performance measure in the absence of coding because, in this case, the errors at the demodulator output are immediately reflected into the digits delivered to the user. This approach was first conceptually extended to the case of modulation plus coding in the presence of hard-decision decoding. In this context, it is commonly accepted that the purpose of coding is to correct the errors at the demodulator output (codes are usually referred to as error-correcting codes!). Therefore, the demodulator symbol error probability is still significant.

But when soft decisions are used, the combined process of modulation and coding needs a different approach. From this viewpoint, the purpose of the modulator–demodulator pair is that of presenting the coder–decoder pair with the best possible discrete input–discrete output (hard decisions or soft decisions) channel or the best possible discrete input–continuous output (unquantized soft decisions) channel for a given available bandwidth and $\mathscr{E}_b/N_0$. The performance measure of this channel is its cutoff rate R_0. The average performance of the coded transmission system is given by (9.132) for block codes and by (9.147) for convolutional codes.

To conclude this discussion, we show how the parameter R_0 can be used to evaluate bounds on the achievable coding gains for a given channel. Equations (9.143) and

† For more details, see Viterbi and Omura (1979), page 314.

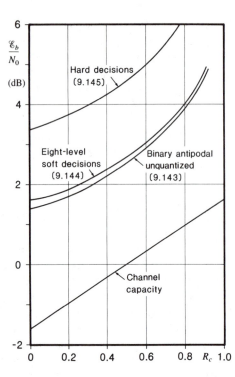

Figure 9.45 Cutoff rate R_0 with hard and soft decisions and binary antipodal modulation over the AWGN channel.

(9.145) are plotted in Fig. 9.45. These curves are obtained by letting $R_c = R_0$ and using (9.87) in (9.145). Let us now comment on these results. We know that ideal uncoded antipodal transmission requires a value of $\mathscr{E}_b/N_0$ of 9.6 dB to achieve an error probability of 10^{-5}. From Fig. 9.45 we see that, for unquantized soft decisions and code rates $R_c = \frac{1}{2}$, we need a value of $\mathscr{E}_b/N_0$ of 2 dB. Therefore, a potential coding gain of the order of 7 dB is available at this rate. Moreover, we can observe that there is not much to be gained by increasing the redundancy of the code, since at zero rate the coding gain is only slightly greater than 8 dB.

Another point that results from Fig. 9.45 is that hard decisions cause a loss of about 2 dB. In the same figure we also have the curve of R_0 for an eight-level quantized receiver with uniform thresholds on the received binary antipodal signal. This curve plots (9.144). The conclusion is that an eight-level quantization presents a negligible degradation with respect to the unquantized case and this explains why this quantization scheme is the most commonly used to implement soft-decision decoding. Finally, in Fig. 9.45 the curve of channel capacity is also plotted. Of course, it provides an absolute limit on the minimum value of $\mathscr{E}_b/N_0$ required to achieve any desired small error probability.

Thus, we can see that for codes of rate $\frac{1}{2}$ an additional coding gain of 2.4 dB is available over the prediction of the R_0 bound. However, coding schemes operating in this region are usually very complex.

9.6 PRACTICAL APPLICATIONS OF CODES

Coding schemes are used not only in sophisticated digital satellite systems but also in many commercial data-transmission equipments. This trend is principally a result of the dramatic decrease in the cost of digital components. Furthermore, in a sophisticated system environment the gain provided by coding can be an attractive and elegant alternative to the upgrading of other system elements (such as the antennas).

Another trend is the gradual blurring of the separation between block and convolutional codes. This separation mainly results from the historic development of the related theory. The real issues concern the performance of the decoding algorithms and their implementation in each particular application. Under this respect, the situation is rapidly evolving also as a consequence of technological developments. Convolutional codes with soft-decision Viterbi decoders are at present an attractive technique in many applications. The storage requirements indicate a constraint length limit of the order of 10. Coding gains of the order of 6 dB have been achieved in practical applications. When larger coding gains and very low bit error probabilities are required, as in deep-space applications, convolutional codes and sequential decoding appear to be the most suitable choice.

In computer-to-computer communications, when large buffers are available, ARQ schemes based on cyclic codes, eventually combined with some FEC capability, are attractive techniques for reasons of performance and flexibility. When the system protocols require the transmission of blocks of data of fixed length, block codes appear most suitable. High performance with lower power can be achieved in this area by using the concatenated coding technique.

When high speeds are needed and reduced complexity is a requirement, suboptimum decoding algorithms in the area of majority-logic (or threshold) decoding are an interesting possibility. A problem of a rather different nature arises when the transmission is performed over constrained-bandwidth channels. In this case the coding scheme should be selected in connection with the signal design in order to obtain the most efficient combination of coding and modulation. Some examples of this approach will be presented in the next section. In conclusion, however, it should be noted that all comparisons in this area are heavily influenced by the present state of the art in digital integrated circuit technology. Advances in this technology could significantly modify these comparisons and the application range.

9.7 SIGNAL-SPACE CODES

When block or convolutional coding is present in the transmission system, the bandwidth of the signal after modulation is wider than that of the uncoded signal for the same source rate and the same modulation scheme. The encoding process requires a bandwidth expansion that is inversely proportional to the code rate. Actually, this bandwidth expansion is traded for a coding gain. Therefore, it could be argued that encoding is a means for trading bandwidth with power, instead of an independent means to improve performance. But the question arises whether codes exist that present significant coding gain without any bandwidth expansion. The answer is found by considering encoding and modulation in a unified signal design approach. The key point of the encoding is the addition of redundancy to the signal. Block and convolutional codes add redundancy in the time domain (more digits transmitted than those emitted by the source). We can add redundancy directly to the signal set. Assume that we have more signals than those strictly required to transmit the source binary digits. Then we can choose signal sequences in such a way as to increase the Euclidean distance among them.

To exploit signal space redundancy, the modulator has memory. Such an encoding process is usually called a *signal-space code*. To be more specific, assume that k information bits per signal must be transmitted. Then 2^k signals are required. If 2^n signals are available, with $n > k$, the rate of the signal-space code is $R_c = k/n$. Indeed, no bandwidth expansion is required, provided that the dimensionality of the signal space is the same for the 2^n signals as for the 2^k signals.

Owing to the modulator's memory, decoding is based on the Viterbi algorithm, which may use soft decisions on the unquantized demodulator outputs. Therefore, the relevant performance parameter for this class of codes is the minimum Euclidean distance $d_{\min}$ between coded sequences of signals.

9.7.1 Ungerboeck Codes

Let us assume that k binary digits must be assigned to a signal belonging to a set of 2^{k+1} signals. The code will have a rate $k/(k + 1)$. The problem is that of mapping each sequence of k message digits into the redundant set of channel signals. Such mapping should maximize the Euclidean distance between sequences of channel signals.

The encoder can be regarded as a system with a finite number of states. We must have 2^k available transitions departing from each state. After selecting a possible trellis diagram, the task is to assign the channel signals to the state transitions in order to achieve the maximum Euclidean distance between signals.

A successful approach follows a rule called *mapping by set partitioning*. It is based on successive partitions of the signal set into subsets of increasing minimum distance. The rule will be better clarified with a simple example. Assume that we want to transmit 2 digits ($k = 2$) per signal and that the signal set has eight signals of the CPSK type. We shall have a rate $\frac{2}{3}$ code. The uncoded transmission would require a signal set with four signals. Therefore, we shall indicate as the reference system the signal set with four signals of the CPSK type for the same bandwidth. Assuming unit-energy signals, the Euclidean minimum distance of this reference system is $d_{\mathrm{ref}} = \sqrt{2}$.

The partitioning of the eight CPSK channel signals into subsets with increasing minimum distance is shown in Fig. 9.46. Notice that the uncoded reference system uses either the subset B_0 or the subset B_1.

Consider now a two-state trellis as in Fig. 9.47. From each state we must have four possible transitions, and for symmetry reasons we choose two *parallel transitions* from each state to the next one. The signals from subset B_0 are assigned to the transitions that originate from state S_1, whereas the signals from subset B_1 are assigned to those of state S_2. Parallel transitions are assigned signals from subsets C_0 or C_1 or C_2 or C_3, respectively. In fact, parallel transitions may imply the occurrence of error events of

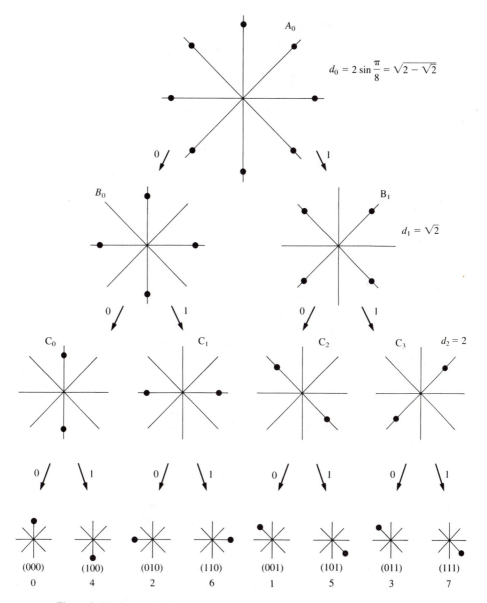

$$d_0 = 2 \sin \frac{\pi}{8} = \sqrt{2 - \sqrt{2}}$$

$$d_1 = \sqrt{2}$$

$$d_2 = 2$$

| (000) | (100) | (010) | (110) | (001) | (101) | (011) | (111) |
| 0 | 4 | 2 | 6 | 1 | 5 | 3 | 7 |

Figure 9.46 Set partitioning of the 8-CPSK signals to be used in Ungerboeck codes.

length 1. Therefore, it is useful that parallel transitions correspond to pairs of signals with the largest possible Euclidean distance.

It can be verified by inspection of Fig. 9.47 that the minimum distance d_{min} on this trellis structure is the distance between the two paths $S_1 S_1 S_1$ and $S_1 S_2 S_1$. Using the notation and results of Fig. 9.46, we get

$$d_{min} = \sqrt{d_0^2 + d_1^2} = 1.608. \tag{9.148}$$

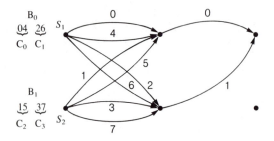

DATA	STATE	
	S_1	S_2
0 0	0 0 0	0 0 1
0 1	0 1 0	0 1 1
1 0	1 0 0	1 0 1
1 1	1 1 0	1 1 1

Figure 9.47 Trellis of a two-state Ungerboeck code of rate $\frac{2}{3}$ using the signals of Fig. 9.46.

Defining an asymptotic coding gain with respect to the uncoded reference system as $10 \log_{10} (d^2_{min}|d^2_{ref})$, this code has an asymptotic coding gain of 1.1 dB. The code table for assigning message digits to signals is given in Fig. 9.47.

The trellis of Fig. 9.47 has only two states. Having this simple example as a guideline, a systematic search for good codes can be implemented. It is expected that increasing the number of states will give higher coding gains.

The following rules for assigning channel signals to the trellis transitions should be applied (mapping by set partitioning). First, when two branches can connect the same two nodes in the trellis (parallel transitions), error events of length 1 can take place. Therefore, parallel transitions must be assigned signals from a subset with the largest possible distance (C_0, C_1, C_2, or C_3).

Consider now error events of length $L > 1$. They will originate from an initial split and a successive merge into the same node, after L steps into the trellis (see Fig. 9.48). With A, B, C, and D denoting subsets of channel signals associated with each branch, and $d(X, Y)$ denoting the minimum distance between a signal in X and one in Y, we see that d^2_{min} will include a term $d^2(A, B)$ and a term $d^2(C, D)$. This implies that, in a good code, the subsets assigned to the same originating state or to the same terminating state (adjacent transitions) must have the largest possible distance (subsets B_0 or B_1). A final rule requires that all the signals be used with the same frequency,

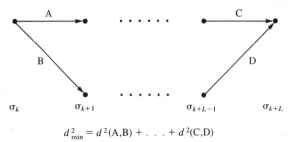

$$d^2_{min} = d^2(A,B) + \ldots + d^2(C,D)$$

Figure 9.48 Error event of length greater than 1.

thus providing a code with a symmetric structure. Although there is no formal proof that assignments of coded signals obeying these rules will provide optimum performance, all the best codes found so far satisfy them.

Notice that in the example of Fig. 9.47 the rule of adjacent transitions could not be satisfied for the transitions terminating in the same node. This explains the poor coding gain achievable with this code. The presence of parallel transitions in these 8-PSK codes limits the achievable minimum distance to $d^2_{min} = 2$, since error events of length 1 can occur.

However, on the other hand, the elimination of parallel transitions reduces the "connectivity" in the trellis and thus allows the extension of minimum length of error events with multiple signal errors. When the number of states is at least four, a trade-off is possible using parallel or distinct transitions. With four states, the trade-off still works in favor of parallel transitions. With eight or more states, only distinct transitions are of interest. Otherwise, the coding gain remains limited to 3 dB.

In Table 9.17, the asymptotic coding gain of the best-known codes for coded 8-PSK, 16-PSK, and 32-PSK are given (see Ungerboeck, 1982, and Benedetto and others, 1987). In all cases, the encoder can be implemented as a (3, 2) convolutional linear encoder followed by a suitable mapper. It can be seen that significant coding gains are achievable by increasing the number N of the states.

TABLE 9.17 Asymptotic coding gains for the best-known signal-space codes for coded 8-PSK, 16-PSK, and 32-PSK. (Ungerboeck, 1982; Benedetto and others, 1987.)

Number of states N	Asymptotic coding gain (dB)		
	8-PSK	16-PSK	32-PSK
4	3.0	3.5	3.5
8	3.6	4.0	4.0
16	4.1	4.4	4.4
32	4.6	5.1	5.1
64	4.8	5.3	5.5
128	5.0		
256	5.4		
512	5.7		

As an example, a possible structure of the encoder for the first code of Table 9.17 is given in Fig. 9.49. The block diagram of the encoder is shown together with the trellis and the state diagram. In the latter, each transition has a label that gives the two information digits and the channel signal described in Fig. 9.46.

Signal-space codes can also be designed for different modulation schemes, particularly multilevel PAM, QASK, and AM-PM (see Problem 9.32). Besides the asymptotic coding gain, it is also important to check the code performance at reasonable values of the signal-to-noise ratio. This can be done by using lower and upper bounds to the symbol error probability. As an example, in Fig. 9.50 upper and lower bounds to the error probability for coded 16-PSK are shown, together with the curve referring to uncoded 8-PSK for comparison (Benedetto and others, 1987). The curves show that the asymptotic gain is almost reached at reasonable values of the error probability, like 10^{-5} or 10^{-6}.

9.7.2 Multi-h Phase Codes

These codes use continuous phase modulated (CPM) signals and are a generalization of the minimum-shift-keying (MSK) modulation (see Section 5.7). Using the notation

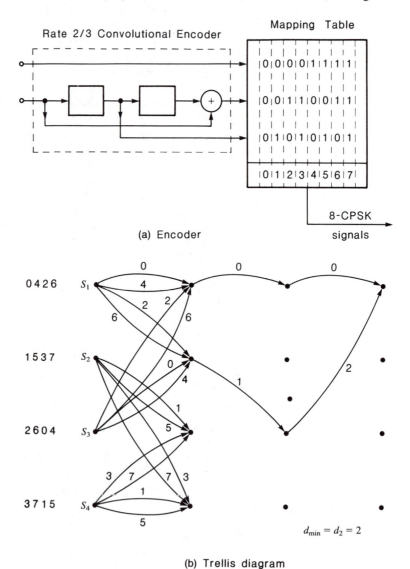

(b) Trellis diagram

Figure 9.49 Rate $\frac{2}{3}$ Ungerboeck code with four states and 8-PSK encoded signals (Table 9.17): (a) Encoder; (b) trellis diagram; (c) state diagram.

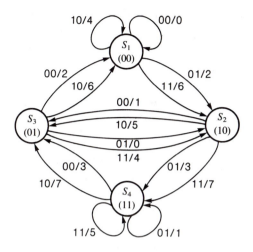

(c) State diagram

Figure 9.49 *(cont.)*

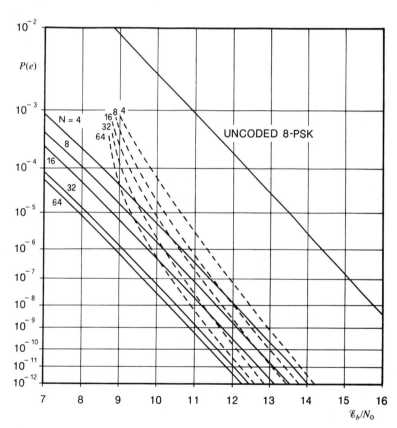

Figure 9.50 Upper and lower bounds to the error probability for coded 16-PSK signals. The curve for uncoded 8-PSK is shown for reference. (Benedetto and others, 1985.)

Sec. 9.7   Signal-Space Codes

of (4.141) and assuming the baseband phase pulse $q(t)$ of the MSK type (Fig. 4.36), let us write the information-carrying phase of the CPM signals as

$$\Phi(t; \xi_k, \sigma_k) = \frac{\pi H_k}{T} \xi_k(t - kT) + \varphi_k, \qquad kT \leq t < (k + 1)T, \qquad (9.149)$$

where ξ_k is the binary random variable that takes values ± 1 representing the binary message digits, and (H_k) is a periodic sequence assuming cyclically the values h_1, h_2, . . . , h_I.

From (9.149) we can see that the phase inside each time interval of duration T grows linearly from the state φ_k to the next state φ_{k+1}. Redundancy is introduced because the modulation index h is not constant as in MSK modulation.

The coding rule for multi-h phase codes is to choose the modulation index H_k cyclically into the set of possible values. That is, h_1 is chosen during the first interval, h_2 during the second, and so on, until all the indexes have been used. For the $(I + 1)$st interval the index h_1 is chosen again. In other words, in each interval of duration T, two frequencies are available for the transmission of the binary digits, and these frequencies are displaced by $\pm h_i/2T$ from the carrier nominal frequency. Phase continuity is maintained at the transition time between adjacent intervals.

When two modulation indexes, h_1 and h_2, are used, there are four possible phase paths, as shown in Fig. 9.51. But owing to the cyclic choice of the modulation index, only a pair of them is used in each time interval. Redundancy is thus introduced in the signal space.

The set of possible modulation indexes should be chosen in such a way as to allow phase paths as far apart as possible to yield a large minimum Euclidean distance. Furthermore, the spectrum of the transmitted signal should maintain the properties of the MSK signal. In this way, coding gains should be achievable without loss in bandwidth efficiency.

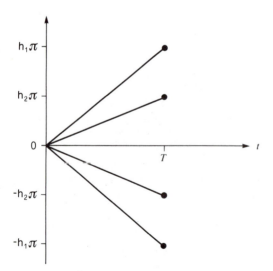

Figure 9.51 Possible phase functions in a multi-h code with two modulation indexes.

The multi-h coding scheme can be described by a phase trellis. If the modulation indexes $\{h_i\}$ include only multiples of $1/m$, where m is an integer, then all phases at the transition times $t = kT$ are multiples of $2\pi/m$. If N is the largest integer such that no pair of phase paths diverging from one state in the trellis merges, the phase code is said to have *constraint length N*. It can be shown that the maximum possible constraint length with I modulation indexes is $(I + 1)$, provided that no two subsets of modulation indexes have the same sum mod I. Let us indicate the modulation indexes as

$$h_i = \frac{L_i}{m}, \qquad m \text{ integer}, \qquad i = 1, 2, \ldots, I, \tag{9.150}$$

where L_i and m are mutually prime. Recalling the discussion of (4.146), the reader can verify that the trellis of a multi-h code exhibits the following properties:

(i) It has a periodic structure, with period T_p given by

$$T_p = \begin{cases} IT & \text{for } L \text{ even}, \\ 2IT, & \text{for } L \text{ odd}, \end{cases} \tag{9.151}$$

where $L = \Sigma_{i=1}^{I} L_i$.

(ii) The number of states is m for L even and $2m$ for L odd.

For practical purposes, the number of states in the code trellis must be kept as small as possible. Furthermore, the modulation indexes should be ratios of small integers; otherwise, the demodulator complexity becomes prohibitive.

Owing to the trellis structure of the code, optimum decoding is accomplished by means of the soft-decision Viterbi algorithm. The performance parameter is the Euclidean minimum distance $d_{\min}$ between pairs of paths in the trellis. Generally, it can be expected that $d_{\min}$ increases with the number of states and with the number I of modulation indexes. However, finding $d_{\min}$ is not trivial and numerical techniques are usually employed.

We shall assume as a reference system for these codes the MSK system that, for the same energy, has a minimum distance $d_{\text{ref}} = 2$. A list of the asymptotic performance of good multi-h codes is given in Table 9.18. Asymptotic coding gains from 2 to 4 dB are possible. Since the codes of Table 9.18 have practically the same spectrum as MSK modulation, the asymptotic coding gains are obtained without any bandwidth expansion.

The coding technique described in this section could be extended in two directions. First, the number of levels could be expanded from binary to multilevel. Second, shaping pulses other than the linear one could be used for the phase transitions between states in the trellis.

9.8 LINE CODES

Until now two different approaches have been considered for encoding the binary digits emitted by the source. In the first (error-correcting codes), redundancy was added directly to the digital sequence. In the second (signal-space codes), redundancy was added to

TABLE 9.18 Asymptotic coding gains for some multi-h codes using I modulation indexes. MSK modulation is used as a reference. (Anderson and Taylor, 1978.)

Number of states	Coding gain (dB)		
	$I = 2$	$I = 3$	$I = 4$
4	1.44	—	—
5	1.86	—	—
6	2.37	—	—
7	2.21	—	—
8	2.49	2.78	—
9	2.38	1.30	—
10	2.58	2.80	—
11	2.52	2.71	—
12	2.65	2.80	—
13	2.60	3.19	—
14		3.13	—
15		2.88	—
16		3.36	3.65
17		3.25	3.02
18			3.07
19			3.44
20			3.78
21			3.29
22			3.88

the available signal set. Another point of view, widely developed in the practice of baseband transmission of digital speech signals, is now explored. The most important requirement for these systems is the real-time operation of the decoder (very small delays allowed). Also, a slight complexity of the demodulation-decoding process was a mandatory design objective, especially in the early days of digital circuit technology. For these reasons, encoding was used mainly *to prevent* to the greatest possible extent the effects of channel impairments. When signal distortion and interferences are the major sources of degradation, besides the additive noise, these impairments depend on the spectral properties of the transmitted signal. In practice, it usually happens that the transfer function of a channel is worse near the band edges. Therefore, a good signal design should concentrate the transmitted power in the middle of the transmission bandwidth. In such a case a smaller distortion should be present in the received signal. To meet this objective, codes can be designed with the aim of shaping the spectrum of the transmitted signal. These codes are called *line codes*.

Besides spectral shaping, other important requirements are introduced when designing these codes. The most important are the reduction of dc wandering, the bandwidth efficiency, and the signal synchronization capabilities. For the latter, consider that the presence of many transitions in the signal sequence is very important for the operation of data-derived symbol synchronizers. When redundancy is introduced in the transmitted signal, error detection can also be achieved. But, unlike in previous codes, this possibility is considered only as a useful by-product of the present encoding approach. We shall consider line codes associated only with baseband transmission, since this was their practical motivation. Their extension to other linear modulation schemes is straightforward.

9.8.1 Binary Line Codes

The easiest way to shape the spectrum of the transmitted signal in a binary signaling scheme is a suitable choice of the signal waveforms. In their simplest implementation these line codes are only a case of signal design.

Some binary formats are shown in Figs. 9.52 and 9.53. The first three codes are simple binary signaling schemes. The *nonreturn-to-zero* (NRZ) code assigns the two antipodal levels to the two binary digits. The *return-to-zero* (RZ) code transmits a one

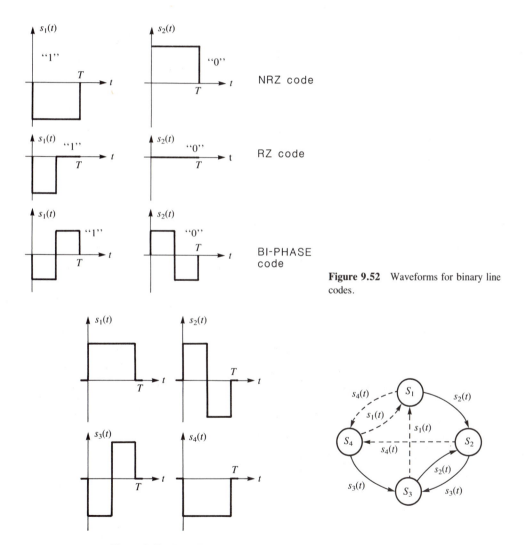

Figure 9.52 Waveforms for binary line codes.

Figure 9.53 Waveforms and state diagram for the encoder of the Miller code.

with a half-symbol pulse and a zero with no pulse. The *Manchester* (or *biphase*) *code* uses again two antipodal signals.

In the case of *Miller code* (or *delay modulation*) the code can be represented through a modulation with memory (Fig. 9.53). Four baseband signals are available. Which one is transmitted depends both on the present digit and on the previous transmitted signal. The encoding rule can be followed on the encoder state diagram of Fig. 9.53. Each state is identified by the label of the previously transmitted signal. Each transition between two states is represented with a solid arrow if a one is to be encoded and with a dashed arrow for a zero. The label on each arrow indicates the generated signal. An example of possible encoded waveforms is shown in Fig. 9.54. The power spectra of these encoded signals can be evaluated with the computational method presented in Section 2.3.1.

For the NRZ code, we get

$$\frac{\mathcal{G}(f)}{\mathcal{E}} = \frac{1}{T}(1 - 2p)^2 \delta(f) + 4p(1 - p)\frac{\sin^2(\pi f T)}{(\pi f T)^2}, \qquad (9.152)$$

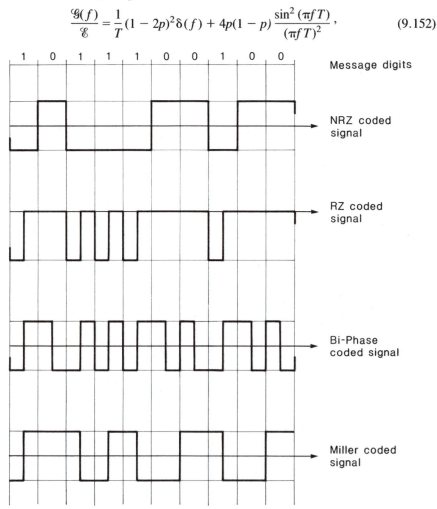

Figure 9.54 An example of encoded waveforms in four binary line codes.

where $\mathscr{E}$ is the energy of each signal, T the signaling period, and p the probability of a zero. When the two binary digits are equally likely, the spectral line at the origin disappears.

For the RZ code, we get

$$\frac{\mathscr{G}(f)}{\mathscr{E}} = \frac{1}{2T}(1-p)^2\delta(f) + \frac{1}{2T}(1-p)^2\sum_{\substack{n=-\infty\\n\neq0}}^{\infty}\left(\frac{2}{\pi n}\right)^2\delta\left(f-\frac{n}{T}\right)$$

$$+\frac{1}{2}p(1-p)\frac{\sin^2(\pi f T/2)}{(\pi f T/2)^2}, \qquad (9.153)$$

where $\mathscr{E}$ is the energy of the nonzero signal. Notice that a discrete component is always present in the spectrum of this coded signal.

For the Manchester code, we get

$$\frac{\mathscr{G}(f)}{\mathscr{E}} = \frac{1}{T}(1-2p)^2\sum_{\substack{n=-\infty\\n\neq0}}^{\infty}\left(\frac{2}{\pi n}\right)^2\delta\left(f-\frac{n}{T}\right) + 4p(1-p)\frac{\sin^4(\pi f T/2)}{(\pi f T/2)^2}. \qquad (9.154)$$

Again, the spectral lines disappear for equally likely binary digits. The case of the Miller code was already considered in Example 2.13.

The power spectra of the NRZ, Manchester, and Miller codes are shown in Fig. 9.55 for the case of equally likely binary digits ($p = 0.5$). Notice that the latter two codes present a spectral minimum at $f = 0$. This property is very important and will be a usual requirement for the design of the codes of the next sections. In the case of baseband transmission, this feature is a countermeasure to poor channel response near dc. Instead, when bandpass transmission is used, the carrier tracking can be achieved more efficiently because a carrier can be inserted at the spectral minimum. When comparing

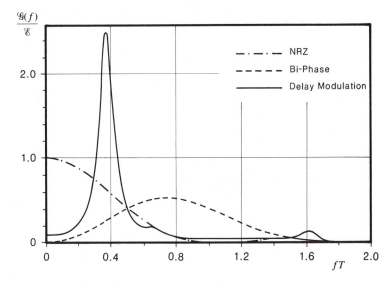

Figure 9.55 Power spectra of binary encoded signals. Only the continuous part of the spectrum is shown for the case of $p = 0.5$.

Manchester and Miller codes, notice that the latter has its power concentrated in a bandwidth that is less than one-half the symbol rate and thus presents a higher bandwidth efficiency. The Manchester code and the Miller code can be easily encoded and decoded.

9.8.2 Correlative Level Encoding

To introduce this encoding technique, let us recall the multilevel baseband PAM, whose transmission scheme is shown in Fig. 9.56. The source binary symbols are converted into an M-level sequence (a_m) of symbols taking values 0, 1, . . . , $(M - 1)$ and corresponding, for instance, to the M antipodal levels used for transmission. If $h(t)$ denotes the response of the overall channel, the line signal $x(t)$ can be represented by

$$x(t) = \sum_i a_i h(t - iT), \tag{9.155}$$

where T is the signaling period. We have $T = T_s \log_2 M$, where $1/T_s$ is the rate of the binary source.

The most significant shortcoming of multilevel modulation schemes is that intersymbol interference (ISI) increases rapidly with the number of signaling levels. Thus, the potentially greater capacity of these systems can hardly be achieved in practice.

The *correlative level encoding* techniques, introduce, deliberately, a controlled amount of ISI over a span of one, two, or more symbols. The net result is a spectral shaping of the transmitted signal at the price of an increased number of transmitted levels with respect to the original sequence. But the spectral shaping should prevent the impairing effects of channel distortions. Notice that redundancy is introduced *amplitude-wise*. That is, there are more signal levels than strictly required. We have a situation similar to that of Ungerboeck codes, but now the redundancy is used to shape the spectrum. In the most general case, the correlative level encoder converts the sequence (a_m) into another sequence (c_m) using the following linear transformation:

$$c_m = \sum_{i=0}^{K-1} g_i a_{m-i}. \tag{9.156}$$

The code is described by the K coefficients $g_0, g_1, . . . , g_{K-1}$, which are usually integers. The size of the alphabet of the coded symbols turns out to be (Kobayashi and Tang, 1971)

$$L = (M - 1) \sum_{i=0}^{K-1} |g_i| + 1. \tag{9.157}$$

Notice that the encoder introduces correlation among the transmitted symbols and this achieves the expected spectral shaping.

An undesired feature of these codes is that the memory introduced by the encoder

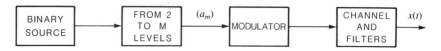

Figure 9.56 Block diagram of the PAM multilevel transmission system.

prevents the possibility of a symbol-by-symbol decoding. In fact, the received symbols are no longer independent. The decoder operation can be formally described by inverting (9.156), from which we get (assuming $g_0 \neq 0$)

$$\hat{a}_m = \frac{1}{g_0} \left[\hat{c}_m - \sum_{i=1}^{K-1} g_i \hat{a}_{m-i} \right], \tag{9.158}$$

where $\hat{c}_m$ is the estimated symbol at the receiver output and the $\hat{a}_m$'s are the reconstructed information symbols. The most important drawback with (9.158) is that the erroneous reception of one L-ary symbol ($\hat{c}_m \neq c_m$) may result in error propagation when it is decoded back into the M-ary sequence ($\hat{a}_m$). A simple example will clarify these first concepts.

Example 9.36 Twinned binary code

This code operates on a binary input ($M = 2$) and is defined by the two ($K = 2$) coefficients $g_0 = 1$, $g_1 = -1$. The coding rule is

$$c_m = a_m - a_{m-1}$$

and the new symbol alphabet is $\{-1, 0, +1\}$ with $L = 3$. The decoding rule is, from (9.158),

$$\hat{a}_m = \hat{c}_m + \hat{a}_{m-1}.$$

The following sequences are an example of encoding and decoding:

$$
\begin{array}{lcccccccccccc}
a_m & 0 & 1 & 0 & 1 & 1 & 1 & 0 & 1 & 0 & 0 & \ldots, \\
c_m & & 1 & -1 & 1 & 0 & 0 & -1 & 1 & -1 & 0 & \ldots, \\
\hat{a}_m & 0 & 1 & 0 & 1 & 1 & 1 & 0 & 1 & 0 & 0 & \ldots.
\end{array}
$$

However, if an error is present in the sequence ($\hat{c}_m$), we can have error propagation in the sequence ($\hat{a}_m$). We can verify this in the following:

$$
\begin{array}{lccccccccc}
\hat{c}_m & & 1 & -1 & 1 & \boxed{-1} & 0 & -1 & 1 & -1 & 0 & \ldots, \\
\hat{a}_m & 0 & 1 & 0 & 1 & 0 & 0 & -1 & 0 & -1 & -1 & \ldots.
\end{array}
$$

Actually, nonpermitted symbols also appear in the sequence ($\hat{a}_m$). This allows error detection. But error propagation is undoubtedly an undesired consequence. □

The error propagation effect can be avoided by *precoding* the M-ary sequence (a_m) before presenting it to the encoder. The operations to be performed are shown in Fig. 9.57. An intermediate sequence (b_m) is generated by the *precoder* according to the following relation:

$$a_m = \sum_{i=0}^{K-1} g_i b_{m-i} \qquad \mathrm{mod}\, M, \tag{9.159}$$

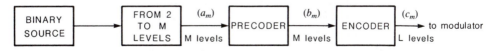

9.57 Block diagram of correlative level encoding.

which means that a_m is obtained by reducing mod M the RHS. Therefore, the actual symbol to be presented to the encoder is obtained from (9.159) as

$$b_m = \frac{1}{g_0}\left[a_m - \sum_{i=1}^{K-1} g_i b_{m-i}\right] \quad \text{mod } M. \tag{9.160}$$

The alphabet size of the precoded symbols b_m remains M, as for the symbols a_m. Moreover, if the symbols of the sequence (a_m) are independent and identically distributed, so are the symbols of the sequence (b_m). The encoder now acts on the sequence (b_m) to generate the sequence (c_m) to be presented to the modulator. We now have

$$c_m = \sum_{i=0}^{K-1} g_i b_{m-i}. \tag{9.161}$$

By comparing (9.161) with (9.159), we can conclude that

$$a_m = c_m \quad \text{mod } M. \tag{9.162}$$

Therefore, the precoder allows the reconstruction of each transmitted symbol only on the basis of the corresponding received L-ary symbol, and thus error propagation is avoided. Let us apply these concepts to the previous example.

Example 9.36 Precoded twinned binary (*continued*)

The equation for the precoder is, from (9.160),

$$b_m = a_m + b_{m-1} \quad \text{mod } 2.$$

And the equation for the encoder becomes $c_m = b_m - b_{m-1}$.
The decoder decides on the basis of the equation

$$\hat{a}_m = \hat{c}_m \quad \text{mod } 2.$$

The following sequences are an example of precoding, encoding, and decoding.

a_m		0	1	0		1	1		1	0	1	0	0	. . . ,
b_m	0	0	1	1		0	1		0	0	1	1	1	. . . ,
c_m		0	1	0	-1	1	-1	0	1	0	0	. . . ,
$\hat{a}_m$		0	1	0		1	1		1	0	1	0	0	. . . .

The reader can immediately verify that no error propagation is possible. □

The code described in the previous example is also known as *bipolar code* or *alternate-mark-inversion* (AMI) code. It represented the first simple example of coded transmission of binary digital information in PCM systems. The line signal is ternary, but each level that could represent $\log_2 3 = 1.58$ bits bears only one bit of information. The binary symbol 0 is represented by no signal on the line, whereas the binary symbol 1 is represented by alternating positive and negative pulses. It is intuitive that this encoding method has the advantage of reducing the dc component of the signal, because a pulse of one polarity is always followed by a pulse of opposite polarity. Moreover, a single transmission error will always give rise to a *bipolar violation*, that is, the presence of a pulse that does not satisfy the bipolar encoding rule. This property does not represent a powerful error-detection scheme. Nevertheless, it allows the monitoring of the quality of the transmission link without the need for any information on the transmitted data. It is the simplest example of error-detection capability caused by the amplitudewise added redundancy. Finally, notice that the transmission of long sequences of zeros results in long periods without timing information. The bipolar code does not present good self-clocking capabilities, and one must eventually resort to *scrambling* the transmitted sequence (see Section 9.2.7). The bipolar code is so important that it is used as a reference when comparing the performance of ternary line codes, as we shall see in the next section. The bipolar code can also be described by the state diagram of Fig. 9.58. As usual, a dashed arrow represents a binary 1 and a solid arrow a 0. The labels $-$, 0, and $+$ are used to indicate the line levels -1, 0, and $+1$, respectively.

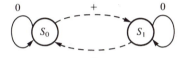

Figure 9.58 State diagram description for the bipolar code (Example 9.36).

We now want to get some insight into the spectral shaping properties of these correlative level codes. Recalling (9.155), the coded line signal is given by

$$x(t) = \sum_i c_i h(t - iT). \tag{9.163}$$

But, using (9.161), we have

$$x(t) = \sum_i \sum_{j=0}^{K-1} g_j b_{i-j} h(t - iT). \tag{9.164}$$

If we let $m = i - j$ and define a new waveform $q(t)$ as

$$q(t) \triangleq \sum_{j=0}^{K-1} g_j h(t - jT), \tag{9.165}$$

we can rewrite (9.164) as

$$x(t) = \sum_m b_m q(t - mT). \tag{9.166}$$

The important conclusion derived from (9.166) is that the encoding process is equivalent to replacing the waveform $h(t)$ with the waveform $q(t)$ for the transmission on the channel. The definition (9.165) can be written in the frequency domain as

$$Q(f) = G(fT)H(f), \tag{9.167}$$

where

$$G(fT) \triangleq \sum_{m=0}^{K-1} g_m e^{-j2\pi mfT} \tag{9.168}$$

is the transfer function of the discrete linear system defined by (9.161). In conclusion, the encoder is a filter with transfer function $G(fT)$. The desired objectives for the spectral shape can be met by assigning suitable constraints in the choice of the code coefficients. As a first example, a code introduces a spectral null at dc if $G(f)$ is zero for $f = 0$. Using (9.168), we get the condition

$$\sum_{m=0}^{K-1} g_m = 0. \tag{9.169}$$

It is immediately verified that the bipolar code (Example 9.36) satisfies this condition.

A spectral null at the Nyquist frequency is obtained when $G(f)$ is zero for $f = 1/2T$. Using again (9.168), we get the condition

$$\sum_{m=0}^{K-1} (-1)^m g_m = 0. \tag{9.170}$$

Let us consider some examples of codes that satisfy these conditions.

Example 9.37 Duobinary code

This code has $K = 2$ and coefficients $g_0 = g_1 = 1$. If the code operates on a binary input ($M = 2$), the encoded symbols are ternary ($L = 3$) and take values in the set $\{0, 1, 2\}$. We usually prefer to transmit antipodal levels. A simple shift in levels gives the desired result. After obtaining from (9.160) the precoded symbols b_m from the binary digits a_m as

$$b_m = a_m - b_{m-1} \qquad \text{mod } 2, \tag{9.171}$$

we convert the binary symbols b_m into binary antipodal (± 1) symbols using the relation

$$b'_m = 2 b_m - 1. \tag{9.172}$$

The duobinary symbols c_m are obtained from (9.161) as

$$c_m = b'_m + b'_{m-1} \tag{9.173}$$

and now belong to the set $\{-2, 0, +2\}$. Finally, the binary information symbols can be recovered from c_m with the relation

$$\hat{a}_m = \frac{\hat{c}_m}{2} + 1 \qquad \text{mod } 2. \tag{9.174}$$

Notice that the receiver can implement this rule by using two thresholds set at -1 and $+1$ and deciding the received signal sample plus noise ($c_m + n_m$) according to the rule

$$\hat{a}_m = \begin{cases} 1, & \text{if } |c_m + n_m| < 1, \\ 0, & \text{if } |c_m + n_m| \geqslant 1. \end{cases} \tag{9.175}$$

The following is an example of possible sequences.

Data sequence	a_m		0	1	1	0	1	0	1	1
Precoded sequence	b_m	0	0	1	0	0	1	1	0	1
Antipodal levels	b'_m	-1	-1	1	-1	-1	1	1	-1	1
Duobinary levels	c_m		-2	0	0	-2	0	2	0	0
Decoded sequence	$\hat{a}_m$		0	1	1	0	1	0	1	1

The extension to the case of a multilevel input is quite straightforward. The reader can verify that (9.171) through (9.174) become for any M, respectively,

$$b_m = a_m - b_{m-1} \quad \mod M,$$
$$b'_m = 2b_m - (M - 1),$$
$$c_m = b'_m + b'_{m-1},$$
$$\hat{a}_m = \frac{\hat{c}_m}{2} + (M - 1) \quad \mod M. \tag{9.176}$$

In this case, the M-ary signal gives $(2M - 1)$ levels that the receiver maps back into the original M-level signal. The following is an example of possible sequences for the case $M = 4$.

Data sequence	a_m		0	1 3	2 3	0	2
Precoded sequence	b_m	0	0	1 2	0 3	1	1
Antipodal levels	b'_m	-3	-3	-1 1	-3 3	-1	-1
Duobinary levels	c_m		-6	-4 0	-2 0	$+2$	-2
Decoded sequence	$\hat{a}_m$		0	1 3	2 3	0	2

To find the spectral shaping introduced by the duobinary code, we obtain from (9.168)

$$G(fT) = 1 + e^{-j2\pi fT} = 2e^{-j\pi fT} \cos(\pi fT). \tag{9.177}$$

It is immediately verified that $G(fT)$ is zero at the Nyquist frequency $f = 1/2T$. The power spectrum of the line signal can be obtained from (9.166), assuming that $h(t)$ is an NRZ signal having the spectrum (9.152). Using (9.177), we obtain

$$\frac{\mathcal{G}(f)}{\mathcal{E}} = 4\left[\frac{\sin(2\pi fT)}{2\pi fT}\right]^2. \tag{9.178}$$

By comparison with (9.152), we conclude that the spectrum of the duobinary sequence (c_m) is compressed in frequency by a factor of 2 relative to the binary NRZ sequence (b'_m). This explains the name duobinary, which means doubling the speed of binary. Finally, it is interesting to notice that the equivalent waveform $q(t)$ of (9.165) for the duobinary code is

$$q(t) = h(t) + h(t - T). \tag{9.179}$$

If we assume for $h(t)$ a perfect Nyquist pulse [(7.22) with $\alpha = 0$], that is,

$$h(t) = \frac{\sin \pi t/T}{\pi t/T}, \tag{9.180}$$

we can see in Fig. 9.59 what $q(t)$ looks like. A full amount of ISI has been introduced at $t = T$. $\square$

The example just shown, and particularly Fig. 9.59, explains the other name used for this type of codes, *partial-response* signaling. In fact, sample points on the received waveform are chosen halfway to a full response. The following example describes another important type of code.

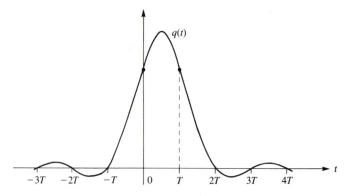

Figure 9.59 Equivalent waveform $q(t)$ for the duobinary code (Example 9.37).

Example 9.38 Modified duobinary code

This code has $K = 3$. The coefficients of the code are $g_0 = 1$, $g_1 = 0$, and $g_2 = -1$. If the code operates on a binary input ($M = 2$), the encoded symbols are ternary ($L = 3$). For an M-ary input, the encoded symbols have ($2M - 1$) levels. The encoding and decoding equations for the general M-ary case are in this case

$$b_m = a_m + b_{m-2} \qquad \text{mod } M,$$
$$b'_m = 2b_m - (M - 1),$$
$$c_m = b'_m - b'_{m-2},$$
$$\hat{a}_m = \frac{\hat{c}_m}{2} \qquad \text{mod } M. \tag{9.181}$$

The following is an example of possible sequences for the case $M = 2$.

Data sequence	a_m			0	1	1	0		1	0	1	1
Precoded sequence	b_m	0	0	0	1	1	1		0	1	1	0
Antipodal levels	b'_m	-1	-1	-1	1	1	1		-1	1	1	-1
Encoded levels	c_m			0	2	2	0		-2	0	2	-2
Decoded sequence	$\hat{a}_m$			0	1	1	0		1	0	1	1

The receiver can implement the decoding rule by using two thresholds set at -1 and $+1$ and deciding on the received sample plus noise ($c_m + n_m$), according to the rule

$$\hat{a}_m = \begin{cases} 0, & \text{if } |c_m + n_m| < 1 \\ 1, & \text{if } |c_m + n_m| \geq 1. \end{cases} \tag{9.182}$$

The spectral shaping introduced by the code is, from (9.168),

$$G(fT) = 1 - e^{j4\pi fT} = 2je^{-j2\pi fT} \sin(2\pi fT). \tag{9.183}$$

It can be verified that $G(fT)$ is zero both at dc and at the Nyquist frequency. It is left as an exercise to the reader to sketch the equivalent waveform $q(t)$ for this code. □

Different classes of *partial-response* signaling schemes can be obtained with a weighted linear superposition of K perfect Nyquist pulses $h(t)$. The equivalent waveform of the code is given by

$$q(t) = g_0 h(t) + g_1 h(t - T) + \cdots + g_{K-1} h[t - (K - 1)T]. \qquad (9.184)$$

An appropriate choice of the coefficients results in a variety of spectral shapes. Some signaling schemes are listed in Table 9.19.

TABLE 9.19 Different signaling schemes of the partial response type

Class	K	g_0	g_1	g_2	g_3	g_4	
I	2	1	1	—	—	—	Duobinary
II	3	1	2	1	—	—	
III	3	2	1	−1	—	—	
IV	3	1	0	−1	—	—	Modified duobinary
V	5	−1	0	2	0	−1	

The performance characteristics of these codes with respect to ideal binary is that they are more suitable for transmitting data on a real channel close to the Nyquist rate because they offer simple means to obtain relatively gentle filter cutoffs and allow different choices that can be advantageous in specific applications. However, there is a price to be paid. Because of the increased number of levels, a larger signal-to-noise ratio is required to obtain the same error probability as for the corresponding PAM system in the presence of white Gaussian noise only. In the following example, the error probability is evaluated for the duobinary code (Example 9.37). However, this penalty is inherent to the symbol-by-symbol detection approach, and it can be partially recovered with more sophisticated decoding techniques. These will be mentioned later.

Example 9.39

Let us take again the duobinary code of Example 9.37. We assume the transmission of an M-level signal according to (9.166). That is,

$$x(t) = \sum_m b'_m q(t - mT), \qquad (9.185)$$

where b'_m takes values in the set $\{-(M - 1), -(M - 3), \ldots, (M - 1)\}$. The Fourier transform of the waveform $q(t)$ of Fig. 9.59 is

$$Q(f) = 2Te^{-j\pi fT} \cos(\pi fT), \qquad |f| \le \frac{1}{2T}. \qquad (9.186)$$

Therefore, its energy is equal to $2T$. The receiver bases its decision on the observation of c_m. If the input levels b'_m are equally likely, it can be verified that the received levels c_m have a triangular probability distribution. That is,

$$P\{c_m = 2i\} = \frac{M - |i|}{M^2},$$

$$i = -(M - 1), -(M - 2), \ldots, 0, 1, \ldots, (M - 1). \qquad (9.187)$$

We assume additive Gaussian noise with power spectral density $N_0/2$. If the receiving filter has an equivalent noise bandwidth $1/2T$, then the variance σ^2 of the received noise samples is $N_0/2T$. The decision thresholds of the receiver will be placed halfway between the signal levels. A decision error occurs whenever the noise moves one signal level to another. Therefore, a tight upper bound on the average symbol error probability is obtained from the union bound

$$P_M(e) \leq \sum_{i=-(M-2)}^{M-2} P\{|r - 2i| > 1 \,|\, c_m = 2i\} P\{c_m = 2i\}$$

$$+ 2\, P\{r - 2(M-1) > 1 \,|\, c_m = 2(M-1)\} P\{c_m = 2(M-1)\}, \tag{9.188}$$

where r represents the sample of the received signal. But it can be immediately verified that

$$P\{|r - 2i| > 1 \,|\, c_m = 2i\} = P\{|r| > 1 \,|\, c_m = 0\}, \tag{9.189}$$

so that

$$P_M(e) \leq \left(1 - \frac{1}{M^2}\right) P\{|r| > 1 \,|\, c_m = 0\}. \tag{9.190}$$

Since

$$P\{|r| > 1 \,|\, c_m = 0\} = erfc\left(\sqrt{\frac{T}{N_0}}\right), \tag{9.191}$$

we have finally

$$P_M(e) \leq \left(1 - \frac{1}{M^2}\right) erfc\left(\sqrt{\frac{T}{N_0}}\right). \tag{9.192}$$

We can express this result in terms of the average energy of the transmitted signal of (9.185). It turns out to be

$$\mathscr{E} = \frac{(M^2 - 1)2T}{3}. \tag{9.193}$$

Therefore, (9.192) becomes

$$P_M(e) \leq \left(1 - \frac{1}{M^2}\right) erfc\left(\sqrt{\frac{3}{2(M^2-1)}\frac{\mathscr{E}}{N_0}}\right). \tag{9.194}$$

Comparing this result with (5.11), we can conclude that the use of the duobinary code, for large values of M, results in a loss of 3 dB in the signal-to-noise ratio with respect to PAM. An improvement over (9.194) can be achieved by equally dividing the partial response function $Q(f)$ between modulator and demodulator, as obtained in (7.27) and (7.28). In this case, the variance of the received noise is given by

$$\sigma^2 = \frac{N_0}{2} \int_{-1/2T}^{1/2T} |Q(f)| \, df = \frac{2N_0}{\pi}. \tag{9.195}$$

Under this hypothesis, the symbol error probability becomes

$$P_M(e) \leq \left(1 - \frac{1}{M^2}\right) erfc\left(\sqrt{\frac{3}{M^2-1}\left(\frac{\pi}{4}\right)^2 \frac{\mathscr{E}}{N_0}}\right). \tag{9.196}$$

The loss in signal-to-noise ratio with respect to ideal M-level PAM is now 2.1 dB. $\square$

Error detection and optimum decoding of correlative level codes

In this section we want to present the correlative level codes in a more general parlance in order to cite some interesting results. However, a lot of details involved are beyond the scope of this book and can be found in the references quoted in the Bibliographical Notes.

The M-level information sequence (a_m) can be represented in polynomial form as

$$A(D) = \sum_{k=0}^{\infty} a_k D^k. \tag{9.197}$$

A correlative level encoder is a discrete linear system whose impulse response can be represented by the polynomial

$$G(D) = \sum_{i=0}^{K-1} g_i D^i. \tag{9.198}$$

The information sequence $A(D)$ is transformed into the precoded sequence $B(D)$ by means of the relation

$$B(D) = \frac{A(D)}{G(D)} \quad \mod M. \tag{9.199}$$

The operation mod M means that the coefficients of the quotient are reduced mod M, where M is the number of levels of the information sequence (a_n). Finally, the correlative level encoder generates the L-level transmitted sequence according to the equation

$$C(D) = G(D)B(D). \tag{9.200}$$

The encoder output $C(D)$ is sent over an AWGN channel, which adds a noise $N(D)$. $Y(D)$ represents the output sequence from the channel. Notice that $N(D)$ and $Y(D)$ are sequences of continuous RVs. In the conventional symbol-by-symbol decoding process, the sequence $Y(D)$ is quantized over L levels, and the quantizer output $\hat{C}(D)$ is equal to $C(D)$ if no errors are introduced by the channel. In this case, the information sequence can be recovered by simply performing a mod M reduction of the sequence $\hat{C}(D)$. The block diagram of a receiver that implements this decoding procedure is shown in Fig. 9.60.

It is our purpose to show how this conventional detector can be replaced by increasingly more complicated devices to achieve either error detection or even error correction of the received sequence. Let us replace the mod M detector of Fig. 9.60

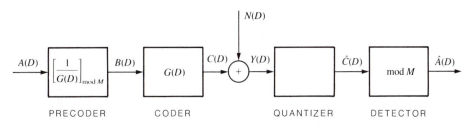

Figure 9.60 Block diagram of the correlative level encoded transmission. The mod-M block is the conventional detector.

with the system of Fig. 9.61. The quantizer output $\hat{C}(D)$ is fed to the inverse filter $1/G(D)$, which is followed by a decoder with transfer function $[G(D)]_{\mathrm{mod}\,M}$. In the absence of errors, the inverse filter output $\hat{B}(D)$ is the same as the precoder output $B(D)$. On the other hand, the decoder performs the inverse operation of the precoder, through the relation

$$\hat{A}(D) = G(D)\hat{B}(D) \qquad \mathrm{mod}\, M. \tag{9.201}$$

The inverse filter and the decoder perform together the equivalent mod M operation of the conventional detector. Therefore, correct recovering of the information symbols is achieved. What is new with this implementation is that the sequence $\hat{B}(D)$ is now available. Since it must contain only the allowable levels, $0, 1, \ldots, M-1$, an error is detected whenever this condition is violated. Practical simplified schemes derived from Fig. 9.61 have been implemented for different codes.

Two different strategies can be adopted in the presence of detected errors. The first is to request retransmission, if possible in the particular application. The second is to monitor the performance of the system by counting the errors and adopting some reasonable action to correct the detected errors. For example, the permitted level closest to the erroneous received one could be chosen. The error-detection procedure just described does not use all the redundancy contained in the coded signal. In fact, the hard decisions taken by the quantizer discard a significant amount of information from the sequence $Y(D)$. Actually, it is clear that the received levels lying close to the thresholds of the quantizer are less reliable than those that are close to the ideal levels. It is thus possible to introduce *ambiguity zones* around the quantizer thresholds, and when the received signal $Y(D)$ falls into these zones, the receiver is temporarily allowed to postpone its decision. Thus, we have a soft quantization of $Y(D)$ on more than L levels, and for some of them (the least reliable) a *null decision* is temporarily taken. The ambiguity symbols are handled by considering a block of received symbols. If only one null decision is present, it is usually possible to recover the correct symbol. This approach is called *ambiguity zone decoding* (AZD). It will be better clarified by the following example.

Example 9.40 AZD of duobinary code

Let us consider the duobinary code of Example 9.37. The encoded symbols belong to the set $\{-2, 0, +2\}$. The correlation introduced by the code introduces the following constraint. Take a block of consecutive symbols with the format given by

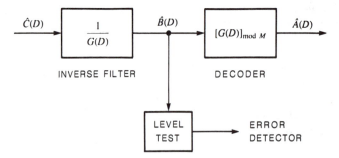

Figure 9.61 Block diagram of the receiver for error detection of correlative level encoded sequences. This receiver replaces the mod-M block of Fig. 9.60.

$$\underbrace{X, 0, 0, \ldots, 0, 0, Y,}_{m \text{ symbols } 0}$$

where X and Y are either -2 or $+2$. Then

$$m \text{ is even if } X = Y,$$
$$m \text{ is odd if } X \neq Y. \tag{9.202}$$

Therefore, an AZD receiver for this code examines blocks of length $(m + 2)$, with the symbols in the first and last position being $+2$ or -2 (extreme values). No extreme value must be present in the intermediate m positions. Consider the situation with a block containing only one null, that is, the block

$$X, \underbrace{0, 0, \ldots 0}_{m_1}, \text{null}, \underbrace{0, 0, \ldots, 0}_{m_2}, Y.$$

Using the rule (9.202), the null can be replaced unambiguously by using the decision table given next, in which $\bar{X}$ is the opposite extreme value from X. That is, if $X = 2$, then $\bar{X} = -2$, and vice versa.

Case	m_1	m_2	Null is
$X = Y$	Even	Even	X
	Even	Odd	0
	Odd	Even	0
	Odd	Odd	$\bar{X}$
$X \neq Y$	Even	Even	0
	Even	Odd	X
	Odd	Even	Y
	Odd	Odd	0

Multiple nulls cannot be replaced unambiguously with simple rules but they can be restored to their most likely values as determined by the null zone in which they fall. An error-detection procedure can check if the restored sequence is permitted. $\square$

A completely different approach for decoding correlative level encoded sequences is based on the following question. Given the received sequence $Y(D)$ of continuous RVs (see Fig. 9.60), how is it feasible to apply the ML approach to obtain the information sequence $A(D)$? The answer to this question comes from the observation that correlative level encoding is equivalent to introducing a controlled amount of ISI into the transmitted waveform. This property is concisely represented in (9.184), which defines, in general, a partial-response signaling scheme. Therefore, the problem of ML decoding of a correlative level encoded sequence is perfectly equivalent to that of recovering a sequence of symbols transmitted over a linear channel with ISI. This problem was solved in Section 7.5, where the ML sequence receiver was shown to be a processor based on the Viterbi algorithm. The encoded sequence corresponds to a path into a trellis that is searched for by the Viterbi processor. To clarify these ideas, the results of Section 7.5 will now be applied to an example.

Example 9.41 Optimum decoding of duobinary code

We consider in this example the duobinary code that was already described in Examples 9.37, 9.39, and 9.40. The M-level information sequence (a_m) is encoded according to (9.176). Using the terminology of Section 7.5, we observe that the encoding scheme has a memory one. The trellis for this code has, at each step k, the states $\sigma_k = (b_{k-1})$ if precoding is used or $\sigma_k = (a_k)$ if precoding is omitted. In both cases there are M possible states at each decoding step. The case of a binary input is illustrated in Fig. 9.62. Both the state and the trellis diagrams are shown for the case of precoded input. In each diagram a continuous transition arrow corresponds to an input symbol 0, whereas a dashed arrow corresponds to an input symbol 1. Moreover, notice that in the trellis diagram the slope of each transition between states corresponds to the transmitted channel symbol c_m. This property is an immediate consequence of the encoding rule expressed in (9.176). The transmitted waveform $q(t)$ for the duobinary code is shown in Fig. 9.59. The samples are $q_0 = 1$ and $q_1 = 1$.

The evaluation of the error probability of the optimum receiver can be performed by using the computations of Example 7.10 with only minor modifications. The final result is in fact the same. So, for transmission without precoding (as in Example 7.10), we get

$$P_M(e) \le M(M-1)\, erfc\left(\sqrt{\frac{2T}{N_0}}\right). \tag{9.203}$$

It was shown in Kobayashi (1971a) that when precoding is adopted the error probability is given by

$$P_M(e) \le 2(M-1)\, erfc\left(\sqrt{\frac{2T}{N_0}}\right). \tag{9.204}$$

Except for the binary case, precoding is also beneficial for the MLD receiver as it was in the conventional symbol-by-symbol detection method. Using the definition (9.193), we can rewrite (9.203) and (9.204) in the form

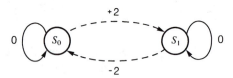

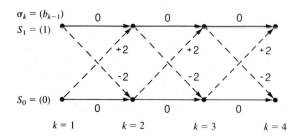

Figure 9.62 State diagram and trellis diagram of a duobinary code with precoding (see Example 9.41).

$$P_M(e) \leq \begin{cases} M(M-1)\, erfc\left(\sqrt{\dfrac{3}{M^2-1}\dfrac{\mathcal{E}}{N_0}}\right), & \text{no precoding,} \\[3mm] 2(M-1)\, erfc\left(\sqrt{\dfrac{3}{M^2-1}\dfrac{\mathcal{E}}{N_0}}\right), & \text{with precoding.} \end{cases} \qquad (9.205)$$

If we compare these results with those for the ideal M-level PAM signaling, we conclude that the loss in noise margin, which was seen to be a disadvantage of correlative level encoding, can be almost completely recovered by using an ML receiver. In fact, the two results are asymptotically equal. The only difference, in favor of the PAM, is the multiplying factors that correspond to fractions of decibels in the ratio $\mathcal{E}/N_0$.

To complete our discussion, we can quote the error probability performance of the AZD receiver derived in Kobayashi and Tang (1971). That is,

$$P_M(e) \leq \frac{3}{2}\left(1 - \frac{1}{M}\right) erfc\left(\sqrt{\frac{12(3 - 2\sqrt{2})}{M^2 - 1}\frac{\mathcal{E}}{N_0}}\right). \qquad (9.206)$$

If we ignore the constant factors and consider only the asymptotic behavior as $\mathcal{E}/N_0 \to \infty$, the decoding error rates for the different receivers for the duobinary code can be given the form

$$P_M(e) \sim erfc\left(\sqrt{\frac{3}{\alpha(M^2 - 1)}\frac{\mathcal{E}}{N_0}}\right) \qquad (9.207)$$

where the parameter α has the following values:

Suboptimum symbol-by-symbol detection (9.194) $\qquad \alpha = 2,$

Optimum symbol-by-symbol detection (9.196) $\qquad \alpha = \left(\dfrac{4}{\pi}\right)^2 = 1.621,$

ML detection (9.205) $\qquad \alpha = 1,$

AZD detection (9.206) $\qquad \alpha = \dfrac{3 + 2\sqrt{2}}{4} = 1.457.$

Although somewhat inferior to the ML receiver, the AZD receiver has the advantage of simpler implementation. In fact, it requires a smaller number of quantization levels, less memory storage, and reduced computational effort. $\square$

The discussion of the previous example can be generalized, as a conclusion, about the performance of correlative level encoding. The existence of redundant levels in the transmitted signal apparently introduces a penalty in the signal-to-noise ratio when compared with an ideal PAM transmission with the same performance. Actually, this is only a feature of the simple conventional detector that processes the received signal on a symbol-by-symbol basis. The desired spectral shaping can be obtained without any practical loss in noise margin if an optimum ML receiver based on the Viterbi algorithm is used in place of the simple conventional detector.

9.8.3 Pseudoternary Line Codes

As previously stated, the bipolar or AMI code is one of the best-known codes for the transmission of multiplex PCM signals on cables. It is a linear ternary code, since it is obtained from a linear combination of the binary information symbols. Other linear ternary codes are the class I (duobinary) and class IV (modified duobinary) partial response codes of Table 9.19. Actually, the name *pseudoternary codes* (PT codes) arose because a three-level line signal is used to represent a binary sequence.

These codes share the property of shaping the spectrum of the transmitted signal but do not achieve any definite self-clocking capability. In fact, long sequences of zero-level signals can cause the loss of timing information. For this reason, these codes usually require a scrambling of the binary sequence to avoid long sequences of zeros. In the attempt to overcome this drawback, a great variety of PT codes was designed using the bipolar code as a reference starting point. These PT codes are less tractable mathematically and more difficult to describe systematically. Nevertheless, they present unique practical advantages. The design goals for these codes can be summarized as follows:

1. To control the shape of the spectrum, especially in the vicinity of the zero frequency. Usual PCM transmission lines require dc removal; as a consequence, their transfer function is poor near the zero frequency.

2. To avoid long sequences of zero-level line signals, in order to render the clock recovery easier.

3. To increase, if possible, the data rate. In fact, the ternary signal can carry up to 1.58 bits per level instead of the 1 bit of the bipolar code.

4. Finally, the redundancy should facilitate error monitoring as in the bipolar code.

The performance of a dc-free code can be expressed in terms of the set of values that the sum of the encoded signal levels can assume. Let us define the *running digital sum* (RDS) of the code as

$$\text{RDS}(N_1, N_2) \triangleq \sum_{i=N_1}^{N_2} c_i \qquad (9.208)$$

where c_i is the encoded signal level at discrete time i. The most common performance index is the *digital sum variation* of the code, defined as

$$\text{DSV} \triangleq \max_{N_1, N_2} \text{RDS}(N_1, N_2) - \min_{N_1, N_2} \text{RDS}(N_1, N_2). \qquad (9.209)$$

It can be shown (e.g., see Justesen, 1982) that the sensitivity of the encoded signal to a cutoff at low frequencies is related to its DSV.

It can be verified that the bipolar code can take only two values of RDS and that its DSV is 1. Actually, this is the smallest possible value of DSV for PT codes.

The PT codes have been classified as *alphabetic* or *nonalphabetic*. The alphabetic codes owe their name to the fact that the binary information symbols are grouped into blocks or *characters* of k symbols. Thus, we have a transmission alphabet of 2^k characters. Each source character is encoded into a line character of n ternary symbols. The encoding is undertaken one character at a time. The encoding rule may or may not be memoryless. Consequently, the encoder has one or more states. An important design goal is that of a decoding algorithm that is independent of the encoder state. One constraint is $2^k \leq 3^n$, that is, $k \leq 1.58\, n$. Small values of k and n are required to avoid excessively complex practical devices. The main feature of these codes is the increase of the data rate with respect to the AMI code. The price paid is that character synchronization (in addition to symbol synchronization) is also required for correct decoding.

The nonalphabetic codes are instead a modification of the AMI code, the aim being the elimination of long sequences of zero-level signals at the price of a slight DSV deterioration. There is no data rate increase. The idea behind these codes is very simple: long sequences of zeros in the bipolar encoded signal are replaced by *filling* sequences that will maintain active the receiver clock. The filling sequence must be recognized by the decoder and eliminated. To this purpose, it contains violations of the bipolar rule that alternates positive and negative pulses. Also in this case, the addition of the filling sequence can be made by a memoryless encoder or by one with memory. What may be important is that the decoder can replace these sequences without considering the encoder state. We shall now describe in some detail the most used nonlinear PT codes.

Nonalphabetic (filled bipolar) PT codes

These codes are derived from the bipolar code by modifying its encoding rule as follows:

1. Binary ones are transmitted by alternating positive and negative pulses.
2. Binary zeros are counted; if there are less than m consecutive of them, they are transmitted as bipolar zero-level signals; if, otherwise, m consecutive binary zeros are counted, they are replaced by an appropriate filling sequence.

These codes are called *modal* if the filling sequence depends on the state of the encoder. They are called *nonmodal* if the choice of the filling sequence is memoryless. All the codes of this class differ for the choice of the filling sequence. In Fig. 9.63, some of these codes are presented and compared with the bipolar code when encoding the same binary data sequence. To clarify the description, the symbol 0 represents a bipolar zero-level signal, the symbol B represents a normal bipolar coded pulse, and the symbol V represents a bipolar coded pulse that violates the alternating bipolar rule.

The first nonmodal code is the *bipolar with 6-zero substitution* (B6ZS) whose 6-symbol filling sequence is chosen to be $B0VB0V$. It can be verified that this code has a DSV of 3. The decoding of the received sequence does not present particular problems since the filling sequence can be easily recognized. Error monitoring can be performed on the recovered bipolar sequence. This nonmodal code can be generalized to the class of BmZS codes, for which m zeros instead of 6 are replaced by suitable filling sequences.

An important class of modal codes is that of the *high-density bipolar* (HDBm) codes, in which sequences of $(m + 1)$ bipolar zeros are replaced by one of the two filling sequences

$$000\ldots0V$$
$$B\underbrace{00\ldots0}_{m-1}V$$

(9.210)

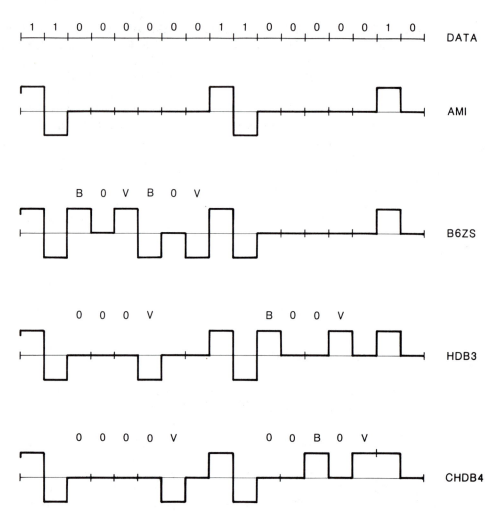

Figure 9.63 Encoding rules of three nonalphabetic PT codes (B6ZS, HDB3, CHDB4) compared with the AMI code when encoding the same binary data sequence. B means a normal alternating bipolar pulse, while V means a bipolar pulse that violates the alternating rule.

according to the rule that the polarity of the violation be always opposite to that of the immediately preceding violation. This is also equivalent to requiring that the number of B pulses between two consecutive violations be always odd. These modal codes have a DSV of only 2, so they present the smallest possible deterioration with respect to the bipolar code. Error monitoring can be performed by checking the alternation in the violation rule. Decoding is not complicated, because each string of $(m + 1)$ received symbols can be checked and set to zero whenever a filling sequence is detected.

An improvement in the decoding rule can be obtained with the *compatible high-density bipolar* (CHDBm) codes, which differ from the HDBm only in the second filling sequence, which is now chosen to be

$$\underbrace{0\,0\,.\,.\,.\,0}_{m-2}\,B0V. \tag{9.211}$$

With this choice, the decoder must always consider only three symbols and set them to zero when the filling sequence is recognized. This feature allows us to use the same decoder for all values of m. The power spectral densities of the codes shown in Fig. 9.63 are presented in Fig. 9.64.

Alphabetic PT codes

We describe here the three alphabetic codes whose line signals for the same binary sequence are shown in Fig. 9.65. The first code is called *pair-selected ternary* (PST) and encodes groups of two binary digits in characters of two ternary symbols. There is no data rate increase, but the code is dc-free and the sequences of zeros are eliminated. The code table is shown in Table 9.20. We can see that only six of the nine possible ternary characters are used. The forbidden characters are 00, $++$, and $--$. The first is discarded in order to eliminate the long sequences of zeros. Instead, the other two are discarded to limit the value of the DSV, which can be verified to be equal to 3.

The encoder has two states. The use of characters with a zero digital sum is unconstrained, whereas the use of characters with an unbalanced digital sum is restricted to an alternate choice between the two states. An easy way to monitor the error performance

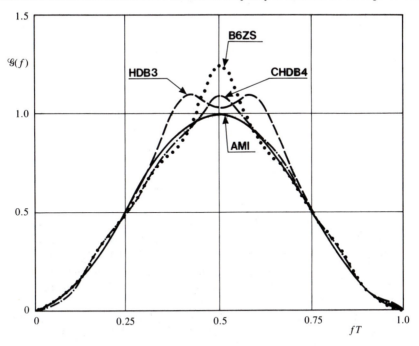

Figure 9.64 Power density spectra of the nonalphabetic PT codes of Fig. 9.63. (Reprinted with permission from Moncalvo and Pellegrini, 1976.)

Sec. 9.8   Line Codes   **519**

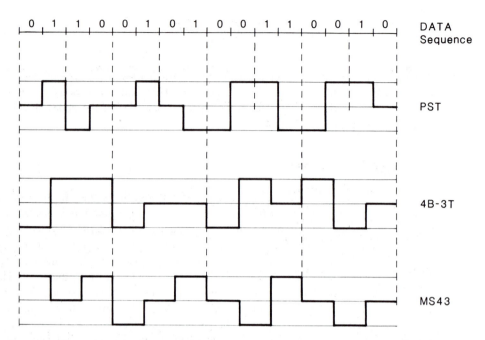

Figure 9.65 Three alphabetic PT codes (PST, 4B-3T, MS43) compared on the basis of the same encoded binary sequence.

TABLE 9.20 Coding table for the PST code

Binary	Ternary	
	State S_1	State S_2
0 0	− +	− +
0 1	0 +	0 −
1 0	+ 0	− 0
1 1	+ −	+ −

of the link consists of obtaining a sequence by replacing with couples of zeros the characters $+ -$ and $- +$ and then monitoring the violation of the bipolarity rule. Also, character synchronization is easy to get by simply checking the absence of the forbidden characters. These are about 30 percent of the total possible characters in the case of misalignment. The PST code was applied in the PCM system, but its design goals are better reached with nonalphabetic codes.

We can conclude that the PST code meets to a certain extent all the design goals required for a PT line code except that of a possible data-rate increase. If one wishes to increase the data rate toward its maximum allowed value while maintaining the elimination of the long sequences of zeros, the best method is to encode 3 binary symbols into characters of 2 ternary levels. The code is called *3B-2T code*. The encoder is memoryless (only one state). The ternary characters are all used, except the character 00. In fact, it is discarded to eliminate the long sequences of zeros. The data rate is 1.5 bits/level, but the penalty paid is that the DSV is infinite. Thus, the 3B-2T code loses control of the spectral shaping in the vicinity of dc.

An interesting step would be to achieve an increase in the data rate while keeping control of the DSV and also to eliminate the long zero sequences. The important goal is obtained with the *4B-3T code* of Table 9.21. Four binary digits are represented with characters of three ternary levels. Thus, the data rate is 1.333 bits/level. The encoder has two states between which it alternates after generating one character. The construction of the code is straightforward.

There are 27 possible ternary characters of length 3 and only one of them (i.e., 000) is discarded to eliminate long zero sequences. The remaining 26 characters are grouped in classes according to the value of their digital sum (DS). The six characters with zero digital sum (DS = 0) are assigned to both encoder states. The remaining 20 characters are grouped into two classes with positive and negative digital sums, respectively. A quick reference to the code table at this point immediately clarifies the encoding rule objectives. The DSV for the 4B-3T code is 7. An increase in the DSV is the penalty associated with an increased data rate. The decoding of a received sequence can be accomplished without memory since there is a one-to-one correspondence from

TABLE 9.21 Coding table for the 4B-3T code

Binary sequence	Ternary characters			
	Positive class state S_1	DS	Negative class state S_2	DS
0 0 0 0	0 + −	0	0 + −	0
0 0 0 1	− 0 +	0	− 0 +	0
0 0 1 0	+ − 0	0	+ − 0	0
0 1 0 0	0 − +	0	0 − +	0
1 0 0 0	+ 0 −	0	+ 0 −	0
0 0 1 1	− + 0	0	− + 0	0
0 1 0 1	+ 0 0	+1	− 0 0	−1
1 0 0 1	0 + 0	+1	0 − 0	−1
1 0 1 0	0 0 +	+1	0 0 −	−1
1 1 0 0	+ + −	+1	− − +	−1
0 1 1 0	− + +	+1	+ − −	−1
1 1 1 0	+ − +	+1	− + −	−1
1 1 0 1	+ + 0	+2	− − 0	−2
1 0 1 1	0 + +	+2	0 − −	−2
0 1 1 1	+ 0 +	+2	− 0 −	−2
1 1 1 1	+ + +	+3	− − −	−3

the ternary characters to the binary ones. Error monitoring and character synchronization are more complex for this code. They can be implemented by checking the compatibility of the received characters with the values of the RDS.

A significant improvement over the 4B-3T code was obtained with the *MS43 code* (Franaszek, 1968), which presents the same properties except for a reduced value of the DSV (from 7 to 5). The code table is shown in Table 9.22. The encoder has three states. The state is defined by the value of the RDS at the end of each ternary character of three symbols. Therefore, encoding is performed under constant monitoring of the RDS and the ternary character is generated according to the Table 9.22. Notice that decoding is still state independent since no ternary character appears twice in different horizontal lines of Table 9.22. Therefore, there is a unique correspondence when going back to binary characters. The character synchronization and error-monitoring capabilities are improved in this code in relation to those of the 4B-3T code. In fact, the value of the RDS when receiving characters with $DS = 2$ or $DS = 3$ must be 4. On the other hand, when receiving characters with $DS = -2$ or $DS = -3$, the value of the RDS must be 1. The importance of the MS43 code lies also in the fact that it was obtained with a theoretic approach based on the code state diagram defined with respect to the design objectives. The power spectral densities of the three codes of Fig. 9.65 are shown together in Fig. 9.66.

TABLE 9.22   Coding table for the MS43 code

Binary sequence	Ternary characters		
	$DS = 1$	$DS = 2$ or 3	$DS = 4$
0 0 0 0	+ + +	− + −	− + −
0 0 0 1	+ + 0	0 0 −	0 0 −
0 0 1 0	+ 0 +	0 − 0	0 − 0
0 1 0 0	0 + +	− 0 0	− 0 0
1 0 0 0	+ − +	+ − +	− − −
0 0 1 1	0 − +	0 − +	0 − +
0 1 0 1	− 0 +	− 0 +	− 0 +
1 0 0 1	0 0 +	0 0 +	− − 0
1 0 1 0	0 + 0	0 + 0	− 0 −
1 1 0 0	+ 0 0	+ 0 0	0 − −
0 1 1 0	− + 0	− + 0	− + 0
1 1 1 0	+ − 0	+ − 0	+ − 0
1 1 0 1	+ 0 −	+ 0 −	+ 0 −
1 0 1 1	0 + −	0 + −	0 + −
0 1 1 1	− + +	− + +	− − +
1 1 1 1	+ + −	+ − −	+ − −

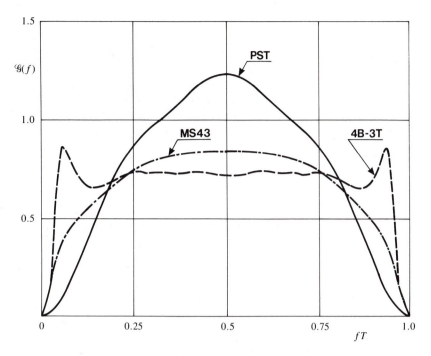

Figure 9.66 Power density spectra of the alphabetic PT codes of Fig. 9.65. (Reprinted with permission from Moncalvo and Pellegrini, 1976.)

BIBLIOGRAPHICAL NOTES

The reader interested in a deeper approach to the theory of both block and convolutional codes can find some excellent books devoted to these subjects. In particular, recommended reading for block codes are the books by Berlekamp (1968), Peterson and Weldon (1972), and MacWilliams and Sloane (1977). A new stimulating approach to the theory of cyclic codes can be found in Blahut (1983). The book by Viterbi and Omura (1979) is a very good reference for both convolutional codes and the random coding bounds for all types of codes on different channels. For the reader interested in the applications and implementation problems of coded transmission, the two books by Clark and Cain (1981) and Lin and Costello (1983) are recommended references. It would be very hard to reference papers on coding theory, besides those motivated by specific details directly in the text. In fact, the literature on the subject is very abundant. A wide bibliography is included in MacWilliams and Sloane (1977). A selected reading in a historical framework can be found in Berlekamp (1974). Concatenated codes were invented by Forney and the most detailed analysis can be found in Forney (1966).

The basic ideas of signal-space coding date back to the early days of information theory. They have been recently revisited by Imai and Hirakawa (1977), Ungerboeck (1982) and Massey (1974, 1984). Tables of good codes for coded 8-PSK, 16-PSK, 32-PSK, and 16-QAM can be found in Ungerboeck (1982), Wilson and others (1984), and Benedetto and others (1987). The performance of Ungerboeck codes over real channels is discussed, for example, in Rhodes and Lebowitz (1981), Taylor and Chan (1981), and Biglieri (1984). Multi-h codes are described in Anderson and Taylor (1978).

Correlative techniques for line codes were proposed by Lender (1963, 1964) and Kretzmer (1966). Null-zone detection of correlative encoded signals can be found in Kobayashi and Tang (1971) and in Smith (1968a). Good survey papers for line codes are Kobayashi (1971a) and Croisier (1970). Our approach was inspired by the latter. An important reference for the theoretical approach to these codes is Justesen (1982). Systematic design methods to find good PT codes are described in Franaszek (1968).

PROBLEMS†

9.1. A (5, 3) block code is defined through the correspondence given in this table:

u_1	u_2	u_3	x_1	x_2	x_3	x_4	x_5
1	1	0	1	0	1	0	1
1	0	1	0	1	0	1	0
0	1	0	0	1	1	0	0

Find the generator matrix of the code.

9.2. Verify that the Hamming distance between two binary sequences is equal to the weight of their mod-2 sum.

9.3. Take the parity check matrix of the extended (8, 4) Hamming code. From (9.13) and (9.20), it is

$$\mathbf{H} = \begin{bmatrix} 1\,1\,1\,0\,1\,0\,0\,0 \\ 0\,1\,1\,1\,0\,1\,0\,0 \\ 1\,1\,0\,1\,0\,0\,1\,0 \\ 1\,1\,1\,1\,1\,1\,1\,1 \end{bmatrix}.$$

(a) Show that the last row, if replaced by the mod-2 sum of all rows, still represents a legitimate parity-check equation.

(b) Verify that the matrix obtained is that of a systematic code equivalent to the original one.

9.4. A (6, 2) linear block code has the following parity-check matrix:

$$\mathbf{H} = \begin{bmatrix} h_1\,1\,0\,0\,0\,1 \\ h_2\,0\,0\,0\,1\,1 \\ h_3\,0\,0\,1\,0\,1 \\ h_4\,0\,1\,1\,1\,0 \end{bmatrix}.$$

(a) Choose the h's in such a way that $d_{min} \geq 3$.

(b) Obtain the generator matrix of the equivalent systematic code and list the four code words.

9.5. Generalize the examples of Problems 9.3 and 9.4 to show that there is always a systematic code equivalent to a given parity-check matrix. *Hint*: In general, the rows of the parity-

† The problems marked with an asterisk should be solved with the aid of a computer.

check matrix form a linearly independent set. Any linear combination of the rows represents a legitimate parity-check equation. The columns can be permuted.

9.6. A systematic (10, 3) linear block code is defined by the following parity-check equations:

$$x_4 + x_1 + x_3 = 0, \qquad x_8 + x_1 + x_2 = 0,$$
$$x_5 + x_3 = 0, \qquad x_9 + x_2 = 0,$$
$$x_6 + x_1 + x_3 = 0, \qquad x_{10} + x_1 + x_2 = 0.$$
$$x_7 + x_2 + x_3 = 0,$$

Find the percentage of error patterns with 1, 2, 3, . . . , 9, 10 errors that can be detected by the code.

9.7. A (5, 2) linear block code is defined by the following table:

$$
\begin{array}{cc}
0 \ 0 & 0 \ 0 \ 0 \ 0 \ 0 \\
0 \ 1 & 0 \ 1 \ 1 \ 0 \ 1 \\
1 \ 0 \quad \Leftrightarrow & 1 \ 0 \ 1 \ 1 \ 1 \\
1 \ 1 & 1 \ 1 \ 0 \ 1 \ 0
\end{array}
$$

(a) Find the generator matrix and the parity-check matrix of the code.
(b) Build the standard array and the decoding table to be used on a BSC.
(c) What is the probability of making errors in decoding a code word?

9.8. Assume that an (n,k) code has minimum distance d.

(a) Prove that every set of $(d - 1)$ or fewer columns of the parity-check matrix $\mathbf{H}$ is linearly independent.
(b) Prove that there exists at least one set of d columns of $\mathbf{H}$ that is linearly dependent.

9.9. Show that the dual of the $(n, 1)$ repetition code is an $(n, n - 1)$ code with $d_{min} = 2$ and with code words always having even weight.

9.10. Given an (n, k) code, it can be shortened to obtain an $(n - 1, k - 1)$ code by simply taking only the code words that have a 0 in the first position, and deleting this 0. Show that the maximal-length (simplex) code $(2^m - 1, m)$ is obtained by shortening the first-order $(2^m, m + 1)$ Reed–Muller code.

9.11. Assume that an (n, k) block code with minimum distance d is used on the binary erasure channel of Example 3.14. Show that it is always possible to decode correctly the received sequence provided that no more than $(d - 1)$ erasures have occurred.

9.12. Consider the following generator matrix of an (8, 5) linear block code:

$$
\mathbf{G} = \begin{bmatrix}
10000 & 111 \\
01000 & 100 \\
00100 & 010 \\
00010 & 001 \\
00001 & 111
\end{bmatrix}.
$$

(a) Show that the code is cyclic, and find both the generator polynomial $g(D)$ and the parity-check polynomial $h(D)$.
(b) Obtain the parity-check matrix $\mathbf{H}$.

9.13. Consider the generator polynomial

$$g(D) = D + 1.$$

(a) Show that it generates a cyclic code of any length.

(b) Obtain the parity-check polynomial $h(D)$, the parity-check matrix **H,** and the generator matrix **G.**

(c) What kind of code is obtained?

9.14. Given the (7, 4) Hamming code generated by the polynomial $g(D) = D^3 + D + 1$, obtain the (7, 3) code generated by

$$g(D) = (D + 1)(D^3 + D + 1).$$

(a) How is it related to the original (7, 4) code?

(b) What is its minimum distance?

(c) Show that the new code can correct all single errors and simultaneously detect all double errors.

(d) Describe an algorithm for correction and detection as in part (c).

9.15. A cyclic code is generated by

$$g(D) = D^8 + D^7 + D^6 + D^4 + 1.$$

(a) Find the length n of the code.

(b) Sketch the encoding circuits with a k or $(n - k)$ shift register.

9.16. Discuss the synthesis of a code capable of correcting single errors and adjacent double errors. Develop an example and compare the numbers n and k with those required for the correction of all double and single errors. *Hint*: Count the required syndromes and construct a suitable parity-check matrix.

9.17. It is desired to build a single-error-correcting (8, 4) linear block code.

(a) Define the code by shortening a cyclic code.

(b) List the code words and find the minimum distance.

(c) Sketch the encoding circuit and verify its behavior with an example.

9.18. Show that the binary cyclic code of length n generated by $g(D)$ has minimum distance at least 3, provided that n is the smallest integer for which $g(D)$ divides $(D^n + 1)$.

9.19. Consider a cyclic code generated by the polynomial $g(D)$ that does not contain $(D + 1)$ as a factor. Suppose that n is odd and show that the vector of all ones is a code word.

9.20. Show that the (7, 4) code generated by $g(D) = D^3 + D + 1$ is the dual of the (7, 3) code generated by $g(D) = D^4 + D^3 + D^2 + 1$.

9.21. Repeat the computations of Example 9.24 for the (15, 11) Hamming code.

9.22. Consider a transmission system that performs error detection over a BSC with transition probability p. Using the weight enumerator $A(z)$, find an exact expression for the probability of undetected errors for the following codes:

(a) Hamming codes.

(b) Extended Hamming codes.

(c) Maximal-length codes.

(d) $(n, 1)$ repetition codes.

(e) $(n, n - 1)$ codes with an overall parity check.

9.23. **(a)** Show that for a linear block code the set of Hamming distances from a given code word to the other $(M - 1)$ code words is the same for all code words. *Hint*: Use Property 3 of Section 9.2.

(b) Prove that for any linear code used on a binary-input symmetric channel with ML decoding we have the *uniform error property*

$$P_w(e) = P_w(e|\mathbf{x}_i), \qquad i = 1, 2, \dots, M.$$

Hint: Write

$$P_w(e|\mathbf{x}_i) = \sum_{j \neq S_i} P(\mathbf{y}_j|\mathbf{x}_i),$$

where S_i is the set of subscripts j of received sequences $\mathbf{y}_j$ that are decoded into the code word $\mathbf{x}_i$, and

$$P(\mathbf{y}|\mathbf{x}_i) = \prod_{k=1}^{n} P(y_k|x_{ik}).$$

Use the symmetry of the channel; that is,

$$P(y_k = 0|x_{ik} = 1) = P(y_k = 1|x_{ik} = 0).$$

9.24. Using the memoryless property of the BSC, that is,

$$P(\mathbf{y}_j|\mathbf{x}_i) = \prod_{k=1}^{n} P(y_{jk}|x_{ik}),$$

derive (9.76) from (9.75).

9.25. Consider the convolutional (3, 1) code with constraint length 3, defined by the generators of (9.99):

$$\mathbf{g}_1 = (111), \qquad \mathbf{g}_2 = (111), \qquad \mathbf{g}_3 = (101).$$

(a) Draw the state diagram of the code.
(b) Obtain the transfer function $T(D, J)$.
(c) Find the free distance d_f of the code.
(d) Evaluate the bit error probability $P_b(e)$ over a BSC with $p = 10^{-3}$ and using the Viterbi decoding algorithm.

9.26. Consider the (2, 1) convolutional code with constraint length 5 and generators

$$\mathbf{g}_1 = (11001), \qquad \mathbf{g}_2 = (11111).$$

(a) Draw the code trellis.
(b) Find the free distance d_f of the code.

9.27. Perform a complete study of the (2, 1) convolutional code of constraint length 2 and generators

$$\mathbf{g}_1 = (10), \qquad \mathbf{g}_2 = (11).$$

9.28. Consider the (2, 1) convolutional code of constraint length 3 and generators

$$\mathbf{g}_1 = (110), \qquad \mathbf{g}_2 = (101).$$

(a) Draw the state diagram of the code.
(b) Verify that the self-loop at state $S_4 = (11)$ does not increase the distance from the all-zero sequence.
(c) The code is catastrophic. Verify with an example.

9.29. A bound on $P_b(e)$ tighter than (9.120) can be obtained for convolutional codes if $P(e_{1d})$ is computed exactly instead of being upper bounded as in (9.113). The all-zero path is assumed to be transmitted, and suppose that the path being compared has distance d. The incorrect path will be selected if there are more than $(d + 1)/2$ errors, with d odd.
(a) Show that in this case

$$P(e_{1d}) = \sum_{i=(d+1)/2}^{d} \binom{d}{i} p^i (1 - p)^{d-i},$$

where p is the transition probability of the BSC.

(b) When d is even, show that

$$P(e_{1d}) = \frac{1}{2} \binom{d}{d/2} p^{d/2}(1-p)^{d/2} + \sum_{i=\frac{d}{2}+1}^{d} \binom{d}{i} p^i(1-p)^{d-i}.$$

(c) Use the preceding results in (9.116) and compare the new bound numerically on an example.

9.30. (Viterbi and Omura, 1979) The bound (9.125) for convolutional codes can be refined and made tighter through the following steps.

(a) Prove, first, the following inequality:

$$erfc\,(\sqrt{x+y}) \le erfc\,(\sqrt{x})e^{-y}, \qquad x \ge 0, \qquad y \ge 0.$$

(b) Since $d \ge d_f$, we may bound (9.122) by

$$P(e_{1d}) \le \frac{1}{2} erfc\left(\sqrt{\frac{d_f R_c \mathscr{E}_b}{N_0}}\right) \cdot e^{-(d-d_f)R_c \mathscr{E}_b/N_0}.$$

(c) Derive the new bound in the form

$$P_b(e) \le \frac{1}{2k_0} erfc\left(\sqrt{\frac{d_f R_c \mathscr{E}_b}{N_0}}\right) e^{d_f R_c \mathscr{E}_b/N_0} \frac{\partial T(D, J)}{\partial J}\bigg|_{J=1,\,D=e^{-(R_c \mathscr{E}_b/N_0)}}.$$

*(d) Compare numerically this bound with (9.125) on an example.

9.31. (Wozencraft and Jacobs, 1965) In this problem we propose to use the method of Section 9.5.2 to derive the cutoff rate R_0 for signaling schemes other than the binary antipodal one that gave the result (9.143).

(a) Assume that M equally spaced antipodal levels $a_1, a_2, \ldots, a_M$ are used in transmission and define

$$d_{ij}^2 \triangleq (a_i - a_j)^2,$$

which is the squared distance between a pair of these levels. Assuming that all levels are equally likely, show that

$$R_0 = -\log_2 \frac{1}{M^2} \sum_{i=1}^{M} \sum_{j=1}^{M} e^{-d_{ij}^2/4N_0}.$$

Verify that this expression reduces to (9.143) for $M = 2$.

(b) A set of M phase-shift waveforms

$$\psi(t; k) \triangleq \sqrt{\frac{2\mathscr{E}}{T}} \sin\left\{2\pi\left(f_0 t + \frac{k}{M}\right)\right\}, \qquad 0 \le t < T, \qquad k = 1, 2, 3, \ldots, M,$$

is used to transmit coded signal in the form

$$s(t) = \sum_{j=0}^{J-1} \psi(t - jT; \xi_j),$$

where ξ_j is a RV taking values in the set $\{1, 2, \ldots, M\}$. These signals are assumed to be equally likely. Show that

$$R_0 = -\frac{1}{2}\log_2 \frac{1}{M}\sum_{k=1}^{M} \exp\left\{-\frac{\mathcal{E}}{N_0}\sin^2\frac{k\pi}{M}\right\}, \qquad \text{for } M \geq 3.$$

(c) Compare numerically these cutoff rates with (9.143).

9.32. Design two Ungerboeck codes of the type of Figs. 9.47 and 9.49, respectively, by using the signals of Fig. P9.1. There are eight signals in the signal set and the code rate is $\frac{2}{3}$. Evaluate the performance in terms of d_{min}^2 and compare with a 4-PSK reference system (uncoded) using the same average energy.

9.33. Evaluate the spectral shaping introduced by the correlative level codes of Table 9.19.

9.34. Show that the results of Example 9.39 are valid also for the modified duobinary code of Example 9.38.

9.35. (Gunn and Lombardi, 1969) Consider a partial-response class IV (see Table 9.19) transmission scheme with an input signal of eight levels ($M = 8$). Let the transmitted levels c_m take values in the set $\{-7, -6, \ldots, 0, 1, \ldots, 6, 7\}$. If b_m are precoded symbols belonging to the set $\{0, 1, 2, \ldots, 7\}$, then show that the following is verified:

$$S_n(J) \triangleq \sum_{j=0}^{J} c_{n+2j} = b_{n+2J} - b_{n-2},$$

for any n and $J \geq 0$. Therefore, the running sum $S_n(J)$ satisfies the relation

$$-b_{n-2} \leq S_n(J) \leq 7 - b_{n-2}.$$

Describe a possible error-detection scheme based on this property.

9.36. Prove the properties of the sequence (b_m) of (9.160) as described in the text just after the formula.

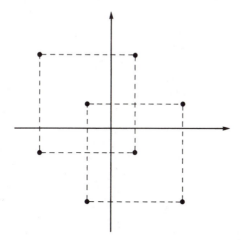

Figure P9.1

CHAPTER 10

Digital Transmission over Nonlinear Channels

In Chapter 6 we considered some of the major impairments affecting digital transmission besides additive Gaussian noise, such as intersymbol interference (ISI), interchannel interference (ICI), and selective fading. They are due to the nonideal characteristics of the linear devices present in the system and to the transmission medium. As we saw, these disturbances can be modeled as terms that sum to the useful signal. This is a direct consequence of the linear assumption made for the components of the system. There are many cases, however, where this assumption is not true, as nonlinear devices significantly contribute to system degradation. One example is encountered in high-speed digital transmission over telephone channels, where nonlinear signal distortion arises principally from inaccuracies in signal companding (compressing–expanding) in telephone transmission. The effect of these nonlinearities on data signals has led to the conjecture (Lucky, 1975) that the error probability performance on data-transmission systems operating at rates greater than 4800 bit/s is almost entirely determined by nonlinear distortion. Another example is the digital satellite link, in which both the earth station and the satellite repeater are equipped with amplifiers operated in a nonlinear region of the input–output characteristics for a better exploitation of the power of the device. The earth station amplifier nonlinearity is often mild, either because it operates some decibels below the saturation point (it is ''backed off''), in a nearly linear region, or because a predistortion linearizer is inserted in the transmitter chain. To the contrary, the satellite amplifier (a traveling wave tube (TWT) or a solid-state device) is driven near to the saturation point and exhibits highly nonlinear characteristics, which must be included in the analysis of the system performance.

In most engineering fields the gap between linear and nonlinear problems is wide. Whereas refined mathematical tools and comprehensive theories are available in the linear case, only very special categories of problems can be analyzed in nonlinear situations. Digital transmission theory is no exception to this rule. No well-established theories exist for the analysis and/or design of a nonlinear digital transmission system. Because of this, the structure of this chapter will be in a sense peculiar with respect to the rest of the book. Most of the topics analyzed in detail before in a linear context will be discussed concisely here, aiming at extending the results, wherever possible, to the nonlinear situations.

Much effort has been devoted to smoothing the differences in the model and in the analysis found in the various approaches to the problems considered. However, the status of the content of this chapter is still that of a work in progress, reflecting in the choice of topics and the analytical tools the preferences of the authors and their research experience. The particular nature of the chapter renders more important the Bibliographical Notes at the end, where reference is made to different approaches to the same problems.

Section 10.1 is devoted to the modeling of a finite-memory nonlinear channel. Then this model is used as a starting point to perform spectral analysis of nonlinear signals (Section 10.2) to derive the optimum symbol-by-symbol linear receiver (Section 10.4) and the optimum maximum likelihood (ML) sequence receiver (Section 10.5). A more specialized model of the nonlinear channel, based on the discrete Volterra series, is introduced in Section 10.3. This model permits the computation of an explicit expression for the received signal in which all the significant contributions (i.e., useful signal, linear and nonlinear interferences, and noise) appear separately. In the same section, this result is used in evaluating the performance of a nonlinear system employing 4-CPSK modulation. Finally, in Section 10.6, the Volterra model is employed to derive the structure of a nonlinear equalizer.

10.1 A MODEL FOR THE NONLINEAR CHANNEL

The model applied to the system analyzed is shown in Fig. 10.1. The source emits a stationary sequence (a_n) of discrete independent RVs of known statistics, one every T seconds; the block labeled "channel" represents the noiseless part of the real channel and includes every device between the source and the receiver (modulator, filters, physical channel, nonlinear devices, etc.); it transforms the discrete-time sequence at its input into a continuous waveform $y(t)$. After the addition of Gaussian noise, the signal enters the receiver, in which it is processed to estimate the transmitted sequence (a_n).

We assume that the channel has a *finite* memory; that is, at any time instant t the channel output $y(t)$ depends only on a finite number, say L, of past source symbols besides the one emitted at time t. Using the definition of shift-register state sequence given in Section 2.2.1, we can define the state σ_n of the channel during the nth interval $((n - 1)T, nT)$ as the n-tuple of the L consecutive symbols a_i:

$$\sigma_n \triangleq (a_{n-1}, a_{n-2}, \ldots, a_{n-L}). \tag{10.1}$$

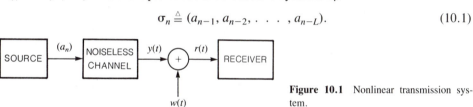

Figure 10.1 Nonlinear transmission system.

The state σ_n then represents the memory of the channel. The emission of a_n from the source forces a transition of the channel state from σ_n to σ_{n+1}. A sequence of states forms a special case of the Markov chain studied in Section 2.2.1 (called the shift-register state sequence).

With the finite memory assumption, we can model the noiseless channel of Fig. 10.1 as in Fig. 10.2, where $h(t - nT; a_n, \sigma_n)$ is the waveform generated in the nth time interval of duration T. The channel output $y(t)$ will then consist of a sum of nonoverlapping waveforms, each defined in an interval of duration T:

$$y(t) \triangleq \sum_n h(t - nT; a_n, \sigma_n). \tag{10.2}$$

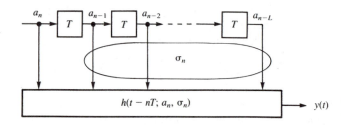

Figure 10.2 Model of the noiseless part of a nonlinear channel with finite memory.

In (10.2), the waveforms $h(t; a_n, \sigma_n)$ are zero outside the interval $(0, T)$ and can assume no more than M^{L+1} different shapes, where M is the cardinality of the set of values assumed by a_n. The waveforms $h(t; \cdot, \cdot)$ will be called *chips*.

Example 10.1

Consider a channel formed by cascading a linear filter with a memoryless nonlinear device like that in Fig. 10.3, used by a system employing a 2-CPSK modulation. The whole channel can be considered as a nonlinear system with memory, in which the memory is due to the linear filter and the nonlinearity to the memoryless nonlinear device. The filter, represented by its impulse response $s(t)$, is a sixth-order Butterworth filter with normalized equivalent noise bandwidth $B_{eq}T$ equal to 1.2. The nonlinear device is a TWT whose AM/AM and AM/PM characteristics are shown in Fig. 10.4 (see Section 10.3 for a detailed explanation of the meaning of these characteristics). From inspection of the impulse response of the filter, a value of $L = 3$ is sufficient to account for the memory of the system. We have then $2^4 = 16$ different chips $h(t; a, \sigma)$ for this system; they are shown in Fig. 10.5.

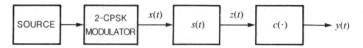

Figure 10.3 Noiseless part of a bandpass nonlinear transmission system using 2-PSK.

The signal $y(t)$ at the output of the channel is then obtained as a sequence of chips chosen according to the values of the sequence (a_n) or, equivalently, (σ_n). An example is shown in Fig. 10.6. □

A final observation is pertinent. This system model can also accommodate a channel encoder (block or trellis encoder) with finite memory.

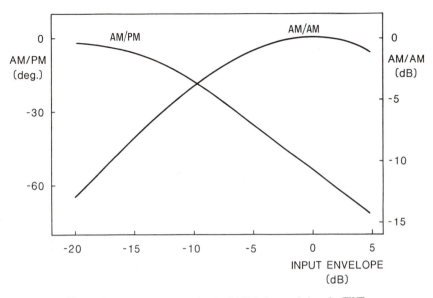

Figure 10.4 Typical AM/AM and AM/PM characteristics of a TWT.

10.2 SPECTRAL ANALYSIS OF NONLINEAR SIGNALS

A general method to evaluate the power spectrum of a random digital signal was presented in Section 2.3. The signal $y(t)$ of (10.2) has the form (2.109), which is the starting point for the derivation of the power spectrum. We can use then the formulas (2.150), (2.151), and (2.152) to compute the spectrum, taking into account the simplification that the state sequence (σ_n) is a shift-register sequence (Section 2.2.1). This reduces the infinite summation in (2.152) to a finite summation ranging from 1 to L, L being the channel memory. The following example shows some results obtained through the application of this method.

Example 10.2

The aim of this example, still referring to Fig. 10.3, is to pictorially describe the effect of the nonlinearity on the power spectrum of a digital signal. To do this, we have computed the power spectra of the signals $x(t)$, the output of an ideal 2-CPSK modulator, $z(t)$, its filtered version, and $y(t)$, the filtered signal passed through the nonlinearity of Fig. 10.4 for different values of the input backoff. Looking at the results shown in Fig. 10.7, the phenomenon of the spectrum spreading due to the TWT appears evident. As a matter of fact, the sidelobes of the CPSK spectrum, attenuated by the filter, are restored by the TWT; also evident is the role played by the input backoff and the consequent trade-off between power efficiency and sidelobes enhancement. This gives rise to ICI in an FDM system. □

10.3 VOLTERRA MODEL FOR BANDPASS NONLINEAR CHANNELS

As previously mentioned, satellite digital transmission represents one of the most important cases of digital communication systems employing a nonlinear channel. It consists of two earth stations (TX and RX in Fig. 10.8), usually far from each other, connected

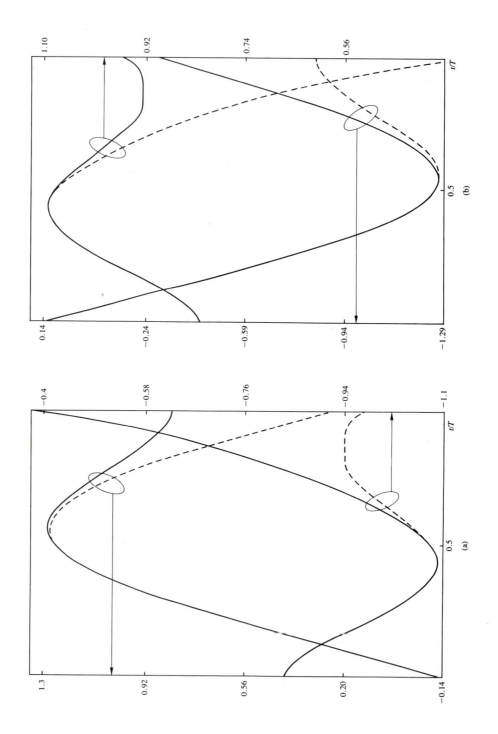

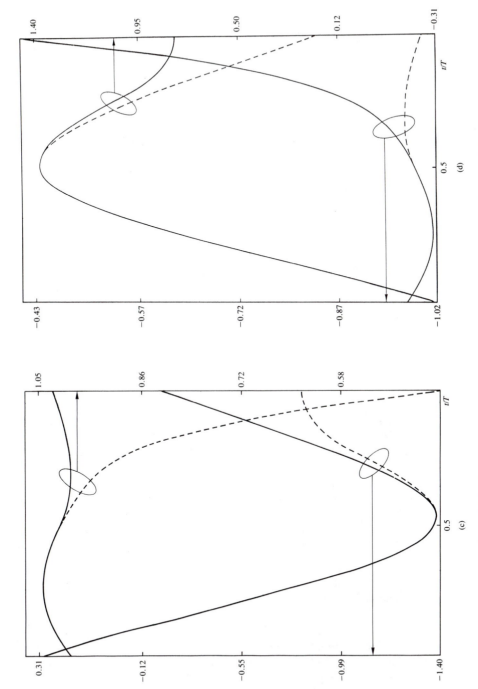

Figure 10.5 T-second duration waveforms ("chips") of the nonlinear system of Example 10.1

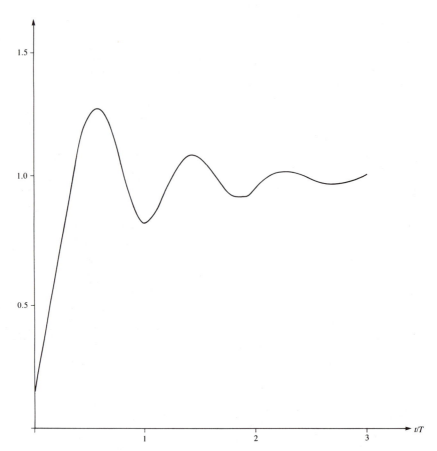

Figure 10.6 Output signal from a noiseless nonlinear channel employed by a 2-PSK modulator.

by a repeater traveling in the sky (satellite) through two radio links (uplink and downlink in the figure).

A block diagram of the system of Fig. 10.8 is shown in Fig. 10.9. The block labeled HPA (high-power amplifier) represents the earth station power amplifier; its input–output power characteristics are nonlinear, of a saturating type like that presented in Fig. 10.4. Although the highest power efficiency is obtained by letting the HPA operate at (or near) the saturation point, it is common practice to operate the HPA a few decibels below saturation in a nearly linear region. This *backing-off* the operation point facilitates the attenuation of the effects caused by nonlinearity (e.g., the spreading of the spectrum of the input signal, which, as seen in Example 10.2, gives rise to ICI).

The TX filter limits the bandwidth of each channel in an FDMA system, whereas the IMUX (input multiplexing) filter limits the amount of uplink noise entering the satellite transponder. The block labeled TWT represents the satellite's on-board amplifier. Owing to the satellite's limited power resources, this amplifier is usually operated at saturation to obtain the maximum efficiency. Typical input–output characteristics of TWTs are shown in Fig. 10.4.

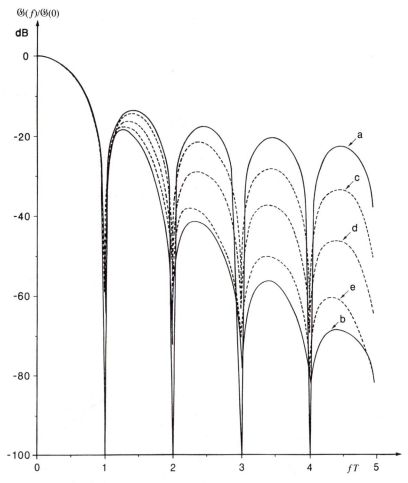

Figure 10.7 Power spectral densities of different signals in the nonlinear transmission system of Example 10.2. Curve a refers to the modulator, curve b to the filter, curve c to the TWT at saturation, and curves d and e to the TWT backed off by 6 and 12 dB, respectively.

Due to bandwidth limitations, the modulated signal at the TWT input does not have a constant envelope. In particular, when transitions between opposite points in the signal space occur (as in CPSK), the envelope may pass through zero. This phenomenon is represented in Fig. 10.10, where the set of the envelope values is drawn in the signal space at the input of the TWT in a typical system situation. These input envelope fluctuations are translated at the TWT output in phase shifts that deteriorate the system performance. The functions of the OMUX (output demultiplexing) filter and RX filter are, respectively, similar to those of TX and IMUX. The receiver is assumed to be a conventional symbol-by-symbol receiver. In this section the receiver optimization is not considered. It will be dealt with later. Our objective is to determine the performance of the system in terms of error probability.

Two main tracks have been followed to assess the error performance of a nonlinear

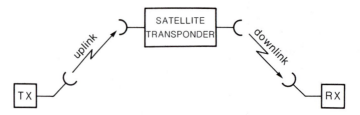

Figure 10.8   Satellite link.

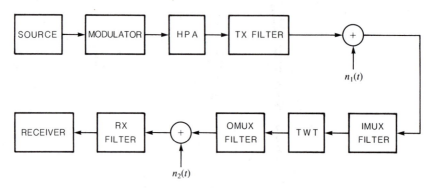

Figure 10.9   Block expansion of a satellite link.

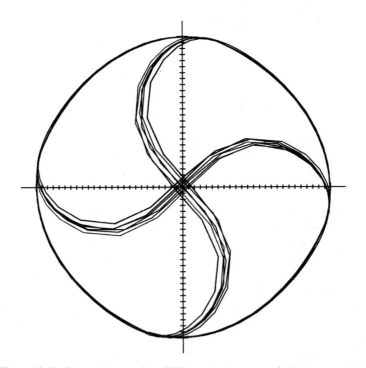

Figure 10.10   Phasor diagram of a 4-PSK signal at the output of a bandpass nonlinear channel obtained by simulation.

satellite link. The first uses simulation tools to analyze a complete model of the system. The second is based on some simplifications of the model such that an analytical approach is feasible. Based on the channel model of Section 10.1, an intermediate approach could be used when the uplink noise is omitted from the system model. This often occurs owing to the higher signal-to-noise ratio in the uplink. The approach consists in evaluating the error probability conditioned on the channel state σ_n, and then averaging over all the possible states by generating the whole set of chips $h(t; \cdot, \cdot)$ in (10.2). This approach stems directly from the direct-enumeration method for computing the error probability in the presence of the ISI described in Section 6.2.1.

In this section we will present a method for the performance analysis based on Volterra series. It renders explicit the dependence of the received signal on the sequence of source symbols (a_n) and allows the application to the nonlinear case of the methods for computing the error probability with ISI as described in Chapter 6.

It is felt that the model based on Volterra series represents a balance between two often conflicting requirements (i.e., faithful representation of the physical system and analytical tractability of the model).

Consider the block diagram of Fig. 10.11. It represents a slightly more detailed version of Fig. 10.9, as the block "receiver" is expanded. This block diagram will be the basis of our analysis. Note that it does not include some types of disturbances that can affect the two radio links of Fig. 10.8, (e.g., fading and interferences from other users of the transmission medium, cochannel and interchannel interference, etc.). Furthermore, carrier and timing recovery is supposed to be ideal. The nonlinear device is a

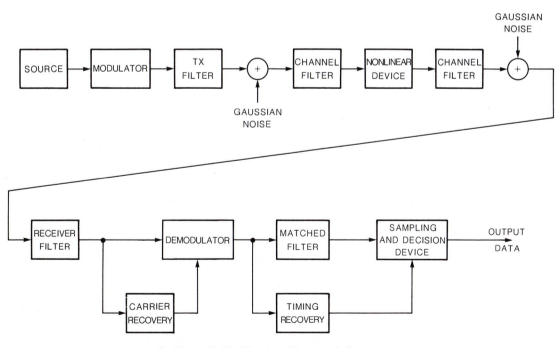

Figure 10.11 Digital satellite transmission system.

Sec. 10.3   Volterra Model for Bandpass Nonlinear Channels   **539**

bandpass memoryless nonlinear system whose input–output characteristics are described by the relationship (see 2.219)

$$y(t) = F[A_x(t)]e^{j\{\varphi_x(t)+\phi[A_x(t)]\}} \tag{10.3}$$

where $y(t)$ is the complex envelope of the output signal and

$$x(t) = A_x(t)e^{j\varphi_x(t)} \tag{10.4}$$

is the complex envelope of the input signal. The nonlinearity is then characterized by two real-valued functions, $F(\cdot)$ and $\phi(\cdot)$, which describe the AM/AM and AM/PM conversion effects. A typical example related to satellite links is shown in Fig. 10.12, where $F(\cdot)$ and $\phi(\cdot)$ represent measured characteristics of a commercial TWT.

Since the whole system represented in Fig. 10.11 is bandpass, we can use the low-pass representation of systems and signals (see Section 6.1), which leads to the block diagram of Fig. 10.13, where

(a_n) is a sequence of discrete iid generally complex RVs.

$x(t)$ is the modulated signal:

$$x(t) = \sum_n a_n\delta(t - nT). \tag{10.5}$$

$s(t)$ is the overall impulse response of the filters preceding the nonlinearity.

$c(\cdot)$ is a complex function that represents the input–output relationships of the nonlinearity:

$$c(\cdot) = F(\cdot)e^{j\phi(\cdot)}. \tag{10.6}$$

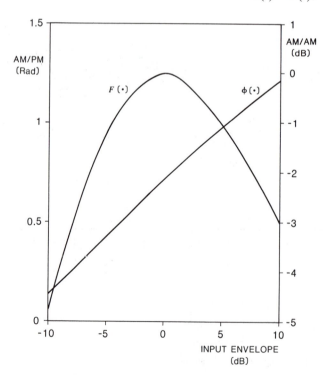

Figure 10.12 Input–output characteristics of the TWT used in Example 10.5.

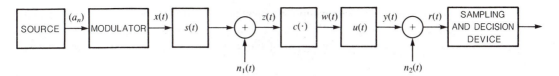

Figure 10.13 Equivalent block diagram of the digital satellite link of Fig. 10.11.

$u(t)$ is the impulse response of a filter that represents the cascade of all linear devices following the nonlinearity.

$n_1(t)$ and $n_2(t)$ are generally complex baseband Gaussian processes, with zero mean and variances σ_1^2 and σ_2^2 (in a satellite link, they represent the uplink and downlink noises, respectively).

The analysis of the system in Fig. 10.13 will follow two steps. First, we will assume that $n_1(t) = 0$ (i.e., no uplink noise is present in the system). Although motivated by the desire for simplicity, this assumption is reasonable for satellite systems in which the greater power of the earth transmitter keeps the signal-to-noise ratio in the uplink higher than in the downlink. Successively, we will include uplink noise in the analysis. As seen in Section 2.4, Eq. (2.235), the relationship between signals $y(t)$ and $x(t)$ in Fig. 10.13 can be expressed as

$$y(t) = \sum_{m=0}^{\infty} \int_{-\infty}^{\infty} \cdots \int_{-\infty}^{\infty} k_{2m+1}(\tau_1, \ldots, \tau_{2m+1})$$

$$\cdot \prod_{i=1}^{m+1} x(t - \tau_i) \prod_{l=m+2}^{2m+1} x^*(t - \tau_l)\, d\tau_1 \ldots d\tau_{2m+1}, \tag{10.7}$$

where $k_{2m+1}(\cdot)$ is the low-pass equivalent Volterra kernel defined in Section 2.4.

Substituting $x(t)$ as in (10.5) into (10.7) yields

$$y(t) = \sum_{m=0}^{\infty} \sum_{n_1=-\infty}^{\infty} \cdots \sum_{n_{2m+1}=-\infty}^{\infty} a_{n_1} \cdots a_{n_{m+1}} a_{n_{m+2}}^* \cdots a_{n_{2m+1}}^*$$

$$\cdot k_{2m+1}(t - n_1 T, \ldots, t - n_{2m+1} T). \tag{10.8}$$

In general, the central problem in using a Volterra approach to the analysis of nonlinear systems with memory consists of estimating the Volterra kernels, which represent a nonparametric characterization of the system. Nevertheless, in our case, the particular structure of the nonlinear system that confines the memory into linear components makes it possible to find a closed-form expression for the kernels.

Consider first a Taylor series expansion, with complex coefficients, of the function $c(\cdot)$ defined in (10.6):

$$c(A) = \sum_{m=0}^{\infty} \gamma_{2m+1} A^{2m+1}. \tag{10.9}$$

The presence in (10.9) of only odd powers of the argument A is a consequence of the bandpass nature of the nonlinearity (see Section 2.4). As shown in Example

Sec. 10.3   Volterra Model for Bandpass Nonlinear Channels   **541**

2.19, the input–output relationship for a bandpass memoryless nonlinear device involves only the complex envelope of the output as a function of the complex envelope of the input. Thus, using (10.9), the output of the nonlinearity (Fig. 10.13) can be expressed as

$$w(t) = \sum_{m=0}^{\infty} \gamma_{2m+1} z^{m+1}(t) z^{*m}(t). \tag{10.10}$$

Since

$$z(t) = \int_{-\infty}^{\infty} s(\tau) x(t - \tau) \, d\tau \tag{10.11}$$

and

$$y(t) = \int_{-\infty}^{\infty} u(\tau) w(t - \tau) \, d\tau, \tag{10.12}$$

we get, through some easy algebra,

$$y(t) = \sum_{m=0}^{\infty} \gamma_{2m+1} \int_{-\infty}^{\infty} \cdots \int_{-\infty}^{\infty} u(\tau) \prod_{r=1}^{m+1} s(\tau_r - \tau) \prod_{s=m+2}^{2m+1} s^*(\tau_s - \tau)$$
$$\cdot \prod_{i=1}^{m+1} x(t - \tau_i) \prod_{l=m+2}^{2m+1} x^*(t - \tau_l) \, d\tau \, d\tau_1 \ldots d\tau_{2m+1}. \tag{10.13}$$

Comparing (10.13) with (10.7), one finds the desired expression for the low-pass equivalent kernels:

$$k_{2m+1}(\tau_1, \ldots, \tau_{2m+1}) = \gamma_{2m+1} \int_{-\infty}^{\infty} u(\tau) \prod_{r=1}^{m+1} s(\tau_r - \tau) \prod_{s=m+2}^{2m+1} s^*(\tau_s - \tau) \, d\tau. \tag{10.14}$$

We now have all the ingredients to find an explicit expression for the sampled received signal entering the decision device in Fig. 10.13. The received signal $r(t)$ is given by

$$r(t) = y(t) + n_2(t), \tag{10.15}$$

where $y(t)$ has been defined in (10.8). Sampling now at $t = t_0$ and defining

$$K_{2m+1}(n_1, \ldots, n_{2m+1}) \triangleq k_{2m+1}(t_0 - n_1 T, \ldots, t_0 - n_{2m+1} T), \tag{10.16}$$

$$N_2 \triangleq n_2(t_0), \tag{10.17}$$

$$R \triangleq r(t_0), \tag{10.18}$$

leads to

$$R = \sum_{m=0}^{\infty} \sum_{n_1=-\infty}^{\infty} \cdots \sum_{n_{2m+1}=-\infty}^{\infty} a_{n_1} \cdots a_{n_{m+1}} a^*_{n_{m+2}} \cdots a^*_{n_{2m+1}}$$
$$\cdot K_{2m+1}(n_1, \ldots, n_{2m+1}) + N_2. \tag{10.19}$$

This result is the starting point for the analysis of a synchronous digital communication system using nonlinear devices. The computation of the discrete kernels $K_{2m+1}(\cdot)$

only involves a convolution integral [see (10.14)], which can be solved using standard numerical techniques like the bilinear z-transform or FFT.

To apply (10.19), we also have to deal with two kinds of infinite summations. The first one, index m, comes from the power series expansion (10.6) of the nonlinearity characteristics. It is usually truncated to some value m_M large enough to accurately represent the function $c(\cdot)$. As an example, in the case of the TWT characteristics shown in Fig. 10.12, $m_M = 3$ allows a reasonable approximation of the curves in the range from zero to a few decibels beyond the saturation point. The second kind of infinite series, indexes n_i, depends on the memory of the linear components of the communication system, that is, $s(t)$ and $u(t)$. As is usual for linear systems, we shall suppose that both $s(t)$ and $u(t)$ have a finite duration. Thus, we can say that each n_i takes value in a finite set of integers. In the following, summations limits will often be omitted. It is intended that the set of values taken by the indexes have finite cardinality.

As seen in Chapter 6, the decision device operates on samples of the in-phase and quadrature demodulated signals, which correspond to the real and imaginary parts of R (i.e., R_P and R_Q). From (10.19), we can extract all the terms containing only the transmitted symbol a_0, which contribute to form what we call the "useful sample" R_0:

$$R_0 \triangleq R_{0P} + jR_{0Q} = a_0 \sum_{m=0}^{m_M} |a_0|^{2m} K_{2m+1}(0, \ldots, 0).$$

This allows one to rewrite (10.19) in the form

$$R = a_0 \sum_{m=0}^{m_M} |a_0|^{2m} K_{2m+1}(0, \ldots, 0) + \sum_{n_1}' a_{n_1} K_1(n_1)$$

$$+ \sum_{m=0}^{m_M} \sum_{n_1} \cdots \sum_{n_{2m+1}} a_{n_1} \cdots a_{n_{m+1}} a^*_{n_{m+2}} \cdots a^*_{n_{2m+1}} \qquad (10.20)$$
$$(n_1, \ldots, n_{2m+1} \neq 0, \ldots, 0)$$

$$\cdot K_{2m+1}(n_1, \ldots, n_{2m+1}) + N_2.$$

Let us define

$$R_{0P} \triangleq \mathcal{R} \left\{ a_0 \sum_{m=0}^{m_M} |a_0|^{2m} K_{2m+1}(0, \ldots, 0) \right\},$$

$$R_{0Q} \triangleq \mathcal{I} \left\{ a_0 \sum_{m=0}^{m_M} |a_0|^{2m} K_{2m+1}(0, \ldots, 0) \right\}, \qquad (10.21)$$

$$R_P \triangleq \mathcal{R} \left\{ \sum_{n_1}' a_{n_1} K_1(n_1) + \sum_{m=0}^{m_M} \sum_{n_1} \cdots \sum_{n_{2m+1}} a_{n_1} \cdots a_{n_{m+1}} a^*_{n_{m+2}} \cdots a^*_{n_{2m+1}} \right.$$
$$(n_1, \ldots, n_{2m+1} \neq 0, \ldots, 0)$$

$$\qquad (10.22)$$

$$\left. \cdot K_{2m+1}(n_1, \ldots, n_{2m+1}) \right\},$$

$$R_Q \triangleq \mathcal{I} \left\{ \sum_{n_1}' a_{n_1} K_1(n_1) + \sum_{m=0}^{m_M} \sum_{n_1} \cdots \sum_{n_{2m+1}} a_{n_1} \cdots a_{n_{m+1}} a^*_{n_{m+2}} \cdots a^*_{n_{2m+1}} \right.$$
$$(n_1, \ldots, n_{2m+1} \neq 0, \ldots, 0)$$

$$\cdot K_{2m+1}(n_1, \ldots, n_{2m+1})\Big\},$$

$$N_{2P} \triangleq \mathfrak{R}\{N_2\},$$
$$N_{2Q} \triangleq \mathfrak{I}\{N_2\},$$
(10.23)

so that (10.20) becomes

$$R = (R_{0P} + R_P + N_{2P}) + j(R_{0Q} + R_Q + N_{2Q}).$$
(10.24)

The structure of (10.20) is quite similar to that of (6.16) and can be viewed as an extension of it. In fact, we recognize in (10.20) the useful signal (first term in RHS), the linear ISI (second term), the nonlinear contribution to ISI (third term), and, finally, the additive Gaussian noise.

As in Chapter 6 for the linear case, we can now apply the methods described in Appendix E to compute the error probability of the system. Those methods are based on the knowledge of a certain number of moments involving the RVs that represent ISI (i.e., R_P and R_Q in our case). With respect to the linear situation, two new factors are present. They give rise to considerable complications. First, it is necessary to compute the discrete Volterra kernels $K_{2m+1}(\cdot)$. This can be achieved with the aforementioned numerical algorithms. Second, the RVs R_P and R_Q cannot be written (as in the linear case) as a sum of independent RVs. The next section is almost entirely devoted to this question. The treatment will deal with M-ary CPSK, this modulation scheme being the most widely used in digital satellite links. However, with minor changes it can be extended to any coherent modulation scheme using a two-dimensional signal constellation.

10.3.1 Error Probability Evaluation for M-ary CPSK Modulation

Let $a_n = e^{j\varphi_n}$ in (10.5), with φ_n assuming equally likely values in the set

$$\left\{\frac{(2k+1)\pi}{M}\right\}_{k=0}^{M-1},$$
(10.25)

which corresponds to M-ary CPSK modulation. The decision device determines the received phase angle ϕ_R:

$$\phi_R = \tan^{-1}\frac{R_{0Q} + R_Q + N_{2Q}}{R_{0P} + R_P + N_{2P}}$$
(10.26)

and decides according to the phase thresholds $2k\pi/M + \Theta$, $k = 0, \ldots, M - 1$, where Θ is a constant phase offset taking into account the value of AM/PM conversion of the TWT at the nominal operating point. This phase conversion has the effect of rotating the signal space of a constant value. This is compensated for by the phase shift Θ of the thresholds.

Following the procedure described in Section 6.2.2 for the linear case, we obtain the following bounds for the error probability, conditioned on the transmitted symbol $a_0 = e^{j\pi/M}$:

$$\max(I_1, I_2) \leqslant P(e|a_0) \leqslant I_1 + I_2,$$
(10.27)

where

$$I_1 \triangleq \frac{1}{2} \int_{\mathcal{L}} erfc \left\{ \frac{\lambda(\Theta + 2\pi/M) + \lambda_0(\Theta + 2\pi/M)}{\sqrt{2}\,\sigma_2} \right\} f_\Lambda[\lambda(\Theta + 2\pi/M)]\, d\lambda, \quad (10.28)$$

$$I_2 \triangleq \frac{1}{2} \int_{\mathcal{L}} erfc \left\{ \frac{-\lambda(\Theta) - \lambda_0(\Theta)}{\sqrt{2}\,\sigma_2} \right\} f_\Lambda[\lambda(\Theta)]\, d\lambda. \quad (10.29)$$

In (10.28) and (10.29), σ_2^2 is the variance of the Gaussian RVs N_{2P}, N_{2Q}; $\mathcal{L}$ and $f_\Lambda(\lambda)$ are the range and pdf of the RV Λ, defined as

$$\Lambda(\beta) \triangleq R_P \sin \beta - R_Q \cos \beta. \quad (10.30)$$

The useful sample in (10.28) and (10.29) is present in λ_0, defined as

$$\lambda_0(\beta) \triangleq R_{0P} \sin \beta - R_{0Q} \cos \beta. \quad (10.31)$$

As in the linear case, the symmetry of the received signals implies that the average error probability coincides with the conditional one. As a matter of fact, the transformations operated by the nonlinear device on the input signal depend on the envelope of the signal itself, which in turn is independent from the transmitted phase.

To evaluate (10.28) and (10.29), we need a few conditional moments of the RV $\Lambda(\beta)$. In the following, we shall derive a recurrent relationship that allows their fast computation.

From (10.30), the conditional moments of Λ can be written as

$$E\{\Lambda^n(\beta)|a_0\} = E\{[R_P \sin \beta - R_Q \cos \beta]^n|a_0\}$$
$$= \sum_{k=0}^{n} \binom{n}{k} E\{R_P^k R_Q^{n-k}|a_0\}(-1)^{n-k} \sin^k \beta \cos^{n-k} \beta \quad (10.32)$$

and the problem of computing the moments of $\Lambda(\beta)$ is reduced to the computation of the joint conditional moments

$$\mu_{kn} \triangleq E\{R_P^k R_Q^{n-k}|a_0\}. \quad (10.33)$$

Define the complex random variable ξ as

$$\xi \triangleq R_P + jR_Q \quad (10.34)$$

so that

$$R_P = \frac{1}{2}(\xi + \xi^*),$$

$$R_Q = \frac{1}{2j}(\xi - \xi^*).$$

Thus we can write (10.33) as

$$\mu_{kn} = \frac{1}{2^n (j)^{n-k}} \sum_{i=0}^{k} \sum_{l=0}^{n-k} \binom{k}{i}\binom{n-k}{l}(-1)^{n-k-l} E\{\xi^{i+l}\xi^{*n-i-l}\}. \quad (10.35)$$

Sec. 10.3   Volterra Model for Bandpass Nonlinear Channels   **545**

Let us get a deeper insight into the structure of the powers of ξ and ξ^*. From (10.22), ξ is given by

$$\xi = \sum_{m=0}^{m_M} \sum_{n_1} \cdots \sum_{\substack{n_{2m+1} \\ (n_1, \ldots, n_{2m+1} \neq 0, \ldots, 0)}} a_{n_1} \cdots a_{n_{m+1}} a^*_{n_{m+2}} \cdots a^*_{n_{2m+1}}$$

$$\cdot K_{2m+1}(n_1, \ldots, n_{2m+1}). \quad (10.36)$$

Also, any power of ξ (and ξ^*) can be considered as output of a Volterra system having ξ (or ξ^*) as input. We shall relate the new Volterra kernels defining the output to the ones that define the input. Let us start with ξ^2. When computing it, the terms $a_{n_1} a_{n_2}$, $a_{n_1} a_{n_2} a_{n_3} a^*_{n_4}$, . . . are multiplied by coefficients obtained as sums of products of Volterra kernels $K_{2m}(\cdot)$ as follows:

$$a_{n_1} a_{n_2} \quad \rightarrow K_1(n_1) K_1(n_2) \triangleq K_2^{(2)}(n_1, n_2),$$

$$a_{n_1} a_{n_2} a_{n_3} a^*_{n_4} \quad \rightarrow K_1(n_1) K_3(n_2, n_3, n_4)$$
$$+ K_3(n_1, n_2, n_4) K_1(n_3) \triangleq K_4^{(2)}(n_1, n_2, n_3, n_4),$$

$$a_{n_1} a_{n_2} a_{n_3} a_{n_4} a^*_{n_5} a^*_{n_6} \rightarrow K_1(n_1) K_5(n_2, n_3, n_4, n_5, n_6) \quad (10.37)$$
$$+ K_3(n_1, n_2, n_5) K_3(n_3, n_4, n_6)$$
$$+ K_5(n_1, n_2, n_3, n_5, n_6) K_1(n_4)$$
$$\triangleq K_6^{(2)}(n_1, n_2, n_3, n_4, n_5, n_6),$$

where the symbol $K_i^{(2)}$ refers to the Volterra coefficient of the ith-order relative to the second power of ξ.

Based on (10.35), we can write ξ^2 as

$$\xi^2 = \sum_{m=0}^{m_M} \sum_{n_1} \cdots \sum_{n_{2+2m}} a_{n_1} \cdots a_{n_{2+m}} \quad (10.38)$$

$$\cdot a^*_{n_{2+m+1}} \cdots a^*_{n_{2+2m}} K_{2+2m}^{(2)}(n_1, \ldots, n_{2+2m}),$$

where $K_{2+2m}^{(2)}(\cdot)$ satisfies the recurrent relationship

$$K_{2+2m}^{(2)}(n_1, \ldots, n_{2+2m})$$

$$= \sum_{i=0}^{m} K_{2i+1}(n_1, \ldots, n_{i+1}, n_{2+m+1}, \ldots, n_{2+m+i}) \quad (10.39)$$

$$\cdot K_{2(m-i)+1}(n_{i+2}, \ldots, n_{2+m}, n_{2+m+i+1}, \ldots, n_{2+2m}).$$

The previous procedure can be extended (see Problem 10.3) to derive the general formulas

$$\xi^l = \sum_{m=0}^{m_M} \sum_{n_1} \cdots \sum_{n_{l+2m}} a_{n_1} \cdots a_{n_{l+m}} a^*_{n_{l+m+1}} \cdots a^*_{n_{l+2m}} K_{l+2m}^{(l)}(n_1, \ldots, n_{l+2m}),$$

$$K_{l+2m}^{(l)}(n_1, \ldots, n_{l+2m})$$

$$= \sum_{i=0}^{m} K_{2i+1}(n_1, \ldots, n_{i+1}, n_{l+m+1}, \ldots, n_{l+m+i}) \quad (10.40)$$

$$\cdot K_{2(m-i)-1+l}^{(l-1)}(n_{i+2}, \ldots, n_{l+m}, n_{l+m+i+1}, \ldots, n_{l+2m}),$$

$$K_m^{(1)}(\cdot) \equiv K_m(\cdot).$$

The reader is invited to challenge his patience in deriving similar relationships for the powers of ξ^* (See Problem 10.4).

Using (10.40) and the corresponding formulas for $(\xi^*)^m$, the averages in RHS of (10.35) can be given the form

$$E\{\xi^l\xi^{*m}|a_0\} = \sum_{k=0}^{m_M}\sum_{i=0}^{m_M}\sum_{n_1}\cdots\sum_{nl+m+2k+2i} E\{a_{n_1}\cdots a_{nl+k+i}a^*_{nl+k+i+1}$$

$$\cdots a^*_{nl+m+2k+2i}|a_0\}\, K^{(l)}_{l+2k}(n_1, \ldots, n_{l+k}, n_{l+k+i+1}, \ldots, n_{l+2k+i}) \qquad (10.41)$$

$$K^{(m)*}_{m+2i}(n_{l+k+1}, \ldots, n_{l+k+i}, n_{l+2k+i+1}, \ldots, n_{l+m+2k+2i}),$$

and the final step toward the knowledge of the moments of Λ [see (10.30)] is the computation of the conditional averages in RHS of (10.41).

Define

$$A_k \triangleq E\{a_n^k\}$$

and remember that the indexes n_i range in a finite set, say $(0, N)$.

Denote by v_m (and v_m^*) the number of indexes of the a_{n_i} (and $a^*_{n_i}$) that take on the values m ($m = 0, 1, \ldots, N$). We have, taking into account that $a_n a_n^* = \exp(\ j\varphi_n)\exp(-j\varphi_n) = 1$,

$$E\{a_{n_1}\cdots a_{nl+k+i}a^*_{nl+k+i+1}\cdots a^*_{nl+m+2k+2i}|a_0\} = a_0^{v_0^-}v_0^*\prod_{i=1}^{N}A_{v_i-v_i^*} \qquad (10.42)$$

and

$$A_k = \begin{cases} 1, & \text{if } k = \pm 2iM, & i = 0, 1, \ldots, \\ -1, & \text{if } k = \pm(2i+1)M, & i = 0, 1, \ldots, \\ 0, & \text{elsewhere.} \end{cases} \qquad (10.43)$$

As an example, we have

$$E\{a_5 a_4 a_0 a_3 a_5 a_0 a_4^* a_4^*|a_0\} = a_0^2 A_{-1}A_1 A_2 = 0.$$

At the end of the section, a step-by-step summary of the whole procedure needed to obtain the error probability may prove useful for applicative purposes.

Step 1. Get the Volterra coefficients $K_{2i+1}(\cdot)$ using, for example, their definition (10.16) and (10.14).

Step 2. Compute the Volterra coefficients of higher order using recurrent relationships similar to (10.40).

Step 3. Compute the moments μ_{kn} defined in (10.35) through (10.41 to 10.43).

Step 4. For a given β, compute the moments of the RV $\Lambda(\beta)$ through (10.32), taking into account (10.33).

Step 5. Compute I_1 and I_2 (and thus the bounds of the error probability) defined in (10.28) and (10.29) using the methods described in Appendix E.

10.3.2 Including Uplink Noise in the Analysis

In general, a wide-sense stationary Gaussian process $n_1(t)$ can be represented in the following form (see Masry, Liu, and Steiglitz, 1968):

$$n_1(t) = \sum_{i=-\infty}^{\infty} \beta_i b_i(t), \qquad (10.44)$$

where $b_i(t)$ are appropriately chosen deterministic functions and (β_i) is a sequence of zero-mean, unit-variance independent Gaussian RVs (possibly complex).

The general representation of (10.44) reduces to the following form:

$$n_1(t) = \sum_{i=-\infty}^{\infty} \beta_i \delta(t - iT) * s(t) \triangleq n_1'(t) * s(t) \qquad (10.45)$$

when (see Problem 10.6) the low-pass equivalent linear systems that limit the uplink noise power can be approximated by an ideal low-pass filter with bandwidth $1/2T$.

If the representation (10.45) is valid, we can use the model of Fig. 10.14, where $n_1'(t)$ is added to the modulated signal before the baseband shaping filter. In this way, the uplink noise is accounted for by simply considering as an input signal $x(t)$, instead of the one given by (10.5), the new signal

$$x(t) = \sum_{i=-\infty}^{\infty} (a_i + \beta_i)\delta(t - iT).$$

The received signal can now be given the same form as (10.19), with the substitution of $(a_n + \beta_n)$ in place of a_n, and the same previous procedure can be applied to the computation of the moments (see Problem 10.7), since the averages to be performed as the final step of the moment computation, that is,

$$E\{(a_{n_1} + \beta_{n_1}) \cdots (a_{nl+k+i} + \beta_{nl+k+i})(a^*_{nl+k+i+1} + \beta^*_{nl+k+i+1})$$
$$\cdots (a^*_{nl+m+2k+2i} + \beta^*_{nl+m+2k+2i})|a_0\}$$

involve products of independent RVs whose moments are known, as in the previous analysis. The interested reader can find all the analytical details in Ajmone Marsan and others (1977, pp. 39–43). The only significant change in the procedure worth mentioning here is that the presence of uplink noise also affects the "useful" signal. In fact, the expression for R_0 in (10.20) now becomes

$$(a_0 + \beta_0) \sum_{m=0}^{m_M} |a_0 + \beta_0|^m K_{2m+1}(0, \ldots, 0).$$

To obtain the error probability, a final average must be computed with respect to β_0. This can be done by using, for instance, a standard Gauss–Hermite numerical quadrature formula (see Example E.1).

Example 10.3

In this example, some results obtained by applying the Volterra series method of analysis will be presented. The system model is that shown in Fig. 10.13. The TWT characteristics were given in Fig. 10.12. The filters $s(t)$ and $n(t)$ are fourth- and second-order Butterworth, respectively. The curves of error probability are given as a function of the parameter

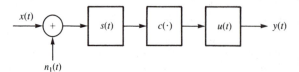

Figure 10.14 Model to include uplink noise.

$$\eta_2 \triangleq 10 \log_{10} \frac{(\mathscr{E}_b)_{\text{sat}}}{N_0},$$

where $(\mathscr{E}_b)_{\text{sat}}$ is the energy per bit at the input of the receiver in correspondence with the TWT operated at saturation point. The sampling instant t_0 was chosen as the one in which the convolution of the impulse responses $s(t)$ and $u(t)$ is maximum, whereas the phase offset Θ chosen was that minimizing the error probability (a good starting point for the optimization is the value corresponding to the AM/PM conversion at saturation).

In Figs. 10.15 to 10.17 the error probability for a quaternary CPSK system is plotted versus the downlink signal-to-noise ratio η_2 for different values of the uplink parameter $\eta_1 \triangleq 10 \log_{10} (1/\sigma_1^2)$. It can be seen that, due to the combined effects of uplink noise and nonlinearity, a "bottoming" effect takes place in the error probability curves. For low values of η_2, it happens that the effect of downlink noise dominates. For large η_2, uplink noise dominates and error probability does not depend on downlink noise power.

By comparing Figs. 10.16 and 10.17, another relevant feature is observed. As a result of an increase of the transmitting filter bandwith from 1.8 to 2.5, the error probability decreases, due to the smaller amount of ISI introduced by the filtering. This is explained by the lack of neighboring channels in the models. Actually, in a multichannel environment, for any increase in the transmitting filter bandwidth a corresponding increase in interference power occurs. By comparing Figs. 10.15 and 10.16, the effects of the presence of nonlinearity can be observed. The most dramatic feature is the increased sensitivity to uplink noise. In fact, when signal plus uplink noise enters a nonlinearity with a saturating characteristic, a signal-suppression effect takes place. In the linear case this does not occur because uplink and downlink noise simply sum up without further corrupting the signal.

Finally, in Fig. 10.18, the sensitivity of the system to offset in timing recovery is shown. The error probability is plotted versus the normalized deviation from the "optimum"

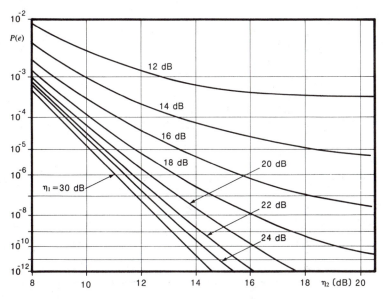

Figure 10.15 Error probability of 4-PSK versus downlink η_2, with uplink η_1 as a parameter. Transmission filter: fourth-order Butterworth with equivalent noise bandwidth $B_{\text{eq}} = 1.8/T$. Receiving filter: second-order Butterworth with $B_{\text{eq}} = 1.1/T$.

Sec. 10.3   Volterra Model for Bandpass Nonlinear Channels   **549**

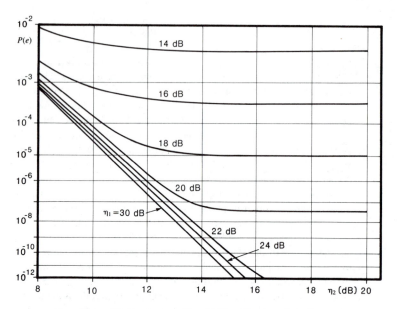

Figure 10.16 Same as Fig. 10.15, without nonlinearity.

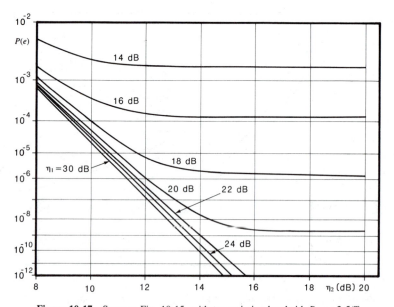

Figure 10.17 Same as Fig. 10.15, with transmission bandwith $B_{eq} = 2.5/T$.

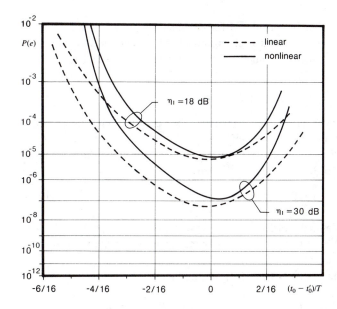

Figure 10.18 Error probability of 4-PSK versus normalized deviation from the "optimum" sampling time t_0', with η_1 as parameter. $\eta_2 = 12.5$ dB. Same channel as in Fig. 10.15.

sampling time t_0'. It is apparent that the presence of the nonlinear device (continuous lines) renders the system behavior more critical to the choice of the sampling instant. $\square$

10.4 OPTIMUM LINEAR RECEIVING FILTER

In the preceding sections, we presented models for the description of nonlinear channels and demonstrated their use to compute certain performance parameters like the power spectrum or the error probability. So far, nonlinearity has been seen as an unwanted source of performance degradation imposed by some particular system demands. Also, the receiver structure was assumed to be the same as in the linear case. In this section and the following, we shall consider the problem of modifying the receiver according to some optimality criterion in order to take into account channel nonlinearity. The analysis is based on the general model introduced in Section 10.1.

Let us consider the symbol-by-symbol receiver shown in Fig. 10.19, in the form of a linear filter followed by a symbol-rate sampler and a memoryless decision device. Here we want to choose the filter impulse response $u(t)$ in such a way that the sample of the received signal $r(t)$ taken at time $t_n = t_0 + nT$ is as close as possible (in the mean-square sense) to the transmitted symbol a_n. In formulas, we aim at finding the $u(t)$ that minimizes the quantity

$$\mathscr{E} \triangleq \mathrm{E}|r(t_n)*u(t_n) - a_n|^2. \tag{10.46}$$

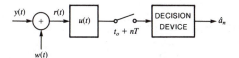

Figure 10.19 Linear symbol-by-symbol receiver.

This optimization has been already considered for the linear case in Section 7.3. Here we shall extend those results to a nonlinear environment. Minimization of $\mathcal{E}$ in (10.46) will be performed in the frequency domain. The received signal (see Fig. 10.1) is given by

$$r(t) = y(t) + w(t), \tag{10.47}$$

where $y(t)$ has been defined in (10.2) and $w(t)$ is a noise process independent of $y(t)$ with power spectral density $G_w(f) > 0$ for all f.

Let us denote with $U(f)$ the transfer function of the receiving filter and with $Y(f)$ the Fourier transform of the signal $y(t)$. We have

$$Y(f) = \sum_n H(f; a_n, \sigma_n)e^{-j2\pi nfT}, \tag{10.48}$$

where $H(f; a_n, \sigma_n)$ is the Fourier transform of the waveform $h(t; a_n, \sigma_n)$. It can assume $N = M^{L+1}$ shapes $\{H(f; i)\}_{i=1}^N$ not necessarily distinct, according to the finite memory assumption made for the channel.

In the frequency domain, (10.46) can be rewritten as

$$\mathcal{E} = \sigma_a^2 + \iint_{-\infty}^{\infty} U^*(f_1)\Gamma_y(f_1, f_2)U(f_2)e^{j2\pi(f_1 - f_2)t_n}df_1 \, df_2 \tag{10.49}$$

$$- 2\Re \int_{-\infty}^{\infty} V(f)U^*(f) \, df + \int_{-\infty}^{\infty} G_w(f)|U(f)|^2 \, df,$$

where the following definitions have been used:

$$\Gamma_y(f_1, f_2) \triangleq E[Y(f_1)Y^*(f_2)], \tag{10.50}$$

$$V(f) \triangleq e^{-j2\pi ft_n}E[a_n Y^*(f)], \tag{10.51}$$

$$\sigma_a^2 \triangleq E|a_n^2|. \tag{10.52}$$

Using the standard variational calculus technique summarized in Appendix C, it can be shown that a necessary and sufficient condition for $U(f)$ to minimize $\mathcal{E}$ is that it be the solution to the integral equation

$$\int_{-\infty}^{\infty} \Gamma_y^*(f, f')U(f')e^{-j2\pi(f-f')t_n}df' + G_w(f)U(f) = V(f). \tag{10.53}$$

Thus far, we have not yet exploited our knowledge of the structure of $Y(f)$, as provided by (10.48). Substituting (10.48) into (10.50) and exploiting the fact that the state sequence is a shift-register sequence, we obtain [see the derivation of (2.116)]

$$\Gamma_y(f_1, f_2) = \frac{1}{T} \sum_{k=-\infty}^{\infty} G_k(f_1)\delta\left(f_1 - f_2 - \frac{k}{T}\right), \tag{10.54}$$

where $G_k(f)$ has been defined as

$$G_k(f) \triangleq \sum_{l=-\infty}^{\infty} E\left\{\{H(f; a_n, \sigma_n)H^*\left(f - \frac{k}{T}; a_{n+l}, \sigma_{n+l}\right)\right\}e^{-j2\pi lfT} \tag{10.55}$$

so that the integral equation (10.53) takes the form

$$\frac{1}{T}\sum_{k=-\infty}^{\infty} G_k^*(f)U\left(f-\frac{k}{T}\right)e^{-j2\pi kt_n/T} + G_w(f)U(f) = V(f). \quad (10.56)$$

Let us now compute the averages involved in the definitions of $V(f)$ and $G_k(f)$:

$$E\{a_n Y^*(f)\} = \sum_{k=-\infty}^{\infty} e^{j2\pi kfT}E\{a_n H^*(f; a_k, \sigma_k)\}. \quad (10.57)$$

The average in RHS of (10.57) can be computed using a method similar to that followed in Section 2.3.1, leading to the result of (2.147). With the same notations, except for the replacement of $s(t; a_n, \sigma_n)$ with $h(t; a_n, \sigma_n)$ (entailing the substitution of the letters s, S with h, H where appropriate), we obtain

$$E\{a_n H^*(f; a_k, \sigma_k)\} = \begin{cases} \mathbf{c}_2^*(f)\mathbf{P}^{k-n-1}\mathbf{c}_3', & k > n, \\ \mathbf{c}_2^*(f)[\mathbf{P}^{1-k+n}]'\mathbf{c}_3', & n \geq k, \end{cases} \quad (10.58)$$

where $\mathbf{c}_2(f)$ has been defined in (2.136), $\mathbf{P}$ [the transition probability matrix of the state sequence (σ_n)] in (2.133), and

$$\mathbf{c}_3 \triangleq \sum_{h=1}^{M} p_h a_h \mathbf{w}\mathbf{E}_h, \quad (10.59)$$

where $p_h = P(a_h)^\dagger$ and $\mathbf{w}$, $\mathbf{E}_h$ are defined in Section 2.3.1. Note that $\mathbf{c}_3$ is a vector whose jth entry is the average of the source symbols a_h that lead the channel to the state S_j, being $\{S_j\}_{j=1}^{ML}$, the set of values assumed by the RV σ_n.

Substitution of (10.58) into (10.57) and of this into (10.51) yields

$$V(f) = e^{-j2\pi ft_0}\mathbf{c}_2^*(f)\mathbf{\Lambda}(f)\mathbf{c}_3' + e^{-j2\pi ft_0}\mathbf{c}_2^*(f)\mathbf{P}^\infty \mathbf{c}_3' \sum_l e^{-j2\pi flT}, \quad (10.60)$$

where

$$\mathbf{\Lambda}(f) \triangleq \sum_{l=1}^{\infty} [\mathbf{P}^{l-1} - \mathbf{P}^\infty]e^{-j2\pi flT} + \sum_{l=0}^{\infty} [\mathbf{P}^{1+l} - \mathbf{P}^\infty]' e^{j2\pi flT}. \quad (10.61)$$

The second term in RHS of (10.60) contains spectral lines at dc and multiples of the symbol rate $1/T$. It disappears when either $\mathbf{P}^\infty\mathbf{c}_3'$ or $\mathbf{c}_2(f)\mathbf{P}^\infty$ is equal to zero. This means that the average value of symbols a_i's or of waveforms $H(f; a_k, \sigma_k)$ is zero. We shall make this assumption in the following; thus we have

$$V(f) = e^{-j2\pi ft_0}[\mathbf{c}_2^*(f)\mathbf{\Lambda}(f)\mathbf{c}_3']. \quad (10.62)$$

Turning our attention to the average in the RHS of (10.55), which defines $G_k(f)$, using a straightforward replica of the algebra leading to (2.147), we obtain

$$E\{H(f; a_n, \sigma_n)H^*(f - k/T; a_{n+1}, \sigma_{n+1})\}$$
$$= \begin{cases} \mathbf{c}_2(f)[\mathbf{P}^{l-1}]'\mathbf{c}_1^\dagger(f - k/T), & l > 0, \\ \mathbf{c}_2(f)\mathbf{P}^{1-l}\mathbf{c}_1^\dagger(f - k/T), & l \leq 0, \end{cases} \quad (10.63)$$

where $\mathbf{c}_1(f)$ was defined in (2.139).

† Here and in (10.59) the subscript h runs over the set of M values assumed by the RV a_n. It might have been more appropriate to employ two different notations (one for the RV and the other for its values). However, we opted for one to avoid an increase in the notations used.

Substitution of (10.63) into (10.55) yields

$$G_k(f) = \mathbf{c}_2(f)\mathbf{\Lambda}(f)\,\mathbf{c}_1^\dagger(f - k/T) + \mathbf{c}_2(f)\mathbf{P}^\infty \mathbf{c}_1^\dagger(f) \sum_{l=0}^{\infty} \cos 2\pi lfT, \qquad (10.64)$$

where $\mathbf{\Lambda}(f)$ was defined in (10.61). With the hypothesis

$$\mathbf{c}_2(f)\mathbf{P}^\infty \mathbf{c}_1^\dagger(f) = 0, \qquad (10.65)$$

we can write finally

$$G_k(f) = \mathbf{c}_2(f)\mathbf{\Lambda}(f)\,\mathbf{c}_1^\dagger\,(f - k/T)\cdot \qquad (10.66)$$

Following the procedure used in Section 7.3 for the linear case, we shall prove that the equation (10.56) admits the solution

$$U_{\mathrm{opt}}(f) = \frac{e^{-j2\pi f t_0}\mathbf{c}_2(f)\boldsymbol{\gamma}(f)}{G_w(f)}, \qquad (10.67)$$

where $\boldsymbol{\gamma}(f)$ is a column M-vector of frequency functions periodic with period $1/T$.

Remembering through the definition (2.136) that $\mathbf{c}_2(f)$ is a vector whose ith component is the average amplitude spectrum of the waveforms available to the modulator when it is in the ith state S_i, the similarity of (10.67) with the result obtained in the linear case (7.38) becomes apparent. As a matter of fact, we have that the optimum receiving filter may be thought of as being composed of a bank of filters matched to the average transmitted waveforms, each one followed by an infinite-length transversal filter.

Let us define

$$\mathbf{b}(f) \triangleq \mathbf{\Lambda}(f)\mathbf{c}_3' \qquad (10.68)$$

so that (10.62) can be rewritten as

$$V(f) = e^{-j2\pi f t_0}\mathbf{c}_2^*(f)\,\mathbf{b}(f). \qquad (10.69)$$

Substitution of (10.66), (10.67), and (10.69) into (10.56) gives

$$\mathbf{c}_2^*(f)\left[\frac{1}{T}\mathbf{\Lambda}^*(f)\sum_k \frac{1}{G_w(f - k/T)}\,\mathbf{c}_1'\,(f - k/T)\,\mathbf{c}_2^*\,(f - k/T) + \mathbf{I}\right]\boldsymbol{\gamma}(f)$$
$$= \mathbf{c}_2^*(f)\mathbf{b}(f). \qquad (10.70)$$

Equivalently, (10.70) can be rewritten in the form

$$\mathbf{c}_2^*(f)\,\mathbf{d}(f) = 0, \qquad (10.71)$$

where $\mathbf{d}(f)$ is the M-vector, periodic with period $1/T$, defined by

$$\mathbf{d}(f) \triangleq \mathbf{A}(f)\,\boldsymbol{\gamma}\,(f) - \mathbf{b}(f) \qquad (10.72)$$

and

$$\mathbf{A}(f) \triangleq \frac{1}{T}\mathbf{\Lambda}^*(f)\sum_{k=-\infty}^{\infty} \frac{1}{G_w(f - k/T)}\,\mathbf{c}_1'\,(f - n/T)\,\mathbf{c}_2^*\,(f - n/T) + \mathbf{I}. \qquad (10.73)$$

The vector $\mathbf{d}(f)$ is now periodic with period $1/T$, so the inverse Fourier transforms of its components, $d_i(t)$, $i = 1, \ldots, M$, have the following form:

$$d_i(t) = \sum_{j=-\infty}^{\infty} \rho_{ij}\delta(t - jT). \tag{10.74}$$

Hence, the time-domain version of (10.71) is

$$\sum_{i=1}^{M} \sum_{j=-\infty}^{\infty} \rho_{ij} c_2^*(t; i) * \delta(t - jT) = \sum_{j=-\infty}^{\infty} \sum_{i=1}^{M} \rho_{ij} c_2^*(t - jT; i) = 0, \tag{10.75}$$

where $c_2(t; i)$ is the inverse Fourier transform of the ith component of $\mathbf{c}_2(f)$.

Thus, if the waveforms $c_2(t; i)$ are linearly independent, (10.75), and hence (10.71), can hold if and only if all the ρ_{ij} are equal to zero. This means that (10.71) admits the only solution $\mathbf{d}(f) = \mathbf{0}$, that is, assuming that $\mathbf{A}(f)$ is nonsingular:

$$\boldsymbol{\gamma}(f) = \mathbf{A}^{-1}(f)\,\mathbf{b}(f). \tag{10.76}$$

Equation (10.76) shows that $\boldsymbol{\gamma}(f)$ is periodic with period $1/T$. This proof can be extended to the case when the waveforms $c_2(t; i)$ are not linearly independent [see Biglieri and others (1984)].

As already noted, the structure of the optimum filter $U_{\text{opt}}(f)$ in (10.67) is that of a bank of matched filters followed by infinite-length transversal filters. This is shown in Fig. 10.20.

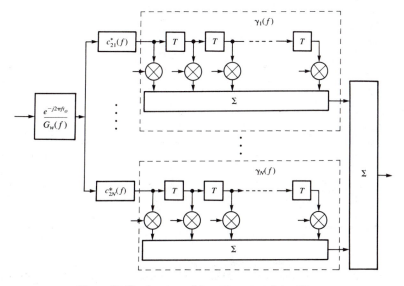

Figure 10.20 Structure of the optimum receiving filter.

Example 10.4

As an example of application, consider a binary CPSK signal transmitted over a nonlinear channel consisting of a fourth-order Butterworth filter cascaded to a nonlinear amplifier (a TWT exhibiting both AM/AM and AM/PM conversion). Figure 10.21 shows the transfer function of the optimum filter for a channel with and without the nonlinearity, for the sake of comparison. The power spectral densities of the received signal are also shown for comparison. The performance of the optimum filter is shown in Fig. 10.22. For comparison, the mean-square error resulting from a second-order Butterworth receiving filter with optimum 3-dB bandwidth is also shown. $\square$

10.5 MAXIMUM-LIKELIHOOD SEQUENCE RECEIVER

In this section we shall derive the structure of the ML sequence receiver for a system using a nonlinear channel. The channel model is still that of Section 10.1 and Figs. 10.1 and 10.2. We suppose that the transmission lasts from 0 to KT. This duration is taken large enough to disregard end effects due to the channel memory (in practice, we shall assume $K \gg L$). The signal at the output of the noiseless part of the channel, represented by (10.2), can thus be written as

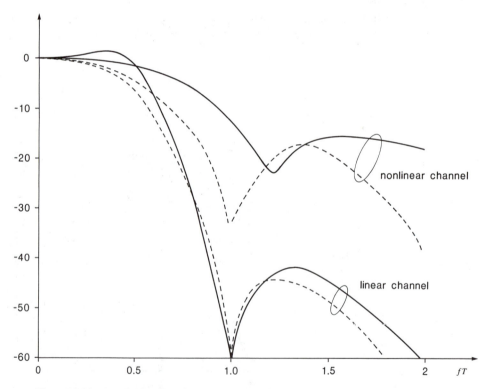

Figure 10.21 Transfer function of the optimum receiving filter for binary CPSK (solid line) and power spectral density (dashed line) of the received signal (binary PSK); fourth-order Butterworth TX filter with $B_{eq} = 1.028/T$; $N_0 = 5.10^{-3}$ W/Hz.

$$y_\mathbf{a}(t) \triangleq \sum_{n=0}^{K-1} h(t - nT; a_n, \sigma_n), \tag{10.77}$$

where the subscript $\mathbf{a}$ denotes the dependence of $y_\mathbf{a}(t)$ on the sequence $\mathbf{a} = (a_0, \ldots, a_{K-1})$ of source symbols that must be estimated from the receiver.

From (10.77) we can see that the entire waveform $y(t)$, $t \in (0, KT)$, is defined by the sequence of states $\sigma_L, \sigma_{L+1}, \ldots, \sigma_K$ or, equivalently, by the sequence of symbols $a_0, a_1, \ldots, a_{K-1}$. Thus we have no more than M^K possible received waveforms in the observation interval, and the ML reception is equivalent to the detection of one out of a finite set of waveforms in additive Gaussian noise. Thus, in principle, the optimum receiver will be made up of a bank of M^K matched filters, one for each possible waveform. The filters' outputs are then sampled at the end of the transmission, and the largest sample is used to select the most likely symbol sequence.

In Chapter 7, solutions were found in the linear case for two major problems arising in the implementation of an ML receiver, (1) the number of matched filters required and (2) the number of comparisons needed to select the largest output. In particular, it was shown that just one matched filter is required to obtain the sufficient statistics for the ML receiver and, also, that the Viterbi algorithm can be used to select the most likely sequence with a complexity that grows only linearly with respect to the message length. In the nonlinear case, the results are different. The second problem will be given a satisfactory solution, still invoking the Viterbi algorithm. With respect to the number of matched filters needed, we shall show that it grows exponentially

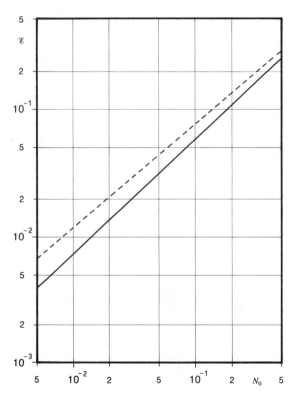

Figure 10.22 Mean-square error with optimum filter (continuous line) and with an optimized second-order Butterworth filter (dashed line). Same system as in Fig. 10.21.

Sec. 10.5   Maximum-Likelihood Sequence Receiver   **557**

with the channel memory L. This is better than the exponential growth with the sequence length K arising from a brute-force approach. But it still makes a practical application of this theory confined to those situations in which the channel memory is very short and/or M is small.

The key to the ML receiver design is the expression of the log-likelihood ratio for the detection of the finite sequence of symbols $\mathbf{a} \triangleq (a_0, a_1, \ldots, a_{K-1})$ based on the observation of the waveform

$$r(t) = \sum_{n=0}^{K-1} h(t - nT; a_n, \sigma_n) + w(t), \qquad t \in (0, KT), \tag{10.78}$$

where $w(t)$ is AWGN. The log-likelihood ratio for $\mathbf{a}$ has the expression (see Section 2.6.3)

$$\lambda_{\mathbf{a}} = \frac{2}{N_0} \mathcal{R} \int_0^{KT} y_{\mathbf{a}}^*(t) r(t) \, dt - \frac{1}{N_0} \int_0^{KT} |y_{\mathbf{a}}(t)|^2 \, dt, \tag{10.79}$$

where $y_{\mathbf{a}}(t)$ is the noiseless waveform defined in (10.77). Using (10.78), (10.79) can be rewritten in the form

$$\lambda_{\mathbf{a}} = \frac{2}{N_0} \mathcal{R} \left\{ \sum_{n=0}^{K-1} \int_0^{KT} h^* (t - nT; a_n, \sigma_n) r(t) \, dt \right\} \tag{10.80}$$

$$- \frac{1}{N_0} \sum_{n=0}^{K-1} \sum_{l=0}^{K-1} \int_0^{KT} h(t - nT; a_n, \sigma_n) h^* (t - lT; a_l, \sigma_l) \, dt.$$

Notice now that $h(\cdot; \cdot)$ has a finite duration T. Thus we have

$$\int_0^{KT} h^* (t - nT; a_n, \sigma_n) r(t) \, dt \tag{10.81}$$

$$= \int_{nT}^{(n+1)T} h^* (t - nT; a_n, \sigma_n) r(t) \, dt, \qquad n = 0, \ldots, K - 1,$$

and

$$\int_0^{KT} h(t - nT; a_n, \sigma_n) h^* (t - lT; a_l, \sigma_l) \, dt \tag{10.82}$$

$$= \begin{cases} 0, & l \neq n, \\ \int_{nT}^{(n+1)T} |h(t - nT; a_n, \sigma_n)|^2 \, dt, & l = n. \end{cases}$$

So, defining

$$Z_n(a_n, \sigma_n) \triangleq \int_{nT}^{(n+1)T} h^* (t - nT; a_n, \sigma_n) r(t) \, dt \tag{10.83}$$

and

$$\mathcal{E}(a_n, \sigma_n) \triangleq \int_0^T |h(t; a_n, \sigma_n)|^2 \, dt, \tag{10.84}$$

we finally get

$$\lambda_{\mathbf{a}} = \frac{2}{N_0} \mathcal{R} \sum_{n=0}^{K-1} Z_n(a_n, \sigma_n) - \frac{1}{N_0} \sum_{n=0}^{K-1} \mathcal{E}_n(a_n, \sigma_n). \tag{10.85}$$

We can observe that:

(i) $\mathcal{E}_n(a_n, \sigma_n)$ is the energy of the waveforms $h(t; a_n, \sigma_n)$;

(ii) $Z_n(a_n, \sigma_n)$ can be obtained as the response, sampled at time $(n + 1)T$, of a filter matched to $h(t; a_n, \sigma_n)$, to a segment of the input $r(t)$ in the interval $(nT, (n + 1)T)$. The number of matched filters required is then equal to M^{L+1}, that is, the number of different values of the pair (a_n, σ_n).

The ML sequence decoding rule requires $\lambda_{\mathbf{a}}$ to be maximized over the set of possible sequences $\mathbf{a}$. Equivalently, multiplying (10.85) by the constant factor N_0 and changing signs, we can say that the ML sequence $\hat{\mathbf{a}}$ is the one that minimizes the quantity

$$l_{\mathbf{a}} \triangleq -2\mathcal{R} \sum_{n=0}^{K-1} Z_n(a_n, \sigma_n) + \sum_{n=0}^{K-1} \mathcal{E}_n(a_n, \sigma_n). \tag{10.86}$$

The ML receiver, besides the addition of the energies $\mathcal{E}_n(\cdot)$, is formed by a bank of filters matched to $h(t; a, \sigma)$ followed by one sampler per branch and by a processor, the ML sequence detector, determining as the most likely transmitted data sequence the one minimizing $l_{\mathbf{a}}$.

Define now the transition between states as

$$\tau_{n+1} \triangleq (\sigma_n, \sigma_{n+1}) \tag{10.87}$$

and observe that there is a one-to-one correspondence between each pair (a_n, σ_n) and τ_n. Thus, write $Z_n(\tau_n)$ and $\mathcal{E}_n(\tau_n)$, so that, defining

$$l_{\mathbf{a}}^{(n)}(\tau_n) \triangleq -2\mathcal{R}(Z_n(\tau_n)) + \mathcal{E}_n(\tau_n), \tag{10.88}$$

we can rewrite (10.86) as

$$l_{\mathbf{a}} = \sum_{n=0}^{K-1} l_{\mathbf{a}}^{(n)}(\tau_n). \tag{10.89}$$

Decomposition of (10.89), together with the fact that the sequence (τ_n) originates from a shift-register sequence, ensures that the Viterbi algorithm (see Appendix F) can be applied to the minimization problem at hand. As in the linear case described in Chapter 7, the ML detection problem reduces to the selection of a path through a trellis whose branches have been associated the values taken by the function $l_{\mathbf{a}}^{(n)}(\tau_n)$, referred to as the metric. The same steps illustrated in Chapter 7 to describe the algorithm can be applied after certain straightforward modifications.

10.5.1 Error Performance

The problem of evaluating the error probability performance of an ML sequence receiver can be conceptually reduced to the general problem examined in Chapter 4. We need only to think of data sequences as points in a signal space in which an error event happens whenever the noise and interferences cause the receiver to decide for a point different from the transmitted one. Of course, the error probability depends on the distribution of the Euclidean distances between all possible pairs of points, and, roughly speaking, they are mainly related to the minimum value d_{min} of these distances.

Usually, finding d_{min} requires exhaustive comparisons of the received signal waveforms corresponding to all possible pairs of symbol sequences. This can be avoided when the structure of points representing the received sequences is completely symmetric in such a way that comparisons of all the sequences with respect to a particular one are representative of the larger set of comparisons that should be needed. In this case we say that the uniform error property applies. A significant example was represented by linear codes, where the error performance is computed using as a reference the all-zero word (block codes) or sequence (convolutional codes).

Unfortunately, in this case, the uniform error property does not hold, so all comparisons are needed. In some cases, like the typical satellite channel with a short memory, the number of sequence pairs that needs to be considered is not very large. Hence, a brute-force computation is feasible. A method to do that can be found in Herrmann (1978). It consists of an algorithm to compute systematically the so-called "chip functions" and "chip distances." It uses these to compute the distances between pairs of received waveforms and then to estimate the symbol error probability.

Here we propose a different approach. We shall derive first an upper bound to the sequence error probability, defined as the probability of choosing as true a transmitted sequence different from the actual one. Then we shall see that the relevant contribution to the error probability, for reasonable values of signal-to-noise ratios, depends on the minimum Euclidean distance between pairs of possible received waveforms. Finally, we shall present an efficient algorithm to compute the minimum Euclidean distance.

Suppose that the sequence $\mathbf{a}_j$ has been transmitted; the probability that the estimated sequence $\hat{\mathbf{a}}$ is different from $\mathbf{a}_j$ can be written as

$$P(e|\mathbf{a}_j) \triangleq P\{\hat{\mathbf{a}} \neq \mathbf{a}_j | \mathbf{a}_j\}. \tag{10.90}$$

Application of the union bound (see Section 4.2.2) allows us to write

$$P(e|\mathbf{a}_j) \leq \sum_{\mathbf{a}_i \neq \mathbf{a}_j} P\{l_{\mathbf{a}_i} < l_{\mathbf{a}_j} | \mathbf{a}_j\} = \frac{1}{2} \sum_{\mathbf{a}_i \neq \mathbf{a}_j} erfc\left(\frac{d(\mathbf{a}_i, \mathbf{a}_j)}{2\sqrt{N_0}}\right), \tag{10.91}$$

where $d(\mathbf{a}_i, \mathbf{a}_j)$ is the Euclidean distance between the two signal sequences obtained at the output of the noiseless part of the channel in correspondence with the symbol sequences $\mathbf{a}_i$ and $\mathbf{a}_j$:

$$d^2(\mathbf{a}_i, \mathbf{a}_j) \triangleq \int_0^{KT} |y_{\mathbf{a}_i}(t) - y_{\mathbf{a}_j}(t)|^2 \, dt. \tag{10.92}$$

The last equality in (10.91) is evident if one thinks of the noiseless received waveforms of duration KT as points in a signal space and then applies the standard formula for the binary error probability between two points at distance $d(\mathbf{a}_i, \mathbf{a}_j)$.

To obtain the average sequence error probability, we need only to average $P(e|\mathbf{a}_j)$ over all the possible symbol sequences $\mathbf{a}_j$ assumed to be equally likely:

$$P(e) = \sum_{\mathbf{a}_j} P(\mathbf{a}_j)P(e|\mathbf{a}_j) \leq \frac{1}{2} \frac{1}{M^K} \sum_{\mathbf{a}_j} \sum_{\mathbf{a}_i \neq \mathbf{a}_j} erfc\left(\frac{d(\mathbf{a}_i, \mathbf{a}_j)}{2\sqrt{N_0}}\right). \tag{10.93}$$

The final step is to consider the discrete finite set of all possible Euclidean distances $d(\mathbf{a}_i, \mathbf{a}_j)$, $\mathbf{a}_i \neq \mathbf{a}_j$ denoted by $D = \{d_l\}$. Denoting by $N(d_l)$ the number of pairs of sequences $(\mathbf{a}_i, \mathbf{a}_j)$ giving rise to noiseless received waveforms at distance d_l, we can rewrite (10.93) as

$$P(e) \leq \frac{1}{2} \frac{1}{M^K} \cdot \sum_{d_l \in D} N(d_l) erfc \left(\frac{d_l}{2\sqrt{N_0}} \right). \tag{10.94}$$

For large values of signal-to-noise ratio, the dominant term of the summation in the RHS of (10.94) is the one containing $d_{min} \triangleq \min_{d_l \in D} d_l$, so that, asymptotically, we can approximate $P(e)$ as

$$P(e) \lesssim \frac{1}{2} \frac{1}{M^K} N(d_{min}) erfc \left(\frac{d_{min}}{2\sqrt{N_0}} \right). \tag{10.95}$$

To compute the sequence error probability (10.94), an extension of the method of the transfer function of directed graphs, used to evaluate the performance of convolutional codes, could be applied (see Viterbi and Omura, 1979, Problem 5.14). However, that approach presents a computational complexity that grows with the fourth power of the number of the states (10.1) and is thus often impractical. We shall content ourselves with the approximation given in (10.95) and describe an efficient algorithm to compute d_{min}.

10.5.2 An Algorithm to Compute d_{min}

Approximation (10.95) to the error probability shows that the minimum distance d_{min} defined in the preceding section plays a fundamental role in assessing the system performance. Besides the method already introduced in Chapter 7, which is valid only in the linear case, we present here an algorithm based on the one described by Saxena (1983) and Mulligan and Wilson (1984) to evaluate d_{min}. Considering the definition of distance (10.92) and the expression (10.77) for $y_\mathbf{a}(t)$, and remembering that $h(t; a, \sigma)$ has a duration of T seconds, we can write

$$
\begin{aligned}
d^2(\mathbf{a}_i, \mathbf{a}_j) &= \sum_{n=0}^{K-1} \int_{nT}^{(n+1)T} |h(t - nT; a_n^{(i)}, \sigma_n^{(i)}) - h(t - nT; a_n^{(j)}, \sigma_n^{(j)})|^2 \, dt \\
&\triangleq \sum_{n=0}^{K-1} d_n^2(a_n^{(i)}, \sigma_n^{(i)}; a_n^{(j)}, \sigma_n^{(j)}),
\end{aligned}
\tag{10.96}
$$

where the meaning of symbols should be obvious.

Understanding the algorithm is facilitated by considering a trellis with $N = M^L$ states $\{S_i\}_{i=1}^N$, which are the values assumed by the RVs σ_n, and with branches labeled with the chips $h(t - nT; a_n, \sigma_n)$, which are uniquely specified by the two states joined by the branch at hand. Thus, a pair of branches specifies one value of $d_n^2(\cdot)$ in (10.96) and the squared distance between two sequences (paths in the trellis) is obtained as a sum of the squared distances between the chips that label the corresponding branches. Finding d_{min} requires the analysis of all the possible pairs of paths in the trellis.

The algorithm is based on the construction and updating, step by step, of an

$M^L \times M^L$ matrix $\mathbf{D}^{(n)} = [\delta_{ij}^{(n)}]$, whose entry $\delta_{ij}^{(n)}$ represents the squared minimum distance between all pairs of paths diverging at the first step from the same initial state (whatever it is) and passing at the nth time instant through the states S_i and S_j. In Fig. 10.23, two such pairs of paths are drawn. Since $\delta_{ij}^{(n)} = \delta_{ji}^{(n)}$, it will be sufficient, at each step, to update only the entries δ_{ij} for $j \geq i$. This updating is done through a recursive algorithm that will be explained next. Note that the entries in the main diagonal of $\mathbf{D}^{(n)}$ represent distances between remerged paths (the "error events" defined in Section 7.5).

Let us now explain the main steps of the algorithm; the figures will refer to the case $M = 2$ for simplicity.

Step 1. For each state S_i, find the M states (the *predecessors*) from which a transition to S_i is possible and construct a table of these predecessors. Set $\delta_{ij} = -1$, for all i and $j \geq i$.

Step 2. For each pair of states (S_i, S_j), find the squared minimum distance between pairs of paths diverging from the same initial state (whatever it is) and reaching the states S_i and S_j in one step. Two of those paths are drawn in Fig. 10.24. Such a distance is $\delta_{ij}^{(1)}$.

Step 3. For both states of the given pair (S_i, S_j), find in the table obtained in step 1 the M predecessors $S_{i_1}, \ldots, S_{i_M}$ and $S_{j_1}, \ldots, S_{j_M}$ (see Fig. 10.25). There exist, in general, M^2 possible paths at the $(n-1)$th time instant that pass, at nth time instant, through the states S_i and S_j. Those paths pass through the pairs

$$(S_{i_1}, S_{j_1}), (S_{i_1}, S_{j_2}), \ldots, (S_{i_1}, S_{j_M}),$$

$$\vdots$$

$$(S_{i_M}, S_{j_1}), (S_{i_M}, S_{j_2}), \ldots, (S_{i_M}, S_{j_M}).$$

The minimum distance between all the paths passing through the pair (S_i, S_j) at the nth step is then computed as

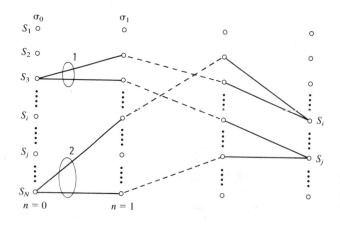

Figure 10.23 Two pairs of paths diverging at $n = 0$ and passing at the same time instant through the states S_i and S_j.

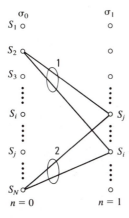

Figure 10.24 Paths starting from different states and reaching the same two states in one step.

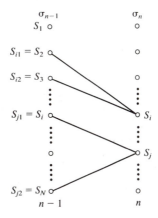

Figure 10.25 Predecessors of states S_i and S_j.

$$\delta_{ij}^{(n)} = \min \{ \delta_{i_1 j_1}^{(n-1)} + d^2(i_1 \to i, j_1 \to j),$$
$$\delta_{i_1 j_2}^{(n-1)} + d^2(i_1 \to i, j_2 \to j),$$
$$\cdot$$
$$\cdot$$
$$\cdot$$
$$\delta_{i_1 j_M}^{(n-1)} + d^2(i_1 \to i, j_M \to j),$$
$$\cdot$$
$$\cdot$$
$$\cdot$$
$$\delta_{i_M j_M}^{(n-1)} + d^2(i_M \to i, j_M \to j)\}. \tag{10.97}$$

In (10.97), the distances $\delta^{(n-1)}$ are known from the preceding step of the algorithm, whereas, for example, $d(i_1 \to i, j_1 \to j)$ denotes the distance between the two chips associated with the transitions $S_{i_1} \to S_i$ and $S_{j_1} \to S_j$. These can be computed at the beginning. When one of the previous distances, say $\delta_{lm}^{(n-1)}$, is equal to -1, the corresponding term in (10.97) disappears, since $\delta_{lm}^{(n-1)} = -1$ means that no pair of paths is allowed passing through S_1 and S_m at the $(n-1)$th step. When $S_i = S_j$, $\delta_{ii}^{(n)}$ represents the squared distance between two paths closed at the nth step on the state S_i (error event). Thus $\delta_{ii}^{(n)}$ is compared with $\delta_{ii}^{(n-1)}$, and, if less than that, it will take its place.

Step 4. If

$$\delta_{ij}^{(n)} < \min_i \delta_{ii}^{(n)} \tag{10.98}$$

for at least one pair (S_i, S_j), set $n = n + 1$ and go to step 3. If not, stop the algorithm and set

$$d_{\min}^2 = \min_i \delta_{ii}^{(n)}. \tag{10.99}$$

Condition (10.98) consists in verifying that all the paths still open at the nth time instant have distances greater than, or equal to, the minimum distance of error events, and guarantees that the latter is the required minimum distance.

The algorithm is quite efficient, both in terms of computer time and complexity. As a matter of fact, at each step the number of pairs of paths to be considered in the computation is constant, equal to M^2. If the algorithm finishes at the kth step, the total number of distances to be computed is at most equal to $kM^2 [N(N + 1)/2]$, where N is the number of states. Also, the amount of memory required by the algorithm is constant and does not grow with k. It requires an array of MN integer entries to store the predecessors of each state and two triangular matrices with $N(N + 1)/2$ real entries to store the old and updated versions of the matrix $\mathbf{D}^{(k)}$.

10.6 IDENTIFICATION AND EQUALIZATION OF NONLINEAR CHANNELS

In the preceding sections we have derived the form of the optimum (in the mean-square sense) linear symbol-by-symbol receiver and of the optimum unconstrained ML sequence receiver. Both receivers require knowledge of the channel. When the channel is not completely known a priori, a preliminary phase, before the transmission of information, must be devoted to channel identification. Moreover, the procedure has to be occasionally repeated as the channel characteristics vary with time. It is the same phenomenon encountered when dealing with adaptive equalization. Here it is complicated by the channel nonlinear behavior.

The conventional linear receiver, possibly equipped with an adaptive equalizer, does not attain the required performance in some applications. A typical example is represented by high-speed (over 10 kbits/s) data transmission over voiceband channels, where nonlinear distortion is one of the most significant factors preventing further increase of the data rate. In this application, the nonlinearity can be modeled as a low-order (second or third) memoryless device embedded in a network where sharp linear filtering operations also take place. Consequently, the overall channel is a nonlinear system with a long memory (L of the order of some tens). This long memory makes the ML sequence receiver impractically complex.

Another example is represented by the previously described satellite channel, where the situation is somewhat reversed. In this case, we have a strong nonlinearity with a short memory (L of the order of a few units). The problem of applying the ML sequence receiver has more to do with the speed of digital integrated circuits, which must cope with the high symbol rate (up to hundreds of Mbits/s), than with the complexity related to the channel memory.

We shall describe in this section a receiver that is intermediate (in terms of complexity and optimality) between the conventional linear receiver and the ML sequence receiver. It can be used indifferently for the discrete channel identification purposes and/or as a nonlinear equalizer. Its structure can easily be made adaptive. The form of the receiver is a nonlinear extension of the tapped-delay-line (TDL) mean-square-error (MSE) equalizer described in Chapter 8. Its nonlinear structure is suggested by the Volterra model of

Section 10.3. As such, it can be used to estimate the parameters of a discrete Volterra model of the channel. The equalizer, which can be seen as a structure-constrained optimum (in the mean-square sense) symbol-by-symbol receiver, will be described in its simpler version. It is intended that some of the refinements to the basic TDL equalizer presented in Chapter 8 (such as decision feedback and distortion cancellation) can be fruitfully applied in this case. The description of the equalizer, as well as the numerical results, refers to a system employing CPSK.

Recalling the expression (10.8) of the signal $y(t)$ (see Fig. 10.13), we can write the samples r_n that form the received sampled sequence (r_n) as

$$r_n = \sum_{m=0}^{\infty} \sum_{n_1} \cdots \sum_{n_{2m+1}} a_{n_1} \cdots a_{n_{m+1}} a^*_{n_{m+2}} \cdots a^*_{n_{2m+1}}$$

$$\cdot K_{2m+1}(n - n_1, \ldots, n - n_{2m+1}) + v_n, \tag{10.100}$$

where v_n is the noise sample.

The structure of (10.100) reflects how the channel output depends both on the channel (through the Volterra coefficients) and on the information symbols. In particular, the symbol structure of PSK modulation ($a_n = e^{j\varphi_n}$) results in insensitivity to certain kinds of nonlinearities. In fact, since $a_n a^*_n = 1$, it is apparent from (10.100) that certain Volterra coefficients K_{2m+1} will contribute to nonlinearities of order less than $2m + 1$. To be more specific, consider first the third-order Volterra coefficients $K_3(n - n_1, n - n_2, n - n_3)$; for $n_1 = n_3$ or $n_2 = n_3$, the channel nonlinearities reflected by these coefficients will not affect a PSK signal, because

$$a_{n_1} a_{n_2} a^*_{n_3} K_3(n - n_1, n - n_2, n - n_3)$$

$$= \begin{cases} a_{n_2} K_3(n - n_1, n - n_2, n - n_3), & \text{if } n_1 = n_3, \\ a_{n_1} K_3(n - n_1, n - n_2, n - n_3), & \text{if } n_2 = n_3, \end{cases} \tag{10.101}$$

and the only contribution is to the linear part of the channel. Similar considerations on the higher-order coefficients show that some of them contribute to the linear part, others only to the third-order nonlinearity, and so on. These considerations can be further pursued if we observe that, for an M-ary CPSK, a_n^M is a constant, which results in a further reduction of sensitivity of PSK to certain nonlinearities. This leads to the noteworthy conclusion that, for CPSK, certain nonlinearities need only a linear compensation, while others affect the signal to a lower degree than other modulation schemes. The overall effect is a further reduction of the number of Volterra coefficients to be taken into account in the channel model.

Example 10.5

As an example of modeling a CPSK nonlinear satellite channel using a Volterra series, consider again the scheme of Fig. 10.13. The transmission filter $s(t)$ includes a rectangular shaping filter and a fourth-order Butterworth with 3-dB bandwidth $1.7/T$. The TWT is described by the characteristics of Fig. 10.12 and is driven at saturation by the PSK symbols. The receiving filter $u(t)$ is a second-order Butterworth with 3-dB bandwidth $1.1/T$. The computed Volterra coefficients for this channel, after deletion of the smallest, are shown in Table 10.1. The thresholds below which we neglected the Volterra coefficients are equal to 0.001 and 0.005 for the linear and nonlinear parts, respectively. After a further reduction

TABLE 10.1 Computed Volterra coefficients for the case of Example 10.5

Linear part		
$K_1(0)$ $\quad=\quad$ 3.4 $\quad+j0.381$	$K_5(0,0,0,0,0)$ $\quad=\quad$ 3.92 $\;-j2.21$	
$K_1(1)$ $\quad=\quad$ 0.052 $+j0.006$	$K_5(1,0,0,0,0)$ $\quad=-0.69$ $\;+j0.388$	
$K_1(2)$ $\quad=-0.048-j0.005$	$K_5(1,0,0,1,0)$ $\quad=\quad$ 0.236 $-j0.133$	
$K_1(3)$ $\quad=\quad$ 0.178 $+j0.020$	$K_5(1,1,0,1,0)$ $\quad=\quad$ 0.070 $-j0.066$	
	$K_5(1,1,0,1,1)$ $\quad=\quad$ 0.074 $-j0.022$	
$K_3(0,0,0)$ $\;=-4.296+j1.741$	$K_5(1,1,1,1,1)$ $\quad=\quad$ 0.039 $-j0.022$	
$K_3(1,0,0)$ $\;=\quad$ 0.388 $-j0.137$	$K_5(2,0,0,0,0)$ $\quad=\quad$ 0.059 $-j0.033$	
$K_3(1,0,1)$ $\;=-0.230+j0.081$	$K_5(2,2,2,2,2)$ $\quad=-0.053+j0.030$	
$K_3(1,1,1)$ $\;=-0.105+j0.037$	$K_5(3,0,0,0,0)$ $\quad=\quad$ 0.349 $-j0.197$	
$K_3(2,0,0)$ $\;=-0.056+j0.020$	$K_5(3,0,0,3,0)$ $\quad=\quad$ 0.118 $-j0.066$	
$K_3(2,2,2)$ $\;=\quad$ 0.074 $-j0.026$	$K_7(0,0,0,0,0,0,0)=-1.14$ $\;+j0.764$	
$K_3(3,0,0)$ $\;=-0.384+j0.136$	$K_7(1,0,0,0,0,0,0)=\quad$ 0.309 $-j0.207$	
$K_3(3,0,3)$ $\;=-0.033+j0.029$	$K_7(1,0,0,0,1,0,0)=-0.106+j0.072$	
	$K_7(3,0,0,0,0,0,0)=-0.107+j0.072$	
	$K_7(3,0,0,0,3,0,0)=-0.043+j0.029$	
Third-order nonlinearities		
$K_3(0,0,1)$ $\quad=\quad$ 0.194 $-j0.068$	$K_5(0,0,0,3,0)$ $\quad=\quad$ 0.233 $-j0.131$	
$K_3(0,0,3)$ $\quad=-0.192+j0.068$	$K_5(1,0,0,0,0)$ $\quad=\quad$ 0.118 $-j0.066$	
$K_3(1,1,0)$ $\quad=-0.115+j0.041$	$K_5(1,1,1,1,0)$ $\quad=\quad$ 0.049 $-j0.028$	
$K_3(3,3,0)$ $\quad=-0.041+j0.015$	$K_5(3,3,0,0,0)$ $\quad=\quad$ 0.059 $-j0.033$	
$K_5(0,0,0,1,0)=-0.460+j0.259$	$K_7(0,0,0,0,1,0,0)=\quad$ 0.231 $-j0.156$	
$K_5(0,0,0,2,0)=\quad$ 0.039 $-j0.022$	$K_7(0,0,0,0,3,0,0)=-0.081+j0.054$	
	$K_7(1,1,0,0,0,0,0)=-0.053+j0.036$	
Fifth-order nonlinearities		
$K_5(0,0,0,1,1)=0.039-j0.022$		

that takes into account the structure of PSK symbols, as previously mentioned, the surviving coefficients are shown in Table 10.2. It is seen that the size of the reduction is relevant. $\square$

The Volterra series representation (10.100) provides us with a basis for representing a general signal processor in the same form. In fact, it seems quite natural to choose, for the general discrete-time processor, the structure suggested by (10.100) after truncating the infinite sums. This processor can be implemented using a TDL, a nonlinear combiner, a number of complex multipliers, and a summing bus, as shown in Fig. 10.26 for the special case $N_1 = N_2 = N_3$.

Thus, assuming a $(2k + 1)$th order equalizer (i.e., a processor with nonlinearities up to the order $2k + 1$), its output sequence (z_n) is related to the input (received) sequence (r_n) by the relationship

$$z_n = \sum_{n_1=0}^{N_1-1} r_{n-n_1} c_1(n_1) + \sum_{n_1=0}^{N_1-1}\sum_{n_2=0}^{N_2-1}\sum_{n_3=0}^{N_3-1} r_{n-n_1} r_{n-n_2} r^*_{n-n_3}$$

(10.102)

$$\cdot \, c_3(n_1, n_2, n_3) + \cdots + \sum_{n_1=0}^{N_1-1} \cdots \sum_{n_{2k+1}=0}^{N_{2k+1}-1} r_{n-n_1} \cdots r^*_{n-n_{2k+1}}$$

$$\cdot \, c_{2k+1}(n_1, \ldots, n_{2k+1}),$$

where N_i represents the number of values assumed by the summation index n_i.

Equation (10.102) shows that the output is related to the input by a finite set of constants $c_1(n_1)$, $c_3(n_1, n_2, n_3)$, $\ldots$, $c_{2k+1}(n_1, \ldots, n_{2k+1})$, which will be referred to hereafter as the *tap weights* of the equalizer, in analogy with the linear case described in Chapter 8. Thus, the design of the equalizer is equivalent to the choice of the tap weights. If we arrange them in the column vector

$$\mathbf{c} = [c_1(0), \ldots, c_1(N_1 - 1), c_3(0, 0, 0), \ldots,$$

$$c_3(N_1 - 1, N_2 - 1, N_3 - 1), \qquad (10.103)$$

$$\ldots, c_{2k+1}(0, \ldots, 0), \ldots, c_{2k+1}(N_1 - 1, \ldots, N_{2k+1} - 1)]'$$

and we define

$$\mathbf{r}_n \triangleq [r_n, r_{n-1}, \ldots, r_{n-N_1+1}, \ldots, r_{n-N_1+1} \cdots r^*_{n-N_{2k+1}+1}]', \qquad (10.104)$$

the input–output relationship for the equalizer can be written in vector form as

$$\mathbf{z}_n = \mathbf{c}' \, \mathbf{r}_n. \qquad (10.105)$$

The input–output relationship (10.105) governing the behavior of the equalizer is linear in the tap-weight vector $\mathbf{c}$; thus, the methods used in the linear case can also be applied

TABLE 10.2 Reduced Volterra coefficients in the case of 4-CPSK modulation (Example 10.5)

Linear part
$K_1(0) = 1.22 + j0.646$ $K_1(1) = 0.063 - j0.001$ $K_1(2) = -0.024 - j0.014$ $K_1(3) = 0.036 + j0.031$
Third-order nonlinearities
$K_3(0, 0, 2) = 0.039 - j0.022$ $K_3(3, 3, 0) = 0.018 - j0.018$ $K_3(0, 0, 1) = -0.035 + j0.035$ $K_3(0, 0, 3) = -0.040 - j0.009$ $K_3(1, 1, 0) = -0.001 - j0.017$
Fifth-order nonlinearities
$K_5(0, 0, 0, 1, 1) = 0.039 - j0.022$

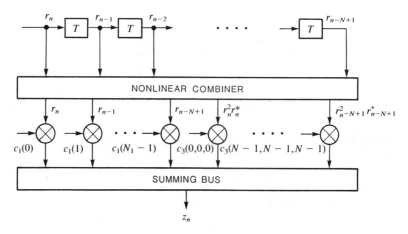

Figure 10.26 Third-order nonlinear equalizer.

here. We want to design the processor so as to get, at its output, a sequence of samples approximating, under an MSE criterion, the sequence of channel input symbols a_{n-D}, where D denotes the allowed delay. Thus, we need to minimize the MSE

$$\mathscr{E} \triangleq E|\mathbf{c}' \, \mathbf{r}_n - a_{n-D}|^2 = \mathbf{c}^\dagger \mathbf{R} \, \mathbf{c} - 2\mathscr{R}[\mathbf{c}^\dagger \mathbf{g}] + E|a_{n-D}|^2 \tag{10.106}$$

with respect to $\mathbf{c}$, where

$$\mathbf{R} \triangleq E\,[\mathbf{r}_n^* \, \mathbf{r}_n'], \tag{10.107}$$

$$\mathbf{g} \triangleq E\,[a_{n-D}\,\mathbf{r}_n^*]. \tag{10.108}$$

The optimum $\mathbf{c}$ can thus be obtained by solving the set of linear equations

$$\mathbf{R} \, \mathbf{c} = \mathbf{g}, \tag{10.109}$$

which admits the solution

$$\mathbf{c}_{\text{opt}} = \mathbf{R}^{-1}\,\mathbf{g} \tag{10.110}$$

provided that $\mathbf{R}$ is positive definite. It is seen from the definition of $\mathbf{R}$ that this condition is fulfilled if, for any arbitrary complex vector $\mathbf{b}$, we have

$$\mathbf{b}^\dagger \mathbf{R} \, \mathbf{b} = E|\mathbf{b}^\dagger \mathbf{r}_n|^2 > 0. \tag{10.111}$$

The RHS of (10.111) can be thought of as the average power of the output of an equalizer with tap weights $\mathbf{b}^*$. This power cannot be zero, due to the presence of the noise added to the samples entering the equalizer. In the absence of noise, (10.111) can only be zero if $\mathbf{b}^\dagger \mathbf{r}_n = 0$ (i.e., the entries of $\mathbf{r}_n$ are linearly dependent).

The solution (10.110) for the optimum tap-weight vector requires knowledge of $\mathbf{R}$ and $\mathbf{g}$. Computation of the averages involved in their definitions (10.107) and (10.108) can be done using the method described in Section 10.3 to evaluate the moments of the RVs representing ISI. In particular, it is possible to evaluate by exhaustive enumeration the part of the averages depending on information symbols and, analytically, the part depending on Gaussian noise. These procedures, of course, allow only off-line computation of $\mathbf{R}$ and $\mathbf{g}$ and are not suitable for the equalizer working in an adaptive mode. However, the algorithms described for adaptive linear equalizers and, in particular, the stochastic-gradient algorithm of Chapter 8 can also be applied fruitfully in this case. Its application leads to the following recursion for the taps updating:

$$\mathbf{c}^{(n+1)} = \mathbf{c}^{(n)} - \alpha(z_n - a_{n-D}) \, \mathbf{r}_n^*. \tag{10.112}$$

Example 10.5 (continued)

Consider again the nonlinear channel of Example 10.5. We want to study the effect of a nonlinear equalizer cascaded to the channel. An important question is worth mentioning at this point. For a given complexity (i.e., a given number of first-, third-, . . . , nth-order coefficients in the TDL), what is the best allocation of those coefficients; that is, how can we choose the range of indexes n_1, n_2, n_3, . . . in (10.102)?

Our experience, which has unfortunately not yet been confirmed by the theory, is that a good use of the allowable complexity consists in allocating the TDL taps such that they make a reasonable copy of the channel Volterra coefficients. In other words, it is convenient to introduce a tap, say $c_3(i, j, k)$, if the corresponding Volterra coefficient $K_3(i, j, k)$ of the channel model has a relevant magnitude. This way of matching the structure of the nonlinear TDL to that of the channel requires previous knowledge of the channel structure. Thus, the first step consists of channel identification using the model of Fig. 10.26. The system behavior has been simulated as in Fig. 10.27. A stochastic-gradient algorithm was used to iteratively modify the coefficients of the nonlinear TDL, with the goal of minimizing $E[|r_n - \hat{r}_n|^2]$, the mean-square difference between the true sample at the channel output and the sample generated by the channel estimator.

The convergence of the identification process is described through a sequence (ξ_k) of running averages of $|r_n - \hat{r}_n|^2$ evaluated over successive blocks of K symbols. If the parameter α is chosen in the field of values allowing the convergence of the algorithm, the sequence (ξ_k) will decrease to the minimum value achievable with the chosen complexity of the channel model.

For the sake of comparison, in Fig. 10.28, (ξ_k) is plotted in the linear case (i.e., when the TWT is not present in the channel and the channel estimator is a linear TDL with 10 taps). In the nonlinear case, we have considered a nonlinear TDL with Volterra coefficients of first, third, and fifth order, allowing a memory of 10 for the linear part, 4 for the third-order part, and 3 for the fifth order. The Volterra coefficients included in the channel estimator are reported in Table 10.3.

In Fig. 10.29 the behavior of (ξ_k) is shown, using $K = 100$. For the curve labeled with Ⓐ, the initial choice of the coefficient α was taken outside the convergence interval of the algorithm. Thus the curve shows an initial divergence that ends when the value of α is suitably reduced. Curves Ⓑ and Ⓒ differ in the choice of the coefficient α governing the updating of nonlinear coefficients. Comparison with the linear case of Fig. 10.28 shows that the MSE settles to a value sensibly lower in the linear case and that the convergence is faster. This happens because of the following:

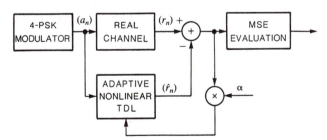

Figure 10.27 Block diagram for the simulation of the channel identifier.

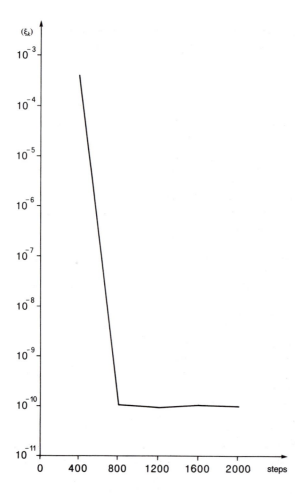

Figure 10.28 Identification of the channel of Example 10.5 in the absence of the TWT (linear channel). The curve presents the behavior of the MSE estimated through a running time average made over blocks of $K = 400$ symbols. The linear equalizer has 10 taps, and $\alpha = 0.1$.

TABLE 10.3 Volterra coefficients considered in the channel estimator

Linear $(N_1 = 10)$	Third order $(N_3 = 22)$		Fifth order $(N_5 = 15)$	
$K_1(0)$	$K_3(0, 0, 1)$,	$K_3(0, 3, 1)$	$K_5(0, 0, 0, 1, 1)$,	$K_5(0, 0, 0, 1, 2)$
$K_1(1)$	$K_3(2, 2, 0)$,	$K_3(0, 0, 2)$	$K_5(0, 0, 0, 2, 2)$,	$K_5(0, 0, 1, 2, 2)$
$K_1(2)$	$K_3(0, 0, 3)$,	$K_3(1, 1, 0)$	$K_5(0, 0, 2, 1, 1)$,	$K_5(0, 1, 1, 2, 2)$
$\vdots$	$K_3(2, 2, 3)$,	$K_3(0, 1, 2)$	$K_5(0, 2, 2, 1, 1)$,	$K_5(1, 1, 1, 0, 0)$
$K_1(9)$	$K_3(1, 1, 2)$,	$K_3(2, 3, 0)$	$K_5(1, 1, 1, 0, 2)$,	$K_5(1, 1, 1, 2, 2)$
	$K_3(0, 1, 3)$,	$K_3(1, 1, 3)$	$K_5(1, 1, 2, 0, 0)$,	$K_5(1, 2, 2, 0, 0)$
	$K_3(2, 3, 1)$,	$K_3(0, 2, 1)$	$K_5(2, 2, 2, 0, 0)$,	$K_5(2, 2, 2, 0, 1)$
	$K_3(1, 2, 0)$,	$K_3(3, 3, 0)$	$K_5(2, 2, 2, 1, 1)$	
	$K_3(0, 2, 3)$,	$K_3(1, 2, 3)$		
	$K_3(3, 3, 1)$,	$K_3(1, 0, 0)$		
	$K_3(3, 3, 2)$,	$K_3(1, 3, 2)$		

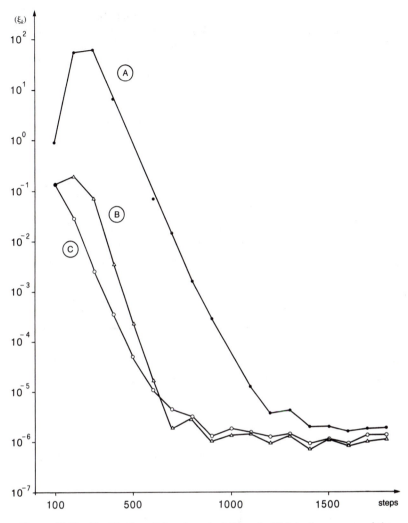

Figure 10.29 Identification of the channel of Example 10.5 in the presence of the TWT (nonlinear channel). The curves present the behavior of the MSE estimated through a running time average made over blocks of $K = 100$ symbols. The curve with black dots refers to a case in which the initial choice of the parameter controlling the gradient algorithm is made outside the convergence interval and then reduced so as to have convergence. The curve with triangles has a value of the parameter controlling the updating of the linear part of the equalizer equal to 0.1, and a value of the parameter controlling the nonlinear part equal to 0.04. The curve with white dots has both parameters equal to 0.01. The Volterra coefficients characterizing the channel are those of Table 10.3.

(i) In the linear case the complexity of the TDL ($N = 10$) is lower. This allows the use of larger α and accelerates the convergence.

(ii) The structure of the estimator in the linear case is much closer to the true channel than in the nonlinear case. In the latter, other nonlinear coefficients should be added to the model in order to decrease the steady-state MSE.

Sec. 10.6   Identification and Equalization of Nonlinear Channels   **571**

Let us now proceed to the design of the equalizer. Having chosen the structure of the nonlinear TDL on the basis of the estimate of the channel, the optimum values of the tap weights $c_i(\cdot)$ of the equalizer can be found through the stochastic gradient algorithm (10.112) aiming at minimizing the MSE $\mathcal{E}$ defined in (10.106). Also, in this case we have used a running average $(\mathcal{E}_k)$ over K symbols, which represents a time average of $|r_n - a_n|^2$.

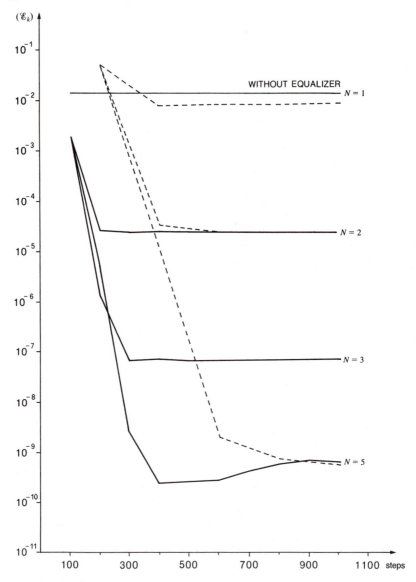

Figure 10.30 Equalization of the channel of Example 10.5 in the absence of the TWT (linear channel). The curves present the behavior of the MSE estimated through a running time average made over blocks of $K = 100$ symbols. The number N of taps of the linear equalizer is a parameter. The solid curves refer to a value of the parameter $\alpha = 0.1$, whereas the dashed curves have $\alpha = 0.05$.

The results relative to $(\mathscr{E}_k)$ in the linear case are shown in Fig. 10.30, for the sake of comparison, for various numbers of the TDL taps. The two sets of curves refer to different values of α and K. The continuous curves have been obtained with $\alpha = 0.1$ and $K = 100$, whereas the dashed ones have $\alpha = 5 \cdot 10^{-2}$ and $K = 200$.

Turning now to the nonlinear case, we can examine the results presented in Fig. 10.31. The set of curves refers to equalizers of increasing complexity cascaded with the nonlinear channel. They suggest the following observations:

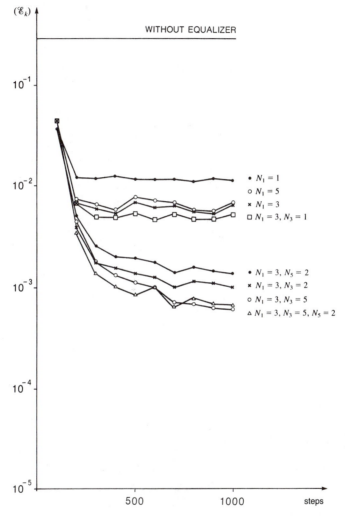

Figure 10.31 Equalization of the channel of Example 10.5 in the presence of the TWT (nonlinear channel). The curves present the behavior of the MSE estimated through a running time average made over blocks of $K = 100$ symbols. The curves marked with ''linear'' refer to a linear equalizer, whereas the ones marked with ''nonlinear'' refer to a nonlinear equalizer. The numbers of linear and nonlinear taps are parameters of the curves. The value of α is equal to 0.1 for the linear taps, and equal to 0.01 for the nonlinear taps. The Volterra coefficients characterizing the channel are those of Table 10.3.

(i) Linear taps cannot compensate for the distortion, since after addition of a certain number of them the MSE will not show any significant reduction.

(ii) The addition of a few nonlinear taps allows a good reduction of the MSE.

This is also shown in Table 10.4, where the equalizer performance is given in terms of η, defined as the ratio between the signal power at TWT saturation and the MSE $\mathscr{E}$ at the output of the equalizer in the absence of noise. $\square$

TABLE 10.4 Signal-to-distortion ratio achievable with linear and nonlinear equalizers in the absence of noise

Equalizer tap weights	η (dB)
$c_1(0)$ to $c_1(1)$	22.5
$c_1(0)$ to $c_1(4)$	23
$c_1(0)$ to $c_1(10)$	23
$c_1(0), c_1(1)$ $c_3(0, 0, 1), c_3(0, 0, 3)$ $c_3(1, 1, 0)$	27
$c_1(0), c_1(1)$ $c_3(0, 0, 1), c_3(0, 0, 3)$ $c_3(1, 1, 0), c_3(3, 3, 0)$ $c_3(0, 0, 2)$	29
$c_1(0), c_1(4)$ $c_3(0, 0, 1), c_3(0, 0, 3)$ $c_3(1, 1, 0), c_3(3, 3, 0)$ $c_3(0, 0, 2), c_5(0, 0, 0, 1, 1)$	37

BIBLIOGRAPHICAL NOTES

Volterra series were first studied by the Italian mathematician Vito Volterra around 1880 as a generalization of the Taylor series of a function. His work in this area is summarized in Volterra (1959). The application of Volterra series to the analysis of nonlinear systems with memory was suggested by Norbert Wiener. Extensive treatments of Volterra series as applied to the description of nonlinear systems can be found in Schetzen (1980) and Rugh (1981). A basic work in this area is represented by the paper of Flake (1963), whereas a recent and good review is found in Schetzen (1981). The modeling and performance evaluation of bandpass nonlinear digital systems follows closely Benedetto, Biglieri, and Daffara (1976, 1979). Discrete Volterra series applied to sampled-data systems are analyzed in Barker and Ambati (1972). Inverse systems of nonlinear systems represented by continuous and discrete Volterra series are investigated in Schetzen (1976) and Wakamatsu (1981). The problem of evaluating the performance of a nonlinear digital transmission system has received considerable attention, particularly as applied to satellite links. The interested reader is invited to scan the extensive reference list in Benedetto, Biglieri, and Daffara (1979).

The derivation of the optimum receiving filter follows that presented in Biglieri, Elia, and Lo Presti (1984). Different approaches have been pursued by Fredricsson (1975) and Mesiya, McLane, and Campbell (1978).

The model and the approach followed in the derivation of the ML sequence receiver are original. Based on an analytical model of the bandpass nonlinearity, Mesiya, McLane, and Campbell (1977) have derived the ML sequence receiver for a binary PSK signal. An ML receiver taking into account also the uplink noise in satellite links has been obtained by Benedetto, Biglieri, and Omura (1981). The Volterra series technique has been applied by Falconer (1978) and, previously, by Thomas (1971) to the design of adaptive nonlinear receivers. The use of orthogonal Volterra series to achieve rapid adaptation in connection with a Volterra series approach is proposed in Biglieri, Gersho, Gitlin, and Lim (1984). Some of the results presented here in connection with nonlinear equalization are original. The treatment follows closely that of Benedetto and Biglieri (1983).

PROBLEMS†

*10.1. Good analytical approximations for typical AM/AM and AM/PM characteristics of TWTs are represented by the expressions

$$F(r) = \frac{\alpha_a r}{1 + \beta_a r^2}, \tag{P1}$$

$$\phi(r) = \frac{\alpha_\phi r^2}{1 + \beta_\phi r^2} \tag{P2}$$

(see Saleh, 1981), where the coefficients α_a, β_a, α_ϕ, and β_ϕ are found by fitting (P1) and (P2) to the experimental data through a minimum mean-square-error procedure. Apply this procedure to approximate the curves of Fig. 10.4 and draw the resulting expressions of $F(r)$ and $\phi(r)$.

*10.2. Consider a channel formed by cascading a second-order Butterworth filter with normalized equivalent noise bandwidth $B_{eq}T = 2.0$ with the TWT of Fig. 10.4. Using a polynomial approximation of the complex TWT characteristics, $F(r) \exp[j\phi(r)]$, with powers up to the seventh, compute the discrete Volterra coefficients (10.16) discarding those smaller than $10^{-3} k_1(t_0)$.

10.3. Derive the recurrent formulas (10.40) for the powers ξ^l of discrete Volterra coefficients.

10.4. Derive the recurrent formulas for the powers $(\xi^*)^m$ of discrete Volterra coefficients using the derivation in Problem 10.3.

10.5. Prove the result (10.42).

10.6. A wide-sense stationary Gaussian process $n(t)$ can be represented as

$$n(t) = \sum_i \beta_i b_i(t), \tag{P3}$$

† The problems marked with an asterisk should be solved with the aid of a computer.

where β_i are unit-variance independent Gaussian RVs and $b_i(t)$ are appropriately chosen deterministic functions. Evaluate the functions $b_i(t)$ in the case of a power spectral density of $n(t)$ equal to G_0 in the interval $(-F, F)$, and zero outside. *Hint*: If $G_n(f)$ is the power spectral density of the process, the functions $b_i(t)$ are obtained as inverse Fourier transforms of $l_i(f)G_n(f)$, where $l_i(f)$ form a complete sequence of orthonormal functions in the Hilbert space with norm

$$\|l\|^2 = \int_{-\infty}^{\infty} |l(f)|^2 G_n(f) \, df. \tag{P4}$$

10.7. Repeat the computations of the moments (10.42) in the case of uplink noise represented as in (10.45).

10.8. Using variational calculus techniques (Appendix C), derive the integral equation (10.53) that minimizes the mean-square error (10.46).

10.9. Compute the average $E\{a_n H(f; a_k, \sigma_k)\}$, obtaining the result (10.58).

10.10. Compute the average $E\{H(f; a_n, \sigma_n)H(f - k/T; a_{n+l}, \sigma_{n+l}\}$, obtaining the result (10.63).

*10.11. Apply the algorithm described in Section 10.5.2 to find $d_{\min}$ for the nonlinear system of Example 10.1.

*10.10. Repeat the simulation of Example 10.5 by setting the 3-dB bandwidth of the transmission filter equal to $2/T$.

APPENDIX A

Useful Formulas and Approximations

A.1 ERROR FUNCTION AND COMPLEMENTARY ERROR FUNCTION

Definitions:

$$erf(x) \triangleq \frac{2}{\sqrt{\pi}} \int_0^x e^{-t^2}\, dt, \tag{A.1}$$

$$erfc(x) \triangleq \frac{2}{\sqrt{\pi}} \int_x^\infty e^{-t^2}\, dt = 1 - erf(x). \tag{A.2}$$

Relation with the normal distribution having mean m and variance σ^2:

$$\frac{1}{\sqrt{2\pi}\sigma} \int_{-\infty}^x e^{-(t-m)^2/2\sigma^2} dt = \frac{1}{2}\left\{1 + erf\left(\frac{x-m}{\sqrt{2}\,\sigma}\right)\right\}. \tag{A.3}$$

The function $erfc(x)$ admits the following asymptotic expansion ($x \to \infty$):

$$erfc(x) \sim \frac{e^{-x^2}}{\sqrt{\pi}\,x}\left\{1 + \sum_{i=1}^\infty (-1)^i \frac{1 \cdot 3 \cdot 5 \cdots (2i-1)}{(2x^2)^i}\right\}. \tag{A.4}$$

Table A.1 shows a comparison, for values of x between 3 and 5, of the approximation (A.4) truncated to its first term with the exact value of $erfc(x)$. Also, relative errors are indicated.

Using relation (A.3) and definitions (A.1) and (A.2), it is straightforward to compute the following probabilities related to the Gaussian RV ξ:

$$P\{\xi < x\} = \frac{1}{2}\left\{1 + erf\left(\frac{x-m}{\sqrt{2}\,\sigma}\right)\right\}, \tag{A.5}$$

$$P\{\xi > x\} = \frac{1}{2}\, erfc\left(\frac{x-m}{\sqrt{2}\,\sigma}\right), \tag{A.6}$$

$$P\{|\xi| < x\} = \frac{1}{2}\left\{erf\left(\frac{x-m}{\sqrt{2}\,\sigma}\right) + erf\left(\frac{x+m}{\sqrt{2}\,\sigma}\right)\right\}, \tag{A.7}$$

$$P\{|\xi| > x\} = \frac{1}{2}\left\{erfc\left(\frac{x-m}{\sqrt{2}\,\sigma}\right) + erfc\left(\frac{x+m}{\sqrt{2}\,\sigma}\right)\right\}. \tag{A.8}$$

TABLE A.1 Comparison between the values of $erfc(x)$ and its approximation

x	$A \triangleq erfc(x)$	$B \triangleq \dfrac{\exp(-x^2)}{\sqrt{\pi}\, x}$	$\dfrac{B - A}{A} \cdot 100$
3.00	0.221E−04	0.232E−04	0.506E+01
3.25	0.430E−05	0.449E−05	0.437E+01
3.50	0.743E−06	0.771E−06	0.380E+01
3.75	0.114E−06	0.118E−06	0.334E+01
4.00	0.154E−07	0.159E−07	0.295E+01
4.25	0.185E−08	0.190E−08	0.263E+01
4.50	0.197E−09	0.201E−09	0.236E+01
4.75	0.185E−10	0.189E−10	0.213E+01
5.00	0.154E−11	0.157E−11	0.193E+01

Figure A.1 gives a useful plot of the function $erfc(x)$.

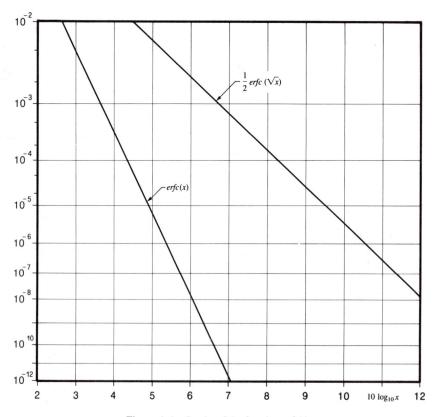

Figure A.1 Graphs of the function $erfc(\cdot)$.

A.2 THE MODIFIED BESSEL FUNCTION $I_0(\cdot)$

Definition:

$$I_0(x) \triangleq \frac{1}{\pi} \int_0^\pi e^{\pm x \cos\vartheta} d\vartheta = \frac{1}{\pi} \int_0^\pi \cosh(x \cos\vartheta) d\vartheta. \tag{A.9}$$

Asymptotic expansion:

$$I_0(x) \sim \frac{e^x}{\sqrt{2\pi x}} \left\{ 1 + \frac{1}{8x} + \frac{9}{2!(8x)^2} + \frac{9 \cdot 25}{3!(8x)^3} + \cdots \right\}. \tag{A.10}$$

A.3 MARCUM Q-FUNCTION AND RELATED INTEGRALS (Schwartz, Bennett, and Stein, 1966)

Definitions and elementary properties:

$$Q(a, b) \triangleq \int_b^\infty \exp\left(-\frac{a^2 + x^2}{2}\right) I_0(ax) x \, dx,$$

$$Q(0, b) = \exp\left(-\frac{b^2}{2}\right), \tag{A.11}$$

$$Q(a, 0) = 1.$$

Asymptotic expansion, valid for $b \gg 1$ and $b \gg b - a$:

$$Q(a, b) \sim \frac{1}{2} \, erfc\left(\frac{b - a}{\sqrt{2}}\right). \tag{A.12}$$

Symmetry and antisymmetry relations:

$$Q(a, b) + Q(b, a) = 1 + \exp\left(-\frac{a^2 + b^2}{2}\right) I_0(ab), \tag{A.13}$$

$$Q(a, a) = \frac{1}{2}[1 + e^{-a^2} I_0(a^2)], \tag{A.14}$$

$$1 + Q(a, b) - Q(b, a) = \frac{b^2 - a^2}{b^2 + a^2} \int_{(a^2+b^2)/2}^\infty e^{-y} I_0\left(\frac{2aby}{a^2 + b^2}\right) dy, \quad b > a > 0. \tag{A.15}$$

Asymptotic expansions, valid for $b \gg 1$, $a \gg 1$, $b \gg b - a > 0$:

$$1 + Q(a, b) - Q(b, a) \sim erfc\left(\frac{b - a}{\sqrt{2}}\right), \tag{A.16}$$

$$Q(a, b) + Q(b, a) - 1 \sim \frac{\exp[-(b - a)^2/2]}{\sqrt{2\pi ab}}, \tag{A.17}$$

$$Q(a, b) \sim \frac{1}{2}\left\{ erfc\left(\frac{b - a}{\sqrt{2}}\right) + \frac{\exp[-(b - a)^2/2]}{\sqrt{2\pi ab}} \right\}. \tag{A.18}$$

For the numerical computation of the Q function, the reader is referred to Brennan and Reed (1965), Pent (1968), and Robertson (1969).

A.4 PROBABILITY THAT ONE RICE-DISTRIBUTED RV EXCEEDS ANOTHER ONE (Schwartz, Bennett, and Stein, 1966)

Given a pair (ξ_1, ξ_2) of independent Rice-distributed RVs with pdf

$$f_{\xi_k}(x_k) = \frac{x_k}{\sigma_k^2} \exp\left(-\frac{a_k^2 + x_k^2}{2\sigma_k^2} \right) I_0\left(\frac{a_k x_k}{\sigma_k^2} \right), \qquad a_k > 0, \quad x_k > 0, \quad k = 1, 2, \tag{A.19}$$

the probability $P\{\xi_2 > \xi_1\}$ that one exceeds the other can be expressed by one of the following equivalent forms:

$$P\{\xi_2 > \xi_1\} = Q(\sqrt{a}, \sqrt{b}) - \frac{v^2}{1 + v^2} \exp\left(-\frac{a+b}{2} \right) I_0(\sqrt{ab}), \tag{A.20}$$

$$P\{\xi_2 > \xi_1\} = \frac{v^2}{1 + v^2} [1 - Q(\sqrt{b}, \sqrt{a})] + \frac{1}{1 + v^2} Q(\sqrt{a}, \sqrt{b}), \tag{A.21}$$

$$P\{\xi_2 > \xi_1\} = \frac{1}{2} [1 - Q(\sqrt{b}, \sqrt{a}) + Q(\sqrt{a}, \sqrt{b})]$$

$$-\frac{1}{2} \frac{v^2 - 1}{v^2 + 1} \exp\left(-\frac{a+b}{2} \right) I_0(\sqrt{ab}), \tag{A.22}$$

where

$$a \triangleq \frac{a_2^2}{\sigma_1^2 + \sigma_2^2}, \qquad b \triangleq \frac{a_1^2}{\sigma_1^2 + \sigma_2^2}, \qquad v \triangleq \frac{\sigma_1}{\sigma_2}.$$

APPENDIX **B**

Some Facts from Matrix Theory

In this appendix we collect together, for ease of reference, some basic results from matrix theory that are needed throughout the book, and in particular in Chapter 8, where extensive use of matrix notations is made. We assume that the reader has had previous exposure to matrix calculus. Thus, we focus our attention mainly on the results that are specifically needed in the text, whereas more elementary material is either skipped or included for reference's sake only.

B.1 BASIC MATRIX OPERATIONS

A real (complex) $N \times M$ matrix is a rectangular array of NM real (complex) numbers, called its *entries*, or *elements*, arranged in N rows and M columns and indexed as follows:

$$\begin{bmatrix} a_{11} & a_{12} & \cdots & a_{1M} \\ a_{21} & a_{22} & \cdots & a_{2M} \\ & & \cdots & \\ a_{N1} & a_{N2} & \cdots & a_{NM} \end{bmatrix}. \tag{B.1}$$

We write $\mathbf{A} = (a_{ij})$ as shorthand for the matrix (B.1). If $N = M$, $\mathbf{A}$ is called a *square matrix*; if $M = 1$, $\mathbf{A}$ is called a *column vector*, and if $N = 1$, $\mathbf{A}$ is called a *row vector*. We shall denote column vectors by using boldface lowercase letters, such as $\mathbf{x}, \mathbf{y}, \ldots$.

Standard operations for matrices are:

(a) Multiplication of $\mathbf{A}$ by the real or complex number c. The result, denoted by $c\mathbf{A}$, is the $N \times M$ matrix with entries ca_{ij}.

(b) Sum of two $N \times M$ matrices $\mathbf{A} = (a_{ij})$ and $\mathbf{B} = (b_{ij})$. The result is the $N \times M$ matrix $\mathbf{C}$ whose entries are $a_{ij} + b_{ij}$. The sum is commutative (i.e., $\mathbf{A} + \mathbf{B} = \mathbf{B} + \mathbf{A}$).

(c) Product of the $N \times K$ matrix $\mathbf{A}$ by the $K \times M$ matrix $\mathbf{B}$. The result is the $N \times M$ matrix $\mathbf{C}$ with entries

$$c_{ij} \triangleq \sum_{k=1}^{K} a_{ik}b_{kj}, \quad i = 1, \ldots, N, \quad j = 1, \ldots, M.$$

The matrix product is not commutative (i.e., in general $\mathbf{AB} \neq \mathbf{BA}$), but it is associative [i.e., $\mathbf{A(BC)} = \mathbf{(AB)C}$] and distributive with respect to the sum [i.e., $\mathbf{A(B + C)} = \mathbf{AB} + \mathbf{AC}$ and $\mathbf{(A + B)C} = \mathbf{AC} + \mathbf{BC}$].

When $\mathbf{AB} = \mathbf{BA}$, the two matrices $\mathbf{A}$ and $\mathbf{B}$ are said to *commute*. The notation $\mathbf{A}^l$ is used to denote the lth power of a square matrix $\mathbf{A}$ (i.e., the product of $\mathbf{A}$ by itself performed $l - 1$ times); that is,

$$\mathbf{A}^l \triangleq \underset{\longleftarrow \, l \, \longrightarrow}{\mathbf{A} \cdot \mathbf{A} \cdots \mathbf{A}}.$$

If we define the *identity matrix* $\mathbf{I} \triangleq (\delta_{ij})$ as the square matrix with elements 0 unless $i = j$, in which case they are equal to 1, multiplication of any square matrix $\mathbf{A}$ by $\mathbf{I}$ gives $\mathbf{A}$ itself, and we can set

$$\mathbf{A}^0 \triangleq \mathbf{I}.$$

(d) Given a square matrix $\mathbf{A}$, there may exist a matrix, which we denote by $\mathbf{A}^{-1}$, such that $\mathbf{AA}^{-1} = \mathbf{A}^{-1}\mathbf{A} = \mathbf{I}$. If $\mathbf{A}^{-1}$ exists, it is called the *inverse* of $\mathbf{A}$, and $\mathbf{A}$ is said to be *nonsingular*. For l a positive integer, we define a negative power of a nonsingular square matrix as follows:

$$\mathbf{A}^{-l} \triangleq (\mathbf{A}^{-1})^l.$$

(e) The *transpose* of the $N \times M$ matrix $\mathbf{A}$ with entries a_{ij} is the matrix with entries a_{ji}, which we denote by $\mathbf{A}'$. If $\mathbf{A}$ is a complex matrix, its *conjugate* $\mathbf{A}^*$ is the matrix with elements a_{ij}^*, and its conjugate transpose $\mathbf{A}^\dagger \triangleq (\mathbf{A}^*)' = (\mathbf{A}')^*$ has entries a_{ji}^*. The following properties hold:

$$(\mathbf{AB})' = \mathbf{B}'\mathbf{A}', \qquad (\mathbf{AB})^\dagger = \mathbf{B}^\dagger\mathbf{A}^\dagger.$$

(f) The *scalar product* of two real column vectors $\mathbf{x}, \mathbf{y}$ is

$$\mathbf{x}'\mathbf{y} \triangleq \sum_{i=1}^{N} x_i y_i = \mathbf{y}'\mathbf{x}. \tag{B.2}$$

If $\mathbf{x}$ and $\mathbf{y}$ are complex, their scalar product is defined as

$$\mathbf{x}^\dagger\mathbf{y} \triangleq \sum_{i=1}^{N} x_i^* y_i = (\mathbf{y}^\dagger\mathbf{x})^*. \tag{B.3}$$

Two vectors are called *orthogonal* if their scalar product is zero.

B.2 NUMBERS ASSOCIATED WITH A MATRIX

(a) The trace

Given a square $N \times N$ matrix $\mathbf{A}$, its *trace* (or *spur*) is the sum of the elements of the *main diagonal* of $\mathbf{A}$:

$$\text{tr}(\mathbf{A}) \triangleq \sum_{i=1}^{N} a_{ii}. \tag{B.4}$$

The trace operation is linear; that is, for any two given numbers α, β and two square $N \times N$ matrices $\mathbf{A}$, $\mathbf{B}$, we have

$$\text{tr}(\alpha\mathbf{A} + \beta\mathbf{B}) = \alpha\,\text{tr}(\mathbf{A}) + \beta\,\text{tr}(\mathbf{B}). \tag{B.5}$$

In general, $\text{tr}(\mathbf{AB}) = \text{tr}(\mathbf{BA})$ even if $\mathbf{AB} \neq \mathbf{BA}$. In particular, the following properties hold:

$$\text{tr}(\mathbf{A}^{-1}\mathbf{BA}) = \text{tr}(\mathbf{B}) \tag{B.6}$$

and, for any $N \times M$ matrix $\mathbf{A}$,

$$\text{tr}(\mathbf{AA}^\dagger) = \sum_{i=1}^{N} \sum_{j=1}^{N} |a_{ij}|^2. \tag{B.7}$$

Notice also that for two column vectors $\mathbf{x}$, $\mathbf{y}$

$$\mathbf{x}^\dagger\mathbf{y} = \text{tr}(\mathbf{x}\,\mathbf{y}^\dagger). \tag{B.8}$$

(b) The determinant

Given an $N \times N$ square matrix $\mathbf{A}$, its *determinant* is the number defined as the sum of the products of the elements in any row of $\mathbf{A}$ with their respective *cofactors* γ_{ij}:

$$\det \mathbf{A} \triangleq \sum_{j=1}^{N} a_{ij}\gamma_{ij}, \qquad \text{for any } i = 1, 2, \ldots, N. \tag{B.9}$$

The cofactor of a_{ij} is defined as $\gamma_{ij} \triangleq (-1)^{i+j}m_{ij}$, where the *minor* m_{ij} is the determinant of the $(N-1) \times (N-1)$ submatrix obtained from $\mathbf{A}$ by removing its ith row and jth column. The determinant has the following properties:

$$\det \mathbf{A} = 0, \qquad \text{if one row of } \mathbf{A} \text{ is zero or } \mathbf{A} \text{ has two equal rows,} \tag{B.10}$$

$$\det \mathbf{A} = \det \mathbf{A}', \tag{B.11}$$

$$\det (\mathbf{AB}) = \det \mathbf{A} \cdot \det \mathbf{B}, \tag{B.12}$$

$$\det \mathbf{A}^{-1} = [\det \mathbf{A}]^{-1}, \tag{B.13}$$

$$\det (c\mathbf{A}) = c^N \cdot \det \mathbf{A}, \qquad \text{for any number } c. \tag{B.14}$$

A matrix is nonsingular if and only if its determinant is nonzero.

(c) The eigenvalues

Given an $N \times N$ square matrix $\mathbf{A}$ and a column vector $\mathbf{u}$ with N entries, consider the set of N linear equations

$$\mathbf{A}\,\mathbf{u} = \lambda\,\mathbf{u}, \tag{B.15}$$

where λ is a constant and the entries of $\mathbf{u}$ are the unknown. There are only N values of λ (not necessarily distinct) such that (B.15) has a nonzero solution. These numbers are called the *eigenvalues* of $\mathbf{A}$, and the corresponding vectors $\mathbf{u}$ the *eigenvectors* associated with them. Note that if $\mathbf{u}$ is an eigenvector associated with the eigenvalue λ then, for any complex number c, $c\mathbf{u}$ is also an eigenvector.

The polynomial $a(\lambda) \triangleq \det (\lambda\mathbf{I} - \mathbf{A})$ in the indeterminate λ is called the *characteristic polynomial* of $\mathbf{A}$. The equation

$$\det (\lambda\mathbf{I} - \mathbf{A}) = 0 \tag{B.16}$$

is the *characteristic equation* of $\mathbf{A}$, and its roots are the eigenvalues of $\mathbf{A}$. The *Cayley–Hamilton theorem* states that every square $N \times N$ matrix $\mathbf{A}$ satisfies its characteristic equation. That is, if the characteristic polynomial of $\mathbf{A}$ is $a(\lambda) = \lambda^N + \alpha_1 \lambda^{N-1} + \cdots + \alpha_N$, then

$$a(\mathbf{A}) \triangleq \mathbf{A}^N + \alpha_1 \mathbf{A}^{N-1} + \cdots + \alpha_N \mathbf{I} = \mathbf{0}, \tag{B.17}$$

where $\mathbf{0}$ is the null matrix (i.e., the matrix all of whose elements are zero). The monic polynomial $\mu(\lambda)$ of lowest degree such that $\mu(\mathbf{A}) = \mathbf{0}$ is called the *minimal polynomial* of $\mathbf{A}$. If $f(x)$ is a polynomial in the indeterminate x, and $\mathbf{u}$ is an eigenvector of $\mathbf{A}$ associated with the eigenvalue λ, then

$$f(\mathbf{A})\,\mathbf{u} = f(\lambda)\mathbf{u}. \tag{B.18}$$

That is, $f(\lambda)$ is an eigenvalue of $f(\mathbf{A})$ and $\mathbf{u}$ is the corresponding eigenvector. The eigenvalues $\lambda_1, \ldots, \lambda_N$ of the $N \times N$ matrix $\mathbf{A}$ have the properties

$$\det (\mathbf{A}) = \prod_{i=1}^{N} \lambda_i \tag{B.19}$$

and

$$\operatorname{tr}(\mathbf{A}) = \sum_{i=1}^{N} \lambda_i. \tag{B.20}$$

From (B.19), it is immediately seen that $\mathbf{A}$ is nonsingular if and only if none of its eigenvalues is zero.

(d) The spectral norm and the spectral radius

Given an $N \times N$ matrix $\mathbf{A}$, its *spectral norm* $\|\mathbf{A}\|$ is the nonnegative number

$$\|\mathbf{A}\| \triangleq \sup_{\mathbf{x} \neq \mathbf{0}} \frac{\|\mathbf{A}\,\mathbf{x}\|}{\|\mathbf{x}\|}, \tag{B.21}$$

where $\mathbf{x}$ is an N-component column vector, and $\|\mathbf{u}\|$ denotes the Euclidean norm of the vector $\mathbf{u}$:

$$\|\mathbf{u}\| \triangleq \sqrt{\sum_{i=1}^{N} |u_i|^2} = \sqrt{\mathbf{u}^\dagger\mathbf{u}}. \tag{B.22}$$

We have

$$\|\mathbf{A}\,\mathbf{B}\| \leq \|\mathbf{A}\| \cdot \|\mathbf{B}\| \tag{B.23}$$

$$\|\mathbf{A}\,\mathbf{x}\| \leq \|\mathbf{A}\| \cdot \|\mathbf{x}\|, \tag{B.24}$$

for any matrix $\mathbf{B}$ and vector $\mathbf{x}$. If λ_i, $i = 1, \ldots, N$, denote the eigenvalues of $\mathbf{A}$, the radius $\rho(\mathbf{A})$ of the smallest disk centered at the origin of the complex plane that includes all these eigenvalues is called the *spectral radius* of $\mathbf{A}$:

$$\rho(\mathbf{A}) \triangleq \max_{1 \leq i \leq N} |\lambda_i|. \tag{B.25}$$

In general, for an arbitrary complex $N \times N$ matrix $\mathbf{A}$, we have

$$\rho(\mathbf{A}) \leq \|\mathbf{A}\| \tag{B.26}$$

and

$$\|\mathbf{A}\| = \sqrt{\rho(\mathbf{A}^\dagger\mathbf{A})}. \tag{B.27}$$

If $\mathbf{A} = \mathbf{A}^\dagger$, then

$$\rho(\mathbf{A}) = \|\mathbf{A}\|. \tag{B.28}$$

(e) Quadratic forms

Given an $N \times N$ square matrix $\mathbf{A}$ and a column vector $\mathbf{x}$ with N entries, we call a *quadratic form* the quantity

$$\mathbf{x}^\dagger\mathbf{A}\mathbf{x} = \sum_{i=1}^{N} \sum_{j=1}^{N} x_i^* a_{ij} x_j. \tag{B.29}$$

B.3 SOME CLASSES OF MATRICES

Let $\mathbf{A}$ be an $N \times N$ square matrix.

(a) $\mathbf{A}$ is called *symmetric* if $\mathbf{A}' = \mathbf{A}$.

(b) $\mathbf{A}$ is called *Hermitian* if $\mathbf{A}^\dagger = \mathbf{A}$.

(c) $\mathbf{A}$ is called *orthogonal* if $\mathbf{A}^{-1} = \mathbf{A}'$.

(d) $\mathbf{A}$ is called *unitary* if $\mathbf{A}^{-1} = \mathbf{A}^{\dagger}$.

(e) $\mathbf{A}$ is called *diagonal* if its entries a_{ij} are zero unless $i = j$. A useful notation for a diagonal matrix is

$$\mathbf{A} = \text{diag}(a_{11}, a_{22}, \ldots, a_{NN}).$$

(f) $\mathbf{A}$ is called *scalar* if $\mathbf{A} = c\mathbf{I}$ for some constant number c; that is, $\mathbf{A}$ is diagonal with equal entries on the main diagonal.

(g) $\mathbf{A}$ is called a *Toeplitz* matrix if its entries a_{ij} satisfy the condition

$$a_{ij} = a_{i-j}. \qquad (B.30)$$

That is, its elements on the same diagonal are equal.

(h) $\mathbf{A}$ is called *circulant* if its rows are all the cyclic shifts of the first one:

$$a_{ij} = a_{(i-j) \bmod N}. \qquad (B.31)$$

(i) $\mathbf{A}$ is called *positive (nonnegative) definite* if all its eigenvalues are positive (nonnegative). Equivalently, $\mathbf{A}$ is positive (nonnegative) definite if and only if for any nonzero column vector $\mathbf{x}$ the quadratic form $\mathbf{x}^{\dagger}\mathbf{A}\mathbf{x}$ is positive (nonnegative).

Example B.1

Let $\mathbf{A}$ be Hermitian. Then the quadratic form $f \triangleq \mathbf{x}^{\dagger}\mathbf{A}\mathbf{x}$ is real. In fact,

$$\begin{aligned} f^* &= (\mathbf{x}^{\dagger}\mathbf{A}\,\mathbf{x})^* \\ &= \mathbf{x}'\mathbf{A}^*\mathbf{x}^* \\ &= (\mathbf{A}^*\mathbf{x}^*)'\mathbf{x} \\ &= \mathbf{x}^{\dagger}\mathbf{A}^{\dagger}\mathbf{x}. \end{aligned} \qquad (B.32)$$

Since $\mathbf{A}^{\dagger} = \mathbf{A}$, this is equal to $\mathbf{x}^{\dagger}\mathbf{A}\mathbf{x} = f$, which shows that f is real. $\square$

Example B.2

Consider the random column vector $\mathbf{x} = [x_1, x_2, \ldots, x_N]'$, and its *correlation matrix*

$$\mathbf{R} \triangleq E[\mathbf{x}\,\mathbf{x}^{\dagger}]. \qquad (B.33)$$

It is easily seen that $\mathbf{R}$ is Hermitian. Also, $\mathbf{R}$ is nonnegative definite; in fact, for any nonzero deterministic column vector $\mathbf{a}$,

$$\begin{aligned} \mathbf{a}^{\dagger}\mathbf{R}\,\mathbf{a} &= \mathbf{a}^{\dagger}E[\mathbf{x}\,\mathbf{x}^{\dagger}]\,\mathbf{a} \\ &= E[\mathbf{a}^{\dagger}\mathbf{x}\,\mathbf{x}^{\dagger}\mathbf{a}] \\ &= E[|\mathbf{a}^{\dagger}\mathbf{x}|^2] \geq 0 \end{aligned} \qquad (B.34)$$

with equality only if $\mathbf{a}^{\dagger}\mathbf{x} = 0$ almost surely; that is, the components of $\mathbf{x}$ are *linearly dependent*.

If $x_1, \ldots, x_N$ are samples taken from a wide-sense stationary discrete-time random process, and we define

$$r_{i-j} \triangleq E[x_{n+i}x^*_{n+j}], \qquad (B.35)$$

it is seen that the entry of $\mathbf{R}$ in the ith row and the jth column is precisely $r_{|i-j|}$. This shows in particular that $\mathbf{R}$ is a Toeplitz matrix.

If $\mathcal{G}(f)$ denotes the discrete Fourier transform of the autocorrelation sequence (r_n), that is $\mathcal{G}(f)$ is the power spectrum of the random process (x_n) [see (2.86)], the following can be shown:

(a) The eigenvalues $\lambda_1, \ldots, \lambda_N$ of $\mathbf{R}$ are samples (not necessarily equidistant) of the function $\mathcal{G}(f)$.

(b) For any function $\gamma(\cdot)$, we have the *Toeplitz distribution theorem* (Grenander and Szegö, 1958):

$$\lim_{N\to\infty} \frac{1}{N} \sum_{i=1}^{N} \gamma(\lambda_i) = \int_{-1/2}^{1/2} \gamma[\mathcal{G}(f)]\, df. \quad \Box \tag{B.36}$$

Example B.3

Let $\mathbf{C}$ be a circulant $N \times N$ matrix of the form

$$\mathbf{C} = \begin{bmatrix} c_0 & c_1 & c_2 & \cdots & c_{N-1} \\ c_{N-1} & c_0 & c_1 & \cdots & c_{N-2} \\ & & \cdots & & \\ c_1 & c_2 & c_3 & \cdots & c_0 \end{bmatrix}. \tag{B.37}$$

Let also $w \triangleq e^{j2\pi/N}$, so that $w^N = 1$. Then the eigenvector associated with the eigenvalue λ_i is

$$\mathbf{u}_i = [w^0\ w^i\ w^{2i} \cdots w^{(N-1)i}]', \tag{B.38}$$

for $i = 0, 1, \ldots, N - 1$. The eigenvalues of $\mathbf{C}$ are

$$\lambda_i = \sum_{l=0}^{N-1} c_l w^{li}, \qquad i = 0, 1, \ldots, N-1, \tag{B.39}$$

and λ_i^* can be interpreted as the value of the Fourier transform of the sequence $c_0, c_1, \ldots, c_{N-1}$, taken at frequency i/N. $\Box$

Example B.4

If $\mathbf{U}$ is a unitary $N \times N$ matrix, and $\mathbf{A}$ is an $N \times N$ arbitrary complex matrix, pre- or postmultiplication of $\mathbf{A}$ by $\mathbf{U}$ does not alter its spectral norm; that is,

$$\|\mathbf{A}\mathbf{U}\| = \|\mathbf{U}\mathbf{A}\| = \|\mathbf{A}\|. \quad \Box \tag{B.40}$$

B.4 CONVERGENCE OF MATRIX SEQUENCES

Consider the sequence $(\mathbf{A}^n)_{n=0}^{\infty}$ of powers of the square matrix $\mathbf{A}$. As $n \to \infty$, for $\mathbf{A}^n$ to tend to the null matrix $\mathbf{0}$ it is necessary and sufficient that the spectral radius of $\mathbf{A}$ be less than 1. Also, as the spectral radius of $\mathbf{A}$ does not exceed its spectral norm, for $\mathbf{A}^n \to \mathbf{0}$ it is sufficient that $\|\mathbf{A}\| < 1$.

Consider now the matrix series

$$\mathbf{I} + \mathbf{A} + \mathbf{A}^2 + \cdots + \mathbf{A}^n + \cdots \tag{B.41}$$

For this series to converge, it is necessary and sufficient that $\mathbf{A}^n \to \mathbf{0}$ as $n \to \infty$. If this holds, the sum of the series equals $(\mathbf{I} - \mathbf{A})^{-1}$.

B.5 THE GRADIENT VECTOR

Let $f(\mathbf{x}) = f(x_1, \ldots, x_N)$ be a differentiable real function of N real arguments. Its *gradient vector*, denoted by ∇f, is the column vector whose N entries are the derivatives $\partial f/\partial x_i$, $i = 1, \ldots, N$. If $x_1, \ldots, x_N$ are complex, that is,

$$x_i = x_i' + jx_i'', \qquad i = 1, \ldots, N, \tag{B.42}$$

the gradient of $f(\mathbf{x})$ is the vector whose components are

$$\frac{\partial f}{\partial x_i'} + j\frac{\partial f}{\partial x_i''}, \qquad i = 1, \ldots, N.$$

Example B.5

If $\mathbf{a}$ denotes a complex column vector, and $f(\mathbf{x}) \triangleq \mathfrak{R}[\mathbf{a}^{\dagger}\mathbf{x}]$, we have

$$\nabla f(\mathbf{x}) = \mathbf{a}. \;\square \tag{B.43}$$

Example B.6

If $\mathbf{A}$ is a Hermitian $N \times N$ matrix, and $f(\mathbf{x}) \triangleq \mathbf{x}^{\dagger}\mathbf{A}\,\mathbf{x}$, we have

$$\nabla f(\mathbf{x}) = 2\mathbf{A}\mathbf{x}. \;\square \tag{B.44}$$

B.6 THE DIAGONAL DECOMPOSITION

Let $\mathbf{A}$ be a Hermitian $N \times N$ matrix with eigenvalues $\lambda_1, \ldots, \lambda_N$. Then $\mathbf{A}$ can be given the following representation:

$$\mathbf{A} = \mathbf{U}\mathbf{\Lambda}\mathbf{U}^{-1}, \tag{B.45}$$

where $\mathbf{\Lambda} \triangleq \text{diag}(\lambda_1, \ldots, \lambda_N)$, and $\mathbf{U}$ is a unitary matrix, so that $\mathbf{U}^{-1} = \mathbf{U}^{\dagger}$. From (B.45) it follows that

$$\mathbf{A}\mathbf{U} = \mathbf{U}\mathbf{\Lambda}, \tag{B.46}$$

which shows that the ith column of $\mathbf{U}$ is the eigenvector of $\mathbf{A}$ corresponding to the eigenvalue λ_i. For any column vector $\mathbf{x}$, the following can be derived from (B.45):

$$\mathbf{x}^{\dagger}\mathbf{A}\,\mathbf{x} = \sum_{i=1}^{N} \lambda_i |y_i|^2, \tag{B.47}$$

where $y_1, \ldots, y_N$ are the components of the vector $\mathbf{y} \triangleq \mathbf{U}^{\dagger}\mathbf{x}$.

BIBLIOGRAPHICAL NOTES

There are many excellent books on matrix theory, and some of them are certainly well known to the reader. The books by Bellman (1968) and Gantmacher (1959) are encyclopedic treatments in which details can be found about any topic one may wish to study in more depth. A modern treatment of matrix theory, with emphasis on numerical computations, is provided by Golub and Van Loan (1983). Faddeev and Faddeeva (1963) and Varga (1962) include treatments of matrix norms and matrix convergence. The most complete reference about Toeplitz matrices is the book by Grenander and Szegö (1958). For a tutorial introductory treatment of Toeplitz matrices and a simple proof of the distribution theorem (B.36), the reader is referred to (Gray, 1971 and 1972). In Athans (1968) one can find a number of formulas about gradient vectors.

APPENDIX C

Variational Techniques and Constrained Optimization

In this appendix, we briefly list some of the optimization theory results used in the book. Our treatment is far from rigorous, because our aim is to describe a technique for constrained optimization rather than to provide a comprehensive development of the underlying theory. The reader interested in more details is referred to (Luenberger, 1969, pp. 171–190) from which our treatment is derived; alternatively, to (Gelfand and Fomin, 1963).

Let R be a function space (technically, it must be a *normed linear* space). Assume that a rule is provided assigning to each function $f \in R$ a complex number $\varphi[f]$. Then φ is called a *functional* on R.

Example C.1

Let $f(x)$ be a continuous function defined on the interval (a,b). We write $f \in C(a, b)$. Then

$$\varphi[f] \triangleq f(x_0), \qquad a \le x_0 \le b,$$

$$\varphi[f] \triangleq \int_a^b w(x) f(x) \, dx, \qquad w \in C(a, b)$$

and

$$\varphi[f] \triangleq \int_a^b f^2(x) \, dx,$$

are functionals on the space $C(a, b)$. $\square$

If φ is a functional on R, and $f, h \in R$, the functional

$$\delta\varphi[f; h] \triangleq \frac{d}{d\alpha} \varphi[f + \alpha h]\Big|_{\alpha=0} \tag{C.1}$$

is called the *Fréchet differential* of φ. The concept of Fréchet differential provides a technique to find the maxima and minima of a functional. We have the following result:

A necessary condition for $\varphi[f]$ to achieve a maximum or minimum value for $f = f_0$ is that $\delta\varphi(f_0; h) = 0$ for all $h \in R$.

In many optimization problems the optimal function is required to satisfy certain constraints. We consider in particular the situation in which a functional φ on R must be optimized under n constraints given in the implicit form $\psi_1[f]=C_1$, $\psi_2[f]=C_2$, . . . , $\psi_n[f] = C_n$, where ψ_1, . . . , ψ_n are functionals on R, and C_1, . . . , C_n are constants. We have the following result:

If $f_0 \in R$ gives a maximum or a minimum of φ subject to the constraints $\psi_i[f]=C_i$, $1 \le i \le n$, and the n functionals $\delta\psi_i[f_0; h]$ are linearly independent, then there are n scalars λ_1, . . . , λ_n which make the Fréchet differential of

$$\varphi[f] + \sum_{i=1}^{n} \lambda_i\psi_i[f] \tag{C.2}$$

to vanish at f_0.

This result provides a rule for finding constrained maxima and minima. The procedure is to form the functional (C.2), to compute its Fréchet differential, and to find the functions f which make it vanish for every h. The values of the "Lagrange multipliers" λ_1, . . . , λ_n can be computed using the constraint equations. Whether the solutions correspond to maxima, or to minima, or neither, can be best determined by a close analysis of the problem at hand.

Example C.2

We want to find the real function $f \in C(0, 1)$ which minimizes the functional

$$\varphi[f] \triangleq \int_0^1 \frac{1}{f^2(x)}\, w(x)\, dx$$

(where $w(x) \in C(0, 1)$) subject to the constraint $\psi[f] = 1$, where

$$\psi[f] \triangleq \int_0^1 f(x)dx.$$

The Fréchet differential of $\varphi[f] + \lambda\psi[f]$ is

$$\frac{d}{d\alpha} \int_0^1 \left\{ \frac{w(x)}{[f(x) + \alpha h(x)]^2} + \lambda[f(x) + \alpha h(x)] \right\} dx \bigg|_{\alpha=0}$$

$$= \int_0^1 \left[\frac{-2w(x)}{f^3(x)} + \lambda \right] h(x)dx.$$

For this functional to be zero for any $h(x)$, we must have

$$f(x) = \left(\frac{2w(x)}{\lambda} \right)^{1/3}.$$

If this result is inserted in the constraint expression, the value of λ can be determined. $\square$

Variational Techniques and Constrained Optimization   Appendix C   **591**

Example C.3

Find the complex function $f(x)$ that minimizes

$$\int_{-\infty}^{\infty} w(x) |f(x)|^2 dx$$

(where $w(x) > 0$ for $-\infty < x < \infty$) subject to the constraint

$$\int_{-\infty}^{\infty} \frac{1}{|f(x)|^2} dx = 1.$$

The relevant Fréchet differential is

$$\frac{d}{d\alpha} \int_{-\infty}^{\infty} \left\{ w(x) [f(x) + \alpha h(x)] f^*(x) + \frac{\lambda}{[f(x) + \alpha h(x)] f^*(x)} \right\} dx \bigg|_{\alpha=0}$$

$$= \int_{-\infty}^{\infty} \left\{ w(x) f^*(x) - \frac{\lambda}{f(x) |f(x)|^2} \right\} h(x) \, dx$$

and this is zero for any $h(x)$ provided that

$$|f(x)|^4 = \frac{\lambda}{w(x)}. \quad \square$$

Transfer Functions
of State Diagrams

A state diagram of a convolutional code such as that of Fig. 9.28 is a *directed graph*. To define a directed graph, we give a set $V = \{v_1, v_2, \ldots\}$ of *vertices* and a subset E of ordered pairs of vertices from V, called *edges*. A graph can be represented drawing a set of points corresponding to the vertices and a set of *arrows* corresponding to each edge of E and connecting two vertices. A *path* in a graph is a sequence of edges and can be denoted by giving the string of subsequent vertices included into the path. In the study of convolutional codes, we are interested in the enumeration of the paths with similar properties.

A simple directed graph is shown in Fig. D.1. There are three vertices, v_1, v_2, and v_3, and four edges (v_1, v_2), (v_1, v_3), (v_2, v_3), and (v_2, v_2). In this graph there are infinitely many paths between v_1 and v_3 because of the loop at the vertex v_2. One path of length 4, is, for instance, $v_1v_2v_2v_2v_3$. Each edge of a graph can be assigned a label. An important quantity, called *transmission* between two vertices, is defined as the sum of the labels of all paths of any length connecting the two vertices. The label of a path is defined as the product of the labels of the edges of the path. For example, the label

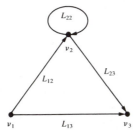

Figure D.1 Example of a directed graph.

of the path $v_1 v_2 v_2 v_2 v_3$ is $L_{12} L_{22}^2 L_{23}$. The transmission between v_1 and v_3 in Fig. D.1 is then given by

$$T(v_1, v_3) = L_{13} + L_{12} L_{23} + L_{12} L_{22} L_{23} + L_{12} L_{22}^2 L_{23} + \cdots$$
$$= L_{13} + L_{12} L_{23}(1 + L_{22} + L_{22}^2 + \cdots) \qquad (D.1)$$
$$= L_{13} + \frac{L_{12} L_{23}}{1 - L_{22}}.$$

Notice that, in deriving (D.1), we have assumed that the labels are real numbers less than 1. Therefore, Fig. D.1 can be replaced by the scheme of Fig. D.2, with the label L'_{13} given by (D.1).

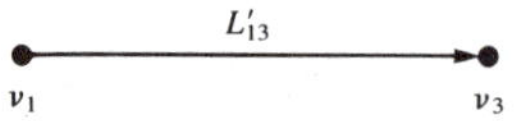

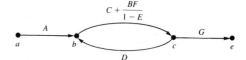

Figure D.2 Reduced graph to compute the transmission $T(v_1, v_3)$.

Given any graph, it is thus possible to compute the transmission between a pair of vertices by removing one by one the intermediate vertices on the graph and by redefining the new labels. As an example, the graph of Fig. D.3 replicates that of Fig. 9.32, with labels A, B, C, . . . , G. Using the result (D.1), the graph of Fig. D.3 can be replaced with that of Fig. D.4, in which the vertex d has been removed. By removing also the vertex b we get the graph of Fig. D.5, and finally the transmission between a and e:

$$T(a, e) = \frac{ACG(1 - E) + ABFG}{1 - E - CD + CDE - BDF}. \qquad (D.2)$$

This result, with suitable substitution of labels, coincides with (9.104).

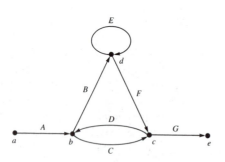

Figure D.3 Directed graph corresponding to the convolutional code state diagram of Fig. 9.32.

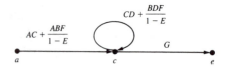

Figure D.4 First step to reduce the graph of Fig. D.3.

Figure D.5 Second step to reduce the graph of Fig. D.3.

There is another technique for evaluating $T(a, e)$ that can be very useful in computations because it is based on a matrix description of the problem. Let us define by x_i the value of the accumulated path labels from the initial state a to the state i, as influenced by all other states. Therefore, the state equations for the graph of Fig. D.3 are

$$x_b = A + Dx_c,$$
$$x_c = Cx_b + Fx_d, \tag{D.3}$$
$$x_d = Bx_b + Ex_d$$
$$x_e = Gx_c.$$

In this approach we have $T(a, e) \triangleq x_e$, and therefore we can solve the system (D.3) and verify that x_e is given again by (D.2).

The set of equations (D.3) can be given a more general and formal expression. Define the two vectors

$$\mathbf{x'} \triangleq (x_b, x_c, x_d, x_e), \tag{D.4}$$
$$\mathbf{x'_0} \triangleq (A, 0, 0, 0)$$

and the state transition matrix $\mathbf{T}$:

$$
\mathbf{T} \triangleq
\begin{array}{cccc}
a & b & c & d
\end{array}
\begin{bmatrix}
0 & D & 0 & 0 \\
C & 0 & F & 0 \\
B & 0 & E & 0 \\
0 & G & 0 & 0
\end{bmatrix}
\begin{array}{c}
a \\
b \\
c \\
d
\end{array}. \tag{D.5}
$$

Using (D.4) and (D.5), the system (D.3) can be rewritten in matrix form as

$$\mathbf{x} = \mathbf{T}\,\mathbf{x} + \mathbf{x}_0. \tag{D.6}$$

The formal solution to this equation can be written as

$$\mathbf{x} = (\mathbf{I} - \mathbf{T})^{-1}\,\mathbf{x}_0, \tag{D.7}$$

or as the matrix power series

$$\mathbf{x} = [\mathbf{I} + \mathbf{T} + \mathbf{T}^2 + \mathbf{T}^3 + \cdots]\mathbf{x}_0. \tag{D.8}$$

Notice that this power series solution is very satisfying when considering the state diagram as being described by a walk into a trellis. Each successive multiplication by $\mathbf{T}$ corresponds to a change in the state vector caused by going one level deeper in the trellis. Notice that the multiplication by $\mathbf{T}$ is very simple since most of its entries are zero. When the number of states is small, the matrix inversion (D.7) is also useful to get directly the result in the form (D.2).

APPENDIX E

Approximate Computation of Averages

The aim of this appendix is to describe some techniques for the evaluation of bounds or numerical approximations to the average $E[g(\xi)]$, where $g(\cdot)$ is an explicitly known deterministic function, and ξ is some random variable whose probability density function is not known explicitly, or is highly complex and hence very difficult to compute exactly. It is assumed that a certain amount of knowledge is available about ξ, expressed by a finite and usually small set of its moments. Also, we shall assume that the range of ξ lies in the interval $[a, b]$, where both a and b are finite unless otherwise stated. The techniques described hereafter are not intended to exhaust the set of possible methods for solving this problem. However, they are general enough to handle a large class of situations and computationally efficient in terms of speed and accuracy. Also, instead of providing a single technique, we describe several, as we advocate that the specific problem handled should determine the technique best suited to it from the viewpoint of computational effort required, accuracy, and applicability.

E.1 SERIES EXPANSION TECHNIQUE

In this section, we shall assume that the function $g(x)$ is analytic at point $x = x_0$. Hence, we can represent it in a neighborhood of x_0 using the Taylor's series expansion:

$$g(x) = g(x_0) + (x - x_0)g'(x_0) + (x - x_0)^2 \frac{g''(x_0)}{2!}$$

$$+ \cdots + (x - x_0)^n \frac{g^{(n)}(x_0)}{n!} + \cdots \quad \text{(E.1)}$$

596

If the radius of convergence of (E.1) is large enough to include the range of the random variable ξ, and we define

$$c_n \triangleq E[\xi - x_0]^n, \tag{E.2}$$

then, averaging termwise the Taylor's series expansion of $g(\xi)$, we get from (E.2)

$$E[g(\xi)] = g(x_0) + c_1 g'(x_0) + c_2 \frac{g''(x_0)}{2!} + \cdots + c_n \frac{g^{(n)}(x_0)}{n!} + \cdots \tag{E.3}$$

It can be seen from (E.3) that $E[g(\xi)]$ can be evaluated on the basis of the knowledge of the sequence of moments $(c_n)_{n=1}^{\infty}$, provided that the series converges. In particular, an approximate value of $E[g(\xi)]$ based on a finite number of moments can be obtained by truncating (E.3):

$$E[g(\xi)] \cong E_N[g(\xi)] \triangleq \sum_{k=0}^{N} c_k \frac{g^{(k)}(x_0)}{k!}, \qquad c_0 = 1. \tag{E.4}$$

The error of this approximation is

$$\begin{aligned} E[g(\xi)] - E_N[g(\xi)] &= \sum_{k=N+1}^{\infty} c_k \frac{g^{(k)}(x_0)}{k!} \\ &= \frac{1}{(N+1)!} E\{(\xi - x_0)^{N+1} f^{(N+1)}[x_0 + \theta(\xi - x_0)]\}, \end{aligned} \tag{E.5}$$

where $0 \le \theta \le 1$. Depending on the specific application characteristics, either one of the second and third terms in (E.5) can be used to obtain a bound on the truncation error. In any case, bounds on the values of the derivatives of $g(\cdot)$ and of the moments (E.2) must be available.

As an example of application of this technique, let us consider the function

$$g(x) = erfc\left(\frac{h + x}{\sqrt{2}\sigma}\right), \tag{E.6}$$

where h and σ are known parameters. This function can be expanded in a Taylor's series in the neighborhood of the origin by observing that (Abramowitz and Stegun, 1972, p. 298)

$$\frac{d^k}{dz^k} erfc(z) = (-1)^k \frac{2}{\sqrt{\pi}} H_{k-1}(z) e^{-z^2}, \tag{E.7}$$

where $H_{k-1}(\cdot)$ is the Hermite polynomial of degree $(k-1)$ (Abramowitz and Stegun, 1972, pp. 773–787). Thus,

$$g^{(k)}(0) = (-1)^k \frac{2}{\sqrt{\pi}(\sqrt{2}\sigma)^k} H_{k-1}\left(\frac{h}{\sqrt{2}\sigma}\right) \exp\left(-\frac{h^2}{2\sigma^2}\right), \tag{E.8}$$

and we finally get

$$\begin{aligned} E[g(\xi)] = erfc\left(\frac{h}{\sqrt{2}\sigma}\right) &+ \frac{2}{\sqrt{\pi}} \exp\left(-\frac{h^2}{2\sigma^2}\right) \\ &\cdot \sum_{k=1}^{\infty} \frac{(-1)^k \mu_k}{(\sqrt{2}\sigma)^k k!} H_{k-1}\left(\frac{h}{\sqrt{2}\sigma}\right), \end{aligned} \tag{E.9}$$

where

$$\mu_k \triangleq E[\xi^k], \qquad k = 1, 2, \ldots, \tag{E.10}$$

are the central moments of the random variable ξ.

The proof that the series (E.9) is convergent, as well as an upper bound on the truncation error, can be obtained by using the following bound on the value of Hermite polynomials (Abramowitz and Stegun, 1972, p. 787):

$$|H_n(z)| \leq \beta\, 2^{n/2}\, \sqrt{n!}\, \exp\!\left(\frac{z^2}{2}\right), \qquad \beta \cong 1.086435, \qquad n = 1, 2, \ldots, \tag{E.11}$$

and the following bound on central moments:

$$|\mu_{k+s}| \leq E[|\xi|^k]\chi^s, \qquad k \geq 0, \qquad s \geq 0, \tag{E.12}$$

where χ denotes the maximum value taken by $|\xi|$. The bound (E.12) can be easily derived under the assumption that ξ is bounded. Using (E.11) and (E.12), we get the inequality (Prabhu, 1971)

$$|R_N[g]| \triangleq |E[g(\xi)] - E_N[g(\xi)]|$$

$$\leq \sqrt{\frac{2}{\pi}}\, \frac{\beta E|\xi^N|}{\sigma^{N+1}}\, \frac{\chi \exp[-h^2/(4\sigma^2)]}{(N+1)\sqrt{N!}} \left[1 - \frac{\chi}{\sigma\sqrt{N+1}}\right]^{-1}, \tag{E.13}$$

which holds provided that

$$\left(\frac{\chi}{\sigma}\right)^2 < N + 1. \tag{E.14}$$

The condition (E.14) can always be met for finite values of χ, provided that N is sufficiently large.

A case of special interest arises when ξ is a symmetric random variable, so that its odd-order central moments are zero:

$$\mu_{2k-1} = 0, \qquad k = 1, 2, \ldots. \tag{E.15}$$

In this case, (E.9) specializes to

$$E[g(\xi)] = erfc\left(\frac{h}{\sqrt{2}\sigma}\right) + \frac{2}{\sqrt{\pi}}\exp\left(-\frac{h^2}{2\sigma^2}\right)$$

$$\cdot \sum_{k=1}^{\infty} \frac{\mu_{2k}}{(\sqrt{2}\sigma)^{2k}(2k)!} H_{2k-1}\left(\frac{h}{\sqrt{2}\sigma}\right) \tag{E.16}$$

and, using the inequality (Abramowitz and Stegun, 1972, p. 787)

$$|H_{2n+1}(z)| \leq |z|\exp\left(\frac{z^2}{2}\right)\frac{(2n+2)!}{(n+1)!}, \qquad n = 0, 1, \ldots, \tag{E.17}$$

the following bound can be derived (Prabhu, 1971):

$$|R_{2N}[g]| \triangleq |E[g(\xi)] - E_{2N}[g(\xi)]|$$

$$\leq \frac{|h|}{2\sqrt{2\pi}\sigma} \exp\left(-\frac{h^2}{4\sigma^2}\right) \frac{\mu_{2N}}{\sigma^{2N+2}} \frac{\chi^2}{(N+1)!}$$ (E.18)

$$\cdot \left[1 - \frac{\chi^2}{\sigma^2(N+2)}\right]^{-1}$$

under the constraint

$$\left(\frac{\chi}{\sigma}\right)^2 < N + 2.$$ (E.19)

It can be seen from (E.13) and (E.18) that the truncation error can be made vanishingly small by taking N, the number of terms retained for computation, sufficiently large. Thus, at first it may seem that the average $E[g(\xi)]$ can be approximated with as great an accuracy as desired. But, in practice, roundoff errors may make it impossible to add up too many terms in the series and still retain a satisfactory accuracy. Notice in particular that in (E.9), and even in (E.16), the terms of the series do not have the same sign. Also, in practice it is virtually impossible to compute a very large set of moments. The inaccuracies in the computations (or in measurements) of the moments increase with their order, so the process of adding more and more terms to the series cannot be extended very far as the computed values of the series become unreliable.

The Taylor's series technique presented so far can be modified by considering a different series expansion for the function $g(\cdot)$. One may want to consider, in lieu of (E.1), a series of the form

$$g(x) = \sum_{n=0}^{\infty} a_n P_n(x), \qquad x \in I,$$ (E.20)

where $P_0(x)$, $P_1(x)$, . . . form a system of polynomials orthonormal in the interval I. Consequently, the coefficients of the series expansion are given by

$$a_n = \int_I g(x) P_n(x)\, dx.$$ (E.21)

If I includes the range of the random variable ξ, by averaging (E.20) termwise we obtain

$$E[g(\xi)] = \sum_{n=0}^{\infty} a_n E[P_n(\xi)],$$ (E.22)

which can in turn be approximated by a finite sum. The computation of (E.22) requires the knowledge of the "generalized moments" $E[P_n(\xi)]$, which can be obtained, for example, as finite linear combinations of the central moments (E.10).

E.2 QUADRATURE APPROXIMATIONS

In this section we shall describe an approximation technique for $E[g(\xi)]$ based on the observation that this average can be formally expressed as an integral:

$$E[g(\xi)] = \int_a^b g(x)f_\xi(x)\,dx, \tag{E.23}$$

where $f_\xi(\cdot)$ denotes the probability density function of the random variable ξ. Having ascertained that the problem of evaluating $E[g(\xi)]$ is indeed equivalent to the computation of an integral, we can resort to numerical techniques developed to compute approximate values of integrals of the form (E.23). The most widely investigated techniques for approximating a definite integral lead to the formula

$$\int_a^b g(x)f_\xi(x)\,dx \cong \sum_{i=1}^{N} w_i g(x_i), \tag{E.24}$$

a linear combination of values of the function $g(\cdot)$. The x_i, $i = 1, 2, \ldots, N$, are called the *abscissas* (or *points* or *nodes*) of the formula, and the w_i, $i = 1, 2, \ldots, N$, are called its *weights* (or *coefficients*). The set of abscissas and weights is usually referred to as a *quadrature rule*. A systematic introduction to the theory of quadrature rules of the form (E.24) is given in Krylov (1962).

The quadrature rule is chosen to render (E.24) as accurate as possible. A first difficulty with this theory arises when one wants to define how to measure the accuracy of a quadrature rule. Since we want the abscissas and weights to be independent of $g(\cdot)$, and hence be the same for all possible such functions, the definition of what is meant by "accuracy" must be made independent of the particular choice of $g(\cdot)$. The classical approach here is to select a number of *probe functions* and constrain the quadrature rule to be exact for these functions. By choosing $g(\cdot)$ to be a polynomial, it is said that the quadrature rule (E.24) has *degree of precision* v if it is exact whenever $g(\cdot)$ is a polynomial of degree $\leq v$ [or, equivalently, whenever $g(x) = 1, x, \ldots, x^v$] and it is not exact for $g(x) = x^{v+1}$.

Once a criterion of goodness for quadrature rules has been defined, the next step is to investigate which are the best quadrature rules and how they can be computed. The answer is provided by the following result from numerical analysis, slightly reformulated to fit our framework (see Krylov, 1962, for more details and a proof):

Given a random variable ξ with range $[a, b]$ and all of whose moments exist, it is possible to define a sequence of polynomials $P_0(x)$, $P_1(x)$, $\ldots$, $\deg P_i(x) = i$, that are orthonormal with respect to ξ; that is,

$$E[P_n(\xi)P_m(\xi)] = \delta_{mn}, \qquad m, n = 0, 1, \ldots. \tag{E.25}$$

Denote by $x_1 < x_2 < \cdots < x_N$ the N roots of the polynomial $P_N(x)$ (they are all real, and lie inside $[a, b]$), and by k_n the coefficient of x^n in the polynomial $P_n(x)$, $n = 0, 1, \ldots$. By defining

$$w_i \triangleq -\frac{k_{N+1}}{k_N}\frac{1}{P_{N+1}(x_i)P'_N(x_i)}, \qquad i = 1, 2, \ldots, N, \tag{E.26}$$

the set $\{x_i, w_i\}_{i=1}^{N}$ is a quadrature rule with degree of precision $2N - 1$. This is the highest degree of precision that can be attained by any quadrature rule with N weights and abscissas.

These quadrature rules are usually called *Gauss quadrature rules* because they were first studied by Gauss. He considered the special case $f_\xi(x) =$ constant.

If $\{x_i, w_i\}_{i=1}^N$ is a Gauss quadrature rule, the error involved in the approximate integration of the function $g(\cdot)$, that is, the difference between the two sides of (E.24), is given by

$$R_N[g] = \frac{1}{(2N)!k_N^2} g^{(2N)}(\eta), \qquad (E.27)$$

provided that $g(\cdot)$ has in $[a, b]$ a continuous derivative of order $2N$; η is a point of $[a, b]$.

Example E.1

For a number of probability density functions $f_\xi(\cdot)$, results concerning Gauss quadrature rules are available in tabular form. For example, if ξ is a Gaussian random variable with mean μ and variance σ^2, the corresponding Gauss quadrature rule is

$$E[g(\xi)] \cong \frac{1}{\sqrt{\pi}} \sum_{i=1}^N w_i g(\sqrt{2}\sigma x_i + \mu), \qquad (E.28)$$

where the x_i, $i = 1, \ldots, N$, are the zeros of the Nth-degree Hermite polynomial. The actual values of w_i and x_i, $i = 1, \ldots, N$, can be found for instance in Abramowitz and Stegun (1972, p. 924). The error is given by

$$R_N[g] = \frac{N!\sigma^{2N}}{(2N)!} g^{(2N)}(\eta), \qquad -\infty < \eta < \infty. \qquad (E.29)$$

It can be seen from this example that the range of the random variable ξ need not be finite for the application of a Gauss quadrature rule to the approximation of $E[g(\xi)]$. $\square$

Computation of Gauss quadrature rules

In the example just shown, the orthogonal polynomials relevant to the computation of the Gauss quadrature rules are well-known classical polynomials. In most instances, however, the probability density function of ξ does not give rise to any polynomial set available in tabular form. In these situations the Gauss quadrature rule must be computed; the relevant fact here is that the set $\{x_i, w_i\}_{i=1}^N$ can be evaluated on the basis of the moments $\mu_1, \mu_2, \ldots, \mu_{2N-1}$ of ξ. In other words, no more than the first $2N - 1$ moments of ξ are necessary to determine explicitly a Gauss quadrature rule with N weights and abscissas.[†]

To see how to undertake this, we use the fact that a Gauss quadrature rule with N weights and abscissas has degree of precision $2N - 1$. Since

$$\mu_k \triangleq E[\xi^k] = \int_a^b x^k f_\xi(x)\, dx, \qquad (E.30)$$

[†] It is often claimed that μ_{2N} is also needed to perform this task; see for instance Golub and Welsch (1969) and Benedetto, de Vincentiis, and Luvison (1973). Actually, the role of μ_{2N} is just that of normalizing the polynomial $P_N(\cdot)$, and its values affect neither the abscissas nor the weights of the quadrature rule (Gautschi, 1970).

for any $0 \leq k \leq 2N - 1$ the Gauss quadrature rule is exact; that is,

$$\mu_k = \sum_{i=1}^{N} w_i x_i^k, \qquad k = 0, 1, \ldots, 2N - 1. \tag{E.31}$$

This system of $2N$ nonlinear equations has the weights and abscissas as unknowns; by solving it, the Gauss quadrature rule can be found.

In general, it is not convenient to solve directly equations (E.31), except for a very few simple cases. A computationally effective technique to determine Gauss quadrature rules based on the moments of ξ has been proposed in Golub and Welsch (1969). The reader is referred to the original paper for details about the computational algorithms. Here it suffices to say that Golub and Welsch's technique consists essentially of two steps: (1) evaluation of the coefficients $(\alpha_n)_{n=1}^{N}$, $(\beta_n)_{n=1}^{N-1}$ of the three-term recurrence relationship satisfied by the polynomials $P_0(x), \ldots, P_N(x)$ orthogonal with respect to the random variable ξ:

$$\beta_n P_n(x) = (x - \alpha_n)P_{n-1}(x) - \beta_{n-1}P_{n-2}(x),$$
$$n = 1, 2, \ldots, N, \quad P_{-1}(x) \equiv 0, \quad P_0(x) \equiv 1, \tag{E.32}$$

and (2) generation of a symmetric tridiagonal matrix whose entries depend on $(\alpha_n)_{n=1}^{N}$, $(\beta_n)_{n=1}^{N-1}$. The weights $(w_n)_{n=1}^{N}$ are found as the first components of the orthonormalized eigenvectors of this matrix, whereas the abscissas $(x_n)_{n=1}^{N}$ are the corresponding eigenvalues. The computations can be performed on the basis of the knowledge of the moments $\mu_1, \ldots, \mu_{2N-1}$. First, the Gram matrix $\mathbf{M}$ of the moments is formed. It has entries

$$(\mathbf{M})_{ij} = \mu_{i+j}, \qquad i, j = 0, \ldots, N.$$

Notice that $\mathbf{M}$ also includes the moment μ_{2N} as an element. Anyhow, as discussed before, the exact value of μ_{2N} is irrelevant; thus, if μ_{2N} is unknown, any value for μ_{2N} will suffice, provided that $\mathbf{M}$ is positive definite. Let $\mathbf{M} = \mathbf{\Gamma}\mathbf{\Gamma}'$ be the Cholesky decomposition of $\mathbf{M}$, where $\mathbf{\Gamma}$ is a lower triangular matrix with positive diagonal entries (see, e.g., Golub and Van Loan, 1983, p. 88). The elements of $\mathbf{\Gamma}$, which can be computed using standard recursive formulas, provide the coefficients of the recursion (E. 32).

Round-off errors in Gauss quadrature rules

In principle, if a sufficiently large number of moments of ξ is available, the error term (E.27) can be made as small as desired by increasing the number N of abscissas and weights of the quadrature rule. In fact, as $N \to \infty$ the RHS of (E.24) converges to the value of the LHS "for almost any conceivable function" $f_\xi(\cdot)$ "which one meets in practice" (Stroud and Secrest, 1966, p. 13). However, this is not true in practice, essentially because the moments of ξ needed in the computation are not known with infinite accuracy. Computational experience shows that the Cholesky decomposition of the moment matrix $\mathbf{M}$ is the crucial step of the algorithm for the computation of Gauss quadrature rules, since $\mathbf{M}$ gets increasingly ill-conditioned with increasing N. Round-off errors may cause the computed $\mathbf{M}$ to be no longer positive definite. Thus its Cholesky decomposition cannot be performed because it implies taking the square root of negative numbers (Luvison and Pirani, 1979). In practice, values of N greater than 7 or 8 can rarely be achieved; the accuracy thus obtained is, however, satisfactory in most situations.

E.3 MOMENT BOUNDS

We have seen in the preceding section that the quadrature rule approach allows $E[g(\xi)]$ to be approximated in the form of a linear combination of values of $g(\cdot)$. This is equivalent to substituting, for the actual probability density function $f_\xi(x)$, a discrete density in the form

$$f_\xi(x) \cong \sum_{i=1}^{N} w_i \delta(x - x_i), \tag{E.33}$$

where $\{w_i, x_i\}_{i=1}^{N}$ are chosen so as to match the first $2N - 1$ moments of ξ according to (E.31).

A more refined approach can be taken by looking for *upper and lower bounds* to $E[g(\xi)]$, still based on the moments of ξ. In particular, we can set the goal of finding bounds to $E[g(\xi)]$ that are in some sense *optimum* (i.e., they cannot be further tightened with the available informations on ξ). The problem can be formulated as follows: given a random variable ξ with range in the finite interval $[a, b]$, whose first M moments μ_1, $\mu_2, \ldots, \mu_M$ are known, we want to find the sharpest upper and lower bounds to the integral

$$E[g(\xi)] = \int_a^b g(x) f_\xi(x)\, dx, \tag{E.34}$$

where $g(\cdot)$ is a known function and $f_\xi(\cdot)$ is the (unknown) probability density function of ξ. To solve this problem, we look at the set of all possible $f_\xi(\cdot)$ whose range is $[a, b]$ and whose first M moments are $\mu_1, \ldots, \mu_M$. Then we compute the maximum and minimum value of (E.34) as $f_\xi(\cdot)$ runs through that set. The bounds obtained are optimum, because it is certain that a pair of random variables exists, say ξ' and ξ'', with range in $[a, b]$ and meeting the lower and the upper bound, respectively, with the equality sign.

This extremal problem can be solved by using a set of results due essentially to the Russian mathematician M. G. Krein (see Krein and Nudel'man, 1977). These results can be summarized as follows.

(a) If the function $g(\cdot)$ has continuous $(M + 3)$th derivative, and $g^{(M+3)}(\cdot)$ is everywhere in $[a, b]$ nonnegative, then the optimum bounds to $E[g(\xi)]$ are in the form

$$\sum_{i=1}^{N'} w_i' g(x_i') \leq E[g(\xi)] \leq \sum_{i=1}^{N''} w_i'' g(x_i''). \tag{E.35}$$

This is equivalent to saying that the two "extremal" probability density functions are discrete, which allows the upper and lower bounds to be written in the form of quadrature rules. If $g^{(M+3)}(\cdot)$ is nonpositive instead of being nonnegative, it suffices to consider $-g(\cdot)$ instead of $g(\cdot)$.

(b) *If M is odd*, then $N' = (M + 1)/2$ and $N'' = (M + 3)/2$. Also, $\{w_i', x_i'\}_{i=1}^{N'}$ is a Gauss quadrature rule, and $\{w_i'', x_i''\}_{i=1}^{N''}$ is the quadrature rule having the maximum degree of precision (i.e., $2N'' + 1$) under the constraints $x_1'' = a$, $x_{N''}'' = b$. *If M is even*, then

$N' = N'' = (M + 2)/2$. Also, $\{w_i', x_i'\}_{i=1}^{N'}$ (respectively, $\{w_i'', x_i''\}_{i=1}^{N''}$) is the quadrature rule having the maximum achievable degree of precision (i.e., $2N'$) under the constraint $x_1' = a$ (respectively, $x_{N''}'' = b$).

A technical condition involved with the derivation of these results requires that the Gram matrix of the moments $\mu_1, \ldots, \mu_M$ be positive definite. For our purpose, a simple sufficient condition is that the cumulative distribution function of ξ have more than $M + 1$ points of increase. If ξ is a continuous random variable, this condition is immediately satisfied. If ξ is a discrete random variable, it means that ξ must take on more than $M + 1$ values. As $M + 1$ is generally a small number, the latter requirement is always satisfied in practice; otherwise, the value of $E[g(\xi)]$ can be evaluated explicitly, with no need to bound it.

Computation of moment bounds

Once the moments $\mu_1, \ldots, \mu_M$ have been computed, in order to use Krein's results explicitly the quadrature rules $\{w_i', x_i'\}_{i=1}^{N'}$ and $\{w_i'', x_i''\}_{i=1}^{N''}$ must be evaluated. From the preceding discussion it will not be surprising that the algorithms for computing the moment bounds (E.35) bear a close resemblance to those developed for computing Gauss quadrature rules. Indeed, the task is still to find abscissas and weights of a quadrature rule achieving the maximum degree of precision, possibly under constraints about the location of one or two abscissas. Several algorithms are available for this computation (Yao and Biglieri, 1980; Omura and Simon, 1980). Here we shall briefly describe one of them, based on the assumption that Golub and Welsch's algorithm

TABLE E.1 Computation of abscissas and weights for moment bounds

	Input moments	Output abscissas and weights
M Odd Lower bound	$\mu_i, \quad i = 1, \ldots, M$	$x_i, w_i, \quad i = 1, \ldots, \dfrac{M+1}{2}$
M Odd Upper bound	$\dfrac{\mu_{i+2} - (a+b)\mu_{i+1} + ab\mu_i}{\mu_2 - (a+b)\mu_1 + ab}, \quad i = 1, \ldots, M-2$	$x_i, w_i, \quad i = 1, \ldots, \dfrac{M-1}{2}$
M Even Lower bound	$\dfrac{\mu_{i+1} - a\mu_i}{\mu_1 - a}, \quad i = 1, \ldots, M-1$	$x_i, w_i, \quad i = 1, \ldots, \dfrac{M}{2}$
M Even Upper bound	$\dfrac{b\mu_i - \mu_{i+1}}{b - \mu_1}, \quad i = 1, \ldots, M-1$	$x_i, w_i, \quad i = 1, \ldots, \dfrac{M}{2}$

described in Section E.3 has been implemented. In particular, we shall show how the known moments of ξ must be modified to use them as inputs to that algorithm and how the weights and abscissas obtained at its output must be modified. These computations are summarized in Fig. E.1 and Table E.1. By using the modified moments v_i of the first column of Table E.1 as an input to Golub and Welsch's (or an equivalent) algorithm, the abscissas and weights of the second column are obtained. Abscissas and weights for moment bounds, to be used in (E.35), are then obtained by performing the operations shown in the third and fourth columns, respectively.

Figure E.1 Summary of the computations to be performed for the evaluation of moment bounds.

These computations will yield tighter bounds as the number M of available moments increases. However, as was discussed in the context of Gauss quadrature rules, M cannot be increased without bound because of computational instabilities. In practice, it is rarely possible to increase M beyond 15 or so. But this gives a sufficient accuracy in most applications.

Example E.2   (Yao and Biglieri, 1980)

As an example of application of theories presented in this section, consider the function $g(\cdot)$ given in (E.6). Using the expression (E.7) for the kth derivative of the error function,

Abscissas for moment bounds	Weights for moment bounds
$x_i' = x_i, \quad i = 1, \ldots, \dfrac{M+1}{2}$	$w_i' = w_i, \quad i = 1, \ldots, \dfrac{M+1}{2}$
$x_1'' = a$	$w_1'' = \dfrac{b - \mu_1 - b \sum\limits_{i=2}^{(M+1)/2} w_i'' + \sum\limits_{i=2}^{(M+1)/2} w_i'' x_i''}{b - a}$
$x_i'' = x_{i-1}, \quad i = 2, \ldots, \dfrac{M+1}{2}$	$w_i'' = \dfrac{-\mu_2 + (a+b)\mu_1 - ab}{(x_i - a)(b - x_i)} w_i, \quad i = 2, \ldots, \dfrac{M+1}{2}$
$x_{(M+3)/2}'' = b$	$w_{(M+3)/2}'' = \dfrac{\mu_1 - a + a \sum\limits_{i=2}^{(M+1)/2} w_i'' - \sum\limits_{i=2}^{(M+1)/2} w_i'' x_i''}{b - a}$
$x_i' = a$	$w_1' = \sum\limits_{i=2}^{(M+2)/2} w_i'$
$x_i' = x_{i-1}, \quad i = 2, \ldots, \dfrac{M+2}{2}$	$w_i' = \dfrac{\mu_1 - a}{x_i' - a} w_i, \quad i = 2, \ldots, \dfrac{M+2}{2}$
$x_i'' = x_i, \quad i = 1, \ldots, \dfrac{M}{2}$	$w_i'' = \dfrac{b - \mu_1}{b - x_i'} w_i, \quad i - 1, \ldots, \dfrac{M}{2}$
$x_{(M+2)/2}'' = b$	$w_{(M+2)/2}'' = 1 - \sum\limits_{i=1}^{M/2} w_i''$

we get

$$g^{(k)}(x) = \left\{ (-1)^k \frac{2}{\sqrt{\pi}(\sqrt{2}\sigma)^k} \exp\left[-\frac{(h+x)^2}{2\sigma^2} \right] \right\} H_{k-1}\left(\frac{h+x}{\sqrt{2}\sigma} \right). \tag{E.36}$$

For a fixed integer k, the sign of the bracketed term in (E.36) does not depend on the value of x. To the contrary, $H_{k-1}[(h+x)/\sqrt{2}\sigma]$, a polynomial having only simple zeros, will change sign whenever its argument crosses a zero value. Hence, $g^{(M+3)}(\cdot)$ is continuous and of the same sign in $[a, b]$ if and only if $[a, b]$ does not contain a root of the equation

$$H_{M+2}\left(\frac{h+x}{\sqrt{2}\sigma} \right) = 0$$

as an interior point. Furthermore, a simple sufficient condition for $g^{(M+3)}(\cdot)$ to be of the same sign in $[a, b]$ is that the largest root of the preceding equation is smaller than a; that is

$$\frac{h+a}{\sqrt{2}\sigma} < z_{max}^{(M+2)}, \tag{E.37}$$

where $z_{max}^{(M+2)}$ is the largest zero of $H_{M+2}(z)$. Table E.2 shows the values of the largest zeros of the Hermite polynomials of degrees 3 to 20.

TABLE E.2 The largest zero of Hermite polynomial $H_k(z)$

k	$z_{max}^{(k)}$
3	1.22474
4	1.65068
5	2.02018
6	2.35060
7	2.65196
8	2.93064
9	3.19099
10	3.43616
11	3.66847
12	3.88972
13	4.10134
14	4.30445
15	4.49999
16	4.68874
17	4.87135
18	5.04836
19	5.22027
20	5.38748

As a simple computation, let $a = -1$, $b = 1$, $M = 3$, and

$$\mu_1 = 0, \qquad \mu_2 = s^2, \qquad \mu_3 = 0.$$

We obtain in this case

$$x_1' = -s, \qquad x_2' = s$$

and

$$w'_1 = w'_2 = \tfrac{1}{2}.$$

For the upper bound we get

$$x''_1 = -1, \qquad x''_2 = 0, \qquad x''_3 = 1$$

and

$$w''_1 = \frac{s^2}{2}, \qquad w''_2 = 1 - s^2, \qquad w''_3 = \frac{s^2}{2},$$

so the moment bounds for $g(\cdot)$ as in (E.6) are

$$\frac{1}{2} erfc\left(\frac{h-s}{\sqrt{2}\sigma}\right) + \frac{1}{2} erfc\left(\frac{h+s}{\sqrt{2}\sigma}\right) \le E\left\{ erfc\left(\frac{h+\xi}{\sqrt{2}\sigma}\right)\right\}$$

$$\le \frac{s^2}{2} erfc\left(\frac{h-1}{\sqrt{2}\sigma}\right) + (1 - s^2)\, erfc\left(\frac{h}{\sqrt{2}\sigma}\right) + \frac{s^2}{2} erfc\left(\frac{h+1}{\sqrt{2}\sigma}\right). \qquad (E.38)$$

These inequalities are valid provided that

$$\frac{h-1}{\sqrt{2}\sigma} \ge 2.02018,$$

which shows that, for $h > 1$, σ must be small enough to make this technique applicable. For $h \le 1$, (E.38) may not be valid. $\square$

E.4 APPROXIMATING THE AVERAGE E [g(ξ, η)]

Before ending this appendix, we shall briefly discuss the problem of evaluating approximations of the average $E[g(\xi, \eta)]$, where $g(\cdot, \cdot)$ is a known deterministic function, and ξ, η are two correlated random variables with range in a region R of the plane. Exact computation of this average requires knowledge of the joint probability density function $f_{\xi,\eta}(x, y)$ of the pair of random variables ξ, η. This may not be available, or the evaluation of the double integral

$$E[g(\xi, \eta)] = \iint_R g(x, y) f_{\xi,\eta}(x, y)\, dx\, dy \qquad (E.39)$$

may be unfeasible. In practice, it is often exceedingly easier to compute a small number of joint moments

$$\mu_{l,m} \triangleq E[\xi^l \eta^m], \qquad l, m = 0, 1, \ldots, M, \qquad (E.40)$$

and use this information to obtain $E[g(\xi, \eta)]$.

The first technique that can be used to this purpose is based on the expansion of $g(\xi, \eta)$ in a Taylor's series. The terms of this series will involve products $\xi^l \eta^m$ ($l, m = 0, 1, \ldots$), so that truncating the series and averaging it termwise will provide the desired approximation.

Another possible technique uses *cubature rules*, a two-dimensional generalization of quadrature rules discussed in Section E.3. With this approach, the approximation of $E[g(\xi, \eta)]$ takes the form

$$E[g(\xi, \eta)] \cong \sum_{i=1}^{N} w_i g(x_i, y_i). \tag{E.41}$$

As a generalization of the one-dimensional case, we say that the cubature rule $\{w_i, x_i, y_i\}_{i=1}^{N}$ has degree of precision v if (E.41) holds with the equality sign whenever $g(x, y)$ is a polynomial in x and y of degree $\leq v$, but not for all polynomials of degree $v + 1$. Unfortunately, construction of cubature rules with maximum degree of precision is, generally, an unsolved problem, and solutions are only available in some special cases. For example, in Mysovskih (1968) a cubature rule with degree of precision 4 and $N = 6$ is derived. This is valid when the region R and the function $f_{\xi,\eta}(\cdot, \cdot)$ are symmetric relative to both coordinate axes; thus, we must have

$$f_{\xi,\eta}(x, y) = f_{\xi,\eta}(-x, y) = f_{\xi,\eta}(x, -y), \qquad (x, y) \in \text{R}. \tag{E.42}$$

With these assumptions, $\mu_{ik} = 0$ if at least one of the numbers i and k is odd. The moments needed for the computation of the cubature rule are then $\mu_{2,0}$, $\mu_{0,2}$, $\mu_{4,0}$, $\mu_{0,4}$, and $\mu_{2,2}$. Under the same symmetry assumptions, a cubature rule with $N = 19$ and degree of precision 9 can be obtained by using the moments $\mu_{2,0}$, $\mu_{0,2}$, $\mu_{4,0}$, $\mu_{0,4}$, $\mu_{2,2}$, $\mu_{2,4}$, $\mu_{4,2}$, $\mu_{0,6}$, $\mu_{6,0}$, $\mu_{2,6}$, $\mu_{6,2}$, $\mu_{8,0}$, $\mu_{0,8}$, and $\mu_{4,4}$ (Piessens and Haegemans, 1975).

If a higher degree of precision is sought or the symmetry requirements are not satisfied, one can resort to "good" cubature rules that can be computed through the joint moments (E.40). Formulas of the type

$$E[g(\xi, \eta)] = \sum_{i=1}^{N_x} \sum_{j=1}^{N_y} w_{ij} g(x_i, y_j) \tag{E.43}$$

with degree of precision $v = \min(N_x - 1, N_y - 1)$ can be found by using the moments $\mu_{h,0}$, $\mu_{0,k}$, $h = 1, \ldots, 2N_x$, $k = 1, \ldots, 2N_y$, and $\mu_{h,k}$, $h = 1, \ldots, N_x - 1$, $k = 1, \ldots, N_y - 1$. Equivalent algorithms for the computation of weights and abscissas in (E.43) were derived in Luvison and Navino (1976) and Omura and Simon (1980).

We conclude by commenting briefly on the important special case in which the two random variables ξ and η are *independent*. In this situation, by using moments of ξ one can construct the Gauss quadrature rule $\{w_i, x_i\}_{i=1}^{N_x}$, and by using moments of η one can similarly obtain the Gauss quadrature rule $\{u_j, y_j\}_{j=1}^{N_y}$. Then it is a simple matter to show that the following cubature rule can be obtained:

$$E[g(\xi, \eta)] = \sum_{i=1}^{N_x} \sum_{j=1}^{N_y} w_i u_j g(x_i, y_j), \tag{E.44}$$

and this has degree of precision $v = \min(2N_x - 1, 2N_y - 1)$.

APPENDIX F

The Viterbi Algorithm

Consider K real scalar functions $\lambda_0(\tau_0)$, $\lambda_1(\tau_1)$, . . . , $\lambda_{K-1}(\tau_{K-1})$, whose arguments τ_0, . . . , τ_{K-1} can take on a finite number of values, and their sum

$$\lambda(\tau_0, \tau_1, \ldots , \tau_{K-1}) \triangleq \sum_{l=0}^{K-1} \lambda_l(\tau_l). \tag{F.1}$$

In this appendix we consider the problem of computing the minimum of $\lambda(\cdot)$, that is, the quantity

$$\mu \triangleq \min_{\{\tau_0,\tau_1,\ldots,\tau_{K-1}\}} \sum_{l=0}^{K-1} \lambda_l(\tau_l). \tag{F.2}$$

Of course, this problem is trivial when the τ_l's are in a sense "independent" (i.e., the set of values that each of them can assume does not depend on the value taken by the other variables). In this case, the solution of the problem is obvious:

$$\mu = \sum_{l=0}^{K-1} \min_{\tau_l} \lambda_l(\tau_l), \tag{F.3}$$

and the minimization of a function of K variables is reduced to the minimization of K functions of one variable. In mathematical parlance, this situation corresponds to the case in which the range of the variables τ_0, τ_1, . . . , τ_{K-1} is the direct product of the ranges of τ_l, $0 \le l \le K-1$.

Instead, let us consider a more general situation in which the range of τ_0, . . . , τ_{K-1} is something different from this direct product. In this case, the value taken on by any of the τ_l affects the range of the remaining variables, and (F.3) cannot be applied.

Example F.1

A tourist wants to travel by car from Los Angeles to Denver in five days. He does not want to drive for more than 350 miles in a day and wants to spend every night in a motel of his favorite chain (say, Motel 60). With these constraints, the routes among which he can choose are summarized in Fig. F.1. The best choice will be the one minimizing the total mileage.

Clearly, the total mileage of a given route is the sum of five terms, where each represents the distance driven in a day. Hence, this shortest-route problem has the form of (F.2), where τ_l is the route followed on the lth day, and $\lambda_l(\tau_l)$ is its length. Also, (F.3) cannot hold, as, for example, the choice to travel from Los Angeles to Blythe on the first day will minimize $\lambda_1(\tau_1)$, but will result in a $\lambda_2(\tau_2)$ value of 228, which is not its minimum value. ☐

When the "trivial" solution (F.3) is not valid, in principle one can solve (F.2) by computing all the possible values for $\lambda(\tau_0, \ldots, \tau_{K-1})$ (which are in a finite number) and choosing the smallest. In certain cases (for instance in Example F.1) this can be done; but here we assume that this task is computationally impractical because of the large number of values to be enumerated. Here we shall look for an algorithm that allows us to avoid the brute-force approach. To this end, we shall first describe a sequential algorithm to compute μ in (F.2) and then investigate under which assumption such an algorithm can be simplified and to what extent.

A sequential minimization algorithm

When the variables $\tau_0, \ldots, \tau_{K-1}$ are not "independent" (in the sense previously stated), if we attempt to minimize $\lambda(\cdot)$ with respect to τ_0 alone, our result will depend on the values of $\tau_1, \ldots, \tau_{K-1}$. This is equivalent to stating that the minimum found

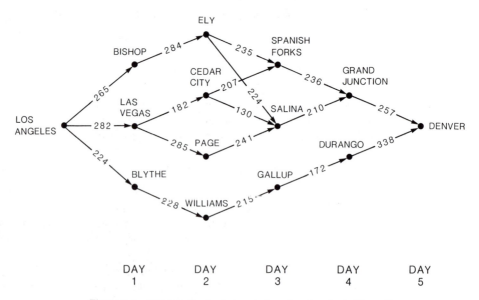

Figure F.1 Which is the shortest route from Los Angeles to Denver?

will be a function of $\tau_1, \ldots, \tau_{K-1}$, say $\mu_1(\tau_1, \ldots, \tau_{K-1})$. Observe also that the minimization of $\lambda(\cdot)$ with respect to τ_0 will involve only the function $\lambda_0(\tau_0)$ and not the remaining terms of the summation in (F.1). The resulting function of $\tau_1, \ldots, \tau_{K-1}$ can be further minimized with respect to τ_1. This operation will only involve $\mu_1(\tau_1, \ldots, \tau_{K-1}) + \lambda_1(\tau_1)$. The result is now a function of $\tau_2, \ldots, \tau_{K-1}$, say $\mu_2(\tau_2, \ldots, \tau_{K-1})$. Repeating this procedure a sufficient number of times, we finish with a function of τ_{K-1} alone, which can be finally minimized to provide us with the desired value μ.

This recursive procedure can be formalized in the following manner. Denoting by $\{\tau_l | \tau_{l+1}, \ldots, \tau_{K-1}\}$, $l = 0, \ldots, K - 2$, the set of values that τ_l is allowed to take on once the values of $\tau_{l+1}, \ldots, \tau_{K-1}$ have been chosen, the procedure involves the following steps:

$$\mu_1(\tau_1, \ldots, \tau_{K-1}) = \min_{\{\tau_0 | \tau_1, \ldots, \tau_{K-1}\}} \lambda_0(\tau_0)$$

$$\mu_l(\tau_l, \ldots, \tau_{K-1}) = \min_{\{\tau_{l-1} | \tau_l, \ldots, \tau_{K-1}\}} [\mu_{l-1}(\tau_{l-1}, \ldots, \tau_{K-1}) + \lambda_{l-1}(\tau_{l-1})],$$

$$l = 2, \ldots, K - 1, \qquad \text{(F.4)}$$

$$\mu = \min_{\tau_{K-1}} \mu_{K-1}(\tau_{K-1})$$

The Viterbi algorithm

Simplifications of the basic algorithm of (F.4) can be derived from the specific structure of the sets $\{\tau_l | \tau_{l+1}, \ldots, \tau_{K-1}\}$. The simplest possible situation arises when, for all $l = 0, \ldots, K - 2$, we have

$$\{\tau_l | \tau_{l+1}, \ldots, \tau_{K-1}\} = \{\tau_l\}; \qquad \text{(F.5)}$$

that is, the values that τ_l can take on are not influenced by those of $\tau_{l+1}, \ldots, \tau_{K-1}$. This is the case where we deal with "independent" variables and we see that (F.4) reduces to (F.3).

A more interesting situation arises when we consider the second simplest case, in which, for all $l = 0, \ldots, K - 2$,

$$\{\tau_l | \tau_{l+1}, \ldots, \tau_{K-1}\} = \{\tau_l | \tau_{l+1}\}; \qquad \text{(F.6)}$$

that is, the values that each τ_l is allowed to take on depend only on τ_{l+1}.

In this case, (F.4) simplifies to

$$\mu_1(\tau_1) = \min_{\{\tau_0 | \tau_1\}} \lambda_0(\tau_0),$$

$$\mu_l(\tau_l) = \min_{\{\tau_{l-1} | \tau_l\}} [\mu_{l-1}(\tau_{l-1}) + \lambda_{l-1}(\tau_{l-1})], \qquad l = 2, \ldots, K - 1, \qquad \text{(F.7)}$$

$$\mu = \min_{\tau_{K-1}} \mu_{K-1}(\tau_{K-1}),$$

which is the celebrated *Viterbi algorithm*.

The Viterbi algorithm can be given an interesting formulation once the minimization problem to be solved has been formulated as the task of finding the shortest route through a graph. To do this, let us describe the general model that in its various forms leads to application of the Viterbi algorithm in this book. Consider an information source generating a finite sequence $\xi_0, \xi_1, \ldots, \xi_{K-1}$ of independent symbols that can take on a finite number M of values. These symbols are fed sequentially to a system whose lth output, say x_l, depends in a deterministic, time-invariant way on the present input and the L previous inputs:

$$x_l = g(\xi_l, \xi_{l-1}, \ldots, \xi_{l-L}). \tag{F.8}$$

The sequence (x_l) can be thought of as generated from a shift register, as shown in Fig. F.2. With this model, it is usual to define the *state* of the shift register at the emission of symbol ξ_l as the vector of the symbols contained in its cells, that is,

$$\sigma_l \triangleq (\xi_{l-1}, \ldots, \xi_{l-L}), \tag{F.9}$$

so we can say that the output x_l depends on the present input ξ_l and on the state σ_l of the shift register:

$$x_l = g(\xi_l, \sigma_l). \tag{F.10}$$

When the source emits the symbol ξ_{l+1}, the shift register is brought to the state $\sigma_{l+1} = (\xi_l, \xi_{l-1}, \ldots, \xi_{l-L+1})$. Now we can define the *transition* between these two states as

$$\tau_{l+1} \triangleq (\sigma_l, \sigma_{l+1}). \tag{F.11}$$

Determination of the range of index l in (F.8) to (F.11) requires some attention, because for $l < L$ and for $l > K$ the function $g(\cdot)$ in (F.8) has a number of arguments

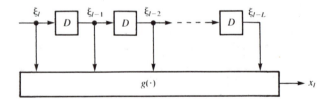

Figure F.2 Generation of a shift-register state sequence.

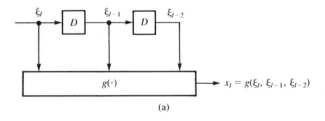

(a)

Figure F.3(a) Example of a shift-register state sequence.

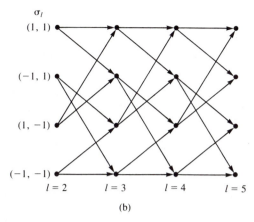

σ_l

(1, 1)

(−1, 1)

(1, −1)

(−1, −1)

$l = 2$ $l = 3$ $l = 4$ $l = 5$

(b)

Figure F.3(b) The associated trellis diagram.

that are not defined. A possible choice is to assume $\xi_l = 0$ for $l < 0$ and $l > K - 1$, and to define the function $g(\cdot)$ accordingly. In this case, l can be assumed to range from 0 to $K + L - 1$. Otherwise, we may want $g(\cdot)$ to be defined only when its $L + 1$ arguments belong to the range of the symbols ξ_l. In this situation, we must assume l in (F.8) to range from L to $K - 1$, which implies in particular that each state can take on M^L values and each transition M^{L+1} values.

Consider now a graphical representation of this process, in the form of a *trellis*. This is a treelike graph with remerging branches, where the nodes on the same vertical line represent distinct states for a given l (this index is commonly called the *time*), and the branches represent transitions from one state to the next. An example should clarify this point.

Example F.2

Consider a shift-register state sequence with $L = 2$, $M = 2$, $\xi_l \in \{-1, 1\}$, and $K = 5$. For $2 < l < 5$, the states σ_l can take on four values: $(\pm 1, \pm 1)$. For $l < 2$ and $l > 5$, we should also consider states including one or two zeros, and assume that $g(\cdot)$ is also defined when one or more of its arguments are zero. In this situation, Fig. F.3b depicts the trellis diagram for $2 \le l \le 5$. Notice that the actual form of $g(\cdot)$ is irrelevant with respect to the structure of the trellis diagram. $\square$

Let us now return to the original minimization problem, to be formulated by way of the use of the trellis diagram defined. It is sufficient to assume that the function to be minimized, the one defined in (F.1), has as its arguments the transitions (F.11) between states, so that the values $\lambda_l(\tau_l)$ can be associated with the branches of the trellis (these are usually called the *lengths* or the *metrics* of the branches). Stated this way, it is easy to see that the set of variables minimizing $\lambda(\cdot)$ corresponds to the minimum-length path through the trellis. Also, it is relatively simple to show that (F.6) holds in this situation. Hence, the Viterbi algorithm can be applied. It suffices to observe that the sequence $\tau_{i+1}, \ldots, \tau_{K-1}$ corresponds to a path in the trellis from σ_i to σ_{K-1}, and that the set of transitions τ_i compatible with such a path is only determined by σ_{i+1}.

The Viterbi Algorithm   Appendix F   **613**

Example F.3

Let us illustrate the application of the Viterbi algorithm to a minimization problem formulated with a trellis diagram. Figure F.4 shows a trellis whose branches are labeled according to their respective lengths. Figure F.5 shows the five steps to be performed to determine the shortest path through this trellis, as briefly described in the following.

With reference to (F.7), the first step in the algorithm is to choose, for each τ_1 (or, equivalently, for each state σ_1), the branch leading to σ_1 and having the minimum length. (This is trivial, because there is only one such branch in our example.) For each value of σ_1, store this shortest path and its length, which corresponds to the value of μ_1 in (F.7). Extend now the paths just selected by one branch. For each state σ_3 select the branch leading to it such that the sum of its length λ_1 plus the value of μ_1 just stored is at a minimum. Store these unique paths, together with their total lengths μ_2.

Extend again these paths by one branch. For each state σ_3 select the branch leading to it such that $\lambda_2 + \mu_2$ is at a minimum and store these minimum paths, together with their lengths. Similar steps should be performed for $l = 4$ and $l = 5$; when $l = 5$ the algorithm is terminated. Then we are left with a single path, the shortest one, together with its length μ. These steps are illustrated in Fig. F.5. $\square$

An interesting fact can be observed from this example. At step 4 (see Fig. F.5d) we see that all the paths selected so far have a common part. In fact, there is a *merge* in that all these paths pass through a single node (the uppermost one for $l = 3$). Clearly, whatever happens from now on will not change anything before this merge. Hence we can deduce that the optimum path will certainly include the first three branches of the path depicted in Fig. F.5d.

Complexity of the Viterbi algorithm

Finally, let us briefly discuss the computational complexity of the Viterbi algorithm. Let N_s denote the maximum number of states in the trellis diagram for any l, and N_t the maximum number of branches leading from the nodes corresponding to time index l to the nodes corresponding to time index $l + 1$, for any l (e.g., $N_s = 4$ and $N_t = 8$ in the trellis diagram of Fig. F.4). As far as memory is concerned, the algorithm requires no more than N_s storage locations, each capable of storing a sequence of paths and their lengths. The computations to perform in each unit of time are no more than N_t additions and N_s comparisons.

As for a shift-register sequence $N_s = M^L$ and $N_t = M^{L+1}$, the complexity of the

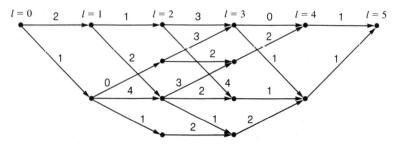

Figure F.4 Trellis labeled with branch lengths.

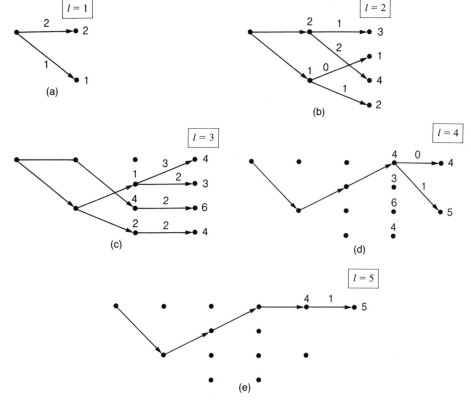

Figure F.5 Finding the minimum path through a trellis.

Viterbi algorithm increases exponentially with the length L of the shift register. Notice that the amount of computations required for the minimization of the function $\lambda(\cdot)$ defined in (F.1) grows *only linearly* with K, whereas the exhaustive enumeration of all the values of $\lambda(\cdot)$ would require a number of computations growing *exponentially* with K.

BIBLIOGRAPHICAL NOTES

The Viterbi algorithm was proposed by Viterbi (1967) as a method for decoding convolutional codes (see Chapter 9). Since then, it has been applied to a number of minimization problems arising in demodulation of digital signals generated by a modulator with memory (see Chapter 4) or in sequence estimation for channels with intersymbol interference (Chapters 7 and 10). A survey of applications of the Viterbi algorithm, as well as a number of details regarding its implementation, can be found in Forney (1973). The connections between Viterbi algorithm and dynamic programming techniques were first recognized by Omura (1969).

References

1. Aaron, M. R., and D. W. Tufts (1966), "Intersymbol interference and error probability," *IEEE Transactions on Information Theory*, vol. IT-12, pp. 24–36.
2. Abend, K., and B. D. Fritchman (1970), "Statistical detection for communication channels with intersymbol interference," *IEEE Proceedings*, vol. 58, pp. 779–785.
3. Abramowitz, M., and I. A. Stegun (eds.) (1972), *Handbook of Mathematical Functions*. New York: Dover Publications.
4. Ajmone Marsan, M., and E. Biglieri (1977), "Power spectra of complex PSK for satellite communications," *Alta Frequenza*, vol. 46, pp. 263–270.
5. Ajmone Marsan, M., S. Benedetto, E. Biglieri, and R. Daffara (1977), *Performance Analysis of a Nonlinear Satellite Channel Using Volterra Series*, Final Report to ESTEC Contract 2328/74 HP/Rider 1, Dipartimento di Elettronica, Politecnico di Torino (Italy).
6. Amoroso, F. (1976), "Pulse and spectrum manipulation in the MSK format," *IEEE Transactions on Communications*, vol. COM-24, pp. 381–384.
7. Amoroso, F. (1980), "The bandwidth of digital data signals," *IEEE Communications Magazine*, vol. 18, pp. 13–24, November.
8. Anderson, J. B., and J. R. Lesh (1981), "Guest editors' prologue," *IEEE Transactions on Communications*, vol. COM-29, pp. 185–186.
9. Anderson, J. B., and D. P. Taylor (1978), "A bandwidth efficient class of signal-space codes," *IEEE Transactions on Information Theory*, vol. IT-24, pp. 703–712.
10. Anderson, R. R., and G. J. Foschini (1975), "The minimum distance for MLSE digital data systems of limited complexity," *IEEE Transactions on Information Theory*, vol. IT-21, pp. 544–551.
11. Andrisano, O. (1982), "On the behaviour of 4Φ-PSK radio relay systems during multipath propagation," *Alta Frequenza*, vol. 51, pp. 112–126.
12. Arens, R. (1957), "Complex processes for envelopes of normal noise," *IEEE Transactions on Information Theory*, vol. IT-3, pp. 203–207.

13. Arsac, J. (1966), *Fourier Transforms and the Theory of Distributions*, translated by Allen Nussbaum and Gretcher C. Heim. Englewood Cliffs, N.J.: Prentice-Hall.
14. Arthurs, E., and H. Dym (1962), "On the optimum detection of digital signals in the presence of white Gaussian noise. A geometric interpretation and a study of three basic data transmission systems," *IRE Transactions on Communications Systems*, vol. CS-10, pp. 336–372.
15. Ash, R. B. (1967), *Information Theory*. New York: Wiley-Interscience.
16. Athans, M. (1968), "The matrix minimum principle," *Information and Control*, vol. 11, pp. 592–606.
17. Aulin, T., and C.-E. W. Sundberg (1981), "Continuous phase modulation—Part I: Full response signaling," *IEEE Transactions on Communications*, vol. COM-29, pp. 126–209.
18. Aulin, T., B. Persson, N. Rydbeck and C.-E. W. Sundberg (1982), *Spectrally Efficient Constant-Amplitude Digital Modulation Schemes for Communication Satellite Applications*, ESA Contract Report, May 1982.
19. Aulin, T., G. Lindell, and C.-E. W. Sundberg (1981), "Selecting smoothing pulses for partial response digital FM," *IEE Proceedings*, vol. 128, pt. F, pp. 237–244.
20. Aulin, T., N. Rydbeck, and C.-E. W. Sundberg (1981), "Continuous phase modulation— Part II: Partial response signaling," *IEEE Transactions on Communications*, vol. COM-29, pp. 210–225.
21. Austin, M. (1967), *Decision-Feedback Equalization for Digital Communication Over Dispersive Channels*, MIT Res. Lab. Electron. Tech. Rep. 461, August 1967.
22. Babler, G. M. (1973), "Selectively faded nondiversity and space diversity narrowband microwave radio channels," *Bell System Technical Journal*, vol. 52, pp. 239–261.
23. Barker, H. A., and S. Ambati (1972), "Nonlinear sampled-data system analysis by multidimensional z-transforms," *IEE Proceedings*, vol. 119, pp. 1407–1413.
24. Barnett, W. T. (1972), "Multipath propagation at 4, 6 and 11 GHz," *Bell System Technical Journal*, vol. 51, pp. 321–361.
25. Beare, C. T. (1978), "The choice of the desired impulse response in combined linear-Viterbi algorithm equalizers," *IEEE Transactions on Communications*, vol. COM-26, pp. 1301–1307.
26. Bedrosian, E. (1962), "The analytic signal representation of modulated waveforms," *IRE Proceedings*, vol. 50, pp. 2071–2076.
27. Bedrosian, E., and S. O. Rice (1971), "The output properties of Volterra systems driven by harmonic and Gaussian inputs," *IEEE Proceedings*, vol. 59, pp. 1688–1707.
28. Belfiore, C. A., and J. H. Park, Jr. (1979), "Decision feedback equalization," *IEEE Proceedings*, vol. 67, pp. 1143–1156.
29. Bellini, S., and G. Tartara (1985), "Efficient discriminator detection of partial response continuous phase modulation," *IEEE Transactions on Communications*, vol. COM-23, pp. 883–886.
30. Bellman, R. E. (1968), *Matrix Analysis*, 2nd ed. New York: McGraw-Hill.
31. Benedetto, S., and E. Biglieri (1974), "On linear receivers for digital transmission systems," *IEEE Transactions on Communications*, vol. COM-22, pp. 1205–1215.
32. Benedetto, S., and E. Biglieri (1983), "Nonlinear equalization of digital satellite channels," *IEEE Journal on Selected Areas in Communications*, vol. SAC-1, pp. 57–62.
33. Benedetto, S., E. Biglieri, and J. K. Omura (1981), "Optimum receivers for nonlinear satellite channels," *5th International Conference on Digital Satellite Communications*, Genova, Italy, March 1981.
34. Benedetto, S., E. Biglieri, and R. Daffara (1976), "Performance of multilevel baseband digital systems in a nonlinear environment," *IEEE Transactions on Communications*, vol. COM-24, pp. 1166–1175.

References

35. Benedetto, S., E. Biglieri, and R. Daffara (1979), "Modeling and performance evaluation of nonlinear satellite links—A Volterra series approach," *IEEE Transactions on Aerospace and Electronic Systems*, vol. AES-15, pp. 494–507.

36. Benedetto, S., E. Biglieri, and V. Castellani (1973), "Combined effects of intersymbol, interchannel, and co-channel interferences in M-ary CPSK systems," *IEEE Transactions on Communications*, vol. COM-21, pp. 997–1008.

37. Benedetto, S., G. De Vincentiis, and A. Luvison (1973), "Error probability in the presence of intersymbol interference and additive noise for multilevel digital signals," *IEEE Transactions on Communications*, vol. COM-21, pp. 181–188.

38. Benedetto, S., M. Ajmone Marsan, G. Albertengo, and E. Giachin (1987), "Combined coding and modulation: Theory and applications," *IEEE Transactions on Information Theory*, to be published.

39. Bennett, W. R., and S. O. Rice (1963), "Spectral density and autocorrelation functions associated with binary FSK," *Bell System Technical Journal*, vol. 42, pp. 2355–2385.

40. Benveniste, A., and M. Goursat (1984), "Blind equalizers," *IEEE Transactions on Communications*, vol. COM-32, pp. 871–883.

41. Berger, T. (1971), *Rate Distortion Theory*. Englewood Cliffs, N.J.: Prentice-Hall.

42. Berger, T., and D. W. Tufts (1967), "Optimum pulse amplitude modulation. Part I: Transmitter–receiver design and bounds from information theory," *IEEE Transactions on Information Theory*, vol. IT-13, pp. 196–208.

43. Berlekamp, E. R. (1968), *Algebraic Coding Theory*. New York: McGraw-Hill.

44. Berlekamp, E. R., ed. (1974), *Key Papers in the Development of Coding Theory*. New York: IEEE Press.

45. Biglieri, E. (1984), "High-level modulation and coding for nonlinear satellite channels," *IEEE Transactions on Communications*, vol. COM-32, pp. 616–626.

46. Biglieri, E., A. Gersho, R. D. Gitlin, and T. L. Lim (1984), "Adaptive cancellation of nonlinear intersymbol interference for voiceband data transmission," *IEEE Journal on Selected Areas in Communications*, vol. SAC-2, pp. 765–777.

47. Biglieri, E., M. Elia, and L. Lo Presti (1984), "Optimal linear receiving filter for digital transmission over nonlinear channels," *GLOBECOM 1984*, Atlanta, Ga., November 1984.

48. Blachman, N. M. (1971), "Detectors, bandpass nonlinearities and their optimization: inversion of the Chebyshev transform," *IEEE Transactions on Information Theory*, vol. IT-17, pp. 398–404.

49. Blachman, N. M. (1982), *Noise and Its Effects on Communication*, 2nd ed. Malabar, Fla: R. E. Krieger Publishing Co.

50. Blahut, R. E. (1983), *Theory and Practice of Error Control Codes*. Reading, Mass.: Addison-Wesley.

51. Blanchard, A. (1976), *Phase-locked Loops*. New York: Wiley.

52. Blanc-Lapierre, A., and R. Fortet (1968), *Theory of Random Functions*, vol. 2. New York: Gordon and Breach.

53. Boutin, N., S. Morissette, and C. Porlier (1982), "Extension of Mueller's theory on optimum pulse shaping for data transmission," *IEE Proceedings*, vol. 129, Pt. F, pp. 255–260.

54. Bracewell, R. N. (1978) *The Fourier Transform and Its Applications*, 2nd ed. New York: McGraw-Hill.

55. Brennan, L. E., and I. S. Reed (1965), "A recursive method of computing the Q function," *IEEE Transactions on Information Theory*, vol. IT-11, pp. 312–313.

56. Butman, S. A., and J. R. Lesh (1977), "The effects of bandpass limiters on n-phase tracking systems," *IEEE Transactions on Communications*, vol. COM-25, pp. 569–576.

57. Cambanis, S., and B. Liu (1970), "On harmonizable stochastic processes," *Information and Control*, vol. 17, pp. 183–202.

58. Cambanis, S., and E. Masry (1971), "On the representation of weakly continuous stochastic processes," *Information Sciences*, vol. 3, pp. 277–290.

59. Campbell, L. L. (1969), "Series expansions for random processes," in *Proceedings of the International Symposium on Probability and Information Theory*, Lecture Notes in Mathematics, no. 89, pp. 77–95. New York: Springer-Verlag.

60. Campopiano, C. N., and B. G. Glazer (1962), "A coherent digital amplitude and phase modulation scheme," *IRE Transactions on Communication Systems*, vol. CS-10, pp. 90–95.

61. Cariolaro, G. L., and G. P. Tronca (1974), "Spectra of block coded digital signals," *IEEE Transactions on Communications*, vol. COM-22, pp. 1555–1564.

62. Cariolaro, G. L., and S. Pupolin (1975), "Moments of correlated digital signals for error probability evaluation," *IEEE Transactions on Information Theory*, vol. IT-21, pp. 558–568.

63. Cariolaro, G. L., G. L. Pierobon, and G. P. Tronca (1983), "Analysis of codes and spectra calculations," *International Journal of Electronics*, vol. 55, pp. 35–79.

64. Castellani, V., L. Lo Presti, and M. Pent (1974a), "Performance of multilevel DCPSK systems in the presence of both interchannel and intersymbol interference," *Electronics Letters*, vol. 10, no. 7, pp. 111–112.

65. Castellani, V., L. Lo Presti, and M. Pent (1974b), "Multilevel DCPSK over real channels," *IEEE International Conference on Communications* (ICC '74), Minneapolis, Minn., June 1974.

66. Chalk, J. H. H. (1950), "The optimum pulse shape for pulse communication," *IEE Proceedings*, vol. 97, pt. III, pp. 88–92.

67. Chang, R. W. (1971), "A new equalizer structure for fast start-up digital communication," *Bell System Technical Journal*, vol. 50, pp. 1969–2001.

68. Chang, R. W., and J. C. Hancock (1966), "On receiver structures for channels having memory," *IEEE Transactions on Information Theory*, vol. IT-12, pp. 463–468.

69. Clark, G. C., and J. B. Cain (1981), *Error-Correction Coding for Digital Communications*. New York: Plenum Press.

70. Corazza, G., and G. Immovilli (1979), "On the effect of a type of modulation pulse shaping in FSK transmission systems with limiter–discriminator detection," *Alta Frequenza*, vol. 48, pp. 449–457.

71. Corazza, G., G. Crippa, and G. Immovilli (1981), "Performance analysis of quaternary CPFSK systems with modulation pulse shaping and limiter-discriminator detection," *Alta Frequenza*, vol. 50, pp. 77–88.

72. Croisier, A. (1970), "Introduction to pseudoternary transmission codes," *IBM Journal of Research and Development*, vol. 14, pp. 354–367.

73. D'Andrea, N. A., and F. Russo (1983), "First-order DPLL's: A survey of a peculiar methodology and some new applications," *Alta Frequenza*, vol. 52, pp. 495–505.

74. Daut, D. G., J. W. Modestino, and L. D. Wismer (1982), "New short constraint length convolutional code construction for selected rational rates," *IEEE Transactions on Information Theory*, vol. IT-28, pp. 794–800.

75. Davenport, W. B., Jr. (1970), *Probability and Random Processes. An Introduction for Applied Scientists and Engineers*. New York: McGraw-Hill.

76. De Buda, R. (1972), "Coherent demodulation of frequency-shift keying with low deviation ratio," *IEEE Transactions on Communications*, vol. COM-20, pp. 429–435.

77. Decina M., and A. Roveri (1987), "Integrated Services Digital Network: Architectures and Protocols," Chapter 2 of K. Feher, *Advanced Digital Communications*. Englewood Cliffs, N.J.: Prentice-Hall.

78. De Jager, F., and C. B. Dekker (1978), "Tamed frequency modulation—a novel method

to achieve spectrum economy in digital transmission," *IEEE Transactions on Communications*, vol. COM-26, pp. 534–542.

79. Devieux, C., and R. Pickholtz (1969), "Adaptive equalization with a second-order algorithm," *Proceedings of the Symposium on Computer Processing in Communications*, Polytechnic Institute of Brooklyn, pp. 665–681. April 8–10, 1969.

80. Di Toro, M. J. (1968), "Communication in time-frequency spread media using adaptive equalization," *IEEE Proceedings*, vol. 56, pp. 1653–1679.

81. Divsalar, D. (1978), *Performance of Mismatched Receivers on Bandlimited Channels*, Ph.D. Dissertation, University of California, Los Angeles.

82. Dugundji, Dj. (1958), "Envelopes and pre-envelopes of real wave-forms," *IRE Transactions on Information Theory*, vol. IT-4, pp. 53–57.

83. Duttweiler, D. L. (1982), "Adaptive filter performance with nonlinearities in the correlation multiplier," *IEEE Transactions on Acoustics, Speech, and Signal Processing*, vol. ASSP-30, pp. 578–586.

84. Dym, H., and H. P. McKean (1972), *Fourier Series and Integrals*. New York: Academic Press.

85. Elia, M. (1983), "Symbol error rate of binary block codes," *Transactions of the 9th Prague Conference on Information Theory, Statistical Decision Functions, Random Processes*, pp. 223–227, Prague, June 1982.

86. Elnoubi, S., and S. C. Gupta (1981), "Error rate performance of noncoherent detection of duobinary coded MSK and TFM in mobile radio communication systems," *IEEE Transactions on Vehicular Technology*, vol. VT-30, pp. 62–76, May 1981.

87. Ericson, T. (1971), "Structure of optimum receiving filters in data transmission systems," *IEEE Transactions on Information Theory*, vol. IT-17, pp. 352–353.

88. Ericson, T. (1973), "Optimum PAM filters are always band limited," *IEEE Transactions on Information Theory*, vol. IT-19, pp. 570–573.

89. Faddeev, D. K., and V. N. Faddeeva (1963), *Computational Methods of Linear Algebra*. San Francisco: W. H. Freeman and Co.

90. Falconer, D. D. (1976), "Jointly adaptive equalization and carrier recovery in two-dimensional digital communication systems," *Bell System Technical Journal*, vol. 55, pp. 317–334.

91. Falconer, D. D. (1978), "Adaptive equalization of channel nonlinearities in QAM data transmission systems," *Bell System Technical Journal*, vol. 57, pp. 2589–2611.

92. Falconer, D. D., and F. R. Magee, Jr. (1973), "Adaptive channel memory truncation for maximum likelihood sequence estimation," *Bell System Technical Journal*, vol. 52, pp. 1541–1562.

93. Falconer, D. D., and L. Ljung (1978), "Application of fast Kalman estimation to adaptive equalization," *IEEE Transactions on Communications*, vol. COM-26, pp. 1439–1446.

94. Fano, R. M. (1961), *Transmission of Information*. Cambridge, Mass.: MIT Press.

95. Fano, R. M. (1963), "A heuristic discussion of probabilistic coding," *IEEE Transactions on Information Theory*, vol. IT-9, pp. 64–74.

96. Feller, W. (1968), *An Introduction to Probability Theory and Its Applications*, 3rd ed. New York: Wiley.

97. Fenderson, G. L., J. W. Parker, P. D. Quigley, S. R. Shepard, and C. A. Siller, Jr. (1984), "Adaptive transversal equalization of multipath propagation for 16-QAM, 90 Mb/s digital radio," *Bell System Technical Journal*, vol. 63, pp. 1447–1463.

98. Flake, R. H. (1963), "Volterra series representation of nonlinear systems," *AIEE Transactions*, vol. 81, pp. 330–335.

99. Forney, G. D., Jr. (1966), *Concatenated Codes*. Cambridge, Mass.: MIT Press.

100. Forney, G. D., Jr. (1970), "Convolutional codes I: Algebraic structure," *IEEE Transactions on Information Theory*, vol. IT-16, pp. 720–738.

101. Forney, G. D., Jr. (1972a), "Lower bounds on error probability in the presence of large intersymbol interference," *IEEE Transactions on Communications*, vol. COM-20, pp. 76–77.

102. Forney, G. D., Jr. (1972b), "Maximum likelihood sequence estimation of digital sequences in the presence of intersymbol interference," *IEEE Transactions on Information Theory*, vol. IT-18, pp. 363–378.

103. Forney, G. D., Jr. (1973), "The Viterbi algorithm," *IEEE Proceedings*, vol. 61, pp. 268–278.

104. Forney, G. D., Jr. (1974), "Convolutional codes III: Sequential decoding," *Information and Control*, vol. 25, pp. 267–297.

105. Forney, G. D., Jr., R. G. Gallager, G. R. Lang, and S. H. Qureshi (1984), "Efficient modulation for band-limited channels," *IEEE Transactions on Selected Areas in Communications*, vol. SAC-2, pp. 632–647.

106. Foschini, G. J. (1975), "Performance bound for maximum-likelihood reception of digital data," *IEEE Transactions on Information Theory*, vol. IT-21, pp. 47–50.

107. Foschini, G. J. (1977), "A reduced state variant of maximum likelihood sequence detection attaining optimum performance for high signal-to-noise ratios," *IEEE Transactions on Information Theory*, vol. IT-23, pp. 605–609.

108. Foschini, G. J. (1985), "Equalizing without altering or detecting data," *AT&T Technical Journal*, vol. 64, pp. 1885–1911.

109. Franaszek, P. A. (1968), "Sequence-state coding for digital transmission," *Bell System Technical Journal*, vol. 47, pp. 143–157.

110. Franks, L. E. (1968), "Further results on Nyquist's problem in pulse transmission," *IEEE Transactions on Communications*, vol. COM-16, pp. 337–340.

111. Franks, L. E. (1969), *Signal Theory*. Englewood Cliffs, N.J.: Prentice-Hall.

112. Franks, L. E. (1980), "Carrier and bit synchronization in data communication: A tutorial review," *IEEE Transactions on Communications*, vol. COM-28, pp. 1107–1120.

113. Franks, L. E. (1983), "Synchronization subsystems: Analysis and design," in K. Feher, *Digital Communications: Satellite/Earth Station Engineering*, Englewood Cliffs, N.J.: Prentice-Hall.

114. Franks, L. E., and J. P. Bubrouski (1974), "Statistical properties of timing jitter in a PAM timing recovery scheme," *IEEE Transactions on Communications*, vol. COM-22, pp. 913–920.

115. Fredricsson, S. (1974), "Optimum transmitting filter in digital PAM systems with a Viterbi detector," *IEEE Transactions on Information Theory*, vol. IT-20, pp. 479–489.

116. Fredricsson, S. (1975), "Optimum receiver filters in digital quadrature phase-shift-keyed systems with a nonlinear repeater," *IEEE Transactions on Communications*, vol. COM-23, pp. 1389–1400.

117. Friedlander, B. (1982), "Lattice filters in adaptive processing," *IEEE Proceedings*, vol. 70, pp. 829–867.

118. Galko, P., and S. Pasupathy (1981), "The mean power spectral density of Markov chain driven signals," *IEEE Transactions on Information Theory*, vol. IT-27, pp. 746–754.

119. Gallager, R. G. (1965), "A simple derivation of the coding theorem and some applications," *IEEE Transactions on Information Theory*, vol. IT-11, pp. 3–18.

120. Gallager, R. G. (1968), *Information Theory and Reliable Communication*. New York: Wiley.

121. Gantmacher, F. R. (1959), *The Theory of Matrices*, Vols. I and II. New York: Chelsea Publishing Co.

122. Gardner, F. M. (1979), *Phaselock Techniques*, 2nd ed. New York: Wiley.

123. Gardner, W. A. (1978), "Stationarizable random processes," *IEEE Transactions on Information Theory*, vol. IT-24, pp. 8–22.

124. Gardner, W. A. (1984), "Learning characteristics of stochastic-gradient descent algorithms: A general study, analysis, and critique," *Signal Processing,* vol. 6, pp. 113–133.

125. Gardner, W. A., and L. E. Franks (1975), "Characterization of cyclostationary random signal processes," *IEEE Transactions on Information Theory*, vol. IT-21, pp. 4–14.

126. Gautschi, W. (1970), "On the construction of Gaussian quadrature rules from modified moments," *Mathematics of Computation*, vol. 24, pp. 245–260.

127. Gelfand, I. M., and S. V. Fomin (1963), *Calculus of Variations*. Englewood Cliffs, N.J.: Prentice-Hall.

128. George, D. A. (1965), "Matched filters for interfering signals," *IEEE Transactions on Information Theory*, vol. IT-11, pp. 153–154.

129. Gersho, A. (1969a), "Adaptive equalization of highly dispersive channels for data transmission," *Bell System Technical Journal*, vol. 48, pp. 55–70.

130. Gersho, A. (1969b), "On combining adaptive equalization and filtering," unpublished paper.

131. Gersho, A., and T. L. Lim (1981), "Adaptive cancellation of intersymbol interference for data transmission," *Bell System Technical Journal*, vol. 60, pp. 1997–2021.

132. Gibby, R. A., and J. W. Smith (1965), "Some extensions of Nyquist's telegraph transmission theory," *Bell System Technical Journal*, vol. 44, pp. 1487–1510.

133. Giger, A. J., and W. T. Barnett (1981), "Effects of multipath propagation on digital radio," *IEEE Transactions on Communications*, vol. COM-29, pp. 1345–1352.

134. Gitlin, R. D., and F. R. Magee, Jr. (1977), "Self-orthogonalizing adaptive equalization algorithms," *IEEE Transactions on Communications*, vol. COM-25, pp. 666–672.

135. Gitlin, R. D., and S. B. Weinstein (1979), "On the required tap-weight precision for digitally implemented, adaptive, mean-squared equalizers," *Bell System Technical Journal*, vol. 58, pp. 301–321.

136. Gitlin, R. D., and S. B. Weinstein (1981), "Fractionally-spaced equalization: An improved digital transversal equalizer," *Bell System Technical Journal*, vol. 60, pp. 275–296.

137. Gitlin, R. D., E. Y. Ho, and J. E. Mazo (1973), "Passband equalization of differentially phase-modulated data signals," *Bell System Technical Journal*, vol. 52, pp. 219–238.

138. Gitlin, R. D., J. E. Mazo, and M. G. Taylor (1973), "On the design of gradient algorithms for digitally implemented adaptive filters," *IEEE Transactions on Circuit Theory*, vol. CT-20, pp. 125–136.

139. Glave, F. E. (1972), "An upper bound on the probability of error due to intersymbol interference for correlated digital signals," *IEEE Transactions on Information Theory*, vol. IT-18, pp. 356–363.

140. Godard, D. (1974), "Channel equalization using a Kalman filter for fast data transmission," *IBM Journal of Research and Development*, vol. 18, pp. 267–273.

141. Godard, D. (1980), "Self-recovering equalization and carrier tracking in two-dimensional data communication systems," *IEEE Transactions on Communications*, vol. COM-28, pp. 1867–1875.

142. Godard, D. (1981), "A 9600 bit/s modem for multipoint communications systems," *Proceedings of NTC'81*, New Orleans, La., pp. B3.3.1–B3.3.5.

143. Goldenberg, L. M., and D. D. Klovsky (1959), "Computer-aided detection of pulse signals," *Trudy LEIS*, vol. 44, pp. 17–26.

144. Golub, G. H., and C. F. Van Loan (1983), *Matrix Computations*. Baltimore, Md.: Johns Hopkins University Press.

145. Golub, G. H., and J. H. Welsch (1969), "Calculation of Gauss quadrature rules," *Mathematics of Computation*, vol. 23, pp. 221–230.

146. Gray, R. M. (1971), *Toeplitz and Circulant Matrices: A Review*, Stanford Electron. Lab. Tech. Rep. 6501–2, June 1971.

147. Gray, R. M. (1972), "On the asymptotic eigenvalue distribution of Toeplitz matrices," *IEEE Transactions on Information Theory*, vol. IT-18, pp. 725–730.

148. Greenstein, L. J., and B. A. Czekaj (1981), "Modeling multipath fading responses using multitone probing signals and polynomial approximation," *Bell System Technical Journal*, vol. 60, pp. 193–214.

149. Greenstein, L. J., and B. A. Czekaj (1982), "Performance comparisons among digital radio techniques subjected to multipath fading," *IEEE Transactions on Communications*, vol. COM-30, pp. 1184–1197.

150. Grenander, U., and G. Szegö (1958), *Toeplitz Forms and Their Applications*. Berkeley, Calif.: University of California Press.

151. Grettenberg, T. L. (1965), "A representation theorem for complex normal processes," *IEEE Transactions on Information Theory*, vol. IT-11, pp. 305–306.

152. Gronemeyer, S. A., and A. L. McBride (1976), "MSK and offset QPSK modulation," *IEEE Transactions on Communications*, vol. COM-24, pp. 809–819.

153. Guidoux, M. L. (1975), "Egalizeur autoadaptatif à double échantillonage appliqué à la transmission de donnés à 9600 bit/seconde" (in French), *L'Onde Electrique*, vol. 55, pp. 9–13.

154. Gunn, J. F., and J. A. Lombardi (1969), "Error detection for partial-response systems," *IEEE Transactions on Communication Technology*, vol. COM-17, pp. 734–736.

155. Gupta, S. C. (1975), "Status of digital phase-locked loops," *IEEE Proceedings*, vol. 63, pp. 291–306.

156. Hänsler, E. (1971), "Some properties of transmission systems with minimum mean-square error," *IEEE Transactions on Communications*, vol. COM-19, pp. 576–579.

157. Hayes, J. F. (1975), "The Viterbi algorithm applied to digital data transmission," *IEEE Communications Magazine*, vol. 13, pp. 15–20, March.

158. Heller, J. A. (1968), "Short constraint length convolutional codes," *JPL Pasadena Space Program Summary 37–54*, vol. 3, pp. 171–174.

159. Heller, J. A. (1975), "Feedback decoding of convolutional codes, in A. J. Viterbi (ed.), *Advances in Communication Systems*, Vol. 4. New York: Academic Press.

160. Helstrom, C. W. (1958), "The resolution of signals in white Gaussian noise," *IRE Proceedings*, vol. 46, pp. 1603–1619.

161. Helstrom, C. W. (1968), *Statistical Theory of Signal Detection*, 2nd ed. Elmsford, N.Y.: Pergamon Press.

162. Herrmann, G. F. (1978), "Performance of maximum-likelihood receiver in the nonlinear satellite channel," *IEEE Transactions on Communications*, vol. COM-26, pp. 373–378.

163. Hestenes, M. R., and E. Stiefel (1952), "Methods of conjugate gradients for solving linear systems," *Journal of Research of the National Bureau of Standards*, vol. 49, pp. 409–436.

164. Hill, F. S., Jr., and M. Blanco (1973), "Random geometric series and intersymbol interference," *IEEE Transactions on Information Theory*, vol. IT-19, pp. 326–335.

165. Ho, E. H., and Y. S. Yeh (1970), "A new approach for evaluating the error probability in the presence of intersymbol interference and additive Gaussian noise," *Bell System Technical Journal*, vol. 49, pp. 2249–2265.

166. Ho, E. H., and Y. S. Yeh (1971), "Error probability of a multilevel digital system with intersymbol interference and Gaussian noise," *Bell System Technical Journal*, vol. 50, pp. 1017–1023.

167. Honig, M. L., and D. G. Messerschmitt (1984), *Adaptive Filters: Structures, Algorithms, and Applications*. Boston: Kluwer Academic Publisher.

168. Huggins, W. H. (1957), "Signal-flow graphs and random signals," *IRE Proceedings*, vol. 45, pp. 74–86.

References **623**

169. Hurd, H. (1969) *An Investigation of Periodically Correlated Stochastic Processes*, Ph.D. Dissertation, Duke University, Durham, N.C.

170. Imai, H., and S. Hirakawa (1977), "A new multilevel coding method using error-correcting codes," *IEEE Transactions on Information Theory*, vol. IT-23, pp. 371–377.

171. Jacobs, I. M. (1974), "Practical applications of coding," *IEEE Transactions on Information Theory*, vol. IT-20, pp. 305–310.

172. Jelinek, F. (1969), "Fast sequential decoding algorithm using a stack," *IBM Journal of Research and Development*, vol. 13, pp. 675–685.

173. Jones, D. S. (1966), *Generalised Functions*. New York: McGraw-Hill.

174. Jones, S. K., R. K. Cavin, III, and W. M. Reed (1982), "Analysis of error-gradient adaptive linear estimators for a class of stationary dependent processes," *IEEE Transactions on Information Theory*, vol. IT-28, pp. 318–329.

175. Justesen, J. (1982), "Information rates and power spectra of digital codes," *IEEE Transactions on Information Theory*, vol. IT-28, pp. 457–472.

176. Kailath, T. (1971), "RKHS approach to detection and estimation problems—Part I: Deterministic signals in Gaussian noise," *IEEE Transactions on Information Theory*, vol. IT-17, pp. 530–549.

177. Kemeny, J. G., and J. L. Snell (1960), *Finite Markov Chains*. New York: Van Nostrand Reinhold.

178. Kettel, E. (1961), "Übertragungssysteme mit idealer Impulsfunktion" (in German), *Archiv der Elektrischen Übertragung*, vol. 15, pp. 207–214.

179. Kettel, E. (1964), "Ein automatisches Optimisator für den Abgleich des Impulsentzerrers in einer Datenübertragung" (in German), *Archiv der Elektrischen Übertragung*, vol. 18, pp. 271–278.

180. Klovsky, D., and B. Nikolaev (1978), *Sequential Transmission of Digital Information in the Presence of Intersymbol Interference*. Moscow (USSR): MIR Publishers.

181. Kobayashi, H. (1971a), "Correlative level coding and maximum-likelihood decoding," *IEEE Transactions on Information Theory*, vol. IT-17, pp. 586–594.

182. Kobayashi, H. (1971b), "A survey of coding schemes for transmission or recording of digital data," *IEEE Transactions on Communication Technology*, vol. COM-19, pp. 1087–1100.

183. Kobayashi, H. (1971c), "Application of Hestenes-Stiefel algorithm to channel equalization," *IEEE International Conference on Communications (ICC'71)*, Montreal, Canada, June 1971.

184. Kobayashi, H., and D. T. Tang (1971), "On decoding of correlative level coding systems with ambiguity zone detection," *IEEE Transactions on Communications Technology*, vol. COM-19, pp. 467–477.

185. Komaki, S. Y., Okamoto, and K. Tajima (1980), "Performance of 16-QAM digital radio system using new space diversity," *IEEE International Conference on Communications (ICC'80)*, Seattle, Washington, June 1980.

186. Kotel'nikov, V. A. (1959), *The Theory of Optimum Noise Immunity*. New York: McGraw-Hill.

187. Krein, M. G., and A. A. Nudel'man (1977), *The Markov Moment Problem and Extremal Problems*, Transl. Math. Monographs, vol. 50. Providence, R.I.: American Mathematical Society.

188. Kretzmer, E. R. (1966), "Generalization of a technique for binary data communication," *IEEE Transactions on Communication Technology*, vol. COM-14, pp. 67–68.

189. Krylov, V. J. (1962), *Approximate Calculation of Integrals*. New York: Macmillan.

190. Larsen, K. J. (1973), "Short convolutional codes with maximal free distance for rates 1/2, 1/3 and 1/4," *IEEE Transactions on Information Theory*, vol. IT-19, pp. 371–372.

191. Lawrence, R. E., and H. Kaufman (1971), "The Kalman filter for the equalization of a

digital communication channel," *IEEE Transactions on Communication Technology*, vol. COM-19, pp. 1137–1141.

192. Leclert, A., and P. Vandamme (1985), "Decision feedback equalization of dispersive radio channels," *IEEE Transactions on Communications*, vol. COM-33, pp. 676–684.

193. Lee, W. U., and F. S. Hill, Jr. (1977), "A maximum likelihood sequence estimator with decision feedback equalization," *IEEE Transactions on Communications*, vol. COM-25, pp. 971–980.

194. Lender, A. (1963), "The duobinary technique for high speed data transmission" *AIEE Transactions on Communications and Electronics*, vol. 82, pp. 214–218.

195. Lender, A. (1964), "Correlative digital communication techniques," *IEEE Transactions on Communications Technology*, pp. 128–135.

196. Lim, T. L., and M. S. Mueller (1980), "Rapid equalizer start-up using least-squares algorithms," *IEEE International Conference on Communications (ICC'80)*, Seattle, Washington, June 1980.

197. Lin, S. (1970), *An Introduction to Error Correcting Codes*, Englewood Cliffs, N.J.: Prentice-Hall.

198. Lin, S., and D. J. Costello, Jr. (1983), *Error Control Coding: Fundamentals and Applications*. Englewood Cliffs, N.J.: Prentice-Hall.

199. Lindell, G., C.-E. W. Sundberg and A. Svensson (1985), "Narrowband coded digital modulation schemes with constant amplitude," *1985 International Tirrenia Workshop on Digital Communications*, Tirrenia, Italy, September 1985. [In E. Biglieri and G. Prati (eds.), *Digital Communications*. Amsterdam: North-Holland, 1986].

200. Lindsey, W. C. (1972), *Synchronous Systems in Communication and Control*. Englewood Cliffs, N.J.: Prentice-Hall.

201. Lindsey, W. C., and C. M. Chie (1981), "A survey of digital phase locked loops," *IEEE Proceedings*, vol. 69, pp. 410–432.

202. Lindsey, W. C., and M. K. Simon (1973), *Telecommunication Systems Engineering*. Englewood Cliffs, N.J.: Prentice-Hall.

203. Loève, M. (1963), *Probability Theory*. Princeton, N.J.: Van Nostrand Reinhold.

204. Lucky, R. W. (1965), "Automatic equalization for digital communication," *Bell System Technical Journal*, vol. 44, pp. 547–588.

205. Lucky, R. W. (1966), "Techniques for adaptive equalization of digital communication," *Bell System Technical Journal*, vol. 45, pp. 255–286.

206. Lucky, R. W. (1975), "Modulation and detection for data transmission on the telephone channel," in J. K. Skwirzynski (ed.), *New Directions in Signal Processing in Communication and Control*. Leiden, Holland: Noordhoff.

207. Lucky, R. W., J. Salz, and E. J. Weldon (1968), *Principles of Data Communications*. New York: McGraw-Hill.

208. Luenberger, D. G. (1969), *Optimization by Vector Space Methods*. New York: Wiley.

209. Lugannani R. (1969), "Intersymbol interference and probability of error in digital systems," *IEEE Transactions on Information Theory*, vol. IT-15, pp. 682–688.

210. Lundgren, C. W., and W. D. Rummler (1979), "Digital radio outage due to selective fading-observation vs. prediction from laboratory simulation," *Bell System Technical Journal*, vol. 58, pp. 1074–1100.

211. Luvison, A., and G. Pirani (1979), "Calculation of error rates in digital communication systems," *Fifth International Symposium on Information Theory*, Tbilisi, USSR, July 1979.

212. Luvison, A., and V. Navino (1976), "Theoretical and experimental development of two-dimensional quadrature rules," *ICCAD International Conference on Numerical Methods in Electrical and Magnetic Field Problems*, Santa Margherita Ligure, Italy, June 1976.

213. Macchi, O., and E. Eweda (1984), "Convergence analysis of self-adaptive equalizers," *IEEE Transactions on Information Theory*, vol. IT-30, pp. 161–176.

214. Macchi, O., and L. Guidoux (1975), "A new equalizer and double sampling equalizer," *Annales des Télécommunications*, vol. 30, pp. 331–338.

215. MacWilliams, F. J., and N. J. A. Sloane (1977), *The Theory of Error Correcting Codes*. Amsterdam: North-Holland.

216. Magee, F. R., Jr., and J. G. Proakis (1973), "Adaptive maximum-likelihood sequence estimation for digital signaling in the presence of intersymbol interference," *IEEE Transactions on Information Theory*, vol. IT-19, pp. 120–124.

217. Makhoul, J. (1978), "A class of all-zero lattice digital filters: Properties and applications," *IEEE Transactions on Acoustics, Speech, and Signal Processing*, vol. ASSP-26, pp. 304–314.

218. Mancianti, M., U. Mengali, and R. Reggianini (1979), "A fast start-up algorithm for channel parameter acquisition in SSB-AM data transmission," *IEEE International Conference on Communications (ICC'79)*, Boston, June 1979.

219. Masamura, T., S. Samejima, Y. Morihiro, and H. Fuketa (1979), "Differential detection of MSK with nonredundant error correction," *IEEE Transactions on Communications*, vol. COM-27, pp. 912–918.

220. Masry, E., B. Liu, and K. Steiglitz (1968), "Series expansions of wide-sense stationary random processes," *IEEE Transactions on Information Theory*, vol. IT-14, pp. 792–796.

221. Massey, J. L. (1963), *Threshold Decoding*. Cambridge, Mass.: MIT Press.

222. Massey, J. L. (1974), "Coding and modulation in digital communications," *1974 International Zürich Seminar on Digital Communications*, Zürich, Switzerland, March 1974.

223. Massey, J. L. (1984), "The how and why of channel coding," *1984 International Zürich Seminar on Digital Communications*, Zürich, Switzerland, March 1986.

224. Massey, J. L., and M. K. Sain (1968), "Inverses of linear sequential circuits," *IEEE Transactions on Computers*, vol. C-17, pp. 330–337.

225. Matthews, J. W. (1973), "Sharp error bounds for intersymbol interference," *IEEE Transactions on Information Theory*, vol. IT-19, pp. 440–447.

226. Matyas, R., and P. J. McLane (1974), "Decision-aided tracking loops for channels with phase jitter and intersymbol interference," *IEEE Transactions on Communications*, vol. COM-22, pp. 1014–1023.

227. Mazo, J. E. (1979), "On the independence theory of equalizer convergence," *Bell System Technical Journal*, vol. 58, pp. 963–993.

228. McEliece, R. J. (1977), *The Theory of Information and Coding*. Reading, Mass.: Addison-Wesley.

229. McLane, P. J. (1980), "A residual intersymbol interference error bound for truncated-state Viterbi detector," *IEEE Transactions on Information Theory*, vol. IT-26, pp. 548–553.

230. McWhirter, J. G., and T. J. Shepherd (1983), "Least-squares lattice algorithms for adaptive channel equalization. A simplified derivation," *IEE Proceedings*, vol. 130, Part F, pp. 532–542.

231. Mengali, U. (1977), "Joint phase and timing acquisition in data transmission," *IEEE Transactions on Communications*, vol. COM-25, pp. 1174–1185.

232. Mengali, U. (1979), *Teoria dei Sistemi di Comunicazione* (in Italian). Pisa: ETS Universitá.

233. Mengali, U. (1983), "A new look at the pulse shaping problem in timing recovery," *1983 International Tirrenia Workshop on Digital Communications*, Tirrenia, Italy.

234. Mesiya, M. F., P. J. McLane, and L. Lorne Campbell (1977), "Maximum-likelihood sequence estimation of digital sequences in the presence of intersymbol interference," *IEEE Transactions on Information Theory*, vol. IT-18, pp. 363–378.

235. Mesiya, M. F., P. J. McLane, and L. Lorne Campbell (1978), "Optimal receiver filters

for BPSK transmission over a bandlimited nonlinear channel," *IEEE Transactions on Communications*, vol. COM-26, pp. 12–22.

236. Messerschmitt, D. G. (1973a), "A geometric theory of intersymbol interference. Part I: Zero-forcing and decision-feedback equalization," *Bell System Technical Journal*, vol. 52, pp. 1483–1519.

237. Messerschmitt, D. G. (1973b), "A geometric theory of intersymbol interference. Part II: Performance of the maximum likelihood detector," *Bell System Technical Journal*, vol. 52, pp. 1521–1539.

238. Messerschmitt, D. G. (1974), "Design of a finite impulse response for the Viterbi algorithm and decision-feedback equalizer," *IEEE International Conference on Communications (ICC'74)*, Minneapolis, June 1974.

239. Meyers, M. H., and L. E. Franks (1980), "Joint carrier phase and symbol timing recovery for PAM systems," *IEEE Transactions on Communications*, vol. COM-28, pp. 1121–1129.

240. Miller, K. S. (1974), *Complex Stochastic Processes*. Reading, Mass.: Addison-Wesley.

241. Moncalvo, A., and G. Pellegrini (1976), "Line Codes for Cable Digital Transmission," (in Italian), *CSELT Technical Reports*, no. 1, pp. 5–20.

242. Moreno, L., and M. Salerno (1983), "Adaptive equalization structures in high capacity digital radio; predicted and observed outage performance," *1983 International Tirrenia Workshop on Digital Communications*, Tirrenia, Italy.

243. Morf, M. (1977), "Ladder forms in estimation and system identification," *IEEE 11th Annual Asilomar Conference on Circuits, Systems, and Computers*, Pacific Grove, Calif., November 1977.

244. Morf, M., A. Vieira, and D. T. Lee (1977), "Ladder forms for identification and speech processing," *Proceedings 1977 IEEE Conference on Decision and Control*, New Orleans, La., December 1977, pp. 1074–1078.

245. Mueller, K. H. (1973), "A new approach to optimum pulse shaping in sampled systems using time-domain filtering," *Bell System Technical Journal*, vol. 52, pp. 723–729.

246. Mueller, K. H., and D. A. Spaulding (1975), "Cyclic equalization. A new rapidly converging equalization technique for synchronous data communication," *Bell System Technical Journal*, vol. 54, pp. 369–406.

247. Mueller, M. S. (1981), "Least-squares algorithms for adaptive equalizers," *Bell System Technical Journal*, vol. 60, pp. 1905–1925.

248. Mueller, M. S., and J. Salz (1981), "A unified theory of data-aided equalization," *Bell System Technical Journal*, vol. 60, pp. 2023–2038.

249. Muilwijk, D. (1981), "Correlative phase shift keying—A class of constant envelope modulation techniques," *IEEE Transactions on Communications*, vol. COM-29, pp. 226–236.

250. Mulligan, M. G., and S. G. Wilson (1984), "An improved algorithm for evaluating trellis phase codes," *IEEE Transactions on Information Theory*, vol. IT-30, pp. 846–851.

251. Mysovskih, I. P. (1968), "On the construction of cubature formulas with the smallest number of nodes" (in Russian), *Doklady Akademii Nauk SSSR*, vol. 178, pp. 1252–1254; English translation in *Soviet Mathematics Doklady*, vol. 9, pp. 277–280.

252. Niessen, C. W., and D. K. Willim (1970), "Adaptive equalizer for pulse transmission," *IEEE Transaction on Communication Technology*, vol. COM-18, pp. 377–395.

253. Nyquist, H. (1928), "Certain topics in telegraph transmission theory," *AIEE Transactions*, vol. 47, pp. 617–644.

254. Odenwalder, J. P. (1970), *Optimal Decoding of Convolutional Codes*, Ph.D. Dissertation, University of California, Los Angeles.

255. Oetting, J. D. (1979), "A comparison of modulation techniques for digital radio," *IEEE Transactions on Communications*, vol. COM-27, pp. 1752–1762.

256. Omura, J. K. (1969), "On the Viterbi decoding algorithm," *IEEE Transactions on Information Theory*, vol. IT-15, pp. 177–179.

257. Omura, J. K. (1971), "Optimal receiver design for convolutional codes and channels with memory via control theoretical concepts," *Information Sciences*, vol. 3, pp. 243–266.

258. Omura, J. K., and M. K. Simon (1980), "Satellite communication performance evaluation: Computational techniques based on moments," *JPL Publication 80–71*, September 15, 1980.

259. Oppenheim, A. V., A. S. Willsky, and I. T. Young, *Signals and Systems*. Englewood Cliffs, N.J.: Prentice-Hall.

260. Paaske, E. (1974), "Short binary convolutional codes with maximal free distance for rates 2/3 and 3/4," *IEEE Transactions on Information Theory*, vol. IT-20, pp. 683–689.

261. Papoulis, A. (1962), *The Fourier Integral and Its Applications*. New York: McGraw-Hill.

262. Papoulis, A. (1965), *Probability, Random Variables, and Stochastic Processes*. New York: McGraw-Hill.

263. Papoulis, A. (1977), *Signal Analysis*. New York: McGraw-Hill.

264. Parzen, E. (1962), *Stochastic Processes*. San Francisco, Calif.: Holden-Day.

265. Pasupathy, S. (1979), "Minimum shift keying: a spectrally efficient modulation," *IEEE Communications Magazine*, vol. 17, pp. 14–22, July.

266. Pent, M. (1968), "Orthogonal polynomial approach for the Marcum Q-function numerical computation," *Electronics Letters*, vol. 4, pp. 563–564.

267. Peterson, W. W., and E. J. Weldon, Jr. (1972), *Error Correcting Codes*, 2nd ed. Cambridge, Mass.: MIT Press.

268. Piessens, R., and A. Haegemans (1975), "Cubature formulas of degree nine for symmetric planar regions," *Mathematics of Computation*, vol. 29, pp. 810–815.

269. Pirani, G., and V. Zingarelli (1984), "Adaptive multiplication-free transversal equalizers for digitally implemented adaptive filters," *IEEE Transactions on Communications*, vol. COM-32, pp. 1025–1033.

270. Prabhu, V. K. (1971), "Some considerations of error bounds in digital systems," *Bell System Technical Journal*, vol. 50, pp. 3127–3151.

271. Price, R. (1972), "Nonlinearly feedback-equalized PAM vs. capacity for noisy filter channels," *IEEE International Conference on Communications (ICC'72)*, Philadelphia, June 1972.

272. Proakis, J. G. (1974), "Channel identification for high speed digital communications, *IEEE Transactions on Automatic Control*, vol. AC-19, pp. 916–922.

273. Proakis, J. G. (1975), "Advances in Equalization for Intersymbol Interference," in A. J. Viterbi (ed.), *Advances in Communication Systems Theory and Applications*, Vol. 4. New York: Academic Press.

274. Proakis, J. G. (1983), *Digital Communications*. New York: McGraw-Hill.

275. Proakis, J. G., and J. H. Miller (1969), "An adaptive receiver for digital signaling through channels with intersymbol interference," *IEEE Transactions on Information Theory*, vol. IT-15, pp. 484–497.

276. Qureshi, S. U. H. (1973a), "An adaptive decision-feedback receiver using maximum-likelihood sequence estimation," *IEEE International Conference on Communications (ICC'73)*, Seattle, Washington, June 1973.

277. Qureshi, S. U. H. (1973b), "Adjustment of the position of the reference tap of an adaptive equalizer," *IEEE Transactions on Communications*, vol. COM-21, pp. 1046–1052.

278. Qureshi, S. U. H. (1977), "Fast start-up equalization with periodic training sequences," *IEEE Transactions on Information Theory*, vol. IT-23, pp. 553–563.

279. Qureshi, S. U. H. (1982), "Adaptive equalization," *IEEE Communications Magazine*, vol. 20, pp. 9–16.

280. Qureshi, S. U. H., and E. E. Newhall (1973), "An adaptive receiver for data transmission

over time-dispersive channels," *IEEE Transactions on Information Theory*, vol. IT-19, pp. 448–457.

281. Qureshi, S. U. H., and G. D. Forney, Jr. (1977), "Performance and properties of a T/2 equalizer," *National Telecommunication Conference (NTC'77)*, Los Angeles, December 1977.

282. Rabiner, L. R., and B. Gold (1975), *Theory and Application of Digital Signal Processing.* Englewood Cliffs, N.J.: Prentice-Hall.

283. Rhodes, S. A. and S. H. Lebowitz (1981), "Performance of coded QPSK for TDMA satellite communications," *5-th International Conference on Digital Satellite Communications*, Genova, Italy, March 1981.

284. Rice, S. O. (1982), "Envelopes of narrow-band signals," *IEEE Proceedings*, vol. 70, pp. 692–699.

285. Robertson, G. H. (1969), "Computation of the noncentral chi-square distribution," *Bell System Technical Journal*, vol. 48, pp. 201–207.

286. Rosenberg, W. J. (1971), *Structural Properties of Convolutional Codes*, Ph.D. Dissertation, University of California, Los Angeles.

287. Rudin, H., Jr. (1967), "Automatic equalization using transversal filters," *IEEE Spectrum*, vol. 4, pp. 53–59.

288. Rugh, W. J. (1981), *Nonlinear System Theory. The Volterra–Wiener Approach.* Baltimore, Md.: Johns Hopkins University Press.

289. Rummler, W. D. (1979), "A new selective fading model: Application to propagation data," *Bell System Technical Journal*, vol. 58, pp. 1037–1071.

290. Rummler, W. D. (1982), "Comparison of calculated and observed performance of digital radio in the presence of interference," *IEEE Transactions on Communications*, vol. COM-30, pp. 1693–1700.

291. Ruthroff, C. L. (1971), "Multiple-path fading on line-of-sight microwave radio systems as a function of path length and frequency," *Bell System Technical Journal*, vol. 50, pp. 2375–2398.

292. Saleh, A. A. M. (1981), "Frequency-independent and frequency-dependent nonlinear models of TWT amplifiers," *IEEE Transactions on Communications*, vol. COM-29, pp. 1715–1720.

293. Saltzberg, B. R. (1968), "Intersymbol interference error bounds with application to ideal bandlimited signaling," *IEEE Transactions on Information Theory*, vol. IT-14, pp. 563–568.

294. Salz, J. (1973), "Optimum mean-square decision-feedback equalization," *Bell System Technical Journal*, vol. 52, pp. 1341–1373.

295. Sato, Y. (1975), "A method of self-recovering equalization for multilevel amplitude-modulation systems," *IEEE Transactions on Communications*, vol. COM-23, pp. 679–682.

296. Satorius, E., and J. Pack (1981), "Application of least squares lattice algorithms to adaptive equalization," *IEEE Transactions on Communications*, vol. COM-29, pp. 136–142.

297. Satorius, E., and S. T. Alexander (1979), "Channel equalization using adaptive lattice algorithms," *IEEE Transactions on Communications*, vol. COM-27, pp. 899–905.

298. Savage, J. E. (1966), "Sequential decoding. The computational problem," *Bell System Technical Journal*, vol. 45, pp. 149–176.

299. Saxena, R. C. P. (1983), *Optimum Encoding in Finite State Coded Modulation*, Report TR83-2, Department of Electrical, Computer and Systems Engineering, Rensselaer Polytechnic Institute, Troy, N.Y.

300. Schetzen, M. (1976), "Theory of pth-order inverses of nonlinear systems," *IEEE Transactions on Circuits and Systems*, vol. CAS-23, pp. 285–291.

301. Schetzen, M. (1980), *The Volterra and Wiener Theories of Nonlinear Systems*. New York: Wiley.

302. Schetzen, M. (1981), "Nonlinear system modeling based on the Wiener theory," *IEEE Proceedings*, vol. 69, pp. 1557–1573.

303. Schichor, E. (1982), "Fast recursive estimation using the lattice structure," *Bell System Technical Journal*, vol. 61, pp. 97–115.

304. Schonfeld, T. J., and M. Schwartz (1971a), "A rapidly converging first-order training algorithm for an adaptive equalizer," *IEEE Transactions on Information Theory*, vol. IT-17, pp. 431–439.

305. Schonfeld, T. J., and M. Schwartz (1971b), "Rapidly converging second-order tracking algorithms for adaptive equalization," *IEEE Transactions on Information Theory*, vol. IT-17, pp. 572–579.

306. Schwartz, M., and L. Shaw (1975), *Signal Processing. Discrete Spectral Analysis, Detection, and Estimation*. New York: McGraw-Hill.

307. Schwartz, M., W. R. Bennett, and S. Stein (1966), *Communication Systems and Techniques*. New York: McGraw-Hill.

308. Shannon, C. E. (1948), "A mathematical theory of communication," *Bell System Technical Journal*, vol. 27, pt. I, pp. 379–423; pt. II, pp. 623–656. (Also reprinted in book form with postscript by W. Weaver. Urbana, Ill.: University of Illinois Press, 1949.)

309. Shimbo, O., and M. I. Celebiler (1971), "The probability of error due to intersymbol interference and Gaussian noise in digital communication systems," *IEEE Transactions on Communication Technology*, vol. COM-19, pp. 113–119.

310. Shnidman, D. A. (1967), "A generalized Nyquist criterion and an optimum linear receiver for a pulse modulation system," *Bell System Technical Journal*, vol. 46, pp. 2163–2177.

311. Simon, M. K. (1976), "A generalization of minimum-shift-keying-type signaling based upon input data symbol pulse shaping," *IEEE Transactions on Communications*, vol. COM-24, pp. 845–856.

312. Sklar, B. (1983), "A structural overview of digital communications—A tutorial review," Part I, *IEEE Communications Magazine*, vol. 21, pp. 4–17, August; Part II, *IEEE Communications Magazine*, vol. 21, pp. 6–21, October.

313. Slepian, D. (1956), "A class of binary signaling alphabets," *Bell System Technical Journal*, vol. 35, pp. 203–234.

314. Slepian, D., ed. (1974), *Key Papers in the Development of Information Theory*. New York: IEEE Press.

315. Slepian, D. (1976), "On bandwidth," *IEEE Proceedings*, vol. 64, pp. 292–300.

316. Smith, J. M. (1977), *Mathematical Modeling and Digital Simulation for Engineers and Scientists*. New York: Wiley.

317. Smith, J. W. (1968a), "A unified view of synchronous data transmission system design," *Bell System Technical Journal*, vol. 47, pp. 273–300.

318. Smith, J. W. (1968b), "Error control in duobinary data systems by means of null-zone detection," *IEEE Transactions on Communication Technology*, vol. COM-16, pp. 825–830.

319. Spaulding, D. A. (1969), "Synthesis of pulse-shaping networks in the time domain," *Bell System Technical Journal*, vol. 48, pp. 2425–2444.

320. Stavroulakis, P., ed. (1980), *Interference Analysis of Communication Systems*. New York: IEEE Press.

321. Stenbit, J. P. (1964), "Table of generators for BCH codes," *IEEE Transactions on Information Theory*, vol. IT-10, pp. 390–391.

322. Stiffler, J. J. (1971), *Theory of Synchronous Communications*. Englewood Cliffs, N.J.: Prentice-Hall.

323. Stroud, A. H., and D. Secrest (1966), *Gaussian Quadrature Formulas*. Englewood Cliffs, N.J.: Prentice-Hall.

324. Symons, F. W., Jr. (1979), "The complex adaptive lattice structure," *IEEE Transactions on Acoustics, Speech, and Signal Processing*, vol. ASSP-27, pp. 292–295.

325. Taylor, D. P., and H. C. Chan (1981), "A simulation study of two bandwidth-efficient modulation techniques," *IEEE Transactions on Communications*, vol. COM-29, pp. 267–275.

326. Thomas, E. J. (1971), "Some considerations on the application of the Volterra representation of nonlinear networks to adaptive echo cancellers," *Bell System Technical Journal*, vol. 50, pp. 2797–2805.

327. Turin, G. L. (1969), *Notes on Digital Communications*. New York: Van Nostrand Reinhold.

328. Ungerboeck, G. (1972), "Theory on the speed of convergence in adaptive equalizers for digital communication," *IBM Journal of Research and Development*, vol. 16, pp. 546–555.

329. Ungerboeck, G. (1974), "Adaptive maximum-likelihood receiver for carrier-modulated data-transmission systems," *IEEE Transactions on Communications*, vol. COM-22, pp. 624–636.

330. Ungerboeck, G. (1976), "Fractional tap-spacing equalizer and consequences for clock recovery in data modems," *IEEE Transactions on Communications*, vol. COM-24, pp. 856–864.

331. Ungerboeck, G. (1982), "Channel coding with multilevel-phase signals," *IEEE Transactions on Information Theory*, vol. IT-28, pp. 55–67.

332. Van Trees, H. L. (1968), *Detection, Estimation, and Modulation Theory*, Vol. 1. New York: Wiley.

333. Varga, R. S. (1962), *Matrix Iterative Analysis*. Englewood Cliffs, N.J.: Prentice-Hall.

334. Vermeulen, F. L., and M. E. Hellman (1974), "Reduced state Viterbi decoders for channels with intersymbol interference," *IEEE International Conference on Communications (ICC'74)*, Minneapolis, June 1974.

335. Vigants, A. (1975), "Space-diversity engineering," *Bell System Technical Journal*, vol. 54, pp. 103–142.

336. Viterbi, A. J. (1965), "Optimum detection and signal selection for partially coherent binary communication," *IEEE Transactions on Information Theory*, vol. IT-11, pp. 239–246.

337. Viterbi, A. J. (1966), *Principles of Coherent Communication*. New York: McGraw-Hill.

338. Viterbi, A. J. (1967), "Error bounds for convolutional codes and an asymptotically optimum decoding algorithm," *IEEE Transactions on Information Theory*, vol. IT-13, pp. 260–269.

339. Viterbi, A. J., and J. K. Omura (1979), *Principles of Digital Communication and Coding*. New York: McGraw-Hill.

340. Volterra, V. (1959), *Theory of Functionals and of Integral and Integro-Differential Equations*. New York: Dover.

341. Wakamatsu, H. (1981), "Inverse systems of nonlinear plant represented by discrete Volterra functional series," *VIII World IFAC Symposium*, Tokyo.

342. Weiner, D. D., and J. F. Spina (1980), *Sinusoidal Analysis and Modeling of Weakly Nonlinear Circuits with Application to Nonlinear Interference Effects*. New York: Van Nostrand Reinhold.

343. Widrow, B., J. M. McCool, M. G. Larimore, and C. R. Johnson, Jr. (1976), "Stationary and nonstationary learning characteristics of the LMS adaptive filter," *IEEE Proceedings*, vol. 64, pp. 1151–1162.

344. Wilson, S. G., H. A. Sleeper, P. J. Schottler, and M. T. Lyons (1984), "Rate 3/4 convolutional coding of 16-PSK: Code design and performance study," *IEEE Transactions on Communications*, vol. COM-32, pp. 1308–1315.

345. Wong, W. C., and L. J. Greenstein (1984), "Multipath fading models and adaptive equalizers in microwave digital radio," *International Conference on Communications (ICC'84)*, Amsterdam, June 1984.

346. Wozencraft, J. M. (1957), "Sequential decoding for reliable communications," *Tech. Report no. 325, RLE MIT*, August 1957.

347. Wozencraft, J. M., and I. M. Jacobs (1965), *Principles of Communication Engineering*. New York: Wiley.

348. Yao, K. (1972), "On minimum average probability of error expression for a binary pulse communication system with intersymbol interference," *IEEE Transactions on Information Theory*, vol. IT-18, pp. 528–531.

349. Yao, K., and E. Biglieri (1980), "Multidimensional moment error bounds for digital communication systems," *IEEE Transactions on Information Theory*, vol. IT-26, pp. 454–464.

350. Yuen, J. H., ed. (1983), *Deep Space Telecommunications Systems Engineering*. New York: Plenum Press.

351. Zadeh, L. A. (1957), "Signal-flow graphs and random signals," *IEEE Proceedings*, vol. 45, pp. 1413–1414.

352. Zigangirov, K. S. (1966), "Some sequential decoding procedures," *Problemy Peredachi Informacii*, vol. 2, pp. 13–25.

Index